NEUROANATOMICAL BASIS OF CLINICAL NEUROLOGY

Orhan Arslan, PhD, DVM
Department of Cell Biology and Anatomy
Finch University of Health Sciences
The Chicago Medical School
Chicago, IL, USA

The Parthenon Publishing Group
International Publishers in Medicine, Science & Technology

NEW YORK LONDON

Library of Congress Cataloging-in-Publication Data

Arslan, Orhan
Neuroanatomical basis of clinical neurology / Orhan Arslan.
p. ; cm.
Includes index.
ISBN 1-85070-578-X (alk. paper)
1. Neuroanatomy. 2. Nervous system--Diseases. I. Title.
[DNLM: 1. Nervous System--anatomy & histology. 2. Nervous System Diseases--pathology. 3. Nervous System Physiology. WL 101 A783n 2001]
QM451 .A77 2001
611.8--dc21

00-048301

British Library Cataloguing in Publication Data

Neuroanatomical basis of clinical neurology
1. Neuroanatomy 2. Nervous system – Diseases 3. Neurology
I. Arslan, Orhan
611.8

ISBN 185070578X

Published in the USA by
The Parthenon Publishing Group Inc.
One Blue Hill Plaza
PO Box 1564, Pearl River
New York 10965, USA

Published in the UK and Europe by
The Parthenon Publishing Group Ltd.
Casterton Hall, Carnforth
Lancs., LA6 2LA, UK

Printed and bound by T. G. Hostench S.A., Spain

Contents

Foreword

The complexity of the central nervous system often overwhelms students attempting to understand how discrete lesions disrupt normal function and cause symptoms. It is for this reason that *Neuroanatomical Basis of Clinical Neurology* fills a real need, by clearly presenting the neuroanatomic mechanisms underlying neurologic diseases.

Dr Orhan Arslan has skillfully integrated clear text and simple diagrams, in order to highlight and clarify the numerous functionally important pathways in the central and peripheral nervous systems. In addition, he has broadened the scope of conventional neuroanatomy texts, which typically emphasize wiring patterns of neurons, to include descriptions of molecular pathways within cells, and the ways in which their perturbation can give rise to disease mechanisms.

This book provides a much-needed link between neurologic disorders and their neuroanatomic origins.

Warren Strittmatter, MD
Professor and Chief
Division of Neurology
Duke University Medical Center
Durham, NC, USA

This book is dedicated to my parents, Inayet and Zübeydeh Arslan, for guiding me through life's journey and to my sister Gülshan and brother Midhat for their everlasting love and inspiration.

Preface

This text reflects my teaching philosophy that neuroanatomy and neurology are mutually interdependent, as understanding of one discipline requires an in-depth knowledge of the other. The concept behind this endeavor was developed through years of teaching experience, tutorials, and reviews of students' feedback at The Chicago Medical School. Discussions among educators and students ascertained the necessity for a functionally and clinically integrated text of neuroanatomy. These ideas and efforts are cohesively synthesized in this uniquely designed text that is easy to understand, yet remains a solid source of information on neuroanatomy and neurology.

This book stands apart from other neuroanatomy texts in that it extensively covers the peripheral and central nervous systems in the context of neurological disorders and conditions. The spinal nerves, plexuses, and associated lesions and disorders are discussed in great detail in conjunction with the central nervous system. Sincere efforts have been made to state most, if not all, of the relevant neuroanatomical and neurological facts.

The book covers both morphological and functional aspects of neuroanatomy. In the morphological part, the developmental aspects, molecular facts and basic neural concepts are explained. These areas are integrated with congenital anomalies and associated conditions. A precise level-by-level description of the spinal cord, brainstem, diencephalon, and brain are documented. Within the cerebral hemispheres, cortical layers, afferent and efferent connections, the white matter, commissures (e.g. corpus callosum) and association fibers (e.g. uncinate fasciculus) along with their lesions and pertinent syndromes (e.g. disconnection syndrome) are exhaustively explored. This is followed by a description of the various forms of cortical dysfunctions (e.g. aphasia, agnosia, and apraxia). In order to maintain continuity, the arterial supply and venous drainage are integrated with the morphological part. Extensive discussion of the cerebral vasculature and emphasis on cerebrovascular accidents emanating from rupture or occlusion of cerebral arteries have been documented. Furthermore, the ventricular system, circulation of the cerebrospinal fluid, brain barriers, and various forms of hydrocephalus are integrated with the description of the brain. A special effort has been made to provide a critical view of the neurotransmitters and their roles in disease processes.

In the context of the peripheral nervous system, great emphasis has been placed on the spinal and cranial nerves, and associated plexuses and nuclei. Much information is given about individual nerve dysfunctions at various levels and comparisons have been provided in an effort to differentiate the sites of these disorders. Spinal and cranial reflexes and their significance in clinical diagnosis are described along with the paths of these peripheral nerves. Neurological testing of each cranial nerve is portrayed in detail. An in-depth approach to the sympathetic and parasympathetic systems is explored, and pertinent clinical conditions are discussed.

In the description of the central nervous system, drawings depicting the morphological features and central connections of the described structures as well as original pictures of brain sections are utilized in an organized schemes. Radiographic images such as CT scans, MRI, and angiograms are added to this textbook to supplement the existing diagrams.

The functional part of this text incorporates a systems approach to neuroanatomical pathways, related lesions, and clinical conditions. A detailed examination of the components of the special sensory systems including the peripheral apparatus of the visual, auditory, and vestibular systems is also provided. Structures in the path of these systems are described and their pertinent clinical significance is carefully studied. This is followed by an organized approach to the pathways that transmit general sensations. In the same fashion the motor system is described, and concerted effort is devoted to the study of upper motor neurons, pathways, and their dysfunctions. A similar discussion of the lower motor neurons is also documented. Finally, examination of the motor system is concluded with an emphasis on the combined lesions of the upper and lower motor neurons. Subsequent to individual coverage of the sensory and motor systems and related deficits, a level-by-level study of the combined lesions of the sensory and motor systems is documented. Numerous, original schematic diagrams have been added to illustrate the neuroanatomical pathways in a simplified manner and summarize their connections.

This book is primarily intended to be used by medical students in their neuroanatomy and/or neuroscience courses. It is equally relevant to students of anatomy, neurobiology, physical therapy, psychology, and other allied health sciences. Additionally, it can be effectively utilized by health practitioners in the fields of neurology, neurosurgery, and rehabilitation medicine.

Orhan Arslan, PhD, DVM

Acknowledgements

The strong support of Richard Hawkins, PhD, President and Chief Executive Officer at the Finch University of Health Sciences/The Chicago Medical School undoubtedly deserves a particular note of gratitude and recognition. His illustrious leadership, vision and enduring inspiration are deeply cherished. I wish to recognize with appreciation the Finch University of Health Sciences and in particular the Department of Cell Biology and Anatomy for providing the resources necessary for the successful conclusion of this academic endeavor. I extend my special thanks to Walid Hindo, MD, Professor of Radiology at The Chicago Medical School for his critical reading, unstinting interest and postive input. I am greatly indebted to Warren Strittmatter, MD, Professor and Chief of Neurology at Duke University Medical Center for his in-depth review of this text, reassuring insight and thoughtful remarks. I am equally gratified by the thorough review and positive comments of James A. Killian, MD, Professor and Vice-Chairman of the Department of Neurology at Baylor School of Medicine. Many thanks to my departmental colleagues, Drs Christopher Brandon, Joseph Dimario, William Frost and Monica Oblinger for their helpful suggestions. I owe a great debt of gratitude and deep-felt appreciation to Talat Arslan, MD, for his significant clinical input and painstaking work on this project.

I also take this opportunity to acknowledge Deborah Rubenstein and Amit Dayal, MS for their artistic expertise and Joseph Nadakapadam for his proficient photographic work. I would like to express my appreciation to David Bloomer, President, Nat Russo, Editor-in-Chief, Dinah Alam, Kate Reeves, Meredith Ross, and the staff at The Parthenon Publishing Group for their unfailing assistance, numerous courtesies, and patience.

Section I

Introduction

Developmental aspects of the nervous system 1

During the third week of gestation, the neural ectoderm forms the neural plate, which eventually becomes the neural tube. The neuroectodermal cells that lie dorsolateral to the neural tube constitute the neural crest cells, giving rise to the autonomic and sensory ganglia. Closure of the neuropores and development of the brain vesicles lead to the formation of various compartments of the central nervous system (CNS). When the neuropores fail to close, a variety of malformations ranging from anencephaly to spina bifida may ensue. Further differentiation of the primitive neural tube leads to the formation of the ependymal, mantle, and marginal layers. Subdivision of the mantle layer into the alar and basal plates serves as a basis upon which the sensory and motor nuclei are localized within the CNS. The neural canal converts into various parts of the ventricular system.

Formation of the neural tube

During the third week of gestation (day 16), the notochord (chorda-mesoderm) induces the development of the neural plate, which is located dorsal to the notochord in the rostral part of the embryonic disc (Figures 1.1, 1.2 & 1.3). Proliferation and differential growth of the neural plate cells, changes in the shape of the cells, stretch posed by the rapidly developing embryo, activities of the microtubules and microfilaments, as well as their intrinsic movements eventually lead to the formation of the neural groove (day 18), which acts as a median hinge region around which the neural folds expand (Figures 1.2 & 1.5). Fusion of the neural folds occurs in the future cervical region at certain sites, gradually expanding in both rostral and caudal directions, leading to the formation of the neural tube (Figure 1.5). Failure of the neural folds to fuse, differentiate, and detach from the surface ectoderm may lead to rachischisis, which is, as described later in this chapter, a group of malformations that assumes a variety of forms depending upon the involved part of the neural tube. The last portions of the neural tube to close are the rostral and caudal neuropores, which maintain connections with the amniotic cavity.

Detachment of the neural tube from the surface ectoderm (future epidermis) and its assumption of a more ventral position are believed to be enhanced by neuronal cell adhesion molecule (NCAM) and N-cadherin synthesized by the neural tube itself. As the neural folds fuse, specialized population of cells in the dorsolateral part of the neural tube forms the neural crest cells (Figures 1.4 & 1.5). Neural crest cells lose their epithelial-specific adhesion molecules and express a new group of cell adhesion molecules such as integrin and laminin. With the help of pseudopodia, which develop from the basal aspects, the neural crest cells are pulled through the basal membrane of the neural tube. Thus, via this process both the surface ectoderm and the basal membrane of the neural tube may guide the migration of the neural crest cells. Migration of these cells occurs in a craniocaudal direction, with the more cephalic cells departing before the closure of the cranial neuropore.

The process that involves formation of the neural plate, floor plate, and neural sulcus, as well as closure of the neuropores with the eventual configuration of the neural tube is known as primary neurulation. Secondary neurulation refers to the development of the caudal neural tube, a process that is initiated by the formation of a solid tube or mass caudal to somite 31 (first or second lumbar somite) and the appearance of ectodermal-lined vacuoles. These vacuoles eventually coalesce and open into the end of the neural tube.

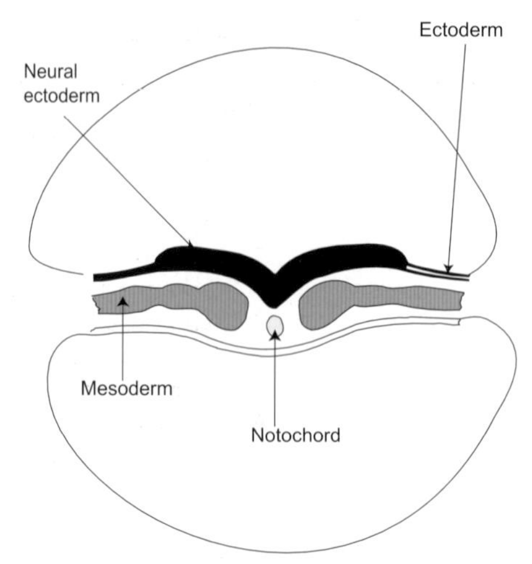

Figure 1.1 Transverse section of the embryo showing the location of the neural ectoderm relative to the notochord and mesoderm

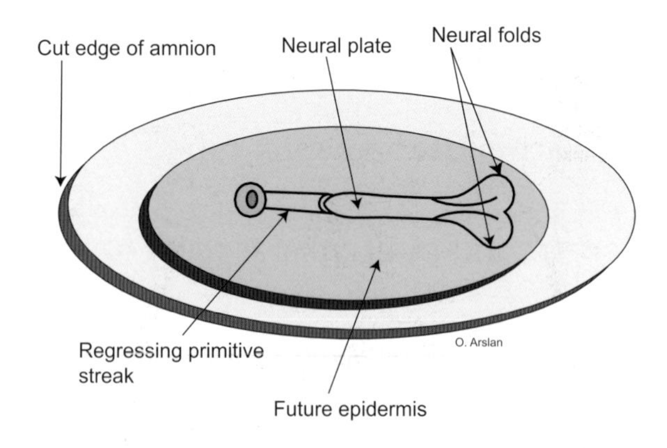

Figure 1.2 Schematic drawing (dorsal view) of the neural plate, neural folds, and their relationships to the future epidermis and primitive streak

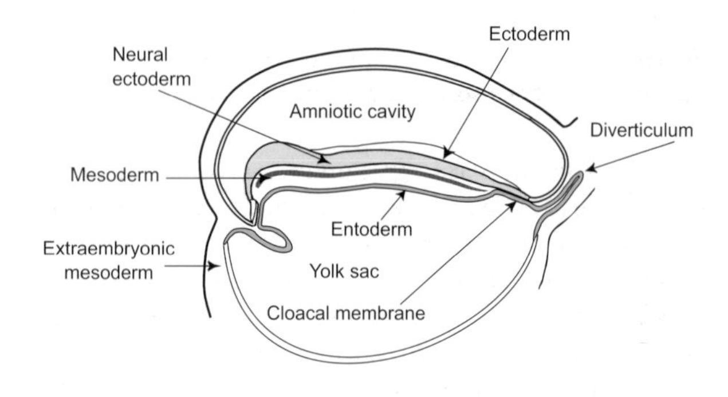

Figure 1.3 Diagram of the neural ectoderm (cephalocaudal) in relationship to the notochord, and intraembryonic coelom. Note the location of the mesodermal tissue

Table 1.1 Derivatives of the neural crest cells

Neuroectodermal cells	*Derivatives*
Neural crest cells	1. Dorsal root ganglia and sensory ganglia of cranial nerves 2. Sympathetic and parasympathetic ganglia 3. Schwann cells 4. Adrenal medulla 5. Melanocytes 6. Neurolemma of the peripheral nerves 7. Cells of the pia and arachnoid mater of the occipital lobe and spinal cord 8. Intraocular muscles 9. Ciliary bodies 10. Carotid bodies

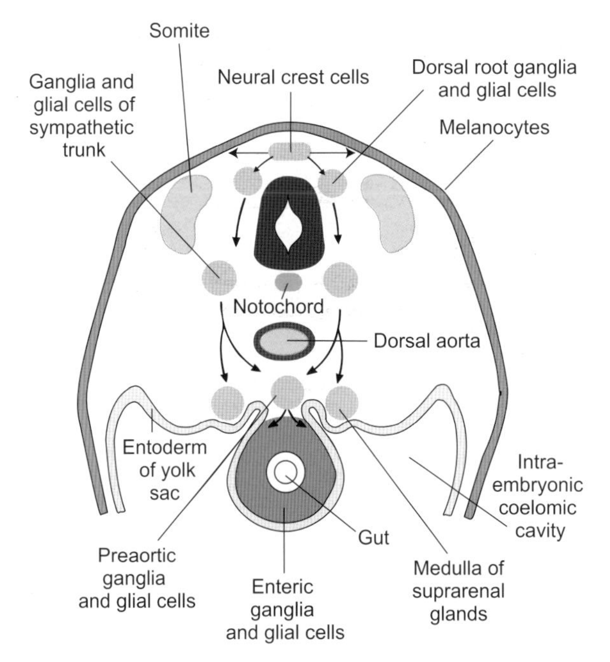

Figure 1.4 Path of migration of the neural crest cells. Note the locations of the somites, notochord, and gut

Later, canalization and final union of the caudal mass with the rest of the neural tube occurs. In 33–35% of embryos the caudal neural tube remains forked.

Neural crest cells contribute to the formation of the sensory ganglia of the dorsal roots, trigeminal, facial, glossopharyngeal and vagus nerves, autonomic ganglia and satellite cells of these ganglia. They also give rise to the Schwann cells, adrenal medulla, melanocytes, auditory nerve, neurolemma of the peripheral nerves, cells of the pia and arachnoid mater of the occipital region and spinal cord, intraocular muscles, ciliary body, and the carotid bodies. The target tissue that receives the migrating cells may determine development of these structures. The target tissue may also influence secretion of catecholamines and acetylcholine by structures of neural crest origin.

Eventually, neural crest cells spread segmentally along the entire length of the neural tube, contributing to the formation of the structures associated with the future peripheral nervous system. Failure of neural crest cells to migrate along the wall of the developing intestinal tract produces congenital loss of parasympathetic ganglia in the Meissner's and Auerbach's plexuses and signs of Hirschsprung's disease (congenital aganglionic megacolon).

Hirschsprung's disease (congenital megacolon) is a congenital anomaly with a male to female ratio of 3:1, and is associated with failure of neural crest cells to migrate. This condition commonly affects the rectum and the sigmoid colon (three-fourths of cases), but rarely involves the entire colon. It may also occur in more proximal locations, depending upon the migratory defect of neural crest cells. It is characterized by impaired peristaltic movement at and beyond the affected part of the colon, followed by bowel stasis, chronic constipation, abdominal distention, hypertrophy, constriction of the aganglionic segment, and dilatation of the colon proximal to the affected area. It usually manifests itself in the newborn by inability to pass meconium, followed by intestinal obstruction. The etiology of this disorder may include mutation of RET receptor tyrosine kinase. RET is a tyrosine kinase that undergoes enzymatic activation and initiates intracellular signaling upon binding the glial-derived neurotrophic factor and the endothelium B receptor.

The posterior neuropore closes at the level of the first or second lumbar somite, which marks the future first or second lumbar spinal segment. *Pax-3, sonic hedgehog,* and *openbrain* genes, as well as folic acid (vitamin B_{12}) and cholesterol play important roles in the closure of this tube. As mentioned earlier, failure of closure of the neuropores produces various forms of dyspharic defects collectively known as rachischisis. The site of closure of the anterior neuropore is represented in the newborn by the lamina terminalis; a vestigial structure located rostral to the hypothalamus.

The neural tube maintains a connection with the amniotic cavity via the anterior and posterior neuropores, which close by the 25th and 27th day, respectively.

Neural tube development induces the formation of somites (a paired segmented division of paraxial mesoderm that develops along the length of the early embryo). The underlying paraxial mesoderm also starts segmentation (day 20). Failure of development of the neural tube may impair development of the surrounding somites (e.g. bone, muscles, and skin) and vice versa. Therefore, malformation of the brain and its coverings may also be accompanied by abnormal ossification of the bony skull. Herniation of brain tissue through a gap in the posterior midline of the skull (cranium bifidum) is known as encephalocele.

Failure of closure of the anterior neuropore results in cranioschisis, an anomaly which is associated with lack of development of the brain (anencephaly) and the skull. Anencephaly may be seen in Meckel (Meckel–Gruber) syndrome, a rare autosomal recessive and fatal condition that is characterized by sloping forehead, posterior meningoencephalocele, polycystic kidney, and polydactyly. It may also be seen in brachydactyly syndrome, an autosomal dominant disorder that exhibits abnormal shortening of the fingers and toes.

Anencephaly (exencephaly) is one of the most severe forms of congenital anomalies that occur between days 18–24 of embryonic development. It is relatively common with an incidence of 1 per 1000 births in the United States. It is most commonly seen in female infants and is usually fatal. The exposed neural tissue undergoes degeneration and becomes converted into a mass of vascular connective tissue, intermixed with masses of degenerated brain and choroid plexus. Due to these defects and the exposure to infectious agents, death usually occurs shortly after delivery. The vault of the skull of the anencephalic fetus fails to form and its base is usually covered with a vascular membrane. The orbits appear shallow and the eyes tend to bulge externally.

The fetus exhibits wide shoulders, short trunk, and a neck, which is commonly absent, giving the impression that the head is stemming directly from the body. Anencephaly may also be associated with absence of vertebral arches, amyelia (absence of the spinal cord) and hydramnios (excess of amniotic fluid) due to the lack of a swallowing reflex. Fifty per cent of anencephalic fetuses are aborted spontaneously. Pregnancy with an anencephalic fetus is usually complicated by delayed onset of labor (most deliveries after 40 weeks' gestation).

Incidence of congenital malformations is approximately 3% in neonates, and CNS anomalies make up approximately one third of all congenital malformations. Thus, the incidence of CNS congenital malformation is around 1%. Manifestations of these congenital defects may vary widely from those occurring incidentally without apparent symptoms to those which are incompatible with life. Malformations may develop as a result of radiation, metabolic disorders, chromosomal abnormalities (e.g. trisomies), chemical exposure (e.g. illicit drugs and alcohol), and infection by viruses (e.g. cytomegalovirus, herpes, rubella), and bacteria (e.g. *Toxoplasma gondii, Treponema pallidum*), etc.

By the end of the fourth week, and following complete separation from the ectodermal surface, the neural tube is composed of a caudal part which becomes the spinal cord and expanded rostral brain vesicles (Figure 1.8) that forms the brain hemispheres and the brainstem. It is important to note that while the most cephalic portion of the neural tube undergoes drastic differentiation, the caudal portion is still forming. At first, the rostral part of the neural tube

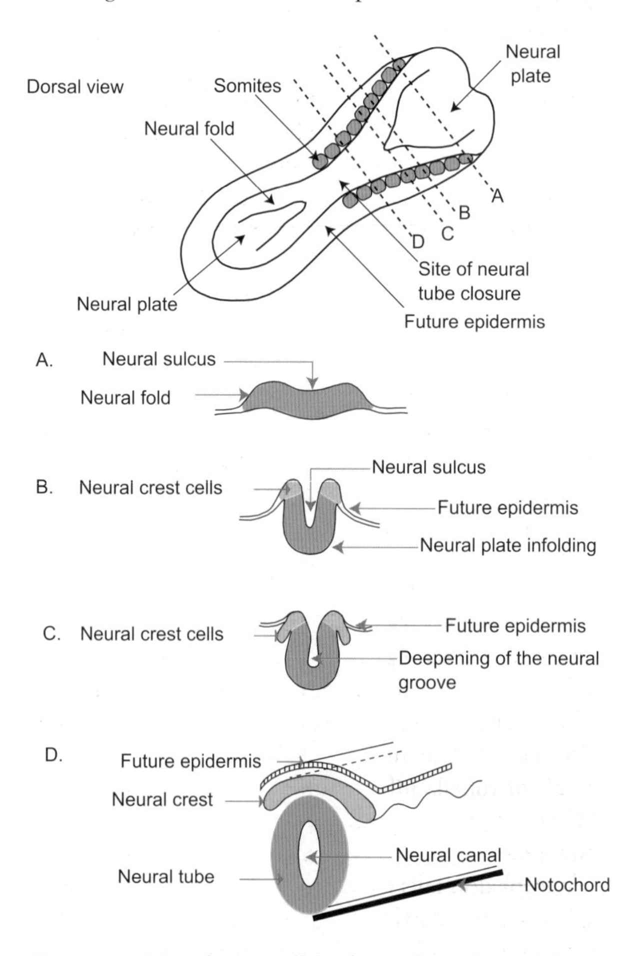

Figure 1.5 Neural crest cells and neural canal. Neural crest cells eventually separate from the newly formed neural tube

Encephalocele, a malformation seen in Klippel–Feil deformity, is commonly associated with fusion of vertebrae, decrease or sometimes increase in the number of vertebrae, and hydrocephalus (enlargement of the brain and skull subsequent to excess of CSF). The prognosis of this condition depends upon the degree of involvement of the brain tissue and associated meninges. Encephalocele should be corrected surgically unless death is imminent or if it concurrently occurs with dextrocardia, laryngomalacia, renal agenesis, or pulmonary hypoplasia. Encephaloceles, which make up 10–20% of all craniospinal malformations, are predominant in female fetuses. The anatomic location of an encephalocele may vary with the geographic regions; in southeast Asia they are more commonly located in the anterior cranial vault, whereas in Europe and the United States they are commonly located in the occiput. The amount of brain tissue within the herniated sac varies and may involve parts of the cerebral hemispheres, cerebellum, and even the brainstem.

Meningoencephalocele occurs when the herniated brain tissue is associated with meningeal coverings. Cranial meningocele refers to the protrusion of the meninges into the sac. The bony defect, which results from incomplete closure of the calvaria, may be confined to the occiput or extend to the arch of the atlas. At birth the infant with encephalocele is usually neurologically normal or may exhibit increased flexor motor tone. Spells of apnea and bradycardia may occur in severe cases of occipital encephaloceles involving the brainstem.

Protrusion of a large amount of brain tissue into the encephalocele may reduce the size of the brain, resulting in microcephaly. Microcephaly may also be due to failure of brain growth, producing a smaller brain with less prominent gyri and a considerably smaller skull than usual. Due to the marked difference between the anterior and posterior ends of an encephalocele, blood vessels may be occluded or ruptured leading to infarction and hemorrhage. Sclerosis in the herniated sac, as well as impairment of CSF circulation and resultant hydrocephalus may also occur. The microcephalic brain may be the result of rubella infection or exposure to radiation. Microcephaly may be associated with holotelencephaly (holoprosencephaly), in which the telencephalon fails to cleave into two cerebral hemispheres and ventricles, but instead forms a single structure with a large single ventricular cavity.

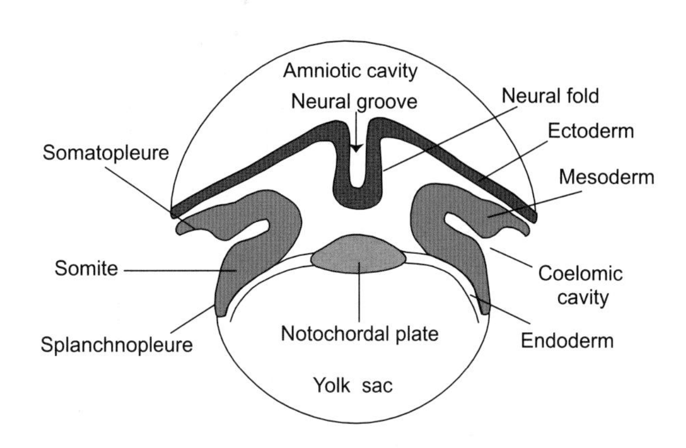

Figure 1.6 Schematic drawing illustrating the neural tube with the dividing sulcus limitans, neural crest cells, and the future epidermis

Macrocephaly, an abnormally large brain and skull, is usually a fatal condition associated with syringomyelia, a disease which exhibits cavitation around the central canal of the spinal cord or lower brainstem (syringobulbia), and may occur in one or both cerebral hemispheres, without any detectable change in intracranial pressure. Macrocephaly may also be seen in individuals with Arnold–Chiari malformation or trauma.

consists of three primary brain vesicles; prosencephalon, mesencephalon, and rhombencephalon (Figure 1.9). Rapid growth of the primary brain vesicles during the fifth week results in the formation of telencephalon, diencephalon, mesencephalon, metencephalon, and the myelencephalon as well as associated flexures. The cephalic flexure lies between mesencephalon and rhombencephalon. The pontine flexure, which develops into the transverse rhombencephalic sulcus, separates the metencephalon and the myelencephalon. Furthermore the cervical flexure lies between the rhombencephalon and the spinal cord.

Persistence of the neural canal within the center of the primitive brain vesicles gives rise to a group of interconnected, fluid-filled cavities that compose the ventricular system. The rhombencephalic vesicle develops into the fourth ventricle; the mesencephalic vesicle becomes the cerebral aqueduct; the diencephalic vesicle converts into the third ventricle; while the telencephalic vesicle develops into the lateral ventricle.

By the middle of the fourth week, the neural tube develops three layers, consisting of the ventricular zone and ependyma (innermost), mantle (intermediate), and marginal (outermost) layers (Figures 1.10 & 1.11). This differentiation commences in the rhombencephalic region and then extends in a craniocaudal direction. The dendritic processes of the mantle neurons (neuroblasts) form the marginal layer (future white matter). In the spinal cord and the rhombencephalon, the ventral part of the mantle layer represents the sites of the motor neurons that appear earlier than the sensory neurons. The mantle layer (future gray matter) consists of a narrow dorsal alar plate and a thick basal plate, separated by the sulcus limitans (Figures 1.10 & 1.11). As it forms the lining of the central canal and ventricular system, the ventricular zone (ependyma) separates the cerebrospinal fluid from the blood vessels of the choroid plexus and neurons of the brain and spinal cord. Mitotic division of the neuroepithelial cells in the ependymal layer forms neuroblasts, which migrate laterally to the mantle layer.

Migration of neuroblasts may be guided by the cytoplasmic extensions of the developing neurons, which are anchored to the pia mater on the outer surface of the CNS. Shortening of these processes may also enable some neurons to migrate. This migration is also dependent upon the radial glial cells, which is clearly evident in the development of the cerebral cortex. In the cerebellum, this migration is seen in the course of the developing granule cells parallel to the surface of the neural tube and along the long processes of the Bergmann glial cells. An adhesion protein molecule known as astrotactin may mediate neuroglial interaction during the migration process. Some investigators propose the possibility of existence of genetically predetermined entities that may guide the migration process.

As the migration from the neuroepithelium or from one of the secondary proliferative zones continues, the differentiated neuronal and glial cell precursors assume either an ellipsoidal configuration (apolar) or present a single process (unipolar). Few apolar or unipolar neurons are retained in the CNS of the vertebrate, but most continue through a bipolar phase. The great majority of neurons develop even more processes and are known as multipolar neurons.

Genetic and molecular aspects of neural development

Expression of certain homeobox genes, at the initial stages of development, may play an important role in the segmentation of the rhombencephalon in which each segment (rhombomeres), a series of eight bilateral protrusions from the rhombencephalic wall, may specify the origin of certain cranial nerves. For example neurons of rhombomere-2 (r2) form the trigeminal ganglion, while r4 forms the geniculate ganglion. Domains of gene expression, for example those for the *HOX-B* genes and the transcription factor Krox20, adjoin rhombomere boundaries. Genes that contain homeoboxes in the human are classified into *HOX-A*, *HOX-B*, *HOX-C*, and HOX-D and are identified by numbers 1–13. These genes may regulate the expression of a number of genes that collectively determine the structure of one body region. *HOX* genes are expressed in the developing rhombomeres and neural crest. The complement of *HOX* genes is thought to form axial and branchial codes that specifies the locations of somites and neurons along the length of the embryo. An interesting feature of specific homeotic (*HOX*) genes is their linear order (colinearity) and their transcription in the cephalocaudal and the 3′ end to 5′ end directions. Since *HOX* gene expression is affected by teratogenes such as retinoic acid, transformation of rhombomeres may occur. Transformation of trigeminal nerve to facial nerve may occur as a result of induced changes of the rhombomeres 2/3 to a 4/5 disposition. Similar changes may be observed upon the application of retinoic acid to other *HOX* codes. Ventro-dorsal patterning of the neural tube may be regulated by the Sonic hedgehog (*Shh*) gene, a product of the notochord and floor plate. *Shh* gene may act by inhibiting the suppression of the expression of *Pax*-3 and dorsalin genes. Non-*HOX* homeobox genes, a separate group of genes, are also involved in developmental patterning of the embryo, but lack the cephalocaudal expression seen in the *HOX* homeobox genes and their role in organogenesis.

Pax genes (*Pax*-1 to *Pax*-9), which belong to the segmentation group of genes, regulate the morphology of the developing embryo. Their protein products are transcription factors and may be implicated in the specification of different regions of the CNS. Pax-3 and Pax-7 are expressed in the alar plate and roof plates as well as the neural crest cells. On the other hand, Pax-5 and Pax-8 are expressed in the intermediate gray columns. Ventricular and basal regions of the neural tube are sites where Pax-6 is expressed. Expression of Pax-1 in the ventromedial region of the somite is induced by several factors such as Shh, notochord and floor plate, whereas expression of Pax-3 and Pax-7 within the dorsolateral region of the somites is induced by the dorsal ectoderm. *Pax* genes may play important roles in certain genetic diseases. Mutated *Pax*-3 and *Pax*-6 may be involved in Waardenburg's syndrome and aniridia, respectively.

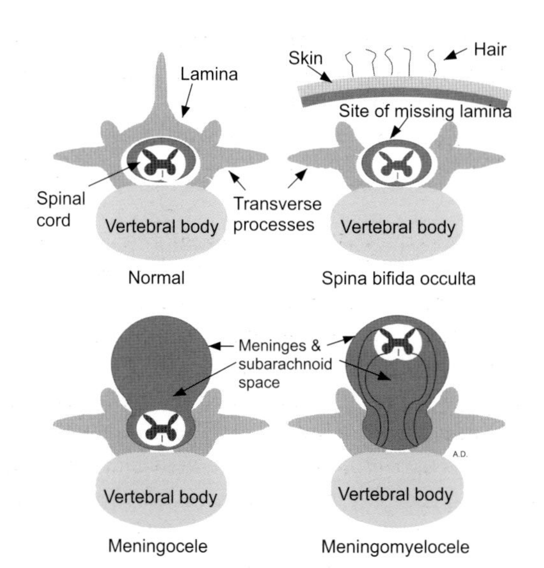

Figure 1.7 Various forms of spina bifida. Note the absence of vertebral lamina and the protrusion of the meninges or meninges and spinal cord tissue

Waardenburg's syndrome is an autosomal dominant condition associated with mutated *Pax*-3. It presents with congenital deafness, wide bridge nose, disorders of pigmentation (white eyelashes, white forelock, and leukoderma), and dystopic canthorum (lateral displacement of the canthi). In aniridia, a rare bilateral hereditary anomaly associated with mutated *Pax*-6, absence or abnormal development of the retina and iris remains a prominent feature.

Extensions of the developing neuronal processes may be governed by many factors that include the development of growth cones, and presence of NCAM, neuroglial adhesion molecules (NgCAM), transiently exposed axonal glycoprotein (TAG-1), actin, extracellular matrix adhesion molecules (E-CAM), and guide-post cells. Growth cones are able to descry the chemical signals (actin and actin-binding proteins) and test the new environment in all directions via filapodia and lamellipodia. Since polymerization of

Spinal dysraphism (defective fusion) refers to a developmental malformation that occurs when the neural tube fails to close completely and the vertebral arches fail to form fully.

Non-closure of the posterior neuropore at day 27, a less frequent anomaly, leads to the formation of spina bifida (myeloschisis) or split spine. It may be ascribed to an excess of vitamin A, teratogenes such as aminopterin, valproic acid and retinoic acid, high concentration of plasma glucose level, trypan blue, or a deficiency of folate. Amniocentesis (obtaining amniotic fluid sample by surgical trans-abdominal perforation) may be useful in detecting myeloschisis. This procedure is based upon the fact that the levels of α-fetoprotein (AFP) at 16–18 weeks of pregnancy in the amniotic fluid are far greater in cases of myeloschisis than in healthy pregnancies. This anomaly is classified into spina bifida occulta and spina bifida cystica (Figure 1.7).

Spina bifida occulta, the least clinically significant form of spina bifida, is a mesodermal (rather than an ectodermal) abnormality, and is rarely associated with neurological dysfunction. It is a closed neural tube defect which is characterized by defective vertebral laminae at certain levels that expose the meninges through a bony gap, without the actual involvement of the spinal cord or meninges. This malformation is usually asymptomatic, unless a developed lipoma or bony process compresses the spinal cord at the affected area, leading to nocturnal enuresis (neurogenic bladder) and retarded development of the lower limb (asymmetrical or sometimes unilateral shortening of one leg and foot). Compression at L5 motor roots may produce calcaneovalgus, which is dorsiflexed, everted, and abducted feet. Equinovarus, plantar-flexed, inverted, and adducted feet may occur as a result of compression of the S1 motor roots. Back pain, impairment of sensations, and sciatica (pain radiating from the back to the leg) may also accompany this condition. Intrauterine ultrasound had not proven useful in the prenatal diagnosis of this condition. Radiographic plain films of the spine in the neonates are also of no use in detecting spina bifida occulta since the vertebral laminae at that time are cartilaginous and radio-lucent, making it very difficult to reveal defective spinous processes. The most common form of spina bifida occulta is the dermoid sinuses and the lipoma-covered defect in spina bifida.

Dermoid sinuses are deep, epithelial lined tracts, sometimes containing hair, that ascend from an external opening over the spine or scalp. They terminate at the intracranial structures and occasionally communicate with the dura mater and subarachnoid space, producing meningitis. Dermoid sinuses are most common in the lumbar or sacral region (65%) and less in the occipital region (30%). These sinuses may go unnoticed and are only detected after repeated bouts of meningitis. Occasionally, brain abscesses occur as a consequence of repeated infections and the patient may present with hypertension, seizures and fever. Lipoma-covered defect, another form of spina bifida occulta, occurs in the sacral or lower lumbar regions, and is very rarely associated with neurological abnormalities.

Spina bifida cystica (spina bifida manifesta, or spina bifida aperta), having an incidence of one per 1000 births, is characterized by a sac-like protrusion of either meninges (meningocele), or a combination of meninges and spinal cord/nerve roots (meningomyelocele).

Meningocele (10% of cases of spina bifida cystica) is not associated with motor or sensory deficits, bowel or bladder dysfunctions. Meningocele is covered by thin easily ruptured meninges and occasionally by skin, and commonly occurs in the lumbosacral portion of the vertebral column. It is frequently associated with myelodysplasia (spinal cord defects), occurring in the majority of cases of spina bifida cystica. A meningomyelocele (90% of cases of spina bifida cystica) is much more common than meningocele and may contain the cauda equina if it occurs in the lumbosacral region, or the spinal cord when the anomaly involves the thoracic or cervical regions.

A lipomyelomeningocele appears when fatty tissue and skin cover the herniated sac. In this case, neurogenic bladder is almost a universal presentation. Meningomyelocele may be accompanied by hydrocephalus, talipes equinovarus (clubfoot, which is plantar flexed, inverted and adducted), sensory and motor deficits, paraplegia, as well as bowel dysfunction. A high percentage (90%) of infants with the meningomyelocele develop hydrocephalus as a part of Arnold–Chiari malformation (syndrome).

Arnold–Chiari malformation (syndrome) is thought to result from fixation of the developing spinal cord in the sac of a meningomyelocele, creating an undue downward strain on the spinal cord and brainstem. This eventually causes displacement of the cerebellum and hindbrain through the foramen magnum into the cervical part of the vertebral canal. Hydrocephalus may be seen in this syndrome as a result of closure of the foramen of Magendie and foramina of Luschka of the fourth ventricle. Blockage of absorption of the cerebrospinal fluid into the dural sinuses, and impairment of the circulation of the fluid around the base and lateral surface of the brain may also contribute to hydrocephalus. Herniation of the cerebellar vermis into the cervical part of the vertebral column may occur as a result of early fusion of its hemispheres, followed by fusion of the neural folds higher in the midbrain. The latter may be associated with stenosis of the cerebral aqueduct (a canal that runs through the mesencephalon). Additionally, an anomalous small posterior cranial fossa may not be of sufficient size to accommodate the growing cerebellum, leading to displacement and herniation.

Arnold–Chiari malformation is divided into four types (I through IV). Type I is seen in young adults as a result of downward displacement of the cerebellar tonsils through the foramen magnum into the cervical part of the vertebral column. It manifests hydrocephalus, dysphagia, dysphonia, syringomyelia, and bilateral cerebellar dysfunctions.

Type II has its onset in neonates and is associated with meningomyelocele and more displacement of the fourth ventricle into the vertebral column. Due to the low level of the spinal cord, the cervical spinal nerve roots assume an ascending course in order to reach their corresponding intervertebral foramina. Patients with this type of anomaly exhibit hydrocephalus, syringomyelia, dysphagia, respiratory stridor, and dysphonia. Type III presents a cyst-like fourth ventricle in the posterior cranial fossa, whereas type IV is characterized by shrinkage of the cerebellum due to hypoplasia. This kind of malformation may exhibit hydrocephalus, dysphagia, dysphonia, syringomyelia, and cerebellar dysfunctions. Telencephalic developmental anomaly may lead to the absence of gyri referred to as lissencephaly or abnormally thick and wide gyri known as pachygyria.

Approximately half of the individuals with spina bifida cystica may also exhibit a partial or complete division of the spinal cord into two symmetrical parts (diastematomyelia). A bony projection, cartilage, or fibrous tissue often separates these two parts. Each part may have its own dural, arachnoidal, and pial coverings, and arterial supply. The part of the spinal cord cranial or caudal to the site of this anomaly remains united. Although the transverse diameter of the vertebral bodies at the affected site may show perceptible widening, the anteroposterior dimension is often narrowed. This anomaly is compatible with life and may not pose any serious neurological complications. Radiographic imaging techniques, myelography, and visible cutaneous manifestations may make diagnosis of this condition possible. Patients with scoliosis may be examined for this type of malformation via myelography.

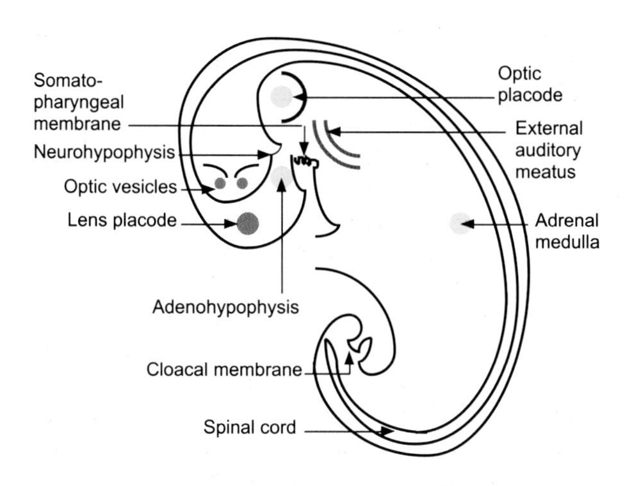

Figure 1.8 Neural tube showing its expanded rostral end with optic vesicles. Observe the location of the adrenal medulla, a derivative of the neural crest cells

actin control to an extent the movement of the growth cones, any substances which limit this process such as fungal toxin cytochalasin B may also hinder their further growth. Additionally, calcium, interaction with other intracellular second messenger systems, and phosphorylation by protein kinase may indirectly affect the direction of neuritic growth by acting upon the actin-binding proteins. Movements of growth cones may also be shaped by NCAM, a molecule that also enhances neuritic fasciculation, laminin, fibronectin and tenascin (cytotactin) which are members of the extracellular matrix adhesion molecules that act via receptor integrins. NgCAM, integrin, and N-cadherin (calcium-dependent molecule) also share important roles in the process of axonal development.

Synaptic connections among developing neurons occur early in development and undergo constant correction and refinement. Non-functional and inappropriate connections may be inhibited by repulsive actions as well as by the mechanical barriers of the surrounding tissue. The time of birth may also determine the ultimate locations and eventually the connections of the neurons.

The initial overproduction of neurons may be controlled by subsequent cell death, which is determined by the available trophic substances in the immediate location. Selectivity of the neurotrophic substances in supporting certain neuronal populations may also determine the fate of certain neurons. Extracellular, intracellular, and transmembranous tyrosine kinase domains remain essential in mediating the effects of neurotrophins. Hormones such as testosterone may also govern the extent of development of certain areas of the CNS.

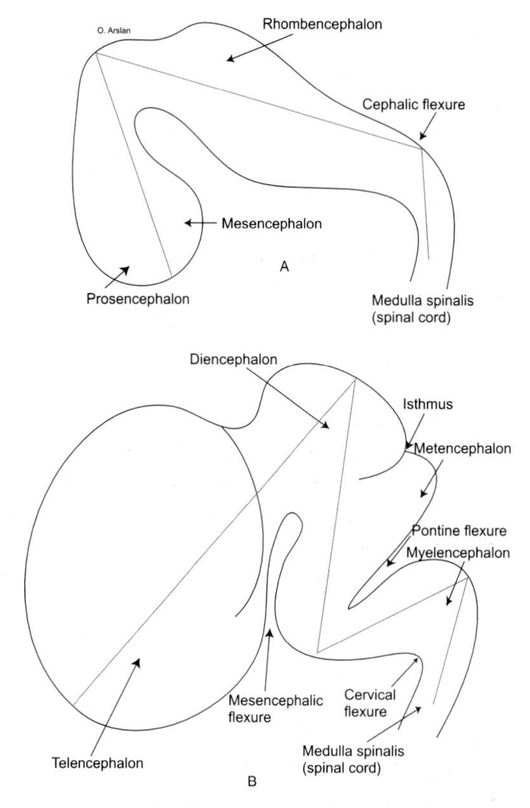

Figure 1.9 Initial differentiation of the rostral neural tube into the primary and secondary brain vesicles. Flexures that separate certain brain vesicles are also illustrated

Spinal cord (myelencephalon)

The spinal cord develops from the caudal portion of the developing neural tube, extending the entire length of the vertebral column (Figures 1.8 & 1.9). During this stage of development the spinal nerves also appear. After the third fetal month the vertebral column outgrows the spinal cord and as the vertebrae

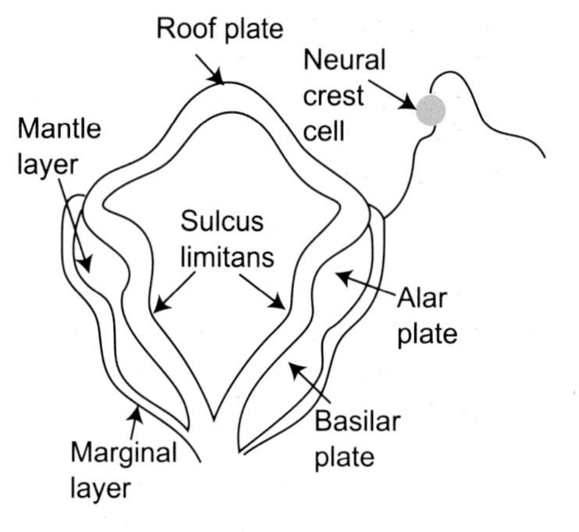

Figure 1.10 A section of the neural tube showing the sulcus limitans, alar and basal plates, the roof and floor plates

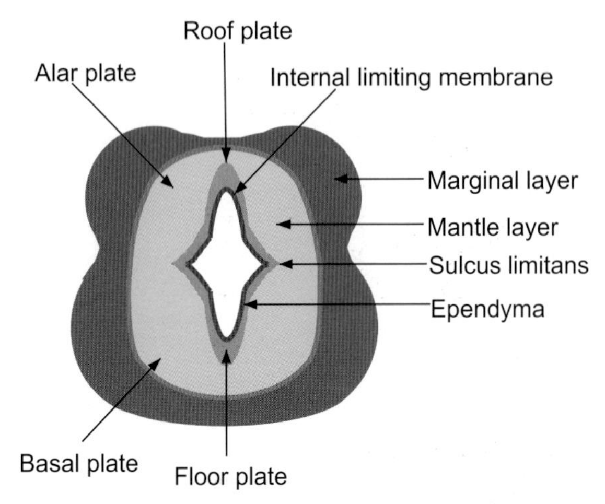

Figure 1.11 Cross-section of the caudal neural tube (future spinal cord). The basal plate forms the ventral horn and the associated motor neurons, whereas the alar plate forms the dorsal horn and sensory neurons of the adult spinal cord

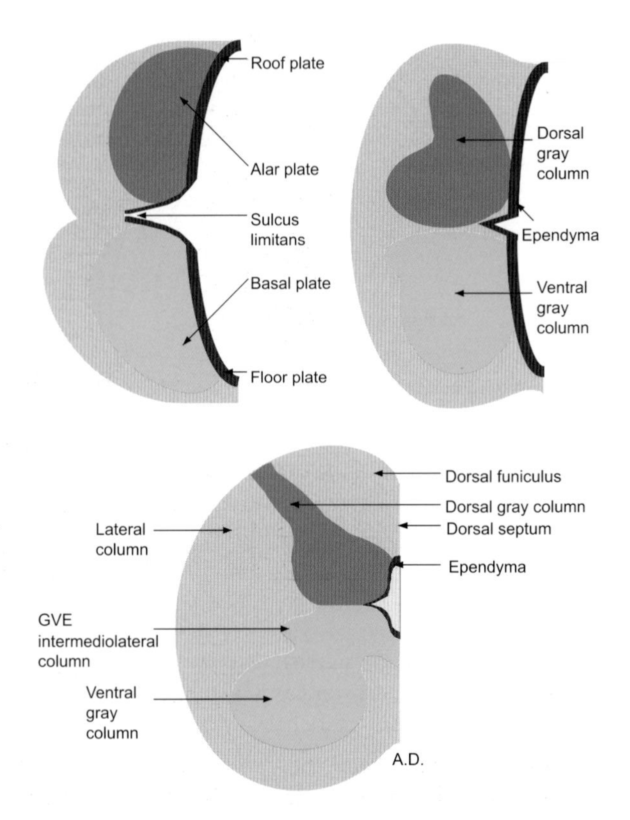

Figure 1.12 Development of the spinal cord. Note the formation of the somatic and autonomic neuronal columns

grow caudally, the dorsal and ventral roots, anchored within their appropriate foramina, pursue a longer course within the vertebral canal. This accounts for the formation of the cauda equina and the final position of the dorsal root ganglia and spinal nerves. By birth, the spinal cord terminates at the level of the third lumbar vertebra. Much of the neural canal obliterates, and the remaining part forms the central canal. The alar plate forms the dorsal horn, which represents the sensory area, while the basal plate, representing the motor area, becomes the future ventral horn of the adult spinal cord. Quite early some of the neuroblasts of the basal plates begin to differentiate into the α motor neurons that extend to form the ventral roots of the spinal nerves. As the basal plates enlarge, they protrude ventrolaterally on each side of the floor plate. They do not fuse in the midline, thus producing the midline ventral median fissure. The roof plate obliterates, and the neural canal reduces in size. The lower end of the spinal cord and the intermediate part of the neural (primitive central) canal, between the level of the second coccygeal and third lumbar vertebra, undergoes necrobiosis (selective death), and degenerate and becomes adherent to the covering pia as the filum terminale. The site of degeneration may, occasionally, give rise to congenital cysts.

Transformation of the dorsal portion of the neural canal, the ependymal cells and their long basal processes give rise to the dorsomedian septum. Obliteration of the floor plate and extension of axons of some neuroblasts across the midline result in the formation of the ventral white commissure. Neuroblasts of the neural tube that develop toward the marginal layer of the cord do so primarily by following a ventrolateral direction. The fasciculus proprius develops from the marginal layer into all three funiculi of the spinal white matter. While the transformation of the neural tube into the spinal cord is progressing, a large mass of axons is added to the cord, which are of neural crest origin.

The cells of the dorsal root ganglia are derived from the neural crest and the neural tube. Their central processes grow into the spinal cord and contribute in great quantity to the white matter of the cord, forming the dorsal columns and the dorsolateral fasciculi. General somatic afferents (generated at or near body surface) and general visceral afferents (generated in or on mucus membranes of visceral structures) lie dorsal to the sulcus limitans and are represented by the dorsal gray and white columns. Conversely, general visceral efferents (autonomic–parasympathetic) and general somatic efferent cell columns (innervate skeletal muscles) lie ventral to the sulcus limitans (Figures 1.11 & 1.12).

It is worth noting that the fibers of the corticospinal tract begin to develop in the ninth week of fetal life and by the twenty-ninth week they achieve their final limit. Those fibers that are destined to the cervical and upper first thoracic segment are in

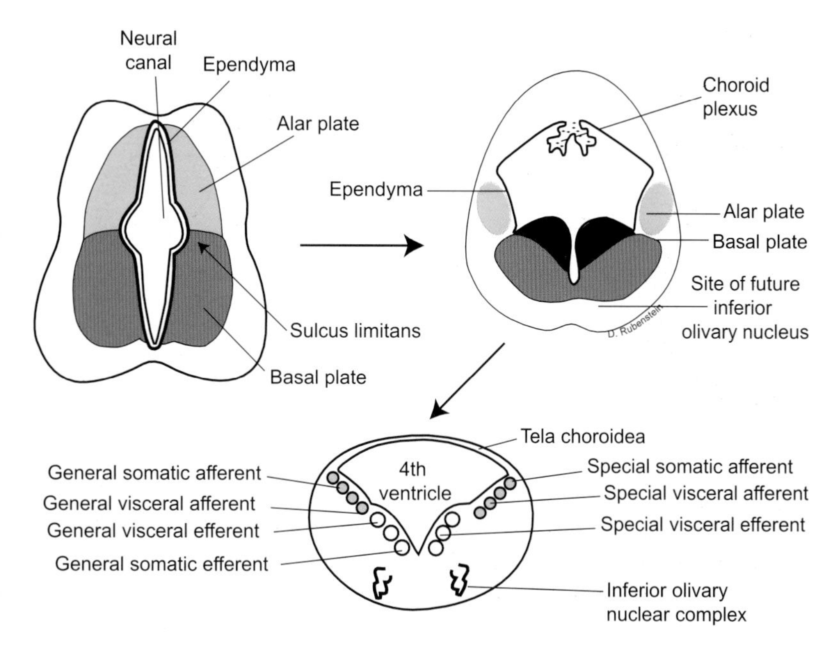

Figure 1.13 Cross-section of the developing medulla. The alar and basal plates are arranged in a mediolateral direction, signifying the locations of the motor and sensory neurons

advance of the fibers that reach the lumbosacral segments, which, in turn, are in advance of the fibers that project to the face.

Myelencephalon

The embryonic myelencephalon forms the caudal part of the rhombencephalon, consisting of alar and basal plates (Figure 1.13). Due to widening of the roof plate, the alar and basal plates eventually assume a more lateral/medial, rather than dorsal/ventral, position. Both the alar and basal plates contribute to the reticular formation, whereas the medullary pyramids remain of telencephalic origin.

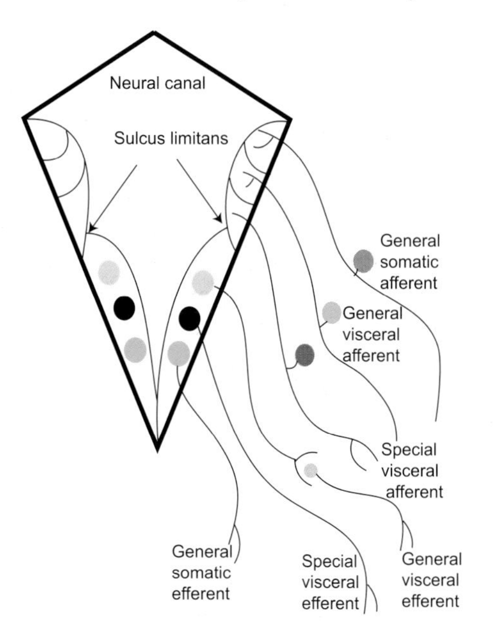

Figure 1.14 Schematic diagram of the somatic and visceral neurons associated with the developing medulla

The alar plate constitutes the sensory part that contributes to the formation of a variety of nuclei (Figure 1.14) which are listed below:

• Vestibular and auditory nuclei contain neurons that subserve special somatic afferents and transmit senses of balance and hearing.

• Spinal trigeminal, gracilis and cuneatus nuclei receive general somatic afferents generated at or near the body surface.

Solitary nucleus contains special and general visceral neurons that subserve taste and visceral sensations, respectively.

• Inferior olivary nuclear complex, a cerebellar relay nucleus.

Similarly, the basal plate (Figure 1.14) gives rise to a group of neurons as shown below:

• Hypoglossal nucleus provides general somatic efferents to the lingual muscles.

• Nucleus ambiguus supplies special visceral efferent fibers to certain muscles of branchial arch origin such as laryngeal, pharyngeal, and palatal muscles.

• The dorsal motor nucleus of vagus and the inferior salivatory nucleus provide general visceral efferents (GVE), which carry parasympathetic fibers.

• The roof plate persists to form the ependyma of the tela choroidea, inferior medullary velum, and the caudal part of the roof of the fourth ventricle. Axons

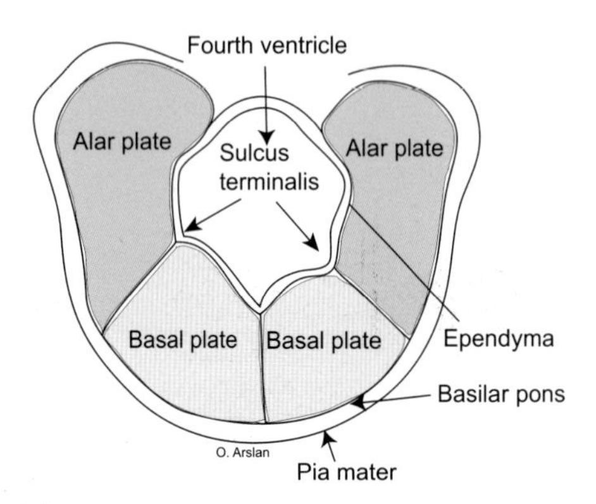

Figure 1.15 Developing pons and associated fourth ventricle

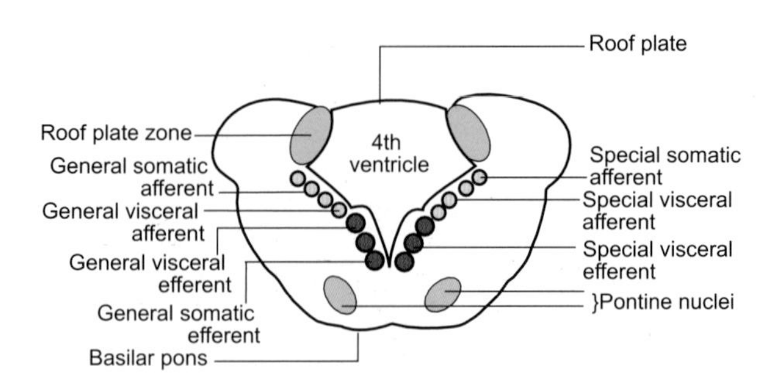

Figure 1.16 A more elaborate diagram of the structures in Figure 1.15. Note the afferent and efferent neurons associated with the basal and alar plates

of the marginal layer are derived from neuronal extensions of the medial lemniscus and the spinothalamic tracts. Attachment of the choroid plexus to the roof of the fourth ventricle is secured by the tela choroidea, which is formed by the ependymal layer of the myelencephalon covered by pia mater; they form the tela choroidea. During the fourth or fifth months of development, the paired foramina Luschka, at lateral recesses of the fourth ventricle, and the single median foramen of Magendie make their appearance.

Metencephalon

The metencephalon differentiates into the pons and cerebellum.

Pons

The basal plate gives rise to primarily efferent nuclei including the abducens, facial, trigeminal, and the superior salivatory nuclei (Figure 1.16).

- Abducens nucleus provides general somatic efferents supplying the lateral rectus.
- Facial and trigeminal nuclei give rise to the special visceral efferents that innervate the facial and masticatory muscles, respectively. These skeletal muscles are of branchial origin.
- Superior salivatory nucleus provides general visceral efferents (parasympathetic fibers) to regulate the glandular secretion of the lacrimal, sublingual and submandibular glands.

Derivatives of the alar plate include somatic and visceral sensory nuclei:

- Vestibular and auditory nuclei transmit impulses of balance and hearing (special somatic afferents).
- Principal sensory nucleus conveys cutaneous impulses from the head (general somatic afferents).
- Solitary nucleus receives both general visceral and special visceral afferents.
- Pontine nuclei are cerebellar relay nuclei.

Cerebellum

The cerebellum is derived from the rhombic lip of the alar plate (Figure 1.17). During the fourth month of development, the posterolateral fissure is the first to appear. The cerebellar cortex develops from the migrating neuroblasts of the external granular layer, which is formed by the germinal cells of the rhombic lip that migrate over the surface of the cortical lip. Both the alar and basal plates contribute to the reticular formation. At about the fifth week of embryonic development, the lateral parts of the alar plates on both sides of the roof of the metencephalon join to form the rhombic lips, which eventually become the cerebellar vermis and hemispheres. The remaining part of the alar plate forms the superior and inferior medullary vela. Some neuroblasts of the mantle layer migrate outward into the marginal layer (towards the surface) to mature and become cerebellar cortical neurons. This migration is guided by the processes of Bergmann glial cells. The group of undifferentiated neuroepithelial cells that move around the rhombic lip region to form the external granular, a superficial layer beneath the pia mater, eventually differentiate into neuroblasts that move inward and mature into the adult granular layer, stellate and basket cells. The young neurons of the superficial part of the mantle layer form the Purkinje and Golgi type II cells. The periventricular neuroblasts that remain at the site of the original mantle layer become the cells of the cerebellar (fastigial, globose and emboliform, and dentate) nuclei.

Mesencephalon

The mesencephalon, morphologically the most primitive of the brain vesicles, contains both basal and alar plates (Figure 1.18). The basal plate of the mesencephalon differentiates into the trochlear and oculomotor nuclei, providing general somatic efferents (GSE) to most extraocular muscles. It also supplies

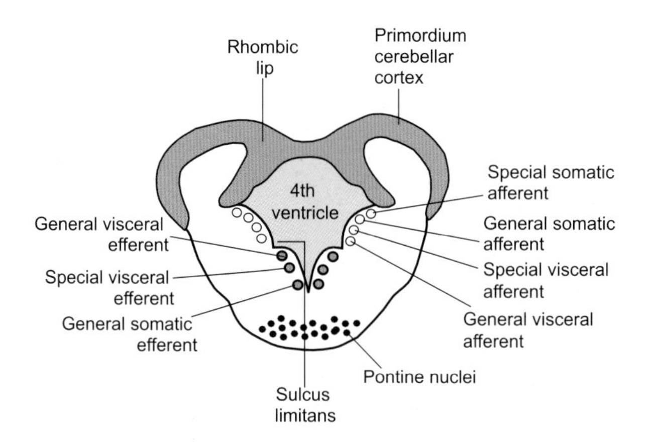

Figure 1.17 Transverse section of the metencephalon. Position of the rhombic lips is also illustrated

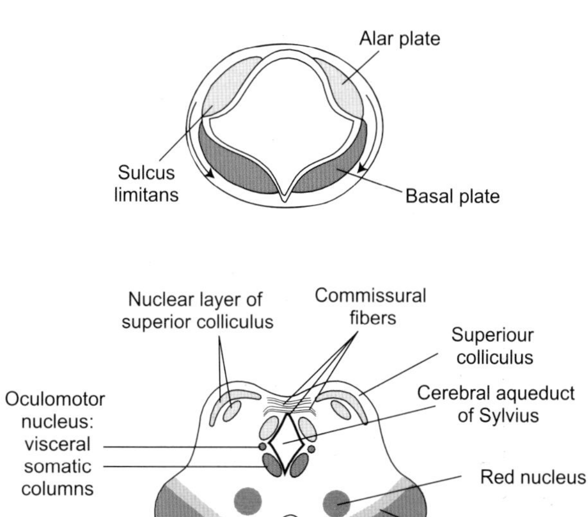

Figure 1.18 The developmental stages of the mesencephalon and derivatives of the alar, basal and roof plates

GVE to the constrictor muscle of the pupil and the ciliary muscle via the Edinger–Westphal nucleus.

Differentiation of the alar plate results in the formation of the superior and inferior colliculi, whereas the corticofugal fibers form the crus cerebri. The substantia nigra, red nucleus, and the reticular formation are probably of mixed origin from neuroblasts of both basal and alar plates. Neural crest cells of the midbrain gives origin to the mesencephalic nucleus. Development of the isthmus (mesencephalo-metencephalic junction) is controlled by fibroblast growth factor 8.

Diencephalon

The diencephalon consists of roof and alar plates but lacks the basal and floor plates. Differentiation of the alar plate here results in the formation of the thalamus, hypothalamus, neurohypophysis, and the infundibulum. A diverticulum of the stomodeum is derived from the Rathke's pouch. Derivatives of the roof plate include the epiphysis cerebri, habenular nuclei and the posterior commissure. The ependyma and vascular mesenchyme of the roof plate give origin to the choroid plexus of the third ventricle.

Telencephalon (Figures 1.9 & 1.19)

The two lateral outpocketings (diverticuli), which arise from the cephalic end of the prosencephalon, form the telencephalon and its main constituents the cerebral hemispheres. These lateral diverticula evaginate from the most rostral end of the neural tube near the primitive interventricular foramen of Monro, and are connected via the midline region known as the telencephalon impar. These diverticula rostrally continue around the foramen of Monro, but caudally remain merged with the lateral walls of the diencephalon. The enormous positive pressure exerted by the accumulated fluid within the neural canal results in the rapid expansion of the brain volume in the early embryo (3–5 days of development). This is aided by the constriction of the neural tube at the base of the brain via the surrounding tissues. The foramen of Monro forms the rostral part of the third ventricle. At the end of the third month the superolateral surface of the cerebral hemisphere shows a slight depression anterior and superior to the temporal lobe. This occurs due to the more modest expansion of this site relative to the adjoining cortical surface. This depression, the lateral cerebral fossa, gradually overlapped by the expanding cortical area converts into the lateral cerebral sulcus (fissure). The floor of this sulcus becomes the insular cortex. Apart from the lateral cerebral and hippocampal sulci the cerebral hemispheres remain smooth until early in the fourth month when the parieto-occipital and calcarine sulci appear. During later stages of development (5th month of prenatal life) the cingulate sulcus followed by (sixth month of prenatal life) the sulci appear on the superolateral and inferior surfaces of the brain. Virtually all sulci become recognizable by the end of the eighth month of development. The ventricular and subventricular parts of the telencephalic lateral diverticula form the ependyma, the cortical neurons,

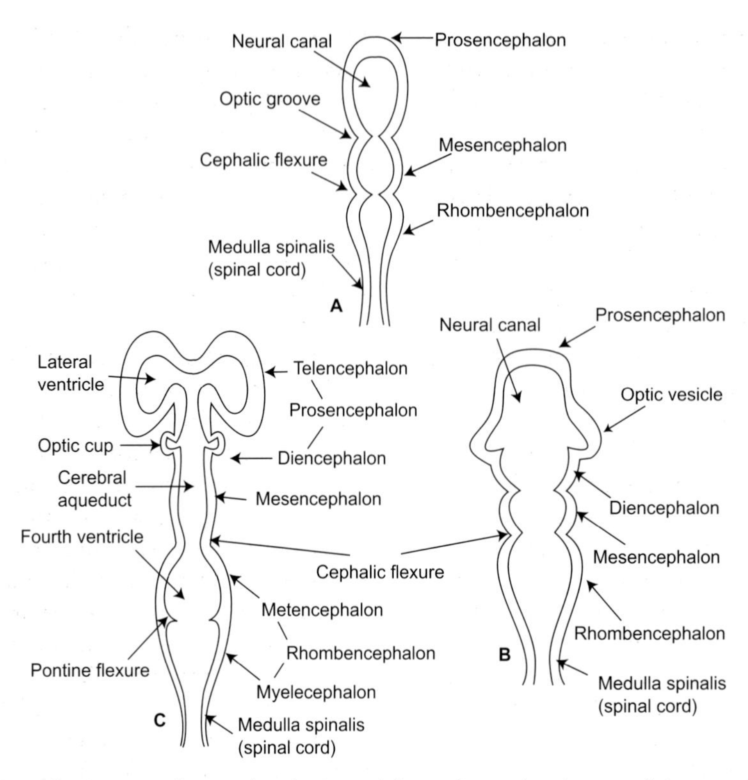

Figure 1.19 Formation of primary and secondary brain vesicles and associated parts of the ventricular system. Letters A, B, and C mark stages of differentiation

Table 1.2 Derivatives of the primary brain vesicles

Primary brain vesicles	*Secondary brain vesicles*	*Neural tube (wall)*	*Neural canal (ventricular system)*
Prosencephalon	Telencephalon	Cerebral hemispheres Rostral part of hypothalamus Rostral part of 3rd ventricle Archicortex, paleocortex, and neorcortex	Lateral ventricles
	Diencephalon	Thalamus, metathalamus Subthalamus, epithalamus & part of hypothalamus	Caudal part of the third ventricle
Mesencephalon	Mesencephalon (midbrain)	Colliculi, tegmentum, cerebral peduncles, cerebral aqueduct	Cerebral aqueduct
Rhombencephalon	Isthmus rhombencephali	Superior medullary velum & superior cerebellar peduncles	Rostral part of fourth ventricle
	Metencephalon	Cerebellum, pons & middle cerebellar peduncles	Middle part of the fourth ventricle
	Myelencephalon	Medulla oblongata	Rostral part of the central canal, caudal part of the fourth ventricle & inferior cerebellar peduncle
Medulla spinalis	Spinal cord	Spinal cord	Central canal

and the glial cells. The intermediate cell layer of the telencephalic diverticula differentiates into the white matter, while the cortical zone differentiates into the various layers of the isocortex.

At the beginning, the wall of the cerebral hemisphere consists of three basic layers that include the inner neuroepithelial, mantle, and marginal layers. Each neuroepithelial cell has a single nucleus and double cytoplasmic extensions. The deep extension reaches the internal limiting laminae, whereas the superficial extension stretches to the external limiting membrane which itself is covered by the pia mater. Attachments of the superficial and deep cytoplasmic extensions are maintained via end-feet that contribute also to the above mentioned membranes or laminae. As the nuclei undergo division, the cytoplasmic processes remain solid. One of the nuclei remains near the ventricular surface while the other migrates within the cytoplasmic extensions to the pial surface. As it reaches the pia mater the cytoplasmic process separates from the original cell and begins to surround the newly formed nucleus. Neuroblasts that maintain position near the pia mater are unipolar with one neuronal extension, which eventually divides into finer processes or dendrites. As the thickness of the cortex increases subsequent to increases in the number of neuroblasts, the unipolar neuroblasts become deeply located. At the same time these neuroblasts begin to form axons that stretch to the ventricular surface and dendrites that extend to the subpial layer. Glioblasts, which differentiate into the astrocytes and oligodendrocytes, are derived from the neuroepithelial cells that line the neural canal when the production of the neuroblasts ceases.

Most cortical neurons follow an 'inside-out' pattern of migration from the ventricular and subventricular zones through the intermediate zones to the cortical plate, allowing the neurons that form at a later stage of development to migrate and maintain an outward position to the neurons that develop earlier. Thus, the recently formed neurons occupy the basal layers of the cortex while the older neurons maintain locations in the superficial layers. In the initial stage of migration the neuroblasts are allowed to proceed to a site between the marginal layer and the white matter. The nuclei of the neuroepithelial cells lie near the ventricle while the cytoplasm elongates to form deep and superficial processes. Some neuroblasts traverse the initial group of migratory neuroblasts to assume a position in the middle third of the mature cortex, whereas others may pursue different courses among the previous group of neuroblasts to reach more superficial positions. This pattern of migration is in line with the radial columnar organization of the cerebral cortex.

The invagination formed by the attachment of the cerebral diverticula to the roof of the diencephalon leads to the formation of the choroid fissure. The latter fissure allows a narrow strip of thin ependymal roof plate, with the accompanying pial covering (tela choroidea) to invaginate into the lateral ventricle. As the temporal lobe develops, the choroidal fissure, with the invaginating tela choroidea and the choroid plexus, continues to increase in length along the medial wall of the developing temporal lobe. Hence, in the adult, the choroid plexus is a continuous structure found in the third ventricle, interventricular foramen, and in the body, trigone and inferior horns of the lateral ventricles. In fetal life, the choroid plexus occupies most of the lateral ventricle, and then gradually decreases in size.

Formation of the anterior commissure (the first commissure to appear during development) is followed by the development of the hippocampal commissure and the corpus callosum (around the fourth month). These commissures are formed by axons that extend from one hemisphere to the other using the embryonic lamina terminalis as a bridge. Initially, the hippocampus appears as a ridge derived from the medial cortical wall.

The basal nuclei are derived from the mantle layer of the telencephalon. Specifically, the caudate and putamen derive from the ganglionic eminences in the ventrolateral part of the telencephalic ventricle, while the pallidum originates from lateral and medial hypothalamic analogs. The corpus striatum assumes a striated appearance following the crossing of fibers that connect the telencephalic vesicles to the diencephalon and brainstem.

Suggested reading

1. Atlas SW, Zimmerman RA, Bilaniuk LT, *et al.* Corpus callosum and limbic system: neuroanatomic MR evaluation of developmental anomalies. *Radiology* 1986;160: 355–62
2. Barsi P, Kenéz J, Solymosi D, *et al.* Hippocampal malrotation with normal corpus callosum: a new entity. *Neuroradiology* 2000;42:339–45
3. Emery JL, Lendon RG. Lipomas of the cauda equina and other fatty tumors related to neurospinal dysraphism. *Dev Med Child Neurol* 1969;11:62–70

4. Epstein NE, Rosenthal RD, Zito J, *et al*. Shunt placement and myelomeningocele repair. Simultaneous vs. sequential shunting. *Childs Nerv Syst* 1985;1:145–7
5. Galliot B, Dollé P, Vigneron M, *et al*. The mouse Hox 1.4 gene: primary structure, evidence for promoter activity and expression during development. *Development* 1989;107:343–59
6. Graham A, Heyman I, Lumsden A. Even-numbered rhombomeres control the apoptotic elimination of neural crest from odd-numbered rhombomeres in the chick hindbrain. *Development* 1993;119:233–45
7. Hensinger RN. Lang JR, MacEwen GD. Klippel-Feil syndrome: A constellation of associated anomalies. *J Bone Joint Surg* 1974;56A:124
8. Hirsch JF, Pierre-Kahn A, Renier D, *et al.* The Dandy-Walker malformation: A review of 40 cases. *J Neurosurg* 1984;61:515–22
9. Hoffman HJ. Comment on Pang D, Dias MS, Ahab-Barmada M. Spinal-cord malformation: part 1: A unified theory of embryogenesis for double spinal-cord malformations. *Neurosurgery* 1992;31:451–80
10. Kuller IA, Globus MS. Fetal therapy. In Brock DJH, Rodeck CH, Ferguson-Smith MA, eds. *Prenatal Diagnosis and Screening.* Edinburgh: Churchill Livingstone, 1992;703–17
11. Lemire RJ. Neural tube defects. *JAMA* 1988;259: 558–62
12. Lorber J, Ward AM. Spina bifida – A vanishing nightmare? *Arch Dis Child* 1985;60:1086–91
13. Moore KL, Persaud TVN. *The Developing Human: Clinically Oriented Embryology*, 6th edn. Philadelphia: W.B. Saunders Co., 1998
14. Norman MG. Malformations of the brain. *J Neuropathol Exp Neurol* 1996;55:133–43
15. Park TS, Hoffman HJ, Hendrick EB, *et al.* Experience with surgical decompression of the Arnold-Chiari malformation in young infants with myelomeningocele. *Neurosurgery* 1983;13:147–52
16. Goldstein S, Reynolds CR. *Handbook of Neurodevelopmental and Genetic Disorders in Children.* New York: Guilford Press, 1999
17. Schoenwolf GC, Smith IL. Mechanisms of neurulation: traditional viewpoint and recent advances. *Development 1990;*109:243–70
18. Serbedzija GN, Fraser SE, Bronner-Fraser M. Pathways of trunk neural crest cell migration in the mouse embryo as revealed by vital dye labelling. *Development* 1990;108:605–12
19. Tassabehji M, Reed AP, Newton VE, *et al.* Waardenburg's syndrome patients have mutations in the human homologue of the Pax-3 paired box gene. *Nature* 1992;355:635–6
20. Wong TT, Lee LS. Membranous occlusion of the foramen of Monro following ventriculoperitoneal shunt insertion: a role for endoscopic foraminoplasty. *Childs Nerv Syst* 2000;16:213–17

Basic elements of the nervous system 2

The nervous system receives, integrates and transmits sensory stimuli. It generates motor activity and coordinates movements. In addition, the nervous system also regulates our emotion and consciousness. In summary, it controls all the activities which preserve the individual and species. These functions are accomplished at cellular levels in the neurons and are enhanced via supportive glial cells. Neurons maintain certain common structural and morphological characteristics that enhance their activities and correlate closely with their functions. Following an injury, these structures may undergo changes or assume different positions in the neurons. Myelin, the covering of certain axons, plays an important role in nerve conduction and exhibits degenerative changes in particular diseases. Glial cells form the skeleton of the nervous system, display several forms and participate in a variety of supportive functions that collectively maintain the optimal environment for neuronal activity.

Neuroglia

Neuroglia are non-excitable supporting cells that outnumber neurons at a ratio of 2:1, forming the skeleton of the central nervous system (CNS). They are of both ectodermal and mesodermal origin and are commonly associated with tumors of the CNS. They have only one type of cell process, do not form synapses, and retain the ability to undergo mitosis. Glial cells provide the optimal milieu for neuronal function by balancing ionic concentration within the extracellular space. They provide nutrients, discard metabolites and cellular debris, and sharpen neuronal signals (preventing cross-talk) or ephapsis by forming the protective myelin sheath. The glial cells mediate the extent of impulse flow, activity of neurons, and frequency of excitation. Thus changes in the glial cell-membrane potentials may occur as a result of the fluctuation in the potassium ion concentration which in turn is affected by level of the generated impulses. They also secrete growth-promoting molecules such as the nerve growth factor, glial-derived neurite-promoting factor (GNPF) which is a protease inhibitor (trypsin, urokinase, and thrombin), and tenascin. Neuronal sprouting may be facilitated by GDN (glial-derived nexin) which prevents the digestion of the extracellular matrix molecules by inhibiting proteases secreted by growth cones. Neuroglia also contribute to the formation of the blood–brain barrier, which selectively permits substances and molecules to enter the CNS. They also allow developing neuroblasts to move to their final destinations. Presence of molecules such as fibronectin, laminin, and cellular adhesion molecules may account for the mechanism by which neurons migrate along processes of certain glial cells and not others. Neuroglial cells are classified as macroglial and microglial cells. Macroglial cells are further subclassified into astrocytes, oligodendrocytes of the CNS, and Schwann cells of the peripheral nervous system.

Macroglia

Macroglia are classified into astrocytes, oligodendrocytes, and ependymal cells.

Astrocytes (star cells) are the largest, the most numerous, and show the most branching among all the glial cells (Figures 2.1, 2.2 & 2.5). They are present in both gray and white matter and possess processes which branch repeatedly in an irregular fashion and assume star-like configurations. Astrocytes establish contacts with the non-synaptic parts of the neurons and form perivascular end feet that extend to the blood capillaries. They retain the capacity to multiply. Astrocytes retrieve glutamate and γ-aminobutyric acid after their release from the nerve endings. They invest most of the synaptic neurons and assume phagocytic function. They also maintain the normal concentration of potassium, which is essential for neuronal activity, by removing and then facilitating its return to the blood. Astrocytes are also considered as the principal glycogen storage site in the CNS. Glycogen breakdown and release of glucose is accomplished by the action of norepinephrine upon β receptor molecules in the astrocytes. During embryonic development precursors of astrocytes (radial glial cells) guide the migration of the developing neurons. The superficial processes of astrocytes extend to the surfaces of the brain and spinal cord to form the external glial limiting membrane and expansions that attach to the pia mater (pia–glial layer). They play a major role in providing a form of scaffolding, or structural support, on which the neurons and their processes are assembled. Astrocytic processes ensheath the initial segments of axons and the bare segments at the nodes of Ranvier.

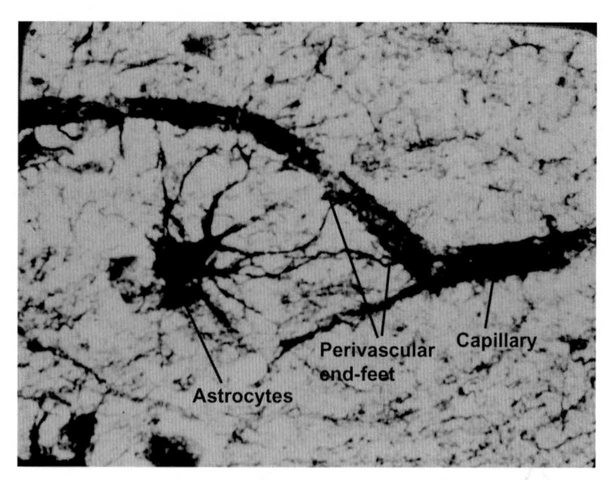

Figure 2.1 Photomicrograph of the astrocytes showing their branches and perivascular end-feet

Astrocytes are classified into protoplasmic and fibrous astrocytes. Protoplasmic astrocytes are located primarily in the gray matter and have shorter processes, while fibrous astrocytes have longer processes and are located primarily in the white matter. Fibrous astrocytes are the scar-forming cells that bridge the gap between severed ends of axons in

Astrocytoma is the most common form of brain tumor (glioma). This tumor may cause increase in intracranial pressure, headache, nausea, and vomiting. Supratentorial glioma may produce a shift of the pineal gland, third ventricle, and anterior cerebral arteries, whereas infratentorial tumors are most likely to produce hydrocephalus.

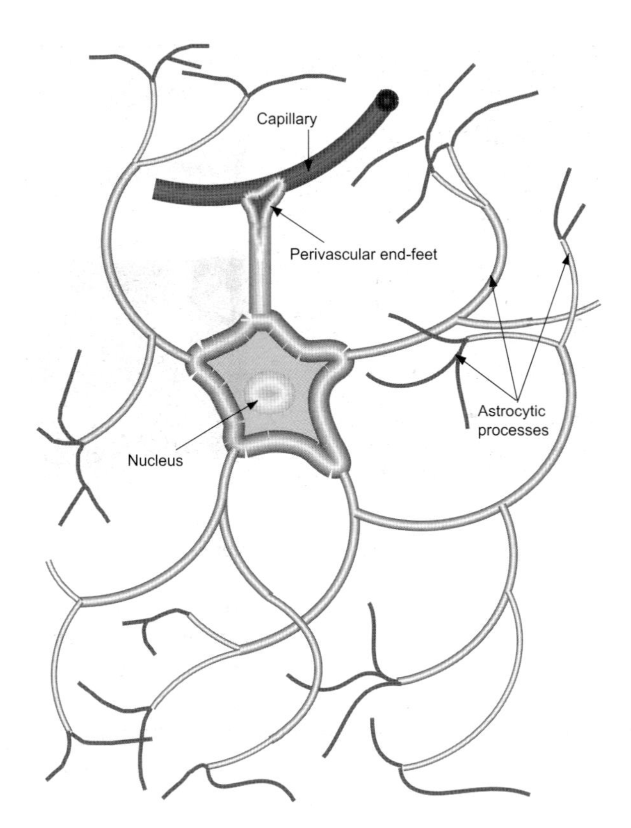

Figure 2.2 Cellular characteristics of the astrocytes and perivascular end-feet

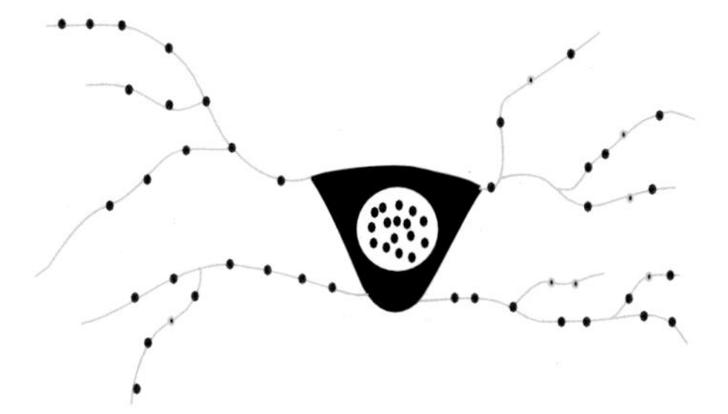

Figure 2.3 This is a simplified diagram of the interfascicular oligodendrocyte with its scanty branches

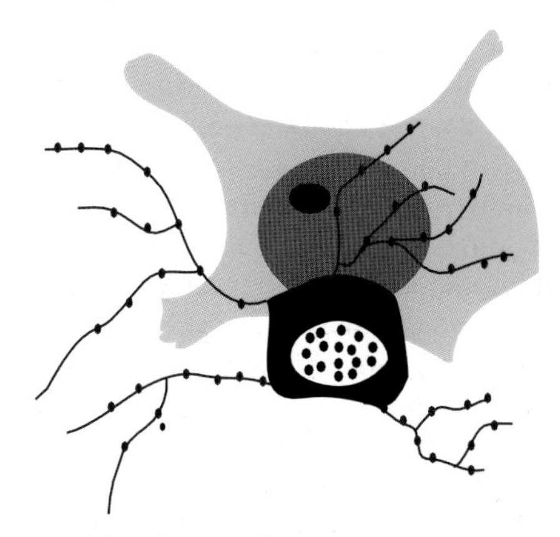

Figure 2.4 This drawing shows a perineuronal oligodendrocyte adjacent to a neuron

pathological conditions involving the CNS. Astrocytes also play a role in the formation of the blood–brain barrier.

Modified astrocytes are classified into Müller cells and Bergmann glial cells. Müller cells of the retina are elongated cell columns that exhibit expanded foot processes forming the inner limiting membrane of the retina. Bergmann glial cells of the cerebellum lie at the Purkinje layer, sending several processes with short side branches that ascend and envelop the Purkinje cell dendrites.

Oligodendrocytes, the myelin-forming glial cells in the white and gray matters, are characterized by relatively few branched processes bearing close resemblance to stellate neurons (Figures 2.3, 2.4 & 2.5). Oligodendrocytes that lie within the white matter, especially of fetal brain and in myelin sheath bundles, and are aligned in rows between nerve fibers, are known as interfascicular oligodendrocytes (Figure 2.3). They are numerous in the fetus and newborn, but rapidly decrease in number (absolutely or relatively) as myelination progresses. Oligodendrocytes that lie closely opposed to neuronal cell bodies in the gray matter are called perineuronal (satellite) oligodendrocytes (Figure 2.4). A few oligodendrocytes that occupy locations near blood capillaries are known as perivascular oligodendrocytes.

Schwann cells are the supporting cells in the peripheral nervous system which are derived from the neural crest cells, forming the capsular (satellite) cells of the dorsal root and autonomic ganglia.

The ependymal cells are arranged as a single layer of epithelial-like cells with variable heights in different regions that form the lining of the ventricular system and central canal of the spinal cord. These cells are the remnants of the embryonic neuroepithelium and maintain their original position after the neuroblasts and glioblasts have migrated into the mantle layer. They vary from columnar to cuboidal to squamous, depending upon their location. The variability of their shape often makes their identification difficult. They have processes which penetrate the brain and extend into the pia mater. Embryonic ependymal cells are ciliated, and some adult cells may retain cilia permanently, producing movements of the cerebrospinal fluid. Modified ependymal cells cover the choroid plexus and play a significant role in the secretion of cerebrospinal fluid. Ependymal cells also have numerous microvilli, exhibiting high oxidative activity. Both cellular structure and chemical reactions reflect the secretory and absorptive functions of these cells. The apices of these ependymal cells are joined by junctional complexes, which are not occluding junctions. Since substances can readily pass between these cells, the brain/CSF interface is not a barrier. The deep basal surface of some adult ependymal cells retains processes that extend for a variable distance from the cell body. Many of the shorter processes are intertwined with a heavy concentration of astrocytic processes forming a subependymal (internal limiting) glial membrane. Specialized ependymal cells may be attenuated and form the lining of the circumventricular organs (medial eminence, subfornical organ, subcommissural organ, and the organ vasculosum of the lamina terminalis).

Ependymal cells with long basal processes that project into the perivascular space that surrounds the underlying capillaries are known as tanycytes. Since, these capillaries are fenestrated, they do not form a blood–brain barrier, allowing substances to pass from the blood and nervous tissue to the CSF via these specialized ependymal cells (tanycytes). These cells are found around the floor of the third ventricle, and in the lining of the median eminence of the hypothalamus, which suggests a possible role of these cells in the secretions of the adenohypophysis.

Microglia

The microglia (Figure 2.6) are small cells whose primary function is phagocytosis of cellular debris associated with pathological processes in the CNS. They probably possess ion-channeled linked P_2 purinoceptors, which may be activated by 5′-adenosine triphosphate in response to injury. Microglia are considered to be a derivative of the angioblastic mesenchyme. Although the consensus dictates that only mature monocytes enter the postnatal brain, the view that monocytes may differentiate into microglial cells has received a large body of support. Microglial cell migration to the nervous tissue occurs through the walls of the parenchymal and meningeal vessels, accounting for the dramatic increase in their numbers at sites of CNS infections. They undergo transformations in and around the site of infection, which include shortening and thickening of cell processes, increase in the volume of the cell body and later retraction of the processes. These changes are followed by complete disappearance of the cellular processes and the conversion of the cell into spherical corpuscular form known as compound granular corpuscles or gutter cells. Promotion of immune response by the microglia and activated T lymphocytes, which enter the brain by crossing the blood–brain barrier, is evidenced in experimentally induced encephalitis. The small microglial cells are more abundant in the gray matter of the CNS and are also found in the retina.

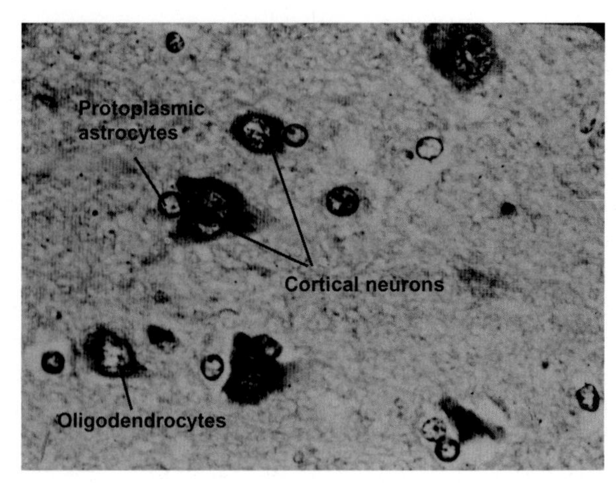

Figure 2.5 Photograph of the various types of astrocytes and oligodendrocytes

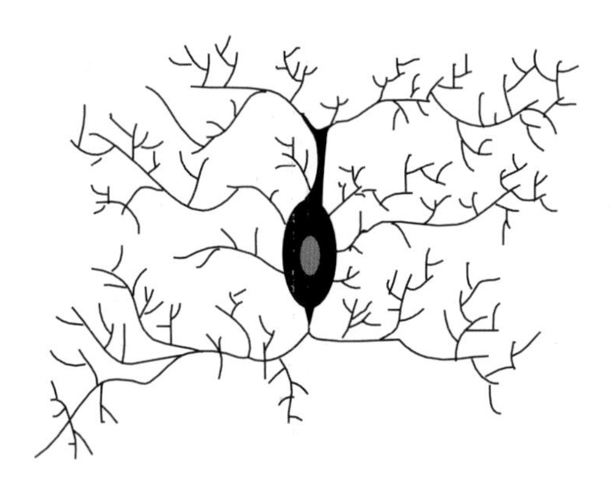

Figure 2.6 Schematic drawing of a microglial cell

Neurons

Neurons form the trophic, genetic, and excitable components of the nervous system, which receive, conduct and transmit nerve impulses. Neurons can be excitatory, inhibitory, sensory, motor or secretory in function. Constant reduction in the number of neurons after birth and the inability of mature neurons to divide represent some of the main characteristics of these cellular entities. A collection of neurons that serve the same overall function and generally share the same efferent and afferent connections is known as a nucleus within the CNS (e.g. trochlear nucleus), and as a ganglion in the peripheral nervous system (e.g. dorsal root ganglion). Although many unique features exist that distinguish neurons from each other, there are common characteristics shared by neurons (Figure 2.7).

The soma (perikaryon) represents the expanded receptive zone of the neuron, consisting of a protoplasmic mass, which surrounds the nucleus. It represents the center for protein synthesis. It harbors the organelles needed for the metabolic functions of the neuron. One (rarely more than one) prominent nucleolus is positioned in the centrally located nucleus. The cytoplasm of the soma contains chromatin bodies (Nissl bodies) which are intracytoplasmic basophilic and RNA-rich masses that are more distinct in the α motor neuron of the spinal cord and in the large neurons of the dorsal root ganglia. Nissl bodies, which extend along dendrites but not axons, are involved in high cellular activity and protein synthesis. These granules begin to disperse or undergo chromatolysis in response to nerve injury or in degenerative conditions. The anatomic location of the cell body has no functional significance. The plasma membrane of the soma, although generally smooth, may possess spinous postsynaptic elevations known as gemules. The soma may engage in axo-somatic, dendro-somatic, and soma-somatic synapses. Within the cytoplasm, neurotubules, neurofilaments, mitochondria, ribosomes, as well as aging pigment which consists of the lipofuscin granules exist. Neurotubules are randomly arranged in the perikaryon but assume longitudinal configuration in the axons and dendrites. At the surface of neuronal soma various enzymes exist, such as adenosine triphosphatase (ATPase), which is activated by sodium and potassium.

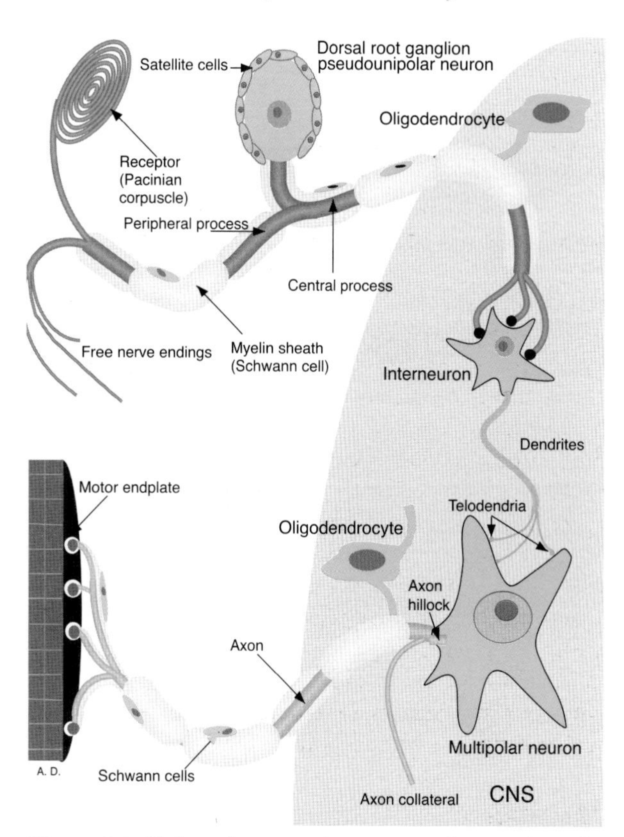

Figure 2.7 Various features of neurons and neuroglia cells. Myelin-forming cells within the CNS and PNS, as well as synaptic connections between these neurons are also shown

Neuronal processes

The dendrites are highly branched processes that originate from the soma and represent the afferent or receptive zone of neurons. They show similar pattern of branching in neurons with similar functions. Dendrites have spines that maximize contact with other neurons, mediating excitatory and inhibitory axo-dendritic as well as dendro-dendritic synapses. They contain microfilaments and microtubules, smooth endoplasmic reticulum, ribosomes and Golgi membranes. More peripheral dendrites, free ribosomes and rough endoplasmic reticulum become progressively sparse and may be entirely lacking. Microtubules and microfilaments are much more conspicuous than in the soma and more regularly aligned along the axis of the dendrite, forming the most striking feature of the dendrites. The microtubules are believed to be involved in the dendritic transport of proteins and mitochondria from the perikaryon to the distal portions of the dendrites. The dendritic transport at a rate of 3mm/h is comparable to some forms of axoplasmic transport. Destruction of the microtubules by drugs, such as colchicine and vinblastine, inhibits this transport. Dendritic transport may also involve viral glycoprotein that is basolaterally targeted. Dendrites contain exclusively the microtubule-associated protein (MAP-2) but do not contain growth-associated protein (GAP-43). For this very reason MAP-2 antibodies are utilized to identify dendrites via immunocytochemical methods.

The axon forms the efferent portion of the neuron, which provides nutrients via the axoplasmic transport. In general, axons are thinner than dendrites, assuming considerable length. Compared to dendrites, axons are more uniform, contain fewer microtubules and more microfilaments, but no ribosomes. Axons are longer than dendrites and may measure up to 6 feet in length, beginning from the axon hillock and giving rise to collaterals that terminate as the telodendria. They provide an avenue for transport of substances to and from the soma. Axons originate from the soma or, less frequently, from the proximal part of dendrites. The axon is divisible into axon hillock, initial segment, axon proper and the telodendria (axonal terminal). The axon hillock is a clearly recognizable elevation that continues with the soma. The relative absence of free ribosomes and rough endoplasmic reticulum is the most obvious feature of the axon hillock. In myelinated axons the initial segment is defined from the axon hillock to the beginning of the myelin sheath. This segment is

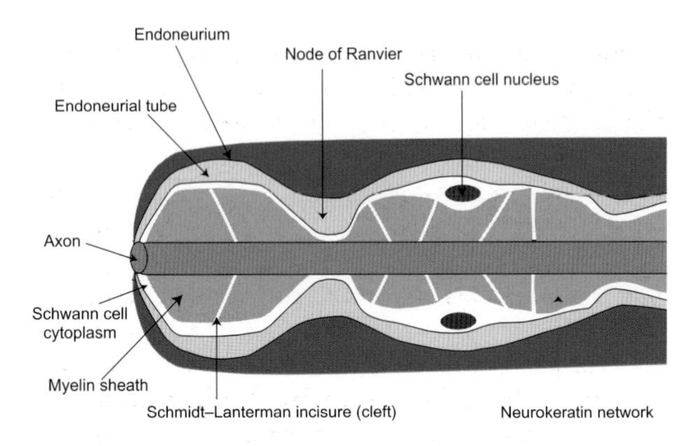

Figure 2.8 Myelin sheath of a peripheral nerve, nodes of Ranvier and associated coverings

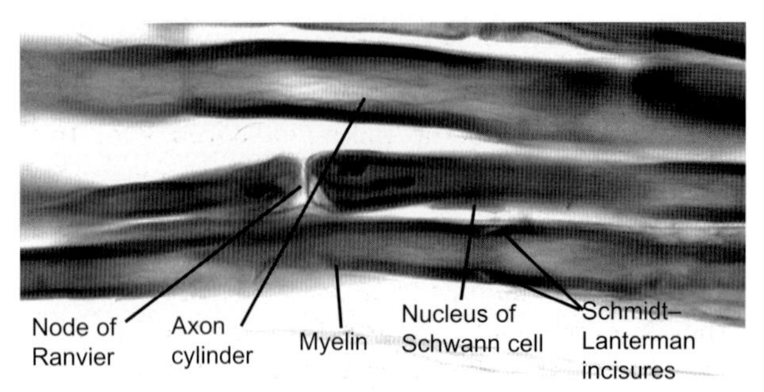

Figure 2.9 This is a longitudinal section of a single myelinated nerve fiber

Neurotropic viruses such as rabies and herpes simplex follow the retrograde direction.

unmyelinated, and maintains inhibitory axo-axonal synapses. It contains some microtubules, neurofilaments and mitochondria, but lacks free ribosomes and rough endoplasmic reticulum. These neurotubules and neurofilaments are gathered into small parallel bundles, connected by electron-dense cross-bridges. Here at the initial segment, the axolemma (plasma membrane bounding the axon) is under-coated (presence of dense-core beneath the plasma membrane) and spike generating. Additionally, spectrin and F-actin (cytoskeletal molecules) are also concentrated at this site, allowing voltage sensitive channels to attach to the plasmalemma. Axonal terminals are initially myelinated, but as they repeatedly branch, myelin sheath will disappear. This will enable these terminals to establish synaptic contacts with neurons in the CNS or with muscle fibers and glands in the PNS. The endings are characterized by tiny swellings known as terminal boutons. Microtubule-associated proteins (MAP) such as tau interconnect axonal microtubules.

Axoplasmic transport within the microtubules may be maintained utilizing protein dynein and kinesin. Neurofilaments are usually found in association with microtubules, as constant components of axons. In the growth cones of the developing axons, filamentous structures finer than neurofilaments exist and are known as microfilaments. These actin filamentous structures facilitate growth and movement and can be inhibited by chemical agents that depolymerize actin. High and low molecular weight proteins of the microfilaments are phosphorylated and antibodies produced for these molecules are used as markers for defining axons and neuronal phenotype. Antibodies raised against GAP-43, a protein contained in the growing and regenerating axons, may be useful for identifying regenerating axons following axonal injury.

Axoplasmic transport is a process that enables proteins, neurotransmitters, mitochondria, and other cellular structures synthesized in the soma or proximal portion of the dendrites to be transported to the axon and axon terminals. This transport may occur in a distal (anterograde) direction toward the axon terminals, while allowing other substances to be transported in the reverse (retrograde) direction from the axon toward the cell body. This process may involve fast, intermediate, and slow phases. The fast phase within the axoplasm may include transport of selected proteins (e.g. molecules carried by the hypothalamo-hypophyseal tract), vesicles, membrane lipids, or enzymes which act on transmitters. This phase of the transport occurs at a speed of 100–400 mm/day, both in anterograde and retrograde directions utilizing smooth endoplasmic reticulum and microtubules. The retrograde component of this phase is formed by the degraded structures within the lysosomes.

The fast phase is energy-dependent and can be inhibited by hypoxia and inhibitors of oxidative phosphorylation, glycolysis, and the citric acid cycle. It has been suggested that proteins that follow the fast axonal transport must either pass through the Golgi complex, or join proteins that do so utilizing clathrin-coated vesicular protein. The intermediate phase transmits mitochondrial proteins at a rate ranging between 15–50 mm/day. The slow phase of the transport utilizes microtubules, microfilaments, and neurofilament proteins, mitochondria, lysosomes, and vesicles, proceeding in the anterograde direction only, at a speed of 0.1–3 mm/day. This phase carries 80% of the substances carried by axoplasmic transport, providing nutrients to the regenerating and mature neurons. The slowest phase deals with the transportation of triplet proteins of tubulin and neurofilaments.

Multiple sclerosis is the most common demyelinating disease of the CNS that leaves the axons relatively preserved. Despite the unknown etiology of this disease, epidemiological, genetic, immunological, and viral factors have been implicated. Decreased suppressor T lymphocytes and increased frequency of a particular kind of HLA are considered some of the immunological abnormalities associated with this disease. Some investigators claim that contraction of measles at an older age and the presence of high measles antibody titer in the CSF may account for the development of this dreadful disease. Europeans and inhabitants of higher altitudes who have lived at least through the age of 15 in cool northern latitudes (above the 37th parallel) may show greater incidence of this disease. It is very rare in Latin America, Japan, and central Africa. Demyelination affects the structures within the CNS (Figure 2.10) with a predilection for the optic nerves, spinal cord, periventricular area, brainstem and cerebellum. This disease is more common in the female and shows higher incidence in first-degree relatives and monozygotic twins. The mean age of onset is 33 years with virtually all cases developing between 15 and 50 years. Although rarely advancing from the onset, this disease produces slowly progressive neurological disorders characterized by relapses and remissions.

The severity of multiple sclerosis increases with time, although improvement accompanied by remission is common. In general, 2–3 years pass before remission. Initial symptoms and signs of this disease are intensified by fever and emotional stress and following infection, trauma, or childbirth. Demyelination involves the optic nerves and the medial longitudinal fasciculus, producing visual deficits and disorders of ocular movements. It also involves the dorsal (posterior) columns and the motor pathways of the spinal cord, causing ataxia and paralysis. Epileptic seizures may occur in some patients. Subsequent to demyelination, scar formation and gliosis may lead to the development of plaques. Lesions of the dorsal columns and anterolateral systems are often symmetrical, causing paresthesia sequentially in the digits, limbs and adjacent parts of the trunk. Locus minoris resistantia, the tendency for relapses in an area of previous activity, may also be seen in this condition.

Selective destruction of the lateral spinothalamic tract may account for the loss of pain and temperature sensations in acute cases. Corticospinal involvement often causes weakness and spasticity along with other signs of upper motor neuron syndromes. Spinal cord lesions can also result, though rarely, in impotence and bladder dysfunction. Men may experience premature or retrograde ejaculation.

The combination of proprioceptive sensory loss, signs of upper motor neuron palsy and cerebellar dysfunctions, disorders of eye movements (nystagmus), and history of visual deficits are all considered diagnostic for this disease. Clinicians consider dysarthria, nystagmus, and tremor (Charcot's triad) as the cardinal signs of this disease. Depression is common in the initial stage and during remission of this disease. Some exhibit euphoria as a sign of relief when the attack subsides, others may live in psychological denial. Overt intellectual impairment is a late sign of this disease due to the fact that the gray matter of the cortex is spared in this disease and demyelination has to be extensive enough to impede the normal cerebral intellectual process. This fact distinguishes multiple sclerosis from Alzheimer's disease and manifestations of cerebrovascular accidents.

Interferon, ACTH, prednisone or other steroid medications combined with physical therapy have proved to be beneficial. This disease should be differentiated from Guillain–Barré syndrome that produces demyelination in the peripheral nervous system, affecting young and middle-aged individuals. Multiple sclerosis may mimic signs of brainstem astrocytoma, neurologic abnormalities of acquired immune deficiency syndrome (AIDS), systemic lupus (which exhibits seizures, stroke, and psychosis), as well as combined system disease (vitamin B_{12} deficiency).

An axon may be myelinated or unmyelinated and ends in the synaptic terminals. Myelin is an insulating cover of cell membrane that encircles axons and is composed of two-thirds lipid and one-third protein (Figures 2.8 & 2.9). The lipid portion is primarily phospholipid (mostly sphingolipid), and to a lesser extent cholesterol in a free form. Interestingly, the macroscopic difference between gray and white matter is attributed to the lipid content of the myelin. Galactocerebrosides, a form of glycoprotein, represents the main component of the myelin protein. Minor lipid species also exist, such as galactosylglycerides, phosphoinositides, and gangliosides. Gangliosides that contain sialic acid (N-acetylneuraminic acid) form 1% of myelin lipid. PNS myelin is LM1 (sialosylparagloboside), whereas GM-4

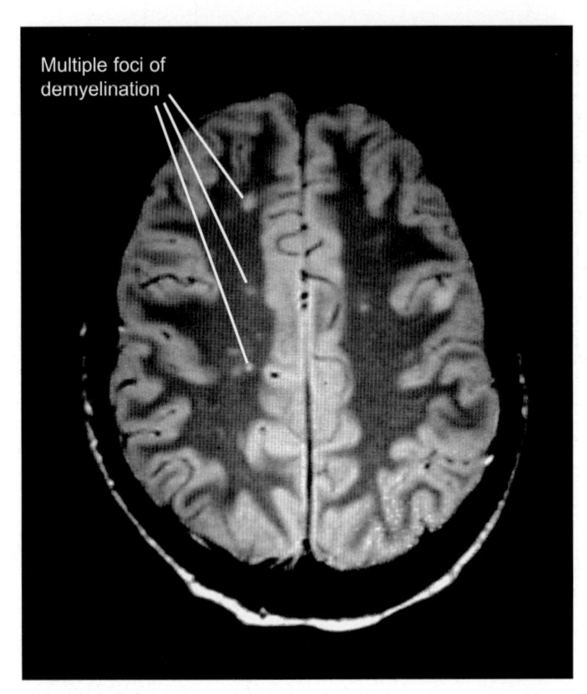

Figure 2.10 This is an MR image of a patient with multiple sclerosis showing multiple foci of demyelination

(sialosylgalactocerebroside) is the main gangloside of CNS myelin. Both PNS and CNS myelin contain, although in low concentration, acidic glycolipids that serve as antigens in myelin.

Myelin allows for substances to be transported between the axon and the myelin forming cells (Schwann cells or oligodendrocytes). It maintains high velocity saltatory nerve conduction, a mode of conduction that proceeds from one node of Ranvier to another in a faster and more energy efficient way. Myelin is not a continuous covering, but rather a series of segments interrupted by nodes of Ranvier. In the peripheral nervous system, each internodal segment represents the territory of one Schwann cell. These nodes are sites of axonal collaterals and bare areas for ion transfer to and from the extracellular space. Extensions of the myelin on both sides of a node of Ranvier are known as paranodal bulbs. These myelin bulbs may lose contact with the axon and undergo degeneration as a result of crush injury. Interruptions within successive layers of myelin are known as Schmidt–Lanterman incisures. Myelin is formed by the oligodendrocytes or Schwann cells during the fourth month of fetal life, and continues into postnatal life.

Myelination is initiated near the soma of neurons and continues toward the axon terminals. It does not cover the axon hillock, dendrites, or axonal terminals. The first step of this process involves surrounding the axon by cytoplasmic membranes of Schwann cells or oligodendrocytes that are detached initially, but later fuse together. The double layer of the Schwann cell plasma membrane forms the meson, which elongates and differentiates into inner and outer parts. Several layers of cell membranes, separated by cytoplasm, surround a given axon. Since myelin formation occurs at a particular site, elongation of the axon requires successive layers of myelin to stretch and cover a larger area of the axon. This results in more layers being concentrated near the center of the internode. When the cytoplasmic and external surfaces of cell membranes come into apposition upon receding of the cytoplasm they form continuous major and minor dense lines, respectively. The minor dense line, also known as the intraperiod line, contains a gap that allows extracellular space to continue with the periaxonal space. This intraperiod gap allows metabolic exchange and serves to accommodate the increasing thickness of the axon by allowing lamellae to slip on one another and thus reduce their numbers. In contrast, oligodendrocytes, the myelin-forming cells in the CNS, are associated with more than one axon and with more than one internodal segment (roughly 15–50 internodes). These multiple associations are maintained by extension of the oligodendrocytes around each axon. Myelination in the CNS begins initially with the vestibular and spinocerebellar tracts. Corticospinal tract and dorsal white column pathways may not be completely myelinated at birth. It should also be remembered that axonal growth and elongation to a destination generally occurs before the migration of oligodendrocytes and formation of myelin.

Unmyelinated axons in the CNS lack any form of ensheathment, whereas unmyelinated axons of the peripheral nervous system are enveloped by Schwann cell cytoplasm. Peripheral axons are lodged in sulci along the surface of Schwann cells. Some Schwann cells in the PNS may encase more than 20 axons through the multiple grooves on their surfaces.

Demyelination may be a primary or secondary process. Primary demyelination affects the thickly myelinated motor fibers and is associated with intact axons as in multiple sclerosis and myelinopathy that affects the thickly myelinated motor fibers of the lower extremity and spares the small sensory fibers. Demyelination secondary to destruction of the axon may be seen in storage diseases and Wallerian degeneration. Incomplete myelination (hypomyelination) occurs in maple syrup urine disease and in phenylketonuria.

In certain diseases such as Refsum's disease and metachromatic leukodystrophy, impairment of α-oxidation and accumulation of phytanic acid lead to

Maple syrup urine disease (branched-chain ketoaciduria) results from anomalies of leucine, valine and isoleucine metabolism as a consequence of a defect of branched chain keto acid decarboxylase. Patients exhibit convulsions, hypertonicity, characteristic odor of urine and perspiration, changes in reflexes, coma, and possible death. Prenatal diagnosis may be possible through enzyme assay of the anomalous metabolites. Acute cases of this disease may be treated by peritoneal and/or hemodialysis.

Phenylketonuria (PKU), a hereditary condition caused by a defect in the phenylalanine decarboxylase, is transmitted as an autosomal recessive trait. It is one the most common aminoacidurias which occurs in one per 20 000 births. This enzymatic defect results in the accumulation of phenylalanine in the blood that may be further metabolized to phenylacetic acid, which is eventually excreted in the urine. Exposure to excessive blood levels of phenylalanine may affect neuronal maturation and myelin formation by desegregation of brain polysomes. It has also been put forward that high concentrations of phenylalanine may inhibit transport of other neutral amino acids across the blood–brain barrier. Others have stated that the inhibitory role of high intracerebral levels of phenylalanine on synaptosomal Na^+–K^+–ATPase activity that regulates the synthesis of neurotransmitters may be responsible for this condition. Newborns with this disease generally do not exhibit clinical manifestations and because of this very reason prenatal screening tests of the amniotic cells and chorionic villi samples are essential for detection of this condition.

Affected infants have lighter skin and eye color, and are not retarded at birth. Eventually, however, patients show signs of mental retardation, seizures, psychoses, extreme hyperactivity, 'musty' body odor and cutaneous rash (eczema). In rare cases PKU may be caused by defects in the metabolism of tetrahydrobiopterin (BH_4), which is the electron donor for the phenylalanine hydroxylase that contributes to the formation of tyrosine and dihydrobiopterin (QH_2). BH_4, which is synthesized from GTP, hydroxylates tyrosine and tryptophan. Since patients are unable to hydroxylate tyrosine or tryptophan, which is mediated by BH_4, restriction of phenylalanine may not prevent the neurological complications from occurring. Patients exhibit convulsions and other severe neurological manifestations.

Gaucher's disease is an autosomal recessive disease, in which the deficiency of the enzyme glucocerebroside results from accumulation of abnormal glucocerebrosides in the reticuloendothelial system. It manifests signs of oculomotor nerve palsy, hepatosplenomegaly, hypertonicity, opisthotonos (a prolonged severe muscular spasm that produces acute arched back), hyperextension of the head and neck, hyperflexion of the arm and hand, tetany, spasticity, and seizures. Individuals with Gaucher's disease may also exhibit an expressionless face, which is described as 'wooden figure'.

Globoid-cell leukodystrophy (Krabbe's disease), a fatal infantile disease caused by deficiency of β-galactosidase and accumulation of galactocerebrosides. Accumulation of an excessive amount of galactocerebrosides leads to disintegration of the myelin in the cerebrum, cerebellum, brainstem, and possibly the spinal cord. This disease is characterized by progressive mental retardation, blindness, convulsion, deafness, signs of pseudobulbar palsy (loss of cranial nerves motor functions due to disruption of cerebral input), and quadriplegia (complete loss of motor functions in the extremities). Rapid cerebral demyelination and presence of globoid cells in the white matter will be seen. Due to generalized rigidity and tonic spasm, the body stiffens and the hand forms a fist, which would be particularly evident when the affected infant is held.

demyelination and production of easily degradable abnormal myelin. Demyelination also occurs in acquired neurometabolic disease (Korsakoff–Wernicke syndrome), due to thiamine deficiency, and in lipid storage (lysosomal) diseases including Gaucher's disease, globoid-cell leukodystrophy, Fabry's disease, Neimann–Pick disease, metachromatic leukodystrophy, Tay–Sachs disease, and Refsum's disease. These genetic disorders are autosomal recessive conditions with the exception of Fabry's disease, a sex-linked abnormality with no ethnic or gender predilection. They are the result of a deficiency of intracellular lysosomal enzymes that regulate the catabolism of sphingolipids. Patients with these disorders carry enzymatic structures in their tissues that are similar to the normal enzymes, but are not capable of degrading lipids.

The diagnosis of these diseases has been made easier following the recognition that antibodies which bind to the affected enzymes may artificially be prepared, and that traces of the above-mentioned non-functional enzymes in the skin fibroblasts, leukocytes, and

Fabry's disease (angiokeratoma corporis diffusum) results from the lack of the α-galactosidase (ceramide trihexosidase) and the accumulation of the glycolipid, dihexosidase and trihexosidase in the autonomic and dorsal root ganglia, myelinated fibers of the brainstem and myocardium. Accumulation of glycolipid in the superior cervical ganglia may be associated with anhydrasis (lack of sweating). Glycolipid accumulation may also occur in the renal tubules and glomeruli, which serves as a diagnostic tool, and also in the hypophysis, skin, eye, and smooth muscles of blood vessels. This fatal condition is an X-linked anomaly, which exhibits corneal opacity, fever, burning pain in the extremities and skin lesions in males. Death occurs as a result of renal failure or disorders associated with vascular hypertension. Ataxia, signs of upper motor neuron palsy, and urinary incontinence are also seen. Demyelination and subsequent loss of thinly myelinated fibers, as a result of the deposited glycolipid in the dorsal root ganglia, may account for the burning pain felt by individuals with this disease. Involvement of the blood vessels may explain the frequency with which cerebrovascular accidents occur in patients with this affliction.

Neimann–Pick disease is a fatal disease of infants, resulting from lack of sphingomyelinase and accumulation of excessive amounts of sphingomyelin in various tissues. However, it is not yet clear whether abnormal breakdown, stereochemical anomaly, or excessive production of sphingomyelin is responsible for the manifestations of this disease. It is thought that presence of this substance is responsible for disintegration of the myelin in the white matter of the brain and brainstem. This disease exhibits pancytopenia (a marked reduction in the number of erythrocytes, leukocytes, and platelets), xanthoma (a benign, fatty, fibrous and yellowish plaque that develops in the subcutaneous tissue, often around tendons), and feeding problems. It may also manifest growth and mental retardation, seizures, deafness, macular irregularity (cherry red macular spots occur in about one-fourth of patients, leading to blindness). Children are cachectic and commonly die between the age of 6 months and 3 years.

Metachromatic leukodystrophy is another fatal disease that results from deficiency of cerebroside sulfatase A and subsequent accumulation of sulfatide in excessive amounts in the myelin. Myelin with its abnormal sulfatide content may stain metachromatically brown upon treatment with acidified cresyl violet. Normally, sulfatide is degraded into cerebroside and inorganic sulfate. Chemically abnormal myelin in the brain, cerebellum, brainstem, spinal cord, and peripheral nerves does not survive and undergoes disintegration. This condition exhibits dementia and progressive paralysis.

Tay–Sachs disease is an autosomal recessive disorder which results from the absence of hexosaminidase A, subsequent to accumulation of GM2 ganglioside in the perikarya of the neurons. Gangliosides are glycolipids normally present in the plasma membrane of neuronal cell bodies. This disease is common in Jewish infants (3–6 months of age) of eastern European origin and French Canadian parents. Patients may exhibit seizures, blindness, laughing spells, abnormal acousticomotor reaction. The latter reaction is characterized by brisk extension of legs and arms followed by clonic jerks of all limbs, neck extension, and startled facial expression in response to sudden sharp noise. Cherry red spots on the macula and progressive intellectual, physical, and neurologic deterioration are additional features of this serious disease. Infants that live longer than 6 months may develop macrocephaly.

Refsum's disease (heredopathis atactica polyneuritiformis) is an autosomal recessive disease, which results from accumulation of phytanic acid (tetramethylated 16 carbon-chain fatty acid) subsequent to a deficiency in the catabolism of fatty acids. It is characterized by demyelination of the spinal nerves, motor neurons of the ventral horn and the dorsal columns. Signs may include polyneuropathy, polyneuritis, nocturnal blindness, deafness, cardiac manifestations, cerebellar deficits, and locomotor ataxia. It may also include retinitis pigmentosa, which refers to the inflammation of the retina associated with progressive loss of retinal response, clumping of pigment, and shrinkage of visual field. Reduction of phytanic acid and amelioration of certain clinical signs may be achieved by dietary restriction of fruit, vegetables, and butter.

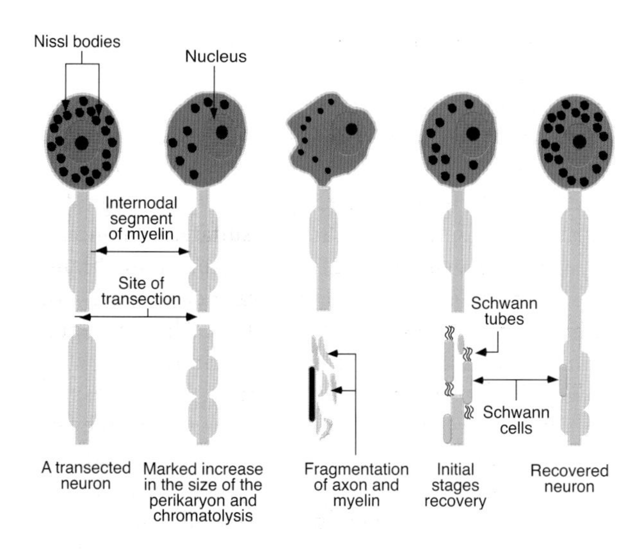

Figure 2.11 Cellular changes and stages of recovery of an injured neuron. Note the changes distal and proximal to axonal injury

amniotic fluid may be detected. Fragmentation of myelin sheath, loss of ability to conduct sensory impulses, impairment of motor function, and trophic changes also occur as the nerve undergoes degeneration (Figure 2.11) following axonal injury. The microscopic alterations (nerve degeneration) in a neuron following damage to its axon may include changes distal to the site of trauma (anterograde degeneration), proximal to the site of damage (retrograde degeneration), or across the axonal terminal into the adjacent neuron (transneuronal degeneration).

Anterograde (Wallerian) degeneration occurs as early as 12–24 hours and includes disintegration of the mitochondria and neurofilaments followed by retraction and fragmentation of the myelin and axon. Cellular debris is later absorbed by the Schwann cells and macrophages in the peripheral nervous system and by the microglia, macrophages, and astrocytes in the CNS. This process is aided by the hydrolytic enzymes of the lysosomes and axonal protease. After two weeks, the cytoplasm of Schwann cells forms tubes or 'guidance tunnels' along the course of the damaged axons, followed by departure of the macrophages. These tubes, known as bands of Bungner, may persist long enough to guide the regrowth of the axon, or may be replaced by the endoneurium. Degenerative changes occur in the terminals earlier if the site of the lesion is adjacent to the synapse.

Retrograde (indirect Wallerian) degeneration is seen in the PNS and CNS, although it does not occur in all neurons. In this type of degeneration, proximal axonal disintegration and break-up of myelin sheath may occur. These changes are accompanied by sealing off of the severed ends by the axolemma, preventing leakage of axoplasm. As the dendrites retract from their synaptic contacts, retraction bulbs are formed at the swollen severed ends of the axons. The soma undergoes chromatolysis, where the Nissl bodies break up near the axon hillock, followed by dissolution of the cytoplasm within three days. Swelling of the soma is accompanied by deviation of the nucleolus into a peripheral eccentric position. Dispersion of the Golgi apparatus is accompanied by an increase in number of the lysosomes, mitochondria, granular endoplasmic reticulum, and free ribosomes. Fully developed neurons may resist the process, allowing slow atrophic changes to occur, whereas poorly developed or immature neurons may die quickly. If regeneration fails death of the affected neuron may eventually occur, which is a most probable outcome in CNS injuries.

The transneuronal (transynaptic) degeneration, on the other hand, occurs in neurons which provide the sole afferents to a neuron that receives axonal injury. It may also be seen in neurons that originally received input from the injured axon. These reactions, which are manifestations of disuse atrophy, extend slowly beyond the synaptic cleft to the adjacent neurons. Thus, the neuronal changes across the synaptic cleft are the result of lack of trophic substances provided to the adjacent neurons by the damaged neurons.

Table 2.1 Metabolic diseases, enzymatic deficiencies and associated metabolites

Disease	*Deficient enzyme*	*Accumulated metabolite*
Gaucher's disease	Glucocerebrosidase	Glucocerebrosides
Globoid leukodystrophy	Galactocerebrosidase	Galactocerebrosides
Fabry's disease	Galactosidase A	Ceramide trihexoside
Neimann–Pick disease	Sphingomyelinase	Sphingomyelin
Metachromatic leukodystrophy	Cerebroside sulfatase A	Sulfatide

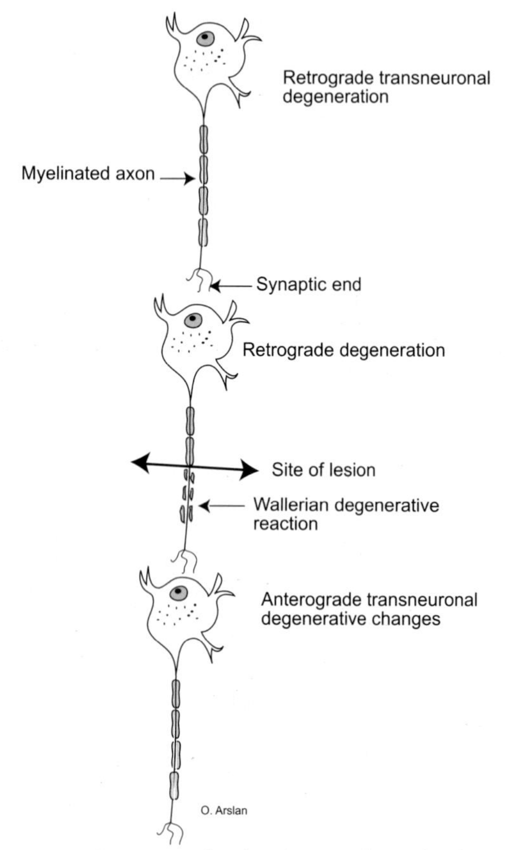

Figure 2.12 Summary of major changes shown in Figure 2.11

Schwannomas and neurofibromas are the most common tumors of the peripheral nervous system, which are represented in von Recklinghausen's neurofibromatosis. This disease exhibits peripheral and central forms. It has an autosomal dominant pattern of inheritance and spontaneous mutations. The gene defect encodes nerve growth factor receptor located on the long arm of chromosome 17. In the peripheral form, patients exhibit *café-au-lait* cutaneous spots in childhood and neurofibromatosis in adults. Mental retardation, epilepsy, spinal deformities, and other tumors such as gliomas and pheochromocytoma may complicate this form of the disease. In the rare central form, multiple meningiomas and Schwannomas occur. It is associated with the loss of specific alleles from chromosomes 22. The possibility of involvement of chromosome 17 does exist.

Regeneration refers to the ability of a neuron to restore function following a traumatic injury. In the peripheral nervous system, regeneration does occur, but is influenced by factors such as the site (the degree of regeneration is inversely proportional to the length of the axon) and the type of injury. Since growth cones (growth projections from the severed axon terminal) fail to properly align with the path of the axon that has undergone a transection injury, regeneration is more difficult than that of crushing injuries in the endoneurium when Schwann cells remain intact. It is noteworthy to add that regeneration is more likely if the site of injury is closer to the target site. The rate of growth of the regenerating axon varies, generally ranging between 3 and 4 mm/day in primates. Signs of regeneration start with the formation of growth cones in the distal end of the proximal segment of the severed axon. These growth extensions, which develop during the first week after the nerve injury, reach the distal segment through guidance tunnels formed by the Schwann cells. These changes are later followed by reconnection with the appropriate target, maturation, which requires recognition, establishment of a functional synapse and myelination as well as increased thickness of the axons. However, neuromas and associated agonizing pain may develop at the ends of the sprouting axons if the distance is long enough not to allow complete approximation of the distal and proximal segments. Axonal sprouting in peripheral nerves occurs when some fibers within the nerve trunk are damaged while the remaining fibers are intact. These sprouts will extend into areas originally innervated by the injured fibers and will restore their function.

Regeneration seems very limited in the CNS, and true growth is almost impossible due to the fact that guidance tunnels are not formed by myelin-forming oligodendrocytes. Additionally, scar and necrotic tissue from trauma or infection may impede the repair process. Growth of axons does not follow a particular pattern to re-establish the connection, thus functional restitution becomes unattainable. The most significant goal of modern rehabilitative medicine is to prevent atrophy of the muscles in individuals with motor neuron diseases. One of the means to achieve this end is to apply electrical stimulation to the affected muscles, preventing denervation hypersensitivity and reducing atrophy.

Neurons are classified according to the chemical nature of the neurotransmitter that they release into cholinergic, adrenergic, noradrenergic, dopaminergic, serotoninergic, GABAergic neurons, etc. Cholinergic neurons release acetylcholine and are commonly found at neuromuscular junctions. Noradrenergic neurons are abundant in the sympathetic ganglia and the reticular formation, whereas adrenergic neurons are found in the adrenal medulla and within the synaptic dense cored vesicles. Dopaminergic neurons are present mainly in the substantia nigra, corpus

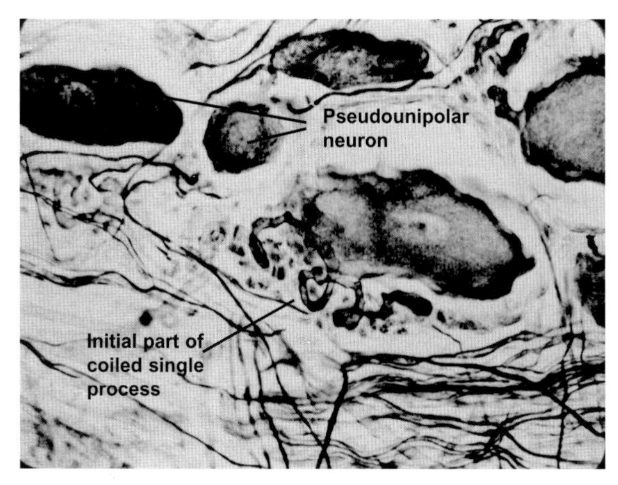

Figure 2.13 Photomicrograph of the pseudounipolar neuron and its initial process

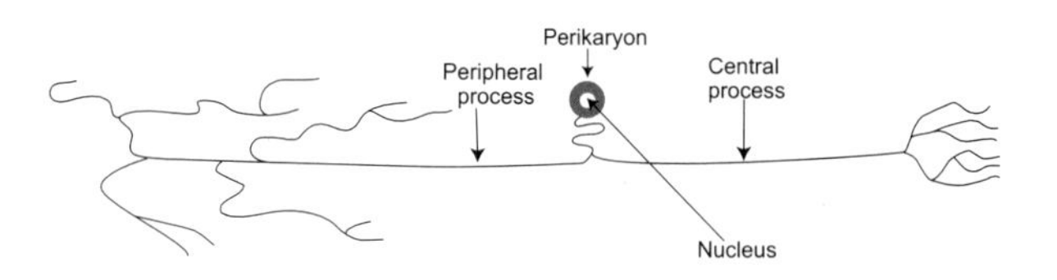

Figure 2.14 Schematic drawing of the pseudounipolar neuron. Cellular elements and associated extensions are also shown

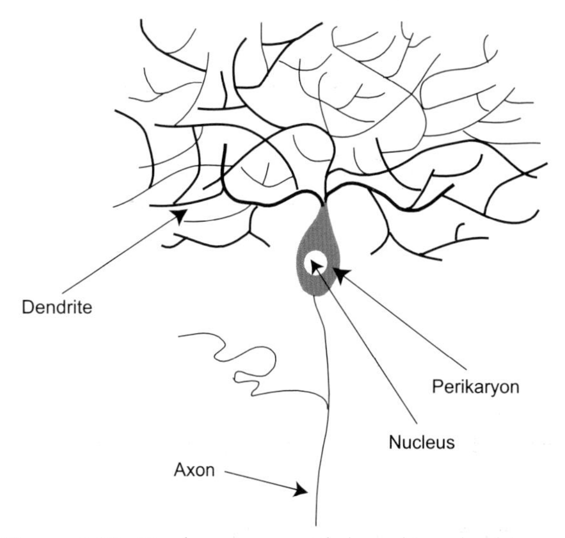

Figure 2.15 Bipolar neuron with branching dendrite and axon

striatum, and cerebral cortex, while serotoninergic neurons occur in the raphe nuclei and in the rounded synaptic vesicles. GABAergic neurons are present in the cerebellar cortex and spinal cord. Neurons may also be classified into pseudounipolar, bipolar, and multipolar neurons.

Unipolar neurons are the most simple class of neurons that exhibit a single extension. This process gives rise to branches, some of which are receptive (dendrites); others function as axons. True unipolar neurons, which are relatively rare in vertebrates, form the dorsal root ganglia, the granule cells of the olfactory system, and the mesencephalic trigeminal nucleus (Figures 2.7, 2.13 & 2.14). Pseudounipolar neurons give off a single process that divides into a peripheral receptive branch (dendrite) and a central extension serving as an axon. Both of these branches maintain structural resemblance to axons.

Bipolar neurons are also a relatively uncommon class of neurons. They are symmetrical cells with ovoid or elongated body and with a single dendritic process and an axon arising from opposite poles. These processes are approximately equal in length. They form the vestibular (Scarpa's) ganglia, spiral (auditory) ganglia, and the retinal bipolar cells (Figure 2.15).

Multipolar neurons (Figures 2.7, 2.16 & 2.17) are the most common types of neurons in the CNS; they form the autonomic ganglia. They possess a single axon with several symmetrically radiating dendrites. Some neurons have multiple axons or lack axons all together. Multipolar neurons can be classified on the basis of dendritic branching pattern and shape of the soma into stellate, pyramidal, fusiform, Purkinje, and glomerular cells.

Stellate (star) cells are found in the spinal cord, reticular formation, and cerebral cortex. They have dendrites of equal lengths (isodendritic) that radiate uniformly in all directions.

Pyramidal cells are multipolar, exhibiting pyramidal-shape soma with basal dendrites and a single apical dendrite that ascend toward the surface of the cerebellar cortex. They are most abundant in the cerebral cortex and hippocampal gyrus.

Fusiform cells are distinguished by their spindle-shaped and flattened soma with dendrites at both ends.

Purkinje cells that form the intermediate layer of the cerebellar cortex have flask-shaped soma with apical tree-like dendritic branches, ascending toward the surface of the cerebellum maximizing synaptic contacts. Purkinje cells are motor neurons that project long axons beyond the area of the soma.

Glomerular cells have a few convoluted dendritic branches and form the mitral and tufted cells of the olfactory bulb. Mitral cells have an inverted

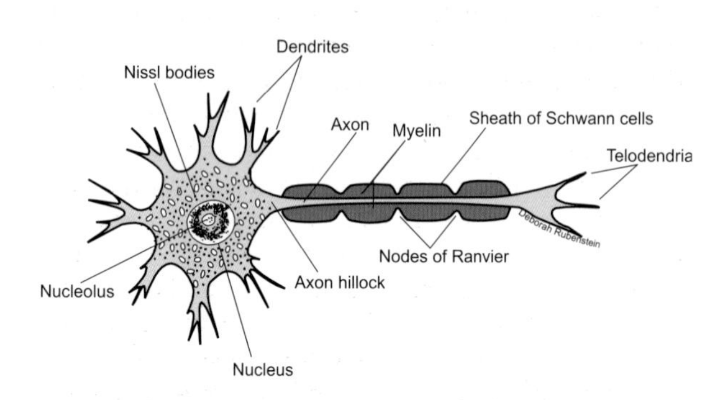

Figure 2.16 Multipolar neuron. Observe the numerous branches of dendrites and a single uniform axon is also shown

cone-shaped dendritic field and soma that resembles a bishop's miter.

Anaxonic cells are abundant in the retina (amacrine cells) and the olfactory bulb, where they are known as granule cells.

On the basis of axonal length, multipolar neurons can also be categorized into Golgi type I, with long axons projecting to distant parts of the CNS; and Golgi type II, possessing short axons that establish contacts with local neighboring neurons. The Golgi type II represents the inhibitory interneurons (such as the periglomular olfactory neurons), which are activated by the ascending sensory pathways and play an important role in lateral inhibition. Neurons without axons, as mentioned earlier, are known as anaxonic, such as the amacrine cells of the retina and granule cells of the olfactory bulb, which establish synapses with parallel neurons.

Neurons can also be classified based on their functional role into somatic motor, somatic sensory, visceral motor, and visceral sensory neurons.

Neuronal communication is maintained through synapses (Figure 2.7), which are specialized junctional complexes formed by the axon terminal of one neuron opposing the dendrites, soma or the axon of another neuron. Synapses represent sites of impulse generation (action potentials) and transmission across a population of neurons within the CNS. Synapse interfaces between neurons provide trophic substances, and act as a 'gate' for controlling impulses. A single axon may establish a synapse with one neuron (e.g. connections of the olivocerebellar fiber with dendrites of the Purkinje neurons). Multisynapses are seen between the parallel fibers of granule cells and the neurons of the molecular layer of the cerebellum. Synaptic glomeruli in the olfactory bulb and the granular layer of the cerebellum consists of an axon that synapses with dendrites of one or more neurons encapsulated by neuroglial cells. In general, synapses consist of presynaptic and postsynaptic components separated by synaptic clefts. Presynaptic processes contain round, granular, or flat vesicles filled with a specific neurotransmitter. Typically round vesicles contain acetylcholine, an excitatory neurotransmitter.

Small granular vesicles with electron-dense cores contain norepinephrine, an excitatory neurotransmitter. Flattened vesicles contain GABA, an inhibitory neurotransmitter. The close relationship between the vesicle morphology and functional synaptic type is evident when considering the association of the flattened synaptic vesicles with symmetrical membrane specializations and spherical vesicles with asymmetrical membrane thickenings. The postsynaptic membrane may be part of a muscle cell or neuron, upon which neurotransmitter molecules bind after crossing the synaptic cleft. The part of the postsynaptic membrane that lies adjacent to the presynaptic membrane is known as the subsynaptic membrane. Synaptolemma is a term that denotes the combined presynaptic and subsynaptic membranes. An increase in the postsynaptic receptor sites may be responsible for the exaggerated response following denervation (denervation hypersensitivity).

Synapses in the CNS differ morphologically and functionally from their counterparts in the peripheral nervous system. They are not always cholinergic (as in the peripheral nervous system), and utilize several excitatory neurotransmitters such as catecholamines (epinephrine, norepinephrine, and dopamine), amino acid neurotransmitters (glutamine, aspartate, cysteine, etc.), serotonin, histamine, enkephalin, etc. Transmission through the central synapses is governed by factors such as diffusion and reabsorption, and may be excitatory or inhibitory (activation drives the membrane potential of the postsynaptic neuron toward or away from its threshold level for firing nerve impulses). Transmission in the peripheral synapses, as in the neuromuscular junction, is generally excitatory, secured by a single presynaptic activation, and is dependent upon the degradation of the neurotransmitters by cholinesterase. Synapses in the CNS occur between one presynaptic ending and several postsynaptic neurons, contrary to the 1:1 synapse ratios in peripheral transmission. The variability and efficiency of transmission and neurotransmitter discharge in the central synapses is dependent upon the number of activated presynaptic endings.

Synapses may also be classified as chemical or electrical. Chemical synapses are slow which involve the release of a neurotransmitter by synaptic vesicles into

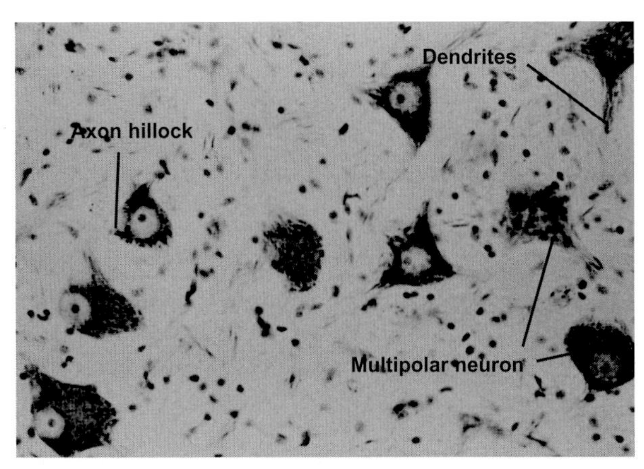

Figure 2.17 These multipolar neurons of the cerebral cortex exhibit an axon with an apical dendrite

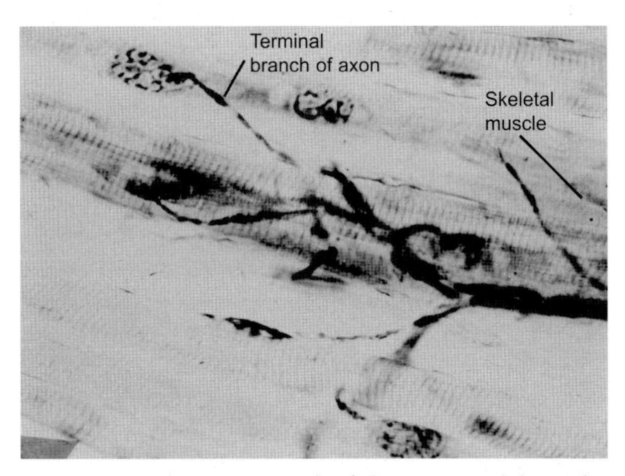

Figure 2.18 Photomicrograph of the motor endplate. The synaptic connection between the terminal axon and skeletal muscle fibers is clearly illustrated

the synaptic cleft, producing changes in the permeability of the postsynaptic membrane. The effect of the neurotransmitter is controlled by local enzymes and/or by reabsorption. Chemical synapses are further categorized on the basis of the utilized neurotransmitter. Cholinergic synapses use acetylcholine, adrenergic synapses utilize epinephrine or norepinephrine, and dopaminergic synapses utilizes dopamine. The electrical synapses exhibit close contact between presynaptic and postsynaptic membranes. Gap junctions enable the nerve impulses to cross directly from one cell to another and act on the postsynaptic membrane. These synapses, which are common in lower vertebrate motor pathways, are similar to the electrical junctions (intercalated discs) of the cardiac muscle cells. Electrical synapses act much more rapidly than chemical synapses.

Classification of synapses may also be based on the morphological characteristics and the type of action they eventually produce, into Gray's Type I and II synapses. Gray's Type I is an excitatory synapse in which the synaptic cleft is wide and the presynaptic and postsynaptic membrane densities are asymmetrical with the subsynaptic zone being thicker than the presynaptic zone. This type of synapse contains a wide variety of neurotransmitters including acetylcholine, glutamate, and hydroxytryptamine. Gray's Type II synapse is an inhibitory synapse in which the synaptic cleft is narrower, and the pre- and postsynaptic membrane densities are symmetrical.

Synapses may also be axodendritic, the most common of which may be symmetrical or asymmetrical. Symmetrical axodendritic synapses predominate near the soma on the larger dendritic trunks. Axosomatic synapses occur on the perikaryon exhibiting both symmetrical and asymmetrical forms. This type of synapse that involves the initial segment of the axon may be inhibitory to cellular discharge. They are commonly symmetrical and may release inhibitory neurotransmitter GABA. Axoaxonic, in general, reduce the amount of neurotransmitter released by the axon and therefore are regarded to mediate presynaptic inhibition. Dendro-somatic and somato-somatic synapses are described in the sympathetic ganglia. Dendro-dendritic synapses are for the most part symmetrical type; however, in the olfactory bulb the dendrites of the mitral cells form asymmetrical synapses with the dendrites of the granule cells.

Neuromuscular junctions (motor endplates) are sites of synaptic contacts between the axons of the peripheral nervous system and the skeletal (striated) muscle fibers (Figure 2.18). This synaptic junction is the site where depolarization of muscle fiber membrane and muscular contraction is initiated. Structurally, each motor endplate consists of presynaptic and postsynaptic membranes. The presynaptic membrane is formed by the plate-like unmyelinated end of a motor axon, with numerous membrane-bound acetylcholine filled vesicles. The postsynaptic membrane, which is formed by muscle cell invagination that corresponds to the presynaptic vesicles, is separated from the extracellular space by the Schwann cells. The synaptic membranes of the motor endplates are separated by synaptic clefts that are larger than the synaptic membranes in the CNS. The release of acetylcholine is dependent upon the frequency of the action potential and the influx of calcium ions. Once released, the acetylcholine diffuses across the synaptic cleft and increases the permeability of the postsynaptic membrane to the sodium and potassium ions, and thus produces depolarization. The endplate potential is local, and its amplitude varies with the distance and the amount of the acetylcholine.

Disease processes, drugs, and exposure to toxins may disrupt chemical transmission. Local anesthetics such as procaine, tetrodotoxin and saxitoxin block the generation of action potentials. Hemicholinium blocks the synthesis of acetylcholine by preventing the re-uptake of choline into the cell. High concentration of magnesium may block the release of acetylcholine and cause paralysis by competing with calcium receptors.

Transmission at the presynaptic level of the neuromuscular junction may be blocked by exotoxin produced by strains of the bacterium *Clostridium botulinum*. This toxin blocks the release of acetylcholine by either binding calcium receptors or preventing entry of calcium ions during the action potential. Botulinum toxin is synthesized in an inactive form, which must be cleaved into heavy and light chains joined by disulfide bridge to become active. Toxicity is initiated by binding of these chains to specific presynaptic receptors. Ultimately the toxic component is discharged from the lysosomes into the cytoplasm of the presynaptic terminals.

Ingestion of clostridial toxin occurs as a result of consuming improperly preserved canned food and vegetables, producing botulism. It may also occur as a result of contamination of a deep penetrating wound with this particular toxin. Symptoms may appear within 2 days, including nausea, vomiting, diplopia, and blurred vision. Paralysis of the extraocular muscles may be accompanied by dilated and unresponsive pupils, dysphagia (difficulty in swallowing), dysarthria (difficulty in speech), and paresis (weakness and not complete paralysis) of muscles of the neck, trunk, and extremities.

Black widow spider venom (α-latrotoxin) is a protein molecule which combines with the presynaptic membrane and allows both sodium and calcium ions to enter the terminal, causing initial massive release of acetylcholine. This is followed by decline and fast depletion of the transmitter that produces initial contraction and painful spasm, rigidity and subsequent paralysis of the associated muscles. Bungarotoxin, venom of *Bungaro multicinctus*, is a protein that inhibits the release of synaptic vesicles from the cholinergic and motor nerve terminals by exhibiting phospholipase A_2 activity. Initially, this toxin produces slight reduction in end-plate potential (EPP) amplitude followed by intensification and then progressive decrease and final complete blockage of the transmitter.

Tetany, another condition that results from abnormalities at the neuromuscular junction, is associated with deficiency of parathormone (hypocalcemia) and vitamin D, and alkalosis due to hyperventilation (high pH increases calcium binding by serum proteins). It is characterized by generalized muscle spasm, tingling sensation in the lips, tongue and digits, and facial muscle spasm. Spasm of the forearm muscles and flexion of the hand at the wrist (Trousseau's sign) may be produced by applying a tourniquet and reducing the blood supply to the forearm. The fingers are pressed together and the thumbs are adducted (obstetrician hand). The lower extremity may be in carpopedal spasm, in which the thigh and knee are extended, whereas the feet are plantar flexed and inverted. In latent tetany, contraction or spasm of the facial muscles may be elicited by tapping the facial nerve trunk (Chvostek's sign). It is treated by injection of calcium that restores normal transmitter release.

Due to development of multiple and highly sensitive ectopic sites, the sensitivity of muscle membrane increases dramatically upon denervation. Denervation hypersensitivity is accompanied by random contraction of individual muscle fibers known as fibrillation. Since these contractions develop individually and are asynchronous, they are not visible through the skin. The postsynaptic membrane contains acetylcholinesterase, an enzyme that limits the duration of action of acetylcholine and curtails the depolarization process by hydrolysis of acetylcholine into choline and acetate.

Antibodies to the (nicotinic) cholinergic receptor may also block transmission at the postsynaptic level of the neuromuscular junction, as in myasthenia gravis (Erb–Goldflam disease). These antibodies bind to the main immunogenic region of the μ-2 subunit receptors and produce cross linkage between cholinergic (nicotinic) receptors and eventually increase the rate of lysosomal degradation and endocytosis. The density of the postsynaptic receptors in this disease may be as low as one-third of normal and the synaptic cleft is widened. Some speculate that mycocytes and thymic muscle-like cells may express AChR on their surface, triggering inflammatory response and subsequent production of cross-reacting antibodies from the thymus to the muscular cholinergic receptors.

Myasthenia gravis (Figure 2.19) is an acquired autoimmune disease of neuromuscular transmission, which is most common in women. It may accompany other systemic autoimmune disorders such as systemic lupus erythematosus and pernicious anemia. This disease usually develops between the second and fourth decades of life. It is characterized by bilateral weakness of the ocular, masticatory, swallowing and muscles of facial expression, as well as muscles of the neck and upper extremity. Clinical signs show varying intensity and occasionally follow an episodic course. Although these signs may be seen occasionally only on one side, weakness of ocular and palpebral muscles remain the first presenting signs.

Ptosis (drooping of the upper eyelid), with lid retraction that cannot be masked by contraction of the frontalis, is the most prominent manifestation of this disease. Repeated forceful opening and closure of the eyes may exacerbate ptosis (Simpson's test). Although pupillary light reflex (characterized by constriction of the pupil of both eyes when light is applied to one eye) remains intact; deficits such as diplopia upon upward gaze and convergence (medial deviation of both eyes), and mydriasis (dilatation of the pupil) are also seen.

Nasal speech, dysphagia, and inability to hold the head up may form additional early signs of this condition. Voluntary muscle weakness is heightened by repetitive maneuvers, inducing dramatic fatigue. Increasing fatigue of individual muscles is more pronounced toward evening. Respiratory muscle involvement in some patients may lead to death (myasthenic crisis). Sensory transmission and tendon reflexes are preserved in this disease. Diagnosis of this disease is done by intravenous injection of a short-acting anticholinesterase agent known as edrophronium chloride. Edrophonium chloride can be used to distinguish between myasthenia gravis and cholinergic crisis. Edrophonium causes temporary relief in patients with myasthenia gravis, but not in patients with cholinergic crisis, which is a diffuse over-activation of the muscarinic receptors and depolarization of nicotinic cholinergic receptors of the skeletal muscles. Over-treatment with anticholinesterase or exacerbation of the disorder may lead to respiratory arrest and possible death.

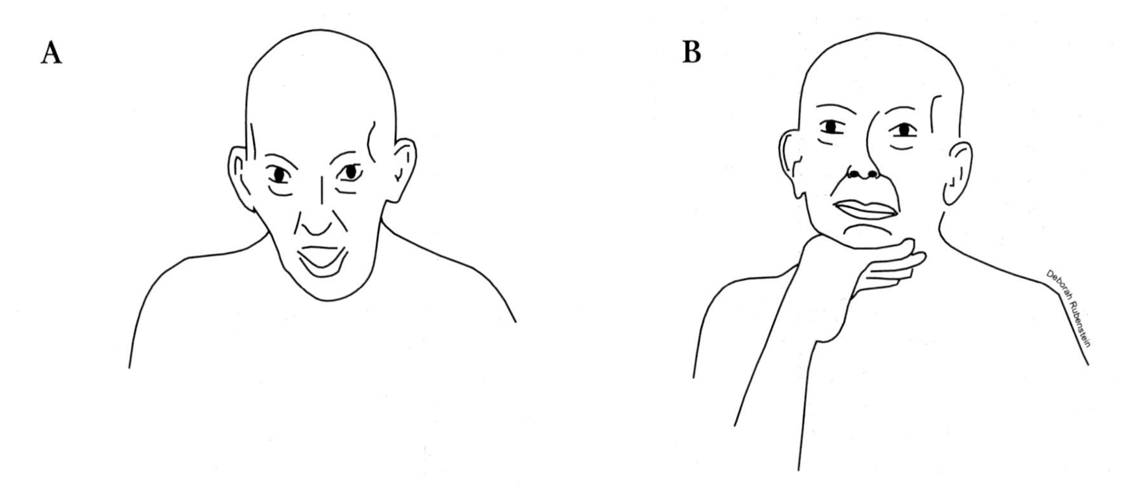

Figure 2.19 These schematic drawings show the manifestations of myasthenia gravis. In (A) observe drooping of the head and jaw. In (B) the patient attempts to hold the head and jaw up with a hand during conversation

Myasthenia gravis should be distinguished from Eaton–Lambert syndrome, which is seen in oat cell carcinoma of the lung, and is associated with pernicious anemia. Antibodies that block the calcium channels essential for the release of acetylcholine cause this syndrome. It manifests itself as weakness in the muscles of the upper extremity, sensory loss, ataxia, and deep tendon areflexia. Sparing of the extraocular muscles is one of the pathognomic features of this disease. It should be noted, however, that the affected muscles show a maximum increase in strength following voluntary exercise (warm-up) which is a unique characteristic of this neurological disorder. The 'warm-up' phenomenon in Eaton–Lambert syndrome is the result of concomitant actions of acetylcholine release and then depletion followed by facilitation of transmitter release by repetitive activities. Autonomic deficits such as sexual dysfunction and dry mouth may also be seen.

Curare drugs (D-tubocurarine) reversibly attach to the postsynaptic membrane, thus preventing any reaction to acetylcholine. D-tubocurarine is a short-acting drug that may be used with local anesthetics to promote muscle relaxation during anesthesia. Its action is terminated by the administration of anticholinesterase. Depolarizing blocking agents such as decamethonium bromide and succinylcholine may mimic acetylcholine at the postsynaptic membrane level. Since these agents are not affected by cholinesterase, they induce prolonged depolarization. Anticholinesterase drugs such as physostigmine and neostigmine prolong the action of the acetylcholine by reversibly inactivating the enzyme. Nerve gas (di-isopropyl fluorophosphate) and organic phosphates irreversibly bind to acetylcholinesterase producing prolonged depolarization, paralysis, and death due to asphyxiation.

α-Neurotoxin is a curare mimetic which is a non-depolarizing blocking agent at the postsynaptic cholinergic receptors. This toxin is produced by snakes of families Elapidae (e.g. cobras, coral snakes, etc) and Hydrophidae (sea snake). α-Neurotoxin consists of the long toxin with 71–74 amino acids and five internal sulfide bonds, and short group with 60–62 amino acids and four internal disulfide bonds. The toxin with the short amino acids exhibits faster binding to the α subunits of the acetylcholine receptor and a reversible dissociation capacity than the long toxin (irreversible binding).

Suggested reading

1. Aguayo AJ, Rasminsky M, Bray GM, *et al.* Degenerative and regenerative responses of injured neurons in the central nervous system of adult mammals. *Phil Trans R Soc* London 1991;331:337–43
2. Atwood HL, Lnenicka GA. Structure and function in synapses: emerging correlations. *Trends Neurosci* 1986; 9:248–50
3. Bhär M, Bonhoeffer F. Perspectives on axonal regeneration in the mammalian CNS. *Trends Neurosci* 1994;17:473–79
4. Colman DR. Functional properties of adhesion molecules in myelin formation. *Curr Opin Neurobiol* 1991;1: 377–81
5. Duchen LW, Gale AN. The motor end plate. In: Swash M, Kennard C, eds. *Scientific Basis of Clinical Neurology*. Edinburgh: Churchill Livingstone, 1985;400–9
6. Fawcett JW, Keynes RJ. Peripheral nerve regeneration. *Annu Rev Neurosci* 1990;13:43–60
7. Hirokawao N. Axonal transport and the cytoskeleton. *Curr Opin Neurobiol* 1993;3:724–31
8. Hökfelt T, Fuxe K, Pernow B, eds. *Progress in Brain Research*, volume 68. Amsterdam: Elsevier, 1986
9. Jessel TM, Kandel ER. Synaptic transmission: A bi-directional and self-modifiable form of cell-cell communication. *Neuron* 1993;10:77–98
10. Kano M. *The Synapse*. Oxford: Oxford University Press, 1994
11. Kelly RB. Storage and release of neurotransmitters. *Neuron* 1993;10:43–53
12. Kennedy RH, Bartley GB, Flanagan JC, *et al.* Treatment of blepharospasm with botulinum toxin. *Mayo Clin Proc* 1989;64:1085–90
13. Lee MK, Cleveland DW. Neurofilament function and dysfunction: involvement in axonal growth and neuronal disease. *Curr Opin Cell Biol* 1994;6:34–40
14. Linden DJ. Long-term synaptic depression in the mammalian brain. *Neuron* 1994;12:457–72
15. Magistretti PJ. Cellular bases of brain energy metabolism and their relevance to functional brain imaging: Evidence for a prominent role of astrocytes. *Cerebr Cortex* 1996;6:50–61
16. Mitchell G. Update on multiple sclerosis therapy. *Med Clin North Am* 1993;77:231–49
17. Nakajima K, Kohsaka S. Functional roles of microglia in the brain. *Neurosci Res* 1993;17:187–203
18. Remahl S, Hildebrand C. Relation between axons and oligodendroglial cells during initial myelination II. The individual axon. *J Neurocytol* 1990;19:883–98
19. Vallee RB, Bloom GS. Mechanisms of fast and slow axonal transport. *Annu Rev Neurosci* 1991;14:59–92
20. Wu L, Saggau P. Presynaptic inhibition of elicited neurotransmitter release. *Trends Neurosci* 1997;20:204–12

Section II

Central nervous system

The central nervous system (CNS) is derived from the neural tube and is comprised of the spinal cord, brainstem, cerebellum, diencephalon, and telencephalon. It initiates and regulates all motor activities and is responsible for the integration and transmission of sensory impulses. In addition, it also controls and regulates all of the activities that preserve the individual and species.

Spinal cord 3

The spinal cord is derived from the caudal part of the neural tube and occupies the upper two thirds of the vertebral column in adults. It stretches between the atlas (upper border of the foramen Magnum) to the intervertebral disc between the first and second lumbar vertebrae. In newborns, it extends to the level of the third lumbar vertebra. The sites of attachment of the 31 pairs of spinal nerves mark the individual segments of the spinal cord. Therefore, each spinal segment is associated with one pair of dorsal and ventral roots. There are eight cervical, 12 thoracic, five lumbar, five sacral, and one coccygeal spinal segments. The spinal cord has a cervical enlargement which corresponds to the fifth cervical through the first thoracic spinal segments (roots of the brachial plexus), and a lumbar enlargement, corresponding to the first lumbar through the third sacral spinal segments (roots of the lumbosacral plexus).

Spinal cord

The spinal cord is the cylindrical part of the central nervous system (CNS), occupying the upper two-thirds of the vertebral column. The lower end of the spinal cord (conus medullaris) shows variation relative to the height of the individual, particularly in females. Flexion and extension of the trunk may also produce relative variation of the lower end of spinal cord. In some individuals the spinal cord may terminate as high as the twelfth thoracic vertebra or may extend as far down to the level of the intervertebral disc between the second and third lumbar vertebrae.

Due to the differential growth of the vertebral column relative to the spinal cord, the spinal cord segments do not always correspond to the vertebral levels. In general the rule of 2 applies to the vertebral levels T1–T10. In other words the injured spinal segments are determined by adding 2 to the level of the affected vertebrae. Spinous processes of T11–T12 vertebrae correspond to the lumbar spinal segments. Accordingly, the cervical spinal nerves exit above their corresponding vertebrae, while the remaining spinal nerves emerge from the vertebral column below the corresponding vertebrae. When the dorsal and ventral roots of the lower lumbar and sacral segments assume a longer course around the conus medullaris to reach the corresponding intervertebral foramina, the cauda equina is formed (Figures 3.1 & 3.2).

The spinal cord is invested by the dura, arachnoid, and the pia mater. The dura mater (pachymeninx), a collagenous tissue, is comprised of an inner meningeal and an outer endosteal layer. The outer endosteal layer forms the periosteum of the vertebral canal and the epineurium (the outermost covering) of the spinal nerves at or slightly beyond the intervertebral foramina. Dural continuation around the filum terminale and the lumbar cistern is known as the dural sac. At the level of the second sacral vertebra the spinal dura joins the filum terminale to attach to the coccyx as the coccygeal ligament. The epidural space (Figures 3.3 & 3.4) contains the internal vertebral plexus, a venous network that maintain connection with the systemic veins.

The arachnoid mater is a loose, irregular, and trabecular layer that is continuous with cranial arachnoid mater. It is generally avascular and surrounds the spinal cord without following the sulci. It is pierced by vessels that supply the pia mater.

The pia mater consists of the epi-pia that contains large vessels and the intima–pial layers. It intimately adheres to the spinal cord, giving rise to the dentate ligaments (Figure 3.3). These ligaments are triangular extensions that extend to the dura, coursing between

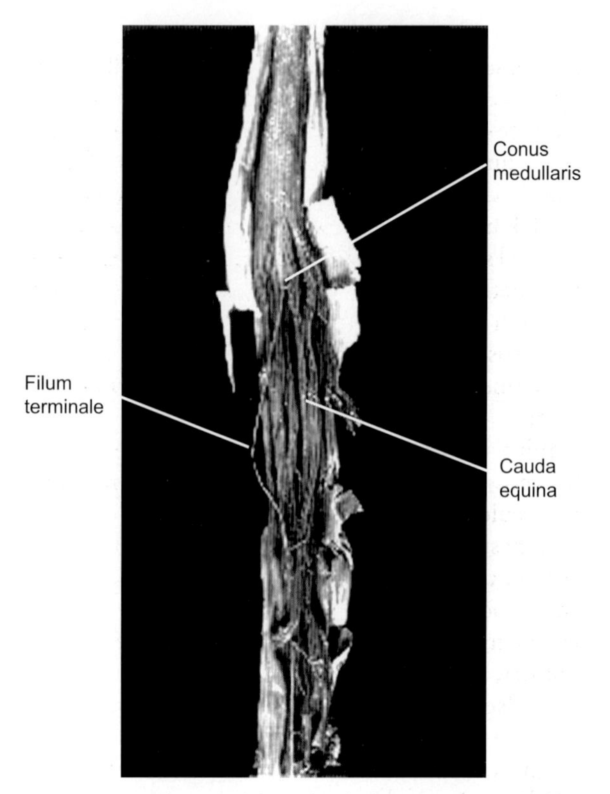

Figure 3.1 The caudal end of the spinal cord, filum terminale, and spinal meninges are shown in this picture

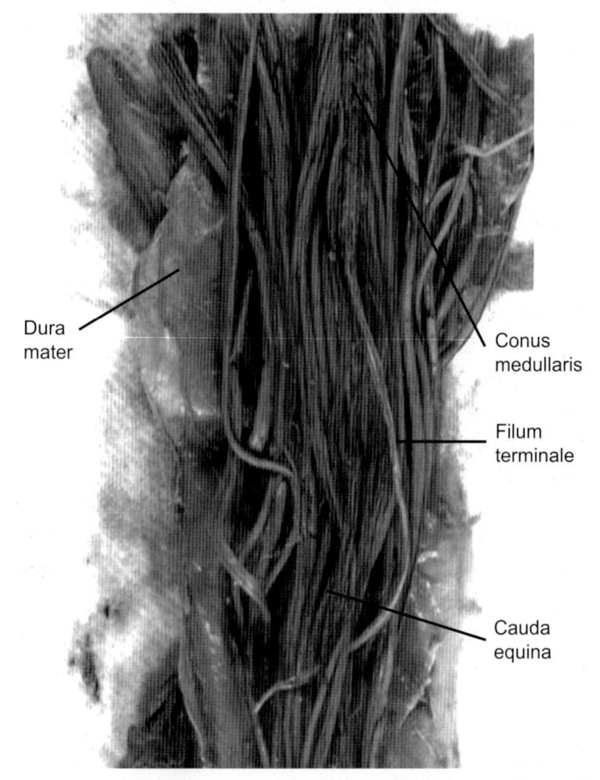

Figure 3.2 A more elaborate picture of the caudal spinal cord. Cauda equina and filum terminale are clearly visible

Pain associated with parturition may be alleviated by injection of local anesthetics into the epidural (extradural) space. Epidural anesthesia is a procedure that involves the administration of local anesthetics via a needle that passes between the L3–L4 lumbar vertebrae. In caudal analgesia the epidural space is reached via the sacral hiatus, utilizing a catheter. Diffusion of the anesthetic solution through the dural coverings of the emerging nerve roots ensures a complete pain blockage and inhibition of perineal reflexes that may unduly prolong the labor.

Spinal epidural abscesses may occur as a result of the posterior spread of infection directly from tuberculous vertebral bodies or during epidural anesthesia. They may also occur as a result of systemic disease and hematogenous spread of *Staphylococcus aureus* infection. Low back pain, which may progress gradually to involve motor, sensory, and/or bowel and urinary incontinence may also be seen in this condition. Infection or abscesses associated with epidural space may be diagnosed via blood and CSF culture as well as CT myelography of the spinal cord.

Hematoma in the epidural space may occur as a result of trauma or spinal diathesis which may partially or completely block the epidural space, producing sudden pain followed by sensory and motor deficits. The dura mater may also be the site of arteriovenous malformation (AVMA) that is commonly seen in the thoracolumbar segments. Dural AVMA may eventually develop into subarachnoid hemorrhage, producing combined upper and lower motor neuron deficits.

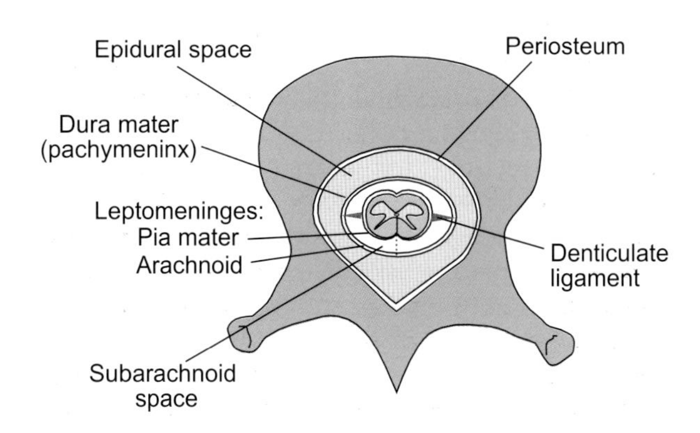

Figure 3.3 Spinal meninges, epidural and subdural spaces, as well as the attachments of the dentate ligaments, are shown

the dorsal and ventral roots. They act as suspensory ligaments for the spinal cord, and extend from the level of the foramen magnum to the level of the first lumbar vertebra. Condensation of the pia mater between the conus medullaris and the second sacral vertebra is known as the filum terminale (internum) (Figures 3.1 & 3.2). Both pia and arachnoid mater form the leptomeninges, and they continue around the spinal nerves as perineurium. The subarachnoid space, between the arachnoid and pia mater, contains the cerebrospinal fluid and spinal arteries and veins. This space shows enlargement around the filum terminale and cauda equina and forms the lumbar cistern.

Blood supply

The vertebral artery (Figure 3.5), the principal arterial source to the cervical segments of the spinal cord, is a branch of the subclavian artery, although occasionally it may arise directly from the aorta, brachiocephalic trunk, or from the thyrocervical trunk. This artery lies anterior to the stellate ganglion, and ascends in the transverse foramina of the upper six cervical vertebrae. In order to gain access to the cranial cavity, it runs through the foramen magnum after piercing the posterior atlanto-occipital membrane. It gives rise to the meningeal, anterior and posterior spinal (Figures 3.6, 3.7 & 3.8), and medullary branches. It joins the vertebral artery of the opposite side to form the basilar artery (Figures 3.5 & 3.7). The vertebral arteries establish anastomosis with the multiple radicular arteries through the spinal branches, external carotid artery through the occipital branch, and subclavian artery through branches of the thyrocervical trunk and occipital artery.

The spinal cord is supplied by the anterior and posterior spinal arteries and by the multiple radicular arteries. The anterior spinal artery is a single vessel that supplies the anterior two-thirds, while the posterior spinal arteries supply the posterior one-third of the spinal cord. On the other hand, the radicular arteries that arise from the neighboring segmental arteries include the ascending cervical, deep cervical, posterior intercostal, lumbar, and lateral sacral arteries. They reach the spinal cord via the intervertebral foramina, following the roots of the spinal nerves, and are considered to be the principal blood supply to the thoracic, lumbar, sacral, and coccygeal spinal segments. Frequently, the radicular arteries are only present on the left side of the thoracic and lumbar spinal segments, and bilaterally in the cervical segments. In 60–65% of individuals one radicular artery (artery of Adamkiewicz or artery of the lumbar

Arachnoiditis (inflammation of the arachnoid matter) may be caused by trauma, invasive imaging procedures (e.g. myelography), or it may remain idiopathic. This condition may produce adhesion of the leptomeninges, obliteration of the subarachnoid space, formation of arachnoid cysts, and possible vascular occlusion.

Arachnoid cysts are congenital outgrowths that assume positions outside (extradural) or inside the dura (intradural). Extradural cysts, common in the thoracic region, may remain asymptomatic or produce compression of the spinal cord and/or roots. Intradural cysts may or may not communicate with the subarachnoid space

The subdural space between the arachnoid mater and the dura mater represents a potential interval, which may accidentally be penetrated during induction of epidural anesthesia, producing toxic effects and eventual spinal cord damage.

Subdural abscess may occur as a result of underlying remote or contiguous infections such as dental, retroperitoneal or tuberculous abscesses. It may also be a spontaneous condition, producing fever, and back pain in the thoracic or lumbar region that radiates to areas of the spinal nerve distributions. Compression of the spinal cord as a result of an abscess may produce paraplegia or quadriplegia. Sensory and/or sphincteric deficits may also occur depending on the site and extent of the abscess.

Subdural hematoma may arise as a result of trauma, anticoagulant therapy, or following lumbar puncture. It produces signs and symptoms similar to subdural abscesses. However, paraplegia and quadriplegia usually occur within minutes to hours in individuals with subdural hematoma, compared to subdural abscess in which the deterioration may take days.

Lumbar puncture (LP), a procedure which is performed to aspirate CSF from the subarachnoid space for the evaluation of signs of meningitis or subarachnoid hemorrhage. It may also be utilized to administer medications to the lumbar cistern. LP or spinal tap is contraindicated in individuals with increased intracranial pressure. This is based upon the fact that a lumbar puncture may precipitate transtentorial herniation by suddenly reducing the pressure in the vertebral canal. The site of puncture in adults is usually between L3–L4 or L4–L5 vertebral interspace, while in infants a much lower level is indicated (L5–S1). In this procedure the skin and interspinous ligaments are anesthetized while the patient lies on his side. The dura and arachnoid mater must be pierced to gain access to the subarachnoid space.

Spinal anesthesia is performed by the injection of anesthetic solution into the lumbar cistern to block the lower thoracic, lumbar, and sacral spinal nerve roots. This procedure is performed when general anesthesia is not desired as in Cesarean section.

Myelography, another procedure that utilizes the lumbar cistern, is used to visualize the vertebral column, spinal cord, and the posterior cranial fossa. A myelographic contrast medium is injected percutaneously via a needle into the lumbar cistern distal to the termination of the spinal cord. The spinal cord and spinal roots become discernible through a series of radiographic images. Since the contrast medium is radiopaque, the spinal cord and the nerve roots may appear radiolucent. For a detailed visualization, CT images may be obtained after contrast injection.

enlargement), generally on the left side, may arise from the lower posterior intercostal arteries or upper lumbar arteries and establish anastomosis with the anterior spinal artery, supplying the lower two thirds of the spinal cord. (Figures 3.6, 3.7 & 3.8).

Venous drainage

The spinal veins form anterior and posterior longitudinal channels. The posterior venous channels drain the posterior half while the anterior venous channels drain the anterior half of the spinal cord. Eventually, these venous channels open into the radicular veins, becoming part of the epidural venous plexus. The epidural venous plexus drains the red bone marrow contained in the vertebral bodies.

Internal organization

Each spinal segment consists of central gray and peripheral white matters that are connected by the corresponding gray and white commissures. The central canal is a tube that pierces the gray commissure of the spinal cord, ascends into the caudal medulla, and continues with the fourth ventricle. This canal does not stretch the entire length of the spinal cord and is frequently obliterated.

Gray matter

The gray matter is a butterfly-shaped area with anterior and posterior horns which are present at all spinal

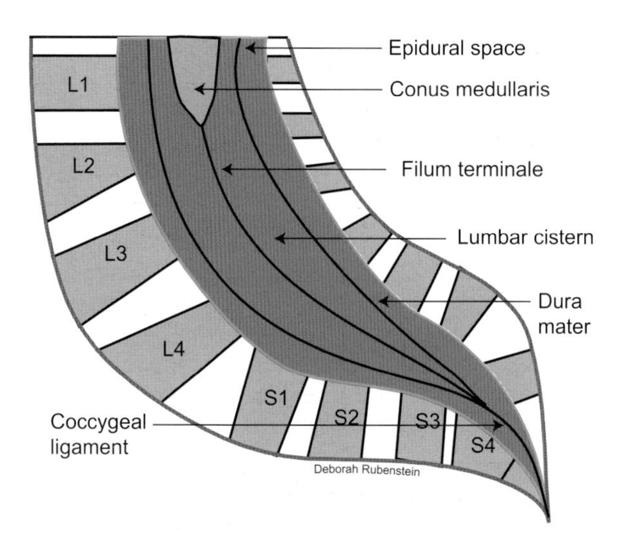

Figure 3.4 Caudal part of the spinal cord showing the meninges, filum terminale, and coccygeal ligament. Corresponding vertebral levels are also shown

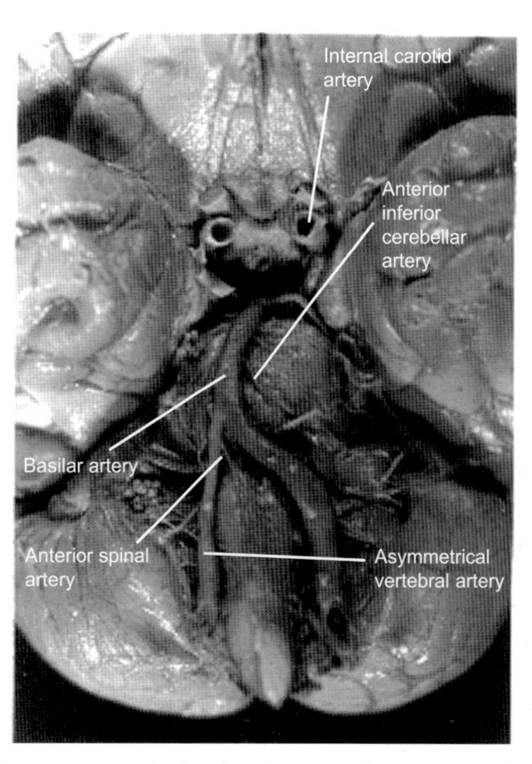

Figure 3.5 Vertebral and anterior spinal arteries are shown on the ventral surface of the medulla

levels. An additional lateral horn that lodges the intermediolateral columns (preganglionic sympathetic neurons) exists in the thoracic and upper two or three lumbar spinal segments. Gray commissures surround the central canal and separate it from the white matter. Most of the spinal cord neurons are small and propriospinal (90%), linking the ventral and dorsal horns within one segment or interconnecting several segments (intersegmental). The intermediate zone between the dorsal and ventral horns is generally formed by medium sized neurons, while the largest neurons occupy the ventral horn.

Based upon the cytoarchitecture of the neuronal cell bodies, the gray matter is classified by Rexed into nine laminae and area or lamina X (Figure 3.9). True lamination is evident in the dorsal horn, and considerable overlap exists among certain laminae.

• Lamina I contains the posteromarginal nucleus, consisting of neurons that display horizontal dendrites in order to maximize their contact with the incoming fibers of the dorsal root. The dorsolateral tract of Lissauer separates this lamina from the surface of the spinal cord.

• Lamina II (substantia gelatinosa) consists of Golgi type II neurons, receiving fibers that carry pain and temperature sensations. Axons of these neurons contribute to the formation of the Lissauer zone (dorsolateral fasciculus). This lamina is the main processing center for nociceptive (noxious) stimuli in the spinal cord.

• Laminae III & IV contain the proper sensory nucleus and occupy a large region of the dorsal horn. This nucleus contributes axons to the lateral spinothalamic tract and receives virtually all sensory modalities carried by the dorsal root.

• Lamina V occupies the neck of the posterior horn and establishes synapses with the corticospinal and rubrospinal tracts. The lateral part of this nucleus is known as the reticular nucleus.

• Lamina VI is present in the spinal cord enlargements and particularly absent in the fourth thoracic through the second lumbar segments.

• Lamina VII forms the intermediate zone, receives fibers from the corticospinal and rubrospinal tracts, and contains the Clarke's, intermediolateral and intermediomedial nuclei. Clarke's nucleus extends from the eighth cervical or first thoracic to the second or third lumbar spinal segments, giving rise to the dorsal spinocerebellar tract. The intermediolateral nucleus occupies the lateral horn between the first thoracic and the second or third lumbar spinal segments, providing preganglionic sympathetic axons. At the second, third, and fourth sacral spinal segments, this nucleus provides preganglionic parasympathetic fibers. The intermediomedial nucleus extends the entire length of the spinal cord and receives visceral afferents.

• Lamina VIII occupies the anterior horn in the spinal cord enlargements, and contains commissural

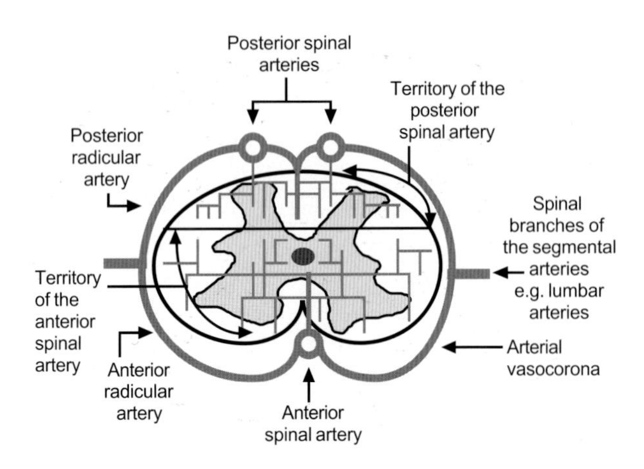

Figure 3.6 Distribution of the anterior and posterior spinal arteries and the formation of the arterial vasocorona

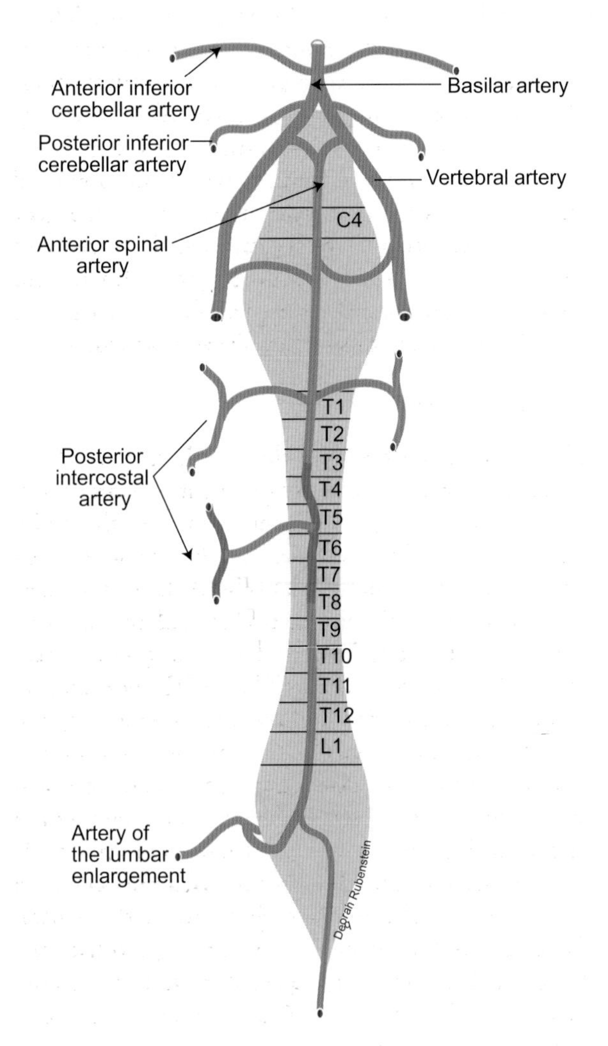

Figure 3.7 Schematic drawing of the anterior spinal artery and its connections to the radicular arteries

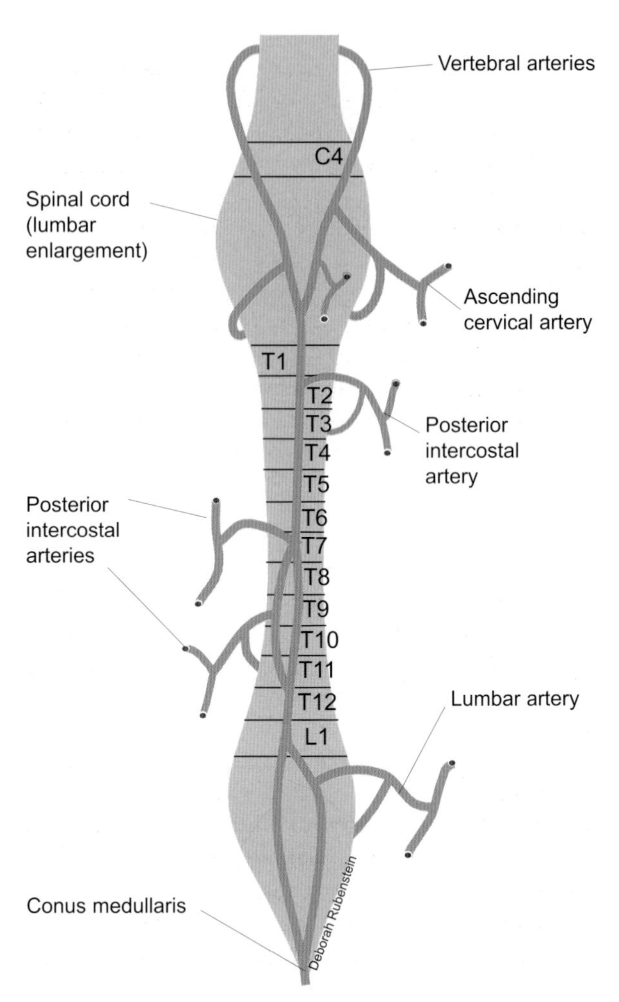

Figure 3.8 Posterior spinal arteries and territory of distribution. Note their connections to the multiple radicular arteries

Spinal segments T1–T4 and L1 are predisposed to infarctions due to the lack of sufficient arterial anastomotic channels and the great distance between the radicular arteries. These watershed infarctions may be seen as a sequel to cardiac arrest, clamping of the aorta, or acute local ischemia. Occlusion of the artery of lumbar enlargement (artery of Adamkiewicz) may produce paraplegia (paralysis of the lower extremities and lower parts of the body), urinary incontinence, and loss of sensation from the lower extremities. Occlusive disease of the anterior spinal artery (Beck's syndrome), as a result of aortic dissecting aneurysm or atheroma, produce combined sensory and motor deficits.

Numerous connections exist at each intervertebral space between the epidural (internal vertebral) venous plexus and systemic veins (superior and inferior vena cava) via the azygos and hemiazygos veins. These connections may serve as a potential route for spread of cancer cells from the thyroid gland and prostate to the vertebral bodies.

The internal vertebral (epidural) plexus affects CSF pressure by forming continuous tamponade of the spinal dural sac. Increased intrathoracic or intra-abdominal pressure (coughing, sneezing, and straining during defecation, or abdominal compression) can thereby increase CSF pressure.

neurons which receive axons of the vestibulospinal, pontine reticulospinal and tectospinal tracts.

- Lamina IX contains α and γ motor neurons that innervate the extrafusal and intrafusal muscle fibers, respectively. The α motor neurons receive excitatory input from the descending pathways and the reflex arcs, and inhibitory input from the propriospinal neurons. Excitatory input far exceeds the inhibitory projections by a ratio of 2:1. They give inhibitory recurrent branches to the interneurons (Renshaw cells), thus facilitating their action. In general, α motor neurons are arranged somatotopically, in which the abductor neurons are located anteriorly, the flexor neurons are positioned posteriorly, and the extensors as well as the adductor neurons maintain intermediate positions. In the lumbosacral segments the neurons for the trunk are medial, the neurons that innervate the foot occupy a lateral position, while neurons for the leg and thigh have intermediate position. In the thoracic segments, lamina IX exhibits a similar somatotopic arrangement whereby the neurons associated with innervation of the abdomen lie medial to the intercostal neurons, and the neurons for the innervation of the back muscles and skin assume an intermediate position. In the cervical segments the neurons for the hand are lateral to the neurons that supply the forearm, whereas trunk neurons are the most medial. Neurons for the arm and shoulder occupy position medial to the forearm and lateral to the trunk neurons. These neurons are classified into tonic and phasic neurons. The tonic α motor neurons innervate the slow, oxidative–glycolytic muscle fibers, exhibiting slow conduction and the ability to readily depolarize. They are inhibited during rapid movement by the Renshaw cells. Phasic neurons display higher threshold and ability to maintain fast conduction, innervating the fast and oxidative–glycolytic muscles. Phasic neurons also send more recurrent branches to the Renshaw cells than the tonic neurons. The γ neurons are located among the α motor neurons, innervating the contractile parts of the muscle spindles. Both α and γ neurons are involved in voluntary movement via the α–γ co-activation and γ loop.

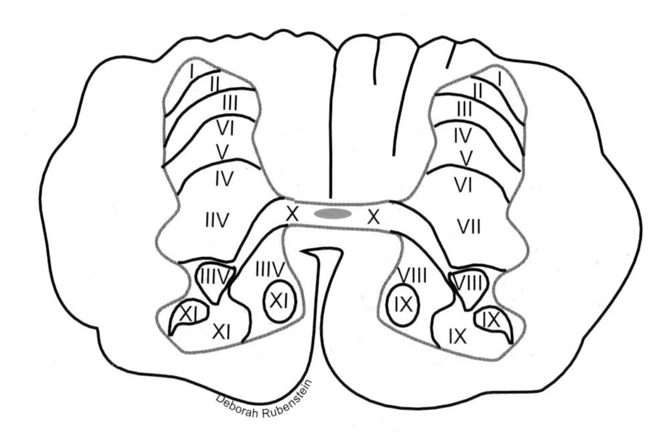

Figure 3.9 Rexed laminae and the cytoarchitecture of the gray columns of the spinal cord

- Lamina X or area X according to some unsupported claims consists of small neurons that form the gray commissures around the central canal. It receives some afferents from the dorsal root fibers and contains neuroglial cells in its ventral part that send cytoplasmic extensions to the adjacent pia mater.

White matter

The white matter occupies the peripheral part of the spinal cord and consists only of neuronal processes. The anterior white commissure connects the white matter on both sides, representing the site of decussation of the lateral and ventral spinothalamic tracts, as well as the ventral spinocerebellar and the anterior corticospinal tracts. The part of the white matter located between the entering fibers of the dorsal roots is known as the dorsal funiculus, containing the dorsal white columns. The part of the white matter that lies between the dorsal and ventral roots on each side is known as the lateral funiculus, containing the lateral corticospinal, rubrospinal and lateral spinothalamic tracts. The area of the white matter between the emerging ventral roots is referred to as the ventral funiculus and contains the ventral spinothalamic, tectospinal, and the reticulospinal tracts as well as the medial longitudinal fasciculus. A tract refers to a group of nerve fibers that have the same origin, destination, course, and function. A fasciculus shares common features of the tract, but the constituent fibers maintain diverse origins.

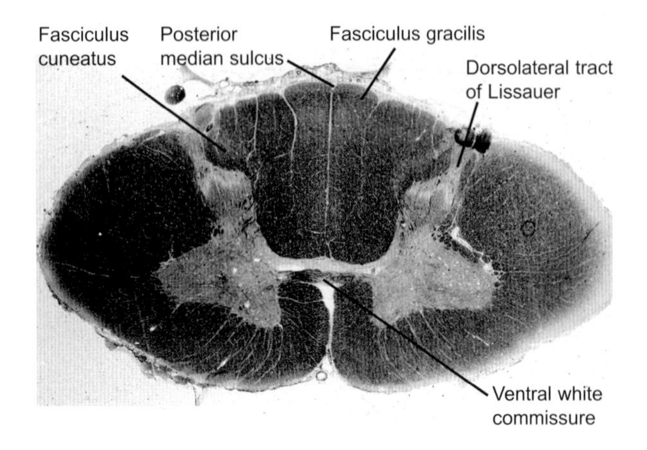

Figure 3.10 Fourth cervical segment. Note the wide transverse diameter and large amount of white and gray matter

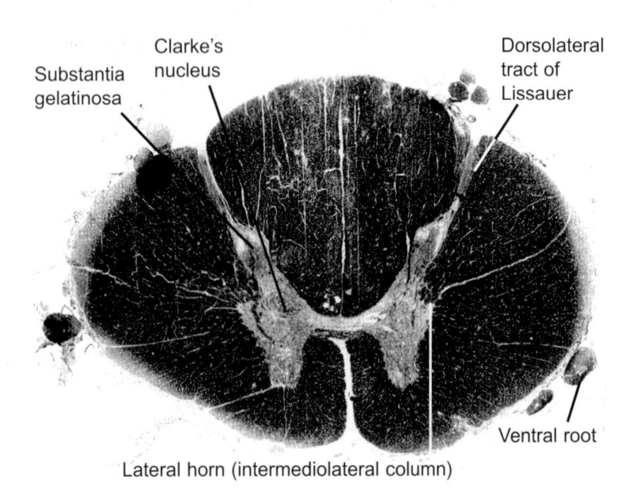

Figure 3.11 Transverse section through the fifth thoracic spinal segment, showing Clarke's nucleus and the intermediolateral column, substantia gelatinosa, and the dorsal tract of Lissauer

Spinal cord segments

The cervical spinal segments are eight in number, are generally large, and have a greater mass of white matter. The dorsal funiculus is divided into a medial gracilis and a lateral cuneatus fasciculi (Figure 3.10).

The thoracic segments (Figure 3.11), which are characterized by a small and distinct lateral horn, contain the intermediolateral cell column which gives rise to the preganglionic sympathetic fibers. The dorsal funiculi of the upper six spinal segments contain the gracilis and cuneatus fasciculi, while the lower six thoracic segments contain only the gracilis fasciculus. Another important feature of the thoracic segments is the presence of Clarke's nuclear column, which is particularly well developed, in the lower two

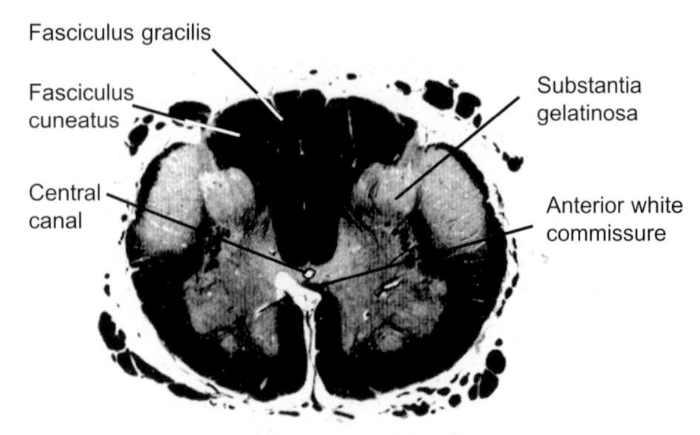

Figure 3.12 Section of the fourth lumbar segment showing the main characteristics of this level

thoracic spinal segments. The axons of this nuclear column form the ipsilateral dorsal spinocerebellar tract that conveys unconscious proprioceptive information from the muscle spindles and Golgi tendons of the lower extremities. Additionally the gray matters of the thoracic segments are tapered in an H-shape.

The axons of this nuclear column form the ipsilateral dorsal spinocerebellar tract that conveys unconscious proprioceptive information from the muscle spindles and Golgi tendons of the lower extremities.

The lumbar segments (Figure 3.12) contain massive amounts of gray matter and relatively less white matter. The upper two lumbar segments contain the continuation of Clarke's nucleus and the intermediolateral columns.

The sacral segments (Figure 3.13) are small compared to other segments and contain large amounts of gray matter. The intermediolateral cell column in the sacral spinal segments provides preganglionic parasympathetic fibers.

Spinal pathways

Ascending tracts

These ascending pathways (Figure 3.14) convey conscious and unconscious sensory information to the higher levels of the CNS. The first order neurons for all ascending tracts from the body are located in the dorsal root ganglion (DRG) of the spinal nerves. The second order neurons are located either in the gray matter of the spinal cord or in the brainstem. The ventral posterolateral (VPL) nucleus of the thalamus constitutes the third order neurons for these pathways. The signals for the information conveyed by the ascending pathways are concerned with the regulation of muscle tone, joint sensation (position sense), vibration, pain and temperature sensations, discriminative tactile sensations, and intersegmental reflexes. These

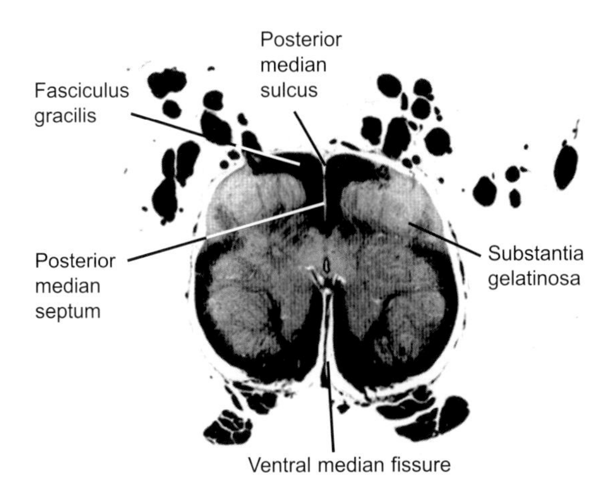

Figure 3.13 Section through the fifth sacral spinal segment

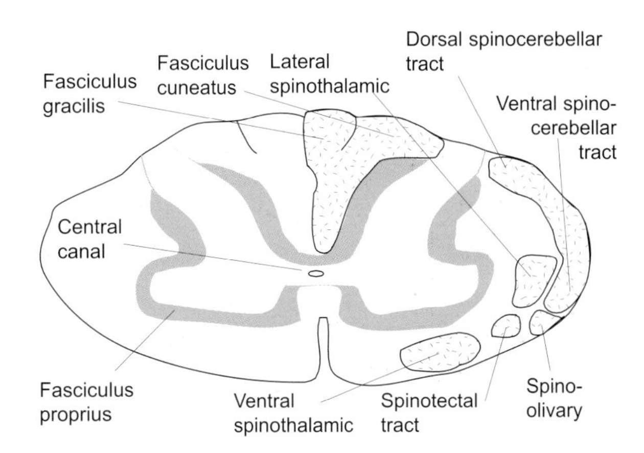

Figure 3.14 Principal ascending pathways within the spinal cord

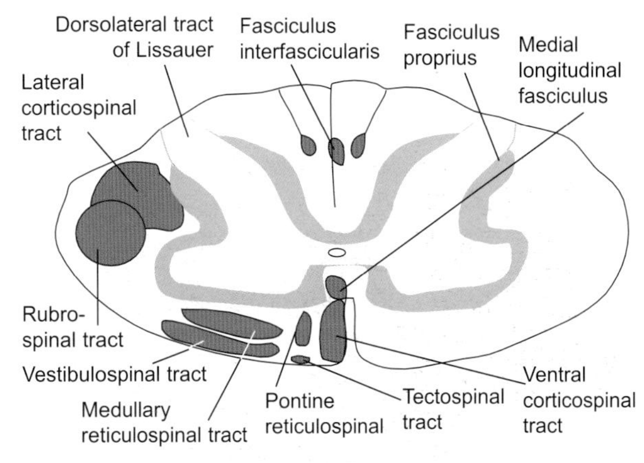

Figure 3.15 Section of the spinal cord showing the descending pathways

pathways may establish monosynaptic connections or utilize an extensive network of neurons and are contained in the funiculi of the spinal cord.

Ascending tracts in the posterior funiculus

The dorsal white columns transmit fine tactile and vibratory sense via the pacinian corpuscles, position and movement sense (kinesthesia) from the muscle spindle. They also convey two point discrimination of simultaneously applied blunt pressure points from the Ruffini corpuscles and stereognosis (ability to recognize form, size, texture and weight of objects) via a variety of receptors.

Ascending tracts in the lateral funiculus

The lateral spinothalamic tract (neospinothalamic), also known as the lateral system, is a contralateral pathway that conveys thermal and painful sensations from somatic and visceral structures. Pain and temperature, received by the free nerve endings, enter the spinal cord via the lateral bundle of the dorsal root.

In the peripheral parts of the lateral funiculus, the dorsal and ventral spinocerebellar tracts are located, carrying unconscious proprioception from the lower extremity to the cerebellum.

Ascending tracts in the ventral funiculus

The ventral spinothalamic tract (paleospinothalamic or anterior system) runs in the ventral funiculus and transmits signals associated with light touch, and possibly tickling, itching and libidinous sensations. Since fine touch and discriminative tactile sensation are primarily carried in the dorsal columns, the clinical significance of this pathway is not clear.

The spino-olivary tract is a contralateral tract, conveying cutaneous information and afferents from Golgi tendon organs to the dorsal and medial accessory olivary nuclei.

The spinoreticular tract is an integral part of the ascending reticular activating system that plays an important role in changing the electrocortical activity of the cerebrum, regulating the state of consciousness and awareness. It establishes a direct link between the spinal cord and the brainstem reticular formation, extending the entire length of the spinal cord. It also contributes to the formation of the spinoreticulothalamic tract.

The spinotectal tract is composed of axons of neurons that are derived from lamina VII of the spinal gray matter. Although the functional significance of this pathway is not clear, its role in modulating the transmission of pain, thermal, and tactile sensation, awaits further study.

Descending tracts

The descending pathways (Figure 3.15) deal with maintenance of posture and balance, control of

Table 3.1 Ascending and descending tracts

Funiculi	*Ascending tracts*	*Descending tracts*
Posterior funiculus	Dorsal white columns	Fasciculus interfascicularis and fasciculus septomarginalis
Lateral funiculus	Lateral spinothalamic and dorsal spinocerebellar tracts	Corticospinal and rubrospinal tracts
Ventral funiculus	Ventral spinothalamic, spino-olivary, spinoreticular and spinotectal tracts	Anterior corticospinal, vestibulospinal, reticulospinal and medial longitudinal fasciculus

visceral and somatic reflex activity, muscle tone, motor activity in general, and modification of the sensory signals. The descending pathways include the corticospinal, rubrospinal, tectospinal, and interstitiospinal tracts. They also include the vestibulospinal and reticulospinal pathways, as well as descending autonomic pathways that are derived from the hypothalamus and the brainstem reticular formation.

Descending tracts in the lateral funiculus

The lateral corticospinal tract is a phylogenetically new pathway that exists in man and other mammals. It continues to develop throughout the first two years of life. This pathway forms the largest crossed component of the corticospinal tract, controlling voluntary motor functions, especially movement associated with the digits.

The rubrospinal tract is a contralateral tract that may be traced from the superior collicular level of the midbrain to the thoracic or lumbosacral segments. This excitatory pathway, which regulates the neurons of the flexor muscles, originates from the magnocellular part of the red nucleus.

Descending tracts in the ventral funiculus

The anterior corticospinal tract represents approximately 10–15% of the corticospinal fibers and travels ipsilaterally in the spinal cord near the anterior median fissure. This pathway influences muscles of the upper extremity and the neck via synapses in the corresponding segments.

The vestibulospinal tracts include the lateral and medial vestibulospinal tracts. The lateral vestibulospinal tract is excitatory and runs the entire length of the spinal cord. The medial vestibulospinal tract is monosynaptically inhibitory and extends to the upper cervical segments.

The reticulospinal tracts comprise the ipsilateral pontine reticulospinal and the predominantly ipsilateral medullary reticulospinal tracts. These tracts convey information received by the reticular formation from the cerebral cortex, cerebellum, cranial nerves, and hypothalamus to the spinal cord.

The medial longitudinal fasciculus is a composite bundle of ascending and descending fibers that originate from vestibular and reticular nuclei. This tract occupies the dorsal portion of the ventral funiculus.

The spinospinal tract (fasciculus proprius) is an intersegmental tract of ascending and descending fibers (crossed and uncrossed), which mediates the intrinsic reflex mechanisms of the spinal cord. It exists in all spinal funiculi, conveying information to higher segments prior to establishing contact with interneurons.

Suggested reading

1. Abdel-Maguid TE, Bowsher D. Alpha and gamma-motoneurons in the adult human spinal cord and somatic cranial nerve nuclei. The significance of dendroarchitectonics studied by the Golgi method. *J Comp Neurol* 1979;186:259-70
2. Appel NM, Lide RP. The intermediolateral cell column of the thoracic spinal cord is comprised of target-specific subnuclei: Evidence from retrograde transport studies and immunohistochemistry. *J Neurosci* 1988; 8:1767–75
3. Batson AJ, Sands J. Regional and segmental characteristics of the human adult spinal cord. *J Anat* 1977;123:797–803
4. Biglioli P, Spirito R, Roberto M, *et al*. The anterior spinal artery: the main arterial supply of the human spinal cord-a preliminary anatomic study. *J Thorac Cardiovasc Surg* 2000;119:376–9
5. Brem SS, Hafler DA, Van Uitert RL, *et al.* Spinal subarachnoid hematoma: A hazard of lumbar puncture resulting in reversible paraplegia. *N Engl J Med* 1981;303:1020–1

6. Brichta AM, Grant G. Cytoarchitectural organization of the spinal cord. In Paxinos G, ed. *The Rat Nervous System*, volume 2. Orlando: Academic Press, 1985;293–300
7. Clark RG. Anatomy of the mammalian spinal cord. In Davidoff RA, ed. *Handbook of the Spinal Cord*, volumes 2/3. New York: Dekker, 1984;1–45
8. Gaffney P, Guthrie JA. Spontaneous spinal epidural haematoma: an unusual cause of neck pain. *J Accid Emerg Med* 2000;17:229–30
9. Heimer L. *The Human Brain and Spinal Cord. Functional Neuroanatomy and Dissection Guide*, 2nd edn. London: Springer-Verlag, 1995
10. Horlocker TT. Complications of spinal and epidural anesthesia. *Anesthesiol Clin North Am* 2000;18: 461–85
11. Kimmel DL. Innervation of spinal dura mater and dura mater of the posterior cranial fossa. *Neurology* 1961;800–9
12. McMenemin IM, Sissons GR, Brownridge P. Accidental subdural catheterization: radiological evidence of a possible mechanism for spinal cord damage. *Br J Anaesth* 1992;69:417–19
13. Rodriguez-Baeza A, Muset-Lara A, Rodriguez-Pazos M, *et al*. The arterial supply of the human spinal cord: a new approach to the arteria radicularis magna of adamkiewicz. *Acta Neurochir* 1991;109:57–62
14. Schneider RC, Crosby EC, Russo RH, *et al*. Traumatic spinal cord syndromes and their management. *Clin Neurosurg* 1972;20:424–92
15. Stein B. Intramedullary spinal cord tumors. *Clin Neurosurg* 1983;30:717–41
16. Tamir E, Mirovsky Y, Halperin N. [Epidural spinal abscess]. *Harefuah* 1999;136:120–2
17. Tekkok IH, Cataltepe K, Tahta K, *et al.* Extradural hematoma after continuous extradural anesthesia. *Br J Anaesth* 1991;67:112–15
18. Tobin WD, Layton DD. The diagnosis and natural history of spinal cord arteriovenous malformations. *Mayo Clin Proc* 1976;51:637–46
19. Ullah M, Salman SS. Localization of the spinal nucleus of the accessory nerve in the rabbit. *J Anat* 1986;145: 97–107
20. Wakabayashi K, Takahashi H. The intermediolateral nucleus and clarke's column in parkinson's disease. *Acta Neuropathol* 1997;94:287–9

Brainstem 4

The brainstem is the infratentorial portion of the central nervous system (CNS), consisting of the medulla, pons and mesencephalon (midbrain). It is connected to the cerebellum via the cerebellar peduncles. The brainstem contains the fourth ventricle, cerebral aqueduct, central canal, and the nuclei of the associated cranial nerves. The reticular formation occupies the central portion of the brainstem, containing the respiratory and cardiovascular centers.

Medulla
- Spinomedullary junction
- Level of the decussation of the internal arcuate fibers
- Midolivary level
- Rostral medulla
- Pontomedullary junction

Pons
- Caudal pons
- Midpons (level of the trigeminal nerve)
- Rostral pons (isthmus)

Mesencephalon (midbrain)
- Inferior collicular level
- Superior collicular level

Medulla

The medulla (Figure 4.1) represents the caudal part of the brainstem, continuing caudally with the spinal cord through the foramen magnum. The rostral end of the medulla is demarcated ventrally by the pontobulbar sulcus, giving passage to the abducens, facial, and vestibulocochlear nerves. Dorsally, it is bounded by a line joining the lateral recesses of the fourth ventricle. The dorsal surface of the rostral half of the medulla (open part) forms the lower part of the rhomboid fossa, while the caudal half (closed part) contains the central canal. The junction of the caudal and rostral medulla is demarcated on the dorsal surface by the obex. The ventral median fissure of the medulla is continuous with the corresponding fissure in the spinal cord, bounded on both sides by the pyramids, and it is obliterated at its caudal part by the decussating fibers of the corticospinal tracts. The olivary eminence lies lateral to the pyramid, and is formed by the underlying inferior olivary nuclear complex. The anterolateral (preolivary) sulcus, which contains the filaments of the hypoglossal nerve, separates the pyramids from the olivary eminence. The post-olivary sulcus is the site of attachment for the glossopharyngeal, vagus and accessory nerves.

Caudally, the posterior surface of the medulla contains the posterior median sulcus and is flanked on both sides by the fasciculi gracilis. The medially located fasciculi gracilis are separated from the laterally located fasciculi cuneatus by the cranial extension of the posterior intermediate septum. These fasciculi terminate in the corresponding tubercles that overlie their respective nuclei. The gracilis and cuneate nuclei receive information concerning conscious proprioception, two-point discrimination, vibratory sense, and discriminative tactile sensation from the upper and lower part of the body. Axons of gracilis and cuneate neurons project to the contralateral ventral posterolateral nucleus of the thalamus via the medial lemniscus. The gracilis fasciculus also acts as a conduit for impulses that originate from stretch receptors and Golgi tendon organs of the lower extremity (unconscious proprioception) and terminates in the Clarke's nucleus. The cuneate fasciculus also contains sensory impulses generated from the stretch receptors and Golgi tendon organs of the upper extremity (unconscious proprioception) en route to the accessory (external) cuneate nucleus. The caudal part of the medulla, between the fasciculus cuneatus and the accessory nerve, contains a prominence produced by the underlying spinal trigeminal tract and nucleus. The central canal runs close to the posterior surface and is continuous with the corresponding canal in the spinal cord caudally and the fourth ventricle rostrally.

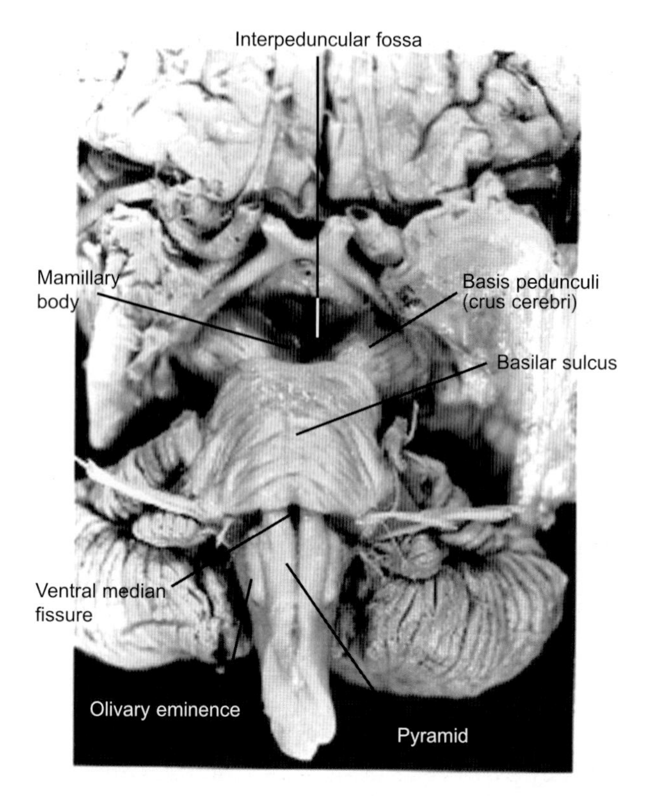

Figure 4.1 Ventral surface of the brainstem and its connections to the cerebellum. Some of the important features of the medulla, pons, and midbrain are shown in this picture

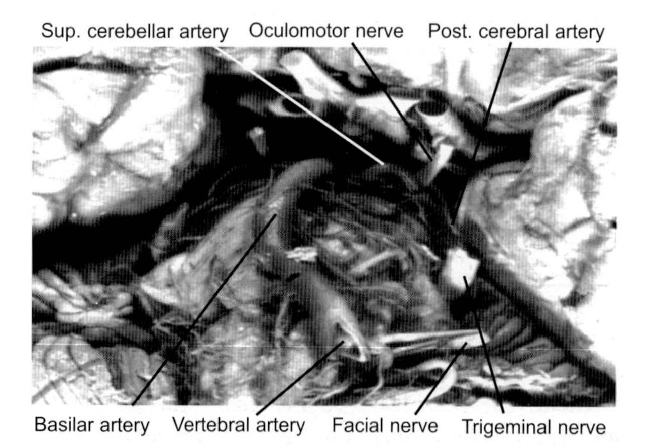

Figure 4.2 Relationship of the ventral surface of the brainstem to the vertebrobasilar arterial system is shown. Most of the associated cranial nerves are also seen. Post., posterior; Sup., superior

The rostral (open) part of the medulla is bounded laterally by the inferior cerebellar peduncles that meet at the obex. It contains the fourth ventricle and forms the lower half of the rhomboid fossa. In this region the rhomboid fossa contains the hypoglossal trigone medially, and the vagal trigone laterally. These trigones overlie the hypoglossal and dorsal motor nucleus of the vagus, respectively. The stria medullaris

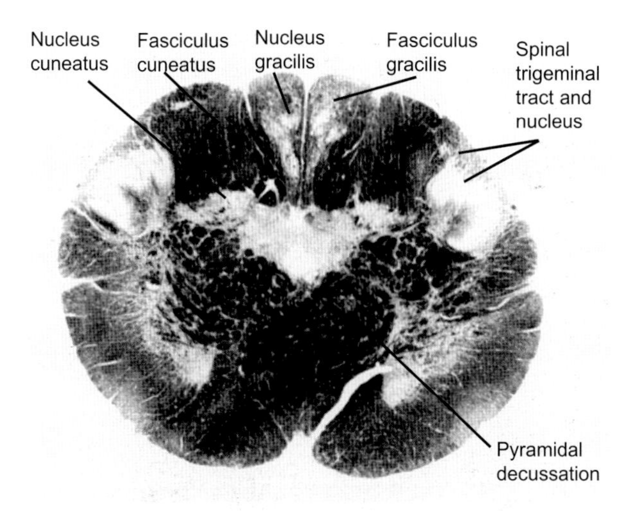

Figure 4.3 Level of the decussation of the corticospinal tracts. The spinal trigeminal tract and nucleus and the dorsal column nuclei are clearly demonstrated

A lesion at the site of pyramidal decussation produces cruciate hemiplegia, which is characterized by paralysis of the ipsilateral upper extremity and contralateral lower extremity.

emanates from the arcuate nuclei, crossing the inferior cerebellar peduncles and the vestibular area of the rhomboid fossa. The vestibular area corresponds to the site of the foramen of Luschka, overlying the medial and inferior vestibular nuclei.

The medial part of the medulla containing the pyramids, medial lemniscus, the hypoglossal nerve and nucleus, and inferior olivary nucleus is supplied by the anterior spinal artery and the bulbar branches of the vertebral artery. The dorsal and lateral parts of the caudal medulla, which contain the associated cuneate, gracilis, dorsal vagal, and solitary nuclei and fasciculi, receive blood supply from the posterior spinal and posterior inferior cerebellar branches of the vertebral artery (Figure 4.2). The latter arterial branch provides blood supply to the lateral medulla and inferior cerebellar peduncle.

Since the pain and thermal fibers terminate in the most caudal portion of the trigeminal nucleus and tract, excision of the spinal trigeminal tract (tracteotomy) may be performed at the level of the medulla to alleviate the intractable pain associated with trigeminal neuralgia, although this is no longer carried out in practice.

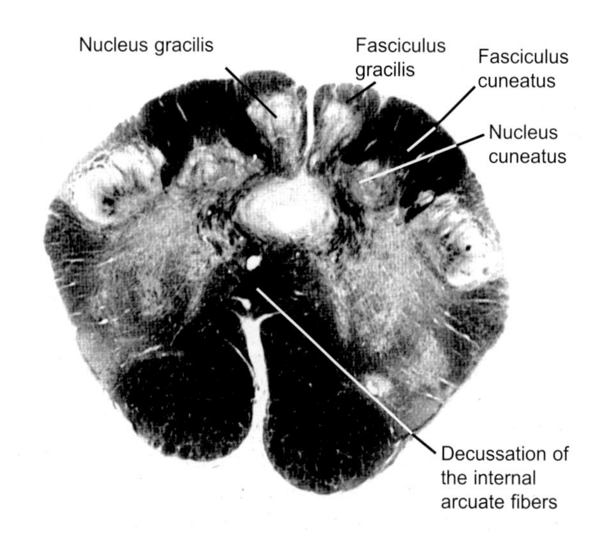

Figure 4.4 Section through the caudal medulla at the level of the sensory decussation. At this level the medial lemnisci are formed. The cuneate and gracilis fasciculi maintain similar positions to Figure 4.3

Venous drainage of the caudal medulla is maintained by the anterior and posterior spinal veins. The rostral medulla drains into the sigmoid, superior or inferior petrosal sinus.

Medullary structures are clearly prominent at three levels and each level may present unique features and contain nuclei or pathways that extend to a more caudal or rostral level. These levels are comprised of the spinomedullary junction, midolivary level, rostral medulla, and the pontomedullary junction.

Spinomedullary junction

The spinomedullary junction (Figure 4.3) marks the gradual transition between the spinal cord and the medulla. It contains, for the most part, structures that extend from the spinal cord with some additional unique neurons, specific to this level of the medulla such as the gracilis and cuneate nuclei. A unique feature of this level is the pyramidal decussation, which marks the site of crossing of the corticospinal fibers.

At a more rostral level, the spinal trigeminal nucleus and tract replace the substantia gelatinosa and the dorsolateral tract of Lissauer, respectively. The spinal trigeminal nucleus extends from the midpons to the upper segments of the spinal cord, representing the rostral extension of the substantia gelatinosa. It receives thermal, painful, and tactile sensations from the head region through branches of the trigeminal, facial, glossopharyngeal, and vagus nerves. The fibers that terminate in this nucleus form the spinal trigeminal tract lateral to the corresponding nucleus. Within the spinal trigeminal nucleus and tract, the

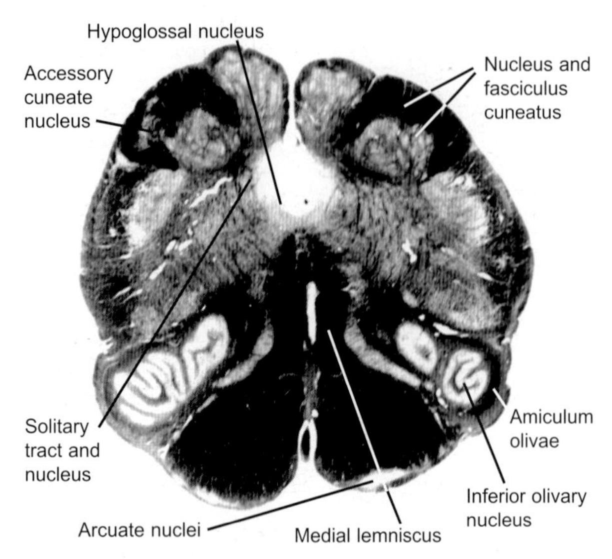

Figure 4.5 Caudal medulla rostral to the level shown in Figure 4.4. The inferior olivary, hypoglossal, solitary, gracilis and cuneatus are clearly evident

The part of the rhomboid fossa immediately rostral to the obex, the area postrema, is a chemoreceptor trigger zone that regulates blood pressure and emesis, responding to apomorphine and certain glycosides. This area occupies the ependyma of the floor of the fourth ventricle. Its activation may account for the emetic action of certain medications. It is rich in dopaminergic receptors, which may account for nausea associated with the administration of high doses of levodopa, and the antiemetic action of the dopamine antagonists. In addition the area postrema, an integral part of the circumventricular organs, lacks the blood–brain barrier. Brain barriers are structures that selectively allow certain substances to enter the CNS and exclude others.

ophthalmic nerve fibers are caudal to the more rostral mandibular nerve fibers, and the maxillary nerve fibers maintain an intermediate position.

The gracilis and cuneate fasciculi are prominent at this level and occupy the corresponding positions to that of the spinal cord. The most significant characteristics of this level are the decussation of the corticospinal tracts and the appearance of the dorsal column nuclei. The crossed fibers represent nearly 85% of the corticospinal tract. The uncrossed fibers form the anterior corticospinal tract, which descends into the anterior funiculus, and later crosses at the anterior white commissure. Some uncrossed fibers pass into the lateral funiculus, forming the anterolateral corticospinal tract.

The medial longitudinal fasciculus (MLF) is displaced laterally by the pyramidal fibers. The gray matter around the central canal is markedly expanded into the reticular formation. The spinocerebellar and spinothalamic tracts occupy the same position as in the spinal cord.

Level of the decussation of the internal arcuate fibers

The gracilis nucleus (Figures 4.4, 4.5 & 4.6) occupies a larger area at this level, stretches rostrally to the level of the obex, and is covered by a thin strip of the gracilis fasciculus. The laterally positioned cuneate nucleus also increases in size and extends more rostrally than the gracilis nucleus, reaching to the midolivary level. Also at this level, the fasciculus cuneatus is greatly reduced. The internal arcuate fibers, axons of the dorsal column neurons, run ventromedially through the reticular formation, decussating in the midline and continuing as the medial lemniscus on the opposite side. This sensory decussation (decussation of the internal arcuate fibers) is a landmark for this level of the medulla. The accessory (lateral) cuneate nucleus initially appears at this level as a lateral appendage to the cuneate nucleus, receiving information from Golgi tendon organs, muscle spindles, and tactile receptors of the upper extremity. It conveys this information to the ipsilateral cerebellum, via the cuneocerebellar tract, and also to the caudal part of the ventral posterolateral nucleus (VPLc) of the thalamus.

The central gray expands into the reticular formation, containing the lateral, dorsal, and ventral reticular nuclei. The spinal trigeminal tract and nucleus occupy a dorsolateral position. The hypoglossal nucleus, the dorsal motor nucleus of vagus, the solitary nucleus, the ambiguus nucleus, the inferior olivary nuclear complex and the arcuate nuclei are first seen at this level. These nuclei will be discussed at the midolivary level.

Midolivary level

The midolivary level (Figures 4.6 & 4.7) contains important medullary structures such as the medial and inferior vestibular, inferior olivary, hypoglossal, solitary, arcuate, and dorsal motor vagal nuclei. Most of the ascending and descending pathways associated with the spinal cord and lower medulla continue at this level. The central canal is converted into the fourth ventricle at this level of medulla, while the fourth ventricle expands to maintain a close relationship to certain medullary nuclei.

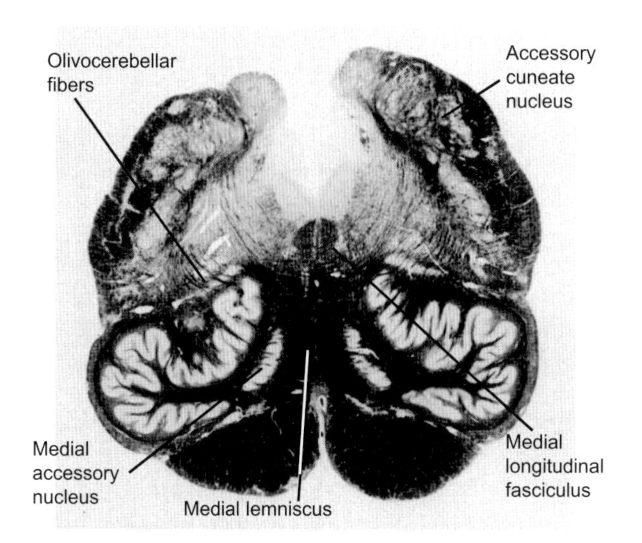

Figure 4.6 The prominent inferior olivary nucleus and expansion of the fourth ventricle are characteristics of this level. The medial longitudinal fasciculus and medial lemniscus maintain similar positions to Figure 4.5

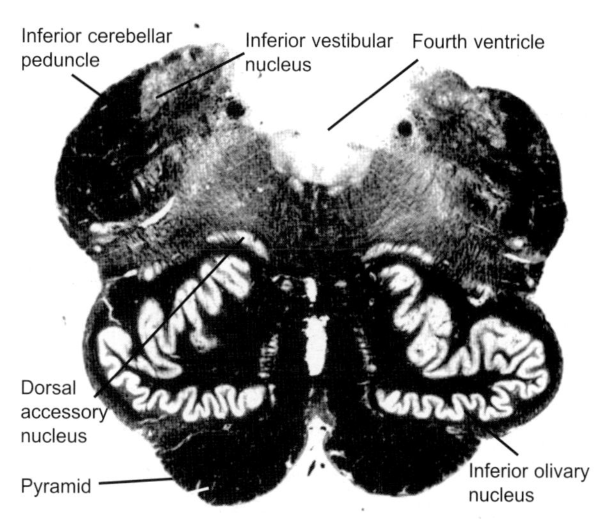

Figure 4.7 Medulla at a more rostral level than in Figure 4.6. Note the great expansion of the inferior olivary and vestibular nuclei, as well as the fourth ventricle

The medial and inferior vestibular nuclei replace the dorsal column nuclei at this level of the medulla. The medial vestibular nucleus (MVN) lies medial to the inferior vestibular nucleus, continuing rostrally with the superior vestibular nucleus. Both the medial and inferior vestibular nuclei receive primary vestibular fibers and convey the information through the juxtarestiform body to the cerebellum. Information received by the medial vestibular nucleus is transmitted to the cervical segments of the spinal cord through the predominantly ipsilateral medial vestibulospinal tract, a component of the MLF. Ascending fibers from the medial and inferior vestibular nuclei also project to the extraocular motor nuclei via the MLF. The medial longitudinal fasciculus contains axons of the vestibular nuclei that project to the extraocular motor nuclei and to the spinal cord.

The inferior vestibular nucleus, the smallest of the vestibular nuclei, extends from the upper end of the nucleus gracilis to the pontomedullary junction. This nucleus is located between the MVN and the inferior cerebellar peduncle. It is characteristically speckled due to crossing of longitudinally oriented primary vestibular fibers. Secondary vestibulocerebellar fibers originate almost exclusively from this nucleus and enter the cerebellum through the juxtarestiform body, terminating in the flocculonodular lobe of the cerebellum.

The arcuate nuclei are cerebellar relay nuclei that occupy a position ventral to the pyramids. These nuclei are derived from the rhombic lip and project to the cerebellum via the inferior cerebellar peduncle. Axons of the arcuate nuclei pursue either a midline course, which decussate in the midline and form the stria medullaris of the fourth ventricle, or a more lateral course superficial to the inferior olivary nucleus, forming the posterior external arcuate fibers. The arcuate nuclei, like the pontine nuclei, receive cortical input and convey this information to the opposite cerebellum.

The inferior olivary nuclear complex, the most characteristic feature of this level, occupies the area above the pyramid. It is the largest nucleus in the medulla, which projects to the cerebellum. It consists of the principal and accessory olivary nuclei. The principal olivary nucleus receives the ipsilateral rubro-olivary component of the central tegmental tract. It also receives bilateral cortico-olivary fibers, projections from the periaqueductal gray matter, crossed fibers from the cerebellar cortex and projections from the accessory oculomotor nuclei. These fibers enclose the inferior olivary nuclear complex as amiculum olivae. The axons of the principal inferior olivary nucleus form the largest component of the inferior cerebellar peduncle. This crossed olivocerebellar tract projects to the lateral parts of the cerebellar cortex. The medial and dorsal accessory olivary nuclei receive impulses from the vestibular, gracilis and cuneate nuclei, and from the spinal cord. They convey this information to the medial portions of the cerebellar cortex.

The hypoglossal nucleus occupies the hypoglossal trigone in the floor of the fourth ventricle. It consists of multipolar neurons that supply general somatic efferents (GSE) to the lingual muscles. This nucleus extends from the level of the inferior olivary nuclear

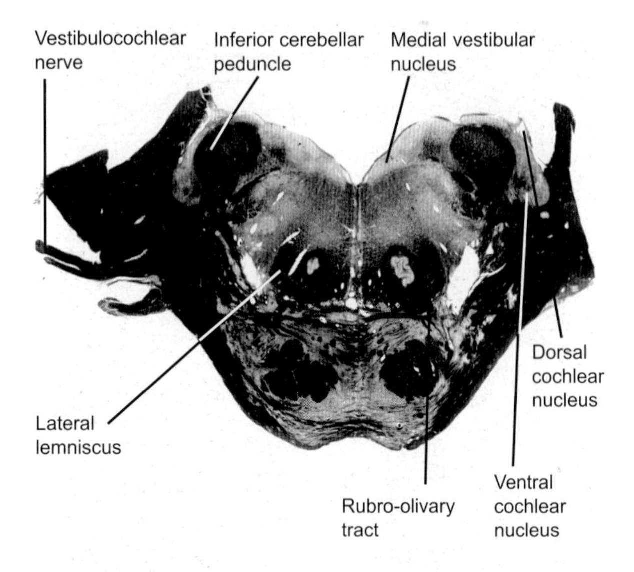

Figure 4.8 Rostral medulla. At this level the inferior olivary nuclei are reduced, the middle cerebellar peduncle is visible, and the cochlear nuclei assume a dorsolateral position to the inferior cerebellar peduncle

complex to the level of the stria medullaris of the fourth ventricle.

The perihypoglossal nuclei are groups of neurons located adjacent to the hypoglossal nucleus. This group includes the nucleus intercalatus, nucleus of Roller, and the nucleus prepositus hypoglossi. The latter nucleus occupies the same position as the hypoglossal nucleus, but at a more rostral level. Through their connection to the extraocular motor nuclei, cerebellum, and the vestibular nuclei, the perihypoglossal nuclei play an important role in controlling eye movements.

The dorsal motor nucleus of the vagus occupies the area dorsolateral to the hypoglossal nucleus, and forms the vagal trigone, a triangular eminence in the fossa rhomboidea. This nuclear column extends from the level of the hypoglossal nucleus both caudally and rostrally, and gives rise to parasympathetic preganglionic (GVE) fibers destined to the thoracic and abdominal viscera.

The nucleus ambiguus lies medial to the lateral reticular nucleus and above the inferior olivary nuclear complex. Neurons of this nucleus give rise to special visceral efferent fibers (SVE), which distribute via the glossopharyngeal and vagus nerves, and general visceral efferents (GVE) via the vagus nerve. The fibers of the ambiguus nucleus join the cranial part of the accessory nerve and travel within branches of the vagus nerve to be distributed to the laryngeal, pharyngeal, and palatal muscles.

The solitary tract is located lateral to the dorsal motor nucleus of the vagus. It is formed by fibers of the vagus, glossopharyngeal, and facial nerves. The solitary nuclear complex surrounds the solitary tract, comprising medial and lateral nuclear groups. The medial portions of the solitary nucleus unite caudal to the obex, forming the commissural nucleus of the vagus nerve. The medial subdivision and the caudal portion of the lateral subdivision of the solitary nucleus receive general visceral afferents (GVA) which convey information from baroreceptors and chemoreceptors, while the dorsal portion of the lateral nucleus receives taste (special visceral afferent) sensation. The solitary nuclear complex conveys general visceral information to the respiratory, cardiovascular and gastrointestinal centers of the brainstem and also to the autonomic neurons in the spinal cord. The part of the solitary nucleus (gustatory subnucleus) which deals with taste sensation projects to the VMN of the thalamus via the solitariothalamic tract.

At this level of the medulla the reticular formation contains neurons for the cardiovascular and respiratory centers. The reticular neurons that remain distinct at this level form the lateral reticular, paramedian reticular, nucleus reticularis gigantocellularis, nucleus reticularis parvocellularis, nucleus raphe obscurus and raphe pallidus. The lateral reticular nucleus lies dorsal to the inferior olivary complex and is primarily a relay nucleus that projects to the cerebellum through the inferior cerebellar peduncle. It receives afferents from the red nucleus, cerebral cortex, and spinal cord. The paramedian nuclei are cerebellar relay nuclei that lie parallel to the median raphe. The nucleus reticularis gigantocellularis occupies the medial zone of the medulla, modulates somatic and visceral motor activity, as well as controls muscle tone through the medullary reticulospinal tract. The nucleus reticularis parvocellularis lies in the lateral (sensory) part of the reticular formation. The nucleus raphe obscurus and raphe pallidus consist of serotoninergic neurons. The less visible inferior salivatory nucleus within the reticular formation of the medulla provides parasympathetic fibers to the glossopharyngeal nerve. These fibers synapse in the otic ganglion and later supply the parotid gland. At this level the inferior cerebellar peduncle becomes discernible and continues throughout the upper medulla, occupying the area lateral to the spinal trigeminal tract and nucleus. It is formed primarily by the cerebellar afferents such as the spinocerebellar tracts, cuneocerebellar, etc. Lateral to the raphe nuclei and dorsal to the pyramids, the MLF occupies a vertical position, containing only descending fibers, which project primarily to the spinal cord.

Rostral medulla

In the rostral medullary level (Figure 4.8) enlargement of the inferior cerebellar peduncle and the appearance of the dorsal and ventral cochlear nuclei, dorsolateral to this peduncle, will be noticed. Expansion of the medial and inferior vestibular nuclei will also be noted at this level. Additionally, the hypoglossal nucleus is replaced at this level by the nucleus prepositus hypoglossi. The reticular formation is expanded and the dorsal motor nucleus of the vagus has disappeared. Similar position to the midolivary level is maintained by the spinal trigeminal nucleus which is crossed by fibers of the glossopharyngeal nerve.

The nucleus raphe magnus contains the main serotoninergic neurons, projecting bilaterally to the spinal cord. This projection courses in the lateral funiculus, inhibiting the spinal neurons that facilitate pain transmission.

Pontomedullary junction

At the pontomedullary junction, the cerebellum and the lateral vestibular nucleus appear for the first time; other structures may maintain similar locations and dimensions, or may undergo reduction relative to lower levels. The cerebellum forms the roof of the fourth ventricle, and the flocculus can be observed. The inferior cerebellar peduncle maintains its large size, and is bounded dorsally by the cochlear nuclei. In addition to the medial and inferior vestibular nuclei, the lateral vestibular (Deiters') nucleus becomes apparent. The medial lemniscus begins to assume a horizontal position and move between the pyramid and the diminishing olivary nuclear complex. There is a specific reduction in the size of the inferior olivary nuclear complex, in particular the accessory olivary nuclei. Furthermore, the nucleus ambiguus, the spinal trigeminal tract and nucleus, and the solitary nucleus maintain their presence at this level. The pyramids begin to disperse, forming fasciculi, as the transition to the basilar pons begins. Additionally, the reticular formation shows expansion. The site of junction of the medulla, cerebellum, and pons forms the cerebellopontine angle, a common site for acoustic neuromas.

Pons

The pons (Figures 4.9 & 4.10) forms the mid portion of the brainstem, and the rostral part of the rhombencephalon. It is connected to the cerebellum via the middle cerebellar peduncle. It is bounded rostrally and caudally by the pontocrural (between the pons and midbrain) and the pontobulbar (between the

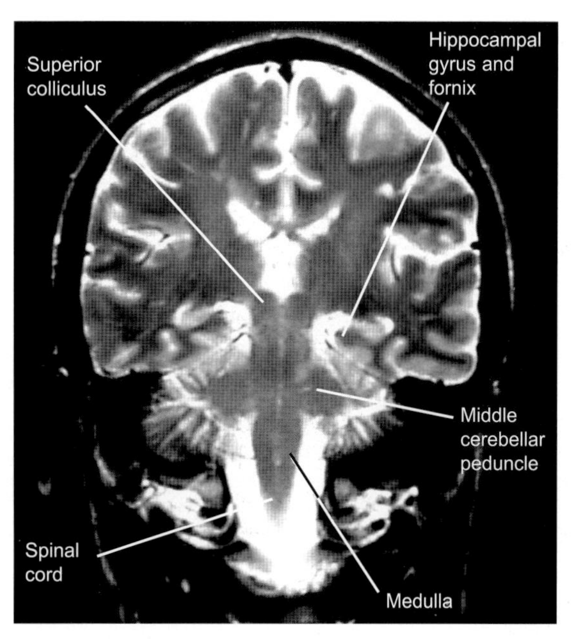

Figure 4.9 MRI scan illustrating the pons and the prominent middle cerebellar peduncle

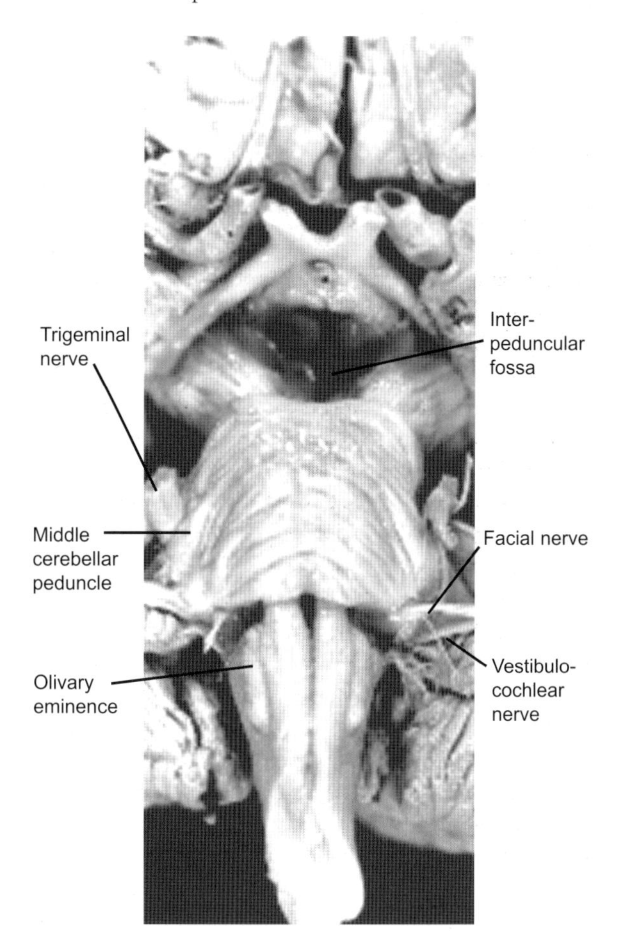

Figure 4.10 Ventral surface of the brainstem. The middle cerebellar peduncle is separated from the pons by the trigeminal nerve and the pontomedullary sulcus marks the exit of the abducens, facial and vestibulocochlear nerves

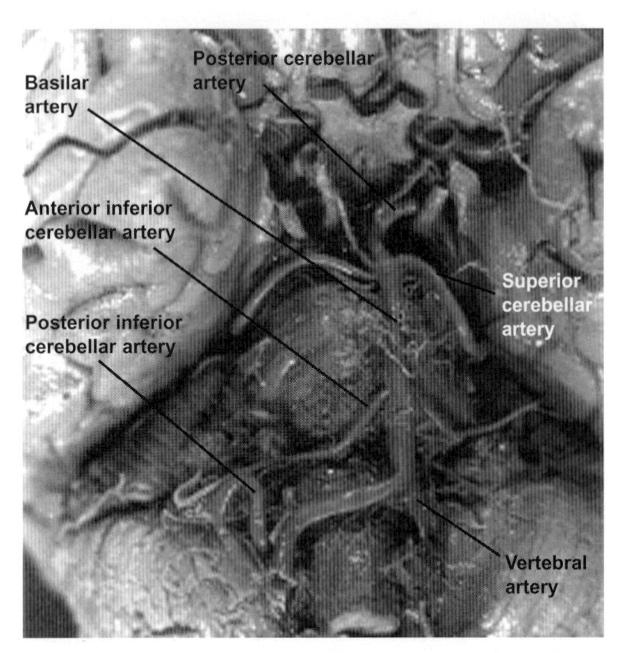

Figure 4.11 Basilar artery on the ventral surface of the pons. The anterior inferior cerebellar, the superior cerebellar and the posterior cerebral arteries are shown. Notice the connection of the vertebrobasilar system to the internal carotid artery via the posterior communicating artery

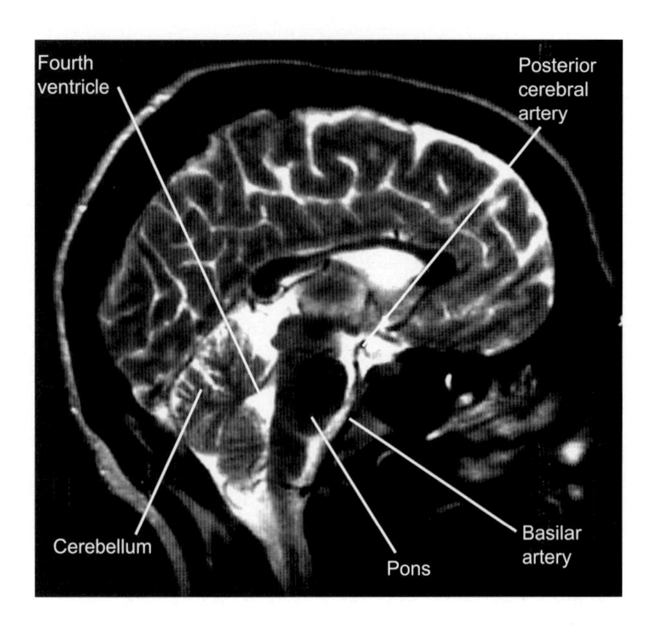

Figure 4.12 MRI scan illustrates the pons ventral to the cerebellum separated by the fourth ventricle. The basilar artery is also identified

medulla and pons) sulci, respectively. Within the pontocerebellar angle, the pontobulbar sulcus gives passage to the abducens nerve in the midline, and the facial and vestibulocochlear nerves more laterally. The dorsal surface of the pons forms the rostral half of the rhomboid fossa, whereas the ventral surface lies adjacent to the basilar part of the occipital bone (clivus) and the dorsum sella of the sphenoid bone. On its ventral surface, the pons is demarcated centrally by the basilar sulcus, which contains the basilar artery. On both sides of the basilar sulcus, the descending cortical motor fibers (corticospinal and corticobulbar tracts) form the pontine protruberances, which are supplied by the pontine branches of the basilar artery.

Ventral to the middle cerebellar peduncle, the trigeminal nerve emerges from the midpons. Transverse section of the pons reveals a dorsal tegmentum and a ventral basilar portion. The tegmentum is the rostral continuation of the medullary reticular formation, while the basilar pons consists of the longitudinal corticospinal and corticobulbar fibers, as well as the transverse pontocerebellar fibers. At the pontobulbar sulcus, the basilar artery is formed by the union of the vertebral arteries, ending in the rostral pons by dividing into the posterior cerebral arteries (Figures 4.11 & 4.12). It supplies the pons, cerebellum, midbrain, temporal, and occipital lobes of the brain. This vessel gives rise to the anterior inferior cerebellar, labyrinthine, pontine, superior cerebellar, and posterior cerebral arteries. Pontine veins open primarily into the sigmoid sinus, or the petrosal sinuses.

Paramedian branches of the basilar artery supply the medial pons, and the short circumferential branches nourish the pontine nuclei, corticospinal and corticobulbar tracts, and some of the trigeminal nuclei. The long circumferential branches of the basilar artery and some branches from the anterior inferior cerebellar artery supply the caudal pontine tegmentum. The superior cerebellar artery supplies the tegmentum of the rostral pons.

Occlusion of the pontine arteries may produce signs and symptoms of locked-in syndrome, a motor disorder that affect the limbs and trunk with the exception of eye movements. For additional information on this condition see Chapter 20.

The pons exhibits certain unique structures and other features that continue caudally with those of the medulla and spinal cord and rostrally with those of the midbrain. All these neuronal structures are best identified at the caudal pons, midpons, and rostral pons.

The presence of the two neuronal populations within the abducens nucleus may account for the distinct difference in the deficits produced by lesions of the abducens nerve versus the abducens nucleus. A lesion that only damages the abducens nerve results in medial strabismus and diplopia, while a damage to the abducens nucleus produces signs of lateral gaze palsy (inability to look to the side of the lesion).

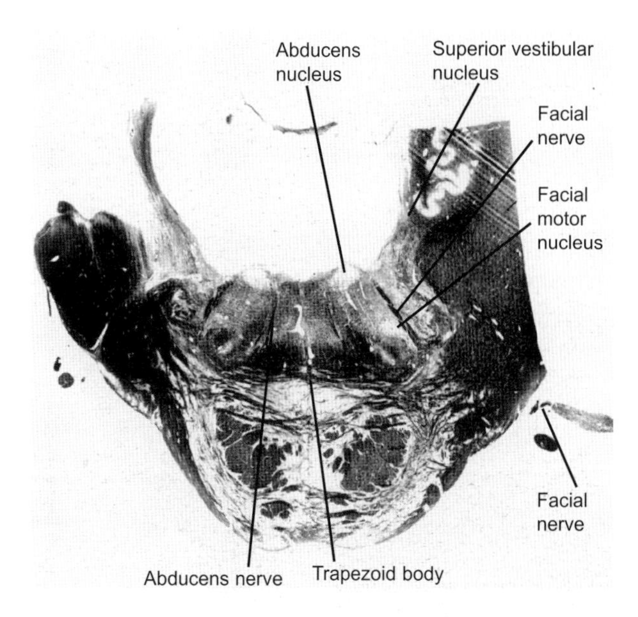

Figure 4.13 Caudal pons and the level of the abducens and facial nuclei. The medial lemniscus assumes a horizontal position and the middle cerebellar peduncle is enlarged. Trapezoid body and superior vestibular nucleus are illustrated

Caudal pons

There are characteristic neuronal elements at the caudal pons, which include the abducens, facial motor, superior vestibular and superior olivary nuclei (Figure 4.13). Appearance of the lateral lemniscus and continuation of the basilar pons are additional features of this level.

The abducens nucleus (GSE) lies deep to the ependyma of the rhomboid fossa, as a component of the facial colliculus. It is surrounded by the motor fibers of the facial nerve. This nucleus is unique among cranial nerve motor nuclei in that it contains two populations of neurons. The larger population is formed by the α motor neurons, which constitutes the abducens nerve, whereas the smaller population of internuclear neurons sends axons through the contralateral MLF to the motor neurons of the oculomotor nucleus, innervating the medial rectus muscle during lateral gaze.

The facial motor nucleus (SVE) occupies the lateral tegmentum, adjacent to the superior olivary nucleus, and gives rise to the motor fibers (SVE) that supply the muscles of facial expression, stapedius, stylohyoid, and the posterior belly of the digastric muscle. These fibers encircle the abducens nucleus (internal genu), form the facial colliculus, and later emerge through the pontocerebellar angle.

The superior salivatory nucleus (GVE) lies adjacent to the caudal end of the facial motor nucleus. It provides preganglionic parasympathetic fibers to the pterygopalatine and submandibular ganglia via the greater petrosal nerve and chorda tympani, respectively. Eventually, the generated impulses enhance lacrimation, salivary secretion, and secretion of mucus glands of the palate, nose and pharynx.

The lateral vestibular (Deiters') nucleus (SSA) consists of multipolar giant neurons, which are the source of the principal vestibulospinal tract. This nucleus is located in the lateral portion of the ventricular floor and extends from the rostral end of the inferior vestibular nucleus to the level of the abducens nucleus. It receives primary vestibular fibers and sends both crossed and uncrossed ascending fibers via the MLF, in a symmetric manner, to the abducens, trochlear, and oculomotor nuclei.

The superior vestibular nucleus (SSA), the most rostral vestibular nucleus, lies inferior to the superior cerebellar peduncle. It receives primary vestibular fibers and sends ipsilateral secondary vestibular fibers through the MLF, predominantly to the trochlear nucleus, and the intermediate and dorsal cell columns of the oculomotor nuclear complex that innervate the inferior oblique and the inferior rectus, respectively.

The ventral and dorsal cochlear nuclei (SSA) are located dorsolateral to the inferior cerebellar peduncle with the ventral cochlear nucleus, the largest, appearing to be continuous with the dorsal cochlear nucleus. The dorsal cochlear nucleus forms the acoustic tubercle in the floor of the fourth ventricle. These cochlear nuclei receive auditory impulses via the central processes of the spiral ganglia and convey the processed auditory information through the acoustic striae to the superior olivary nuclei and trapezoid nuclei and eventually to the inferior colliculi via the lateral lemniscus.

The trapezoid body is formed mainly by the decussating ventral acoustic striae from the ventral cochlear, superior olivary, and trapezoid nuclei. This bundle of fibers occupies a midline position ventral to the medial lemniscus.

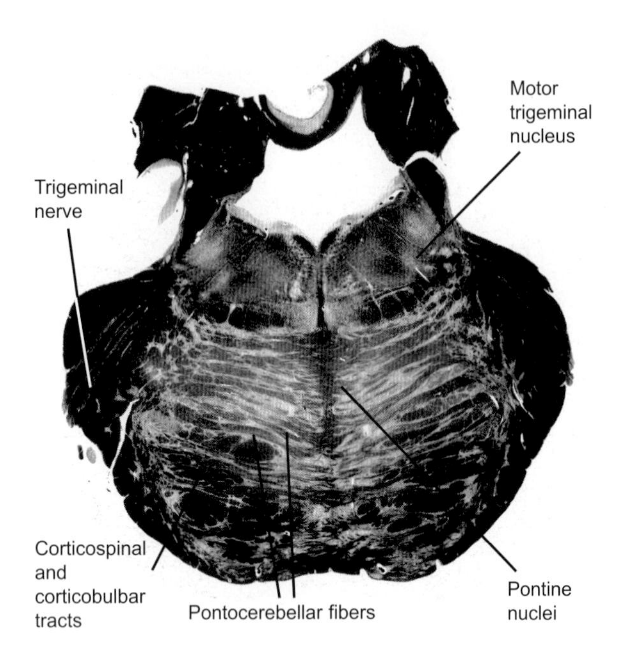

Figure 4.14 Section of the pons at the level of the trigeminal nerve. Corticospinal and corticobulbar tracts and the pontine nuclei occupy the entire basilar pons

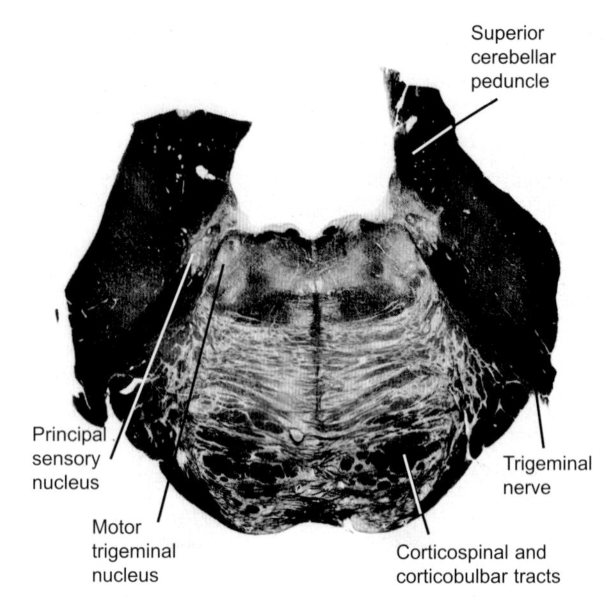

Figure 4.15 Section of the pons at the level of the trigeminal nerve showing the motor and principal sensory nuclei

The superior olivary nucleus is an auditory relay nucleus, which extends from the level of the facial motor nucleus to the level of the trigeminal motor nucleus. It receives collaterals from the acoustic striae of both sides. This nucleus contributes fibers to the trapezoid body and the lateral lemniscus.

The lateral lemniscus represents the main ascending auditory pathway and is located lateral to the superior olivary nucleus, containing the corresponding nucleus that receives secondary and tertiary auditory fibers.

The spinal trigeminal nucleus lies lateral to the facial motor nucleus and ventral to the lateral vestibular nucleus. It continues rostrally with the chief sensory nucleus of the trigeminal nerve. This nucleus receives pain and temperature sensations (GSA) from the head region through branches of the trigeminal, facial, glossopharyngeal, and vagus nerves. It provides axons to the ventral trigeminal tract (lemniscus). Descending fibers from the trigeminal ganglion form the spinal trigeminal tract lateral to the corresponding nucleus.

The inferior cerebellar peduncle reaches its maximum size at this level and connects the medulla to the cerebellum, transmitting most of the cerebellar afferent fibers from the spinal cord. It consists of a medial portion (juxtarestiform body) and a lateral portion (restiform body).

The MLF contains ascending vestibulo-ocular fibers and axons of the internuclear neurons of the abducens nucleus en route to the ventral cell column of the oculomotor nuclear complex. It also contains fibers of the interstitiospinal, medial vestibulospinal and pontine reticulospinal tracts.

The reticular formation at this level of the pons contains the nucleus reticularis pontis caudalis that contributes to the pontine reticulospinal tract. The area of the reticular formation adjacent to the abducens nuclei has been given special consideration due to its involvement in lateral gaze palsy. This region, the parapontine reticular formation (PPRF), is thought to contain the center for lateral gaze. Ischemic degeneration of the parapontine reticular formation may occur as a result of occlusion of the basilar artery.

In the basilar part of the pons descending corticofugal fibers such as the corticospinal, corticobulbar, and the pontocerebellar fibers lie. Pontocerebellar fibers arise from the pontine nuclei, constituting an indirect pathway from the cerebral cortex to the cerebellum through the middle cerebellar peduncle. As it descends through the pons towards the medulla, the central tegmental tract deviates laterally, occupying the center of the pontine tegmentum.

The middle cerebellar peduncle (Figure 4.9), the largest cerebellar peduncle, may remain visible at this level, consisting exclusively of the pontocerebellar fibers.

The medial lemniscus occupies a horizontal position in the ventral part of the tegmentum, dorsal to the trapezoid body.

Midpons (level of the trigeminal nerve)

The midpontine level (Figures 4.14 & 4.15) is characterized by the presence of the trigeminal nerve fibers, principal sensory, mesencephalic trigeminal, and motor trigeminal nuclei. The distinct appearance of the superior cerebellar peduncle and visibility of the middle cerebellar peduncle are additional features of this pontine level.

The principal (chief) trigeminal sensory (Figures 4.14 & 4.15) nucleus lies lateral to the fibers of the trigeminal nerve and is concerned with pressure and tactile sensations. In this nucleus, the mandibular fibers occupy a dorsal position to the maxillary fibers which assume an intermediate position, whereas the ophthalmic fibers occupy a ventral position in this nucleus. The ascending fibers from the ventral part of this nucleus contribute to the ventral trigeminal tract or lemniscus, while the dorsal part comprises the dorsal trigeminal tract or lemniscus. The most rostral part of the spinal trigeminal nucleus (pars oralis) merges with the principal sensory nucleus.

The motor trigeminal nucleus lies medial to entering fibers of the trigeminal nerve, and provides innervation through its mandibular division (V3) to the muscles of mastication, tensor tympani, tensor palatini, and the anterior belly of the digastric muscle. The motor trigeminal nucleus mediates the jaw jerk reflex and receives bilateral corticobulbar fibers (via interneurons).

The mesencephalic nucleus is the only sensory nucleus which contains unipolar neurons that are retained within the CNS. This nucleus extends from the level of the trigeminal nerve to the caudal midbrain. It primarily conveys proprioceptive impulses from the muscles of mastication to the motor trigeminal nucleus, mediating the monosynaptic 'jaw jerk' reflex. It also receives input from the facial and ocular muscles, temporomandibular joint, and the peridontium. In that regard it may be considered homologous to some degree to the Clarke's nucleus of the spinal cord. The mesencephalic tract is formed by processes of the corresponding nucleus and provides collaterals to the motor root of the trigeminal nerve. The mesencephalic tract is formed by processes of the corresponding nucleus, providing collaterals to the motor root of the trigeminal nerve.

- The superior cerebellar peduncle lies dorsolateral to the fourth ventricle and contains cerebellar efferents that have not yet undergone decussation.
- The middle cerebellar peduncle is pierced by the trigeminal nerve, losing its connection to the cerebellum at this level.
- The nucleus reticularis pontis oralis is part of the medial reticular zone that contributes to the pontine reticulospinal tract.
- The superior central nucleus is also seen at this level of the pons.
- The basilar pons expands at this level to contain essentially the same structures seen in the caudal pons.
- The fourth ventricle is narrower and is bounded dorsally by the superior medullary velum.

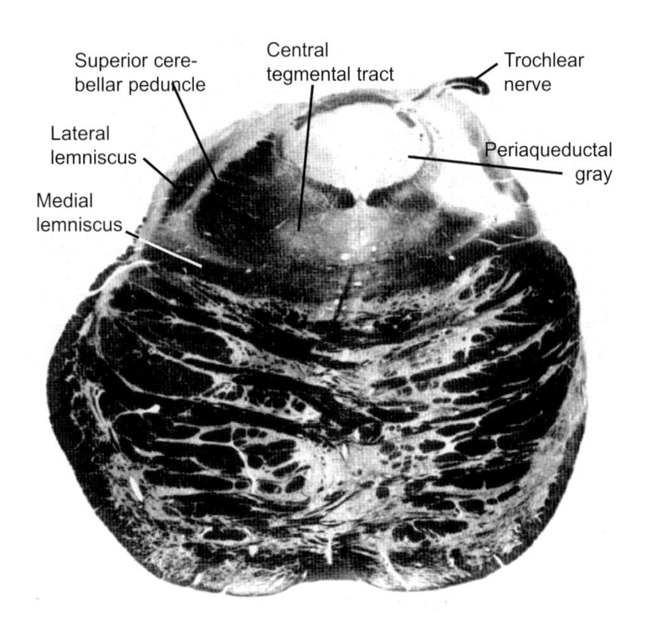

Figure 4.16 Rostral pons (isthmus). Note the decussation of the trochlear nerves in the superior medullary velum and crossing of the superior cerebellar peduncles within the tegmentum

Rostral pons (isthmus)

At the level of the isthmus (Figure 4.16) the fourth ventricle terminates and the cerebral aqueduct begins. It is the site of decussation of the trochlear nerve fibers and the appearance of the pigmented locus ceruleus. The basilar pons assumes its greatest size and the lemniscal triad occupy a dorsal position in the pontine tegmentum.

At more rostral levels, the central gray matter expands around the fourth ventricle and becomes the periaqueductal gray. At the same time expansion of the basilar part of the pons and narrowing of the tegmental portion occurs. Additionally, the medially

Since the superior medullary velum contains the decussating fibers of the trochlear nerves, a lesion of this area may produce bilateral trochlear nerve palsy (superior medullary velum syndrome).

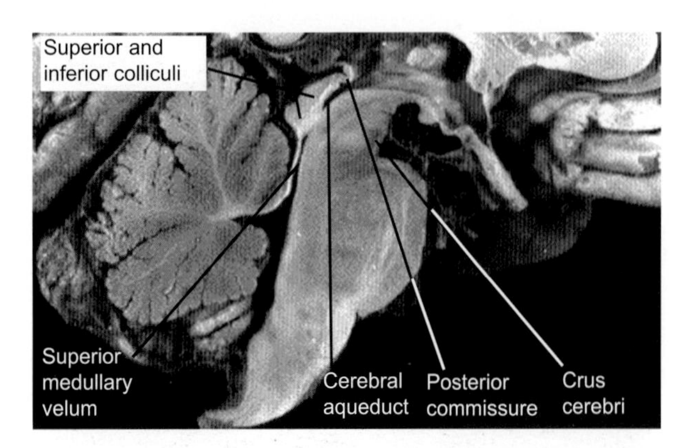

Figure 4.17 Midsagittal section of the brainstem. The midbrain with the associated cerebral aqueduct and tectum are clearly visible

directed fibers of the superior cerebellar peduncles begin entering the tegmentum. The locus ceruleus and the trigeminal mesencephalic nucleus occupy basically the same position.

Dorsolateral to the superior cerebellar peduncle the lemniscal trigone occupies the lateral part of the tegmentum. This trigone is formed by the lateral lemniscus dorsally, the medial lemniscus ventrally, and by the spinal lemniscus assuming an intermediate position. The trigeminal lemniscus appears dorsal to the medial lemniscus and the central tegmental tract becomes a very prominent tegmental structure.

The reticular formation at this level of the pons consists of a number of nuclei, which include reticulotegmental, superior central (median), and the dorsal raphe nuclei.

- The reticulotegmental nucleus is one of the pontine paramedian reticular nuclei that receives input from the dentate nuclei of the cerebellum and the cerebral cortex. This nucleus projects back to the contralateral cerebellar cortex and the ipsilateral vermis.
- The superior central (median) nucleus lies dorsal to the reticulotegmental nucleus. It sends impulses to the superior colliculus, pretectum, hippocampal formation, and the mamillary bodies.
- The dorsal raphe nucleus projects to the lateral geniculate nucleus, the neostriatum, the substantia nigra, the pyriform lobe, and the olfactory bulb.

Both the superior central and dorsal raphe nuclei consist of serotoninergic neurons which project via the medial forebrain bundle to the hypothalamus which in turn projects to the substantia nigra, intralaminar thalamic nuclei, and the septal area.

Since the medulla and pons contribute to the formation of the fourth ventricle and share important relationships with this cavity, a brief account of this ventricle will serve a useful purpose.

The fourth ventricle lies dorsal to the pons and medulla and is rostrally continuous with the cerebral aqueduct and caudally with the central canal. Lateral extensions of this ventricle dorsal to the inferior cerebellar peduncles are known as the lateral recesses, which terminate in the lateral foramina of Luschka. Within the caudal part of the fossa rhomboidea (the floor of the fourth ventricle), the hypoglossal and vagal trigones are seen. These trigones overlie the hypoglossal and dorsal motor nucleus of vagus, respectively. Pneumotaxic and emetic centers, which lack a blood–brain barrier, are represented in area postrema, a small site near the vagal trigone and on both sides of the obex. The rostral half of the rhomboid fossa contains the facial colliculus and the pigmented locus ceruleus. The facial colliculus is formed by the facial nerve encircling the abducens nucleus. Noradrenergic neurons are concentrated in the locus ceruleus, a pigmented nucleus, which extends from the rostral pons to the caudal midbrain. Another site in the fossa rhomboidea is the vestibular area which overlies the vestibular nuclei and corresponds to the site of the lateral recesses. It is crossed on both sides by fibers of the stria medullaris of the fourth ventricle, which emanate from the arcuate nuclei. Examination of the roof of the fourth ventricle reveals a cranial part formed by the superior medullary velum and superior cerebellar peduncles, and a caudal part mainly formed by the inferior medullary velum containing the foramen of Magendie. Circulation of CSF between the subarachnoid space (cerebellomedulary cistern) and the fourth ventricle is maintained via the foramina Luschka and Magendie.

Mesencephalon (midbrain)

The midbrain (Figures 4.17, 4.18, 4.19 & 4.20) represents the shortest part of the brainstem and is derived from the unmodified third brain vesicle. It is bounded rostrally by an imaginary line that passes behind the posterior commissure and the mamillary bodies, and caudally by another line that connects the pontocrural sulcus and the posterior borders of the inferior colliculi. The midbrain consists of the tectum, tegmentum, and the basis pedunculi. Each half of the midbrain, excluding the tectum, is known as the cerebral peduncle. The tectum (quadrigeminal plate) consists of four rounded eminences connected by commissures. The larger and more grayish upper pair of eminences is known as the superior colliculi, while the smaller lower pair is termed the inferior colliculi.

For descriptive purposes the midbrain (Figures 4.18, 4.19 & 4.20) may be classified into superior and

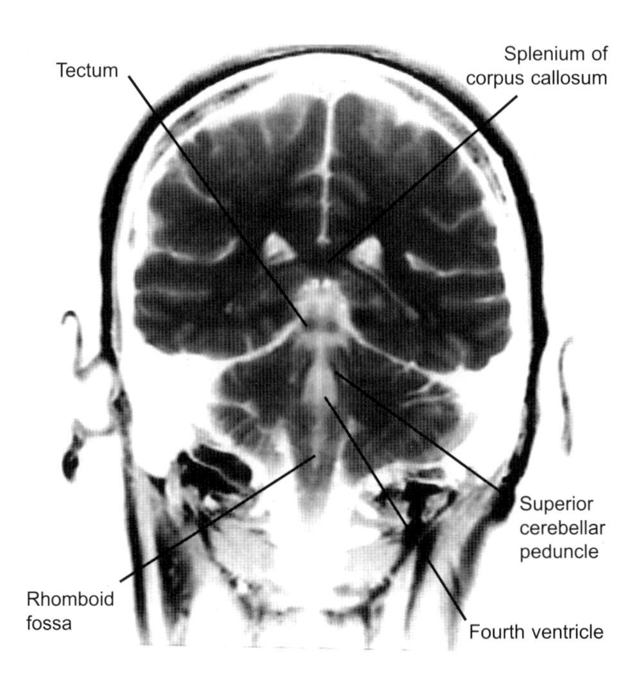

Figure 4.18 This MRI scan illustrates the tectum and connection of the midbrain to the cerebellum via the superior cerebellar peduncle

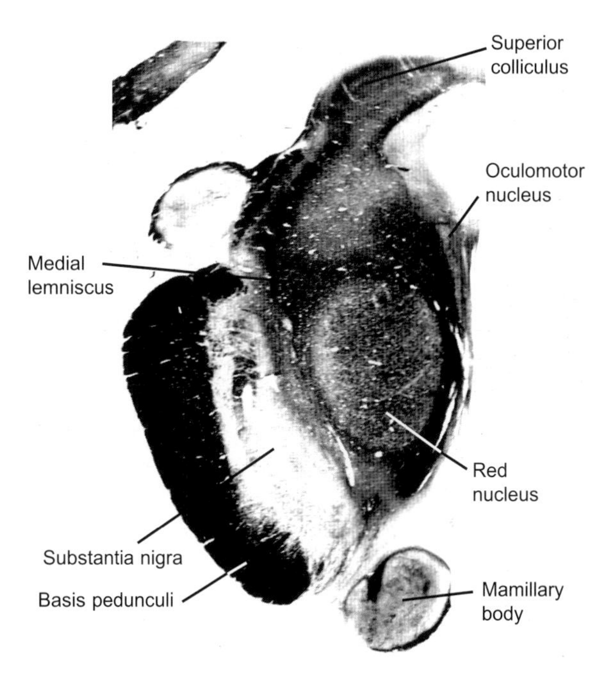

Figure 4.19 Section of the left half of the midbrain at the level of the superior colliculus. This photograph illustrates the tectum, tegmentum, substantia nigra, and the crus cerebri

inferior collicular levels. These two levels contain common features and structures that include the cerebral aqueduct, substantia nigra, crus cerebri, and the trigeminal, spinal (anterolateral system), and medial lemnisci.

The cerebral aqueduct interconnects the third and fourth ventricles, separating the tectum from the tegmentum of the midbrain. The periaqueductal gray matter surrounds this duct, containing the accessory oculomotor nuclei and the dorsal tegmental nucleus. This area gives rise to an inhibitory descending pathway for painful stimuli.

The substantia nigra is the largest pigmented nucleus of the midbrain, lying dorsal to the crus cerebri. It consists of pars reticularis and pars compacta. The pars compacta is the main source of the inhibitory neurotransmitter dopamine, forming with the pars lateralis the group A9 of Dahlström and Fuxe. Group A9 together with the retrorubral nucleus (group A8) constitutes the principal dopaminergic neurons of the midbrain. Group A10 (paranigral nucleus) interconnects the pars compacta of the substantia nigra on both sides. Acetylcholine is also contained in these neurons. The substantia nigra receives input from the striatum, globus pallidus, and the subthalamic nucleus. It projects to the striatum, reticular formation, tectum, pedunculopontine nucleus, and the thalamus. Nigral projection to the tectum arises from the pars reticulata, influencing coordination of head and eye movements. In addition to dopamine, the substantia nigra contains a high concentration of substance P, serotonin and glutamic acid decarboxylase (GAD).

> Obstruction of the cerebral aqueduct may occur congenitally, resulting in a non-communicating hydrocephalus.

> A lesion of the substantia nigra is responsible for signs and symptoms of Parkinson's disease.

The crus cerebri is the most ventral portion of the cerebral peduncles, containing in its medial two thirds the corticospinal and corticobulbar fibers. Within the crus cerebri the frontopontine fibers are medial, while the parietopontine, temporopontine and occipitopontine fibers assume a more lateral position. These peduncles form the boundaries of the interpeduncular fossa that gives passage to the oculomotor nerve. The floor of this fossa forms the posterior perforated substance, allowing passage of the central branches of the posterior cerebral arteries.

The trigeminal lemnisci lie dorsal to the medial lemniscus, conveying pain, temperature, and tactile

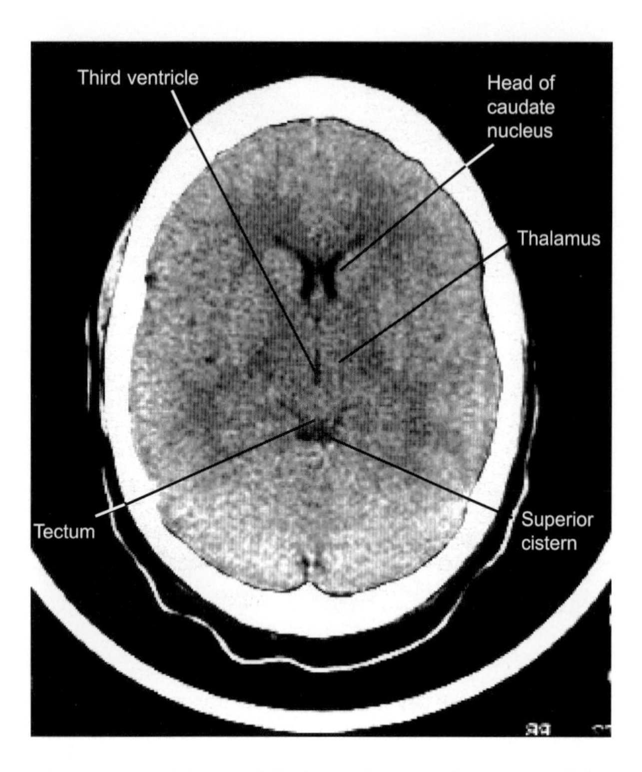

Figure 4.20 CT scan of the brain showing the tectum of the midbrain in relationship to the third ventricle and thalamus

sensations to the ventral posteromedial nucleus of the thalamus from the face and head region. The ventral trigeminal lemniscus, a crossed pathway derived from the spinal trigeminal nucleus and the ventral portion of the principal sensory nucleus, terminates in the ventral posteromedial (VPM) nucleus of thalamus. Neuronal axons of the dorsal portion of the principal sensory nucleus form the ipsilateral dorsal trigeminal lemniscus.

The spinal lemniscus (anterolateral system) is located dorsal to the lateral part of the substantia nigra, bounded dorsolaterally by the lateral lemniscus and ventromedially by the medial lemniscus. It consists primarily of fibers of the lateral spinothalamic tract.

The medial lemniscus is the principal pathway that conveys conscious proprioception, fine tactile sensation, and vibratory sense to the thalamus. It assumes a horizontal position ventral to the trigeminal and spinal lemnisci.

The medial longitudinal fasciculus at this level consists of ascending vestibulo-ocular fibers and axons of abducens internuclear neurons, projecting to the trochlear and oculomotor nuclei.

The central tegmental tract, a composite bundle of fibers, lies dorsolateral to the red nucleus and lateral to the MLF. The descending components of this pathway originate from the red nucleus, periaqueductal gray, thalamus, and the tegmentum of the midbrain en route to the inferior olivary nucleus where it forms the amiculum olive. This pathway regulates intrareticular conduction through its short ascending fibers. It also conveys cortical motor input to the contralateral cerebellum through the inferior olivary nucleus, and eventually to the red nucleus and basal nuclei (a collection of subcortical nuclei embedded in the white matter of the cerebral hemispheres) via cerebellar projections. The central tegmental tract is also considered to be the main ascending pathway for the reticular formation, conveying impulses to the subthalamus and the intralaminar thalamic nuclei.

The dorsal longitudinal fasciculus, a predominantly ipsilateral tract ventrolateral to the cerebral aqueduct, contains ascending and descending pathways. It connects the hypothalamus to the Edinger–Westphal, solitary, salivatory, facial, hypoglossal, and ambiguus nuclei, as well as to the tectum.

The midbrain is supplied by branches of the posterior cerebral, posterior communicating, anterior choroidal, and the superior cerebellar arteries. The paramedian branches of the posterior cerebral and posterior communicating arteries supply midline structures including the oculomotor nuclei, medial longitudinal fasciculi, and medial parts of the substantia nigra. The lateral tegmentum, medial lemnisci, spinal lemnisci, substantia nigra, and crus cerebri are supplied by the circumferential branches of the posterior cerebral and superior cerebellar arteries. The long circumferential branches of the posterior cerebral artery and small branches from the superior cerebellar artery supply the tectum. Venous blood of the midbrain terminates in the internal cerebral vein or the great cerebral vein of Galen.

Inferior collicular level

The inferior collicular level (Figures 4.22 & 4.23) contains the inferior colliculi, trochlear nucleus, tegmental nuclei, locus ceruleus, lateral lemniscus, and pedunculopontine nucleus.

Lesions involving the trochlear nerve result in extorsion, impairment of downward movement of the affected eye and vertical diplopia, which increases on downward gaze. Patients compensate for this deficit and ameliorate diplopia by tilting the head to the contralateral side (Bielschowsky sign). Infants with this deficit may develop torticollis (spasmodic contracture of the sternocleidomastoid muscle) as a result of continuous tilting of the head. This condition is also discussed in detail in Chapter 21.

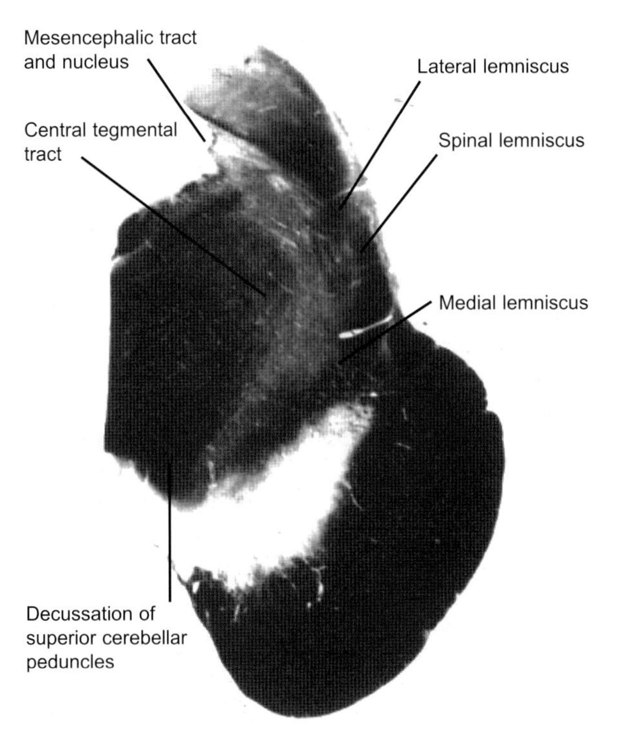

Figure 4.21 Image of the right half of the midbrain. Mesencephalic trigeminal nucleus and lemniscal triad (medial, lateral, and spinal lemnisci)

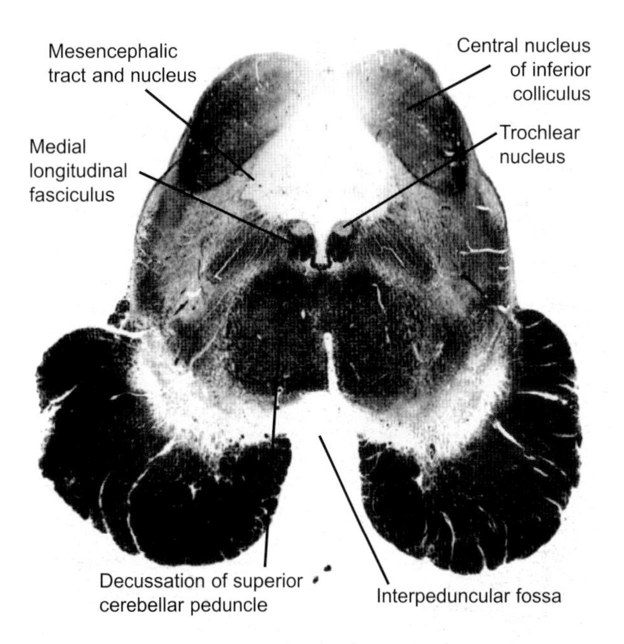

Figure 4.22 Section of the midbrain at the level of the inferior colliculus illustrating the trochlear nucleus and the site of decussation of the superior cerebellar peduncles

The inferior colliculus is connected to the medial geniculate body via the inferior brachium, representing the auditory reflex center. It consists of a large central nucleus, responding to binaural impulses, and a pericentral nucleus concerned with the ipsilateral auditory impulses. The principal afferent to the inferior colliculus is the lateral lemniscus that conveys auditory impulses to the medial geniculate body via the brachium of the inferior colliculus.

The trochlear nucleus is a round nucleus embedded within the medial longitudinal fasciculus, lying ventral to the periaqueductal gray matter. Its axons, which constitute the trochlear nerve, travel dorsally and decussate completely in the superior medullary velum. These fibers emerge from the dorsal surface of the pons, immediately below the inferior colliculi, innervating the superior oblique muscle.

The dorsal tegmental (supratrochlear) nucleus is located between the trochlear nuclei, receiving input from the mamillary bodies and the interpeduncular nucleus through the mamillotegmental tract. Projections from this nucleus ascend within the dorsal longitudinal fasciculus to be distributed to the limbic system nuclei of the diencephalon and telencephalon.

The ventral tegmental nucleus is considered to be a continuation of the superior central nucleus of the pons. Both the dorsal and ventral tegmental nuclei convey impulses to the lateral hypothalamus, the preoptic area, and the mamillary bodies via the dorsal longitudinal fasciculus, mamillary peduncle, and the medial forebrain bundle.

The dorsal raphe nucleus, which lies adjacent to the dorsal tegmental nucleus, synthesizes and transports serotonin.

The interpeduncular nucleus is located dorsal to the interpeduncular fossa receiving the habenulopeduncular tract.

The pedunculopontine nucleus, which lies in the lateral tegmentum ventral to the inferior colliculus, modulates the activities of the nigral and pallidal neurons through its connections to the cerebral cortex, globus pallidus, and the substantia nigra. Its compact part is mainly cholinergic that projects to the thalamus, regulating locomotion. This nucleus is crossed by fibers of the superior cerebellar peduncle.

The parabigeminal nucleus contains a collection of cholinergic neurons that are located lateral to the lateral lemniscus and ventrolateral to the inferior colliculus. These cholinergic neurons regulate both moving and stationary visual stimuli via their bilateral connections to the superior colliculus.

The locus ceruleus appears at the rostral pons medial to the trigeminal mesencephalic nucleus. It is a pigmented bluish nucleus which may readily be

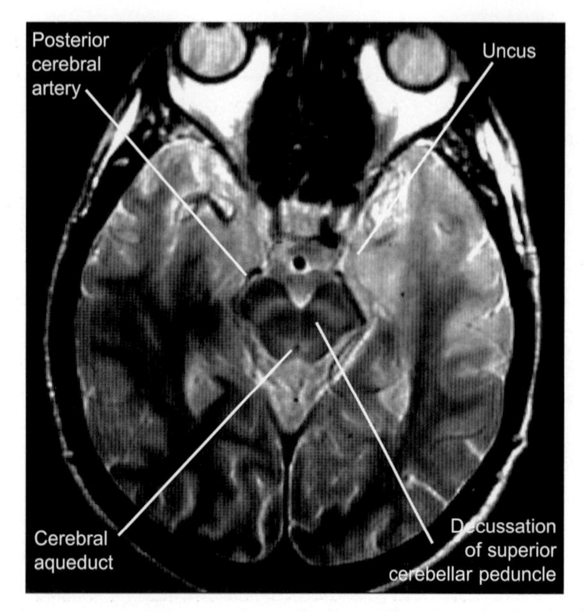

Figure 4.23 MRI scan through the midbrain at the level of the inferior colliculus. This image illustrates the relationship of the midbrain to the uncus and posterior cerebral arteries

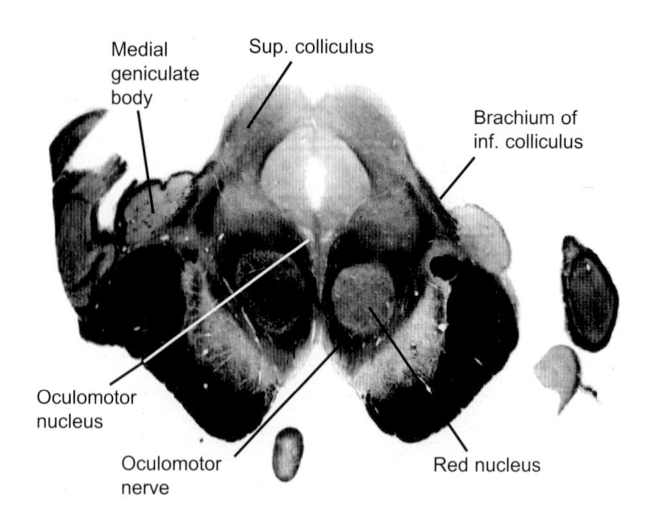

Figure 4.24 Section of the midbrain at its junction with diencephalon. The superior colliculus, oculomotor nucleus and nerve, and the red nucleus are principal features of this level

identified on a gross brain. This nucleus synthesizes and transports norepinephrine to the midbrain, cerebellum, medulla, spinal cord, diencephalon, and telencephalon. The spinal projection of the locus ceruleus descends in the lateral funiculus and exerts direct inhibitory influences upon the neurons that form the lateral spinothalamic tract. These projections use α receptors and not the opiate receptors, as is the case with the raphe-spinal tract. Noradrenergic neurons of the locus ceruleus project via the medial forebrain bundle, stria medullaris, and the mamillary peduncle to the cerebral cortex, hypothalamus, midbrain tegmentum, and the telencephalon. Intralaminar thalamic nuclei receive profuse projections from the locus ceruleus. Fibers of the locus ceruleus also terminate on small cerebral vessels and capillaries accounting for the possible role of this nucleus in the regulation of the cerebral blood flow. Locus ceruleus also regulates the functions of the preganglionic sympathetic neurons of the spinal cord and modulates cerebellar activities. Additionally, this nucleus is involved in the reinforcement mechanism essential for learning, and REM sleep (see also the reticular formation, Chapter 5).

The superior cerebellar peduncles complete their decussation within the tegmentum. These peduncles represent the main cerebellar output to the ventral lateral nucleus of the thalamus and the red nucleus.

The lateral lemniscus is the principal ascending auditory pathway, which occupies the dorsolateral part of the midbrain, terminating in the inferior colliculus.

Superior collicular level

In addition to the common characteristics discussed earlier the midbrain at this level contains the superior colliculus, red nucleus, and the oculomotor nucleus (Figures 4.19, 4.20 & 4.24).

The superior colliculus (Figures 4.19 & 4.24) consists of alternate gray and white laminae, constituting the visual reflex and vertical gaze centers. It is connected to the lateral geniculate body and partly to the optic tracts via the superior brachium. The superior colliculus lies ventral to the posterior commissure and the pineal gland, a relationship that bears clinical significance in Parinaud's syndrome in which vertical gaze palsy is associated with a tumor of the pineal gland. Movements of the head and neck toward visual and auditory stimuli are mediated by projections of the superior colliculi to the brainstem and spinal cord via the tectobulbar and tectospinal tracts, respectively. In addition to its role in visual reflexes, the superior colliculus also receives general somatic afferents through the spinotectal tract, a pathway that is incorporated within the spinal lemniscus.

The red nucleus (Figures 4.19 & 4.24), a highly vascularized nucleus in the center of the midbrain tegmentum, is encircled by fibers of the superior cerebellar peduncle. Medial to the red nucleus the habenulopeduncular tract (fasciculus retroflexus) descends, projecting to the interpeduncular nucleus. It is crossed by fibers of the oculomotor nerve en route to the interpeduncular fossa. It has a caudal magnocellular and a rostral parvocellular part. The magnocellular part gives rise to the contralateral rubrospinal tract, which controls flexor muscle tone, delivering cerebellar input to the upper three cervical spinal cord segments. Fibers of the rubrospinal tract are incorporated within the lateral corticospinal tract, maintaining identical termination sites. Cerebral cortical input to the red nucleus is conveyed via the corticorubral fibers, while cerebellar input emanating from the contralateral dentate, globose and emboliform nuclei are carried via the middle cerebellar peduncle. Information received by the red nucleus is delivered to laminae V through VII of the spinal cord. The parvocellular part of the red nucleus form the rubro-olivary tract that projects to the ipsilateral inferior olivary nucleus within the central tegmental tract. Due to massive connections of the inferior olivary nucleus to the cerebellum, the rubro-olivary tract forms part of a cerebellar feedback loop.

The oculomotor nuclear complex has a 'V' shaped configuration and is located medial to the MLF. It consists of somatic and visceral columns. The somatic columns (GSE) provide innervation to the extraocular muscles (with the exception of the lateral rectus and superior oblique) and levator palpebrae muscles, while the visceral columns (Edinger–Westphal nucleus) provide presynaptic parasympathetic fibers (GVE) to the ciliary ganglia which eventually innervate the constrictor pupillae and ciliary muscles.

Suggested reading

1. Bandler R, Shipley MT. Columnar organization in the midbrain periaqueductal gray: modules for emotional expression? *Trends Neurosci* 1994;17:379–89
2. Dunnett SB, Annett LE. Nigral transplants in primate models of parkinsonism. In Lindvall O, Bjarklund A, Widner H, eds. *Intracerebral Transplantation in Movement Disorders. Experimental Basis and Clinical Experiences.* Amsterdam: Elsevier Science Publishers, 1991:27–51
3. Glendenning DS, Vierck CJ. Lack of proprioceptive deficit after dorsal column lesions in monkeys. *Neurology* 1993;43:363–6
4. Godefroy JN, Thiesson D, Pollin B, *et al.* Reciprocal connections between the red nucleus and the trigeminal nuclei: a retrograde and anterograde tracing study, *Physiol Res* 1998;47:489–500
5. Hopkins DA. The dorsal motor nucleus of the vagus nerve and the nucleus ambiguus: structure and connections. In Hainsworth R, McWilliam PN, Mary DASG, eds. *Cardiogenic Reflexes.* Oxford: Oxford University Press, 1987;185–203
6. Kudo M, Sakurai H, Kurokawa K, *et al.* Neurogenesis in the superior olivary complex in the rat. *Hear Res* 2000; 139:144–52
7. Leichnetz GR, Smith DJ, Spencer RF. Cortical projections to the paramedian tegmental and basilar pons in the monkey. *J Comp Neurol* 1984;228:388–408
8. Lynch JC, Hoover JE, Strick PL. Input to the primate frontal eye field from the substantia nigra, superior colliculus and dentate nucleus demonstrated by transneuronal transport. *Exp Brain Res* 1994;100:181–6
9. McClung JR, Goldberg SJ. Organization of motoneurons in the dorsal hypoglossal nucleus that innervate the retrusor muscles of the tongue in the rat. *Anat Rec* 1999; 254:222–30
10. McKitrick DJ, Calaresu FR. Reciprocal connection between nucleus ambiguus and caudal ventrolateral medulla. *Brain Res* 1997;770:213–20
11. Nakaso K, Nakayasu H, Isoe K, *et al.* A case of dissecting aneurysm of the basilar artery presented as superior pons type of Foville's syndrome. *Rinsho Shinkeigaku* 1995;35:1040–3
12. Narai H, Manabe Y, Deguchi K, *et al.* Isolated abducens nerve palsy caused by vascular compression. *Urology* 2000;55:453–4
13. Nguyen L, Spencer RF. Abducens internuclear and ascending tract of deiters inputs to medial rectus motoneurons in the cat oculomotor nucleus: neurotransmitters. *J Comp Neurol* 1999;411:73–86
14. Ostrowska A, Zimny R, Zguczynski L, *et al.* The neurons of origin of non-cortical afferent connections to the oculomotor nucleus. A retrograde labeling study in the rabbit. *Arch Ital Biol* 1991;129:239–58
15. Panneton WM, Loewy AD. Projections of the carotid sinus nerve to the nucleus of the solitary tract in the cat. *Brain Res* 1980;191:239–44
16. Pinganaud G, Bernat I, Buisseret P, *et al.* Trigeminal projections to hypoglossal and facial motor nuclei in the rat. *J Comp Neurol* 1999;415:91–104
17. Voogd J, Bigarè F. Topographic distribution of olivary and corticonuclear fibers in the cerebellum: a review. In Courville J, de Mountigny I, y Latha RE, eds. *The Inferior Olivary Nucleus.* New York: Raven Press, 1980:207–34
18. Wang SF, Spencer RF. Morphology and soma-dendritic distribution of synaptic endings from the rostral interstitial nucleus of the medial longitudinal fasciculus (riMLF) on motoneurons in the oculomotor and trochlear nuclei in the cat. *J Comp Neurol* 1996;366: 149–62
19. Webster HH, Jones BE. Neurotoxic lesions of the dorsolateral pontomesencephalic tegmentum-cholinergic cell area in the cat. Effects upon sleep-waking states. *Brain Res* 1988;458:285–302
20. Welsh JP, Chang B, Menaker ME, *et al.* Removal of the inferior olive abolishes myoclonic seizures associated with a loss, of olivary serotonin. *Neuroscience* 1998;82: 879–97

Reticular formation 5

The reticular formation is considered to be one of the oldest functional units in the central nervous system (CNS). It occupies the central core of the brainstem, extending rostrally to include the midline, intralaminar and reticular thalamic nuclei, and the zona incerta of the subthalamus. Reticular neurons mediate local reflexes, receiving collaterals from the ascending and descending pathways with the exception of the medial lemniscus. The reticular formation is bounded ventromedially by the pyramidal tracts and the medial lemnisci, and dorsolaterally by the secondary sensory pathways. Regulation of both somatic and visceral (autonomic) motor functions and modulation of the electrocortical activities are maintained by massive reticular connections to the autonomic centers in the brain and the spinal cord. Additional functions of the reticular formation include control of emotional expression, pain transmission, and regulation of reflex activities associated with the cranial nerves.

General organization of the reticular formation

General organization of the reticular formation

The reticular formation consists of deeply localized and poorly identified nuclear groups, which are particularly scattered in the brainstem. It contains centers that generate motor activities (e.g. walking and running), and regulate conjugate eye movements, and respiratory and cardiovascular activities. It also contains centers for expiration, inspiration, vomiting, and deglutition, and is also concerned with the regulation of blood pressure. Reticular neurons are classified into median raphe, paramedian, and medial and lateral nuclear columns.

The raphe nuclei (Figure 5.1) synthesize and transport serotonin to various areas of the CNS through ascending and descending projections. Dahlström and Fuxe grouped these serotoninergic neurons into nine clusters. Serotonin, the neurotransmitter for raphe nuclei, is involved in the slow (NREM) phase of sleep, behavioral regulation, and inhibition of pain transmission. Raphe neurons also contain substance P. Raphe neurons within the pons project inhibitory impulses to the parapontine reticular formation (PPRF) producing rapid eye movement (REM) during sleep. They also send inhibitory fibers to the nuclei reticularis pontine oralis and caudalis, which, upon release from inhibition, elicit involuntary movements of the trunk and limbs. Inhibitory input to the pontine neurons that project to the lateral geniculate body may be responsible for the pontine–geniculate–occipital spikes and associated visual-natured REM sleep. It has been suggested that the latter phenomenon and associated dreams may be suppressed by certain medications or upon consumption of dairy products that contain tryptophan. Lesions of the raphe nuclei produce prolonged insomnia. This group of nuclei includes the raphe pallidus and raphe obscurus, raphe magnus, dorsal raphe nucleus, and superior central nucleus.

The nucleus raphe pallidus is located in the pons, projecting to laminae I, II and V of the dorsal horns of all spinal segments, as well as to the intermediolateral columns. The spinal projections of the raphe pallidus convey pain-controlling input from the periaqueductal gray matter to the spinal cord.

Nucleus raphe obscurus lies in the pons and forms the inhibitory intermediate raphe spinal projection to the spinal cord, modulating the sympathetic neurons of the intermediolateral column and thus regulating the cardiovascular function.

The nucleus raphe magnus is located in the rostral medulla and caudal pons and contains B3 neurons, maintaining bilateral projections to the spinal cord via the Lissauer tract, and to the posterior part of the spinal trigeminal nucleus within the spinal trigeminal tract. These spinal projections descend in the lateral funiculus and may inhibit the neurons which transmit painful stimuli, acting as an additional endogenous analgesic pathway. The serotoninergic component of the spinal projection, which represents 20% of all these fibers, establishes excitatory synaptic contacts with the enkephalinergic neurons and inhibitory contacts with the lateral spinothalamic tract neurons of the substantia gelatinosa. These synaptic contacts at spinal levels produce stimulus-bound profound analgesia, a common phenomenon seen upon stimulation of the periaqueductal and periventricular gray matter. They are also thought to enhance motor response to nociceptive stimuli via fight and flight response.

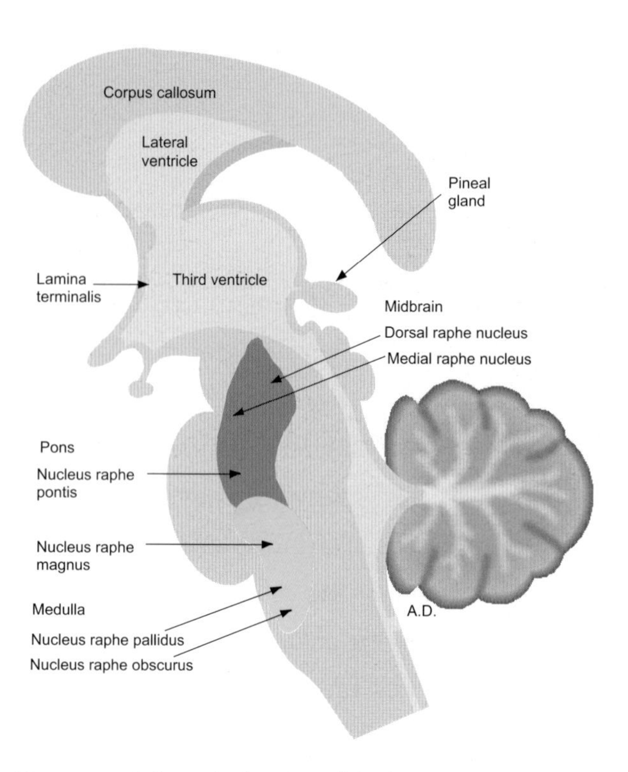

Figure 5.1 Schematic drawing of the brainstem illustrating the midline raphe nuclei of the reticular formation

The dorsal nucleus of the raphe is located in the midbrain and at the pontomesencephalic junction. This nucleus corresponds to group B7, conveying impulses to the hypothalamus, putamen, caudate nucleus, amygdala, intralaminar thalamic nuclei, and septal region via the medial forebrain bundle. It also projects to the superior colliculus, mamillary bodies, hippocampal formation, substantia nigra, locus ceruleus, olfactory cortex, and the lateral geniculate body. Reciprocal connections between the dorsal nucleus of raphe and limbic system are documented.

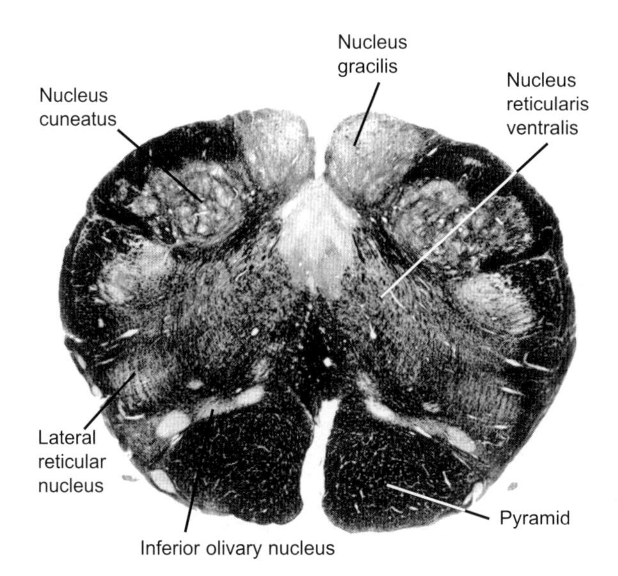

Figure 5.2 Section of the caudal medulla at the level of decussation of the internal arcuate fibers. The lateral reticular nucleus is dorsolateral to the inferior olivary nucleus and the nucleus reticularis ventralis is centrally located

Ascending fibers from this nucleus distribute via the dorsal longitudinal fasciculus and the medial forebrain bundle to the mamillary body, ventromedial hypothalamus, habenula, preoptic area, and suprachiasmatic nucleus.

The superior central nucleus (Figure 5.5) is located at the pontomedullary junction, corresponding to groups B6 and B8 of Dahlström and Fuxe. It projects diffusely to all areas of the cerebral cortex and specific regions of the cerebellar cortex. It also maintains reciprocal connections with the limbic system, projecting via the medial forebrain bundle and dorsal longitudinal fasciculus to areas that overlap with the dorsal raphe nuclear projections.

The paramedian nuclei (Figure 5.3) lie parallel to the medial longitudinal fasciculus (MLF) and the medial lemniscus, and consist of nuclei that maintain reciprocal connections with the cerebellum. The reticulotegmental, a paramedian nucleus at the rostral pons, maintains connections with the cerebellum and the globus pallidus.

The medial reticular nuclei form the efferent zone of the reticular formation which receives afferents from the spinal cord and collaterals from the spinothalamic tracts, as well as the cochlear, vestibular and trigeminal nerves. They consist primarily of the nucleus reticularis gigantocellularis (Figure 5.3) in the medulla and the nucleus reticularis pontine caudalis and oralis (Figure 5.4 & 5.5) in the pons. These reticular nuclei convey information received from the cerebral cortex, vestibular nuclei, cerebellum, spinal cord, and the lateral zone of the reticular formation to the spinal cord through the inhibitory medullary reticulospinal tract and excitatory pontine reticulospinal tract. Both reticulospinal tracts control posture (sitting and standing) and automatic movements (walking), and exert powerful influences upon the axial and proximal appendicular muscles. Both tracts function in conjunction with the vestibulospinal tract. The pontine reticulospinal is a massive uncrossed tract that descends through the medial longitudinal fasciculus and anterior funiculus, terminating in laminae VII, VIII, and IX. It activates extensor α motor neurons via the γ loop. The input of the basal nuclei to the pontine reticular nuclei is evident in the postural disorders exhibited by patients with Parkinson's disease. On the other hand, the medullary reticulospinal tract descends in the lateral funiculi, terminating in laminae VII, VIII, and IX of both sides and laminae IV, V, and VI on the ipsilateral side. This tract synapses upon the flexor α motor neurons.

The medial reticular nuclei regulate eye movements through projections via shorter fibers to the extraocular motor nuclei and sensory nuclei of the brainstem. These reticular neurons in the rostral pons and midbrain, which project via ascending fibers of the central tegmental tract to the intralaminar thalamic nuclei and hypothalamus, receive multiple sensory input. This reticular input to the intralaminar thalamic nuclei is utilized to produce alterations in the electrocortical activity and sleep–wake cycle, and affect our consciousness by projecting to diffuse areas of the cerebral cortex via the central tegmental tract. The latter pathway, a component of the ascending reticular activating system (ARAS), is responsible for changing the level of consciousness. The ARAS, a tegmental polysnaptic network is not only associated with consciousness, but also with memory, emotion, drive, and motivation. It has extrinsic and intrinsic elements; the extrinsic element consists of neurons in the medulla and pons that respond to stimulation generated by the cranial and spinal nerves, without being involved in the sleep–wake cycle. By contrast, the intrinsic element is represented in the mesencephalic neurons, which exhibit cyclic (e.g. diurnal) activity related to the projection of the anterior hypothalamic and suprachiasmatic areas to the

Damage to the reticular formation at the level of the rostral pons and caudal medulla may lead to coma or akinetic mutism (coma vigil). An EEG similar to the slow phase of sleep with no appreciable change in the autonomic and somatomotor reflexes or eye movements are characteristics of this condition.

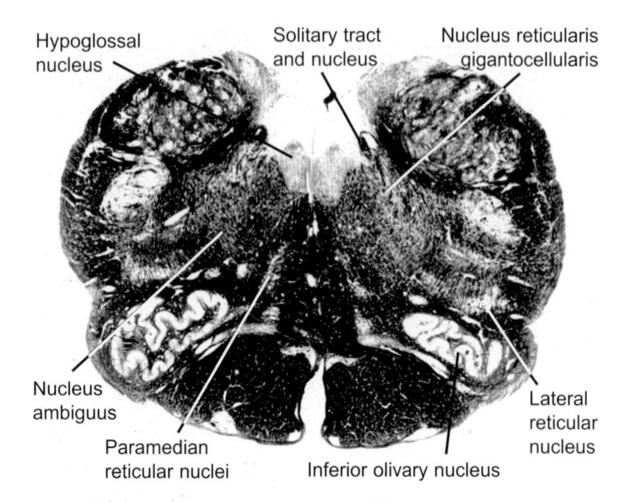

Figure 5.3 Section of the medulla at the midolivary level. The paramedian reticular nuclei lie parallel to the midline and lateral to the raphe nuclei. The nucleus reticularis gigantocellularis represents the medial reticular zone and the source of the medullary reticulospinal tract

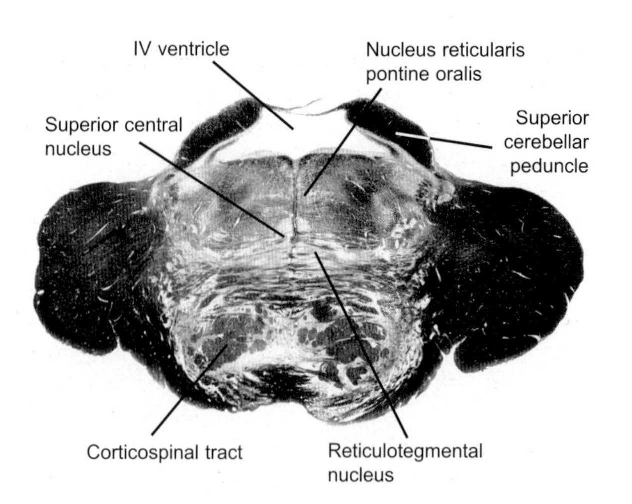

Figure 5.5 Section of the pons at the level of the trigeminal nerve. The locations of the superior central and reticulotegmental nuclei, as well as the nucleus reticularis pontine oralis are illustrated

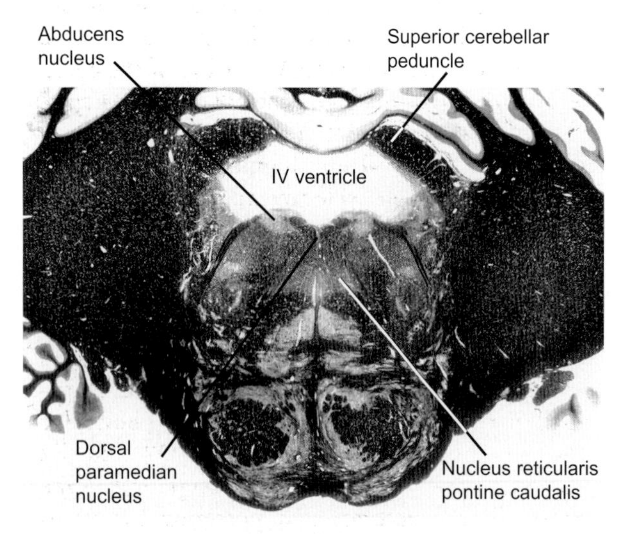

Figure 5.4 In this section of the pons at the level of the abducens nucleus, the nucleus reticularis pontine caudalis is shown superior to the medial lemnisci

midbrain via the medial forebrain bundle.

The lateral reticular zone represents the sensory reticular area that receives input from the ascending sensory pathways as well as the cerebral cortex. In the medulla, this zone contains the nucleus reticularis parvocellularis and nucleus reticularis lateralis. In the rostral pons and caudal midbrain, it contains the reticulotegmental, Kölliker–Fuse, medial parabrachial, and pedunculopontine nuclei. In the caudal medulla it lies medial to the solitary nucleus, nucleus ambiguus and dorsal motor nucleus of vagus, containing at this level the adrenergic C2 and noradrenergic A2 group of reticular neurons. In general, they are considered to be precerebellar relay nuclei, which receive input from a variety of sources such as the cerebral cortex, spinal cord and vestibular nuclei and maintain reciprocal connections with the cerebellum. The lateral reticular zone also contains L & M micturition centers, noradrenergic A, adrenergic C, and chlolinergic Ch neurons which are scattered in the medulla and upper pons. The noradrenergic neurons are classified into A1, A2, A4, A5, A6 and A7 groups (A3 is absent in humans). Adrenergic neurons are categorized into C1 and C2, while cholinergic neurons are divided into Ch5–Ch6. C2 and A2 groups are located in the medulla in close proximity to the nucleus ambiguus. In the lateral tegmental area of the pons noradrenergic cell groups A1, A2, A4, A5, A6 and A7, adrenergic groups C1–C2, and cholinergic groups Ch5–Ch6 are visible. A2, A4, A5 and C1 cellular groups are located ventrolaterally in the pons near the facial motor nucleus. A4 extends along the medial surface of the superior cerebellar peduncle. A5, in conjunction with C1, may act to regulate vasomotor (blood pressure, caliber of the vessels, heart rate etc.) activities. A1 and A7 are located in the lateral pontine tegmentum. A1, A2, A5 and A7 influence the activities of the amygdala, septal nucleus, hypothalamus, bed nucleus of stria terminalis, and diagonal band of Broca via the central tegmental tract (CTT) and the medial forebrain bundle (MFB).

The nucleus reticularis parvocellularis is located between the nucleus reticularis gigantocellularis and the spinal trigeminal tract and nucleus. It contains

The locus ceruleus is believed to have a role in the rapid eye movement phase (REM) of sleep and in the control of cortical activity through the intralaminar thalamic nuclei. Rapid conjugate eye movements and increase in intracranial pressure as well as heart and respiratory rates characterize REM sleep, which begins ninety minutes after sleep onset in the early evening. It also involves increase in temperature, nocturnal tumescence, hypotonia, and very brief episodes of facial and limb movement. In this stage, the muscles remain paretic and flaccid, and the deep tendon reflexes can not be elicited. It represents 20–25% of total sleep time, consisting of brief episodes (3–4 each night) of rapid eye movements (each episode may last 5–30 minutes) that follow the predominant cycles of non-rapid eye movement (NREM) phase which may last up to an hour and half. In the REM phase muscle movement and tone in the head, trunk and extremities are absent (flaccid) and their tendons remains areflexic. Angina and cluster headache may also occur during this phase. Since autonomic activities occur in a motionless individual, REM phase is also known as paradoxical sleep. These autonomic activities include increase in pulse and blood pressure, metabolism and blood flow in the brain.

NREM sleep represents 75–80% of nocturnal sleep in adults and 50% of the newborn. In this phase, the eye movements are slow and muscles maintain tone with normal deep tendon reflexes. Autonomic activities are minimal, although secretion of growth hormone and prolactin increases during this phase. Sleepwalking (somnambulism), bedwetting (enuresis), night terrors, and seizures may occur in NREM sleep. NREM sleep is divided into four stages that are distinguished by the slower and higher voltage EEG patterns and depth of unconsciousness:

• Stage 1 lasts only for 1 minute or up to 7 minutes. Sleep is easily discontinued during this stage (maintains low arousal threshold), e.g. softly calling one's name, closing door, etc. The length and percentage of this stage may increase when the sleep is disrupted.

• Stage 2 manifests sleep spindles or K complexes on EEG. Arousal in stage 2 may be produced by a more intense stimulus. As this progresses, there is a gradual appearance of high voltage slow activity on EEG, which eventually meets the criteria for stage 3.

• Stage 3 presents with high voltage (75 uV) and slow wave (2 cycles per minute) activity accounting for more than 20% but less than 50% of the EEG activity. This stage usually lasts only a few minutes in the first cycle and is transitional to stage 4.

• Stage 4 is identified when the high voltage slow wave activity is more than 50% of the EEG activity, and generally lasts 20–40 minutes in the first cycle.

Stages 3 and 4 are often referred to as slow wave sleep, δ sleep or deep sleep. It is the most restorative sleep. Rising to a lighter NREM sleep stage is usually associated with a series of body movements. NREM and REM sleep continue to cycle throughout the night, with REM sleep episodes generally becoming longer, Stage 3 and 4 sleep occupy less time in the second cycle and may disappear altogether from later cycles as stage 2 sleep expands to occupy the NREM portion of the cycle. On average across the night, the NREM–REM cycle is approximately 90–110 minutes. In the young adult, slow wave sleep dominates the NREM portion of the sleep cycle toward the beginning of the night, while REM sleep tends to be greatest in the last third of the night.

In addition to serotonin's role in NREM sleep and norepinephrine's role in REM sleep, numerous other substances are also involved in the sleep mechanism such as the hypothalamic δ sleep-inducing peptide, corticotropin-releasing hormone (suppresses the slow-wave sleep), D2 and E2 prostaglandins (act upon the sleep and wake centers in the anterior and posterior hypothalamus, respectively). Lesion of the locus ceruleus suppresses the deep phase of sleep. In contrast, long term arousal may result from increased secretion of the catecholamines from this nucleus. The balanced opposing effects of serotoninergic and noradrenergic neurons regulate the sleep states.

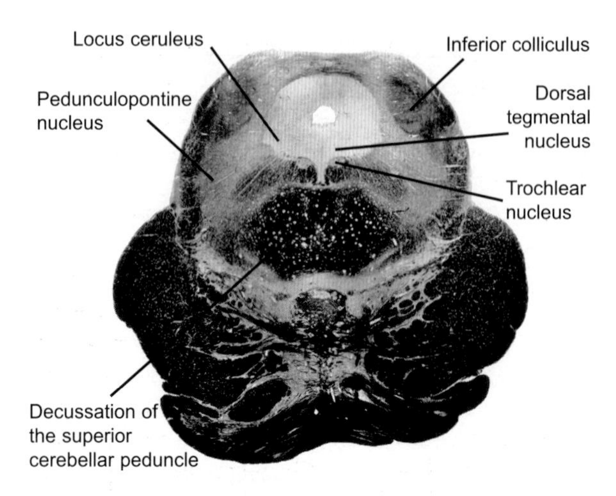

Figure 5.6 Section of the pons at the level of the trochlear nucleus. The pigmented locus ceruleus is located ventrolateral to the cerebral aqueduct within the periaqueductal gray matter. The pedunculopontine nucleus is dorsolateral to the superior cerebellar peduncle

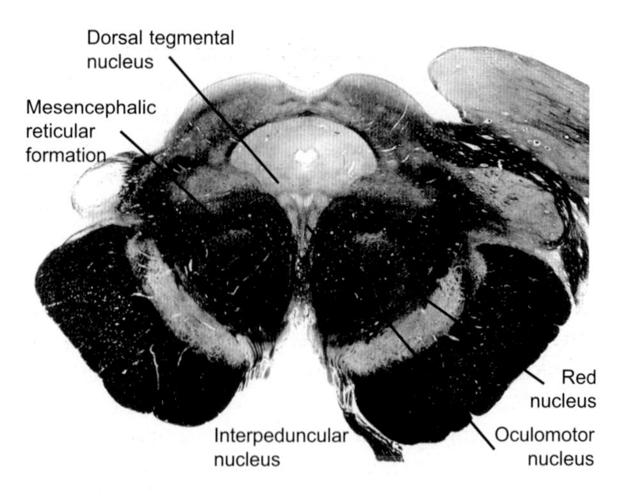

Figure 5.7 Section of the midbrain at the level of the superior colliculus. The mesencephalic reticular formation occupies the center of the tegmentum lateral to the red nucleus. The dorsal tegmental nucleus is located dorsal to the oculomotor nucleus, whereas the interpeduncular nucleus is dorsal to the interpeduncular fossa

noradrenergic A2 and adrenergic C2 cellular groups. The connections of the hypoglossal to the trigeminal and facial motor nuclei are maintained by the parvocellular reticular nucleus. Afferents to this nucleus arise from the contralateral red nucleus and cerebral cortex.

The lateral reticular nucleus (Figures 5.2 & 5.3) occupies the lateral medulla, projecting to the cerebellum via the inferior cerebellar peduncle.

The reticulotegmental nucleus (Figure 5.5) is located in the center of the ventral pontine tegmentum at the level of the trigeminal nerve and isthmus. It maintains reciprocal connections to the cerebellum, receiving afferents via the superior cerebellar peduncle, and projecting back to the cerebellum through the middle cerebellar peduncle.

The Kölliker–Fuse nucleus is located medial and ventral to the superior cerebellar peduncle. It acts as a pneumotaxic center by projecting to the inspiratory and expiratory centers in the medullary and pontine reticular formation. These centers convey the input to the preganglionic sympathetic neurons of the upper three or four thoracic spinal segments, phrenic nucleus, and the intercostal neurons.

The medial parabrachial nucleus is located medial and ventral to the superior cerebellar peduncle, establishing reciprocal connections with the insular cortex, amygdala, and hypothalamus. This nucleus projects to the pontine micturition M center.

The pedunculopontine nucleus (Figure 5.6), which is classified as group Ch5, is comprised of cholinergic neurons that lie dorsolateral to the superior cerebellar peduncle at the caudal mesencephalon. It continues with a more caudal cholinergic cell group (Ch6) in the pontine central gray matter. This nucleus is crossed by fibers of the superior cerebellar peduncle, and receives input from the globus pallidus, substantia nigra, and primary motor cortex. It projects primarily to substantia nigra.

The locus ceruleus (noradrenergic group A6) and nucleus subceruleus are pigmented noradrenergic neurons (Figure 5.6) contained in the rostral pons which merge caudally with group A4. These nuclei play an important role in the regulation of paradoxical (REM) phase of sleep and control of sensory neurons. The locus ceruleus is known to project monosynaptically to the cerebral cortex, spinal cord (intermediolateral lateral column), hippocampal formation, septal area, and diencephalon. It also projects to structures in the telencephalon and the entorhinal area (secondary olfactory cortex) via the stria medullaris thalami, stria terminalis, fornix and the medial forebrain bundle.

Narcolepsy is a condition in which patients enter into brief, often unpredictable, and irresistible episodes of sleep during waking hours (when conducting activities that require constant attention). Sleep attacks during a single day may range between one to as many as 20 and each attack may last from a few minutes to as much as a few hours. It has been estimated that up to 50% of these individuals suffer from memory loss. Some narcoleptics may experience automatic behaviors that may include rapid blinking, repetitive motor activities and irrelevant speech.

This condition starts in the second decade of life between adolescence and 25 years and may follow traumatic emotional situations. Excessive daytime sleepiness, preceded by a feeling of overwhelming fatigue may be an initial sign of narcolepsy. Presence of excessive daytime somnolence, cataplexy, sleep (hypnagogic and hypnopompic) hallucinations and sleep paralysis, represent the narcolepsy tetrad that is essential for the diagnosis of narcolepsy. Most narcoleptics present with cataplexy, an excitement, anger, or fear-induced condition.

Cataplexy usually occurs months to years following the initial onset of sleep attacks. Visual and/or auditory hallucinations may be detected in individuals with prolonged cataplectic attacks. It is characterized by sudden loss of muscle tone which may lead to sudden collapse of the patient to the ground. Hypnagogic hallucinations refers to the hallucinations that occur prior to falling asleep whereas hypnopompic hallucinations occur upon awakening.

Sleep paralysis is similar to cataplexy but refers to the paralysis that may be experienced by the patient slightly before the onset of sleep or shortly after awakening. The individual is aware of the paralysis and often describes the struggle to get up. This paralysis is temporary and may be terminated spontaneously or after mild sensory stimulation.

The gold standard for diagnosis of narcolepsy is sleep study. In this diagnostic method the patient is allowed adequate amount of sleep approximately 10 days prior to the study. This is followed by a multiple sleep latency test (MSLT) in which the patient is asked to nap every 2 hours throughout the day while the time to fall asleep (sleep latency) is measured. Narcoleptics usually require less than 4 minutes to fall asleep, a considerably less amount of time compared to normal individuals. They also usually enter REM sleep within minutes of the onset of sleep. Occurrence of more than two episodes of sleep onset REM period (SOREMPs) during the MSLT is virtually diagnostic of this disorder. This condition is commonly known to be associated with human leukocyte antigen (HLA) DR2 on the short arm of chromosome 6. Although the inheritance pattern is believed to be autosomal dominant, environmental factors should also be considered in the assessment of this condition. First-degree relatives of narcoleptics may exhibit excessive daytime somnolence and have much greater incidence of narcolepsy. Treatment of this disorder may be accomplished by stimulants such as methylphenidate, which is thought to increase the release of norepinephrine and thus decrease REM sleep. Prolonged administration of this medication may cause insomnia, irritability, psychosis, and possible tolerance.

Centrally acting $\alpha 1$ adrenergic agonists such as modafinil, and selegiline, a monoamine oxidase B inhibitor, although less effective than methylphenidate, may be used as second-line agents. Tricyclic antidepressants (TCA) and serotonin reuptake inhibitors such as femoxetine and fluoxetine may be used for treatment of cataplexy. Amphetamine, methylphenidate may be used as treatment of narcolepsy. Since cataplexy, hypnogogic (sleep) hallucination, and sleep paralysis occur in REM sleep, tricyclic antidepressant and monoamine oxidase (MAO) inhibitors which suppress REM are used as therapeutic measures for these disorders. γ-Hydroxybutyrate and behavioral modification may be useful in the treatment of cataplectic episodes.

Suggested reading

1. Aldrich MS. Narcolepsy. *Neurology* 1992;42(Suppl 6):34–43
2. Baghdoyan HA, Lydic R. M2 muscarinic receptor subtype in the feline medial pontine reticular formation modulates the amount of rapid eye movement sleep. *Sleep* 1999;22:835–47
3. Dahlström A, Fuxe K. Evidence for the existence of monamine-containing neurons in the central nervous system. *Acta Physiol Scand* 1964;232:1–55
4. Dahlström A, Fuxe K. Evidence for the existence of monoamine neurons in the central nervous system. II. Experimentally induced changes in the intraneuronal amine levels of bulbospinal neuron systems. *Acta Physiol Scand* 1965;247:1–36

5. Flegontova VV. The topography and cellular organization of the reticular formation of the thoracic part of the spinal cord in the cat. *Neurosci Behav Physiol* 2000;30: 111–13
6. Gerrits N, Voogd J. The projection of the nucleus reticularis tegmenti pontis and adjacent regions of the pontine nuclei to the central cerebellar nuclei in the cat. *J Comp Neurol* 1987;258:52–69
7. Guilleminault C, Pelayo R. Narcolepsy in children: a practical guide to its diagnosis, treatment and follow-up. *Paediatr Drugs* 2000;2:1–9
8. Hobson JA. Sleep and dreaming: Induction and mediation of REM sleep by cholinergic mechanisms. *Curr Opin Neurobiol* 1992;2:759–63
9. Kishi E, Ootsuka Y, Terui N. Different cardiovascular neuron groups in the ventral reticular formation of the rostral medulla in rabbits: single neurone studies. *J Autonom Nerv Syst* 2000;79:74–83
10. Mancia M. One possible function of sleep: To produce dreams. *Behav Brain Res* 1995;69:203–6
11. Martin GF, Hostege G, Mettler WR. Reticular formation of the pons and medulla. In Paxinos G, ed. *The Human Nervous System.* San Diego: Academic Press, 1990:203–20
12. McGinty D, Szymusiak R. Keeping cool: A hypothesis about the mechanisms and functions of slow-wave sleep. *Trends Neurosci* 1990;13:480–7
13. Nakayama J, Miura M, Honda M, *et al.* Linkage of human narcolepsy with HLA association to chromosome 4p13-q21. *Genomics* 2000;65:84–6
14. Sasaki S, Isa T, Naito K. Effects of lesion of pontomedullary reticular formation on visually triggered vertical and oblique head orienting movements in alert cats. *Neurosci Lett* 1999;265:13–16
15. Siegel JM, Rogawski MA. A function of REM sleep: Regulation of noradrenergic receptor sensitivity. *Brain Res Rev* 1999;13:213–33
16. Swadling C. Narcolepsy is a rare yet serious disorder. *Nurs Times* 1999;95:41
17. Valentino RJ, Curtis AL, Page ME, *et al.* Activation of the locus ceruleus brain noradrenergic system during stress: circuitry, consequences, and regulation. *Adv Pharmacol* 1998;42:781–4
18. Waitzman DM, Silakov VL, DePalma-Bowles S, Ayers AS. Effects of reversible inactivation of the primate mesencephalic reticular formation. II. Hypometric vertical saccades. *J Neurophysiol* 2000;83:2285–99
19. Webster HH, Jones BE. Neurotoxic lesions of the dorsolateral pontomesencephalic tegmentum-cholinergic cell area in the cat. II. Effects upon sleep-waking states. *Brain Res* 1988;458:285–302
20. Wilson VJ. Vestibulospinal reflexes and the reticular formation. *Prog Brain Res* 1993;97:211–17

Cerebellum 6

The cerebellum is derived from the rhombic lip of the alar plate and constitutes an important part of the rhombencephalon. It is located in the posterior cranial fossa, covered by the tentorium cerebelli, and separated by the fourth ventricle from the dorsal surfaces of the pons and medulla. The tentorium cerebelli is lodged between the cerebellum and the occipital lobes of the brain. The cerebellum coordinates (synchronizes) voluntary motor activity and eye movements, regulates phonation, maintains equilibrium, and processes exteroceptive impulses.

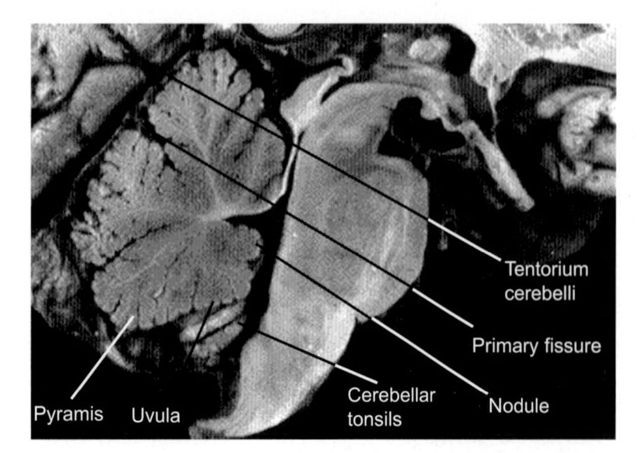

Figure 6.1 Midsagittal section of the brainstem and cerebellum. The fourth ventricle separates the cerebellum from the pons and medulla. The components of the cerebellar vermis are illustrated

Morphologic and histologic features of the cerebellum

The study of the cerebellum reveals a corpus cerebelli with inputs from the spinal cord, pontine and trigeminal nuclei, and a flocculonodular lobe that maintains primary connections to the vestibular nuclei. The corpus cerebelli may be subdivided into regions that receive spinal and pontine inputs. The cerebellum lies caudal and inferior to the occipital lobes and dorsal to the pons and midbrain (Figures 6.1 & 6.2). It is connected to the midbrain, pons and medulla via the superior, middle and inferior cerebellar peduncles, respectively. It comprises two lateral hemispheres, which are connected by the midline vermis. Vermis and cerebellar hemispheres are separated by the paravermal sulcus. Posteriorly these hemispheres are separated, though incompletely, by the falx cerebelli, contained in the posterior cerebellar notch. This deep notch continues inferiorly with the cerebellar vallecula, a median fossa between the two hemispheres. Each cerebellar hemisphere consists of a peripheral cortical gray matter, which is thrown into lamina or cerebellar folia, and a central white medullary substance, containing the cerebellar nuclei. Such an arrangement is also present in the cerebral cortex. In contrast, the spinal gray and white matter assume a reverse arrangement.

The fourth ventricle penetrates the white matter as a transverse gap, the fastigium, marking the junction of the superior and inferior medullary vela (white matter bands that extends rostrally and caudally in the roof of the fourth ventricle). The cerebellar hemispheres are divided into lobes and lobules by the primary, horizontal, pre- and postpyramidal,

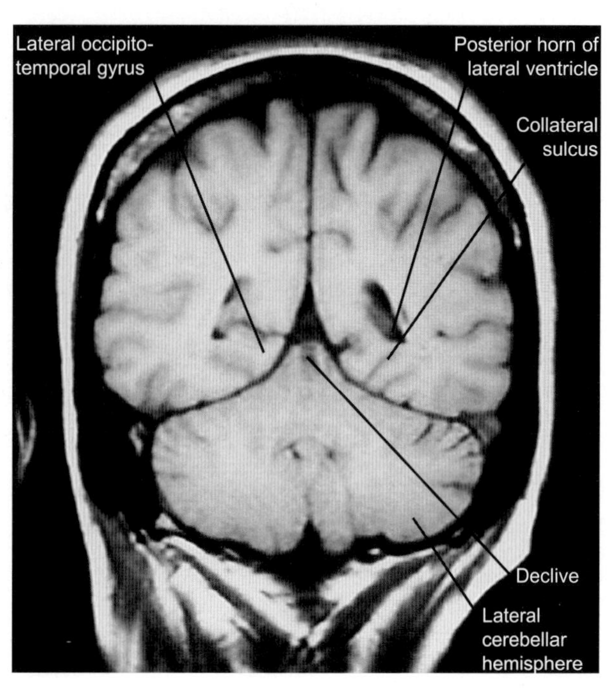

Figure 6.2 MRI scan of the brain through the occipitotemporal gyri. The cerebellum lies inferior to the occipital lobe and is separated by the tentorium cerebelli

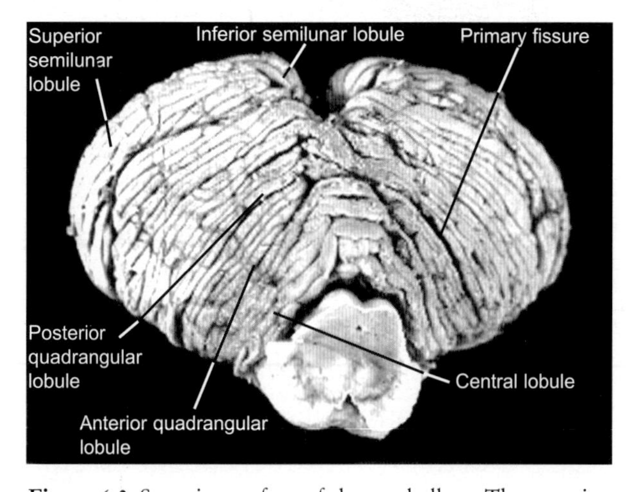

Figure 6.3 Superior surface of the cerebellum. The superior vermis and the corresponding parts of the cortical hemispheric areas are shown

postlingual and postcentral, and posterolateral fissures (Figures 6.3, 6.4 & 6.5). Developmentally, the posterolateral fissure is the first fissure to appear, marking the caudal boundary of the flocculonodular lobe. The primary fissure serves as a landmark, separating the rostral anterior lobe from the more caudal posterior lobe. The cerebellum further divides into superior and inferior surfaces by the horizontal fissure, which also divides the ansiform lobule (cerebellar cortical areas that correspond to the folium and

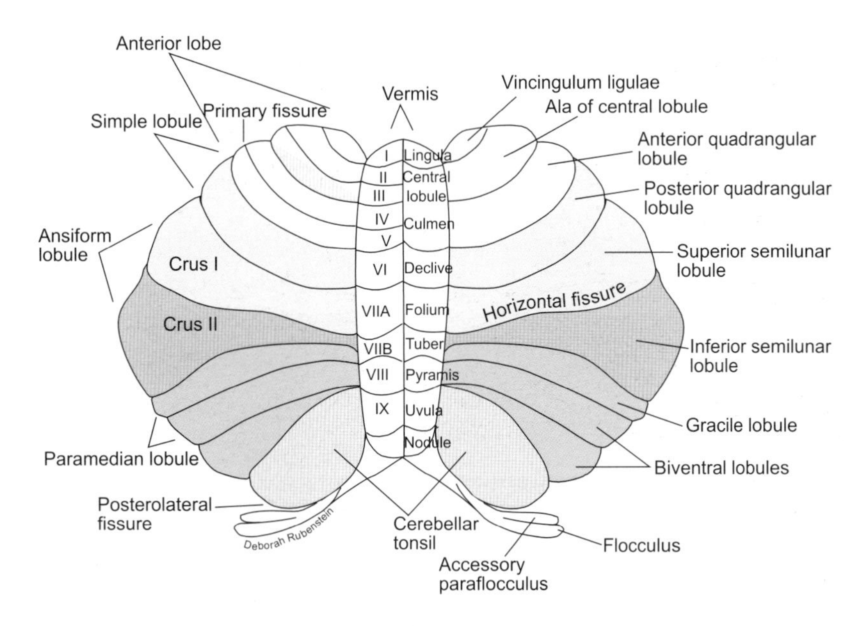

Figure 6.4 Various parts of the vermis and associated cortical parts are shown in this diagram

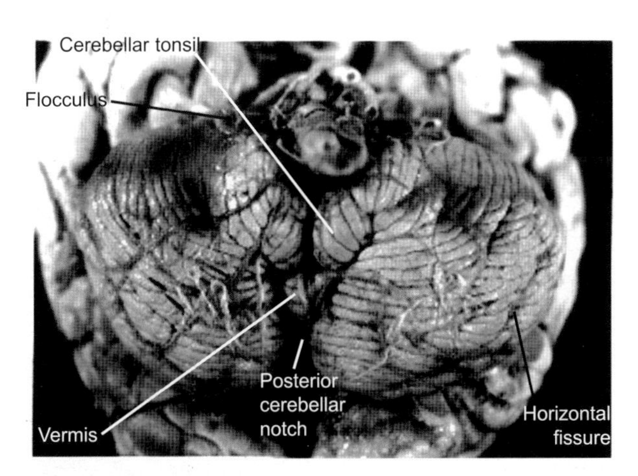

Figure 6.5 Inferior surface of the cerebellum. Note the branches of the anterior and posterior inferior cerebellar arteries on this surface. The posterior cerebellar notch is occupied by the falx cerebelli

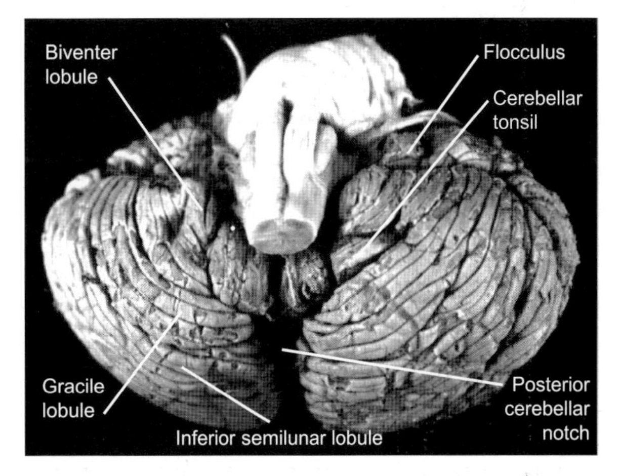

Figure 6.6 Inferior view of the cerebellum. Cerebellar tonsils and flocculus are clearly visible. Biventeral, gracile, and inferior semilunar lobules are illustrated

tuber vermis) into superior and inferior semilunar lobules. The midline vermis, which interconnects the cerebellar hemispheres, is divided into segments. Each vermal segment expands into the cerebellar hemispheres, with the exception of the lingula (Bolk's nomenclature). For example the central lobule of the vermis corresponds to the ala of the cortex, the culmen to the quadrangular lobule, declive to the posterior quadrangular (simple) lobule, whereas both the folium and tuber vermis correspond to the ansiform (superior and inferior semulunar lobules) and biventral lobules of the cerebellar cortex.

Similarly, the pyramis corresponds to the biventral lobules, and the uvula is associated with the cerebellar tonsils and the medial belly of the biventral lobule (Figures 6.4, 6.5 & 6.6). The flocculi remain attached to the midline nodule, forming the flocculonodular lobe. The prepyramidal fissure bounds the tuber vermis caudally and the postpyramidal fissure separates the uvula from the pyramid.

Larsell classified the vermis in a simplified pattern in which each hemispheric lobule can be related to one of the vermian parts. Lobules I–V of Larsell constitute the anterior lobe, lobule VI extends to the cortex as the simple lobule, whereas lobule VII corresponds to the ansiform lobule. In the same manner, lobule VIII has a hemispheric extension represented in the paramedian lobule (a combination of the gracile

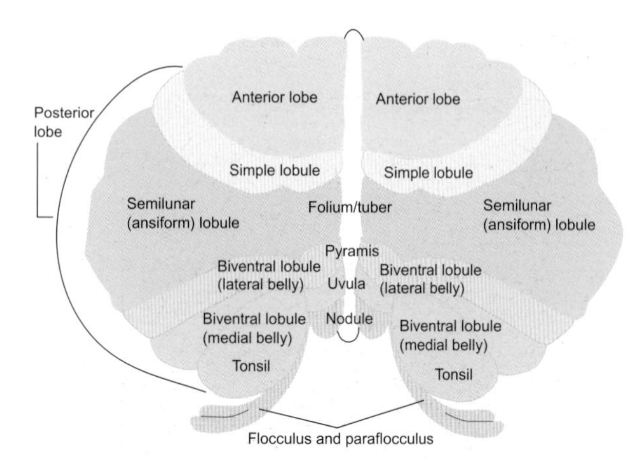

Figure 6.7 Various lobes of the cerebellum and associated fissures. Note that the morphologic divisions may correspond to a great extent to the functional classification

lobule and the lateral belly of the biventral lobule). The paraflocculus is shared by the vermian lobules IX and X. Extension of lobule X is the flocculus.

Voogd divided the cerebellar cortex into longitudinal zones that maintain discrete connections to the cerebellar nuclei. On the basis of this mapping, the efferent fibers from the cerebellar cortex to the cerebellar nuclei appear to arise in cortical zones that belong to the same compartment as the nuclear region to which they project.

The vermis and its cortical expansions are sites of somatotopic representation of the body. As a case in point, the head and face are represented in the simple lobule, the leg in the central lobule, whereas the arm occupies a distinct area in the culmen.

Blood flow to the cerebellum is maintained by the superior cerebellar, anterior and posterior inferior cerebellar arteries.

The superior cerebellar artery supplies the superior portion of the cerebellum, middle and inferior cerebellar peduncles, superior medullary velum, and the choroid plexus of the fourth ventricle. This vessel originates from the basilar artery (an artery formed at the pontobulbar sulcus by union of the vertebral arteries).

The anterior inferior cerebellar artery (AICA) supplies the anterolateral part of the inferior surface of the cerebellum, pyramids, tuber vermis, dentate nucleus, white matter of the cerebellum, and most of the tegmentum of the caudal pons. AICA arises from the lower part of the basilar artery and forms a loop into the internal acoustic meatus, where it frequently gives rise to the labyrinthine artery.

The posterior inferior cerebellar artery (PICA) supplies the cerebellar hemispheres and the inferior

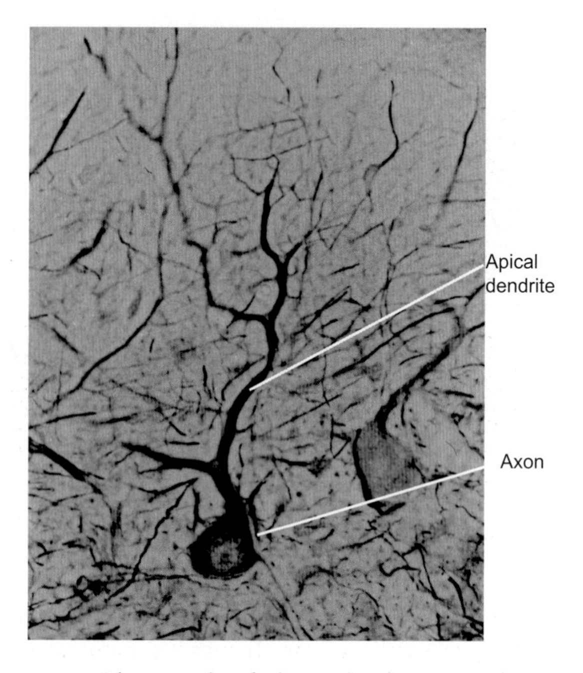

Figure 6.8 Photograph of the Golgi neuron with its prominent apical dendrite

vermis, uvula, nodule, cerebellar tonsils, choroid plexus of the fourth ventricle, and dentate nuclei.

Cerebellar veins are classified into superior, inferior, and lateral veins. The superior cerebellar veins drain into the great cerebral vein of Galen, the inferior cerebellar veins open into the straight sinus, while the lateral veins drain into the transverse, superior and inferior petrosal sinuses.

Phylogenetically the cerebellum ranges from the oldest to the most recent, consisting, in that order, of the archicerebellum, paleocerebellum, and the neocerebellum, respectively. The archicerebellum, which corresponds to the flocculonodular lobe, represents the oldest and most primitive part of the cerebellum. The paleocerebellum, the second oldest cerebellar component, consists of the anterior lobe and part of the anterior vermis. Most of the cerebellar hemispheres (posterior lobe) and part of the posterior vermis are the latest to appear phylogenetically, forming the neocerebellum. Functionally the cerebellum is classified into the vestibulocerebellum, spinocerebellum, and the pontocerebellum. The vestibulocerebellum is comprised of the flocculonodular lobe and part of the uvula. It has reciprocal connections with the vestibular nuclei, maintaining equilibrium and mediating vestibulo-ocular reflex. The inter-relationship between the vestibular nuclei, oculomotor system, and the cerebellum plays an important role in the coordination of voluntary eye movements. The spinocerebellum includes most of

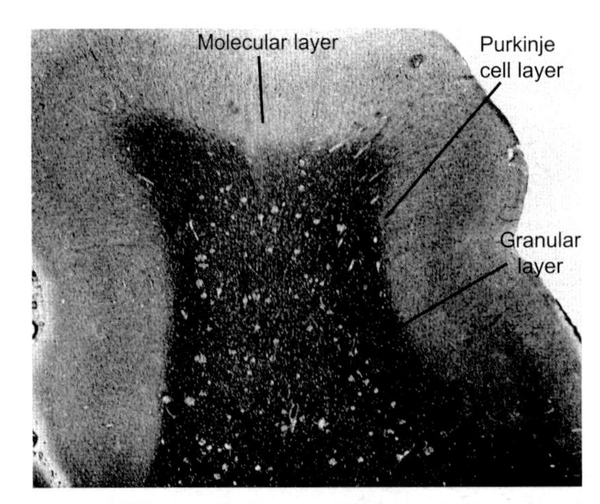

Figure 6.9 The cerebellar cortical layers. Note the single Purkinje layer between the outer molecular and inner granular layer

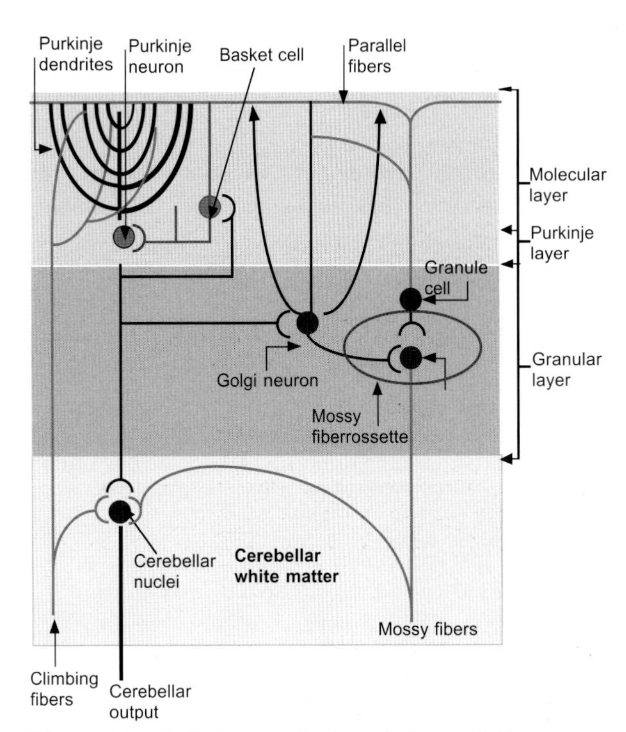

Figure 6.10 Cellular organization of the cerebellar cortex. The mossy fibers, cerebellar glomeruli, and climbing fibers are illustrated

the anterior lobe, pyramis, and corresponding cortical parts (biventral lobules). It receives input from the spinal cord and mesencephalic trigeminal nucleus and deals with propulsive movements (e.g. walking and swimming).

The flocculonodular lobe regulates activities that govern posture and movement through its connections to the brainstem and spinal cord. The posterior lobe with the exception of the simple lobule, gracile and biventral lobule and the corresponding vermal parts (declive and pyramis), comprises the pontocerebellum. This lobe receives cerebral cortical input via the pontine nuclei. It coordinates fine movements through its connections to the cerebral cortex.

The cerebellar cortex consists of granular, Purkinje, and molecular layers. These layers contain interneurons (granule, Golgi, basket and stellate cells) which are functionally inhibitory (Figures 6.8, 6.9 & 6.10). The granular layer (innermost layer) consists of small, densely packed granule cells, Golgi neurons, and mossy fiber rosettes. Granule cells give rise to axons that ascend into the molecular layer through the Purkinje layer, bifurcating into T-shaped fibers that run parallel to the direction of the cerebellar folia (parallel fibers). NMDA receptors, which are voltage dependent and produce slow depolarization and opening of the calcium channels in the postsynaptic neurons, are the predominant ionotropic glutamate receptors in the immature granule cells. However, AMP (α-amino-3-hydroxy-5-methyl-4-isolaxone propionic acid) and high affinity kainate types of non-NMDA glutamate receptors mediate fast excitatory transmission. It must be realized that receptors mediating cholinergic (nicotinic and muscarinic) receptors also exist in the granule cells. During their course, axons of the granule cells establish extensive contacts with the successive Purkinje cells like telephone poles. This transmission between the parallel fibers of the granule cells and the Purkinje dendrites are mediated by metabotropic receptors (type mGluR2 and GluR7) which are coupled to the phosphoinositide hydrolysis second messenger system and by the ionotropic glutamate receptors of the AMP type. Transmission between Purkinje and parallel fibers may be blocked by adenosine that binds to A1-adenosine receptors of the parallel fibers. Desensitization of Purkinje cell AMP and reduction of the synaptic transmission may occur as a result of simultaneous activation of the parallel and climbing fibers, a process known as long-term depression (LTD).

LTD is dependent upon an increase in intracellular calcium and on cyclic guanosine 3',5'-monophosphate (cGMP)-dependent protein kinase in Purkinje cells. Since synthesis of cyclic AMP is catalyzed by soluble guanylate cyclase (present in all cerebellar cells), which is activated by NO synthetase, inhibition of production of cGMP by NO may abolish LTD. Cyclic GMP is located in Bergmann glial cells and astrocytes. Changes in the circuitry of the cerebellum during learning and adaptation processes may be attributed to the LTD. Thus, LTD and adenosine are the two factors by which transmission between granule and Purkinje cells are blocked. Climbing fiber activity determines the extent of release of adenosine.

The excitatory input mediated by the parallel fibers and incoming mossy fibers results in repetitive firing of the Purkinje neurons with the conventional type of action potentials.

Golgi neurons (Figure 6.8), which are also located in this layer, have somas that are lodged in the outer portions of the granular layer, and dendrites that extend into the molecular layer. Their dendrites and axonal branches extend in a radially symmetrical pattern, in exact contrast to the pattern of branching of other inhibitory neurons and Purkinje cells. Golgi cells also establish inhibitory axo-somatic contacts with the mossy and parallel fibers of granule cells in the mossy synaptic glomeruli. Golgi cells contain GABA (especially $\alpha6$ subunit) and glycine, a feature that distinguish these cells from the basket and stellate cells. The predominance of the $\alpha6$ subunit in the Golgi neuron versus the granule cells may account for the scarcity of benzodiazepine-sensitive $GABA_A$ receptors in the granular layer. Some Golgi cells may also contain enkephalin and somatostatin, as well as choline acetyltransferase (CHAT) that catalyzes the synthesis of acetylcholine. However, calcium binding proteins are strikingly lacking in these cells. Since most Golgi cells contain the mGluR2 type of metatropic glutamate receptor that is dependent upon cyclic adenosine monophosphate, inhibition of AMP and subsequently Golgi cells may result from stimulation of the glutamatergic parallel or mossy fibers. Mossy fibers, which constitute the bulk of the cerebellar afferents (non-olivary fibers), terminate in the granular layer as mossy rosettes (Figure 6.10). Sites of linkages between mossy rosettes, axons of the Golgi neurons, and dendrites of the granule cells occur in the synaptic glomeruli.

The Purkinje cell layer (Figures 6.9 & 6.10) consists of a single row of flask-shaped neurons with massive and flattened dendritic trees, establishing contact with the climbing fibers and the axons of the granule cells. Subunits of $GABA_A$, which are present in the cerebellar neurons, include $\alpha1$, α_3, β_2 and μ_2. Combinations of α_1, β_2, and m2 subunits display high affinity binding for benzodiazepine ligands. Subunit α_1 is present in the soma of Purkinje cells, opposite the terminals of the basket cells, whereas the $\alpha3$ subunit occupies the proximal dendritic tree where the stellate cell terminates. Elements of the second messenger system that mobilize calcium from subsurface cisterns (e.g. receptor for $InsP_3$ or inositol 1,4,5-triphosphate) and protein kinase C are contained in the Purkinje neurons. Calcium binding proteins such as calmodulin, calbindin, and parvalbumin are also located in these cells. In fact, these elements are functionally related to the metabotropic receptors

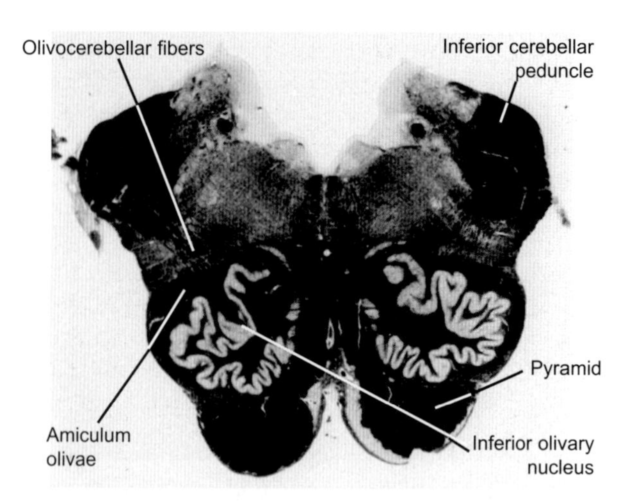

Figure 6.11 Section through the medulla at the midolivary level. The massive fibers of the olivocerebellar tract represent the main component of the inferior cerebellar peduncle

that mediate synaptic transmission between parallel fibers and Purkinje cells. All these factors may play important roles in the increase of calcium concentration in Purkinje cells. Purkinje axons project to the cerebellar nuclei in a mediolateral direction, axons of vermal neurons project to the fastigial nucleus, paravermal neurons to the globose and emboliform nuclei, whereas the lateral hemispheric neurons project to the dentate nucleus and the lateral vestibular nuclei. Thus, the cerebellar excitatory input must overcome the tonic inhibitory impulses generated by the Purkinje cells upon the cerebellar nuclei. These projections, which represent the inhibitory corticonucleocerebellar tract, are arranged in (a)symmetrical longitudinal bands that reflect the Zebrin-positive cells, a group of proteins that are contained in certain populations of Purkinje neurons.

The molecular layer (Figure 6.10) is a largely acellular outer layer, containing the dendrites of the Purkinje neurons, climbing fibers, axons of the granule cells, stellate cells, and the basket cells. Stellate and basket cells have similar structures and their dendrites extend toward the surface and their axons run transversely to the folia and parallel to the dendrites of the Purkinje cells. They receive excitatory input from the parallel fibers, and their soma receives collaterals from the Purkinje neurons, as well as parallel and mossy fibers. These cells, which contain NMDA receptors, mediate the parallel fiber-Purkinje inhibition. Large epithelial (Bergmann) glial cells and their radiating branches and processes that surround all cerebellar cortical neurons are also present in this layer.

Bergmann cells are associated with signaling processes, and are the primary source of cGMP in the cerebellar cortex. These cells, among all cerebellar neurons, uniquely contain α_2 subunit of the $GABA_A$. The enzyme, which is involved in the hydrolytic cleavage of 5′-nucleotide monophosphates and formation of the adenosine, also resides in these cells. Among other features of Bergmann cells are their possession of kainate receptors with specific ionic arrangement. The release of homocysteic acid, a putative amino acid neurotransmitter, by these cells is dependent upon the climbing fibers. They are capable of glutamate uptake and its conversion into glutamine, a process that enable the synthesis of glutamate by glutamatergic terminals. These terminals form the external limiting membrane.

Cerebellar nuclei

The cerebellar nuclei are embedded in the white matter, consisting of the fastigial, globose and emboliform, and the dentate nuclei. Neurons of these nuclei contain ionotropic glutamate receptors of the NMDA and non-NMDA types, as well as GABAergic ($GABA_A$ α_1-β_2 or 3-μ_2) and glycine receptors.

The fastigial nucleus is the most medial and phylogenetically the oldest cerebellar nucleus. It is located near the apex of the fourth ventricle where the superior and inferior medullary vela joins. It receives input from the vermis, and collaterals of cerebellar cortical afferents, and afferents from the medial and inferior vestibular nuclei, locus ceruleus, and the reticulotegmental nucleus. It projects to the vestibular nuclei and the reticular formation.

The globose and emboliform nuclei are intermediate in phylogeny and location, receiving input from the paravermal cortex, reticulotegmental nucleus, medial and dorsal accessory olivary nuclei, and collaterals from other cerebellar afferents.

Secondary trans-synaptic degeneration and atrophy of the inferior olivary nucleus may occur as a result of cerebellar cortical degeneration involving the superior vermis. Olivopontocerebellar atrophy, as the name indicates, is associated with degeneration of the inferior olivary nucleus, pontine nuclei, and cerebellar cortex. This condition exhibits cerebellar ataxia, signs of upper motor neuron palsy, involuntary movements, autonomic disorders, and mental retardation. A lesion which disrupts the circuitry between the dentate nucleus and the inferior olivary nucleus may result in palatal myoclonus, a continuous rhythmic contraction of the posterior pharyngeal muscles that resembles tremor.

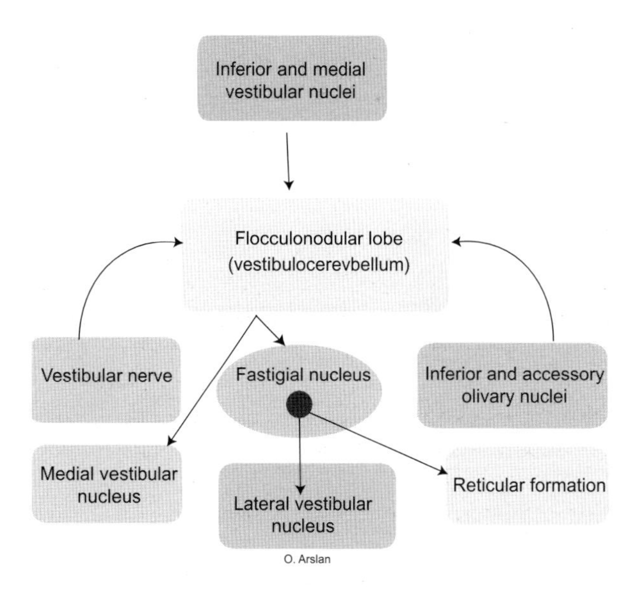

Figure 6.12 Schematic drawing of the projections of the vestibular nerve (primary vestibulocerebellar tract) and the vestibular nuclei (secondary vestibulocerebellar tract) to the cerebellum

The dentate nucleus is the most lateral nucleus and developmentally is the most recent. Most cerebellar afferents, cortical fibers from the lateral hemispheric zone, afferents from the pontine nuclei, trigeminal sensory nuclei, reticulotegmental nucleus, inferior olivary nucleus, locus ceruleus, and the raphe nuclei terminate in this nucleus. It forms the main cerebellar output which projects through the superior cerebellar peduncle to the red nucleus and the ventral lateral nucleus of the thalamus.

Cerebellar afferents

Climbing fibers

The climbing fibers (Figure 6.10), which utilize L-glutamate as a neurotransmitter, primarily represent the olivocerebellar fibers that establish synaptic contacts with the Purkinje dendrite. In addition to the L-glutamate or L-aspartate, peptides are also contained in certain subpopulations of climbing fibers. Corticotropin-releasing factor (CRF) is also contained in these fibers. A single axon of the inferior olivary nucleus establishes excitatory connections with about a dozen Purkinje neurons and each Purkinje neuron in turn receives only one climbing fiber. The dendrites of the Purkinje neurons are entirely entwined by the climbing fiber, making roughly over two hundreds synaptic contacts. A single climbing fiber may evoke excitatory postsynaptic potential which maintains an amplitude greater than 25 mV (complex spikes), and greatly

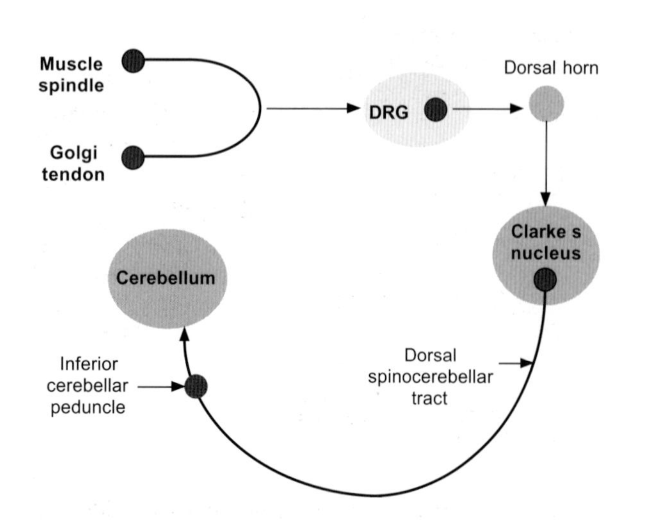

Figure 6.13 Clarke's nucleus and its projection to the cerebellum through the dorsal spinocerebellar tract. DRG, dorsal root ganglion

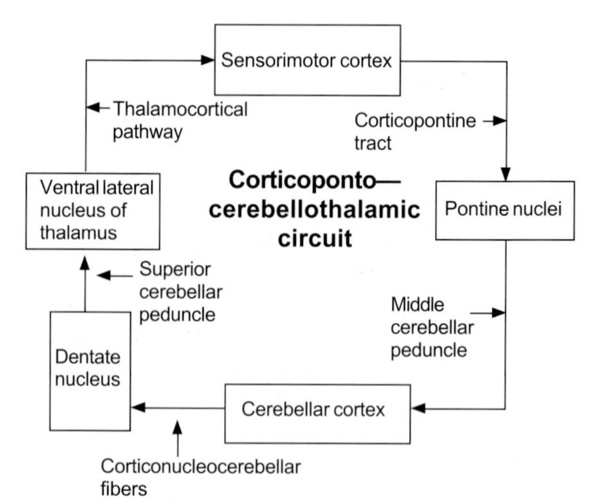

Figure 6.15 This circuit is mediated via the connections of the pontine nuclei which project to the cerebellar cortex. Series of neurons are involved, as a closed circuit loop, in the transmission of impulses between the cerebellum, thalamus, and motor cerebral cortex

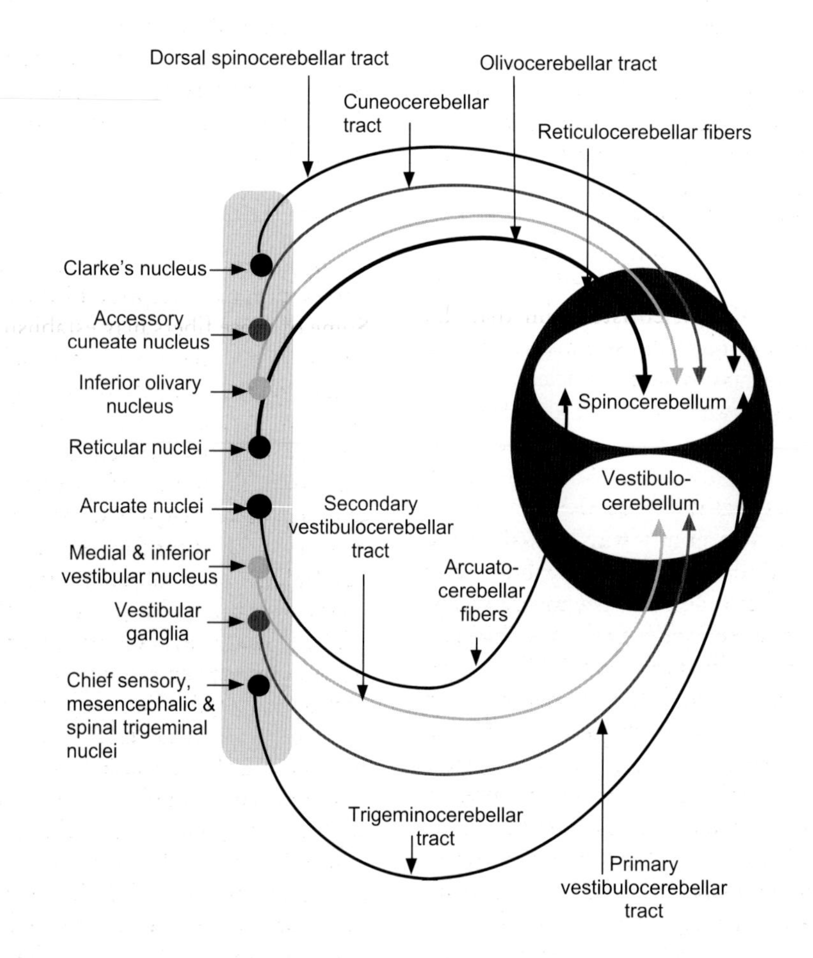

Figure 6.14 The major afferents to the cerebellum via the inferior cerebellar peduncle. Note the bulk of afferents come from the inferior olivary nucleus

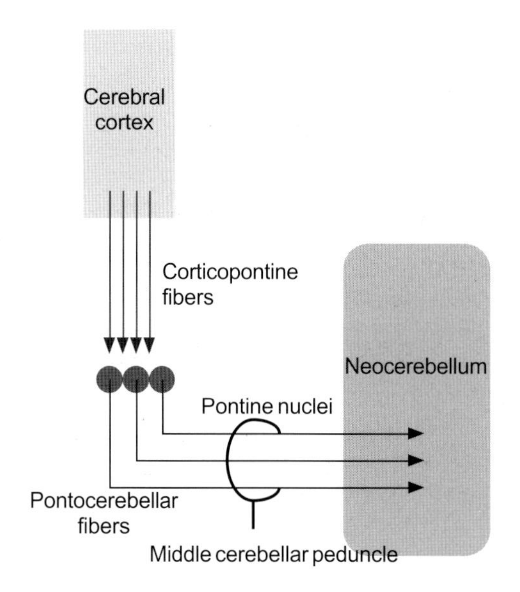

Figure 6.16 The neuronal components of the middle cerebellar peduncle. Contralateral axons of the pontine nuclei form this peduncle

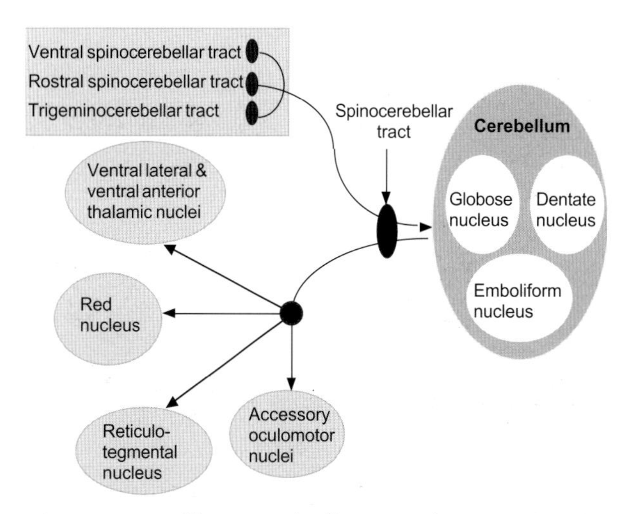

Figure 6.17 Afferents and efferents pathways within the superior cerebellar peduncle. The cerebellar projection to the thalamus constitutes the principal tract within this peduncle. Note that some fibers also terminate in the red nucleus, reticulotegmental and accessory oculomotor nuclei

exceeds the Purkinje cell threshold. Due to this very reason a single impulse in the climbing fibers always elicits an action potential in more than ten Purkinje cells. Thus, the action potential of a single climbing fiber is not a graded response, but an all-or-none action. Thus, the Purkinje cells respond to the generated action potential with complex spikes, in contrast to the simple spikes evoked by the T-parallel fibers of the granule cells. According to some investigators the climbing fiber is primarily concerned with fast, ballistic movements. Other scientists claim that the climbing-fiber system reflects the summation of the inhibitory and excitatory synaptic activity at any instant time. Some also theorize that the signals in the climbing fibers are meant to ascertain error in executing a motor activity (e.g. during learning stages), which suggest that the frequency of firing of the climbing fibers is not in any way related to the direction or the speed of a motor activity, but rather to the disturbances of that activity. Glutamate receptors at the synaptic connections between the Purkinje neurons and the climbing fibers are non-NMDA type, which are responsible for the influx of calcium ions into the dendritic tree of the Purkinje neurons. The olivocerebellar tract (Figures 6.10, 6.11 & 6.14), a contralateral pathway which arises from the inferior and the accessory olivary nuclei, represents the major climbing-fiber system. It crosses the medullary reticular formation to be distributed to the opposite vermal and cerebellar hemispheres via the inferior cerebellar peduncle. The accessory olivary nuclei convey joint, tactile, visual and vestibular impulses to the cerebellum from certain nuclei of the medulla. The inferior olivary nucleus sends information to the cerebellum, which is received from the spinal cord via the spino-olivary pathway, motor cortex, periaqueductal gray, and accessory oculomotor nuclei to the cerebellum.

Mossy fibers

Mossy fibers refer to all afferents of the cerebellar cortex with the exception of the olivocerebellar tract. Some of these fibers may establish excitatory synaptic contacts with the cerebellar nuclei, basket, Golgi, and stellate neurons. These fibers may be concerned with slow and tonic motor activity. L-glutamate is the primary neurotransmitter in mossy fibers with the exception of the secondary vestibulocerebellar fibers that utilize acetylcholine. Mossy fibers include the primary and secondary vestibulocerebellar, spinocerebellar, reticulocerebellar, pontocerebellar, tectocerebellar, and trigeminocerebellar tracts.

The primary vestibulocerebellar fibers (Figures 6.12 & 6.14) are the central processes of the neurons of the vestibular ganglia. These fibers run within the juxtarestiform part of the inferior cerebellar peduncle and terminate in the vestibulocerebellum of the same side.

Secondary vestibulocerebellar fibers (Figures 6.12 & 6.19) originate from the medial and inferior vestibular nuclei. These fibers run within the juxtarestiform part of the inferior cerebellar peduncle and terminate in the vestibulocerebellum on both sides.

The spinocerebellar tracts (Figures 6.13 & 6.14), which are represented in two principal pathways,

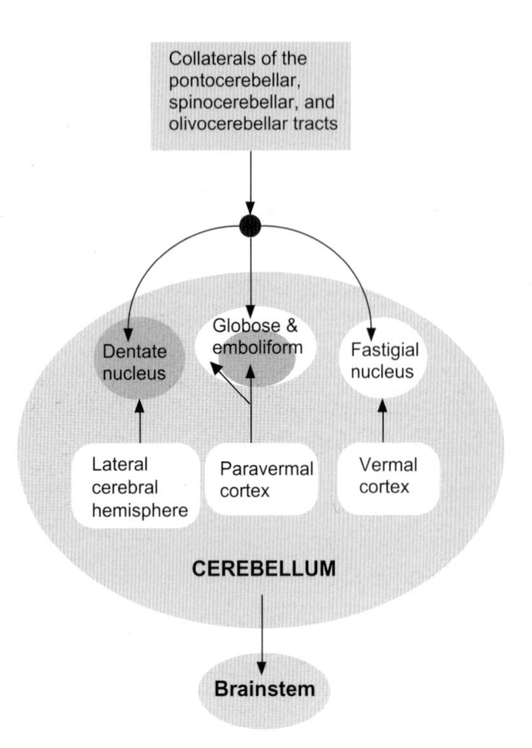

Figure 6.18 Somatotopic projections of the Purkinje cell axons to the cerebellar nuclei. These nuclei also process information received via collaterals of some mossy and climbing fibers. Purkinje axons also convey information to the brainstem

carry proprioceptive, stretch, and tactile sensations, terminating in the spinocerebellum (tactile input mainly terminate ipsilaterally in the simple lobule).

The dorsal spinocerebellar tract represents the axons of the Clarke's column, which extends in the thoracic and upper lumbar segments of the spinal cord. It is an ipsilateral tract that enters the spinocerebellum through the inferior cerebellar peduncle, carrying proprioceptive, tactile, and pressure impulses from individual muscles and joints of the lower extremity and lower half of the trunk. The upper limb equivalent of this tract is the cuneocerebellar tract that is derived from the accessory cuneate nucleus and terminates ipsilaterally in the pontocerebellum and spinocerebellum via the inferior cerebellar peduncle.

The ventral spinocerebellar tract (Figure 6.17) is derived from the intermediate gray columns and the border cells of the anterior horn cells of the thoracolumbar and sacral segments. Information conveyed by this crossed tract originates from the whole lower extremity, reaching the cerebellum through the superior cerebellar peduncle. The fibers of this tract cross again within this peduncle and terminate in the ipsilateral spinocerebellum. Lamina VII of the cervical enlargement gives rise to the rostral spinocerebellar, an ipsilateral, tract, which represents the upper limb equivalent of the ventral spinocerebellar tract. It enters the cerebellum through the inferior and superior cerebellar peduncles, to be distributed to the anterior lobe of the cerebellum. The ventral and rostral spinocerebellar tracts act jointly as a relay center reflecting the neuronal activities in the descending motor pathways.

The reticulocerebellar tract (Figures 6.14, 6.17 & 6.20) originates from the pontine reticulotegmental, and the medullary lateral and paramedian reticular nuclei. The lateral reticular nucleus projects bilaterally via the inferior cerebellar peduncle to the vermis of the spinocerebellum, fastigial and emboliform nuclei. This projection conveys information from all levels of the spinal cord (spinoreticular) that initially establishes synaptic contacts with the neurons of the lateral cervical nucleus and later projects to the medullary reticular formation, and eventually to the cerebellum. The paramedian nuclei, which receive fibers from the interstitial nucleus, tectum, spinal cord and cerebral cortex, send efferents to the entire cerebellum with the exception of the paraflocculus. The reticulotegmental nucleus projects to the anterior lobe, simple lobule, the folium and tuber vermis via the middle cerebellar peduncle. Some fibers also terminate in the fastigial, globose, and dentate nuclei.

The pontocerebellar tract (Figures 6.16 & 6.22) delivers the cortical inputs from the ipsilateral motor, visual, and auditory cortices to the contralateral pontocerebellum (some to the ipsilateral pontocerebellum) by way of the middle cerebellar peduncle. It comprises, by far, the most massive afferent system which passes through the anterior and posterior limbs of the internal capsule (a massive bundle of fibers, which consists of afferent and efferent fibers, connecting the cerebral cortex to subcortical centers, as well as the spinal cord). It then runs through the basis pedunculi of the midbrain and enters the basilar pons, establishing synapses with the pontine nuclei.

The arcuatocerebellar fibers (Figure 6.14) are derived from the arcuate nuclei which are located ventral to the pyramids of the medulla. These fibers pursue two distinct courses, a superficial ventral and a deeper dorsal course. The superficial (ventral external arcuate) fibers run along the lateral and ventral surfaces of the medulla, entering the cerebellum via the inferior cerebellar peduncle. Fibers that follow a dorsal course (posterior external arcuate fibers) run near the midline of the floor of the fourth ventricle and extend laterally as the stria medullaris of the fourth ventricle, entering the cerebellum via the inferior cerebellar peduncle. The sites of termination of the arcuatocerebellar fibers include the flocculonodular and the pontocerebellum.

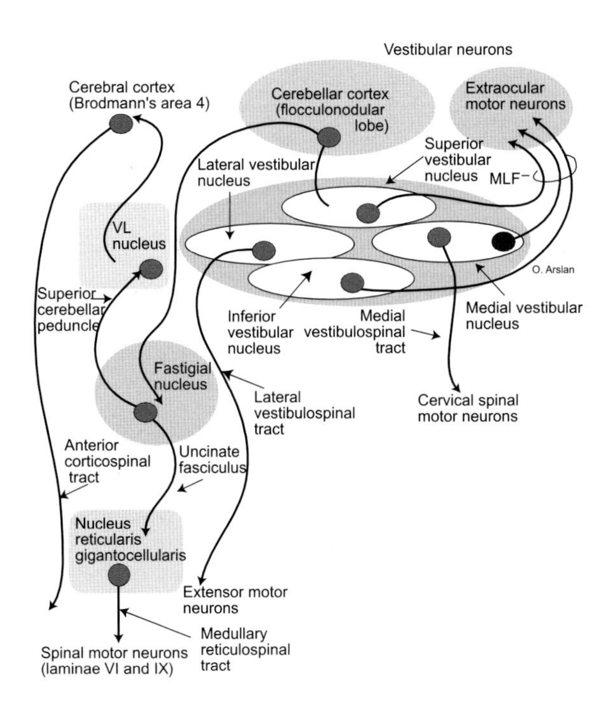

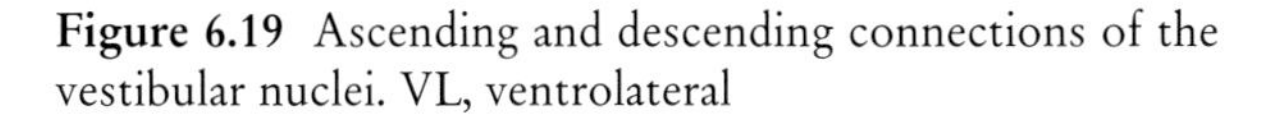

Figure 6.19 Ascending and descending connections of the vestibular nuclei. VL, ventrolateral

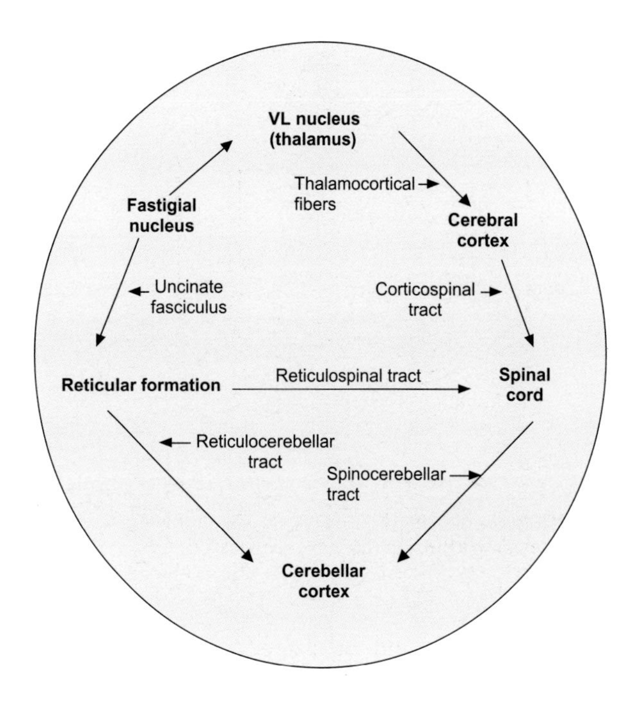

Figure 6.20 Efferent and afferent fibers of the reticular formation within the cerebellar feedback loop. VL, ventrolateral

The tectocerebellar fibers convey auditory and visual information to the pontine nuclei and then to the spinocerebellum, through the superior cerebellar peduncle.

The trigeminocerebellar tract (Figure 6.14) originates from the mesencephalic nucleus of the trigeminal nerve, conveying proprioceptive impulses from the muscles of mastication and muscles of facial expression to the contralateral simple lobule and rostral vermis (spinocerebellum) via the superior cerebellar peduncle. Some trigeminal fibers also originate from the principal sensory and spinal trigeminal nuclei, which receive tactile sensation, and convey this sensation to the anterior lobe via the inferior cerebellar peduncle.

Aminergic cerebellar afferents include fibers from the locus ceruleus and raphe nuclei that project to the cerebellar cortex via the superior and inferior cerebellar peduncles. Axons of the locus ceruleus form a network in the molecular layer where they increase the GABA-mediated inhibition of Purkinje cells which is produced by the cerebellar interneurons (stellate and basket cells). They also enhance the release of glutamate from the parallel fibers onto the Purkinje cells, sharpening the signals and reducing the background activity. Serotoninergic fibers emanate mainly from the medullary reticular formation. Noradrenergic projections to the cerebellum inhibit Purkinje cells by β-adrenergic receptor-mediated inhibition of adenylate cyclase in the Purkinje cells. Dopaminergic neurons that project to the cerebellum gain origin from the ventral tegmentum and may act upon the D_2 and D_3 receptors of the molecular layer. Some cholinergic fibers may also be found in the Purkinje cell layer.

Cerebellar efferents

Some cerebellar efferents arise from the cerebellar cortex and project to the vestibular nuclei and the deep cerebellar nuclei as the corticonucleocerebellar fibers. Others originate from the cerebellar nuclei and project to the thalamus and the brainstem reticular formation.

The corticonucleocerebellar (Figure 6.18) fibers represent the axons of the Purkinje neurons of the cerebellar cortex that project to the vestibular and cerebellar nuclei. Neuronal axons of the vermal cortex extend to the vestibular and fastigial nuclei with a certain degree of specificity. In general, Purkinje neurons of the nodule and uvula project to the cerebellar nuclei and to all of the vestibular nuclei with the exception of the lateral vestibular nucleus, while that of the paleocerebellum projects to the lateral and inferior vestibular nuclei. There are A and B parallel zones of Purkinje cells in the vermis of the anterior lobe and simple lobule that project to the rostral part of the

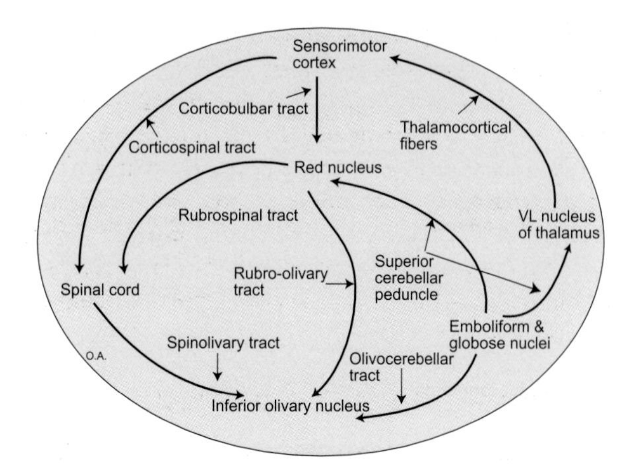

Figure 6.21 Schematic drawing of the connections of the red nucleus within a cerebellar feedback circuit

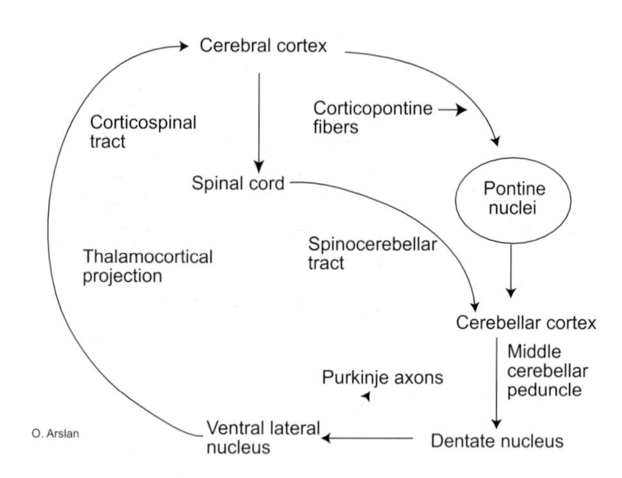

Figure 6.22 The corticocerebrocerebellar circuit and role of the pontine nuclei in mediating coordinated motor activity initiated by the cerebral cortex

fastigial nucleus and the lateral vestibular nucleus, respectively. The caudal part of the fastigial nucleus receives input from the folium and tuber vermis that in itself receive a visual input. This visual information is used to calibrate saccadic eye movements. Pyramis also projects to the same areas that receive input from the anterior lobe. Uvular projections are more far reaching to the interposed and dentate nuclei. Intermediate (paravermal) zones which include C1, C2, and C3 project to the interposed nucleus. C1 and C3 zones send fibers to the emboliform nucleus, whereas C2 projects primarily to the globose nucleus. Purkinje neurons of the lateral cerebellar cortex form D1 and D2 parallel zones that send axons to the caudolateral and rostromedial parts of the dentate nucleus, respectively.

The dentatorubrothalamic tract is formed by axons of the dentate nucleus, which runs within the superior cerebellar peduncle, and terminates in the ventral lateral and ventral posterolateral thalamic nuclei. These terminations are specific and do not show overlap with the pallidal termination. The thalamic nuclei that receive cerebellar output project to the motor cortex via the thalamocortical radiation. This connection enables the cerebellum to exert influence over motor activity. Collaterals of this projection are also given to the red nucleus, oculomotor nucleus, and associated accessory oculomotor nuclei (interstitial nucleus of Cajal and nucleus of Darkschewitch). Efferents from the globose and emboliform nuclei project to the contralateral ventral posterolateral, intralaminar, and ventral lateral thalamic nuclei via the superior cerebellar peduncle. Collaterals of these fibers terminate in the red nucleus.

The fastigiovestibular pathway (Figure 6.19) is an excitatory pathway derived from the fastigial nucleus that projects to the ipsilateral vestibular nuclei via the juxta-restiform body. It also projects to the lateral and inferior vestibular nuclei, and to the nucleus reticularis gigantocellularis of the contralateral medulla via the uncinate fasciculus (hook bundle of Russel). There are fastigial nuclear projections to the ventral lateral and ventral posterolateral thalamic nuclei. Cerebellar influences upon the motor activity may be mediated via the fastigial projection to the ventral lateral nucleus of thalamus, the primary motor cortex, and the corticospinal tract. The lateral vestibular nucleus conveys this information to the spinal cord, regulating the motor activity of the antigravity muscles. Fastigial projection to the nucleus reticularis gigantocellularis allows the fastigial nucleus to influence motor activity via the medullary reticulospinal tract. This connection serves as an additional route by which the fastigial nucleus regulates motor activity.

Cerebellar circuits

The cerebellar functions are performed by series of closed-circuit pathways, which indicate the complexities of cerebellar connections and the role each circuit and associated centers play in the programming, sequencing, grading, and ultimately coordinating motor activities.

Cerebellovestibular circuit

The cerebellovestibular circuit (Figure 6.19) enables the vestibular nuclei to project to the spinal cord via the excitatory lateral vestibulospinal tract and the inhibitory medial vestibulospinal tract. Through this feedback loop information received from the vestibular receptors is also conveyed to the cerebellum via the primary and secondary vestibulocerebellar tracts.

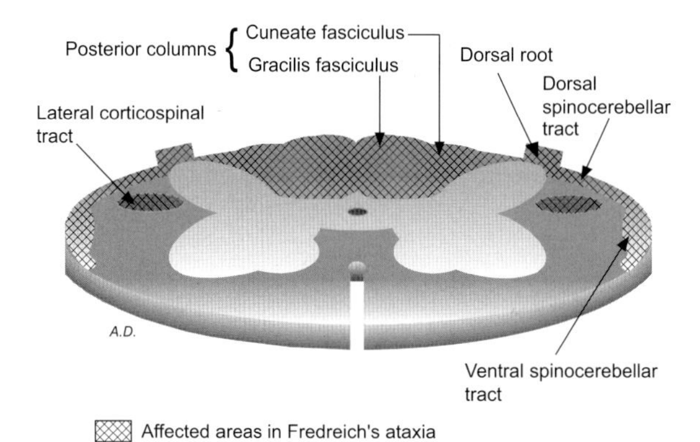

Figure 6.23 Section of the spinal cord showing the lesions associated with Fredreich's ataxia. The dorsal columns, lateral corticospinal and spinocerebellar tracts are affected

Vestibular projections to the flocculonodular lobe eventually influence the activities of the ventral lateral nucleus of thalamus and the motor cerebral cortex via their connections to the fastigial nucleus. Purkinje neurons of the nodule and flocculus project to the medial and inferior vestibular nuclei that receive input from the vestibular nerve. Since these vestibular nuclei project to the motor nuclei that govern eye muscles, the flocculus may play an important role in smooth pursuit movement of the eye by suppressing the vestibulo-ocular reflex.

Part of this circuit is represented by the ascending vestibulocular fibers to extraocular motor nuclei, which run in the medial longitudinal fasciculus, enabling vestibular impulses to coordinate eye movements and fixate gaze.

Reticulocerebellar circuit

Information received by the reticular formation, which originate from diverse sources such as the fastigial nucleus, cerebellar cortex, and spinal cord, is processed and send back to the cerebellum directly via the inferior cerebellar peduncle (as a component of the reticulo-cerebellar tract; Figure 6.20)). It is also sent indirectly via the reticulospinal tract to the spinal cord that projects back to the cerebellum by the spinocerebellar tracts. The spinal cord also conveys impulses to the cerebellum, which are generated in the cerebral cortex and delivered via the corticospinal tract to spinal cord segments.

Rubrocerebellar circuit

The red nucleus (Figure 6.21) receives input from the ipsilateral cerebral cortex and the contralateral cerebellar nuclei. Through its connection to the inferior olivary nucleus (via the rubro-olivary tract), the red nucleus influences activities of the cerebellar cortex through the massive olivocerebellar fibers. Thus, the combination of corticorubral fibers, spinal projections of the red nucleus (rubrospinal tract), spino-olivary fibers (conveying multisensory input to the inferior olivary nucleus from the spinal cord), inferior olivary nucleus itself and the olivocerebellar fibers, cerebellar nuclei, and finally the red nucleus serves to complete this feedback circuit.

Corticocerebrocerebellar circuit

The corticocerebrocerebellar circuit (Figure 6.22) is mediated by the diffuse projections of the cerebral cortex to the pontine nuclei, which sequentially project to the cerebellar cortex via the middle cerebellar peduncle, completing the cerebrocerebellar tract. The cerebellar cortex, via its projection to the dentate nucleus (corticonucleocerebellar tract), acts upon neurons that influence the ventral lateral thalamic nucleus (dentatorubrothalamic pathway), which eventually affect the cerebral motor cortex. Thus, voluntary movements initiated by the motor cerebral cortex are modulated by this feedback loop.

Intracerebellar circuit

The intracerebellar feedback loop (Figure 6.10) utilizes the inhibitory connections of granule cells with the Golgi neurons and the reciprocal inhibition exerted by the Golgi neurons (negative feedback) which is exemplified by the cerebellar glomeruli. Granule cells also exert feed-forward inhibition upon the Purkinje cells through connections to other interneurons such as the basket and stellate cells. Within this feedback circuit collaterals of Purkinje neurons project to the inhibitory cerebellar interneurons which sequentially send fibers back to the same Purkinje neurons or other interneurons.

Functional and clinical considerations

The cerebellum maintains balance and posture by gradual modulation of muscle tension and by maintaining the orderly sequence of muscular contractions. It also plays an important role in the timing of movements. These complex activities are accomplished by integrating information received from certain areas of the cerebral cortex that are involved in the planning and command aspects of movements with the sensory feedback arising in the periphery during the course of movement.

The variations in severity of the signs and symptoms of cerebellar dysfunction depend upon the extent of the lesion and duration of the insult. These manifestations are usually seen ipsilaterally and as a constellation of deficits. They emerge as signs of release from the inhibition exerted on intact structures by the cerebellum. Cerebellar deficits may occur as a result of direct compression or invasion of cerebellar tissue by a developing mass, ischemia, tumors or hemorrhage of the posterior cranial fossa and subsequent obstruction of the cerebrospinal fluid pathway. Developing mass may also produce secondary effects upon other areas of the cerebellum by pressure or compression of the vessels. It should be understood, however, that pure cerebellar deficits produced in experimental animals are rarely encountered in man. Most patients exhibit a combination of gait and postural disturbances (ataxia), asynergy, hypotonia, visual disturbances, vertigo, dementia, headache, nausea and vomiting.

Posture and gait abnormalities (ataxia) resemble drunken gait, which is broad-based, irregular, and staggering. These deficits include tendency to fall (patients become apprehensive and frightened to stand), limb ataxia (past pointing of the extremities), and difficulty in walking in a straight line. Inability to stand may require the patient to seek support and attempt to alleviate the situation by constant adjustment of the extremities and head (titubation). Patients keep the feet too widely apart or too closely together, with the head and body deviated toward the side of the lesion during walking.

Asynergy refers to the lack of coordinated action between muscle groups or movements which normally maintain the proper degree of harmony, smooth and accurate sequencing. Asynergistic muscles lack synchronous activity, skill, and speed. Lack of proper sequence and grouping of muscles which are associated with successive components of a motor activity may produce movement which is decomposed and broken down into puppet-like acts. Asynergy of the muscles of the mouth, pharynx and larynx may lead to disturbance of the mechanism that regulates breathing and phonation (dysarthria). This disturbance produces a peculiar form of speech, scanning (telegraphic) or staccato speech, which is a slow, slurred, explosive, and ataxic speech with prolonged intervals between syllables, and wrong pauses. Asynergy may also be manifested in the form of dysmetria, adiadochokinesis, hyperkinesia, and rebound phenomenon of Holmes.

Dysmetria is the inability to gauge the distance, range, speed, or power of movement. Over-shooting (hypermetria) or under-shooting (hypometria) of the intended target may occur as a result of lack of appreciation of distance or range. Individuals may perform the act slowly or very rapidly with minimal or maximal power.

Adiadochokinesis, the inability to perform alternate successive pronation and supination of the forearm, opening and closing of the fists, tapping the finger or properly execute the finger-to-nose or heel-to-shin tests. In the finger-to-nose test, the examiner asks the patient to put his finger on his nose; the finger begins to oscillate gradually and then violently as it approaches the nose. In the heel-to-shin test the patient, while in supine position, is asked to touch the knee of one leg with heel of the other extremity and then move the heel downward in front of the shin to the ankle joint.

Hyperkinesia is seen in the form of non-rhythmic, jerky, irregular, uncontrollable, coarse, to-and-fro movements of the limbs (kinetic or intentional tremor) during the course of a movement, or upon command as in finger-to-nose and heel-to-shin tests. The amplitude of the tremor increases as the intended target is approached.

Rebound phenomenon of Holmes refers to the lack of normal checks of agonist and antagonist muscles, and tendency to overshoot the target than stopping smoothly. For example, sudden release of the flexed arm against resistance by the examiner may cause the released arm to strike the patient's face due to the delay in contraction of the triceps brachii, a muscle that ordinarily is responsible for arrest of overflexion.

Hypotonia (reduced muscle tone) results in weak, flabby and fatigued muscles (asthenia). The affected muscles may not resist passive movements of the joints into extreme degrees of flexion or extension. The contraction and relaxation phases of movements become slow, delaying the initiation of voluntary movements. When the patient is asked to outstretch his or her forearms, the outstretched forearm on the affected side assumes a pronated position and generally maintains a higher position than the limb on the contralateral side.

Other deficits include vertigo (sense of rotation of self or environment), visual and ocular disorders, hyporeflexia, macrographia, dementia, headache, nausea, and vomiting. Cerebellar disease usually produces anteroposterior or side-to-side movements of the environment, and sense of instability during walking.

It is important to note that occlusion of the subclavian artery, medial to the origin of the vertebral artery, by arteriosclerotic plaques, may also produce vertigo, a weaker pulse and lower pressure in the upper extremity relative to the lower extremity of the affected side. This results from diversion of blood flow from the vertebral artery on the (intact) opposite side, which maintains a higher blood pressure, to the vertebral artery on the occluded side, with a lower blood pressure, and subsequently to the subclavian artery distal to the site of occlusion. Diversion of blood to the subclavian artery (stealing) from the vertebral artery is generally exacerbated during physical effort (increased metabolic demand), leading to manifestations of cerebellar ischemia which include vertigo and dizziness. On this basis, the described condition is known as subclavian steal syndrome.

Occlusion or stenosis of the proximal part of the vertebral artery, however, may produce, although rarely, transient ischemic attack (TIA), and vertigo at rest (to be distinguished from vertigo associated with exertion, which is seen in subclavian steal syndrome). Other deficits also seen such as diplopia (double vision), oscillopsia (sensation of oscillation of the viewed object), numbness and hemiparesis. Occlusion or stenosis of the vertebral artery on one side rarely slows the blood flow to the brainstem or cerebellum if efficient collateral circulation in the neck as well as symmetry of vertebral arteries is maintained. Stenosis of the basilar artery may cause symptoms, which vary from vertigo, nausea, dysarthria (difficulty in speech), hemianesthesia, and paresis of conjugate eye movements, to dysphagia, diplopia, occipital headache, and vertical and horizontal nystagmus.

Ocular and visual disorders are the result of disruption of the cerebellovestibulo-ocular reflexes, including nystagmus, ocular dysmetria, disturbances of conjugate gaze, and diplopia. Nystagmus is a rhythmic oscillation of one or both eyes at rest or with ocular movements. Patients with cerebellar dysfunction cannot maintain gaze away from the midline (rest) position, and attempts to do so may result in slow movements of the eyes towards the center (slow component of nystagmus). The rapid corrective movement in the direction of the gaze is considered the fast component of cerebellar nystagmus. The direction as well as amplitude of its fast phase decreases with sustained deviation of the eyes towards the target. Following return of the eyes toward the midline, the nystagmus resumes with the fast phase away from the midline. This transient oscillation of the eyes upon gazing toward a target, which is associated with blurred vision, may result from ocular dysmetria. Blurred vision that improves by closing one eye is a sign of dysconjugate gaze subsequent to cerebellar dysfunction. Inflammatory and degenerative diseases of the cerebellum may reduce visual acuity or result in transient or permanent blindness. Permanent blindness is seen in certain familial diseases associated with degeneration of the spinocerebellar tracts. Diplopia, the most common ocular manifestation, may be constant or transient deficit. Smooth pursuit movements are also impaired, forcing the patient to track moving objects by compensatory saccades.

Dementia may result from obstructive hydrocephalus, which compresses the enlarged brain against the bony rigid wall of the skull. Obstructive hydrocephalus develops from compression of the fourth ventricle by a cerebellar mass or hemorrhage within the posterior cranial fossa. Impairment of memory and transient confusion are also common features of cerebellar degenerative diseases.

Headache, the most common manifestation of cerebellar dysfunction, is typically severe, persistent, dull pain of the occipital or frontal region which shows no lateralization and remains unresponsive to conventional analgesics. Postural changes may exacerbate headache. Frontal headache is a manifestation of deviation of the tentorium cerebelli, which is innervated by the recurrent meningeal branch of the ophthalmic nerve, a branch that also supplies the skin of the forehead.

Nausea and vomiting are most severe in the morning and may last for months. Postural changes may attenuate or reduce the severity of nausea. Nausea and vomiting may be abrupt dependent upon the extent and degree of progression of cerebellar deficits. With associated weight loss, these deficits may obscure the true etiology and may lead the physician to undertake extensive abdominal evaluation. Nausea and vomiting may result from compression or irritation of the emetic center in the brainstem.

Neocerebellar lesions involve major parts of the cerebellar hemispheres and the corresponding parts of the posterior vermis. Signs associated with the appendicular muscles include spooning of the hand (hyperextension of the fingers) and intention tremor, which may be unilateral or bilateral, and is noted during movement. The patient may exhibit mild ataxic (broad based and unsteady) gait, hypotonia, tendency to fall toward the affected side, and asynergy (which includes dysmetria, adiadochokinesis, and rebound phenomenon). Rebound phenomenon is characterized by uncontrollable oscillation of the outstretched arm up and down upon sudden release of pressure by the examiner. Scanning (telegraphic) speech, a form of dysarthria, which is characterized by slurred, labored, garbled, hesitating and monotonous speech, with inappropriate pauses may also be observed. Handwriting may be affected in the same manner, showing characteristically letters larger than normal (macrographia). Nystagmus, a late common sign, occurs as a result of destruction of the cerebellar connections to the vestibular nuclei. Horizontal nystagmus, which is seen in neocerebellar lesions, is commonly associated with impairment of tracking movements, and becomes markedly visible upon gazing to the side of the lesion.

Paleocerebellar lesions are rare and affect the anterior lobe. However, increased extensor muscle tone and postural reflexes accompanied by truncal ataxia (nodding movements of the head and trunk), and signs of decerebrate rigidity (due to involvement of the brainstem) can also be seen in this type of lesion. Impairment of gait with relative preservation of the upper extremity are some additional signs of this condition.

Archicerebellar lesions target the flocculonodular lobe, producing deficits identical to midline (vermal) lesions. These lesions may occur as a result of medulloblastoma, a childhood malignant tumor arising in the roof of the fourth ventricle which occurs between 5 and 10 years of age, multiple sclerosis, chronic alcoholism, tumor or vascular disease. Deficits usually include truncal ataxia (staggering gait and unsteady posture while standing) and positional nystagmus without appendicular ataxia (ataxia of limb movement). Truncal ataxia necessitates constant support due to the inability of the patient to maintain standing position. Midline lesions of the cerebellum restricted to the lingula, superior medullary velum, and the superior cerebellar peduncle also produce bilateral trochlear nerve palsy, nystagmus, and ipsilateral tremor of the corresponding limb.

Cerebellar lesions may occur in multiple sclerosis, acoustic neuroma, Benedikt's syndrome, cerebellar herniation, Foix's syndrome, Nothnagel's syndrome, cerebellar aplasia, and Friedreich's ataxia.

Multiple sclerosis, as described earlier, is a multifocal demyelinating disease of the CNS. It produces lesions in the brain, brainstem, and cerebellum. Patients exhibit an unsteady gait, truncal ataxia, intention tremor and slurred speech. The most severe form of cerebellar ataxia occurs in MS, where the slightest attempt to move the trunk or limbs results in a violent and uncontrollable ataxic tremor. This is usually due to involvement of the dentatorubrothalamic tract and the adjacent structures in the tegmentum of the midbrain.

Cerebellar hemorrhage produces headache, vertigo and ataxia, and may also lead to ipsilateral conjugate gaze disturbances.

Acoustic neuromas, a cerebellopontine angle tumor, may also produce cerebellar dysfunctions by compressing the flocculus and the middle cerebellar peduncle. This tumor, although commonly benign, is associated with widening of the internal acoustic meatus.

Benedikt's syndrome is a condition that results from destruction of the oculomotor nerve, red nucleus, medial lemniscus, and possibly the spinothalamic tracts. It exhibits signs of ipsilateral oculomotor palsy, contralateral cerebellar dysfunctions, loss of pain, thermal, and position sensations on the opposite side (also described with the motor system and cranial nerves, Chapters 20 and 11). Occlusion of the superior cerebellar artery may produce degeneration of the superior cerebellar peduncle, and the spinal and trigeminal lemnisci, producing some of the signs of Benedikt's syndrome. This vascular occlusion results in ipsilateral atonia, ataxia, asthenia, bradyteleokinesis (hesitancy and slowness in completion of a movement), and loss of pain and temperature sensation on the contralateral face and body.

Herniation of the cerebellum refers to the bilateral or unilateral downward displacement of the cerebellar tonsils that occurs as a result of a tumor of the posterior cranial fossa or frontal lobe, or from the pressure exerted by edematous brain. The paraflocculus and possibly part of the temporal lobe may herniate through the foramen magnum into the cervical part of the vertebral column. Cerebellar herniation is characterized by occipital headache, tonic spasmodic contracture of the neck (nuchal rigidity) and back muscles (cerebellar fits), fixation of the head toward the lesion side, as well as extension and medial rotation of the extremities. These symptoms are followed later by progressive loss of consciousness as a result of destruction of the ascending reticular activating system. Vasomotor changes such as reduced cardiac contraction and feeble pulse, numbness in the upper extremity, and difficulty of swallowing may all be attributed to compression of the structures and nuclei in the brainstem. Mechanical displacement of the fourth ventricle in this herniation may eventually cause compression of the medullary respiratory center, and death ensues following respiratory arrest. A tumor that causes upward deviation of the cerebellum may lead to compression of the fourth ventricle and obstruction of the foramina of Magendie and Luschka, followed by hydrocephalus. These deficits may occur in Dandy–Walker syndrome.

Dandy–Walker syndrome is caused by failure of development of the cerebellar vermis, thinning of the posterior cranial fossa and the formation of a cyst-like midline structure that replaces the fourth ventricle. In patients with this condition, the cervical spinal nerve roots assume an ascending course in order to reach their corresponding foramina. This syndrome may also be associated with agenesis (failure of development) of the corpus callosum.

Foix's syndrome is a manifestation of a lesion involving the red nucleus and the superior cerebellar peduncle, sparing the third cranial nerve. Patients with this condition may exhibit cerebellar ataxia and hemichorea on the contralateral side. However, if the lesion involves the superior cerebellar peduncles at the point of their decussation, cerebellar ataxia will be seen bilaterally.

Nothnagel's syndrome results from an expanding lesion of the tectum that impinges upon and exerts downward pressure upon the superior vermis of the cerebellum. Cerebellar symptoms and paralysis of the extraocular muscles ipsilaterally are the distinguishing features of this condition.

Frontal lobe tumors may produce cerebellar deficits contralateral to the side of the tumor as a result of increased intracranial pressure and possible compression of the corticopontine fibers. Manifestations of mental disorders appear long before any cerebellar signs.

Cerebellar aplasia is a congenital malformation in which part of the cerebellar hemisphere or most of the cerebellum does not develop, and is usually associated with anomalies of the contralateral inferior olive. Neonates with this condition exhibit intention tremor and other motor dysfunctions associated with standing and walking. Cerebellar aplasia is also seen in individuals with encephalocele and in Klippel–Feil syndrome.

Friedreich's ataxia (Figure 6.23) is an autosomal recessive disorder resulting from degeneration of the spinocerebellar tracts, dorsal white columns, Clarke's column, dorsal roots, and dorsal root ganglia. The onset of this disease is usually before the end of puberty, exhibiting gait ataxia of upper and lower extremities, muscular weakness, areflexia, and loss of joint sensation from the lower extremity. Patients may also manifest pes cavus (high arched foot and clawing of the toes), scoliosis (exaggerated lateral curvature of the spine), and cardiomyopathy. Nystagmus, vertigo and hearing loss may also be seen in this disease. Essential tremor, if present, may be a minor manifestation. Refsum's disease and abetalipoproteinemia (Bassen–Kornzweig syndrome) share some features with Friedreich's ataxia.

Suggested reading

1. Amarenco P, Hauw JJ. Cerebellar infarction in the territory of the anterior and inferior cerebellar artery. *Brain* 1990;113:139–55
2. Audinat E, Gähwiler BH, Knopfel T. Excitatory synaptic potentials in neurons of the deep nuclei in olivo-cerebellar slice cultures. *Neuroscience* 1992;49: 903–11
3. Bear MF, Malenka RC. Synaptic plasticity: LTP and LTD. *Curr Opin Neurobiol* 1994;4:389–99
4. Chen S, Hillman DE. Colocalization of neurotransmitters in the deep cerebellar nuclei. *J Neurocytol* 1993;22: 81–91
5. Chumas PD, Armstrong DC, Drake JM, *et al.* Tonsillar herniation: the rule rather than the exception after lumboperitoneal shunting in the pediatric population. *J Neurosurg* 1993;78:568–73
6. Dietrichs E, Walberg F. Cerebellar nuclear afferents – where do they originate? A re-evaluation of the projections from some lower brainstem nuclei. *Anat Embryol* 1987;177:165–72
7. Dow RS. Cerebellar syndromes including vermis and hemispheric syndromes. In Vinken PJ, Bruyn GW, eds. *Handbook of Clinical Neurology*, volume 2. Amsterdam: North Holland Publishing, 1969
8. Fortier PA, Smith AM, Kalaska JF. Comparison of cerebellar and motor cortex activity during reaching: directional tuning and response variability. *J Neurophysiol* 1993;69:1136–49
9. Garthwaite J, Brodbelt AR. Synaptic activation of N-methyl-D-aspartate and non-N-methyl-D-aspartate receptors in the mossy fiber pathway in adult and immature rat cerebellar slices. *Neuroscience* 1989;29:401–12
10. Kawamura K, Hashikawa T. Projections from the pontine nuclei proper and reticular tegmental nucleus on to the cerebellar cortex in the cat: an autoradiographic study. *J Comp Neurol* 1981;201:395–413
11. King JS, Cummings SL, Bishop GA. Peptides in cerebellar circuits. *Prog Neurobiol* 1992;39:423–42
12. Larsell O, Jansen J. *The Comparative Anatomy and Histology of the Cerebellum. The Human Cerebellum,*

Cerebellar Connections, and the Cerebellar Cortex. Minneapolis: University of Minnesota Press, 1972
13. Leiner HC, Leiner AL, Dow RS. The human cerebro-cerebellar system: its computing, cognitive, and language skills. *Behav Brain Res* 1991;44:113–28
14. Llinás R, Welsh JP. On the cerebellum and motor learning. *Curr Opin Neurobiol* 1993;3:958–65
15. Montarolo PG, Palestim M, Strata P. The inhibitory effect of olivocerebellar input on the cerebellar Purkinje cells in the rat. *J Physiol* 1982;332:187–202
16. Morard M, deTribolet N. Traumatic aneurysm of the posterior inferior cerebellar artery: case report. *Neurosurgery* 1991;29:438–41
17. Raymond JL, Lisberger SG, Mauk MD. The cerebellum: a neuronal learning machine? *Science* 1996;272:1126–31
18. Reisser C, Schukriecht HF. The anterior inferior cerebellar artery in the internal auditory canal. *Laryngoscope* 1991;101:761–6
19. Welsh JP, Harvey JA. The role of the cerebellum in voluntary and reflexive movements: history and current status. In Llinás R, Sotelo C, eds. *The Cerebellum Revisited*. London: Springer, 1992:301–34
20. Zhang N, Ottersen OP. In search of the identity of the cerebellar climbing fiber transmitter: immunocytochemical studies in rats. *Can J Neurol Sci* 1993;20(Suppl 3):S36–S42

Diencephalon 7

The diencephalon is bounded by the lamina terminalis rostrally and the posterior border of the mamillary bodies caudally. It is divided by the hypothalamic sulcus, which extends between the interventricular foramen of Monro and the cerebral aqueduct, into dorsal and ventral portions. The dorsal portion consists of the thalamus and epithalamus, while the ventral portion is comprised of the hypothalamus and subthalamus. The third ventricle separates the diencephalon from the corresponding part on the opposite side.

Characteristics of the diencephalon

Thalamus
- Ventral nuclear group
- Lateral group of thalamic nuclei
- Medial group
- Intralaminar nuclei
- Midline nuclei
- Ventral thalamus

Hypothalamus
- Hypothalamic afferents
- Hypothalamic efferents

Pituitary gland (hypophysis cerebri)

Epithalamus

Subthalamus

Characteristics of the diencephalon

The diencephalon, an important part of the central nervous system (CNS) rostral to the brainstem, consists of the thalamus, epithalamus, hypothalamus, and subthalamus. The thalamus and epithalamus are separated from the hypothalamus and subthalamus by the hypothalamic sulcus. The latter sulcus extends between the interventricular foramen of Monro and the cerebral aqueduct. The diencephalon lies between the cerebral hemispheres, caudal to the lamina terminalis. It contains the third ventricle (Figure 7.1), which is bounded superiorly by the ependyma and the pia mater that join together to form the tela choroidea. This ventricle stretches from the lamina terminalis rostrally to the cerebral aqueduct caudally. It is connected to the lateral ventricle via the interventricular foramen of Monro and to the fourth ventricle through the cerebral (Sylvian) aqueduct. It extends into the pineal stalk as the pineal recess and to the area superior to the optic chiasma as the optic recess. It is frequently crossed by fibers of the interthalamic adhesion (massa intermedia), and also extends into the infundibulum as a funnel-shaped infundibular recess. The cerebrospinal fluid within this ventricle is secreted by the choroid plexus, which is attached to the tela choroidea and is supplied by the posterior choroidal artery.

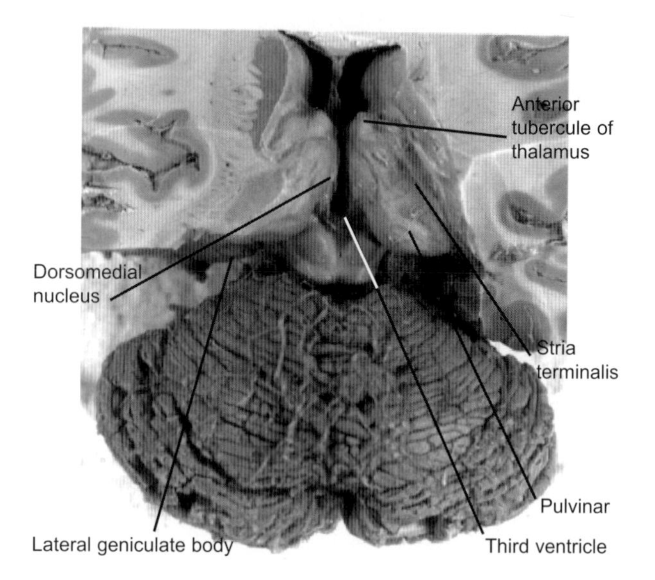

Figure 7.1 Dorsal surface of the diencephalon. Thalami are located on both sides of the third ventricle

Thalamus

The thalamus is the largest component of the diencephalon, which lies superior to the hypothalamic sulcus. It is separated from the caudate nucleus by the genu of the internal capsule. It forms the floor of the central part of the lateral ventricle, the posterior boundary of the interventricular foramen of Monro, and part of the lateral wall of the third ventricle (Figures 7.1 & 7.2 & 7.3). Both thalami may be interconnected in the midline by an intermediate mass (interthalamic adhesion) which extends behind the interventricular foramen of Monro.

It receives direct sensory impulses of all modalities, with the exception of the olfactory sensation. Olfactory information, combined with some other sensations, is conveyed to the olfactory cortex, amygdala, and the mamillary body. Impulses that arise from the cerebellum and the basal nuclei are also integrated in the thalamus. These impulses are then projected to the motor and premotor cortices. To the same extent, visceral activities are also influenced by the thalamic connections to the hypothalamus and cingulate gyrus. In general, the thalamus integrates and modifies the sensory and motor inputs and selectively tunes up the output signals in a manner most efficient for

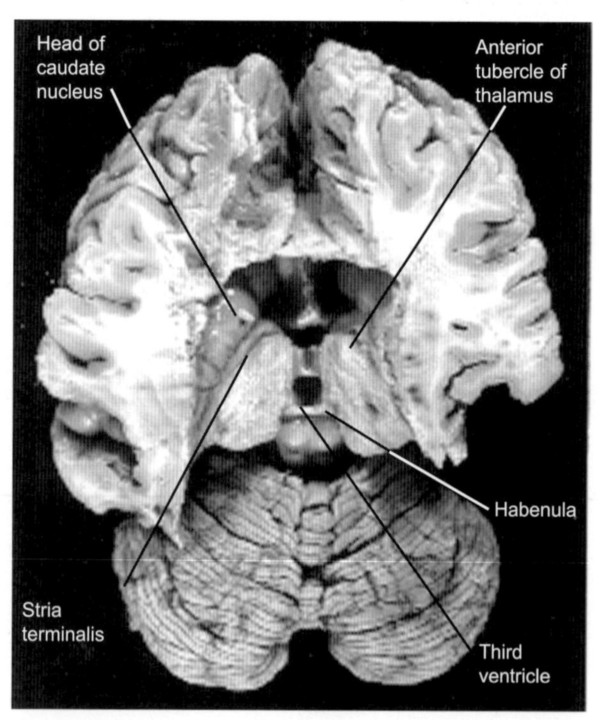

Figure 7.2 Dorsal surface of the diencephalon in relation to the basal nuclei. Note the habenular commissure connecting the habenular nuclei of the epithalamus

stimulation of the cerebral cortex. Conscious awareness of pain, crude touch, temperature, and pressure sensations may occur at thalamic level. The thalamus also processes information that influences the electrocortical activity in the sleep–wake cycle through the ascending reticular activating system. Additionally, the thalamus is also involved in the modification of

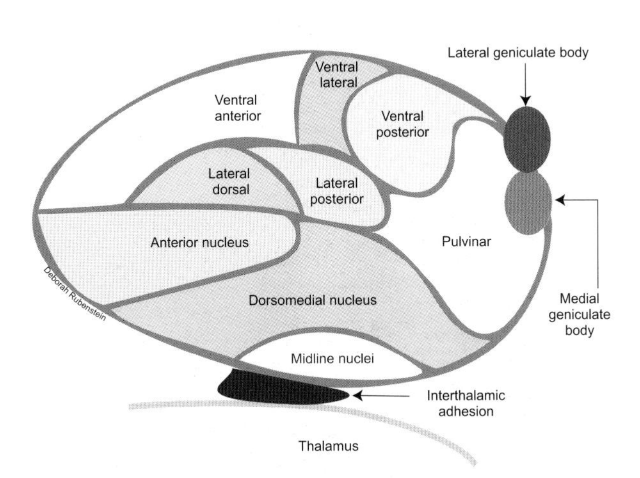

Figure 7.3 Three-dimensional diagram of the thalamus. Major nuclear groups are demarcated by the internal medullary lamina. Midline connection of thalami (interthalamic adhesion) is also illustrated

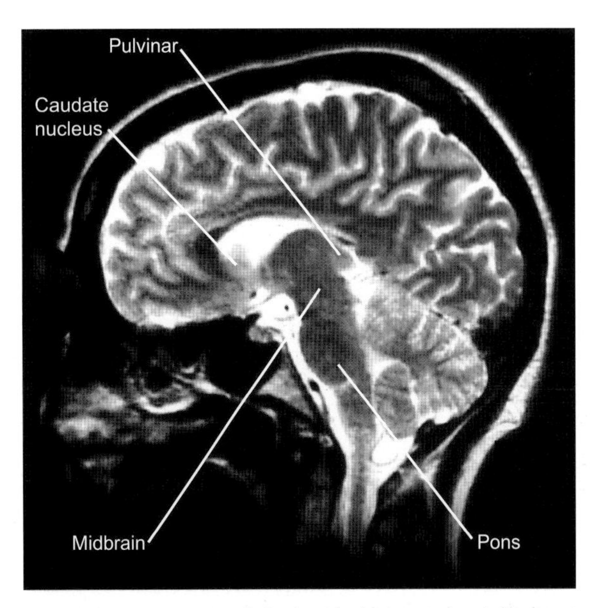

Figure 7.4 MRI scan of the brain showing the thalamus in relation to the caudate nucleus, midbrain, and corpus callosum. The posterior cerebral artery (main source of blood supply to the thalamus) is indicated

the affective component of behavior via its connections to the limbic system.

With respect to the blood supply of the thalamus, the posterior cerebral artery plays an important role (Figures 7.4 & 7.28). As the terminal branch of the basilar artery, it supplies the midbrain and portions of the occipital and temporal lobes, giving rise to the posterior choroidal artery and central branches. These central branches include the posteromedial (thalamoperforating) and the posterolateral (thalamogeniculate) arteries. The latter branch may also arise from the posterior communicating artery, which emanates from the internal carotid artery. Anteromedially, the thalamus is supplied by the thalamoperforating arteries, while the caudal parts of the thalamus, including the pulvinar and geniculate bodies, are supplied by the thalamogeniculate arteries. Numerous branches from the posterior choroidal and posterior communicating arteries supply the superior and inferior parts of the thalamus, respectively. Occlusion of the terminal part of the basilar artery, supplying the diencephalon and midbrain, may result in pupillary dilatation or constriction, loss of light reflex, vertical gaze, and short-term memory, hallucination, agitation and coma or hypersomnolence.

The thalamus has dorsal and medial surfaces that are separated by the stria medullaris thalami. The dorsal thalamus is divided into medial and lateral nuclear groups by the internal medullary lamina (Figure 7.3). The diverging limbs of the internal medullary lamina surround the anterior nucleus of thalamus, forming

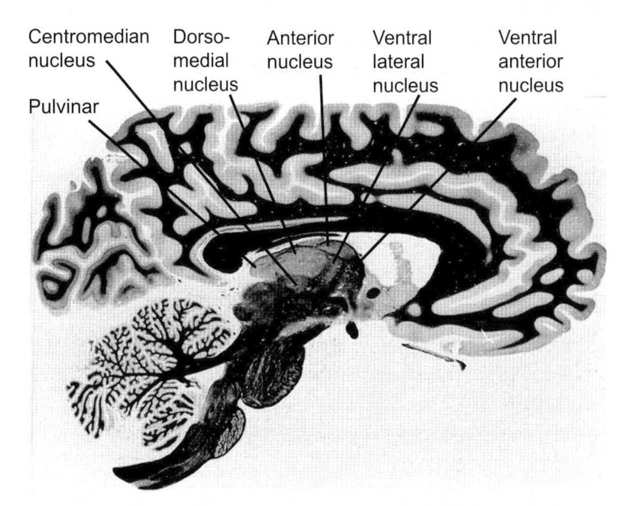

Figure 7.5 Midsagittal photograph of the brain. Anterior, ventral anterior, and dorsomedial nuclei, as well as the pulvinar are illustrated

the anterior tubercule. The ventral portion consists of the caudally located medial and lateral geniculate bodies (metathalamus), ventral anterior, ventral lateral and the ventral posterior nuclear groups. The ventral posterior nuclear complex comprises the ventral posterolateral and ventral posteromedial (arcuate) nuclei. The dorsal portion incorporates the lateral dosal and the lateral posterior nuclear groups, and the pulvinar. Along the periventricular gray matter of the third ventricle, a group of midline thalamic nuclei exist, which are more prominent in the interthalamic adhesion (massa intermedia). Another group of thalamic nuclei, the intralaminar nuclei, is located in the substance of the internal medullary lamina. The reticular nucleus of the thalamus, which is considered to be a continuation of the zona incerta of the subthalamus, is located in the ventral thalamus, lateral to the external medullary lamina. The latter separates the thalamus from the internal capsule. All thalamic nuclei maintain reciprocal connections with the cerebral cortex. On the basis of their connections to the cerebral cortex and ascending pathways, the thalamic nuclei have customarily been classified as specific and non-specific nuclei. The latter is further subdivided into relay nuclei and association nuclei. However, the assumption upon which this classification rests remains tenable. It is clear that both specific, dense projections exist; all cortical areas receive more than one such type of thalamic input. It is equally clear that diffuse, non-specific projections do not originate from a single, discrete group of thalamic nuclei, and many 'specific' nuclei may also convey 'non-specific' projections to widespread corticla areas.

The anterior nucleus (Figures 7.1, 7.2, 7.3, & 7.5) is bounded by the bifurcating limbs of the internal medullary lamina, forming a prominent swelling at the anterior pole of the thalamus (anterior tubercle). It also constitutes the posterior boundary of the foramen of Monro. Profuse reciprocal connections between this nucleus and the mamillary body (via the mamillothalamic tract), as well as the hippocampal formation (via the fornix) do exist. This nucleus consists of anteroventral, anteromedial, and anterodorsal subnuclei. The anteroventral and the anteromedial subnuclei receive input from the ipsilateral medial mamillary nucleus, whereas the anterodorsal subnucleus receives bilateral afferents from the lateral mamillary nuclei and the midbrain reticular formation. Projections from the anterodorsal and anteromedial subnuclei are destined to the anterior and middle portions of the cingulate gyrus.

Due to its close association with the limbic system circuitry, the anterior nucleus is considered to be an essential element in the short-term memory. The significance of this fact is illustrated in the anterograde amnesia observed in lesions of the mamillothalamic tract, a pathway that terminates in the anterior nucleus. This type of amnesia is seen in Korsakoff's syndrome, which results from a thiamine deficiency (see the limbic system, Chapter 17).

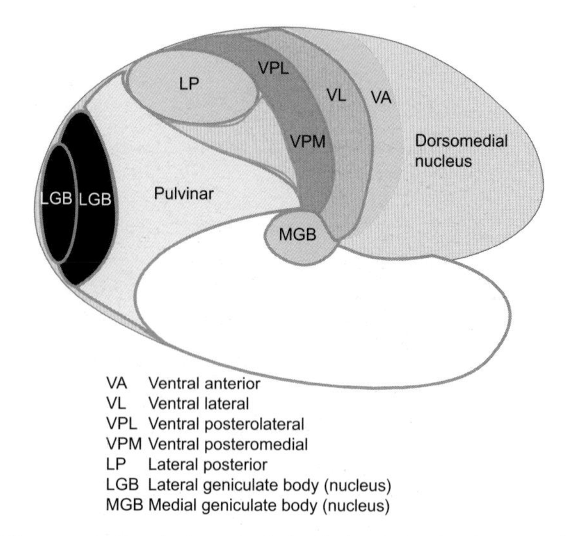

Figure 7.6 Cortical areas of the thalamic projections on the lateral surface of the cerebral hemisphere. These projections are specific sites where certain sensory and/or motor impulses are integrated

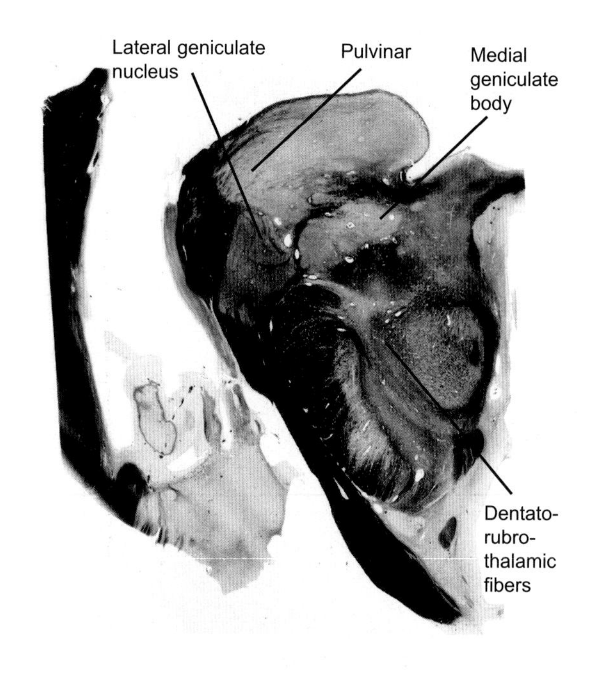

Figure 7.7 Section through the thalamomesencephalic junction. Metathalamus (lateral and medial geniculate bodies) are shown. The pulvinar overlies the geniculate bodies, forming the main component of the posterior thalamus

The cingulate gyrus also maintains reciprocal connections to the various divisions of the anterior thalamic nucleus. Additionally, the anterior nucleus interconnects with other thalamic nuclei of the same and opposite sides. It is an integral component of the limbic system, providing a linkage between the thalamus, hippocampus, hypothalamus, and the

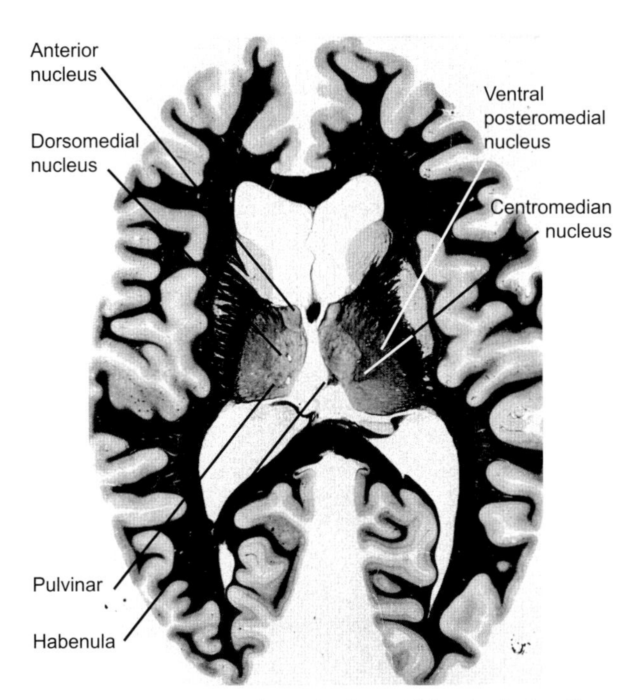

Figure 7.8 Section through the mid-level of the diencephalon. The centromedian (intralaminar) and ventral posteromedial nuclei as well as the pulvinar are clearly shown

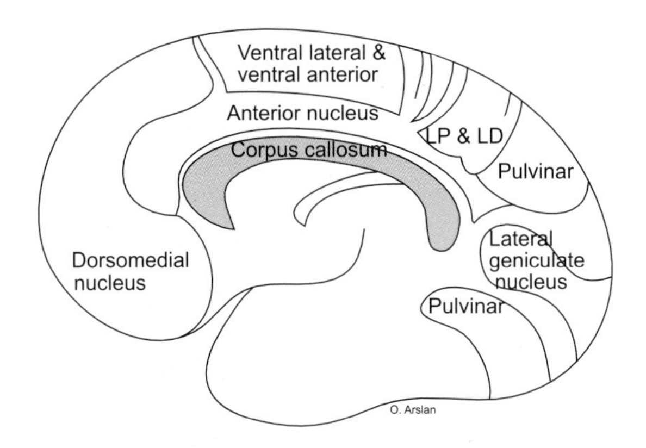

Figure 7.9 Schematic diagram of the medial surface of the brain illustrating cortical projections of thalamic nuclei. LD, lateral dorsal; LP, lateral posterior

cingulate gyrus. Constructing a balance between instinctive and volitional behavior is thought to be an important function of this nucleus.

Ventral nuclear group

The ventral nuclear group comprises the metathalamus, ventral posterior nuclear complex, the ventral anterior nucleus, and the ventral lateral nucleus. The metathalamus consists of the lateral and medial geniculate bodies.

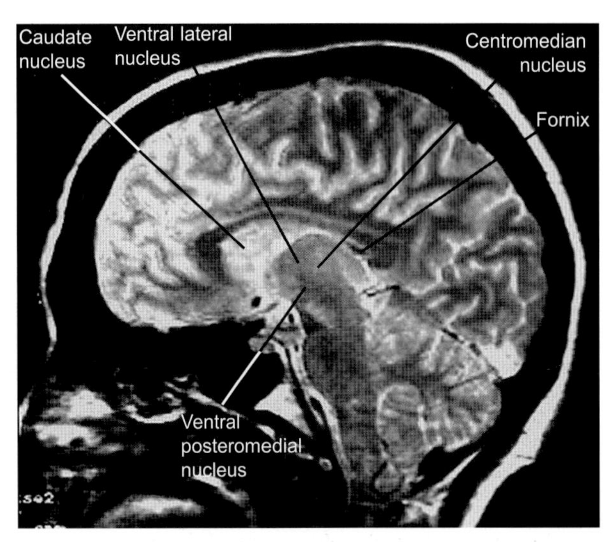

Figure 7.10 MRI scan of the brain showing the ventral lateral, centromedian, ventral posteromedial nuclei

The lateral geniculate nucleus (body) is a visual relay nucleus which grossly resembles Napoleon's hat and lies lateral and rostral to the medial geniculate (Figures 7.1, 7.3, 7.7, 7.9 & 7.21). It consists of six laminae. Laminae 1 & 2 form the magnocellular part, while laminae 3, 4, 5, and 6 comprise the parvocellular part. The neurons of the magnocellular part are only sensitive to the black and white colors, respond quickly, and possess a high-resolution capacity. The parvocellular neurons project primarily to Brodmann's area 17, are responsive to colors, less sensitive to low-contrast stimuli, and remain slow in reacting to visual stimuli. Laminae 1, 4, and 6 receive visual impulses from the crossed fibers of the optic tract whereas the second, third, and fifth laminae only receive fibers from the ipsilateral optic tract. The inferior visual quadrant is represented in the medial portion, whereas the superior quadrant is represented in the lateral part of the lateral geniculate body. Fibers from the macula project to the caudal part of the lateral geniculate body. The lateral geniculate body has reciprocal connections with the primary visual cortex (Brodmann's area 17).

The efferent fibers from this nucleus run within the retrolenticular part of the internal capsule, forming the geniculocalcarine tract (optic radiation). The lower fibers of the optic radiation course within the temporal lobe and loop backward (Meyer's loop) to join the more dorsal fibers. These fibers terminate primarily within lamina IV of the striate cortex (Brodmann's area 17), parastriate (Brodmann's area 18), and the peristriate (Brodmann's area 19) cortices.

The medial geniculate body (MGB) is an auditory relay nucleus ventrolateral to the thalamus, separated

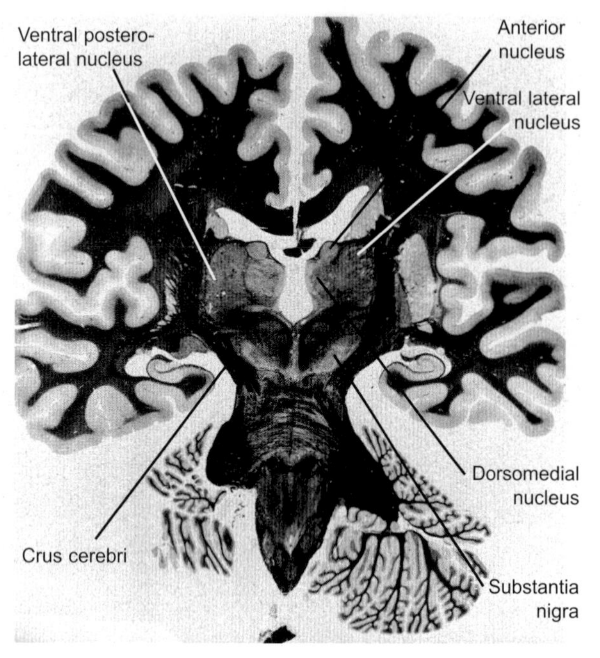

Figure 7.11 Coronal section of thalamus through the ventral posterolateral thalamic nucleus. In this view the anterior, dorsomedial, and ventral lateral nuclei are seen

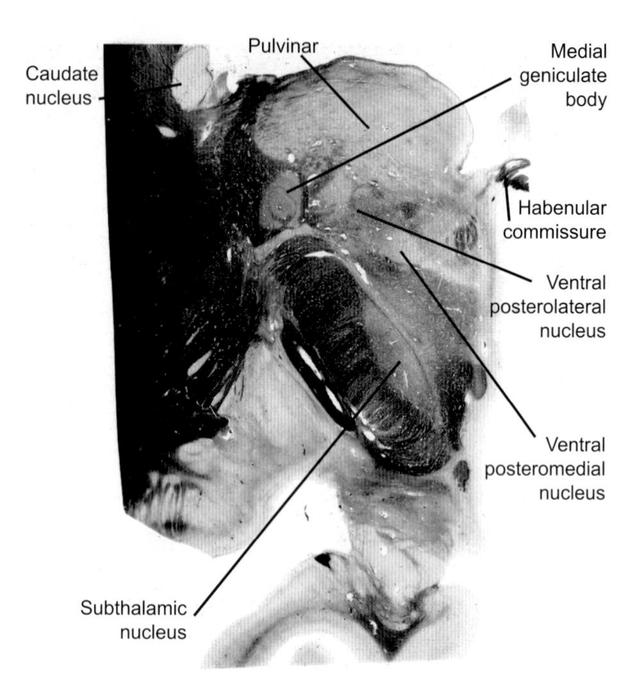

Figure 7.12 Section of the thalamus through the habenular commissure. Note the distinct ventral posterolateral and ventral posteromedial nuclei as well as the pulvinar

from the pulvinar by the brachium of the superior colliculus (Figures 7.3, 7.6, 7.7, 7.18 and 7.23). It consists of medial, ventral and dorsal nuclei. The medial (magnocellular) nucleus receives fibers from the inferior colliculus and the deep part of the superior colliculus, indicating the possible role of this nucleus in the mediation of modalities other than sound. It projects diffusely to lamina VI of the auditory, insular and opercular cortices. The dorsal (posterior) nucleus overlies the ventral nucleus, receives afferents from the pericentral nucleus of the inferior colliculus and from other auditory relay nuclei. A broad range of frequencies is regulated in the dorsal nucleus, accounting for the lack of tonotopic organization. Projection of the dorsal nucleus is limited to the secondary auditory cortex (Brodmann's area 22). Neurons of the ventral nucleus receive afferents from the inferior colliculus of the same side via the brachium of the inferior brachium. It is interesting to note that the ventral nucleus exhibits tonotopic arrangement in which low frequencies project laterally whereas high-pitched sounds are conveyed medially. The ventral nucleus terminates primarily in layer IV of the primary auditory cortex. Commissural neurons do not exist between the medial geniculate bodies.

The ventral posterior nucleus (VPN) (Figures 7.3, 7.6, 7.11 & 7.12) consists of the ventral posterolateral (VPL), the ventral posteromedial (VPM), and the ventral posterior inferior nucleus (VPI).

The ventral posterolateral nucleus (Figures 7.3, 7.6, 7.11 & 7.12) conveys impulses received via the spinothalamic tract, solitariothalamic tract, spinocervicothalamic tract, and the medial lemniscus to the dorsal and the intermediate regions of the postcentral gyrus, and to the secondary somatosensory cortex. Due to the considerable difference in the density of the peripheral innervation of different body regions, many more neurons tend to respond to stimulation of the hand than the trunk. Similarly, the distorted mapping of the body in this nucleus also reflects the difference in innervation density. Within this nucleus, the cervical fibers terminate medially, the thoracic and lumbar fibers end dorsally, and the sacral fibers are positioned laterally. Neurons of the VPL are modality specific and are unaffected by anesthesia.

The ventral posteromedial nucleus (VPM) is also known as the arcuate nucleus, because of its crescent shape (Figures 7.3, 7.6, 7.8, 7.10 & 7.12). It lies lateral to the centromedian nucleus and consists of a medial parvocellular part, which receives gustatory impulses through the ipsilateral solitariothalamic tract, and a lateral principal part, which receives general sensation (tactile, thermal) from the head region. The general sensations from the head region ascend via the crossed ventral trigeminal tract and the uncrossed dorsal trigeminal tract.

The ventral trigeminal tract originates from the spinal trigeminal nucleus and the ventral part of the principal sensory nucleus. The uncrossed dorsal

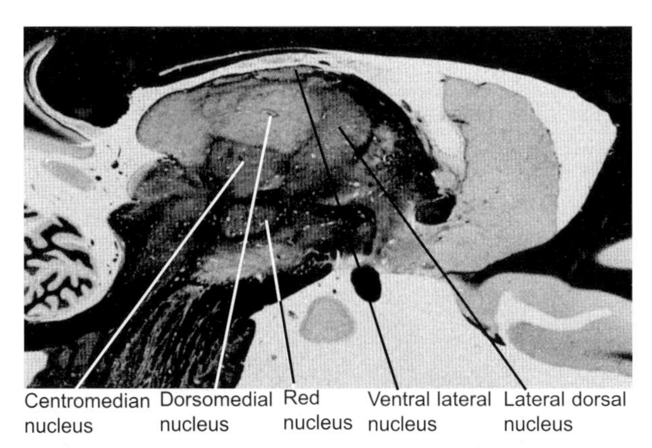

Figure 7.13 Sagittal view of the thalamus showing ventral lateral, dorsomedial, centromedian, and ventral anterior nuclei

trigeminal tract is derived from the dorsal part of the principal sensory nucleus. Information received by the VPM is conveyed to the lower part of the post-central gyrus and to the secondary somatosensory cortex via the thalamocortical fibers. VPL and VPM projections to the sensory cortex are contained in the posterior limb of the internal capsule.

The ventral posterior inferior nucleus (VPI) lies in close proximity to the thalamic fasciculus and the reticular nucleus of the thalamus. It receives terminals of the ascending vestibular fibers, which bypass the medial longitudinal fasciculus, delivering this information bilaterally to the vestibular cortical center, which lies adjacent to the facial region in the primary sensory cortex. Neurons of this nucleus respond to deep stimuli, particularly tapping.

The posterior thalamic zone (PTZ) is a nuclear complex, which is located dorsal to the medial geniculate body and medial to the pulvinar. It is comprised of the posterior nucleus, which is continuous with the ventral posterior inferior nucleus, the suprageniculate nucleus, and the nucleus limitans (which lies between the pulvinar and the pretectum). This nuclear complex has a broad range of connections, and it is neither place nor modality specific. In particular, the posterior nucleus of the PTZ receives pain and nociceptive stimuli via the lateral spinothalamic tract. It also receives tactile, vibratory and auditory impulses. The posterior thalamic nuclei project principally to the secondary somatosensory cortex of both cerebral hemispheres, particularly area IV.

The ventral anterior nucleus (VA) is bounded anteriorly and laterally by the reticular nucleus (Figures 7.3, 7.5, 7.6 & 7.13)). It forms the anterior pole of the ventral nuclear group. It has rich connections with midline and intralaminar nuclei as well as the reticular nucleus. It is interesting to note that no fibers have

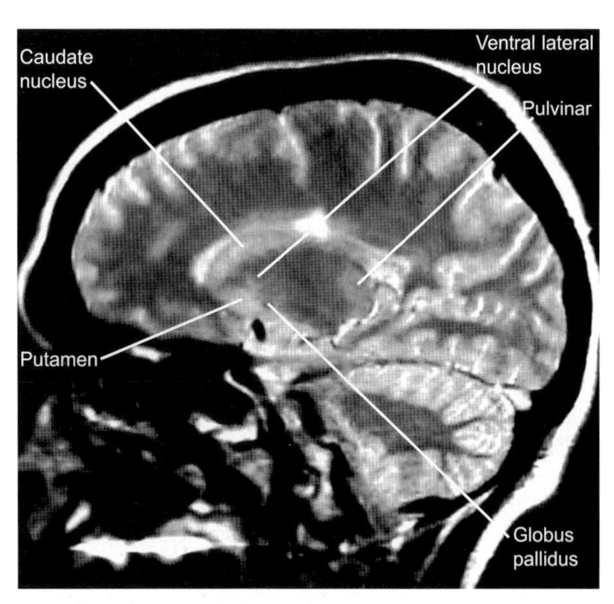

Figure 7.14 MRI scan illustrating the ventral lateral nucleus and pulvinar

been traced from the VA nucleus to the contralateral thalamus or to the striatum. The mamillothalamic tract crosses the VA nucleus. The magnocellular part of this nucleus receives input from the midbrain reticular formation, midline and intralaminar nuclei, and projecting to the orbitofrontal cortex and Brodmann's area 8. Due to its linkage to the intralaminar nuclei, wide areas of the cerebral cortex can be activated and desynchronization of the electrocortical activity can be achieved by stimulation of this nucleus. The ventral anterior nucleus as well as the orbitofrontal cortex is involved in the physiological phenomenon known as the recruiting response. The substantia nigra also projects to the magnocellular part of the nucleus, running parallel to the mamillothalamic tract. On the other hand, the principal part of this nucleus conveys input generated by the globus pallidus (via the thalamic fasciculus) and contralateral cerebellar nuclei (via the superior cerebellar peduncle) to the premotor cortex (Brodmann's area 6). The connections of the VA nucleus to the premotor cortex, globus pallidus, and substantia nigra (areas thought to be involved in dyskinetic movements) may explain the reduction or abolition of tremor seen in Parkinsonism.

The ventral lateral nucleus (VL) (Figures 7.3, 7.6, 7.10, 7.11, 7.13 & 7.14), also known as the ventral intermediate nucleus, occupies the area posterior to the ventral anterior nucleus. It is divided into a rostral (oral part), a posterior (caudal part), and a medial part. The afferents to the ventral lateral nucleus are derived from the contralateral dentate nucleus (via the superior cerebellar peduncle), ipsilateral red nucleus,

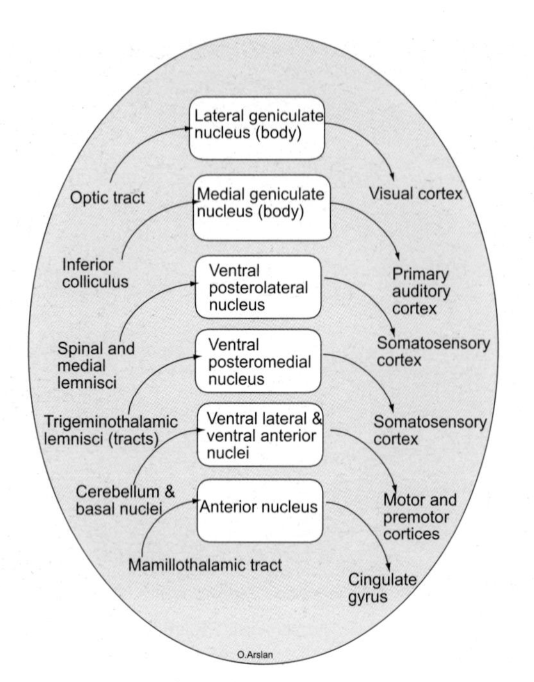

Figure 7.15 Summary of the main afferent and efferent connections of the thalamic relay nuclei

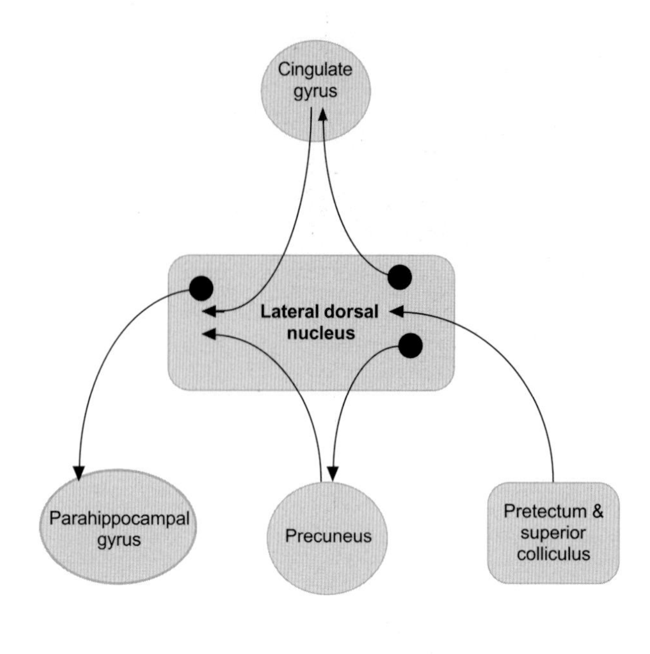

Figure 7.16 Schematic drawing of the afferent and efferent connections of the lateral dorsal nucleus

contralateral globus pallidus (through the thalamic fasciculus), the pars reticulata of the substantia nigra, and also from the precentral gyrus and premotor cortices. The afferents from the substantia nigra terminate in the medial part, while the remaining afferents terminate in the oral and caudal parts of the VL nucleus. Projections of the VL nucleus to the precentral gyrus are somatotopically arranged and establish monosynaptic connections with the neuron of this gyrus. The VL nucleus is part of the link between the cerebellum and the motor cortex, the basal nuclei and the cerebral cortex. Therefore, the cerebellum and the basal nuclei influence motor activity through their projections to the VL nucleus. Surgical ablation of the VL nucleus may be performed in order to relieve the tremor and rigidity associated with Parkinsonism.

Association nuclei, which project to the cortical association areas, but do not receive direct input from the ascending pathways, include the lateral dorsal, lateral posterior and dorsomedial nuclei, and the pulvinar.

Lateral group of thalamic nuclei

The lateral group of thalamic nuclei includes the lateral dorsal nucleus (LD), lateral posterior nucleus, and pulvinar. The lateral dorsal nucleus (Figures 7.3, 7.9, 7.13, 7.16 & 7.18) represents the rostral extension of the dorsal group thalamic nuclei. It lies posterior to the anterior nucleus, and receives input from the superior colliculus and pretectal area. Reciprocal connections exist between this nucleus and the cingulate, and the percuneate gyri. The LD nucleus also projects to the posterior part of the parahippocampal gyrus.

The lateral posterior nucleus (LP) (Figures 7.3, 7.9, 7.17 & 7.18) lies caudal to the lateral dorsal nucleus. The geniculate bodies, as well as neurons of the ventral group thalamic nuclei, convey information to this nucleus. The parietal and occipital lobes (Brodmann's areas 5 & 7) have reciprocal connections with this nucleus. However, no known connections exist between the lateral dorsal nucleus and the primary sensory, visual, or auditory cortices.

The pulvinar (Figures 7.1, 7.2, 7.3. 7.7, 7.8, 7.12, 7.13 & 7.19) is the largest thalamic nucleus and phylogenetically is the most recent. It overlies the geniculate bodies, separated from them by the brachium of the superior colliculus. It has multisensory functions, receiving input from the intralaminar nuclei, geniculate bodies and the superficial layers of the superior colliculus, as well as reciprocal connections with the temporal, parietal, and occipital lobes. Wernicke's sensory speech center (Brodmann's area 22) maintains a rich connection with this nucleus. Visual input from the retina also reaches the pulvinar via the lateral geniculate body and the superior colliculus. Visual impulses from the inferior and lateral parts of the pulvinar, which projects to the supragranular layers of the primary visual cortex, and to layers I, III, and IV of the secondary visual cortex (areas 18 and 19),

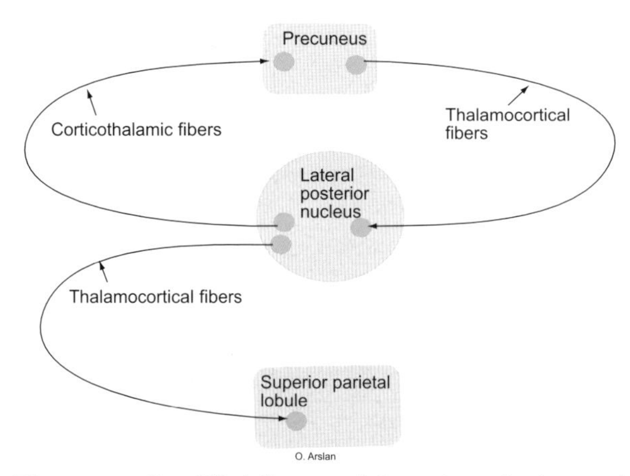

Figure 7.17 Simplified diagram of the main projections and afferent of the lateral posterior thalamic nucleus

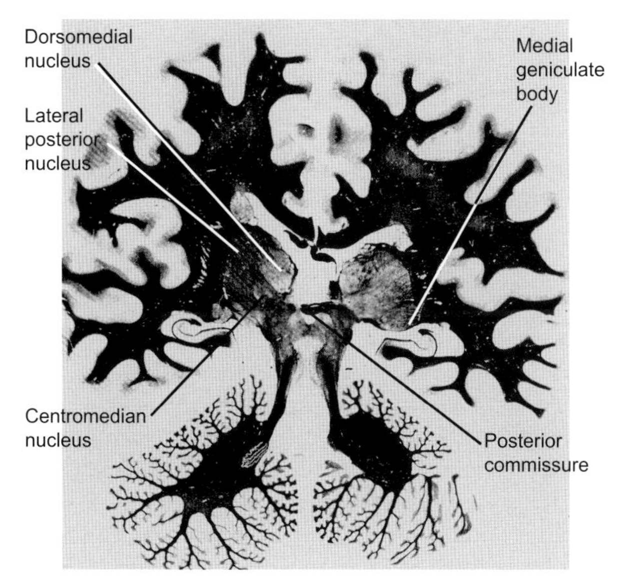

Figure 7.18 This section through the posterior commissure illustrates the prominent dorsal medial, lateral posterior, and centromedian nuclei and the medial geniculate body

constitute the extrageniculate visual pathway, linking the retina, tectum and visual cortex. The medial part has primary reciprocal connections with the posterior parietal cortex. Based on these connections, the pulvinar is implicated in visual and oculomotor control, as well as pain and speech modulation.

Medial group

The medial group of thalamic nuclei consists of a single large dorsomedial nucleus. The dorsomedial nucleus (Figures 7.1, 7.3, 7.5, 7.8, 7.11, 7.13, 7.18, 7.19, 7.23 & 7.24) is located between the internal medullary lamina and the third ventricle. It is considered a relay nucleus for transmission of impulses to

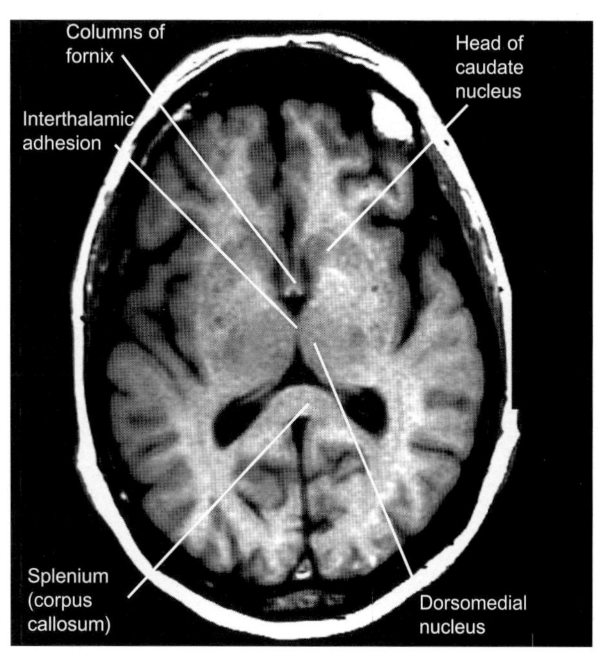

Figure 7.19 An inverted MRI scan of the brain showing the dorsomedial nucleus and the interthalamic adhesion where midline nuclei are located

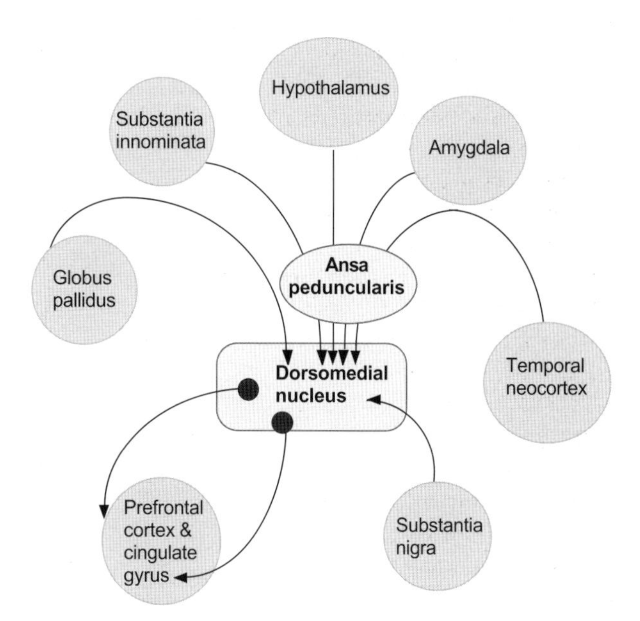

Figure 7.20 Schematic diagram of the afferents of the dorsomedial nucleus of the thalamus which contribute to the formation of the ansa peduncularis

the hypothalamus. It has extensive connections with the intralaminar and midline thalamic nuclei. This nucleus consists of magnocellular, parvocellular, and paralaminar parts. Through its connections, the magnocellular part integrates impulses received through the ansa peduncularis from the amygdala, lateral hypothalamus, diagonal band of Broca,

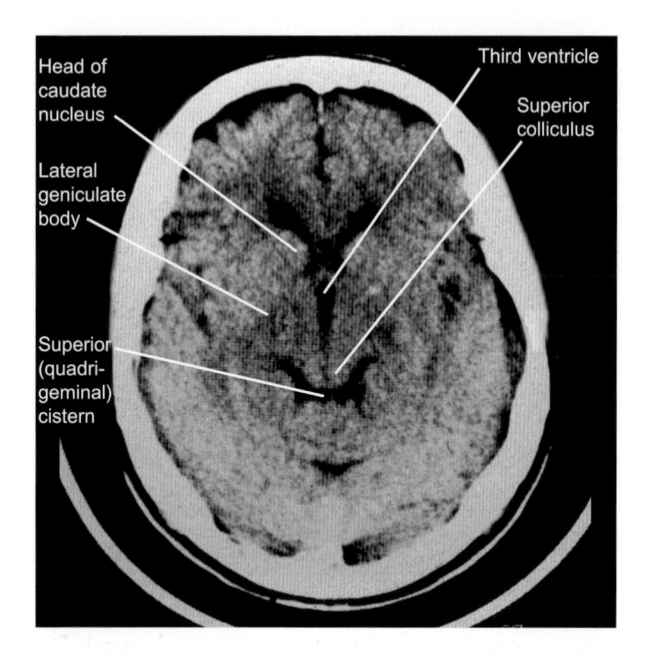

Figure 7.21 This CT scan shows the lateral geniculate body and its location rostral and lateral to the superior colliculus

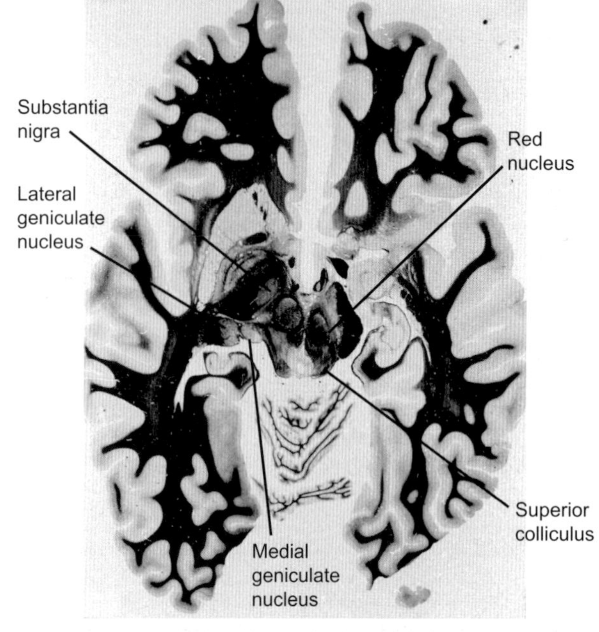

Figure 7.22 Section through diencephalo-mesencephalic junction. The lateral and medial geniculate bodies are prominently illustrated

orbitofrontal cortex, and the substantia innominata. The ansa peduncularis is comprised of the inferior thalamic peduncle and the interconnecting fibers between the amygdala and the preoptic area. The parvocellular part establishes reciprocal connections with the prefrontal cortex (Brodmann's areas 9, 10, 12 & 13), and receives input from the globus pallidus. The paralaminar part maintains reciprocal connections to the premotor cortex (Brodmann's areas 6 and 8) and also receives input from the pars reticulata of the substantia nigra.

Intralaminar nuclei

The intralaminar thalamic nuclei are embedded in the internal medullary lamina. These include the centromedian (Figures 7.3, 7.8, 7.10, 7.13, 7.18 & 7.23), parafascicular, central lateral and central medial nuclei. They receive afferents from the globus pallidus, striatum, and dentate nuclei of the cerebellum, pedunculopontine nucleus and terminals and collaterals from the spinal, medial and trigeminal lemnisci. Ascending reticular fibers, which are derived from the nucleus reticularis gigantocellularis, nucleus reticularis ventralis and dorsalis of the medulla and nucleus reticularis pontis, form the principal afferents to the intralaminar nuclei. They project ipsilaterally to the centromedian-parafascicular nuclear (CM-PF) complex and to the paracentral and central nuclei.

The spinoreticulothalamic (paleospinothalamic) tract projects bilaterally to the intralaminar nuclei. Despite the diffuse cortical projections of the intralaminar nuclei, this connection is not reciprocal. The centromedian is the largest nucleus in this group, which together with the medially located parafascicular nucleus, forms the CM-PF nuclear complex. This nuclear complex sends fibers to the putamen and the substantia nigra. Other smaller intralaminar nuclei project to the caudate nucleus. The projection of the precentral gyrus to the centromedian nucleus is predominantly ipsilateral. Some of these fibers run within the internal capsule and crus cerebri. Due to its connection to the striatum (caudate and putamen), motor cortex and indirectly to the globus pallidus, the centromedian nucleus serves as an important element of a feedback circuit between these areas. The habenulopeduncular tract (fasciculus retroflexus) divides the parafascicular nucleus into medial and lateral divisions. The PF nucleus conveys the input received from the premotor cortex (Brodmann's areas 6, 8 and 9) to the putamen. Desynchronization of electrocortical activity is attributed to the diffuse connections of the intralaminar nuclei to all areas of the cerebral cortex and to their extensive input from the reticular formation. Both the intralaminar thalamic nuclei and the specific thalamic nuclei project to the cerebral cortex, giving off collaterals to the reticular nucleus.

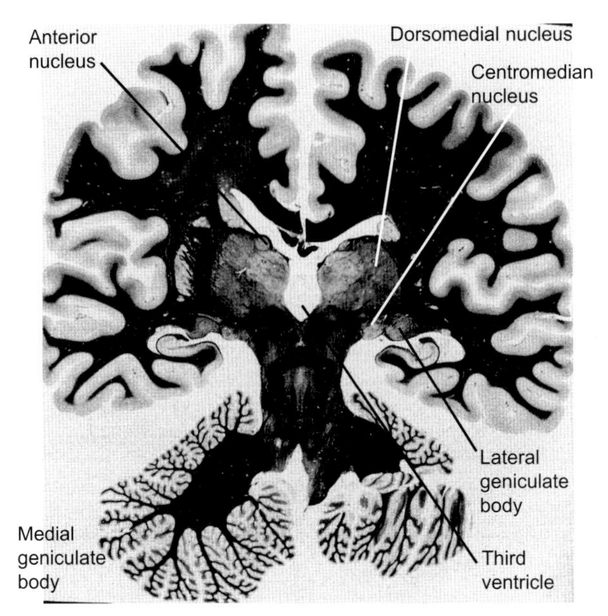

Figure 7.23 Section of the thalamus through the anterior nucleus. Dorsomedial, centromedian, and ventral postero-lateral nuclei are illustrated. The geniculate bodies are also visible

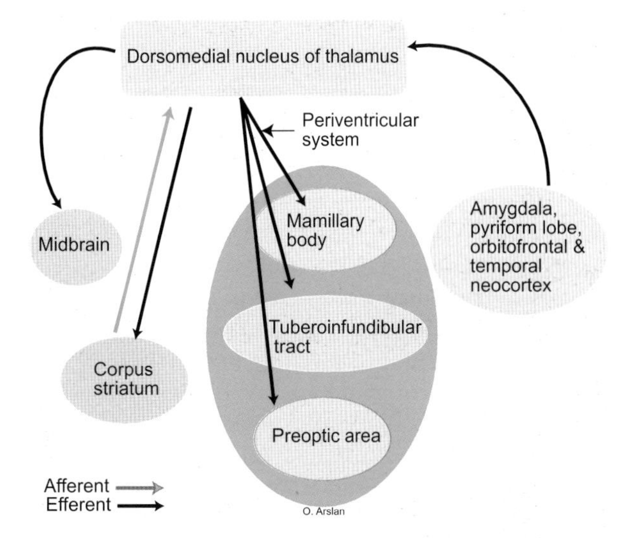

Figure 7.24 Schematic diagram of the afferent and efferent fibers of the dorsomedial nucleus of the thalamus. Note that the connection to the corpus striatum is bilateral. The projections of the dorsomedial nucleus to the mamillary body, tuberoinfundibular region, and preoptic area comprise the periventricular system

The dorsomedial nucleus is associated with the affective qualities of behavior and somatic sensation through its connections with the hypothalamus and the prefrontal cortex. It has no connection with any particular sensory nucleus. Removal of this nucleus and/or a prefrontal lobectomy in an individual with frontal lobe tumor may also modify or reduce the emotional stress associated with chronic pain. Patients who have undergone this surgical procedure report feeling pain without being distressed. This is identical to the perception of pain that patients report when given narcotics. Reduction in anxiety, aggressive behavior or obsessive thinking may result from destruction of this nucleus. Some degree of amnesia and confusion may develop later.

Midline nuclei

The midline nuclei (Figure 7.3) are located, when present, within the interthalamic adhesion (massa intermedia) and in the periventricular area of the third ventricle. These nuclei comprise the paleothalamus (relatively new on phylogenetic basis), consisting of anterior and posterior paraventricular nuclei, as well as the rhomboidal, reuniens, central, and paratenial nuclei. These nuclei receive afferents from the reticular formation, corpus striatum, cerebellum, hypothalamus, and spinothalamic tracts. The main projections of these nuclei reach the cingulate gyrus, amygdala, entorhinal cortex, and the prepyriform cortex.

Ventral thalamus

The principal nuclear group of the ventral thalamus is the reticular nucleus. The reticular nucleus structurally resembles and is continuous with the zona incerta. It lies between the external medullary lamina and the posterior limb of the internal capsule. All areas of the cerebral cortex as well as the midbrain reticular formation, and the globus pallidus convey impulses to the reticular nucleus. This nucleus does not project to the cerebral cortex, but it does influence the activity of other thalamic neurons that project to the cerebral cortex.

Hypothalamus

The hypothalamus (Figure 7.25, 7.26, 7.27 & 7.28) lies posterior to the optic chiasma, ventral to the thalamus and between the third ventricle and the subthalamus. It is connected ventrally to the pituitary gland, and is divided into lateral and medial parts by the fornix. The hypothalamus is comprised of mamillary bodies, tuber cinereum, medial eminence, and infundibulum.

The hypothalamus may also be classified into the preoptic, supraoptic, tuberal and mamillary areas. The

Thalamic syndrome (Dejerine–Roussy syndrome or thalamic apoplexy) may result from occlusion of the posterior choroidal, the posterior cerebral, and the thalamogeniculate arteries. Hemiparesis on the contralateral side may be seen initially as a result of edema that compresses the corticospinal tracts as they descend through the posterior limb of the internal capsule. Ataxia and clumsiness in the affected limb may occur and is often associated with episodes of severe pain following a transient anesthesia (loss of all sensations) contralaterally. In general, it is attributed to a loss of joint sensation and a loss of the ability to appreciate movement. At the onset of the syndrome, complete contralateral hemianesthesia and hemianalgesia occur. Pain, temperature, and crude touch gradually return; however, positional sense and stereognosis are lost permanently. Sensation in the face may or may not be affected, depending upon the involvement of the VPM nucleus of the thalamus. Visual deficits, such as homonymous hemianopsia, may also be detected if the lateral geniculate body is damaged. Hypersensitivity to slight superficial stimuli, which results in severe pain, is common in this condition. Thalamic pain is a diffuse, burning and lingering type of sensation that may be elicited by mild stimuli of touch, pressure, and vibration, and aggravated by emotional conditions. A pinprick may cause agonizing and intolerable pain. Pressure from one's clothing or sound from a musical instrument may produce tremendous discomfort. These sensations occur in response to any stimulus; however, pain threshold is increased and pain sensation is often prolonged and may be elicited by a stronger stimulus.

Occasionally a condition known as thalamic hand may be observed on the contralateral side. It is characterized by flexion of the digits at the metacarpophalangeal joints and extension at the distal phalangeal joints, along with flexion and pronation of the wrist. General hypotonia results in a constant need to re-adjust posture. Thalamic hematomas may cause prominent sensory deficits contralaterally. Due to the caudal location of the corticospinal fibers, relative to the hematoma, muscular weakness is less likely to occur. However, choreiform movements and ataxia may be seen in the contralateral limb. Ocular deviation, in which one eye exhibits a lower position than the opposite eye at rest and maintain an exaggerated convergence (pseudo abducens nerve palsy) and an inability to look upward are also observed. These ocular changes, which result from compression of the adjacent tectal and pretectal areas by a thalamic hematoma, give the appearance of an individual staring at the bridge of his nose. Hematoma of the left thalamus is generally associated with fluent aphasia. A right-sided thalamic hematoma is accompanied by anosognosia and left sided visual neglect. Due to involvement of the intralaminar thalamic nuclei and the central tegmental tract, thalamic hemorrhage may initially be associated with decreased consciousness.

preoptic area is located rostral to the optic chiasma and ventral to the anterior commissure, while the supraoptic area lies above the optic chiasma and includes the supraoptic, paraventricular, and suprachiasmatic nuclei. The preoptic area is the site of termination of many fibers that carry neuromediators such as angiotensin II, δ sleep-inducing peptide, dynorphin, enkephalin, endorphin, galanin, γ-aminobutyric acid (GABA), glycine, substance P, vasoactive intestinal peptide (VIP), serotonin, acetylcholine, adrenaline, dopamine, ACTH, luteinizing hormone-releasing hormone, etc.

The magnocellular secretory neurons of the supraoptic and paraventricular nuclei of the hypothalamus secrete antidiuretic hormone (ADH) and oxytocin. These hormones are synthesized on the ribosomes of the cell body and packaged in Golgi complexes. Both hormones are peptides linked to precursor protein, neurophysin. Exocytosis of the secretory granules into the pericapillary spaces occurs following depolarization of the neuroendocrine cells. ADH changes the permeability of the distal convoluted and collecting tubules, thus increasing the reabsorption of water into the bloodstream and counteracting dehydration. ADH neurons also secrete other peptides such as dynorphin, galanin, and amino acids histidine and isoleucine. In mammals, ADH neurons receive input from the median preoptic area, and from the noradrenergic neurons of the brainstem that carry cardiovascular input. They are inhibited by GABA and excited by angiotensin II, α_2 adrenergic, glutamate, and acetylcholine. Oxytocin is secreted in conjunction with small amounts of enkephalins, galanin, and dynorphin. Oxytocin-secreting neurons receive afferents from the uterine cervix, vagina (Ferguson reflex), and nipple for milk secretion. Thus, oxytocin causes an increase in the contraction of the smooth muscles of the mammary gland and the uterus. Both ADH and oxytocin have overlapping functions and are delivered to the posterior lobe of the pituitary gland through the supraoptico-paraventriculohypophysial tract.

Table 7.1 Summary of connections of thalamic nuclei

Afferents	*Thalamic nuclei*	*Efferents*
Medial lemniscus	Ventral posterolateral (VPL)	Somesthetic cortex (postcentral gyrus)
Spinothalamic tract	Ventral posterolateral (VPL)	Somesthetic cortex (postcentral gyrus)
Trigeminal lemnisci	Ventral posteromedial (VPM)	Somesthetic cortex (postcentral gyrus)
Inferior colliculus	Medial geniculate body	Auditory cortex
Optic tract & visual cortex	Lateral geniculate body	Visual cortex (Brodmann's area 17)
Hypothalamus, prefrontal cortex	Dorsomedial nucleus	Prefrontal cortex
Mammillothalamic tract & cingulate gyrus	Anterior	Cingulate cortex
Globus pallidus, motor cortex & cerebellum	Ventral lateral	Primary motor cortex (precentral gyrus)
Substantia nigra, dentate nucleus & premotor cortex	Ventral anterior	Premotor cortex
Hypothalamus	Lateral dorsal	Cingulate and precuneate gyri
Parietal & occipital cortices	Lateral posterior	Parietal and occipital cortices
Intralaminar nuclei & tectum	Pulvinar	Temporal, parietal, and geniculate bodies
Ascending reticular activating system	Intralaminar	Putamen, substantia nigra, globus pallidus, and lateral spinothalamic tracts
Reticular formation, globus pallidus & spinothalamic tracts	Midline	Cingulate gyrus, amygdala, cerebellum, hypothalamus and cortex
Cerebral cortex, globus pallidus & reticular formation	Reticular	Thalamic nuclei projecting to midbrain

The suprachiasmatic nucleus lies postero-superior to the optic chiasma, in close proximity to the third ventricle. It receives bilateral input from the retina, mediating the biological clock. This nucleus which has ventrolateral and dorsomedial subdivisions, is involved in the regulation of body temperature, hormonal concentrations of plasma, renal secretion, sleep–wake cycle, and day–night cycle in motor activity. The ventrolateral part of the nucleus consists of neurons that remain immunoreactive for VIP, receiving additional afferents from the midbrain raphe and the lateral geniculate body. The glutamatergic retinal afferents, which mediate the rhythm to the light–dark cycle, are received by the neurons of the ventrolateral subdivision of the nucleus. The receiving neurons respond to the onset and offset of light and intensity of light, by changing their firing rate. One of the striking characteristics of the retinally-related neurons of the suprachiasmatic nucleus is their ability to express a specific complement of subunits for the N-methyl-D-aspartate (NMDA) receptors. It has also been suggested that the expression of c-*fos* genes within the suprachiasmatic nucleus may be induced by the application of light during the night that reset the biological clock. On the other hand, the dorsomedial subdivision contains parvocellular neurons that are immunoreactive to arginine vasopressin. Neuronal axons of the suprachiasmatic nucleus project to the paraventricular, tuberal, and ventromedial hypothalamic nuclei. The projections of the suprachiasmatic nucleus to the reticular formation that eventually affect the activities of the sympathetic neurons and secretion of melatonin from the pineal gland may be mediated via the paraventricular nucleus. This nucleus contains vasopressin, VIP, and neurotensin. Vasopressin neurons in the suprachiasmatic nucleus show marked reduction in Alzheimer's disease. One of the most important functions of the suprachiasmatic nucleus is the ability to mediate circadian rhythm, a mechanism that remains active even in the tissue grafts of fetal hypothalamus that contain the suprachiasmatic neurons.

It is thought that deterioration of the suprachiasmatic nucleus leads to the disruption of the circadian rhythms and disturbances in the sleep–wake cycle. Disturbance during sleep may predispose affected

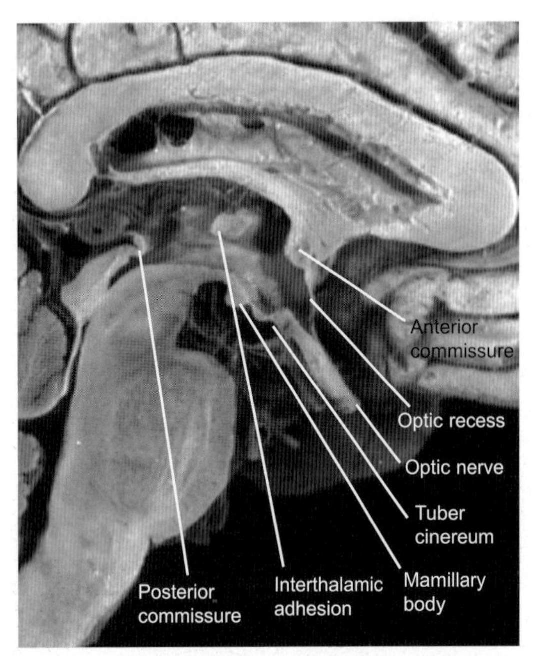

Figure 7.25 Midsagittal section of the brain illustrating the components and boundaries of the hypothalamus

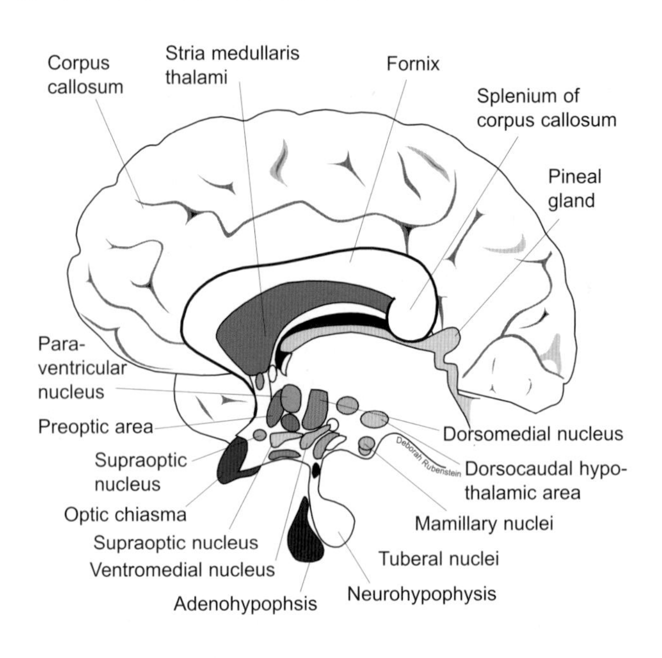

Figure 7.27 Hypothalamic nuclei and associated areas

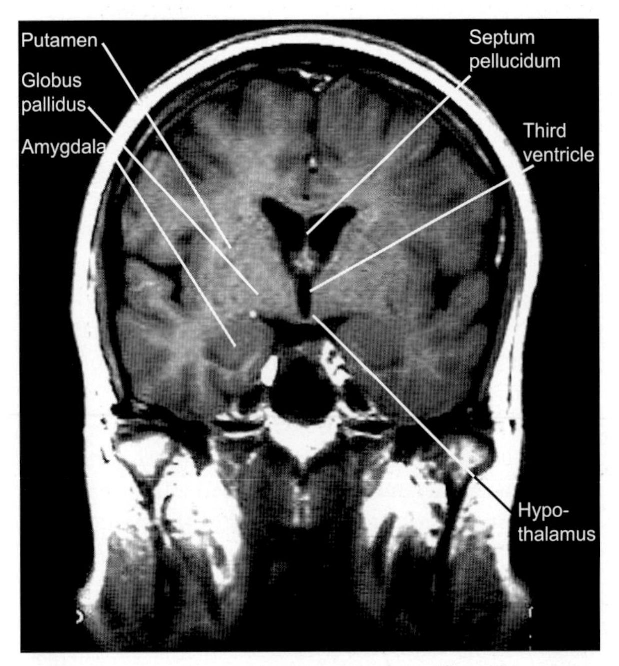

Figure 7.26 An inverted MRI scan of the brain. The hypothalamus, as indicated, forms the floor of the third ventricle

individuals to the development of sundowner syndrome, a condition that is characterized by increased confusion, vocalization, sleep apnea, restlessness, agitation and pacing in the early evening, and dementia.

On each side, the tuberal area extends between the infundibulum and the mamillary bodies, containing the tuberal, dorsomedial, and ventromedial nuclei. The parvocellular neurons of the tuberal nuclei of the hypothalamus secrete the hormone regulating factors (HRF), which are delivered to the median eminence (upper portion of the infundibulum) via the tubero-infundibular tract. The hypophysial portal system then carries these substances from the median eminence to the anterior lobe of the pituitary where they may enhance or inhibit the release of the hormones from the anterior lobe of the pituitary gland. The tuberal nuclei contain histamine, galanin, GABA, cholinesterase, but not choline acetyltransferase.

The lateral hypothalamic region, which lies lateral to the column of the fornix and mammilothalamic tract, contains the lateral hypothalamic nucleus. The mamillary bodies (MBs) are located posterior to the tuber cinereum, anterior to the interpeduncular fossa, and rostral to the anterior perforated substance. They consist of the medial and lateral mamillary nuclei. The mamillary area consists of the medial and lateral mamillary nuclei. Histamine, GABA, and galanin are the main contents of the medial mamillary nucleus.

The hypothalamus integrates visceral and endocrine functions. It is an important component of the limbic system through which emotions gain expression. It contains sympathetic (posterolateral) and parasympathetic (anteromedial) centers, and centers for thermo-regulation (anterior center provides the mechanism for heat dissipation whereas the posterior center mediates activities that produce and conserve heat). Feeding centers, including the hunger center (lateral hypothalamic nucleus) and the satiety center (ventromedial nucleus), as well as centers for sleep–wake cycle (suprachiasmatic nucleus and anterior hypothalamus), memory and behavioral regulation (ventromedial nucleus) are all contained in the hypothalamus. Behavioral regulation may encompass fear, rage, pleasure, sexual attitude, and reproduction. Stimulation of the anteromedial part of the hypothalamus increases gastrointestinal motility and bladder contractions. It also decreases the heart rate, and produces peripheral vasodilatation. Stimulation of the posteromedial part produces mydriasis (dilation of

> Unilateral obstruction of the internal carotid artery is usually asymptomatic, unless it is acute, due to the fact that blood will flow from the opposite side via the arterial circle of Willis. Branches of the arterial circle of Willis are classified into anteromedial, anterolateral, posteromedial, and posterolateral arterial groups.

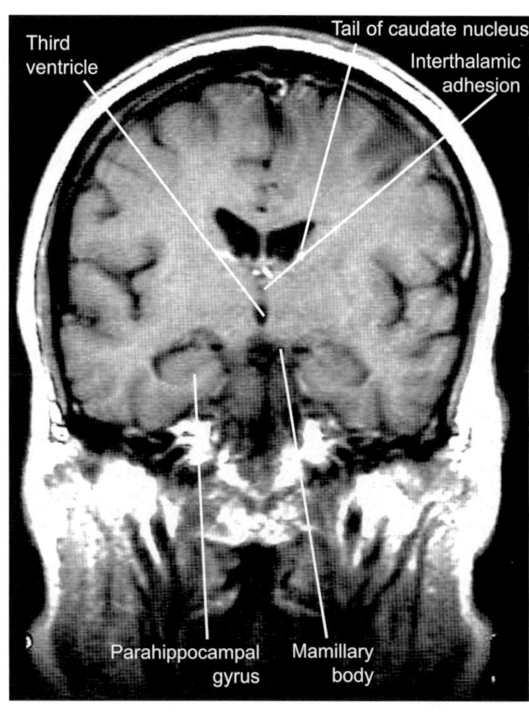

Figure 7.28 In this inverted MRI scan, the prominent mamillary region of the hypothalamus is illustrated

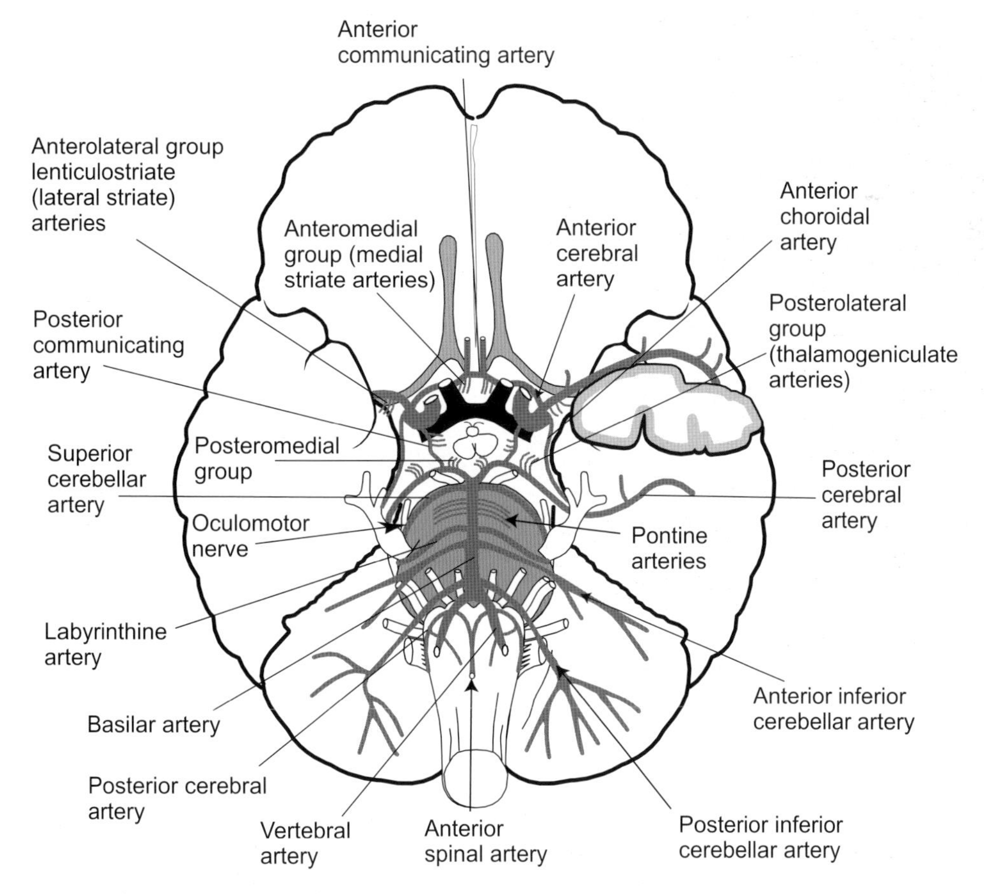

Figure 7.29 Diagram of the arterial circle of Willis, which encircles the mamillary bodies, tuber cinereum, and optic chiasma. Observe the main branches that emanate from the contributing arteries of this circle

the pupil) increased heart rate, and peripheral vasoconstriction.

The posteromedial (thalamo-perforating) branches of the posterior cerebral artery primarily supply the hypothalamus, including the mamillary bodies, infundibulum, tuber cinereum, and pituitary gland. The mamillary bodies, tuber cinereum, and optic chiasma are surrounded by the arterial circle of Willis (Figure 7.29), an arterial chamber formed by branches of the internal carotid and vertebral arteries, providing shunting between the vertebrobasilar system and the carotid systems. It does not always show symmetry as one or more branches may undergo hypoplasia. The specific arterial branches involved in this circle are the anterior cerebral, internal carotid, anterior and posterior communicating, and the posterior cerebral arteries.

Hypothalamic afferents

Hypothalamic afferents emerge from the hippocampal formation, septal area, amygdala, zona incerta, subthalamic nucleus, tegmental nuclei, periaqueductal gray matter, and the retina. The hypothalamus also receives noradrenergic input from the locus ceruleus, serotoninergic input from raphe nuclei, and cholinergic afferents from the ventral tegmental nucleus (Figures 7.30 & 7.31). It also receives fibers from the nucleus ambiguus, solitary nucleus, superior colliculus, Edinger–Westphal and hypoglossal nuclei.

The hippocampal formation projects to the hypothalamus via the fornix, a robust bundle of fibers that emanate from the hippocampal gyrus. As it approaches the anterior commissure, the fornix divides into the pre-commissural column, which terminates in the lateral hypothalamic nucleus and post-commissural column, projecting to the medial mamillary nucleus (Figure 9.1).

The septal area and midbrain reticular formation also project to the hypothalamus via the medial forebrain bundle.

The amygdala sends fibers to the hypothalamus via the stria terminalis and the ventral amygdalofugal fibers. The stria terminalis originates from the corticomedial nucleus and terminates in the medial hypothalamic region. Input to the lateral hypothalamic region emanates from the basolateral nucleus and is carried by the ventral amygdalofugal fibers.

The tegmental nuclei project to the lateral mamillary nucleus via the mamillary peduncle, conveying cholinergic input.

The periaqueductal gray matter of the midbrain projects to the medial hypothalamic region through the dorsal longitudinal fasciculus. The nucleus ambiguus, solitary nucleus, superior colliculus, Edinger–Westphal and hypoglossal nuclei also project to the hypothalamus via the dorsal longitudinal fasciculus.

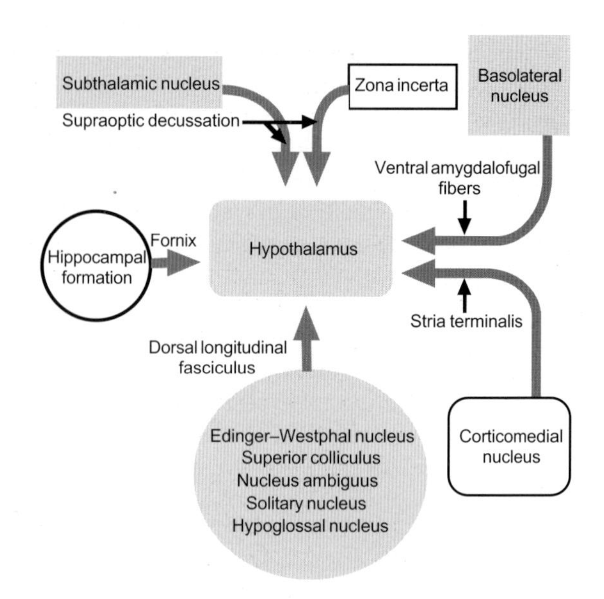

Figure 7.30 Schematic drawing of the afferent and efferent fibers of the hypothalamus. Some of these connections, e.g. from the hippocampal formation, are bilateral

The retina sends projections to the suprachiasmatic nucleus of the hypothalamus via the retinohypothalamic pathway. This connection is utilized in the control of activities of the pineal gland (diurnal regulation).

Locus ceruleus (adrenergic), raphe nuclei (serotoninergic), and the mesolimbic dopaminergic neurons also project to the hypothalamus. Adrenergic group A11 projects to the medial hypothalamic nuclei, whereas groups A13 and A14 convey impulses to the dorsal and rostral hypothalamus.

Hypothalamic efferents

The diverse functions of the hypothalamus are maintained through its multiple projections to the anterior nucleus of thalamus, midbrain reticular formation, pituitary gland, and the spinal cord (Figure 7.31).

The principal mamillary fasciculus emanates from the mamillary body and divides into the mamillothalamic, which projects to the anterior nucleus of the thalamus, and mamillotegmental tract, terminating in the tegmental nuclei of the midbrain. Mammillary projections to the anterior nucleus of the thalamus reach the cingulate gyrus via the cingulum to the hippocampal gyrus and then back to the mamillary body via the fornix. This constitutes the Papez circuit of emotion, an element thought to be essential in regulating emotion.

Lesions of the hypothalamus produce a multitude of symptoms depending upon the location and the structures involved. To produce hypothalamic dysfunction, a lesion must be bilateral. One of the most common sources of hypothalamic dysfunction is craniopharyngioma, a tumor of Rathke's pouch. Posterior hypothalamic lesions are more likely to injure the major pathways associated with the hypothalamus. Bilateral damage to the medial forebrain bundle may result in deep coma, whereas lesions of the dorsal longitudinal fasciculus may disrupt the heat production and heat-dissipating mechanisms. Destruction of the supraoptic nuclei or supraopticohypophysial tract may produce diabetes insipidus, a condition which is characterized by polydipsia (excessive drinking of water) and polyuria (copious urination). Involvement of the tuberal nuclei may lead to cessation of the hormone regulating factors and resultant sexual dystrophy. Destruction of the satiety center (ventromedial nucleus) produces obesity as a result of hyperphagia (excessive eating), whereas damage to the feeding center results in anorexia. Obesity may be observed in cases where the ventromedial nucleus is damaged. Destruction of the satiety center produces hyperphagia (excessive eating), whereas damage to the feeding center results in anorexia.

Hypothermia and hyperthermia may result from destruction of the anterior and posterior nuclei of the hypothalamus, respectively. Selective destruction of the posterior hypothalamic region causes poikilothermia (a condition in which body temperature varies with the environmental temperature). Bilateral destruction of the medial region of the hypothalamus produces violent behavior in previously docile individuals. Destruction of the posterior hypothalamus may result in hypersomnia (excessive sleeping).

Lesions of the intermediate hypothalamic area that destroy the mamillary bodies, fornix, and the stria terminalis may produce signs of Korsakoff's psychosis. Marked anterograde amnesia (short-term memory loss) and preservation of intermediate and long-term memories characterize this syndrome. Consciousness usually is not altered, but affected individuals have a tendency to fabricate when responding to questions (compensatory confabulation). This syndrome is seen in chronic alcohol abuse associated with thiamine deficiency.

Hypothalamic tumors are often slow-growing, achieving a large size prior to the appearance of symptoms. Signs such as hydrocephalus, focal cerebral dysfunction, and hypopituitarism, are often seen. Slow-growing tumors produce dementia, disturbances of food intake, and endocrine dysfunctions. Acute destructive processes of the hypothalamus may lead to coma or to autonomic disturbances. Diseases which affect the hypothalamus and pituitary gland may have both endocrine and non-endocrine manifestations.

The hypothalamohypophyseal tracts comprise the supraoptichypophysial and the tuberinfundibular tracts. The supraoptichypophysial tract delivers ADH and oxytocin from the magnocellular neurons of the supraoptic and paraventricular nuclei to the neurohypophysis, whereas the tuberoinfundibular tract transmits hormone-regulating factors from parvocellular neurons of the tuberal nuclei to the adenohypophysis through the hypophysial portal system. Although the tuberal nuclei are the major source of the dopaminergic hormone regulating factors to the adenohypophysis, the preoptic, periventricular, dorsomedial, and ventromedial hypothalamic nuclei also contribute to this system. Diffuse projections also arise from the tuberal nuclei to areas of cerebral cortex, and to areas that contains cholinesterase that typically undergo neurofibrillary degeneration in Alzheimer's disease.

The dorsal longitudinal fasciculus contains fibers, largely uncrossed, that project from medial and periventricular areas of the hypothalamus to the central gray of the midbrain, accessory oculomotor, salivatory nuclei, solitary, and facial nuclei transmitting impulses between neurons of the reticular formation and the hypothalamus. Some cholinergic fibers are described as descending from the hypothalamus to the cerebellum via the superior cerebellar peduncle. Serotoninergic fibers also ascend from the raphe nuclei of the brainstem within this pathway to the hypothalamus.

Direct hypothalamospinal projections to the intermediolateral columns of the spinal cord emanate from the paraventricular and dorsomedial nuclei. The ventromedial nucleus projects via the medial forebrain bundle to the midbrain reticular formation, central nucleus of amygdala, bed nucleus of stria terminalis, and the basal nucleus of Mynert. The septal area and amygdala maintain reciprocal connections with the hypothalamus.

The hypothalamus shares embryological, morphological, and functional characteristics with the pituitary gland. The hormonal secretions of the pituitary gland are dependent upon the functional

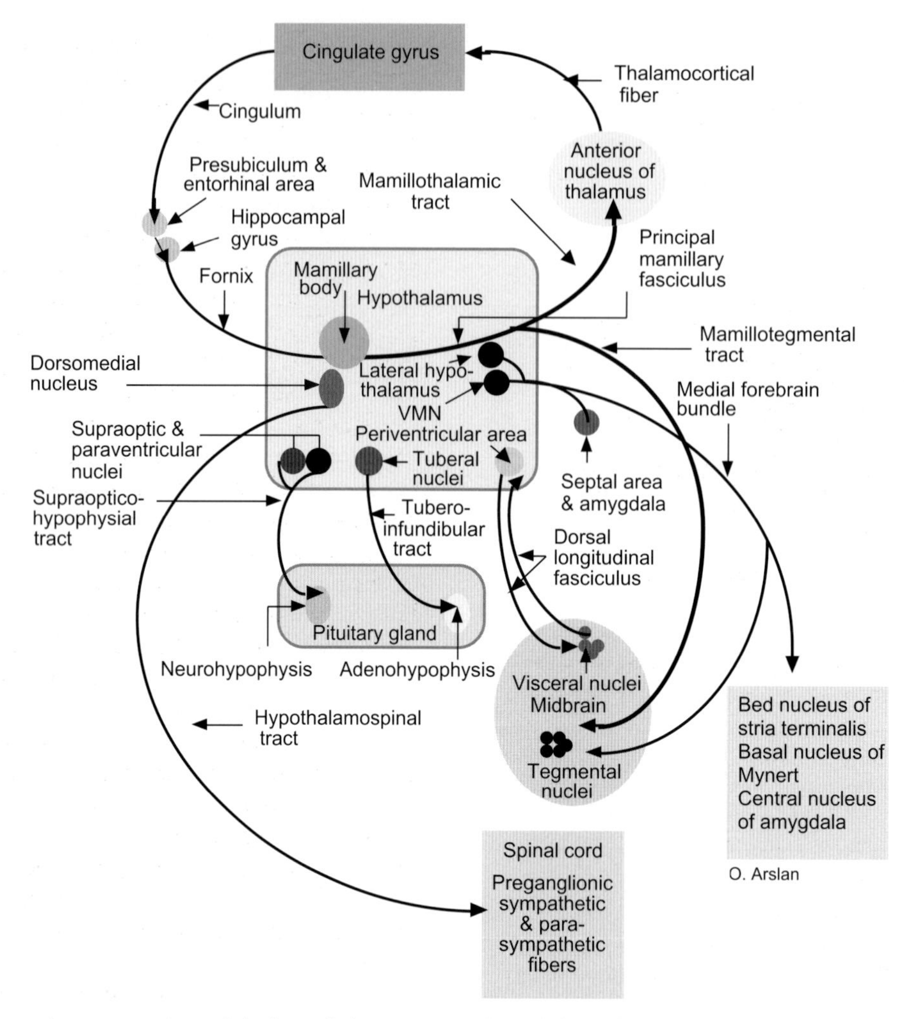

Figure 7.31 The various connections of the hypothalamus, some of which form feedback loops such as Papez circuit of emotion. VMN, ventromedial nucleus

integrity of the hypothalamus. This is accomplished through the hypothalamic input to the medial eminence via the tuberoinfundibular tract which eventually affects the secretion of the adenohypophysis, and by the hypothalamic neurons that secret the ADH and oxytocin and stored in the neurohypophysis.

Pituitary gland (hypophysis cerebri)

The pituitary gland (Figure 7.32) develops from Rathke's pouch (an ectodermal outpocketing of the stomodeum) and the infundibulum (a downward extension of the diencephalon). Rathke's pouch forms the adenohypophysis, the intermediate lobe, and the pars tuberalis.

The diencephalic (neural ectodermal) part of the pituitary gland contains neuroglia, and receives fibers from the hypothalamus. It consists of the adenohypophysis and the neurohypophysis.

Remnants of the Rathke's pouch may give rise to craniopharyngioma, a common tumor in children which extends dorsally, and involves the third ventricle, producing dwarfism, visual disturbances, and erosion of sella turcica.

Enlargement of the sella turcica, subsequent to development of a pituitary tumor, may be detected radiographically. The position of the pituitary gland above the sphenoidal sinus may also be utilized in surgical removal of pituitary tumors via a trans-sphenoidal approach. Additionally, the location of the adenohypophysis posterior and inferior to the optic chiasma may account for disruption of the nasal retinal fibers as a sequel to an adenoma, and the development of bitemporal heteronymous hemianopsia (tunnel vision).

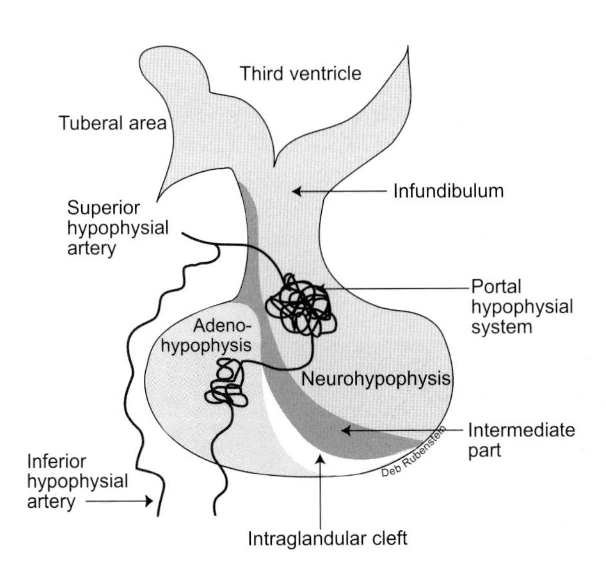

Figure 7.32 The adenohypophysis, neurohypophysis, infundibulum, and the intermediate part of the pituitary gland. The formation of the portal hypophysial system, an important link between the hypothalamus and adenohypophysis is illustrated

The cavernous sinuses flank this gland, which lies in the hypophysial fossa of the sella turcica. It is covered partly by the diaphragma sella, and is connected to the hypothalamus by the infundibulum.

The pituitary gland is supplied by the superior and inferior hypophysial branches of the internal carotid artery. The superior hypophysial artery supplies the tuberal region, the infundibular stalk, optic chiasma, and the medial eminence. The inferior hypophysial artery forms an arterial ring around the infundibular stem and provides blood supply to the lower infundibulum and the posterior lobe. These arteries establish numerous anastomoses and branch repeatedly, terminating into capillaries and capillary sinusoids in the medial eminence and the infundibular stem. These capillary sinusoids are the beginning of the hypophysial portal system, which conveys blood to the epithelial tissue of the anterior lobe via primary and secondary capillary venous plexuses. The portal hypophysial system forms two capillary beds, which drain the venous blood of the pituitary gland, carrying the hormone regulating factors (HRF) from the medial eminence of the hypothalamus to the anterior lobe of the pituitary gland.

The adenohypophysis comprises the anterior lobe, intermediate part, and tuberal region. It secretes somatotropin (growth hormone), thyrotropin, prolactin or luteotropin, adrenocorticotropin,

Pituitary gland dysfunctions are associated with a variety of pathological conditions and syndromes. Some of these conditions primarily affect the adenohypophysis, while others may be confined to the neurohypophysis.

Sheehan's syndrome, is a condition that exhibits persistent amenorrhea, asthenia (muscle weakness), visual disorders, episodes of hypotension, and altered consciousness. It is associated with hypopituitarism of the anterior lobe, and is observed in individuals with postpartum hemorrhage and spasm of the infundibular arteries.

Adenoma of the anterior pituitary produces a variety of symptoms, depending upon the type of tumor, which may include acromegaly and/or gigantism (enlargement of the face, hands, and feet), impotence, amenorrhea (cessation of menses), galactorrhea, or Cushing's disease. The latter is characterized by moon facies, obesity with prominent fat pad in the neck and shoulder, hypertension, renal calculi, and irregular menses.

Pituitary tumors most often arise from the anterior lobe and are classified, according to the degree of their endocrine activity, into endocrine-active and endocrine-inactive tumors.

Endocrine-active tumors are for the most part microadenomas, and may be detected by radiographic imaging of the sella turcica. They may increase secretion of many hormones such as growth hormone, producing acromegaly or gigantism. On the other hand, some tumors may result in under-secretion of hormones, producing dwarfisim (abnormally short body stature) in childhood among other deficits.

Endocrine-inactive tumors, by contrast, become clinically significant only upon enlargement, thus compressing the adjacent nerves or brain tissue. These tumors may produce headache, visual disturbances, and extraocular motor palsies. Surgical removal may be accomplished by a trans-sphenoidal approach; however, extensive subfrontal or parasellar expansion of the tumor may require a transfrontal approach. Tumors that grow laterally or wedge under the optic nerve may necessitate craniotomy.

Lesions of the posterior lobe may occur in skull fractures, suprasellar and intrasellar tumors, tuberculosis, vascular disease or aneurysms. These lesions produce temporary signs of diabetes insipidus (due to lack of vasopressin), a condition, which is characterized by copious and dilute urination and excessive thirst. The onset is generally sudden and nocturia (excessive urination at night) is a presenting symptom in most patients.

luteinizing hormone (interstitial-cell-stimulating hormone), follicle-stimulating hormone, and melanocyte-stimulating hormone. The hormone regulating factors of the hypothalamic tuberal nuclei regulate the adenohypophysis. The tuberal nuclei project to the medial eminence through the tubero-infundibular tract and regulate the glandular function of the adenohypophysis via the portal hypophysial vessels.

The neurohypophysis is comprised of the posterior lobe and the infundibular stem, and acts as a storage depot for vasopressin and oxytocin. Approximately half the volume of the neurohypophysis consists of axonal swellings; the largest are the Herring bodies, which may be as large as erythrocytes. Herring bodies provide a source of secretory granules and have longer life span (more than 2 weeks). Vasopressin, also known as antidiuretic hormone (ADH), enhances water reabsorption by the distal convoluted tubules of the kidneys, whereas oxytocin causes milk ejection from lactating mammary glands and contraction of the uterine muscles. These hormones are secreted by the paraventricular and the supraoptic nuclei of the hypothalamus. During suckling, stimulation of the sensory nerve endings in the female nipple and areola activates the spinoreticular fibers, which subsequently relay the generated impulses to the supraoptic and paraventricular nuclei via the dorsal longitudinal fasciculus. The infundibulum primarily consists of fibers of the hypothalamohypophysial tracts that project to the neurohypophysis.

Epithalamus

The epithalamus occupies the dorsolateral part of the diencephalon and consists of the pineal body (epiphysis cerebri), stria medullaris thalami, habenula and the posterior commissure (Figures 7.33, 7.34 & 7.35).

The pineal gland (Figures 7.33, 7.34 & 7.35) is an endocrine gland, which is located rostral to the superior colliculi and posterior commissure, and inferior to the splenium of the corpus callosum. It is connected to the habenula and the posterior commissure via the laminae of the pineal stalk. The latter contains the pineal recess of the third ventricle. This gland secretes melanocyte-stimulating hormone (lipotropic hormone), which plays an analgesic role due to its endorphin content. It also secretes indolamines such as melatonin and associated enzymes (N-acetyltransferase and hydroxyindole-O-methyltransferase which synthesizes melatonin from serotonin) that show sensitivity to the variations in diurnal light and circadian rhythms. It is also believed to influence the secretion of the gonadotropic hormones, maintaining a regulatory role in reproductive development. Through these secretions the pineal gland may exert a regulatory influence, modifying the activity of the pituitary, adrenal, and parathyroid glands, as well as gonads. Darkness activates the secretion of melatonin, inhibiting sexual development through a series of neuronal chains in the retina, hypothalamus, reticular formation, and the spinal cord. Retinal input to the suprachiasmatic nucleus of the hypothalamus is eventually conveyed to the reticular formation. Activation of the reticular formation produces excitatory effects on the sympathetic neurons of the intermediolateral column of the upper two thoracic spinal segments, via the reticulospinal tracts. Sympathetic neurons convey the secretomotor impulses from the reticular formation to the pineal gland through the nervi Conarii (sympathetic

Bilateral lesions of the suprachiasmatic nuclei may abolish the rhythmic activities of N-acetyltransferase and produce low levels of hydroxyindole-O-methyltransferase activity, resulting in disruption of the circadian rhythms associated with the sleep–wake cycle and spontaneous motor activities that pertain to drinking and feeding.

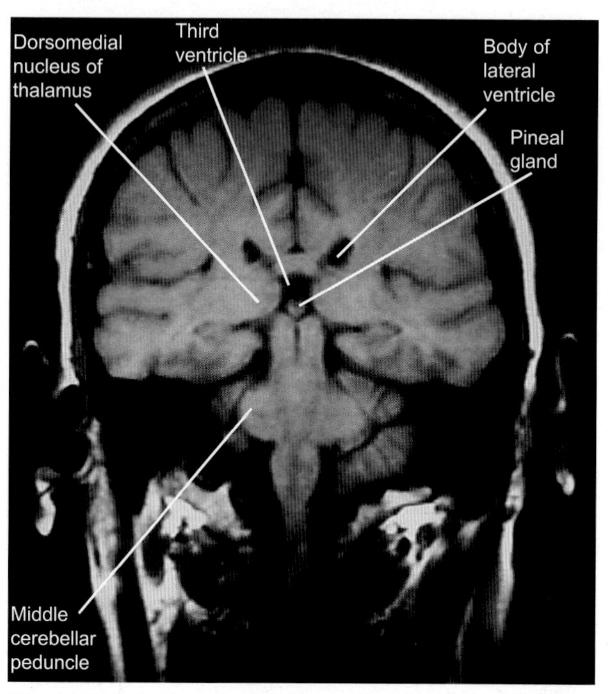

Figure 7.33 An inverted MRI scan showing the epithalamus with its main components the pineal gland and habenula. Observe their relationship to the third ventricle

The most common malignant tumor of the pineal gland is germinoma (atypical teratoma), which is frequently seen in young males. Pinealoma (pineal gland tumor) may compress the tectum and the posterior commissure, producing signs of Parinaud's syndrome, which is characterized by bilateral vertical gaze palsy, hydrocephalus, and loss of pupillary light reflex. Cysts or tumors associated with the pineal gland may compress the hypothalamus, resulting in obesity and hypogonadism.

Pineal (sand) concretions or corpora arenacea in the astrocytes were considered as an important radiographic site for detection of brain shift associated space-occupying intracranial mass on the contralateral side. Deviation of the gland from a midline position may be considered significant. However, it must also be remembered that due to the relative large size of the right cerebral hemisphere, a normal pineal gland may slightly deviate to the left side.

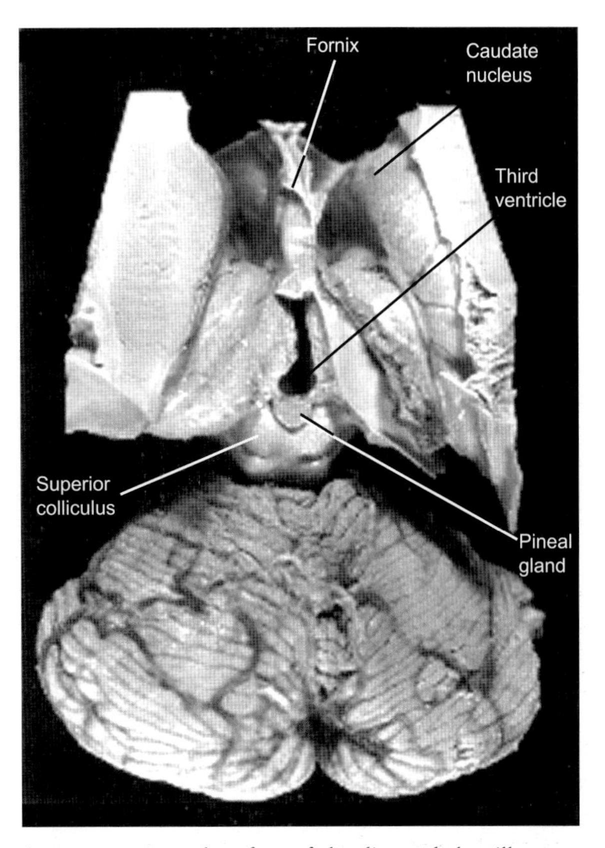

Figure 7.34 Dorsal surface of the diencephalon illustrates the epithalamus and its relationship to the third ventricle and tectum

postganglionic fibers), resulting in release of catecholamines from the pinealocytes and subsequent receptor-mediated (β-adrenergic) increase of cyclic adenosine monophosphate (cAMP).

The pineal gland also secretes norepinephrine and contains significant concentrations of hypothalamic peptides such as luteinizing hormone-releasing hormone (LHRH), thyrotropin-releasing factor and somatostatin. Pinealocytes also contains tryptophan hydroxylase and aromatic amino acid decarboxylase, which are involved in the synthesis of serotonin. Increase of cAMP evokes augmentation in serotonin N-acetyltransferase activity that is followed by increased pineal glandular activity and production of melatonin, and eventual inhibition of reproductive development.

The stria medullaris thalami consists of axons that originate from the septal area, runs dorsomedial to the thalamus, and terminates in the habenular nuclei of both sides. It also contains afferents to the habenular nuclei from the septal area that receives input from the amygdala, hippocampal formation, and the anterior perforated substance via the septal area.

The habenulae (Figures 7.33, 7.34, 7.36 & 7.37) are two triangular eminences which are comprised of the medial and lateral habenular nuclei, representing the sites of convergence of major limbic system pathways. The lateral habenular nucleus receives afferents from the globus pallidus, substantia innominata, tectum, lateral hypothalamus, prepyriform cortex, lateral preoptic area, basal nucleus of Mynert, midbrain raphe nuclei, olfactory tubercle, and the pars compacta substantia nigra. Input from the prepyriform cortex, tectum, nucleus basalis of Mynert, septal and the lateral preoptic area travels within the stria medullaris thalami. The stria medullaris thalami contains also fibers that transport neuromediators such as acetylcholine, norepinephrine, serotonin, GABA, LHRH, somatostatin, vasopressin, and oxytocin. The lateral habenular nucleus projects back to the substantia nigra and also to the midbrain reticular formation, and the hypothalamus. The medial habenular nucleus, the smallest component of the habenular nuclear complex, receives fibers from the serotoninergic neurons of the midbrain reticular formation and from the septofimbrial nucleus. The latter nucleus receives input from the amygdala and the hippocampal formation. Some adrenergic fibers also project to the medial habenular nucleus from the superior cervical ganglion of the sympathetic trunk. The output from the habenular nuclear complex primarily emanates from the medial habenular, a cholinergic nucleus that projects to the interpeduncular nuclei of the midbrain via the habenulopeduncular tract (fasciculus retroflexus). This pathway enables

The posterior commissure may be damaged by pineal tumors. However, no known symptoms are associated with lesions of this commissure, though impairment of visual tracking movement has been reported.

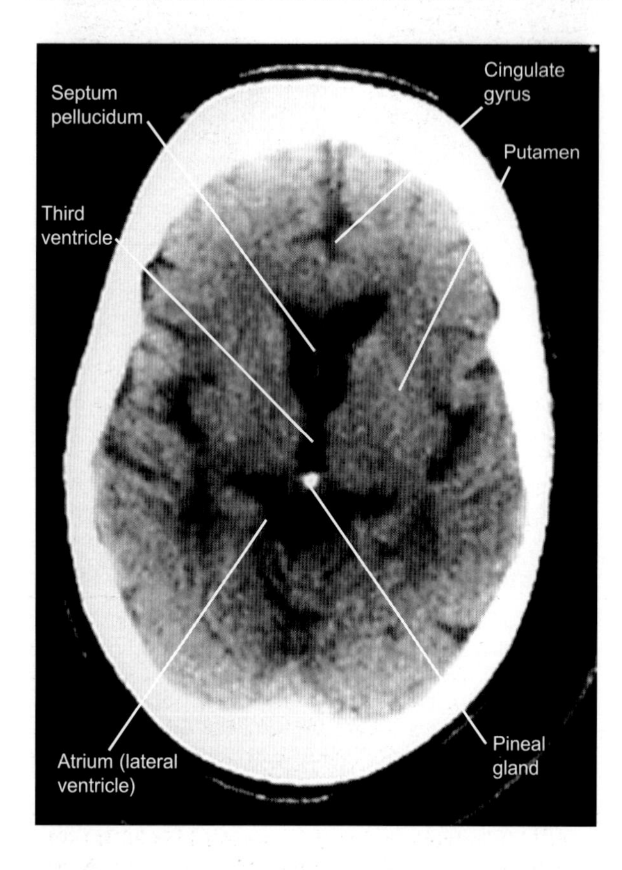

Figure 7.35 This computed tomography scan shows the pineal gland with calcareous concretion (brain sand) and the third ventricle.

the habenula to exert influences upon the gastric and salivatory secretions as well as the preganglionic neurons of the spinal cord via the tecto-tegmento-spinal and the dorsal longitudinal fasciculi. The possible role of the habenula in the regulation of sleep has been suggested by some investigators. Extensive metabolic, thermal, and endocrine disturbances may accompany damages to the habenular nuclei.

The posterior commissure consists of fibers which cross the midline inferior to the pineal gland and dorsal to the superior colliculi and cerebral aqueduct. Accessory oculomotor and pretectal nuclei project to the corresponding structures on the contralateral side through this commissure.

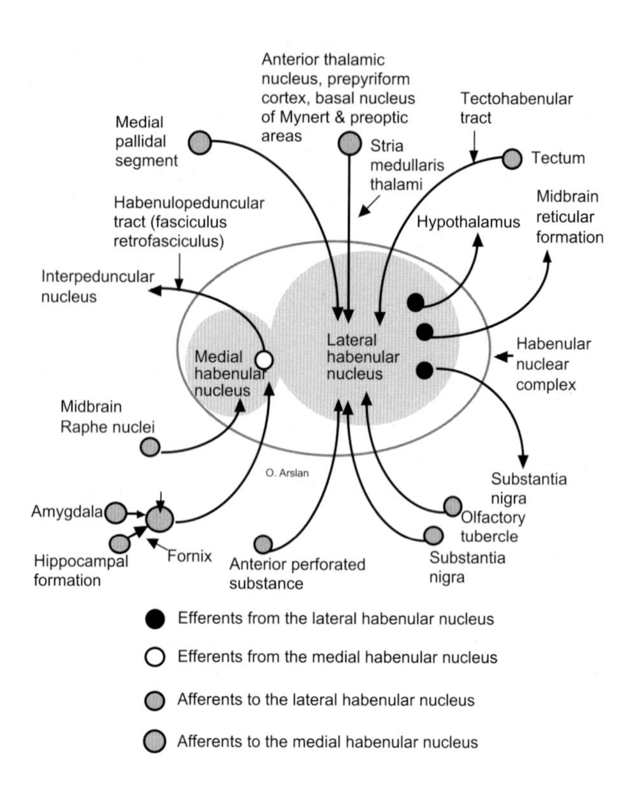

Figure 7.36 Schematic diagram of the principal afferents and efferents of the habenula

Subthalamus

The subthalamus is a small area of the diencephalon that lies ventral to the hypothalamic sulcus and lateral and caudal to the hypothalamus. It contains the subthalamic nucleus, zona incerta, cranial portions of the red nucleus, and substantia nigra. It also contains the spinal and medial lemnisci, and efferent fibers from the globus pallidus (ansa lenticularis, lenticular, thalamic and subthalamic fasciculi) and cerebellum which are destined to the thalamus. In mammals it also contains the entopeduncular and prerubral nuclei. It continues caudally with the tegmentum of the midbrain, separated from the globus pallidus by the internal capsule (Figure 7.36).

The subthalamic nucleus (Figures 7.7, 7.12 & 7.37) is a biconvex nucleus which lies medial to the internal capsule, caudally overlying the substantia nigra. It receives input from the lateral segment of the globus pallidus, reticular formation and the motor and

Lesions of the subthalamic nucleus produce violent, uncontrollable movements of the contralateral extremities, a condition known as hemiballismus, which is described in Chapter 21.

prefrontal cortices. It projects to both segments of the globus pallidus via the subthalamic fasciculus and to the pars reticulata of the substantia nigra. It is also connected to the ipsilateral red nucleus, mesencephalic reticular formation, and zona incerta. The subthalamic nucleus integrates motor activities through its connections with the basal nuclei, substantia nigra, and tegmentum of the midbrain. It is thought to have an inhibitory influence upon the globus pallidus.

The zona incerta is a thin layer of gray matter ventral to the thalamic fasciculus that extends with the reticular nucleus of the thalamus and the midbrain reticular formation. It contains dopaminergic neurons and receives cholinergic afferents from the midbrain tegmentum.

The prerubral and entopeduncular nuclei are located ventral to the zona incerta, and adjacent to the posterior limb of the internal capsule. It receives fibers from the globus pallidus, which are destined to the midbrain reticular formation. These nuclei project through the central tegmental tract to the inferior olivary nucleus.

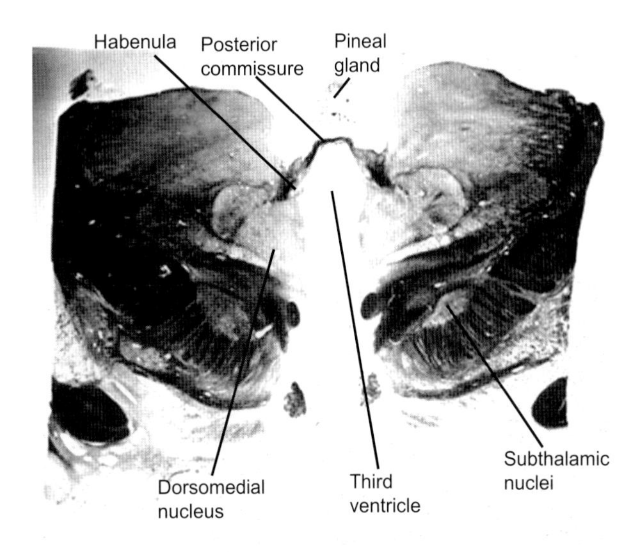

Figure 7.37 Section through the posterior commissure showing the pineal gland, habenula, and certain parts of the subthalamus

Suggested reading

1. Benabid AL, Benazzouz A, Limousin P, *et al.* Dyskinesias and the subthalamic nucleus. *Ann Neurol* 2000;47(Suppl 1):S189–92
2. Braak H, Braak E. Anatomy of the human hypothalamus (chiasmatic and tuberal region). *Prog Brain Res* 1992;93:3–16
3. Cardoso ER, Peterson EW. Pituitary apoplexy: a review. *Neurosurgery* 1994;14:363–73
4. Daikoku S. Functional anatomy of the hypothalamic hypophyseal system: neuroendocrine control mechanism. *Kaibogaku Zasshi J Anat* 1993;68:288–304
5. Dermon CR, Barbas H. Contralateral thalamic projections predominantly reach transitional cortices in the rhesus monkey. *J Comp Neurol* 1994;344:508–31
6. Ghika-Schmid F, Bogousslavsky J. The acute behavioral syndrome of anterior thalamic infarction: a prospective study of 12 cases. *Ann Neurol* 2000;48:220–7
7. Groenewegen HJ, Berendse HW. The specificity of the "nonspecific" midline and intralaminar thalamic nuclei. *Trends Neurosci* 1994;17:52–7
8. Guridi J, Obeso JA. The role of the subthalamic nucleus in the origin of hemiballism and parkinsonism: new surgical perspectives, *Adv Neurol* 1997;74:235–47
9. Imai K. Bilateral simultaneous thalamic hemorrhages – case report. *Neurol Med Chir (Tokyo)* 2000;40:369–71
10. Kao YF, Shih PY, Chen WH. An unusual concomitant tremor and myoclonus after a contralateral infarct at thalamus and subthalamic nucleus. *Kaohsiung J Med Sci* 1999;15:562–6
11. Krout KE, Loewy AD. Periaqueductal gray matter projections to midline and intralaminar thalamic nuclei of the rat. *J Comp Neurol* 2000;424:111–41
12. Mai JK, Kedziora O, Teckhaus L, Sofroniew MV. Evidence for subdivisions in the human suprachiasmatic nucleus. *J Comp Neurol* 1991;305:508–25
13. Moore RY, Silver R. Suprachiasmatic nucleus organization. *Chronobiol Int* 1998;15:475–87
14. Power BD, Mitrofanis J. Specificity of projection among cells of the zona incerta. *J Neurocytol* 1999;28:481–93
15. Sato F, Parent M, Levesque M, Parent A. Axonal branching pattern of neurons of the subthalamic nucleus in primates. *J Comp Neurol* 2000;424:142–52
16. Sperlágh B, Maglóczky Z, Vizi ES, Freund TF. The triangular septal nucleus as the major source of ATP release in the rat habenula: a combined neurochemical and morphological study. *Neuroscience* 1998;86: 1195–207
17. Tremblay N, Bushnell MC, Duncan GH. Thalamic VPM nucleus in the behaving monkey. II. Response to air-puff stimulation during discrimination and attention tasks. *J Neurophysiol* 1993;69:753–63
18. Van den Pol AN, Dudek FE. Cellular communication in the circadian clock, the suprachiasmatic nucleus. *Neuroscience* 1993;56:793–811
19. Wittkowski WH, Schulze-Bonhage AH, Bockers TM. The pars tuberalis of the hypophysis: a modulator of the pars distalis? *Acta Endocrinol* 1992;126:285–90
20. Yañez J, Anadón R. Afferent and efferent connections of the habenula in the rainbow trout (Oncorhynchus mykiss): an indocarbocyanine dye (DiI) study. *J Comp Neurol* 1996;372:529–43

Telencephalon 8

The telencephalon is comprised of the cerebral hemispheres and the basal nuclei, containing the lateral ventricles. It is a derivative of the lateral diverticula, which are interconnected by the median telencephalic impar. The cerebral hemispheres are comprised of the frontal, parietal, temporal, occipital, limbic, and the central (insular cortex) lobes. Examination of the cerebral cortex reveals an outer gray cortical area and an inner white matter. Fasciculi cross the white matter connecting areas within the same and opposite hemisphere. Blood supply of the cerebral hemisphere is secured by the carotid and vertebrobasilar systems.

Cerebral hemispheres

The cerebral hemispheres are mirror image duplicates that occupy the cranial cavity and are interconnected by the corpus callosum. The corpus callosum, which lies ventral and partly caudal to the anterior cerebral vessels and the falx cerebri, consists of the rostrum, genu, trunk, and splenium. The sagittal (interhemispheric) sulcus that partially separates the cerebral hemispheres, contains the falx cerebri. Both cerebral hemispheres are composed of an outer gray matter thrown into folds (gyri) and an inner white matter, containing the basal nuclei. Each cerebral hemisphere is divided into the frontal, parietal, temporal, occipital, central (insular cortex), and limbic lobes via sulci and fissures (Figures 8.1, 8.2, 8.6, 8.7, 8.8 & 8.9). The frontal, parietal, temporal and the occipital lobes are interconnected by the genu, trunk, and the splenium of the corpus callosum, respectively. The central sulcus separates the frontal and parietal lobes and contains the rolandic branch of the middle cerebral artery. The lateral cerebral (Sylvian) fissure or sulcus begins in the Sylvian fossa, contains the middle cerebral artery and demarcates the temporal lobe from the frontal and parietal lobes.

The lateral cerebral fissure gives rise to anterior, ascending, and posterior rami that divide the inferior frontal gyrus into orbital, angular and opercular parts. The angular and opercular parts form the Broca's motor speech center. The insular cortex forms the floor of this fissure.

Each hemisphere contains the lateral ventricle. This ventricle extends in a rostrocaudal direction from the frontal lobe to the occipital lobe, and inferiorly to the temporal lobe. The lateral ventricle on each side consists of an anterior horn that continues with the body (central part) and posterior and inferior horns. The lateral ventricles are separated via the septum pellucidum, communicating with the third ventricle via the interventricular foramen of Monro. Within the frontal lobe, the anterior horn extends inferior and posterior to the corpus callosum and superior and lateral to the caudate nucleus. The anterior horns are separated by the septum pellucidum. The floor of the central part (body) is formed by the thalamus and caudate nucleus, which are separated by the stria terminalis and the thalamostriate vein. The central part contains the choroid plexus and continues with the posterior horn in the occipital lobe and inferior horn in the temporal lobe. The posterior horn stretches in the occipital lobe medial to the calcarine fissure which forms the calcar avis, a visible prominence on its medial wall. The lateral wall and the roof of this horn are formed by the tapetal fibers of the

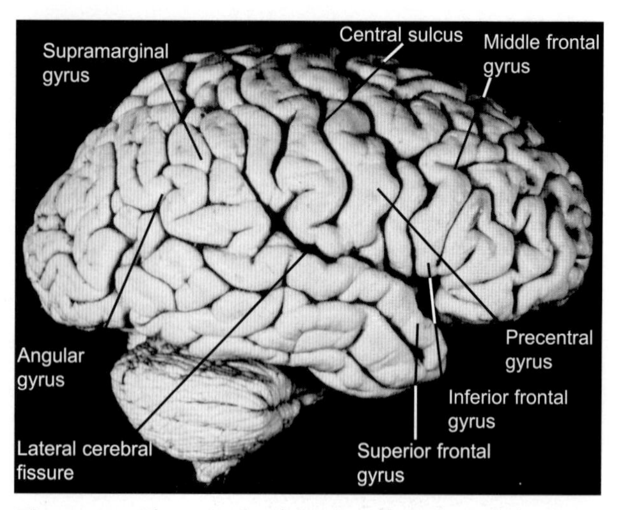

Figure 8.1 Photograph of the lateral surface of the brain showing prominent gyri. Lateral cerebral fissure and central sulcus are also shown

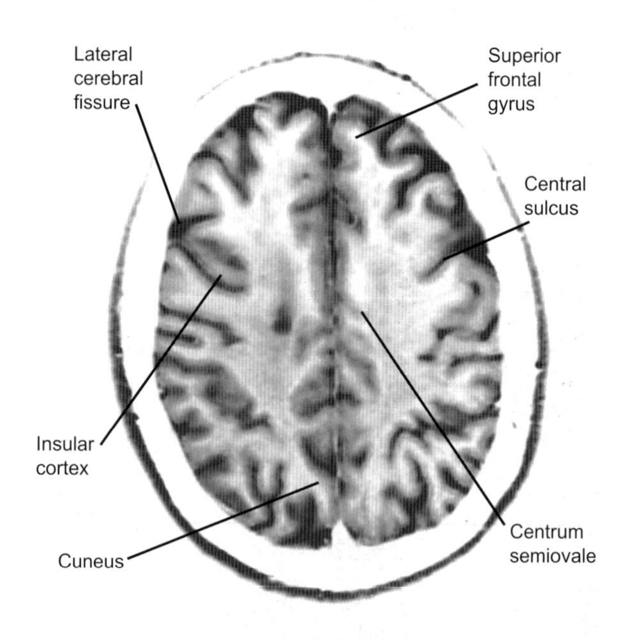

Figure 8.2 In this inverted MRI scan cerebral cortex and central white matter are shown

trunk of the corpus callosum. The transition zone between the posterior and inferior horns is termed the collateral trigone (Figures 8.3, 8.4, & 8.5).

The caudate nucleus is contained in the floor of the posterior horn and the roof of the inferior horn. The amygdala, which is attached to the tail of the caudate nucleus, lies immediately rostral to the tip of the inferior horn. The inferior horn curves downwards and ends near the temporal pole, containing in its floor the collateral eminence (formed by the collateral

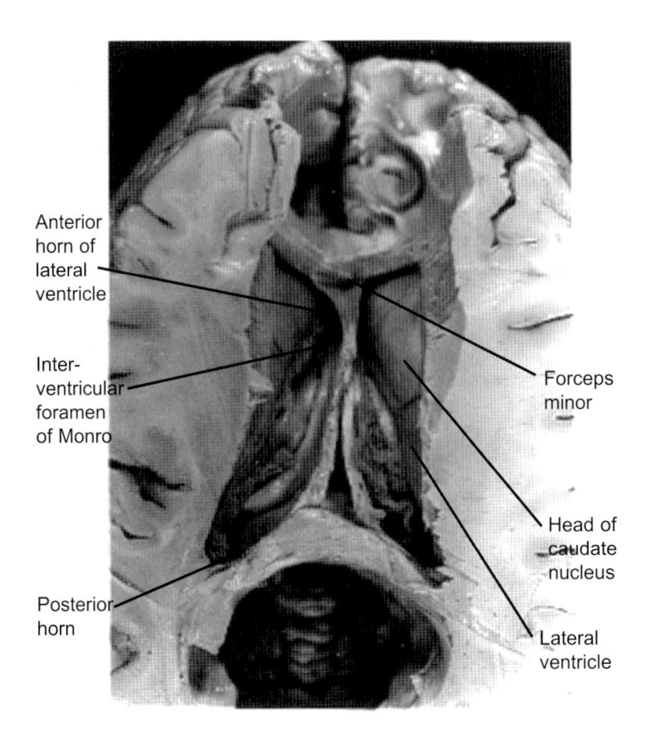

Figure 8.3 Horizontal section of the brain illustrating the boundaries of the anterior and posterior horns of the lateral ventricle and its central part. The connection of the lateral and third ventricles is maintained via the interventricular foramen of Monro

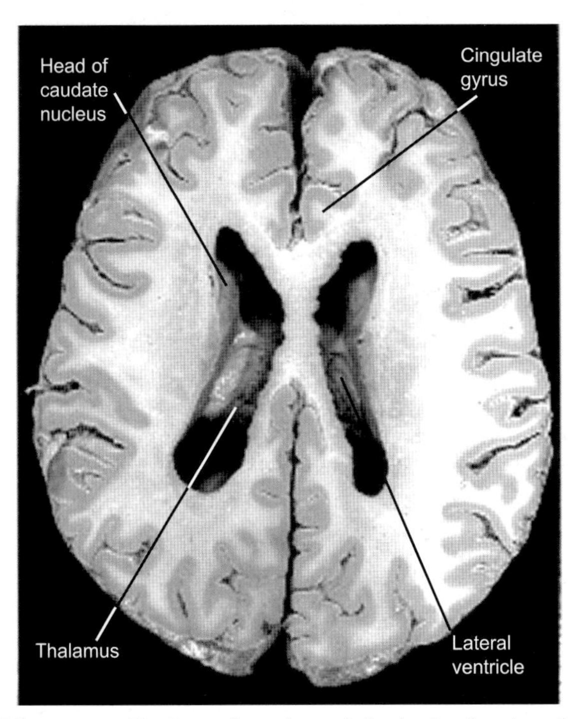

Figure 8.4 Horizontal section of the brain showing the location of the caudate nucleus and thalamus in relation to the anterior horn and central part of the lateral ventricle. The transition between the atrium and inferior horn of the lateral ventricle is visible

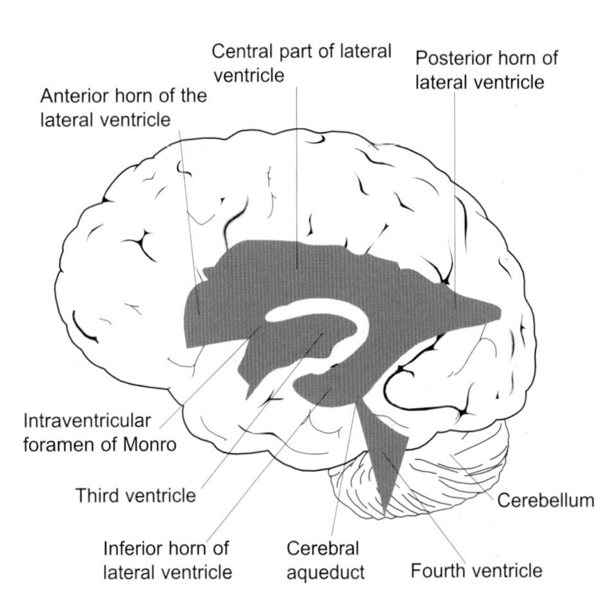

Figure 8.5 Diagram of a lateral view of the ventricular system

sulcus), hippocampal gyrus, fimbria of the fornix, dentate gyrus, and the choroid plexus. The roof and the lateral wall of this horn are formed by the tapetal fibers of the trunk of the corpus callosum. This plexus is supplied by the anterior and posterior choroidal arteries that originate from the internal carotid and posterior cerebral arteries, respectively.

Frontal lobe

The frontal lobe (Figures 8.1, 8.2, 8.7, 8.8 & 8.9) lies rostral to the central sulcus, superior to the lateral cerebral (Sylvian) fissure. It occupies the anterior cranial fossa, superior to the olfactory bulb and tract, as well the cribriform plate of the ethmoid bone. It contains the superior and inferior frontal sulci, which separate the superior, middle and inferior frontal gyri. Medially, the cingulate sulcus separates the medial part of the superior frontal gyrus (medial frontal gyrus) from the cingulate gyrus. The orbital gyri on the inferior surface of the frontal lobe are demarcated from the rectus gyrus by the olfactory sulcus that contains the olfactory tract. This lobe contains the primary motor center (Brodmann's area 4) in the

> Damage to the motor cortex, as a result of trauma, tumor, hemorrhage or thrombosis of the middle cerebral artery produces signs of upper motor neuron (UMN) palsy which are confined to the upper extremity, trunk, and head region (see Chapter 20).

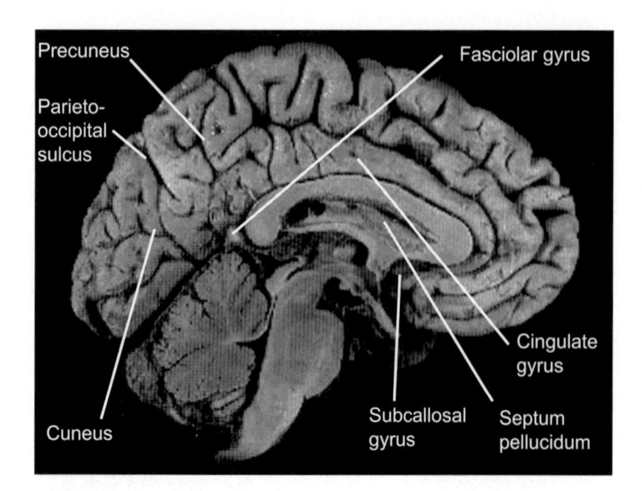

Figure 8.6 Midsagittal section of the brain showing the medial surface of the frontal, parietal, and occipital lobes

Frontal lobe seizures may activate the frontal eyefield and produce conjugate eye movement to the opposite side. Damage to this area, however, produces tonic deviation of both eyes to the side of the lesion (as if the patient were pointing to the affected side). This is due to the fact that a lesion in one hemisphere leads to the unopposed action of the contralateral frontal eyefield. It is interesting to note that this condition gradually attenuates, and in a matter of hours or days, movement of the eyes as far as the midline becomes possible again. However, free gaze movement will be regained much later, with nystagmus being the final diagnostic indicator of the lesion.

precentral gyrus, and the frontal eyefield (Brodmann's area 8) in the middle frontal gyrus which is responsible for conjugate deviation of the both eyes to the opposite side. It also contains Broca's motor speech center (Brodmann's areas 44 and 45) in the inferior frontal gyrus. Within Broca's motor speech center, linguistic rules and grammar (syntax) and the template for phonation are formed, and speech, melody and rhythm are regulated. The prefrontal cortex occupies the area rostral to premotor cortex.

As discussed earlier, numerous gyri and cortical areas constitute the frontal lobe, which include the precentral, superior, middle, and inferior frontal gyri, premotor, prefrontal, and supplementary cortices.

The precentral gyrus (Brodmann's area 4) lies between the central and the precentral sulci and constitutes the primary motor cortex for the entire

Ablation of the motor and premotor cortices leads to flaccid paralysis, for unknown reasons. Isolated lesion of Brodmann's area 6 results in motor apraxia (inability to perform familiar motor activity in the absence of any detectable motor or sensory deficits).

Prefrontal lobectomy is a surgical procedure, involving bilateral removal of the prefrontal cortices, which is used rarely in the treatment of frontal lobe tumors. In the past, it was utilized to modify the behavior of psychotic individuals, and to relieve intractable chronic pain unresponsive to conventional analgesics. Patients who have undergone this type of operation no longer complain of pain and appear to be oblivious to it. They exhibit emotional lability and superficial affect. Alterations in personality, disposition, drive and outlook are also prominent. Patients become disinhibited, lack initiative, and become less creative. They display irresponsible attitude and seem indifferent, with no changes in recent memory and intelligence. Lobectomized patients show a lack of restraint, an absence of hostility and boast. Reasoning, logical thinking, and problem-solving abilities are all impaired. Considerable loss of drive and ambition is observed.

Recent studies conducted on patients with depressive diseases indicate that transcranial electromagnetic stimulation of the prefrontal cortex, in which bouts of magnetic waves are passed through the brain, is effective in the treatment of certain depressive illnesses.

body, with the exception of the lower extremity. A distorted somatotopic organization of the body in this gyrus is known as the motor homunculus. Neurons of the precentral gyrus contribute to the formation of the corticospinal and corticobulbar fibers and maintain reciprocal connections with the ventral lateral nucleus of the thalamus.

Above the superior frontal sulcus lies the superior frontal gyrus that continues medially with the medial frontal gyrus.

The middle frontal gyrus, as mentioned earlier, contains the frontal motor eyefield (Brodmann's area 8) which is not regulated by the visual stimuli.

The inferior frontal gyrus is subdivided into orbital,

Since the prefrontal cortex is supplied by the orbitofrontal branches of the middle cerebral artery, infarction in the territory of the orbitofrontal branch may produce depressive syndromes which may not respond to tricyclic antidepressants.

Unilateral lesion of Brodmann's area 6 produces contralateral transient pathological grasp reflex, while bilateral lesion causes hypertonia in the muscles of the upper extremity. Paralysis or paresis is not seen in these lesions.

Frontal lobe hematoma may occur as a result of rupture of the middle cerebral artery or its branches, producing abulia, which is characterized by a lack or impairment of verbal spontaneity and initiative as well as the inability to perform volitional acts or make decisions. Patients appear sedentary and withdrawn. Expansion of the hematoma to involve the frontal eyefield may produce tonic conjugate deviation of both eyes toward the side of lesion. It usually resolves after a few days as the intact contralateral frontal gaze center compensates for the deficit. Involvement of the precentral gyrus in this hematoma produces paralysis of the opposite half of the body.

Occlusion of the middle cerebral artery may produce coma and forced deviation of the eyes toward the occluded side when the frontal eyefield is involved. Occlusion of the anterior cerebral artery proximal to the callosomarginal branch may cause a large infarction on the medial surface of the frontal lobe, resulting in intellectual deterioration, apraxia, visible primitive reflexes (such as sucking reflex), incontinence, and possible aphasia. Although rare, bilateral occlusion of the anterior cerebral arteries may occur when both arteries arise from a common stem, producing significant infarctions in both frontal lobes. Patients with this extensive lesion exhibit a variety of symptoms which include paralysis or paresis of both legs, urinary and bowel incontinence, behavioral changes, development of primitive reflexes (e.g. grasp reflex and sucking reflex), and abulia (inability to perform volitional acts or make decisions). Frontal lobe lesions may also induce bowing shift of the anterior cerebral artery, and Brun's ataxia, a broad-based, short-stepped gait, increasing the risk of the affected individuals for falls.

Many pathological reflexes may be observed in individuals with frontal lobe lesions such as grasp, snout, suck, and palmomental reflexes.

The grasp reflex is an involuntary tonic grasp response, associated with slow flexion of the fingers. Attempts by the examiner to withdraw the grasped fingers will only augment the patient's grasp. It may be elicited by touching or stroking the anterior surface of the wrist, the center of the palmar surface of the hand and digits, and between the thumb and index finger. The snout reflex is characterized by puckering and protrusion of the lips upon light percussion of the middle upper lip. The suck reflex is an involuntary sucking movement of the lips, produced by a blunt object stroked across the lips. The palmomental reflex exhibits contraction of the mentalis and the orbicularis oris muscles upon striking the palm of the infant's hand with a sharp object.

triangular and opercular parts. The orbital part is anterior to the anterior ramus; the triangular part lies between the anterior and ascending rami, whereas the opercular part is lodged posterior to the ascending ramus of the lateral cerebral fissure. As mentioned earlier the opercular and triangular parts of this gyrus collectively form Broca's motor speech center.

The premotor cortex is composed of Brodmann's areas 6 and 8. It occupies the area immediately anterior to the motor cortex, which includes part of the middle and frontal gyri. Most of the afferents to this cortical area are derived from the ventral anterior thalamic nucleus. Stimulation of Brodmann's area 6 produces motor activities, which are characterized by attentive or orientative movements such as turning the head and eyes.

The prefrontal cortex is rostral to the premotor cortex, a well-developed area in the human. It includes Brodmann's areas 9 and 10 and the orbital gyri (Brodmann's area 46). This cortical area has reciprocal connections with the dorsomedial nucleus of the thalamus. It is also connected to the temporal cortex via the uncinate fasciculus and to the parietal and occipito-temporal association cortices. The prefrontal cortex is important in the establishment of emotional responses, programming and intellectual functions.

The supplementary motor area (Brodmann's area 6) primarily refers to the medial frontal gyrus, which forms the medial extension of the superior frontal gyrus above the cingulate gyrus. This area exhibits a small motor homunculus that functions independently from the primary motor cortex. Stimulation of this region results in bilateral synergistic movements of a postural nature of both the axial and appendicular musculature, and rapid uncoordinated movements, as well as complex pattern of motor activities.

Parietal lobe

The parietal lobe is bounded by a line which connects the pre-occipital notch to the upper end of the parieto-occipital sulcus (Figures 8.7 & 8.8). On the lateral surface, this lobe consists of the postcentral gyrus, (Brodmann's area 3, 1, 2) and the superior and inferior parietal lobules. Medially, it contains the precuneus and the posterior part of the paracentral lobule. This lobe is concerned with mimicking (imitation of speech or action), pantomimicking actions (imitation of gestures without words) or copying.

The postcentral gyrus contains neurons which represent the primary sensory cortex (Brodmann's areas 3, 1, 2). The main input to this area arises from the ventral posterior nuclei of the thalamus (VPL and VPM), a somatotopically arranged projection in the form of a sensory homunculus (lips, tongue, and thumb have a larger representation). The body's distorted and disproportionate representation in this gyrus is also based upon the relative densities of the neuronal populations (sensory homunculus). Brodmann's area 3 responds to joint sensations, Brodmann's area 1 responds to cutaneous stimuli, and

> Parietal lobe hematoma may produce loss of general sensations on the opposite side of the body. When the dominant hemisphere is involved, signs of neglect on the contralateral half of the visual field may also occur. Frequently, parietal lobe diseases may mimic certain manifestations of frontal lobe diseases.

> The site of termination of the fibers that emanate from the postcentral gyrus may explain the deficit in kinesthetic perception that follows upper motor neuron lesions.

> Ablation of the postcentral gyrus results in the inability to appreciate texture, weight, and slight changes in temperature. The ability to recognize painful, tactile, pressure and proprioceptive sensations may not be significantly impaired, although localization of noxious and tactile stimuli may be severely affected.

> A lesion of the angular gyrus in the dominant hemisphere may produce visual agnosia, in which the ability to read and comprehend written word (alexia) and copy (agraphia) are lost. Bilateral damage to the angular gyri may result in the loss of ability to judge the location of objects and their relationship to each other in space.

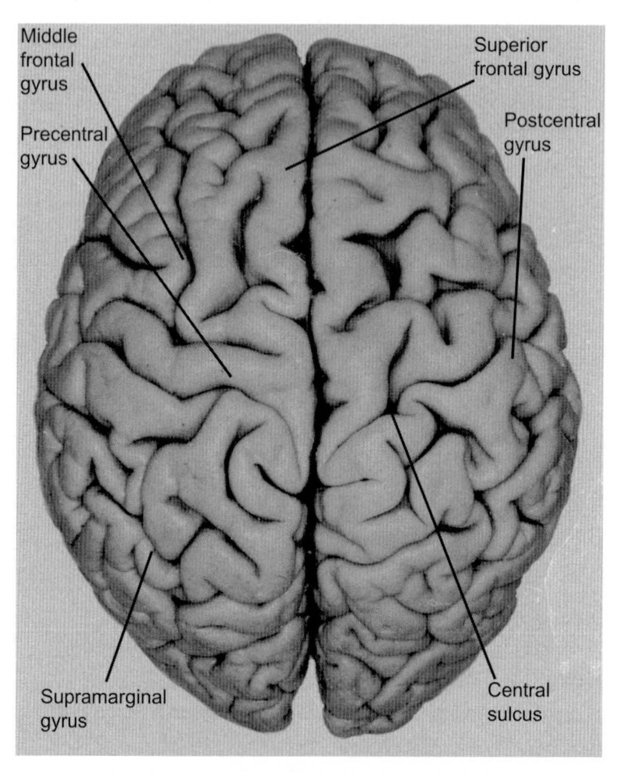

Figure 8.7 Dorsal surface of the brain with the precentral, postcentral, and supramarginal gyri clearly demarcated

both Brodmann's areas 1 and 2 are excited by stimuli from joints. There are neurons in this gyrus that evoke a motor response upon stimulation. Some fibers of the postcentral gyrus descend in conjunction with the pyramidal tract and terminate in the gracilis and cuneatus nuclei.

The secondary somatosensory area is located at the base of the postcentral gyrus. However, certain parts of the body (e.g. face, mouth and throat) are not represented in this area. Lesions of this area do not result in any recognizable deficit.

The superior parietal lobule is represented by Brodmann's areas 5 and 7. Damage to this lobule in the dominant hemisphere may impair the ability to recognize certain parts of one's body on the contralateral side. Initial hypotonia may also be observed. The inferior parietal lobule, which consists

Damage to the paracentral lobule may result from occlusion of the anterior cerebral artery, producing spastic paralysis of the muscles of the contralateral lower extremity, and bowel and urinary incontinence.

Damage to the anterior portion of the parietal lobe results in Dejerine's cortical sensory (Verger–Dejerine) syndrome, which is characterized by contralateral loss of discriminative sensory modalities (joint sensation and stereognosis), including the face. Pain and temperature sensations remain intact, but the ability to localize these sensory impulses may be affected.

of the supramarginal and angular gyri, is separated from the superior parietal lobule by the interparietal sulcus.

The supramarginal gyrus (Figure 8.7) encircles the posterior end of the lateral cerebral sulcus, and is designated as Brodmann's area 40.

The angular gyrus (Figure 8.1) designated as Brodmann's area 39, surrounds the terminal part of the superior temporal sulcus. Due to its massive connections with the association cortices of the auditory, visual and superior parietal lobe, the angular gyrus acts as a nodal point of integration for these modalities of sensations into symbols. It also signals the premotor cortex in the dominant hemisphere to initiate action via the superior longitudinal (arcuate) fasciculus within the same hemisphere and via the corpus callosum in the non-dominant hemisphere. On the left side, it converts the written word to its auditory equivalent (graphemes to phonemes).

The precuneus is bounded by the marginal branch of the cingulate sulcus rostrally and the parieto-occipital sulcus caudally.

The paracentral lobule is formed by the medial extension of the precentral and postcentral gyri. It contains neurons which are responsible for the motor and sensory innervation of the lower extremity and the regulation of certain physiological functions such as defecation and micturition.

Destruction of the supramarginal gyrus in the dominant hemisphere may disrupt its connection to other association cortices, resulting in tactile and proprioceptive agnosia or astereognosis, a condition that refers to the inability to recognize texture, form, and size by palpation, apraxia, left–right disorientation, and disturbances of body image.

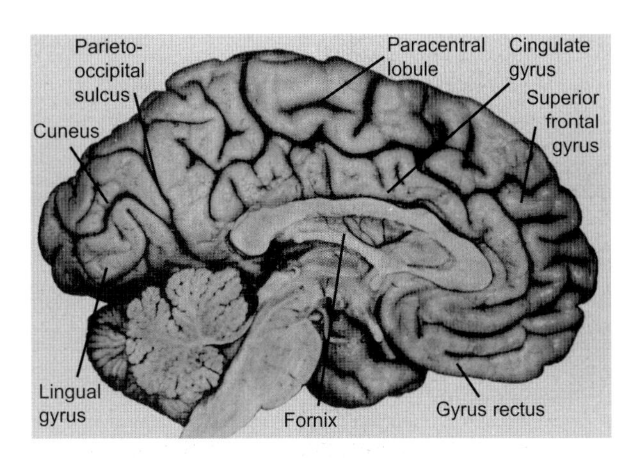

Figure 8.8 In this photograph of the medial surface of the brain, the individual gyri of the frontal, parietal, and occipital lobes are shown. The parieto-occipital and calcarine fissures are prominent

Temporal lobe

The temporal lobe (Figures 8.1 & 8.9) lies inferior to the lateral cerebral fissure, rostral to the pre-occipital notch, and inferior and caudal to the frontal lobe. It contains the inferior horn of the lateral ventricle, the hippocampal and dentate gyri, fimbria of the fornix, and the amygdala. The lateral surface of this lobe consists of the superior, middle, and the inferior temporal gyri, which are separated by the temporal sulci. The superior temporal sulcus runs between the superior and middle temporal gyri, and is surrounded caudally by the angular gyrus. Separation of the middle and inferior temporal gyri is maintained by the inferior temporal sulcus. Additional gyri exist on the inferior surface of the temporal lobe, which include the occipito-temporal and the parahippocampal gyri. Medial to the superior temporal gyrus, the transverse gyri of Heschl (Brodmann's areas 41 and 42) make their appearance, constituting the primary auditory cortex.

In the dominant hemisphere, the posterior part of the superior temporal gyrus contains the secondary

Temporal lobe hematoma may produce herniation and brainstem compression, and may lead to stupor and possible death.

Selective destruction of parts of areas 41 and 42 produces sigoma contralaterally, a hearing deficit, which is comparable to scotoma associated with lesions of the visual pathway.

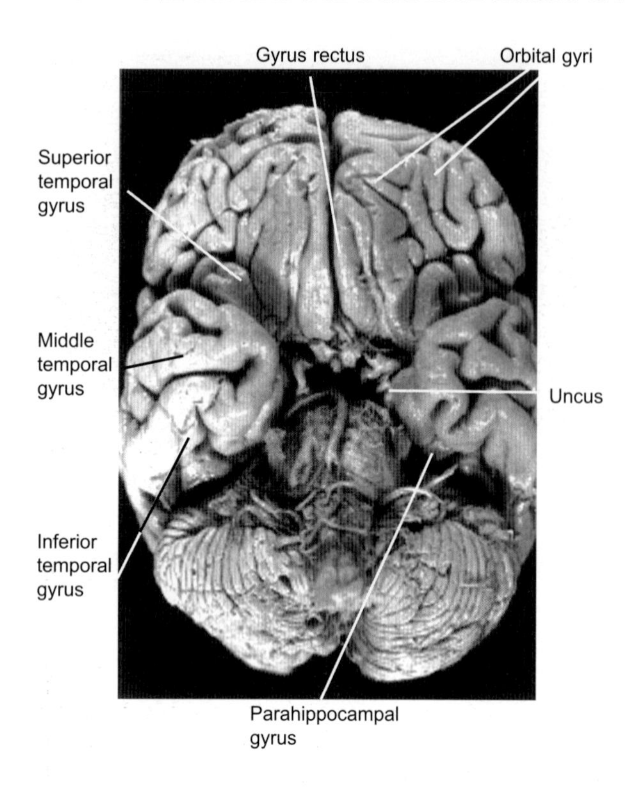

Figure 8.9 Inferior surface of the brain illustrating the gyri associated with the frontal and temporal lobes

auditory cortex (Brodmann's areas 22, 51 and 52) or Wernicke's zone. Wernicke's zone is the site where templates for phonemes (speech sounds such as 'f' or 'ph' and words) are linked to primitive auditory sensations received from the primary auditory cortex. These templates are also linked to the auditory, visual and olfactory systems, as well as to other sensory modalities in the angular gyrus. Due to these connections, Wernicke's area can regulate auditory perception and store visual imagery and verbal comprehension. Aside from the superior temporal gyrus, other gyri also exist on the lateral surface of the temporal lobe. Between the superior and inferior temporal sulci lies the middle temporal gyrus. The inferior temporal gyrus (Brodmann's area 20) lies lateral to the occipitotemporal sulcus, making an appearance on both the inferior and lateral surfaces of this lobe. This cortical area is considered a higher visual association zone, which receives in its posterior part major input from the occipitotemporal cortex, representing the contralateral visual field.

A hematoma that damages the superior frontal gyrus in the dominant (usually the left) hemisphere may produce receptive aphasia (inability to comprehend spoken language).

Uncal herniation is a condition in which the uncus as well as the medial portion of the temporal lobe are forced to protrude, usually as a result of supratentorial mass (hematoma) into the middle cranial fossa, over the sharp margin of the tentorial notch. The mass often causes a substantial rise in the intracranial pressure and displacement of the uncus into and below the tentorial (Kernohan's) notch. The swollen uncus may compress the midbrain, oculomotor nerve, crus cerebri, and the posterior cerebral artery. Patients with uncal hernia exhibit signs of ipsilateral or bilateral oculomotor palsy, which include mydriasis (pupillary dilatation) and lateral strabismus (lateral deviation of the eye), followed by downward and lateral deviation of the eye, and eventual external ophthalmoplegia.

Irregularities in respiration ranging from Cheyne–Stokes breathing to respiratory arrest also occur. As in subdural hematomas, reduced states of consciousness and coma develop rapidly, as a result of compression of the ascending reticular activating system following oculomotor palsy. Signs of ipsilateral upper motor neuron paralysis, followed by decerebrate rigidity as a result of destruction of the inhibitory corticospinal tract in the crus cerebri and unopposed action of the excitatory pathways, are eventually concluded in total flaccidity.

Uncinate fits are seizures that produce involuntary movements of the mouth and tongue and the perception of noxious hallucinatory odors which may be accompanied by irrational fears of the surroundings. They may occur as a result of a lesion, infarct or tumor affecting the uncus, amygdala and possibly the gustatory area. Due to involvement of the temporal lobe, cognitive functions, such as memory, orientation and attention, may also be impaired.

Pick's disease is another condition that selectively produces atrophy of the temporal and frontal lobes, sparing the posterior two-thirds of the superior temporal gyrus. It may also involve the occipitotemporal region (the silent cortical area) which is concerned with the storage of memories from the visual and auditory systems. Damage to this wide array of cortical areas may result in epileptic seizures combined with amnesia, auditory hallucinations, and the *déjà vu* phenomenon.

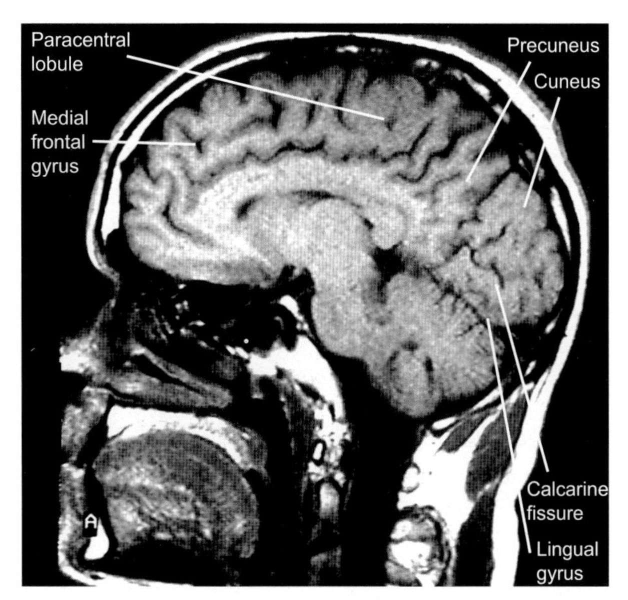

Figure 8.10 An inverted MRI scan of the medial surface of the brain showing some of the associated gyri

Immediately lateral to the parahippocampal (fusiform) gyrus, the occipitotemporal gyrus makes its course. The parahippocampal gyrus (Figure 8.9) represents the inferior portion of the limbic lobe, expanding rostrally and medially into the uncus, a small tongue-like projection between the collateral and hippocampal sulci. This projection constitutes the primary olfactory center (Brodmann's area 34) which receives and processes olfactory information received from the lateral olfactory stria. Uncus and olfactory pathways are considered integral parts of the limbic system.

Occipital lobe

The occipital lobe (Figures 8.6, 8.8 & 8.10) lies caudal to an imaginary line that connects the pre-occipital notch to the upper end of the parieto-occipital fissure. On its lateral surface the superior and inferior occipital gyri are separated by the lateral occipital sulcus. Caudal to the lateral occipital sulcus, the descending occipital gyrus makes its appearance. Medially, this lobe is formed by the cuneus and lingual gyrus. The cuneus, which is bounded rostrally by the parieto-occipital sulcus and caudally by the calcarine fissures, receives visual impulses from the lower quadrant of the opposite visual field. Inferior to the calcarine fissure and medial to the collateral sulcus lies the lingual gyrus that receives input from the upper quadrant of the opposite visual field.

Rostral to the occipital pole, the lunate sulcus runs vertically between the striate and peristriate cortices. The lunate sulcus, which contains the parastriate

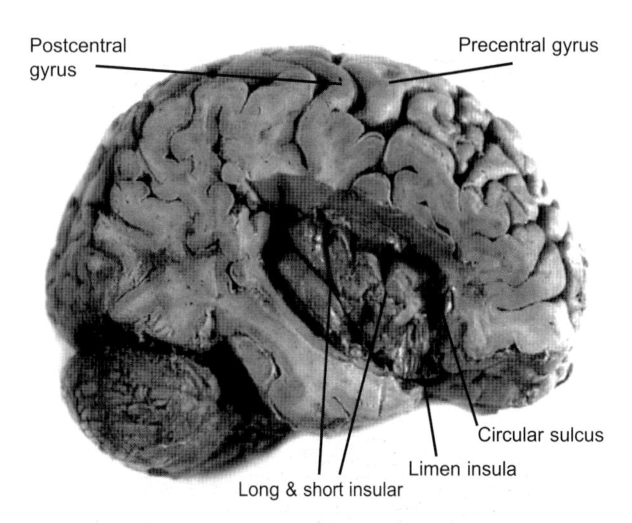

Figure 8.11 In this picture, the lateral surface of the brain is dissected to delineate clearly the boundaries of the insular gyri

cortex, is crossed at its upper and lower ends by the superior and inferior polar sulci, receptively. Enclosed between the polar sulci is the macular area of the primary visual (striate) cortex.

Insular cortex (central lobe)

The insular cortex (Figure 8.11) or central lobe consists of long and short insular gyri, forming the floor of the lateral cerebral fossa. It continues anteriorly with the anterior perforated substance and is surrounded by an incomplete circular sulcus. This sulcus is deficient rostrally and inferiorly where the limen insula is located. Continuation of the insular cortex is marked by the opercular areas of the frontal, parietal and temporal lobes. It receives input from the ventral posterior nucleus, medial geniculate body, dorsomedial thalamic nucleus, pulvinar and intralaminar nuclei, maintaining ipsilateral connections with the primary and secondary somatosensory cortex, inferior parietal lobule, and the orbitofrontal cortex. It has been suggested that the insular cortex plays an important modulatory role in the perception and recognition of fine touch, auditory impulses, and taste, and is thought to be associated with language function.

Limbic lobe

The limbic lobe consists of the subcallosal, paraterminal (septal area), cingulate, and the fasciolar gyri (Figures 8.8, 8.9, 8.10 & 8.12). A rostral region of this lobe, which lies between the lamina terminalis and the posterior olfactory sulcus, is termed the subcallosal (para-olfactory) gyrus. It should also be noted that the paraterminal (precommissural septum) gyrus

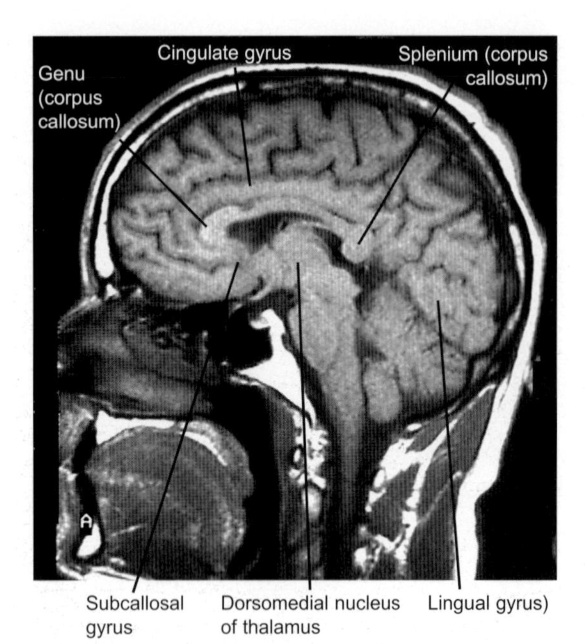

Figure 8.12 Inverted midsagittal MRI of the brain illustrating some components of the limbic lobe that surrounds the corpus callosum as well as adjacent structures

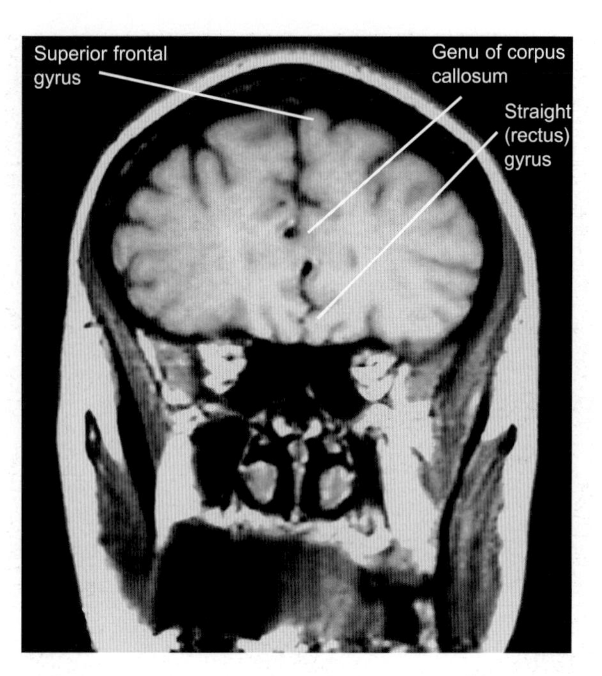

Figure 8.13 MRI scan through the genu of the corpus callosum. The frontal lobes are connected by the forceps minor, an extension of the genu of the corpus callosum

A unilateral cerebral mass may compress the cingulate gyrus of the contralateral hemisphere against the falx cerebri and produce cingulate or subfalcial herniation, a serious condition that may end in death.

receives input from the olfactory system via the medial olfactory stria; hippocampal gyrus through the precommissural column of the fornix; and the lateral hypothalamus via the medial forebrain bundle (MFB). It also receives impulses from the tegmental nuclei via the mamillary peduncle. Aside from its projection to the tegmental nuclei of the midbrain through the MFB, the paraterminal gyrus also sends fibers to the habenular nuclei contained in the stria medullaris thalami. It is also connected through the diagonal band of Broca to the periamygdaloid cortex.

The cingulate gyrus (Figures 8.8, 8.10 & 8.12) is part of the limbic lobe, maintaining reciprocal connections with the anterior nucleus of the thalamus. Through this thalamic nucleus, the hypothalamus influences the activities of the cingulate gyrus. This gyrus projects to the entorhinal cortex, and influences the hypothalamus through fibers of the fornix. Due to these diverse connections somatic and visceral responses may be elicited by stimulation of the anterior part of the cingulate gyrus. Stimulation of the posterior cingulate gyrus elicits pleasurable reactions.

The fasciolar (retrosplenial) gyrus connects the cingulate and the induseum griseum to the dentate gyrus of the hippocampal formation.

Midline interconnection of the cerebral hemispheres is maintained primarily by the corpus callosum (Figures 8.8, 8.12 & 8.13). This commissural structure, covered by the ventricular ependyma, forms the roof of the lateral ventricle. Superior to the corpus callosum lies the anterior cerebral vessels and the falx cerebri. It consists of the rostrum, genu, trunk and splenium. Fibers of the genu (forceps minor), which lies between the rostrum and trunk, connect the frontal lobes. Apart from its small size, the rostrum extends from the genu to the lamina terminalis; its superior surface is attached to the septum pellucidum. Wide cortical areas of the hemispheres are connected via the trunk, whereas the splenium, the thickest part of this commissure, forms the base of the longitudinal sagittal cerebral fissure, connecting the occipital lobes (forceps major). The splenium protrudes into the posterior horn of the lateral ventricle as the bulb of the posterior horn. The callosal trunk is covered by the induseum griseum (supracallosal gyrus), a thin lamina of gray matter which is continuous with the cingulate, fasciolar, and the paraterminal gyri, containing the medial and lateral longitudinal striae of Lancisi. Fibers of the trunk and splenium form the tapetum, which constitutes the roof and the lateral walls of the posterior and inferior horns of the lateral ventricle.

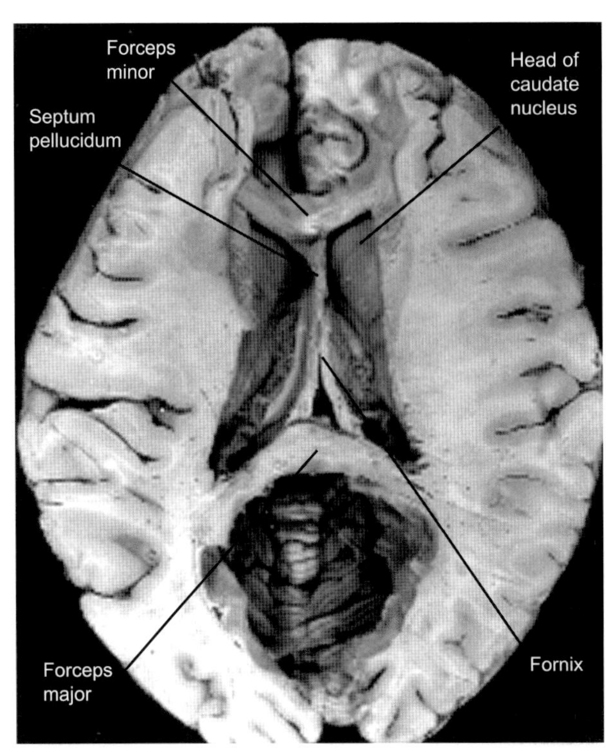

Figure 8.14 In this horizontal section the septum pellucidum and its relation to the fornix is illustrated

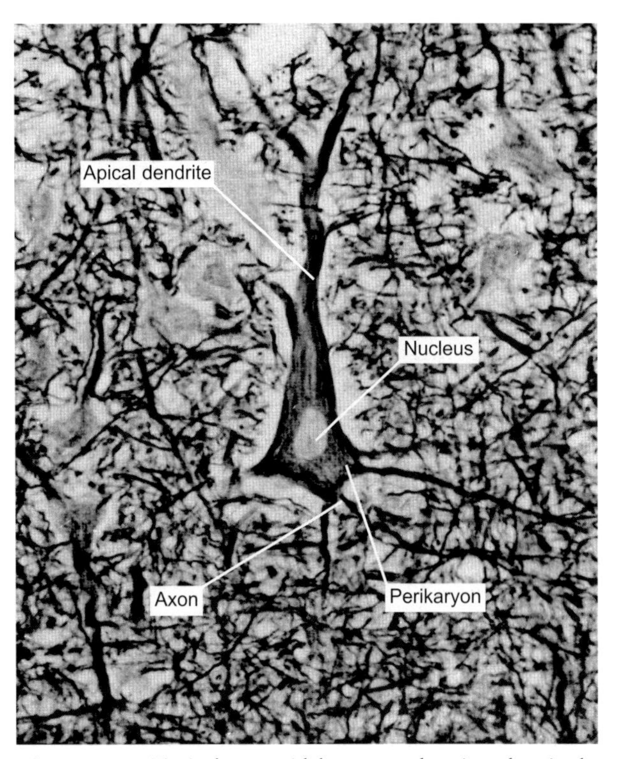

Figure 8.15 Typical pyramidal neuron showing the single axon and a distinct apical dendrite

A vertical partition, the septum pellucidum (supracommissural septum), extends between the corpus callosum and the fornix, forming the medial wall of the anterior and central parts of the corpus callosum. This septum (Figures 8.12 & 8. 14) consists of two laminae of fibers, sparse gray matter and neuroglia. Note that the fornix (Figures 8.6, 8.8, 8.12 & 8.14), a robust bundle at the inferior border of the septum pellucidum, is formed by the axons of the pyramidal layer of the hippocampal gyrus. This bundle starts as the alveus, converges into the fimbria, which ascends toward the splenium of the corpus callosum, and more rostrally above the thalamus as the crura of the fornix. These crura, interconnected by the hippocampal commissure, unite rostrally to form the body of the fornix which attaches to the corpus callosum. Eventually the body of the fornix divides into precommissural and postcommissural columns by the fibers of the anterior commissure. The precommissural column terminates in the lateral hypothalamus, while the postcommissural column projects primarily to the mamillary nuclei.

Cerebral cortex (gray matter)

In the human, the cerebral cortex (pallium) is the most highly evolved portion of the CNS and in particular the cerebral hemisphere. It consists of a thin shell of a large number of neuronal cell bodies, unmyelinated nerve fibers, glial cells, and blood vessels. The presence of the neuronal cell bodies and extensive capillary network are responsible for the gray appearance of the cerebral cortex. It is well documented that the cerebral cortex maintains crucial roles in the perception, fine discrimination, integration of various modalities of sensation, and in the regulation of visceral and somatic motor activities. It contains afferent (thalamocortical), efferent (projection), commissural, and association fibers. In order to increase the surface area of the brain, the cortex is thrown into convolutions (gyri) separated by sulci.

On a phylogenetic basis, the cerebral cortex is divided into archipallium, paleopallium, and neopallium. The archipallium (oldest cortex) is comprised of the hippocampal, dentate, and fasciolar gyri, and the subiculum. The hippocampal gyrus (cornu ammonis) is superior to the subiculum and the parahippocampal gyrus. It lies in the floor of the temporal horn of the lateral ventricle and forms the pes hippocampi, a rostral swelling which is covered by the ependymal layer. As mentioned earlier, the axons of the hippocampal neurons form the alveus, which continues with the fimbria of the fornix. The dentate gyrus, which lies between the hippocampal gyrus (cornu ammonis) superiorly and the subiculum

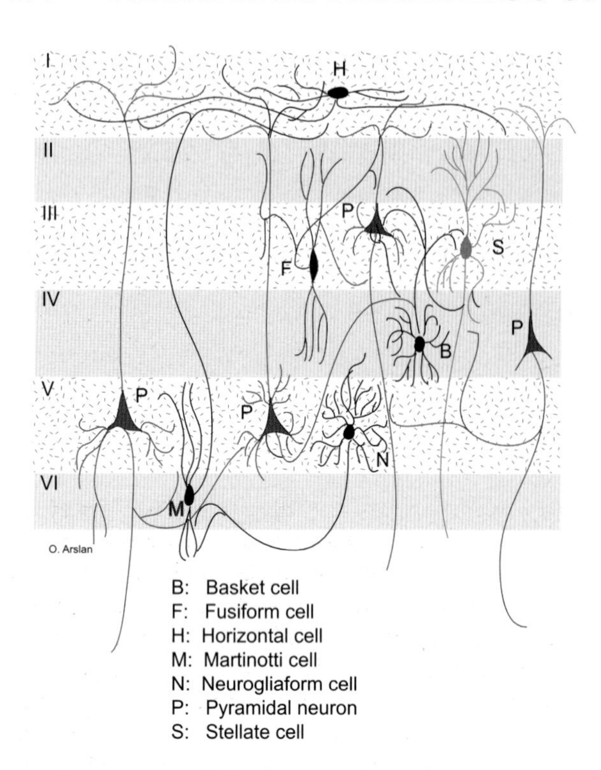

Figure 8.16 Schematic representation of the various neurons associated with the cerebral cortex

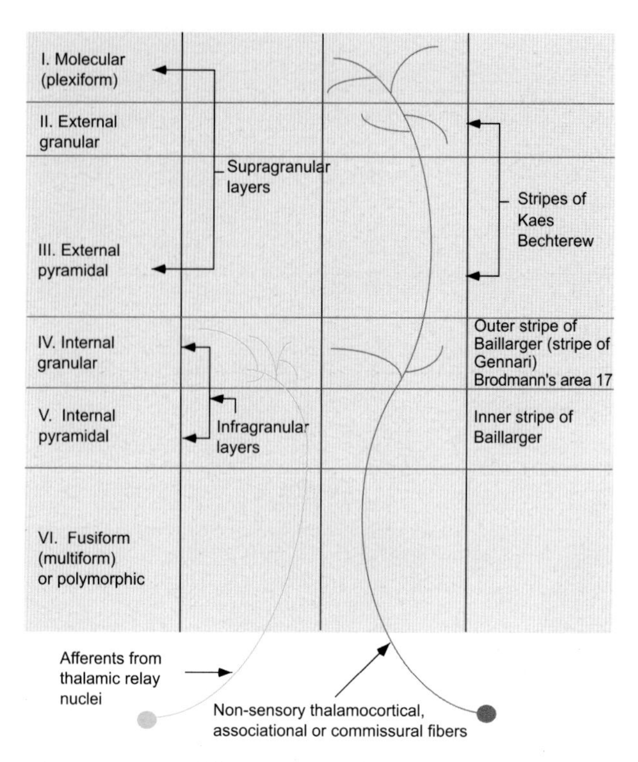

Figure 8.17 The six-layered isocortex. Note the supragranular and infragranular layers and the sites of termination of the specific thalamic afferent and non-sensory thalamocortical fibers

inferiorly, is separated from the subiculum by the hippocampal sulcus. A transitional zone between the six-layered cortex of the parahippocampal gyrus and the three-layered cortex of the cornu ammonis is termed the subiculum (see also the limbic system).

The paleopallium (old cortex) includes the olfactory cortex and the pyriform lobe, which are integral parts of the limbic system. Both the archipallium and paleopallium comprise the allocortex or heterogenetic cortex, consisting of three layers. Most of the cerebral cortex, approximately 90%, constitutes the neopallium (neocortex, or isocortex), which is formed by six distinct layers (homogenetic cortex). By the seventh month of intrauterine development, the six layers of the homogenetic cortical regions become distinct.

The cerebral cortex consists of pyramidal and non-pyramidal neurons such as the stellate (granule), fusiform, horizontal cells of Cajal, and cells of Martinotti, which are arranged in horizontal and vertical layers. The pyramidal cells (Figures 8.15 & 8.16) have apical dendrites, which run perpendicular to the cortical surface, and basal dendrites that branch locally parallel to the cortical layer. Axons of the pyramidal neurons travel to the subcortical areas as projection fibers or course within the white matter as association fibers. These axons may give rise to recurrent collaterals, and represent the primary output of the cerebral cortex. Betz cells are giant pyramidal cells, occupying the precentral gyrus. There are numerous granule cells, which function as interneurons, with their axons and many dendrites concentrated in lamina IV of the cerebral cortex (Golgi type II). Fusiform cells (Figure 8.16) occupy mainly the deep cortical layers, particularly layer V, and possess axons that form projection fibers. Horizontal cells of Cajal (Figure 8.16) have axons that are confined to the superficial layers of cerebral cortex, whereas Martinotti cells (Figure 8.16) are scattered diffusely throughout the cortical layers.

Neuronal cytoarchitecture of the homotypical isocortex reveals six layers or laminae (Figures 8.16, 8.17 & 8.18):

I. The molecular layer consists of horizontal cells of Cajal, dendrites of the pyramidal neurons, and axons of Martinotti cells.

II. The external granular layer is a receptive layer, consisting of small pyramidal and granule cells. Neurons of this layer project to the molecular layer and to deeper cortical layers in order to mediate intracortical circuits.

III. The external pyramidal layer primarily contains pyramidal neurons, projecting to the white matter as

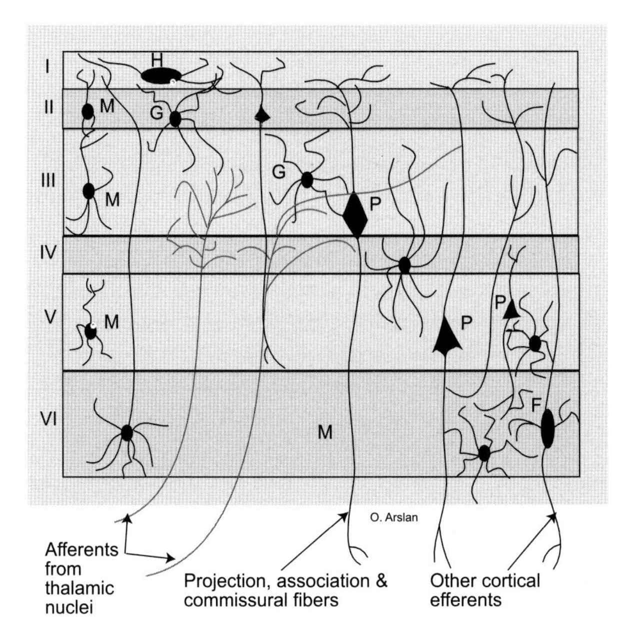

Figure 8.18 Cortical layers with their distinct neuronal population, showing areas of distribution of the afferents derived from the specific thalamic nuclei and areas of distribution of the commissural and association fibers

association fibers and to the opposite hemisphere as commissural fibers. This layer also contains granule and Martinotti cells, as well as the horizontal band of Kaes–Bekhterev (distinctive stripes in layers II and III). Laminae I, II, and III are concerned with associative and receptive functions.

IV. The internal granular layer receives all of the sensory projections from the thalamic relay nuclei via the thalamocortical fibers (e.g. optic radiation, auditory radiation, and primary sensory systems). It consists of densely packed stellate cells, containing the external myelinated bands of Baillarger. These outer and inner bands are formed by tangential thalamocortical fibers and may be visible to the naked eye. The outer band lies in lamina IV and is produced by afferents from the specific thalamic relay nuclei. The outer band is conspicuous in the striate cortex as the strip of Gennari, while the inner band is formed by the basal dendrites and the myelinated collaterals of the large pyramidal (Betz) cells. In the visual cortex the internal granular layer consists almost exclusively of simple cells which respond to stimuli from only one eye (but not both). The cortical layers that lie superficial and deep to layer IV respond to visual inputs from both sides. The infragranular (V and VI) layers are the first to be formed during embryological development. Subsequent cell migration through the infragranular layers allows the neurons to form more superficial layers.

V. The internal pyramidal (ganglionic) layer contains the largest pyramidal (Betz) neurons, giving rise to the corticospinal and corticobulbar fibers. It is pierced by dendrites and axons from other layers, including association and commissural fibers.

VI. The multiform layer contains small pyramidal and Martinotti cells, giving rise to projection fibers to the thalamus. It is crossed by commissural and association fibers. Cortical neurons of this layer project fibers, which maintain a feedback loop with the thalamic nuclei.

The relative thickness of the pyramidal and granular layers may be used as a basis to classify the cerebral cortex into five distinct areas. This classification grades the cortex from a purely motor to a purely sensory cortex and from the thickest to the thinnest. These cortical layers also vary from a layer that contains the least number of granule cells to a layer that consists mostly of granule cells. According to this classification the cerebral cortex is divided into agranular, frontal, parietal, polar, and granular cortex.

The agranular (motor) cortex is the thickest type that lacks or contains only a few granule cells in layers II and IV. It is exemplified in the heterotypical motor cortex (Brodmann's area 4), partly in the premotor cortex (Brodmann's areas 6 and 8), and paracentral lobule.

The frontal type of cortex is a homotypical cortex with a very thin granular layer, which is represented in the superior frontal, postcentral, and the inferior temporal gyri, superior parietal lobule, precuneus, and in parts of the middle and inferior frontal gyri.

Examination of the parietal type of cortex reveals six distinct layers with a particularly thin pyramidal layer, which is evident in the prefrontal cortex, inferior parietal lobule, and the superior temporal and occipitotemporal gyri.

In the polar cortex, which is represented in the frontal and occipital poles, a well-developed granular layer is evident.

The granular (konicortex), the thinnest of all cerebral cortices, contains a granular layer that achieved maximum development, and is symbolized in the homotypical cortices of cuneus, lingual, parahippocampal, and postcentral gyri as well as the transverse gyri of Heschl.

The cerebral cortex also exhibits vertical lamination, which represents the functional units of the cerebral cortex, extending through all cellular elements. This arrangement is absent in the frontal cortex, but distinctly evident in the parietal, occipital, and temporal cortices. The neurons of the vertical columns in the sensory cortex establish contacts with each other via interneurons (Golgi type II cells). Each column receives impulses from the same receptors, is stimulated by the same modality of sensation, and discharges for the same duration, maintaining an identical temporal latency. The isocortex (neopallium) is also divided into sensory, motor, and association cortices. The sensory cortex is further subdivided into primary, secondary, and tertiary sensory cortices. The motor cortex is classified into primary motor, premotor, and supplementary motor cortices. On the other hand the association cortices include parts of the parietal, temporal, and occipital cortices.

Sensory cortex

The sensory cortex deals with the perception and recognition of sensory stimuli. It imparts unique characteristics to sensations, enabling their identification on the basis of both comparative and temporal (spatial) relationships. It includes primary and secondary sensory cortices.

Primary sensory cortex

The primary sensory cortices are modality and place specific, receiving information from the specific thalamic nuclei. Depending upon the modality, sensations from body, projections from the visual fields and auditory spectrum are represented topographically in the contralateral primary sensory cortices. They include the somesthetic cortex (Brodmann's areas 3, 1, & 2), visual or striate (Brodmann's area 17), auditory (Brodmann's areas 41 & 42), gustatory, and vestibular cortices.

The primary somesthetic cortex (Brodmann's areas 3, 1, and 2) subserves general somatic afferents (deep and superficial), occupying the postcentral gyrus (Figure 8.19 & 8.32). It consists of three cytoarchitecturally distinct cortical stripes. Area 3 receives tactile sensation, and area 1 forms the apex of the postcentral gyrus, receiving deep and superficial sensation. Area 2 lies in the posterior surface of the postcentral gyrus, deals with deep sensation, and receives collaterals from the other two areas. The primary somesthetic cortex receives projections from the ventral posterolateral and the ventral posteromedial thalamic nuclei. Pain and thermal sensations are only minimally represented. The distorted representation of the body in this cortex is known as the sensory homunculus. This homunculus is formed according to the innervation density of the body part.

The primary visual (striate) cortex, representing Brodmann's area 17, lies on the banks of the calcarine fissure of the occipital lobe (Figure 8.23). It receives information from the third, fourth, fifth and the sixth layers of the lateral geniculate body. There is a visuotopic representation in which the peripheral part of the contralateral visual field is represented rostrally, while the macular visual field is delineated caudally in the occipital pole. No commissural fibers connect the striate cortices of both hemispheres. This cortical area receives blood supply from the internal carotid artery and the vertebrobasilar system via the middle and posterior cerebral arteries, respectively. This cortex consists of vertical columns that discharge more or less as a unit, maintaining a topographical representation. It contains simple cells in the internal granular layer that respond to edges, rectangles of light, and bars presented in a particular receptive field axis of orientation to one eye. They possess 'on' and 'off' (inhibitory surrounding) centers. Complex cells have no 'on' and 'off' centers, spread in many layers of the striate cortex, and receive input from the simple cells. However, they do respond constantly to moving stimuli from both eyes. The vertical columns of the striate cortex may also be viewed as ocular dominance and orientation columns. Ocular dominance columns receive visual input from one eye only, are divided into alternate, independent, and super-imposed stripes. Each column receives visual information from either the left or right eye. Development of these columns requires visual input. Deprivation of the binocular vision, as may be experienced in individuals with severe strabismus, may lead to unequal development of the ocular dominance columns and potential blindness. Orientation columns are much smaller than the dominance columns, are genetically determined, present at birth, and respond to a slit of light at a certain axis of orientation.

The primary auditory cortex (Brodmann's areas 41, 42) lies in the medial surface of the superior temporal gyrus (Figure 8.19). It is represented by the transverse gyri of Heschl, receiving the auditory radiation via the sublenticular part of the posterior limb of the internal capsule. Auditory radiation is formed by the axons of

> Interestingly, anesthetics have a far greater effect on the secondary sensory area than on the primary sensory cortex.

> Due to the bilateral representation of the auditory impulses, a unilateral lesion of the auditory cortex produces impairment of hearing on the contralateral side, but not total loss, with some degree of hearing loss on the ipsilateral side.

Table 8.1 Primary sensory cortex

Cortical areas	*Major afferents*	*Major efferents*	*Function*	*Deficits*
Postcentral gyrus, Brodmann's areas 3, 1 & 2	Ventral posterolateral & ventral postero-medial nuclei, secondary somato-sensory cortex along the upper bank of the lateral cerebral fissure, claustrum, basal nucleus of Meynert & locus ceruleus	Superior parietal lobule (Brodmann's areas 5, 7 & 6); ventral postero-medial and ventral posterolateral nuclei & corticopontine fibers	Perception of somesthetic sensations in a somatotopic manner	Loss of discriminative sensations such as position of body parts, weight differences, texture and two-point discrimination, and slight reduction in pain & temperature perception
Inferior part of the postcentral gyrus adjacent to insular cortex	Visceral afferents	Not well defined	Integration of visceral sensations	Not well documented
Striate cortex,- Brodmann's area 17 (cuneate & lingual gyri)	Lateral geniculate nucleus	Lateral geniculate body & Brodmann's area 18 spatial representation	Primary center for visual perception	Contralateral homonymous hemianopsia (commonly with macular sparing), stimulation produces flash of light & visual hallucination
Transverse gyri of Heschl (Brodmann's areas 41 & 42)	Medial geniculate body	Wernicke's zone & medial geniculate body	Perception of auditory impulses and tonotopic representation	Due to bilaterality of auditory projections no noticeable auditory deficits will be detected
Olfactory cortex (pyriform lobe)	Olfactory bulb via olfactory tract & striae	Dorsomedial nucleus, orbitofrontal, insular cortex, amygdala, entorhinal cortex & hypothalamus	Perception of olfactory impulses	Ablation of the olfactory cortex produces anosmia, while irritation, as in uncinate fits, produces olfactory aura that precedes seizures

the ventral nucleus of the medial geniculate neurons. It should be noted that impulses reaching the primary auditory cortex emanate from the auditory receptors of both sides with contralateral predominance. In this cortex a distinct tonotopic representation is exhibited in which higher frequencies are received medially and caudally, while lower frequencies occupy lateral and rostral areas.

The gustatory cortex (Brodmann's area 43), confined to the parietal operculum, receives afferents from the ventral posteromedial thalamic nucleus.

The vestibular cortex is not a distinct cortical center, since the vestibular input is intermingled with other sensations. It is thought to occupy the distal part of the primary sensory cortex, a cortical area adjacent to Brodmann's area 2. The thalamic nuclei which receives somatosensory (VPL and VPI) and motor impulses (VL) also receive vestibular input. This overlap may play a role in the regulation of conscious awareness of spatial relationships at the level of the thalamus.

Secondary sensory cortex

The secondary sensory cortices surround the primary sensory cortices, occupy a smaller area than the primary cortices and receive input from the intralaminar and midline thalamic nuclei. Their topographic representation is either a mirror image or an inverted image, relative to the one perceived by the primary cortices.

The secondary somesthetic area is primarily associated with noxious and painful stimuli. It occupies the superior lip of the lateral cerebral (Sylvian) fissure, distal to the postcentral gyrus. Large and diverse receptive areas convey a variety of sensory impulses to this cortex. Impulses are conveyed bilaterally, with a unilateral predominance, from the posterior thalamic zone and the ventral posterolateral thalamic nuclei. This cortex exhibits a distorted somatotopic arrangement in which the facial region lies adjacent to the corresponding area in the primary sensory cortex.

The secondary visual cortex (Brodmann's areas 18 and 19) surrounds the striate cortex and receives information from the primary visual cortex (Brodmann's area 17) and the pulvinar. The visual impulses, which reach the superficial layers of the superior colliculus, project to the inferior and lateral part of the pulvinar. These impulses eventually terminate in the secondary visual cortex, constituting the extrageniculate visual pathway. Secondary visual cortices are connected to subserve visual memory functions and other components of vision. In the peristriate cortex (Brodmann's area 18) an inverted visual field receptive topography exists, as compared to the striate cortex. It receives input from Brodmann's area 17, the pulvinar, and the lateral geniculate nucleus. It projects to the peristriate cortex of the opposite hemisphere. This cortical area is essential for visual depth perception (stereoscopic vision). The unique bilateral representation of the visual image is achieved by the interhemispheric connections of the peristriate cortices through the corpus callosum. This bilaterality, which is not mediated by the lateral geniculate nucleus, ensures that no gap exists between the single image generated by both eyes. The parastriate cortex (Brodmann's area 19) surrounds Brodmann's area 18, maintaining identical retinotopic representation.

The secondary auditory cortex (Wernicke's zone-Brodmann's area 22) surrounds the primary auditory cortex, and receives afferents from the dorsal and medial subnuclei of the medial geniculate nucleus. This cortical area maintains reciprocal connections with the opposite hemisphere.

Motor cortex

The motor cortex is also known as agranular cortex because of the masking (attenuation) of the granular layers, particularly the inner granular layer. It occupies most of the frontal lobe and exerts control over the axial and appendicular muscles. It has a number of subclassifications, which include the primary and supplementary motor cortices, as well as premotor area and motor eyefield.

Contralateral flaccid palsy, followed by a gradual spasticity, Babinski's sign, increased deep tendon reflexes, and clonus are the prominent (upper motor) signs of lesions of the primary motor cortex. Recovery from such injury is only moderate.

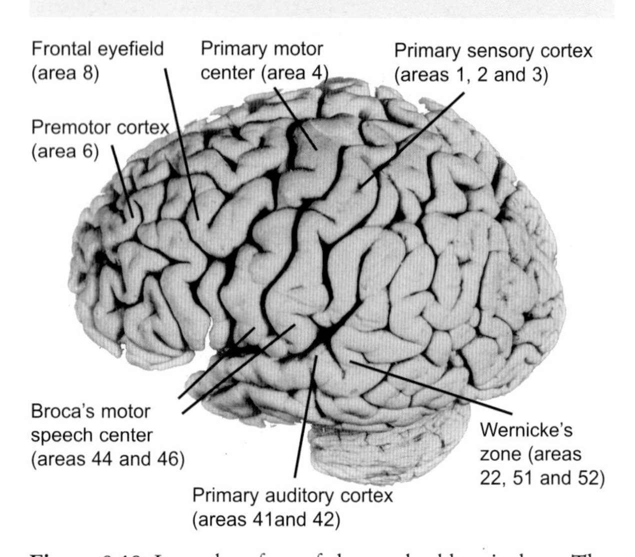

Figure 8.19 Lateral surface of the cerebral hemisphere. The main gyri of the frontal temporal and parietal lobes are indicated. Speech areas, frontal eyefield, and auditory cortex are shown

The primary motor cortex (Brodmann's area 4), represented by the precentral gyrus and part of the paracentral lobule, contains the giant pyramidal cells of Betz that project to the lumbosacral segments of the spinal cord (Figures 8.19 & 8.32). Like the primary sensory cortex, the body is also represented here in a distorted fashion and arranged according to the relative innervation density (motor homunculus). In this homunculus, the foot, leg, and the thigh occupy the anterior part of the paracentral lobule, whereas the gluteal region, trunk, upper extremity, followed by the hand, digits, and the head in a descending fashion, on the precentral gyrus. On the lower end of this homunculus, the tongue, muscles of mastication, and the larynx are designated. A brief glance at the homunculus reveals disproportionately large areas for the hand and especially the thumb, as well as the face. Ablation of the precentral gyrus results in spastic palsy (increased muscle tone in the antigravity muscles) on the contralateral side. Approximately up to 30% of corticospinal tract fibers arise from the primary motor cortex. The connection of the primary and secondary somatosensory cortices to the primary motor cortex enables the ventral posterolateral nucleus to convey information to the motor cortex.

Table 8.2 Motor cortex

Cortical areas	*Major afferents*	*Major efferents*	*Function*	*Deficits*
Precentral & postcentral gyri, and paracentral lobule	Premotor (Brodmann's area 6), postcentral gyrus (areas 3, 1, 2), contralateral (Brodmann's area 4, and Brodmann's areas 5 & 7), ventral lateral & ventral anterior nuclei	Corticospinal, corticopontine, corticotegmental, corticothalamic to ventral lateral and ventral anterior nuclei	Provide motor control to the α motor neurons & regulate reflexes and muscle tone	Contralateral spastic palsy, Babinski's sign, hyperreflexia, clonus & loss of superficial abdominal and cremasteric reflexes
Premotor cortex (Brodmann's area 6)	Precentral gyrus, postcentral gyrus, superior parietal lobule (areas 5 & 7), ventral anterior and ventral lateral nuclei	Corticostriate, ventral anterior & ventral lateral nuclei	Motor for skeletal muscles	Stimulation of this area produces generalized movements. When area 4 is destroyed it mimics the activities of area 4
Frontal eyefield (Brodmann's area 8)	Motor, premotor & visual cortices	Other parts of the cortex	Voluntary eye movements conjugately to the opposite side	Both eyes conjugately move to the side of the lesion
Supplementary motor cortex	Postcentral gyrus (areas 3, 1, 2), precentral gyrus (area 4),VL & VL nuclei	Area 4, striatum, spinal cord via the corticospinal tract	Controls contraction of the postural muscles	Ablation produces spastic contracture of flexor muscles & grasp reflexes

Specific projections to lamina V of the primary motor cortex arise from the ventrolateral (VL) nucleus of thalamus. Other cortical areas such as Brodmann's areas 5 and 6 also project to the motor cortex. Thus, the VL nucleus conveys the input received from the cerebellar nuclei to the motor cortex.

The supplementary motor cortex (Brodmann's areas 8 & 9), a duplication of the primary motor cortex which occupies the medial frontal gyrus (medial part of the superior frontal gyrus), may overlap Brodmann's areas 4 and 6. It mediates contraction of the postural muscles on both sides. It also plays an important role in the planning and initiation of movements. It becomes active even if the intended movement did not occur.

> Ablation of the supplementary motor cortex produces an increase in flexor muscle tone, leading to spasmodic contracture and pathologic grasp reflexes on both sides. The face and upper limb is represented anteriorly, lower limb posteriorly, and the trunk occupies a more inferior position in this cortex.

The premotor cortex (Brodmann's area 6) occupies part of the superior frontal gyrus and maintains functional and topographical representations similar to the primary motor cortex (Figures 8.19 & 8.32). The motor programs in this cortex regulate the activities which are essential for any motor activity such as the rhythm and strength of contraction of the muscles. The frontal motor eyefield (Brodmann's area 8) controls conjugate eye movement to the opposite side (Figure 8.19).

Association cortices includes cortical areas that are located between visual, auditory, and somatosensory cortices which integrate generated auditory, visual, gustatory, and general sensory impulses. This integration serves a variety of functions, including recognition of shape, form, texture of objects, awareness of body image, relationships of body parts to each other and their location. These cortical areas also regulate the conscious awareness of body scheme, physical being,

Table 8.3 Association cortex

Cortical areas	*Major afferents*	*Major efferents*	*Function*	*Deficits*
Superior parietal lobule (Brodmann's areas 5 & 7)	Lateral posterior, lateral dorsal nuclei; Brodmann's area 3, 1, 2 (postcentral gyrus	Corticotegmental & corticopontine pathways	Integration of general sensory and visual information	Contralateral astereognosis & inability to recognize writing on skin although ability to feel and appreciate general weight & temperature remains unchallenged
Brodmann's areas 18 & 19	Brodmann's area 17 and pulvinar	Midbrain tegmentum, tectum, pontine nuclei & pulvinar	Interpretation of visual information & regulation of optokinetics & accommodation reflexes	Inability to recognize an object in the opposite field of vision. Deficits in optokinetic & accommodation reflexes. Stimulation of this area produces visual hallucinations
Wernicke's speech center (Brodmann's area 22)	Primary auditory cortex (Brodmann's areas 41 & 42) & other cortical areas	Other association cortices	Integrates auditory & visual information	Sensory aphasia may result from unilateral lesion, but is most pronounced when the damage is bilateral
Prefrontal cortex	Dorsomedial nucleus & other cortical areas	Pontine nuclei, dorsomedial nucleus & other cortical areas	Deals with personality, drive, emotion, intellect Affective component of pain	Marked changes in personality, behavior & judgement, most evident when lesions are bilateral. Stimulation produces changes in blood pressure, respiratory rate, gastric motility
Cingulate gyrus	Anterior nucleus of thalamus	Entorhinal cortex	Autonomic manifestations & pain perception	Vascular changes & changes in temperature regulation
Broca's speech center (Brodmann's areas 44 & 45)	Wernicke's zone & other cortical areas	Contralateral Broca's center & motor nuclei of cranial nerves	Phonation	Broca's aphasia
Inferior parietal lobule (Brodmann's areas 39 & 40)	Pulvinar & secondary visual cortex	Prefrontal & premotor and insular cortices	Spatial & three-dimensional perception	Inability to draw, place blocks, orient spatially or identify contralateral body parts

and recognition and comprehension of language symbols. They may also be involved in planning of motor functions and modulation of sensory impulses. Association cortices encompass the superior parietal lobule (Brodmann's areas 5 & 7), inferior parietal lobule which comprises the supramarginal (Brodmann's area 40) and the angular gyri (Brodmann's area 39), the posterior part of the superior temporal gyrus (Wernicke's zone – Brodmann's area 22), and the secondary visual cortex (Brodmann's areas 18 & 19).

Cortical afferents

Cortical afferents are derived primarily from the spinal cord, cerebellum, and basal nuclei. They project to the cerebral cortex via the thalamocortical radiations (peduncles). The cortical afferents that originate from the spinal cord include the dorsal column-medial lemniscus and the spinal lemniscus. These afferents convey impulses to the sensory cortex via thalamocortical fibers that emanate from the ventral posterolateral nucleus. Other afferents from the spinal trigeminal and principal sensory nuclei are conveyed through the trigeminal lenmnisci to the ventral posteromedial thalamic nucleus. These fibers also project to the sensory cortex via thalamocortical fibers. Afferents, which carry motor information, are conveyed to the motor and premotor cortices from the cerebellum and basal nuclei via neuronal axons of the ventral lateral and ventral anterior thalamic nuclei. The limbic lobe and prefrontal cortex receive afferents, subserving emotion, behavior, memory and mood, from the anterior and dorsomedial nuclei of the thalamus. Visual and auditory cortical afferents, derived from the lateral and medial geniculate nuclei, are carried by the optic and auditory radiations, respectively. The sleep–wake cycle is partly regulated by cortical afferents from the intralaminar thalamic nuclei within the ascending reticular activating system. Noradrenergic afferents that arise from the locus ceruleus project to all cortical areas and laminae (especially to lamina I) via the central tegmental tract and internal capsule, bypassing the thalamus. These noradrenergic fibers may inhibit the moderately active cortical neurons and enhance the signal-to-background ratio. They do not affect the cortical sensory neurons.

An important and consistent feature of the thalamocortical fibers is their organization into the superior, inferior, anterior, and posterior peduncles. The superior peduncle contains fibers that connect the ventral posterior nucleus to the postcentral gyrus, and the ventral anterior nucleus to the premotor

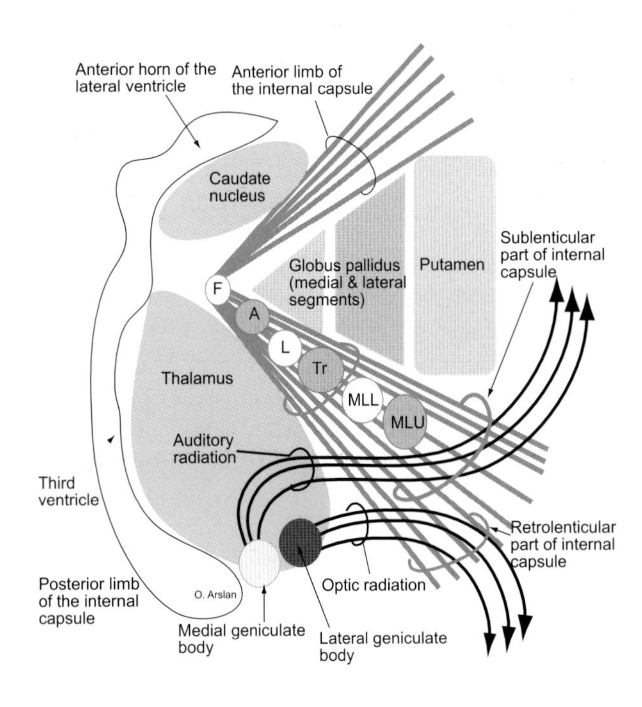

Figure 8.20 Topographic location of the internal capsule and its main constituents. Note the massive number of fibers crossing through the posterior limb of the internal capsule. F, face (corticobulbar fibers); A, arm; L, lower extremity; Tr, trunk (corticospinal fibers); MLU, medial lemniscus fibers from the upper extremity; MLL, medial lemniscus fibers from the lower extremity

cortex. It also contains fibers that connect the ventral lateral nucleus to the motor cortex, and the lateral dorsal and lateral posterior nuclei to association cortices of the parietal lobe. The inferior peduncle contains the auditory radiations, projecting from the medial geniculate nucleus to the transverse gyri of Heschl via the sublenticular part of the internal capsule. The anterior peduncle contains fibers connecting the dorsomedial and anterior thalamic nuclei to the cingulate and prefrontal cortices. The posterior peduncle runs in the retrolenticular part of the internal capsule, connecting the lateral geniculate nucleus to the primary visual cortex of the occipital lobe.

Cortical efferents

Massive cortical efferents project to the spinal cord, brainstem, and the thalamus. Some of these fibers are destined to the spinal cord, while others terminate in a subcortical area (e.g. thalamus, brainstem, etc.)

A prominent projection of the cortex is represented in the corticospinal tract (described in detail later with the upper motor neurons) which originates from motor, premotor, and somesthetic areas of the cerebral cortex and descends in the ventral part of the brainstem. Most fibers of the corticospinal tract decussate at the level of the caudal medulla, forming the lateral corticospinal tract, while the remaining ipsilateral fibers constitute the anterior and anterolateral corticospinal tracts.

The corticobulbar tract, also described with upper motor neurons, is a cortical pathway that acts upon the motor nuclei of the cranial nerves.

The corticothalamic tracts form part of the reciprocal pathway that connects certain areas of the cerebral cortex to the thalamic nuclei. These corticofugal pathways include projections from the primary motor cortex (Brodmann's area 4) to the ventral lateral and centromedian nuclei, as well as projections from the prefrontal cortex to the dorsomedial nucleus. They also encompass projections from the cingulate gyrus to the anterior nucleus and from the premotor cortex (Brodmann's area 6) to the ventral anterior and ventral lateral nuclei. In addition, cortical projections from premotor and motor cortical areas to the intralaminar nuclei, and from the primary sensory cortex to the ventral posterolateral and ventral posteromedial nuclei, are also included. Projections from the primary auditory and visual cortices to the medial and lateral geniculate nuclei respectively constitute additional corticothalamic pathways. It is interesting to note that the thalamic reticular nucleus receives afferents from all areas of the cortex with no reciprocal projections to these areas.

Fibers of the corticopontine tract that project to the pontine nuclei maintain diverse origin from the pyramidal neurons in layer V of the primary motor and prefrontal cortices (frontopontine), primary sensory cortex (parietopontine), temporal lobe (temporopontine), and the primary visual cortex (occipitopontine). These fibers show a somatotopic arrangement in the crus cerebri of the midbrain, in which the frontopontine fibers occupy a medial position to the parieto-occipitotemporopontine fibers. The corticopontine tract, the ipsilateral component of the corticopontocerebellar pathway, enables the cerebral cortex to influence cerebellar activity.

A third group of fibers form the corticoreticular tract which originates from the motor, premotor, visual, and auditory cortices and terminates in the nucleus reticularis gigantocellularis of the medulla and the nucleus reticularis pontis oralis of the pons. This pathway may be partially responsible for signs of decerebrate rigidity, which is seen upon transection of the brainstem at the intercollicular level.

Another cortical projection, the corticotectal tract, consists of fibers from the secondary visual and primary auditory cortices as well as the frontal eyefield that project to the tectum. Visual tracking movements are regulated by fibers from the secondary visual cortex (Brodmann's areas 18 & 19) that project to the superficial layers of the ipsilateral superior colliculus. Control of saccadic eye movements is maintained by the projections of frontal eyefield (Brodmann's area 8) to the middle layers of the superior colliculus. In the same manner, movements of the eye and the head toward auditory stimuli is mediated by the projections of the primary auditory cortex to the deep layers of the superior colliculus.

The corticorubral tract is a component of the corticorubrospinal tract, which originates from the motor and premotor cortices of both hemispheres and descends to terminate somatotopically in all areas of the red nucleus. Note that the ipsilateral corticorubral fibers from the motor cortex mainly terminate in the magnocellular part of the red nucleus, whereas the contralateral fibers from the premotor and motor cortices terminate in the parvocellular part of the red nucleus.

Finally, fibers from all areas of the cerebral cortex form the corticostriate which project bilaterally to all regions of the caudate and putamen. Bilateral corticostriate fibers are primarily derived from the motor, premotor, and sensory cortices, and are somatotopically arranged. In order for the contralateral fibers to reach the striatum, they cross in the corpus callosum and pursue their course in the subcallosal fasciculus. In particular, the caudate nucleus, a component of the striatum, receives input primarily from the prefrontal cortex.

Cortical afferents accompanied by cortical efferents are contained in the internal capsule, a V-shaped bundle which consists of anterior limb, genu, and posterior limb with retro- and sublenticular parts (Figures 8.20, 8.21 & 8.24). The corticothalamic and thalamocortical projections form the bulk of the internal capsule. An important aspect of the internal capsule is the somatotopic arrangement of its fibers. For instance, the anterior limb contains fibers which maintain reciprocal connections with the frontal lobe, the genu consists of corticobulbar and corticospinal, whereas the posterior limb (lenticulothalamic) is much more massive consisting of fibers of the corticospinal and frontopontine pathways, and the superior thalamic radiation (spinothalamic tracts and medial lemniscus). It also contains the optic and auditory radiation in the retrolenticular and sublenticular parts, respectively. Due to this extensive concentration of sensory and motor fibers within the posterior limb, occlusion or hemorrhage associated with a small arterial branch that supplies the posterior limb may lead to profound sensory and/or motor deficits. Visual and auditory deficits may also be observed if the retrolenticular and sublenticular portions are involved. Other deficits, such as spastic paralysis of the muscles of mastication, muscles of facial expression, and palatal muscles may be observed if the corticobulbar fibers are involved.

Diverse arterial sources contribute to the blood supply of the internal capsule. In particular, the posterior communicating artery, which is most commonly hypoplastic, supplies the genu and a portion of the posterior limb of the internal capsule. Additional blood supply to the genu of the internal capsule may be derived from the internal carotid artery. The medial striate artery (recurrent artery of Heubner), which arises from the anterior cerebral artery, and sometimes from the middle cerebral artery, pierces the anterior perforated substance to supply the rostral part of the anterior limb. The anterior choroidal artery, a branch of the internal carotid artery, runs above the uncus along the optic tract, and enters the inferior horn of the lateral ventricle via the choroidal fissure, to supply the posterior limb, including the retrolenticular part. The lateral striate (lenticulostriate) artery supplies the anterior limb and the dorsal part of the posterior limb of the internal capsule.

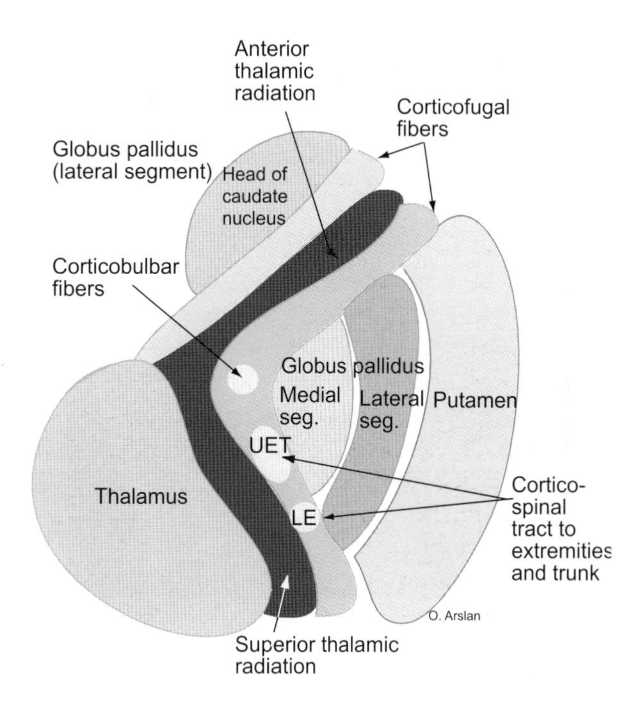

Figure 8.21 The internal capsule. Note its location between the thalamus, caudate nucleus, and the lentiform nucleus. seg., segment

Cerebral white matter

The white matter of the cerebral hemispheres lies deep to the gray matter of the cerebral cortex, consisting of nerve fibers and glial cells. It contains the basal nuclei and the central branches of the cerebral arteries. The fibers that course within the white matter are classified into commissural, association, and projection (Figure 8.22).

Occlusion of the anterior cerebral artery proximal to the origin of the recurrent artery of Heubner may produce no detectable deficit if the collateral circulation from the corresponding artery of the opposite side is maintained. However, a lack of efficient anastomosis may result in infarction of the anterior limb of the internal capsule and destruction of the cortico-ponto-cerebellar tract, producing frontal dystaxia (partial ataxia in which the patient exhibits difficulty in controlling voluntary movements.

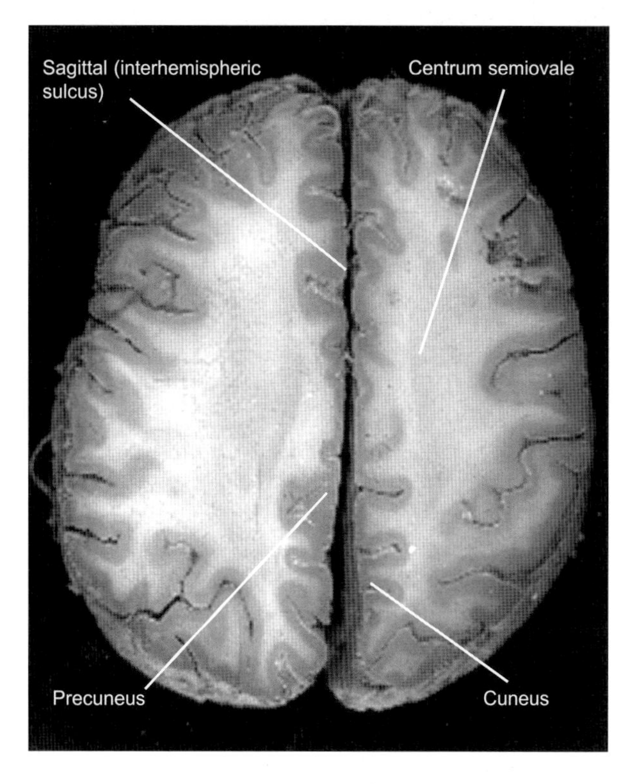

Figure 8.22 Horizontal section of the brain indicates the distinction between the central white matter and the peripheral cortical areas

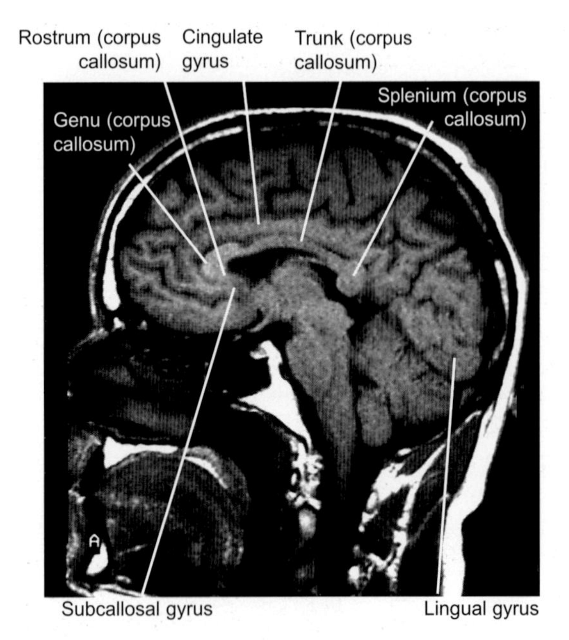

Figure 8.23 Inverted MRI showing components of the corpus callosum and their relationships to individual cerebral lobes, gyri and the septum pellucidum

Commissural fibers

The commissural fibers interconnect identical or non-identical areas of the two cerebral hemispheres, forming the corpus callosum, anterior commissure, and the hippocampal commissure. Interhemispheric communication that mediates learning process is accomplished to a great extent by the corpus callosum, the largest of all commissures, which interconnects the cerebral hemispheres (Figures 8.23 & 8.25). However, exceptions exist in regard to this fact, for instance, the striate (primary visual) cortex and the hand area of the cerebral cortex do not project commissural fibers through the corpus callosum.

The anterior commissure (Figures 8.24 & 8.26) is embedded in the upper part of the lamina terminalis, superior to the optic chiasma, resembling the handle of a bicycle. It divides the fornix into precommissural and postcommissural columns. This commissure exhibits an anterior smaller portion, which connects the olfactory tracts to the anterior perforated substances of both sides, and a posterior larger portion, interconnecting the parahippocampal and the middle and inferior temporal gyri, as well as the amygdaloid nuclei. In addition, the anterior commissure interconnects the anterior olfactory nuclei, diagonal bands of Broca, olfactory tubercles, prepyriform cortices, entorhinal areas, nucleus accumbens septi, bed nuclei of stria terminalis, and the frontal lobes.

Disconnection syndrome (split brain) encompasses a constellation of symptoms, which result from interruption of the interhemispheric commissures or intrahemispheric connections. This syndrome may be produced by a midsagittal section of the corpus callosum, a procedure commonly used in the past as a treatment for intractable epilepsy. Infarction of the pericallosal branch of the anterior cerebral artery may result in similar deficits, producing two hemispheres that function independently. This functional independence involves perception, cognition, mnemonic, learned, and volitional activities. Therefore, information generated in the non-dominant hemisphere is expressed only in a non-verbal manner and could not be expressed in writing or speech, as the dominant hemisphere has a major role in linguistic expression. In individuals with divided corpus callosum, most voluntary daily activities, native intelligence, memory, verbal reasoning, and temperament are not generally affected. Presence of bilateral sensory representation and the compensatory development of bilateral motor representation may explain intactness of these functions.

Various forms of agnosia may also be associated with lesions of the corpus callosum. Owing to interruption of the callosal fibers and the fact that information is not transferred from the right hemisphere to the speech center in the left hemisphere, patients may exhibit an inability to name objects presented into the left visual field or placed in the left hand. Affected individuals are also unable to carry out verbal commands with the left hand, but are capable of naming objects presented into the right visual field or hand. Thus the inability to match an object in one hand (or seen with one eye) with one placed in the other hand (or seen with the other eye) becomes evident. These observations indicate that information received or generated by the non-dominant hemisphere could not be expressed verbally or in writing, and that visual information is not transferred between the two hemispheres. Comprehension of spoken and written languages is not affected due to bilateral representation.

Another smaller, yet important posterior commissure exists which lies caudal to the pineal gland and rostral to the superior colliculi. It contains fibers, which arise from the habenular and Darkschewitsch nuclei, interstitial nucleus of Cajal, and the nucleus of the posterior

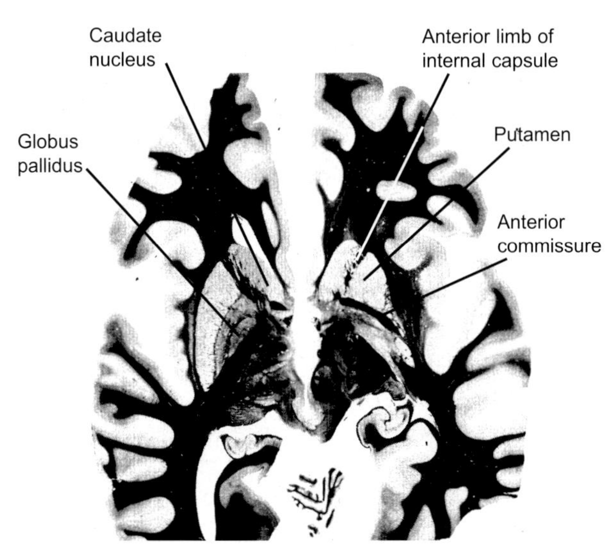

Figure 8.24 The course of the anterior commissure as a midline structure and its continuation inferior to the lentiform nucleus is illustrated

Occlusion of the posterior cerebral artery may result in disruption of the splenium of the corpus callosum, preventing transfer of information from the right to the left visual cortices, a process essential for the visual recognition of objects. Therefore, patients with this type of lesion are able to read, but unable to write or name colors. In normal individuals, visual stimuli elicit non-verbal associations (tactile, taste or smell) which are transmitted across the intact anterior part of the corpus callosum.

Destruction of the anterior part of the corpus callosum (e.g. due to infarction of the anterior cerebral artery (anterior cerebral artery syndrome), distal to the origin of the anterior communicating artery, prevents verbal information to be conveyed from language centers of the left hemisphere to the appropriate centers of the right (mute) hemisphere. This may produce left arm apraxia in which the patient is unable to identify numbers or letters written by a blunt object on the affected left extremity, or perform movements using the left arm and leg upon verbal or written commands, although normal spontaneous movements are maintained. Due to infarction of the rostral part of the corpus callosum, a patient with left arm apraxia is also unable to name an object placed in the left hand, or write or print with the left hand. Damage to the corpus callosum may also occur as a result of excessive consumption of red wine (Marchiafava–Bignami syndrome), producing subtle and variable manifestations.

commissure. It also contains fibers from the superior colliculi and the pretectal nuclei.

Additional interhemispheric linkage is secured by the hippocampal commissure, which extends between the crura of the fornix and hippocampal gyri, inferior to the splenium of the corpus callosum.

Association fibers

Association fibers establish linkage between areas within the same cerebral hemispheres, consisting of short and long fasciculi, with the short fasciculi interconnecting adjacent gyri within the same cerebral hemisphere. Long fasciculi, which include the cingulum, uncinate, superior longitudinal, inferior

Complete failure of the corpus callosum to develop, agenesis of the corpus callosum, may occur during the fourth to the twelfth week of development. Agenesis of the corpus callosum, a congenital anomaly, is accompanied by a partial or complete absence of the cingulate gyrus and septum pellucidum, or the appearance of ipsilateral longitudinal fibers, which include some that fail to cross the midline. This form of agenesis is not commonly associated with neurological deficits, although mild mental retardation, seizure, and motor deficits may be seen. Another form of this condition is associated with development of small, multiple gyri (micropolygyria) and heterotopias of the gray matter. In general, agenesis of the corpus callosum is associated with inherited metabolic disorders (pyruvate dehydrogenase deficiency, glutaric aciduria type II), and with chromosomal abnormalities as in Dandy–Walker, Aicardi's, and Cogan syndromes.

In Dandy–Walker syndrome, the cerebellar vermis fails to develop and the fourth ventricle is replaced by a cyst-like midline structure. Patients with this condition exhibit thinning of walls of the posterior cranial fossa and the roots of the cervical spinal nerves assume an ascending course in order to reach their corresponding foramina. As mentioned earlier, this syndrome may also be associated with agenesis (failure of development) of the corpus callosum and related manifestations.

Both Aicardi's syndrome, which is characterized by seizure, microcephaly, and vertebral defects such as hemivertebra, and Cogan syndrome that manifests keratitis (inflammation of the cornea) and vestibulo-auditory disorders, may show signs of agenesis of the corpus callosum.

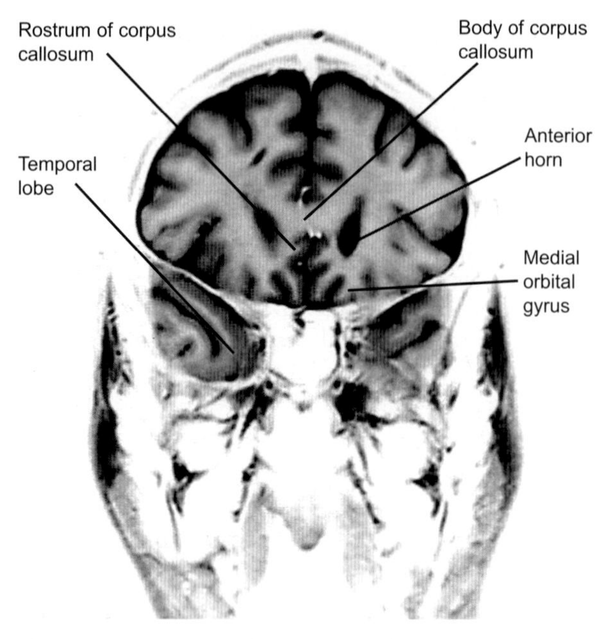

Figure 8.25 An inverted MRI scan (coronal section) of the brain through the anterior horn of the lateral ventricle. Observe the body and rostrum of the corpus callosum, individual cerebral lobes, gyri and the septum pellucidum

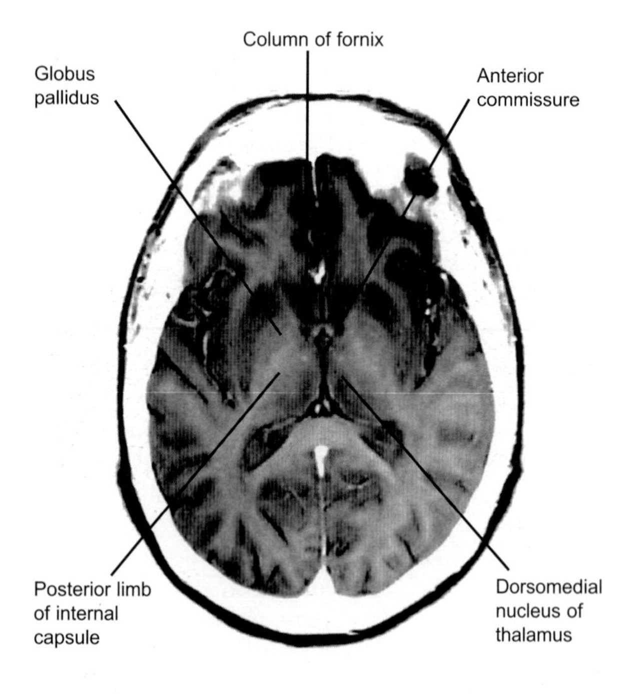

Figure 8.26 An inverted MRI scan (horizontal view) of the brain through the frontoparietal operculum. Note the curved midline anterior commissure rostral to the columns of the fornix. The splenium of the corpus callosum and the posterior horn of the lateral ventricle are also visible

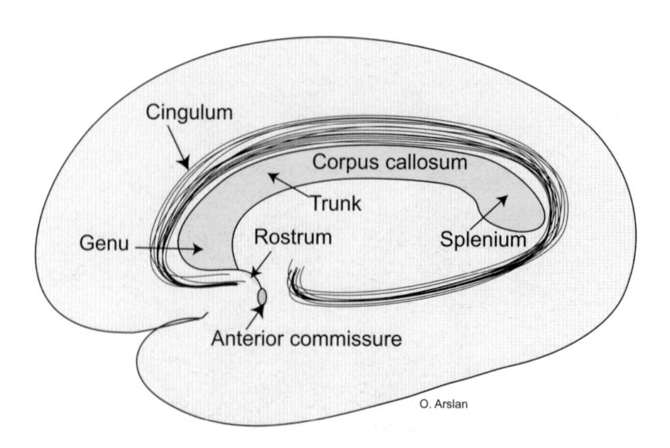

Figure 8.27 The cingulum within the cingulate gyrus. Observe its relationship to the corpus callosum

longitudinal, superior and inferior occipitofrontal fasciculi, connect distant areas of the cerebral hemispheres.

Structures that comprise the limbic system such as the subcallosal and paraterminal gyri rostrally and the parahippocampal and adjacent temporal gyri caudally are interconnected by the cingulum, a curved bundle of association fibers that lies deep to the cingulate gyrus and follow its course (Figures 8.27 & 8.28). On the other hand, Broca's speech center is connected to the rostral parts of the temporal gyri via the uncinate fasciculus, which enables input generated/or integrated in the temporal lobe to affect articulation (Figures 8.29, 8.30 & 8.31).

A group of large association fibers that course lateral to the corona radiata and the internal capsule form the superior longitudinal (arcuate) fasciculus, connecting the rostral parts of the frontal lobe to the secondary visual cortex (Brodmann's area 18, 19) and the parietal and temporal gyri (Figures 8.28, 8.30 & 8.31).

Another group of association fibers forms the inferior longitudinal fasciculus (Figure 8.28) which extends from the secondary visual cortex to the inferotemporal cortex, coursing lateral to the occipital horn of the lateral ventricle, and is separated by the optic radiation and tapetum.

Frontal and occipital lobes are also connected by the superior occipitofrontal (subcallosal) fasciculus that lies inferior and lateral to the corpus callosum. This fasciculus is separated from the superior longitudinal fasciculus by the corona radiata.

An additional bundle, the inferior occipitofrontal fasciculus, connects the frontal to the occipital lobe, and runs within the temporal lobe proximal to the uncinate fasciculus. It has been suggested that the uncinate fasciculus is part of the inferior occipitofrontal fasciculus (Figure 8.31).

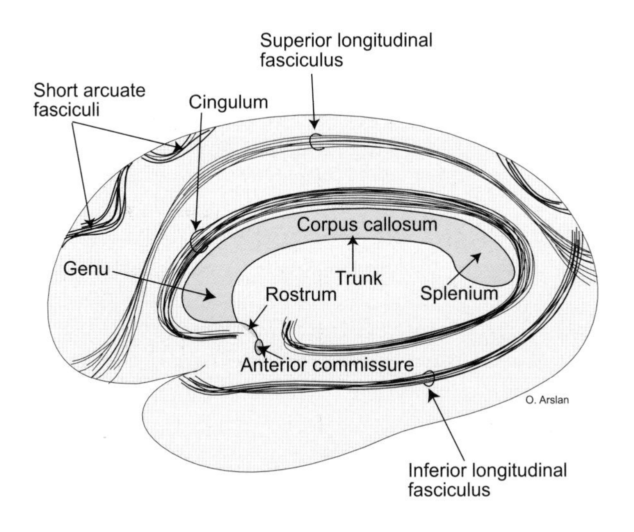

Figure 8.28 Midsagittal drawing of the brain. The superior and inferior longitudinal fasciculi are illustrated

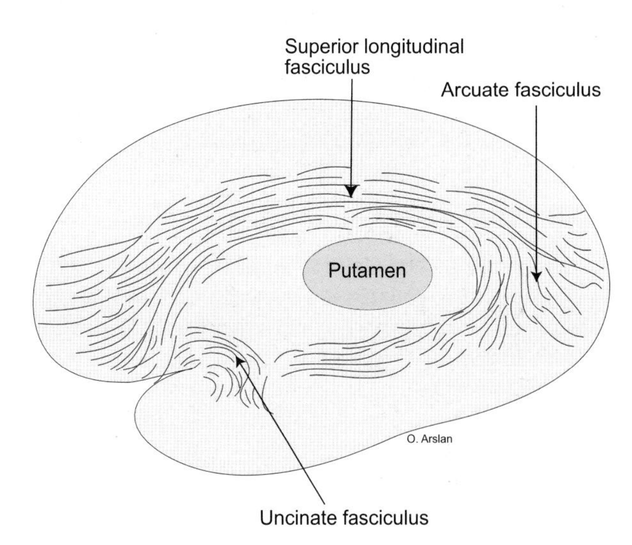

Figure 8.30 Schematic drawing of some of the association fibers within the cerebral hemisphere

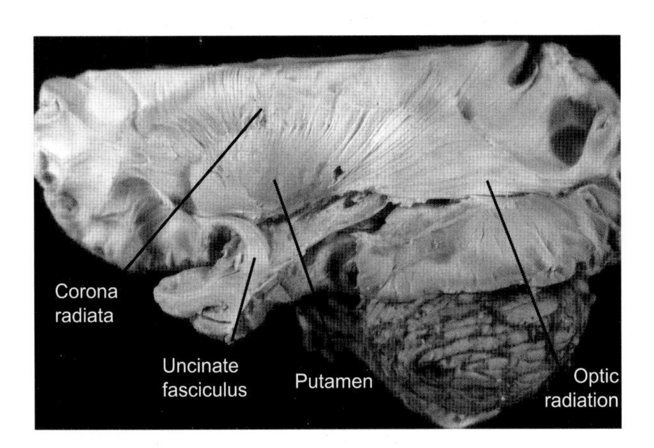

Figure 8.29 The lateral surface of the brain after removal of the frontal and temporal gyri. The fibers of the corona radiata are shown

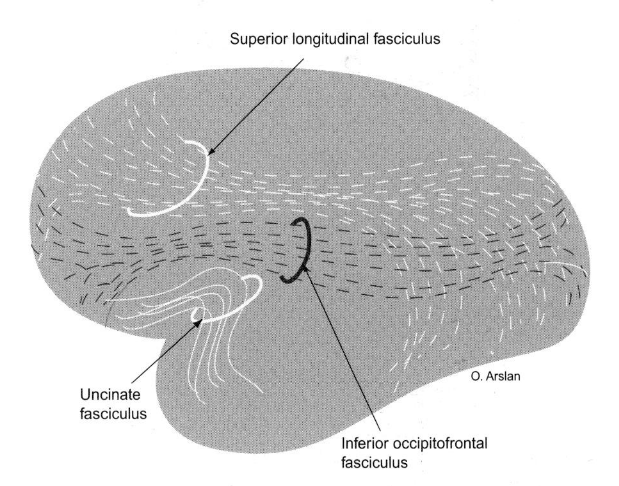

Figure 8.31 Diagram of the inferior occipitofrontal, superior longitudinal, and uncinate fasciculi are illustrated to emphasize their relative locations

Bilateral disruption of the superior longitudinal fasciculi may produce visual agnosia, which is characterized by the inability to name an object or describe its function.

Projection fibers

Projection fibers are corticofugal axons which extend from the cerebral cortex to the subcortical nuclei, brainstem, and the spinal cord. They form the corona radiata that runs between the superior longitudinal and the superior occipitofrontal fasciculi. They run within the internal capsule and may descend to terminate in the spinal cord or other subcortical areas. These projection fibers include the corticospinal, corticobulbar, corticostriate, corticopontine, etc.

Cerebral dysfunctions

Cerebral lesions occur in a variety of diseases and conditions such as Alzheimer's disease, post-encephalitis, Parkinsonism, and pseudobulbar palsy (associated with destruction of the corticobulbar tracts), and progressive supranuclear palsy. These diseases may produce speech and language disorders that may include speech derangement as in Alzheimer's disease, palilalia (compulsive repetition of a phrase with increasing speed and decreasing volume) as in progressive supranuclear palsy, and aphasia. Other cerebral lesions produce apraxia (inability to perform familiar motor activity without sensory or motor damage) or agnosia (inability to recognize objects and symbols despite intactness of sensory pathways). Cortical dysfunctions may also produce dementia, and seizures which will be discussed later in this chapter.

Neuronal sequence in verbal identification

The following structures and/or centers are involved in an orderly manner in naming a visual object:

Object → optic nerve & tract → lateral geniculate nucleus → Brodmann's area 17 of the visual cortex → Brodmann's area 18 → angular gyrus (Brodmann's area 39) → Wernicke's area (area 22) → pattern is formed → arcuate fasciculus → Broca's motor speech center (Brodmann's areas 44 & 45) → facial area of the motor cortex → lingual, palatal and laryngeal muscles via the corresponding cranial nerves.

Cerebral dominance

Despite the similarity in the morphologic features of the two cerebral hemispheres and the symmetrical projections of the sensory pathways, each hemisphere maintains specialization in certain higher cortical functions. In most individuals, the posterior part of the superior temporal gyrus expands and shows greater length on the left cerebral hemisphere. A minority of brains exhibit this feature in the right hemisphere.

Identification of objects and comprehension of language may be accomplished by the right (mute, non-dominant or creative) hemisphere, utilizing visual and tactile information. This hemisphere also integrates visual impulses with spatial information and motor activities, as in drawing, and mediates musical tones, facial recognition, construction, and other non-verbal activities. Thus, the non-dominant hemisphere is holistic, concerned with spatial perception, gesturing that accompanies speech (prosody), and recognition of familiar objects. In other words, this hemisphere is holistically creative, lacks details, and defies rules and logic.

Aphasia is the inability to comprehend the spoken and written language, or express thoughts via words despite the fact that sensory systems, mechanisms of articulation and the associated structures (Figures 8.32 & 8.33) are intact. This disorder should not be confused with other speech abnormalities that occur in cerebellar dysfunctions, hypoglossal or vagus nerve damage (changes in speech tone) or upper motor neuron palsy. Generally, the severity of aphasia depends upon the site of cortical damage and duration of the disorder. Although complete recovery does not always occur, the earlier the onset, the better the chance of recovery. Speech areas are comprised of the angular and opercular parts the inferior frontal gyrus (Broca's motor speech center), posterior part of the superior temporal gyrus (Wernicke's receptive speech center), and the fibers that connect these centers in the ipsilateral and contralateral hemisphere. The following disorders involve cognitive functions (prepositional features) of language in the dominant hemisphere (usually the left) hemisphere.

Broca's (non-fluent, expressive, motor, verbal) aphasia (Figures 8.32 & 8.33) is seen in individuals with lesions of Broca's speech center (Brodmann's areas 44 and 45). This condition involves disorders of speech and written language. It is characterized by difficulty initiating or repeating words, restricted vocabulary and grammar, and labored and awkward speech with a tendency to delete adverbs and adjectives as well as connecting words (telegraphic

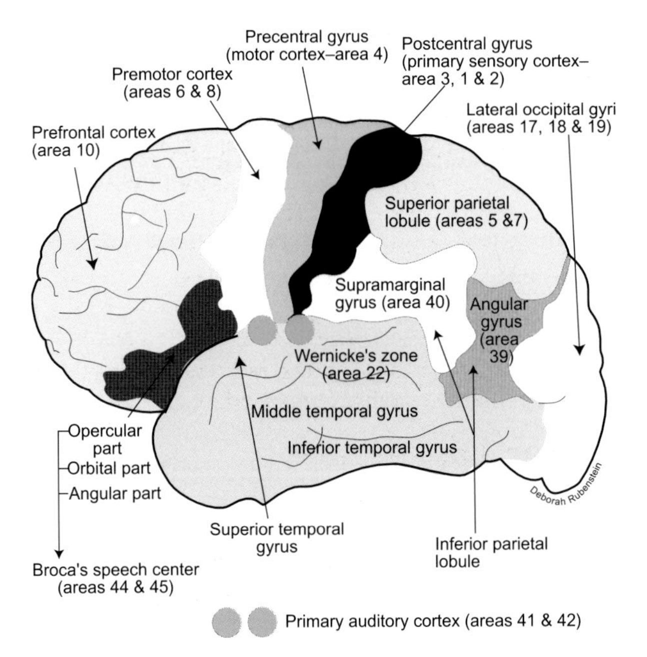

Figure 8.32 Prominent centers and gyri associated with the frontal, parietal, and temporal lobes. The Broca's speech center and Wernicke's zone are distinctly illustrated

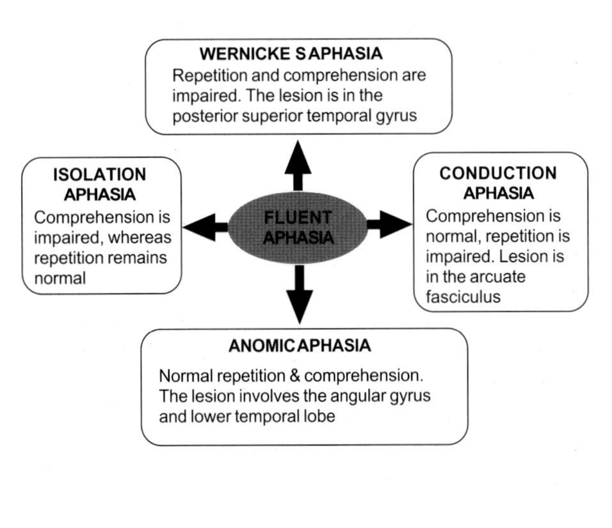

Figure 8.33 Schematic representation of various forms of fluent aphasia

speech). Although comprehension of the spoken and written language is usually preserved, writing is severely affected. Due to disruption of the ipsilateral corticospinal, corticobulbar and optic radiation, Broca's aphasia often is associated with right-sided hemiplegia, supranuclear facial (lower facial) palsy, and right sided homonymous hemianopsia. Tongue and lip movements are usually impaired. Involvement of the frontal eyefield may result in conjugate deviation of both eyes to the left side.

Wernicke's (receptive, fluent, sensory, syntactic, acoustic) aphasia results from destruction of the association cortex that occupies the posterior portion of the superior temporal gyrus (Brodmann's area 22) of the dominant hemisphere (Figures 8.32 & 8.33). Since Wernicke's area integrates verbal memory, visual and sound patterns with the correct word phonemes that are essential elements for reading and writing, destruction of this area may impair the comprehension of spoken language, ability to read or write, and find appropriate words (circumlocution). Patients also exhibit enormous verbal output, and a tendency to enhance speech by irrelevant word substitution such as 'fen for pen' known as verbal paraphasia. Sound transpositions (phonemic paraphasias) or substitution of phoneme (literal aphasia), and increased rate and pressure of speech (logorrhea), as well as unwillingness to terminate speech may also be observed. Although the grammar of the spoken language is not correct, the free usage of a variety of tenses in an unusual combination gives the impression of a speech with proper syntax. For an examiner who does not speak the patient's language, observing any speech dysfunctions may not be an easy task. Since patients remain unaware of the deficit, their speech conveys little meaning. For example, patients with Wernicke's aphasia have trouble repeating statements.

Patients with Wernicke's aphasia may be labeled psychotic because of the similar nature of their speech to a thought disorder of frontal lobe origin. Patients may tend to develop a mania-like psychosis characterized by hyperactivity, rapid speech, euphoria or irritable mood. Some patients with Wernicke's aphasia may exhibit superior quadranopsia (one-fourth blindness), indicating involvement of the geniculocalcarine pathway (Meyer's loop) as it courses through the temporal lobe. Individuals with Wernicke's aphasia may also exhibit a striking lack of concern (a finding that is not seen in individuals with Broca's aphasia) that may be replaced by paranoid behavior.

Anomic (semantic, amnesic) aphasia (Figure 8.33) may occur at the end of Wernicke's aphasia or may be present as a distinct disorder. It is characterized by the inability to find words with relatively intact comprehension. Individuals with this condition do not exhibit paraphasia, but lack substantive words in the speech. It is associated with alexia and agraphia, and occasionally with right superior quadranopsia. It can be caused by an injury to the parieto-occipital cortex, which may extend to involve the angular gyrus in the left dominant hemisphere. This condition may be an early language disturbance detected with expanding brain tumors.

A lesion that disrupts the arcuate fasciculus connecting Wernicke's zone to Broca's speech center causes conduction (central) aphasia (Figure 8.33). The main deficit in this type of aphasia is the inability to repeat words, choose and sequence phonemes. Comprehension of silent reading remains intact, but the ability to read aloud is lost. Reading aloud requires intactness of the association visual cortex, splenium of the corpus callosum, angular gyrus, Wernicke's area, Broca's speech area, arcuate fasciculus, and the motor cortex. Patients clearly become aware of their language deficit, especially when they entangle a key word. Some degree of apraxia may also be observed in the limb and facial muscles.

Transcortical aphasia (Figure 8.33) occurs as a result of compromise of the blood supply to the watershed areas surrounding the speech centers in the frontal and parietal lobes. These areas are supplied by the middle, anterior, and posterior cerebral arteries.

Transcortical sensory aphasia (isolated speech area syndrome) is a rare type of fluent aphasia which may result from a cerebrovascular accident in the watershed areas of the parieto-occipital cortex, and at the junctions of the middle, anterior, and posterior cerebral arteries. This vascular lesion disrupts the connection between Broca's and Wernicke's centers and other parts of the brain. Patients are unable to start conversation, and the response to a question generally contains confabulatory, automatic, irrelevant and repetitious paraphasia. Repetition may take the form of echoing (parrot-like), word phrases, and melodies (echolalia). Although reading and writing abilities are abolished, articulation of memorized materials remains intact.

Transcortical motor (dynamic) aphasia (Figure 8.33) is a form of non-fluent aphasia (speech is reduced to less than 50 words per minute) in which repetition is unaffected, and the speech is grammatically accurate. Inability to initiate conversation is the primary deficit, although comprehension of sounds and written language remains functional. Anoxia or multiple infarctions that disrupts the connections of Broca's area to the rest of the frontal lobe may be responsible for this disorder.

Transcortical motor and sensory aphasias are produced by lesions of the area surrounding the lateral cerebral fissure. Automatic repetition of words (echolalia) is the main speech-related function performed by these types of individuals.

Global aphasia results from a lesion which destroys Broca's and Wernicke's centers, and the connecting arcuate fasciculus. There is loss of ability to comprehend, articulate, read, write, and name a viewed object. It is usually associated with right homonymous hemianopsia, right hemiplegia, and right hemianesthesia.

Aphemia (subcortical motor aphasia) is characterized by the inability to imitate, repeat or produce sounds, with preservation of the ability to read and write. Auditory comprehension and word finding also remain intact. This condition may result from destruction of the output from Broca's speech center.

Subcortical sensory aphasia (pure word deafness) results from destruction of the primary auditory cortex and the transcallosal fibers that carry information from the non-dominant hemisphere. Since Heschl's gyrus is damaged on the left (dominant) hemisphere and no auditory impulses are able to reach Wernicke's zone from the opposite hemisphere, comprehension and repetition of the spoken word are not possible.

Alexia with agraphia is seen in individuals with a lesion involving the angular gyrus in the inferior parietal lobule. It is characterized by impairment of the ability to read and write, and inability to comprehend symbols and words. Comprehension of sounds and the ability to articulate are not affected. Due to the proximity of the angular gyrus to the temporal lobe, as well as other areas of the parietal lobe, additional deficits may also be seen including anomic aphasia, loss of right–left recognition and ability to identify fingers.

Pure alexia or alexia without agraphia (pure word blindness) is a rare condition that occurs as a result of disruption of the visual association cortices (Brodmann's areas 18 & 19) and the splenial fibers that convey visual information to the visual cortex and angular gyrus of the dominant hemisphere. It may be caused by occlusion of the left posterior cerebral artery. Patients exhibit right homonymous hemianopsia due to destruction of the left visual cortex. Disruption of the transcortical splenial fibers may result in the inability of patients to recognize words or letters, even their own. Since visually presented objects excite a variety of sensory systems in the mute hemisphere (tactile, taste, and olfactory) that are conveyed by the intact callosal fibers, naming these objects is possible. Generally, aphasics develop alexia because of the difficulty in comprehending the meaning of words, conversion of grapheme to phoneme (Wernicke's aphasia), or in formulating grammar (Broca's aphasia).

Pure (aphasic) agraphia is a rare condition seen in individuals with lesions of the angular gyrus. Reading remains intact, but writing and spelling is severely affected. It may also be produced by a lesion of the motor association cortex of the frontal lobe.

Disconnection syndromes, as discussed earlier, represent a constellation of deficits seen in complete transection of the corpus callosum. Patients with these syndromes may exhibit agraphia in the left hand, but not the right, as well as apraxia (ability to execute oral commands or perform familiar tasks is lost in the absence of sensory or motor deficits).

Hemi-optic aphasia is characterized by the inability to name objects seen in the left visual field, while maintaining the ability to recognize these objects via the left hand which is guided by the right hemisphere. This syndrome is seen in individuals with bilateral epilepsy and subsequent to surgical transection of the corpus.

Tactile aphasia is a condition in which the patient is unable to identify objects placed in the left hand, but is able to do so when the object is placed in the right hand. As the right non-dominant hemisphere, which has lost its connection to the left hemisphere (due to disruption of the callosal fibers), is responsible for the recognition of an object, identification of an object placed in the left hand will not be possible.

In 95% of males and 80% of females, the dominant hemisphere is the left. It is less creative and designed to carry out sequential analysis. It is conceived to comprehend spoken and written languages and to express thoughts into words. Additionally, sequencing of phonemic and syntactical characteristics of language, mathematical calculations, analytical functions, and fine skilled motor activities are regulated by the left hemisphere. Right-handed individuals (dextrals) constitute 80% of the population, 10% are left-handed, and the other 10% of the population are ambidextrous. In approximately 90–97% of individuals who use primarily their right hands, the left hemisphere is dominant for language. The other 3–10% of right-handed persons have the speech center in the right hemisphere. In 60–65% of left-handed individuals (sinistrals) the speech center is located in the left hemisphere; 20–25% have the speech center in right hemisphere; and in 15–20% of the population, the speech center is bilateral. In 60% of ambidextrous individuals the speech center is located in the left hemisphere, in 10% of this population it is positioned in the right hemisphere, and in 30% of these individuals it lies in both hemispheres. Recovery from aphasia in sinistrals is more complete than dextrals. Positron emission tomography (PET) may be used to detect the increased blood flow into the dominant hemisphere.

Aprosodia is a condition in which the affective component of language, such as musical rhythm, facial expression, and comprehension of these gestures is lost or impaired. These disturbances accompany lesions of the non-dominant hemisphere (usually the right). A lesion that involves the inferior frontal gyrus of the right hemisphere (mirror image of Broca's area) produces speech which is monotonous, agestural (lacks gestures), and lacks melody, timing, and other affective content (motor aprosodia). These individuals tend to have difficulty in conveying tune, singing, and understanding emotional reactions. They exhibit hemiplegia and lower facial palsy on the left side. A lesion which destroys the posterior part of the superior temporal gyrus in the right hemisphere (mirror image of the Wernicke's zone) may result in the inability to recognize and comprehend the emotional content of the spoken language (sensory aprosodia). The emotional gestures produced will not be appropriate for the occasion or the content of the speech.

Alexithymia, frequently seen in patients with psychosomatic disorders, is characterized by the inability to express emotion through words. Disruption of the connection between the affective (right) hemisphere and the expressive left hemisphere may account for this deficit.

Basal nuclei

The basal nuclei are embedded in the white matter of the cerebral hemispheres. Together with the substantia nigra, red and subthalamic nuclei, reticular formation, and the claustrum, they comprise the subcortical motor system. This system is concerned with stereotyped movements, suppression of cortically induced movements, regulation of posture, and adjustment of muscle tone. The basal nuclei influence the motor activity by projecting to specific nuclei of the thalamus which in turn deliver the received impulses to the cerebral motor and premotor cortices. Specifically, the basal nuclei consist of the corpus striatum and amygdala. The amygdala maintains connections with the striatum, thalamus, cerebral cortex, and to the structures that constitute the limbic system (see Chapter 17). The corpus striatum comprises the globus pallidus, caudate nucleus, and the putamen. Both the caudate nucleus and putamen form the neostriatum, representing the afferent portion of the basal nuclei, whereas the lentiform nucleus refers to both the putamen and the globus pallidus. For further discussion of the basal nuclei, see Chapter 21.

Blood supply of the cerebral hemispheres

Most of the neurological disorders, whether reversible or irreversible, are the result of vascular diseases or accidents. Knowledge of anatomy of the cerebral vessels is of utmost clinical importance, as it is essential in performing and interpreting angiographic imaging, understanding the deficits associated with vascular accidents, and in developing treatment plans. The blood supply to the brain tissue is maintained by the carotid and vertebrobasilar arterial systems. There are specific features of the capillaries associated with these two arterial systems. For instance, these capillaries, are composed of non-fenestrated continuous endothelial cells, connected by tight junctions and encased by perivascular end feet of astrocytes, possessing large number of mitochondria and pinocytotic vesicles. Approximately

Apraxia is a disorder characterized by the inability to execute learned voluntary functions without any detectable motor or sensory deficits, mental disability or comprehension deficits. It is classified into kinetic, ideomotor, parietal, callosal, sympathetic, ideational, constructional, and buccofacial apraxia.

Kinetic apraxia is characterized by the inability to perform fine, skilled movements in one extremity, often due to a lesion of the contralateral primary motor cortex (Brodmann's area 4).

Ideomotor apraxia is a disorder in which patients exhibit an inability to perform a given task upon command, despite retaining aptness to execute acts automatically such as opening or closing the eyes. This deficit may be categorized into parietal, callosal, and sympathetic apraxia.

Parietal apraxia is due to destruction of the arcuate fasciculus that establishes connection between the motor centers in the frontal lobe and the centers that formulate motor activities in the parietal lobe. This form of apraxia is bilateral and may be associated with conduction aphasia.

Callosal apraxia, as the name indicates, is produced by disruption of the genu of the corpus callosum that connects the premotor areas of both hemispheres. Thus, information generated in the upper extremity area of the premotor cortex of the left frontal lobe would not reach the corresponding area of the right frontal lobe. Since motor activity is conveyed via the corticospinal tract (which is crossed), this deficit will be seen in the left arm.

Sympathetic apraxia is another form of ideomotor apraxia, resulting from damage to the premotor area of the left frontal lobe, an area adjacent to Broca's speech center. As discussed earlier, the premotor area of the left lobe provides motor commands to the corresponding area on the right side via the corpus callosum. Since the corticospinal tract is partly derived from the premotor area, the generated motor impulses will be conveyed to the spinal levels on the contralateral side. Therefore, destruction of the left premotor area produces a paralyzed right limb and an apractic left limb (in sympathy for the affected right limb). Patients with this lesion may also exhibit similar deficits in the buccofacial muscles, and possible aphasia due to proximity of the lesion to Broca's center in the left hemisphere.

Ideational apraxia is characterized by the inability to perform a complex task or series of acts in a purposeful manner and in a proper sequence. It is caused by a lesion in the parietal lobe of the dominant hemisphere or as a result of a diffuse brain disorder (dementia). It stems from the loss of ability to appreciate and formulate the idea necessary to carry out a complex task, although individual acts within the task can be executed without any difficulty.

Agnosia refers to the failure to recognize and understand the symbolic significance of sensory stimuli, despite the presence of the learned skill, intactness of the sensory receptors and pathways, and the absence of mental disorders or dementia. This deficit is commonly associated with disruption of the connection of the primary, secondary, or tertiary sensory cortices with the association cortical areas that store memories for the stimulus. Agnosias are classified into visual, auditory and tactile agnosia, simultanagnosia, anosognosia, Babinski's agnosia, reduplicative paramnesia, asymbolia, and apractognosia.

Visual agnosia is the inability to recognize objects by vision, while retaining the capability to identify the same objects with other sensory modalities. Individuals with this condition do not have defects in the visual apparatus or pathway, and can recognize people. This disability may change in severity from time to time. Lesions are generally confined to the visual association cortex of the temporal and parietal lobes and the connecting fibers to this cortex. Visual agnosia, a primary characteristic of Klüver–Bucy syndrome, may take the form of alexia, protopagnosia, facial and finger agnosia, graphagnosia, color, auditory, and tactile agnosia.

Alexia is a form of visual agnosia that is characterized by the inability to read as a consequence of failure to recognize the written words.

Protopagnosia is a disability resulting from bilateral destruction of the occipitotemporal gyri. Patients with this condition are unable to identify familiar people, or they may identify them as impostors (Capgras syndrome), or as strangers (reverse Fergoli syndrome), or as strangers mistaken for familiar people (Fergoli syndrome). Patients also are unable to identify familiar objects removed from their common visual context.

Facial agnosia is the inability to identify people by their faces, although identification of these individuals is possible via their voice, demeanor, or dress.

Finger agnosia (anomia) is a deficit characterized by the incapacity to identify and name fingers. This dysfunction is often seen in Gertsmann syndrome, and is associated with anarithmetria (loss of arithmetic concept), agraphia or dysgraphia, and left–right disorientation of the body parts and objects. This syndrome is a combination of agnosia and apraxia, and is produced by a lesion of the angular and supramarginal gyri of the dominant cerebral hemisphere.

Graphagnosia refers to the inability to recognize letters or numbers written on the palm.

Color agnosia refers to inability to identify the color of an object, despite retaining the capacity to match cards of different colors. It is not a sex-linked deficit and lesions commonly affect the inferomedial parts of the occipital and temporal lobes. Patients with this deficit do not have color blindness.

Auditory agnosia occurs as a result of bilateral or unilateral destruction of the Wernicke's zone in the dominant hemisphere, and is characterized by the inability to recognize familiar sounds with no detectable deficits in the auditory system or pathway.

Tactile agnosia is characterized by the inability to recognize objects through touch without impairment of the tactile receptors, spinal nerves, or ascending sensory pathways. The primary lesion is in the supramarginal gyrus of the dominant hemisphere (left), and to a lesser degree in the postcentral gyrus. Astereognosis refers to the loss of ability to recognize the texture, form and shape of an object by tactile sensation (e.g. inability to recognize a coin placed in the palm of the hand).

Simultanagnosia (spelling dyslexia) refers to the inability to recognize sensory stimuli received simultaneously and the inability to read except for the shortest words. This condition is a manifestation of a lesion of Brodmann's area 18, or dysfunctions in scanning of the visual images. It may be associated with visual deficits.

Anosognosia (asomatognosia), an important part of 'hemispatial neglect', is characterized by indifference to or rationalization of the symptoms, or denial of an illness of serious nature. Patients appear inattentive and experience allocheira, which refers to an individual's experience of a stimulus applied to one side of the body and being felt on the opposite side. Allocheira results from lesion of the superior parietal lobule of the right (non-dominant) hemisphere.

Babinski's agnosia refers to the neglect or even the denial of existence of a paralyzed limb in a paralytic patient.

Dementia refers to a progressive, diffuse, and multifocal decline in the intellectual and cognitive abilities that impairs daily functioning in the presence of normal consciousness. These include loss of memory accompanied by dysfunction in at least one other mental function such as memory, language, emotion or behavior and cognition (judgement, abstract thinking, etc.) This may be classified into cortical and subcortical dementia.

Cortical dementia may be seen in Alzheimer's disease (discussed in Chapter 16) which accounts for approximately 50% of cases of dementia. Cortical dementia also occurs in multiple sclerosis (discussed in Chapter 1), normal pressure hydrocephalus (discussed later in this chapter), Pick's disease, neurosyphilis, lyme disease, subacute sclerosing panencephalitis, Creutzfeldt–Jakob disease, and in individuals with frontal and temporoparietal lobe lesions. Cortical dementia is manifested in the inability to recall events, but patients remain conscious and fully aware of their intellectual dysfunction.

Pick's disease (lobar atrophy or sclerosis) is an extremely rare condition that shows severe signs of frontal or temporal lobe dysfunctions. It is a slowly progressive disease that initially manifests behavioral disorders, lack of insight, and poor mental functions. These are followed by loss of retentive memory, language impairment, and appearance of primitive reflexes such as grasp and sucking reflexes. Dementia in this disease has a slow presenile course that resembles Alzheimer's dementia.

In neurosyphilis, which is caused by the spirochete *Treponema pallidum*, dementia develops as a result of brain infection and is seen as a late manifestation. It may be preceded by amnesia and certain personality changes. Patients with syphilitic dementia may also exhibit paresis, dysarthria, tremor, Argyll Robertson pupils, and locomotor ataxia.

Dementia pugilistica may occur as a result of repeated head trauma or cerebral concussions in professional boxers' 'punch-drunk state'.

Lyme disease, which is caused by the spirochete *Borrelia burgdorferi*, may manifest dementia in the second stage of the disease weeks or months after the onset of infection. Dementia may be detected in association with sleep and emotional disorders, slowing of memory, irritability, facial nerve palsy, and some degree of poor concentration.

Cortical dementia may also occur in subacute sclerosing panencephalitis (SSPE), a particularly fatal disease that affects children and young adults before the age of 20, months or years after a measles attack. The probable etiology is altered rubeola virus. It initially manifests mood disorders, insomnia, hallucinations, lack of concentration, which is eventually followed by myoclonic jerks, and dementia. Patients may also exhibit choreiform movements, rigidity, dysphagia, and cortical blindness.

Creutzfeldt–Jakob disease is a progressive neurodegenerative condition that manifests cortical dementia, and commonly assumes sporadic or familial forms. In the sporadic form, the transmission of the disease may occur as a result of contaminated neurosurgical instruments, injection of human growth hormone, infected corneal transplant, and handling of cadaveric dura mater. Patients with the sporadic form may also show elevated levels of 14-3-3 brain protein in the CSF. Myoclonic jerking and periodic EEG complexes (sharp wave pattern superimposed on slow background rhythm) accompany dementia. Patients may also develop sleep disorder, ataxia, hemiparesis, aphasia, and hemianopsia. The familial form is thought to be an autosomal dominant inheritance condition with point mutations, deletions, or insertions in the coding sequence of the gene for the prion protein (PrP) on the short arm of chromosome 20. The most common mutation that produces the clinical picture of this form of the disease is at codon 200. It has an earlier onset and its course is more protracted. The EEG changes are often missing and the 14-3-3 protein is not detected in the CSF, as is the case with the sporadic form. Different phenotypic manifestations may develop as a result of these mutations. Some may exhibit cerebellar ataxia, spastic paresis, and dementia at a later stage. Progressive insomnia, dementia, and dysautomonia that prove to be fatal, may also be seen in association with a mutation in the gene for PrP in this form of the disease.

Subcortical dementia, which is associated with lesions of the brainstem, cerebellum, or basal nuclei, may be observed in individuals with Parkinson's and Huntington's diseases (which are described in Chapter 21), AIDS patients, and in progressive supranuclear palsy. In this type of dementia speech dysfunction, motor deficits, and generalized slowing of cognitive function (mental akinesia) are prominent.

AIDS dementia occurs in the late stages of the disease subsequent to meningitis and encephalitis, whereas progressive supranuclear palsy also known as 'Steele–Richardson–Olszewski syndrome' exhibits parkinsonian manifestations including impairment of voluntary eye movements, prolongation of thought processes, irritability, and apathy.

Seizures are irregular, brief, sudden, excessive and repeated activation and depolarization of neurons within the brain that produce hypersynchrony, where large cell group fires together, creating an electrical storm. The location of neuronal activity and the length of activation determine the eventual manifestations, including disturbance of sensation and consciousness, convulsive motor activity, or some combination of these disorders. Increased neuronal excitability may lead to neuronal synchrony and recruitment as well as accumulation of extracellular potassium or decline of postsynaptic GABAergic inhibition. Seizures are viewed in two categories: partial and generalized.

Partial seizures manifest abnormal movements or sensations and/or stereotyped behavioral patterns on one side. These focal seizures result from localized lesions such as scars, tumors or arteriovenous malformations, head trauma, infection, prenatal injury, fever, hypo-/hyperglycemia, stroke, alcohol withdrawal, or congenital or metabolic disorders. They either remain localized or spread to adjacent cortical areas depending on the extent of glutamate mediated excitation combined with the decrease in GABAergic inhibition. Partial seizures are classified into simple and complex partial seizures.

A simple partial seizure is characterized by the relative localization of the abnormal discharge in the brain, usually to one hemisphere. It may arise as a result of activation of foci in the primary motor, premotor, supplementary motor, or prefrontal cortices. Auras are very common and may be the only manifestation in this condition. As it involves the motor, sensory, and autonomic systems, consciousness usually remains unaffected.

Partial motor seizures may be manifested in turning the head and eyes to the contralateral side. Jacksonian 'march' seizure is a motor seizure in which rhythmic and clonic twitching starts on the contralateral hand and 'marches' up the arm, to the face, and down the leg in seconds, followed commonly by unconsciousness. However, it may be localized, affecting the foot, thumb, or mouth angle on the contralateral side.

Partial sensory seizure is associated with a variety of sensations depending upon the selective involvement of certain parts of the sensory homunculus. These sensations may include epigastric rising sensation, tingling in the lips, fingers, or toes that spreads to adjacent areas. It may also be associated with vertigo, olfactory hallucinations, or visual disorders such as sensations of darkness and light flashes.

Complex partial seizure occurs infrequently and irregularly, and is marked by impaired but not loss of consciousness, exhibiting widely varied clinical characteristics. Automatic behaviors such as chewing, lip smacking, fumbling with clothes, scratching genitalia, thrashing of arms or legs, or loss of postural tone may be seen in the complex partial seizure. An 'aura' may precede this type of seizure. Dyscognitive states, which include increased familiarity (*déjà vu*) or unfamiliarity (*jamais vu*), autonomic responses, exhaustion, and illusions, may also be observed. However, these disorders are often more indicative of anxiety attacks than seizures.

Generalized seizures, which result from involvement of both cerebral hemispheres, mostly with no apparent structural damage, are inherited for the most part. Many of the generalized seizures begin during childhood or adolescence. The two main categories of generalized seizures are the convulsive and non-convulsive forms. The common convulsive type is the tonic–clonic (formerly known as grand mal) seizure, whereas the non-convulsive forms include absence seizures (petit mal). Convulsive type may also include less common varieties such as purely tonic, atonic or clonic generalized seizure. Non-convulsive seizures also encompass atypical absence seizures. In summary, convulsive generalized seizures include tonic–clonic seizures, status epilepticus, tonic, atonic and clonic seizures.

The tonic–clonic seizures (grand mal epilepsy) occur in 4–10% of all cases of epilepsy and is viewed in two types: a) awakening clonic–tonic–clonic seizures, which are provoked by sleep deprivation, excessive fatigue, and alcohol consumption; b) tonic–clonic seizures, which occur both during sleep and wake periods, and maintains better remission rate following drug therapy than the first type. In general, patients with tonic–clonic seizures may or may not sense the approach of convulsions (prodrome phase) in the forms of apathy or ecstacy, movement of the head, abdominal pain, headache, pallor or redness of the face, or unusual sensations. These experiences (aura), which last for few seconds, may represent an impending simple partial seizure. In the ictal phase, which lasts 10–30 seconds, sudden loss of consciousness and fall to the ground, a brief period of flexion of the trunk and elbow, followed by a longer extension of the back and neck, jaw clamping, and stiffness of the limbs may occur. Suspension of breathing, dilatation of the pupil, cyanosis, and possible urinary incontinence are also observed. This tonic stage is followed by a clonic phase in which mild generalized tremor is followed by violent flexor spasm, facial grimaces and tongue biting. At this stage increase of blood pressure, pulse rate, salivary secretion, and sweating occur.

When the clonic phase terminates, movements cease, and the patient remains apneic, lethargic, confused, and often falls asleep (postictal phase). The EEG shows characteristic initial desynchronization, lasting for a few seconds, followed by a 10-second period of 10Hz spikes. In the clonic phase, the spikes become mixed with slow waves and then the EEG shows a polyspike wave pattern. The EEG tracing becomes nearly flat when all movements have ceased.

Status epilepticus refers to the condition in which seizures last longer than 30 minutes or follow one another in a rapid fashion. A new wave of seizures begins before the previous one has ceased, with no recovery of consciousness or behavioral function. This condition is commonly caused by abrupt withdrawal of anticonvulsant medications, high fever, metabolic disorder, or cerebral lesions. There are two types of status epilepticus: convulsive and non-convulsive. Non-convulsive type affects behavior but does not produce tonic or clonic movements, which may be evident in the continuous lethargic and clouded state that a patient experiences with a prolonged absence seizure or a series of complex partial seizures. The convulsive type may be life threatening due to the sequence of events that may include circulatory collapse and hyperthermia.

Repeated nature of epileptic seizures may be explained experimentally on the basis of kindling theory. Induction of seizure activity in experimental animals increases in intensity subsequent to the increase in stimulus repetitions. Therefore, an intense seizure can be produced, after a long period (months), by the application of a relatively mild electrical stimulus. It has been suggested that prolonged electrical stimulation triggers a series of events in neural circuitry that enables repeated seizure activity to be produced by an appropriate stimulus. This theory may be similar to long-term potentiation (LTP) in which a brief period of intense activity may elicit a persistent change in the synaptic property of the neuron.

Tonic seizures are characterized by sudden bilateral tonic (stiffening) muscle contractions of the entire body, accompanied by altered consciousness, without being followed by a clonic phase. They are brief and range from a few seconds to a minute and occur more commonly during sleep. This condition is usually caused by cerebral lesion and can occur at any age.

Atonic seizures exhibit several forms, but is classically characterized by loss of postural muscle tone and responsiveness, and fall of the patient to the ground. Loss of muscle tone may be less severe, producing nodding of the head or drooping of the eyelids. These seizures occur at any age as brief episodes with sudden onset followed by immediate recovery.

Clonic seizures are characterized by generalized clonic (jerking) movements that are not preceded by a tonic phase. These movements are often asymmetric and irregular, and although rare, they usually occur in children.

Non-convulsive generalized seizures comprise both absence and atypical absence seizures.

Absence seizures are brief episodes (2–15 seconds) of staring (simple absence), accompanied by impairment of awareness and responsiveness. During the episode patients may briefly stop talking or responding. In a more complex absence seizure blinking or synchronized mouth or hand movements accompany the staring episode. They occur suddenly and leave the patient postictally alert and responsive. These seizures may be so brief that patients themselves are sometimes unaware of them and to the observer it appears as a moment of absent-mindedness or day dreaming. They usually begin between the age of 4 and 14, and often resolve by the age of 18. Approximately 50% of individuals with this type of seizure develop generalized tonic–clonic convulsions by the end of the second decade.

Atypical absence seizures usually start and end gradually. They are not produced by hyperventilation and often last more than ten seconds. Like the typical absence, they also begin at an early age. In contrast to the typical absence seizures, these episodes are often seen in individuals with mental retardation and other neurologic disorders. In this condition, staring is accompanied by partial diminution in responsiveness (i.e. able to respond to questions and remember events). Blinking or jerking movements and tonic or atonic seizures may also occur.

15% of cardiac output, equivalent to 750 ml per minute, is directed to the brain. The gray matter of cerebral cortex receives blood far greater than the white matter. The difference between the arterial and venous pressure may determine the cerebral blood flow. In addition, intrinsic factors such as the condition of the cerebral vessels, changes in the tension of carbon dioxide, and alteration in pH will also affect the blood flow to the brain. Due to the small size of frontal lobes in infants, the cerebral vessels are concentrated within the Sylvian triangle.

As mentioned earlier, the branches of the internal carotid and basilar arteries supply the brain. The internal carotid artery (Figures 8.34, 8.35, 8.38, 8.39 & 8.40) arises from the common carotid artery at the level of the upper border of the thyroid cartilage, coursing in the neck in close proximity to the sympathetic trunk. This vessel supplies the rostral two thirds of the brain hemispheres, diencephalon, nasal cavity, forehead, eye and orbit. Initially it ascends in the neck within the carotid sheath (cervical part), entering the carotid canal to gain access to the cranial cavity. Within the carotid canal (petrous part), it lies anterior to the cochlea and tympanic cavity, separated by a thin bony lamella (Figure 8.36). This petrous part courses near the trigeminal ganglion, separated from it by the roof of the carotid canal. During its intracranial course the internal carotid artery passes over the cartilaginous plate of foramen lacerum and continue through the carotid sulcus to the cavernous sinus. In this sinus (cavernous part), it reposes adjacent to the abducens, oculomotor, trochlear, ophthalmic, and maxillary nerves, giving rise to the cavernous, meningeal, and hypophysial arteries. As it leaves the sinus, it turns anteriorly, ventral to the optic nerve and dorsal to the oculomotor nerve. At this point (cerebral part), it gives rise to the ophthalmic, anterior cerebral, middle cerebral, posterior communicating, and anterior choroidal arteries.

The cavernous and the cerebral parts of the internal carotid artery are collectively known as the carotid siphon (Figure 8.34 & 8.36). Occlusion of one or both internal carotid arteries (at the level of the carotid siphon) activates the collateral circulation between the meningeal branches of the external and the internal carotid arteries. Collateral circulation also develops between branches of the lenticulostriate arteries. As a result of this anastomosis one cerebral hemisphere is supplied with blood through a network of vessels (transdural anastomosis) resembling the 'rete mirabile' of lower mammals. This anastomosis may be detected in Moya-Moya syndrome via angiography as a 'puff of smoke'.

The Wada test, in which sodium amobarbital is injected into each internal carotid artery, may be utilized to establish dominance of the cerebral hemisphere. Brief aphasia, which accompanies injection into the internal carotid artery, may determine the dominant hemisphere.

Many arterial branches arise from the cerebral part of the internal carotid artery, including the ophthalmic, anterior cerebral, middle cerebral, anterior choroidal, and the posterior communicating arteries. As the first branch of the internal carotid artery, the ophthalamic artery pierces the optic nerve and supplies the four quadrants of the retina through its central retinal branch (an end artery that lacks any anastomosis), and the choroid and the sclera via the ciliary branches. It also supplies the orbit, nasal cavity, ethmoidal sinuses, dura mater, and the scalp (Figure 8.35).

One of the terminal branches of the internal carotid artery is the anterior cerebral artery (ACA) that runs above the corpus callosum, coursing on the medial surface of the brain (Figures 8.34, 8.35, 8.37, 8.38, 8.39 & 8.42). It is connected to the corresponding artery of the opposite side via the anterior communicating artery. It supplies the medial surface of the brain hemisphere, including the cingulate, medial frontal gyrus, upper parts of the precentral and postcentral gyri, and the paracentral lobule. Orbital and frontopolar branches supply the orbitofrontal cortex, and the rostral and medial parts of the frontal lobe, respectively. Within the cingulate sulcus, the cingulate branch makes its appearance, supplying the corresponding gyrus. Additionally, the callosomarginal branch supplies the paracentral lobule, whereas the pericallosal artery, a terminal branch of the anterior cerebral artery, supplies the precuneus, assuming a typical 'sausage' configuration in syphilitic endarteritis. There are smaller yet important branches that arise from the ACA such as the medial striate (recurrent artery of Heubner) which supplies head of the caudate nucleus, putamen, and anterior limb of the internal capsule. The latter branch is absent in about 3% of the population.

Most of the lateral surface of the brain is supplied by branches of the middle cerebral artery (Figures 8.34, 8.35, 8.38, 8.39 & 8.40) which travels in the lateral cerebral fissure and across the insular cortex. These areas of distribution include the precentral, postcentral, angular, supramarginal, superior temporal, middle frontal, and the inferior frontal gyri, as well as the transverse gyri of Heschl. Numerous branches arise from the middle cerebral artery including the anterior temporal, orbitofrontal,

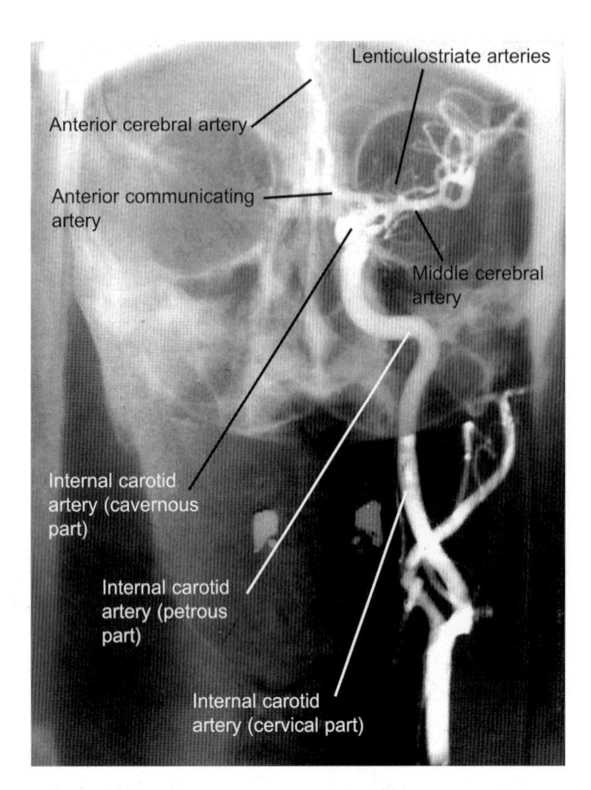

Figure 8.34 Anteroposterior arteriogram showing the common carotid artery, and various parts of the internal carotid artery and its branches. The course of the internal carotid artery within the carotid sheath, petrous temporal bone, and cavernous sinus are illustrated

Vascular disorders produce dysfunctions as a sequel to ischemia or hemorrhage. Cerebral emboli may occur at anytime, but are especially common during daytime. They may originate from the pulmonary veins, cardiac valves or chambers, and from plaques in the aortic arch or its branches. Ulceration and occlusion increase the likelihood of stenosis in the affected vessels.

prerolandic, rolandic, anterior and posterior parietal, angular, posterior temporal, and lenticulostriate branches. Brodmann's area 8 (frontal eyefield) which controls horizontal saccadic eye movements toward the contralateral visual field, lies within the vascular domain of the ascending frontal branch 'candelabra' of this artery, a branch formed by contribution of the prerolandic, rolandic and the anterior parietal branches. The middle cerebral artery also gives off the lateral striate (lenticulostriate or Charcot's artery of cerebral hemorrhage), and possibly the medial striate artery, which enter the anterior perforated substance and supply the globus pallidus, putamen, the anterior limb of the internal capsule.

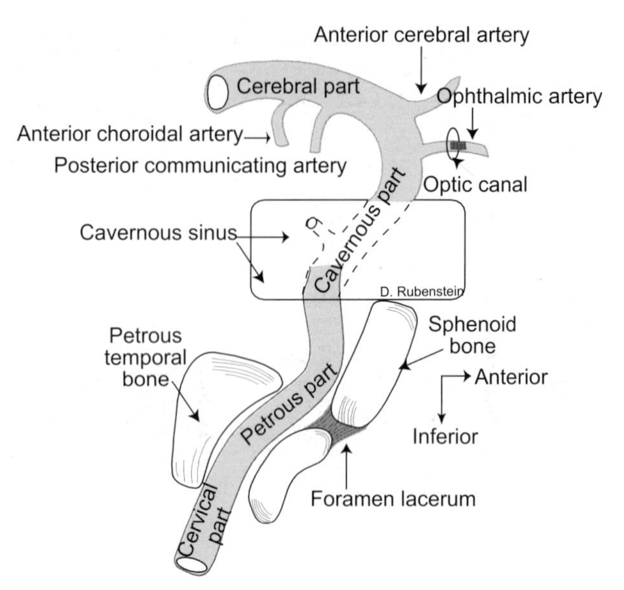

Figure 8.35 In this schematic drawing the course of the internal carotid artery and principal branches are indicated

In addition to cerebral branches, the internal carotid artery gives rise to the anterior choroidal artery (Figure 8.35) which supplies the choroid plexus of the inferior horn of the lateral ventricle, hippocampal and dentate gyri, internal capsule, and the optic tract.

Another vessel that arises from the internal carotid artery and contributes to the arterial circle of Willis is the posterior communicating artery, which provides blood supply to the hypothalamus, optic tract, and tuber cinereum. It runs superior to the oculomotor nerve, a relationship that bears clinical significance. An aneurysm that develops in this artery may produce signs of oculomotor palsy (see Chapter 11).

The posterior cerebral arteries (Figures 8.37, 8.39 & 8.41) which commonly emanate from the basilar artery also supply the brain hemispheres. However, in 25% of individuals one of the posterior cerebral arteries may arise from the internal carotid artery, replacing the posterior communicating branch. In 5% of individuals the posterior cerebral arteries on both sides gain origin from the internal carotid arteries.

After providing blood to the midbrain and portions of the occipital and temporal lobes, the posterior cerebral artery divides into posterior temporal and internal occipital branches. The latter branch gives rise to the calcarine and parieto-occipital arteries. The calcarine artery runs within the calcarine fissure and supplies the primary visual cortex, establishing anastomosis with the middle cerebral artery. Aside from its distribution to the cuneus and precuneus, the parieto-occipital branch also supplies the

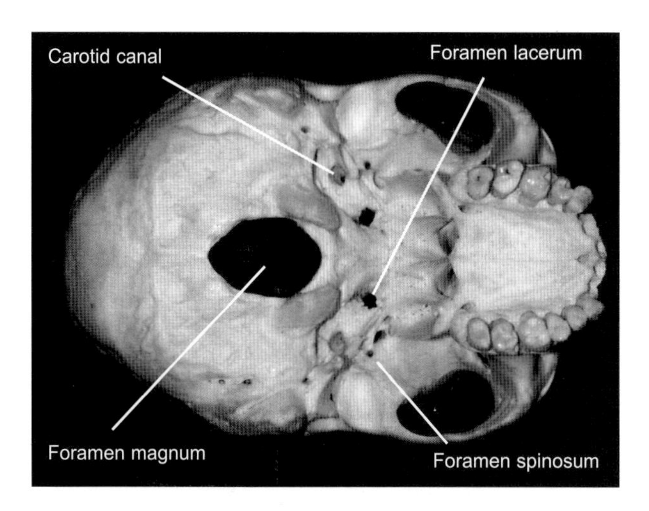

Figure 8.36 Inferior surface of the cranial base. The carotid canal and foramen lacerum are illustrated

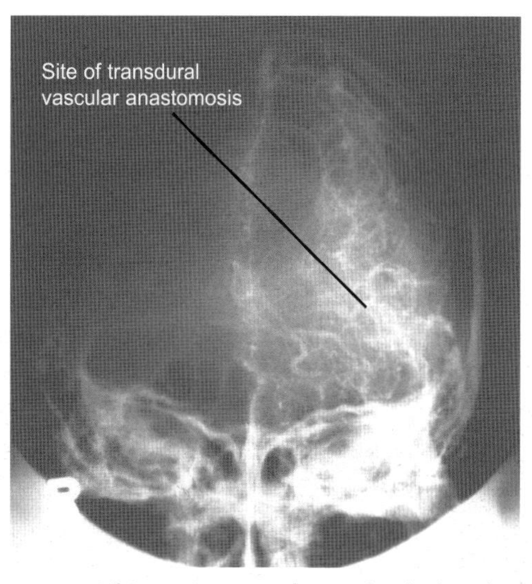

Figure 8.37 This angiogram of an individual with Moya-Moya syndrome illustrates the site of trandural anastomosis

splenium of the corpus callosum (Figures 8.15 & 8.16). There are additional branches from the posterior cerebral artery including the thalamogeniculate and thalamoperforating arteries that supply the thalamus. Macular sparing in individuals with cerebrovascular accident, involving either the middle or posterior cerebral arteries, may be attributed to the rich vascular anastomosis around the occipital pole between these two vessels.

Venous drainage of the cerebral hemispheres

Venous drainage of the brain is maintained by the superficial and deep cerebral veins. The superficial set of veins (Figure 8.43 & 8.44) include the superior and inferior cerebral, superficial middle cerebral, and the anastomotic (connecting) veins. Venous blood from the superior, upper lateral and medial surfaces of the cerebral hemispheres is drained via numerous superior cerebral veins that open into the superior sagittal sinus.

It is important to note that the anterior group of superior cerebral veins open at a right angle to the superior sagittal sinus while the posterior group assumes a more oblique direction and against the current created by the blood in the dural sinuses. The inferior cerebral veins drain the orbital, frontal, and temporal gyri and open into the transverse sinus or the cavernous sinus, whereas the superficial middle cerebral vein runs across the lateral cerebral fissure to open into the cavernous or sphenoparietal sinus, draining the area around the lateral fissure. Two anastomotic veins interconnect the above-mentioned superficial cerebral veins. These connecting veins are comprised of the superior anastomotic vein of Trolard connecting the superior cerebral and superficial middle cerebral veins to the superior sagittal sinus, and the inferior anastomotic vein of Labbe that connects the middle cerebral vein to the transverse sinus.

Examination of the deep cerebral veins (Figures 8.43, 8.44, 8.45, 8.46 & 8.47) reveals the internal cerebral veins, great cerebral vein of Galen, basal and occipital veins. Near the foramen of Monro, the internal cerebral vein is formed by the choroidal, septal, epithalamic, lateral ventricular, and thalamostriate veins. Each internal cerebral vein runs medial to the thalamus on the dorsal aspect of the third

Moya-Moya syndrome (Figure 8.37) is a rare condition which is seen particularly in children and young adults and commonly in females. It was originally described in Japan and later in the west in individuals suffering from occlusive or stenotic diseases of the major branches of the arterial circle of Willis, and in association with neurofibromatosis, sickle cell anemia, retinitis pigmentosa, Down's syndrome, and Fanconi's anemia. Individuals with this condition exhibit convulsions associated with subarachnoid hemorrhage, alternating hemiplegia, and other variable neurological disorders. Ischemic events are commonly observed in the younger individuals, whereas subarachnoid or cerebral hemorrhage may be seen in the adults.

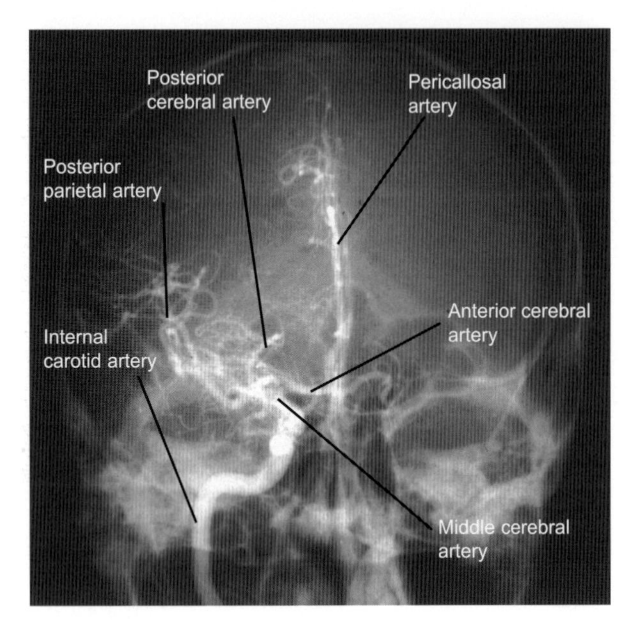

Figure 8.38 Another angiogram of the internal carotid artery showing branches of the middle and anterior cerebral arteries. A branch of the posterior circulation (posterior cerebral artery) is also shown

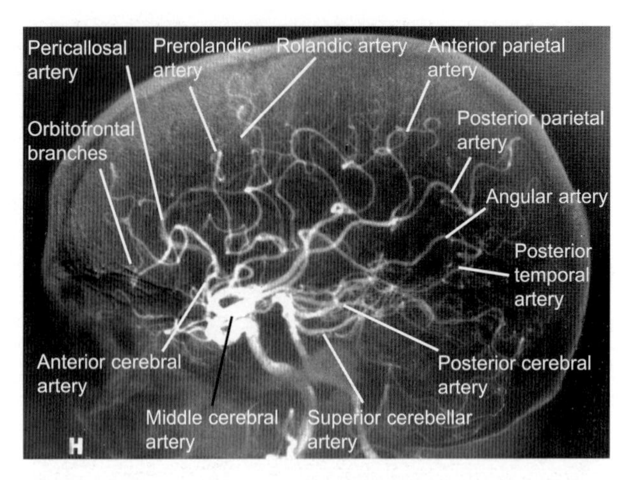

Figure 8.39 In this angiogram branches of the internal carotid artery and branches of the middle cerebral artery are shown. Some branches of the posterior circulation are also visible

ventricle, and courses inferior to the splenium of corpus callosum.

- The choroidal vein drains the venous blood of the hippocampal gyrus, fornix, corpus callosum, and the choroid plexus of the lateral ventricle.
- The septal vein drains the septum pellucidum and corpus callosum.
- The epithalamic vein receives venous blood from the habenula and pineal gland.
- The lateral ventricular vein receives venous tributaries from the caudal thalamus, parahippocampal gyrus, and the lateral ventricle.
- The thalamostriate (terminal) vein courses caudally between the caudate nucleus and thalamus, draining both these structures.

The great cerebral vein of Galen (Figures 8.43, 8.44, 8.45, 8.46 & 8.47) is formed caudal to the pineal gland by the union of the two internal cerebral veins. This vein is located inferior to the splenium and within the cisterna ambiens (a dilatation of the subarachnoid space superior to the cerebellum). It joins the inferior sagittal sinus to open into the straight sinus at the junction of the tentorium cerebelli and the falx cerebri. At this site, it receives the occipital, posterior callosal veins, and the basal vein of Rosenthal.

The occipital vein drains the venous blood of the occipital and the temporal lobes.

The posterior callosal vein carries the venous blood from the caudal portion of the corpus callosum to the great cerebral vein of Galen. Near the anterior perforated substance, the basal vein (of Rosenthal) is formed by contributions from the deep middle cerebral, anterior cerebral, inferior striate, and occipital veins.

- The deep middle cerebral vein drains the insular cortex and follows the lateral cerebral sulcus.
- The anterior cerebral vein drains the medial surface of the brain accompanied by the anterior cerebral artery.
- The inferior striate vein exits from the anterior perforated substance and drains the inferior surface of striatum.

Tumors of the temporal lobe may displace the middle cerebral artery upward. Also, an interesting correlation exists between the extensive area of distribution of this artery and the increase in the vulnerablity to infection. Lacunar infarcts and possible death due to rupture of the lenticulostriate artery may occur in hypertensive individuals. However, these lacunar infarcts do not usually cause headache, and transient ischemic attacks prior to the infarction are infrequent.

When the posterior cerebral arteries are derived from the internal carotid artery, transient ischemic attack (TIA) as a result of occlusion of the internal carotid artery may also involve areas of distribution of the vertebral arteries.

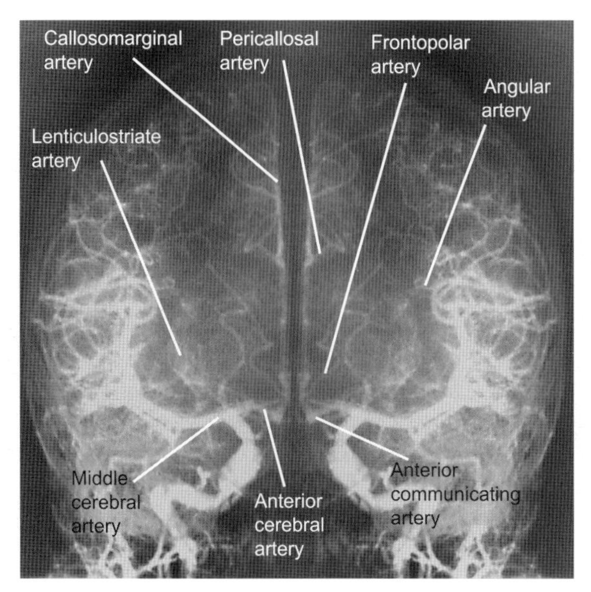

Figure 8.40 Angiogram of the internal carotid artery and its main cerebral branches. Branches of the middle cerebral artery are clearly visible

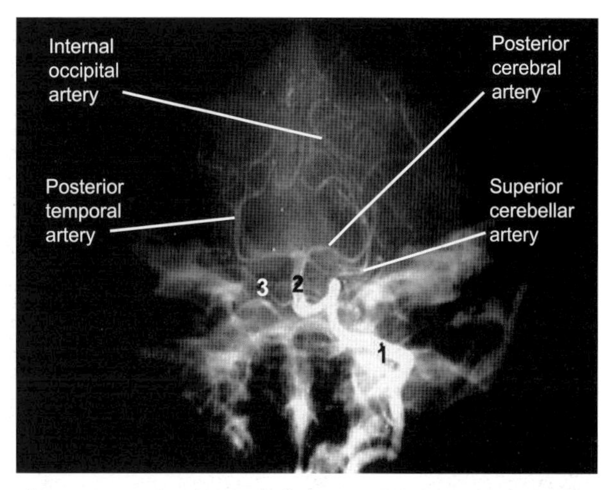

Figure 8.41 Angiogram of the posterior cerebral circulation. The vertebral artery and its branches are illustrated. 1. Vertebral artery; 2. basilar artery; 3. anterior inferior cerebellar artery

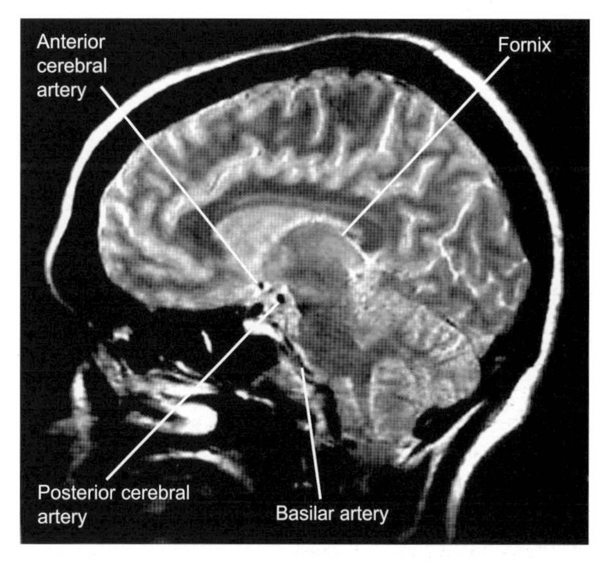

Figure 8.42 An MRI of the brain in the sagittal plane. Note the main branches of the anterior and posterior cerebral circulation

Meninges

The brain is enveloped by the meninges that form the coverings for the brain, contribute to the formation of the dural sinuses, brain barriers, and the cerebrospinal fluid (CSF). These coverings send partitions that separate the cerebral hemispheres from each other and from the cerebellum and the brainstem. Meninges consist of the dura mater (pachymeninx), the arachnoid and pia mater (leptomeninges). Since meninges continue with the epineurium and perineurium of peripheral nerves, meningitis may produces irritation of the meninges around the brain, spinal cord, and peripheral nerves.

The dura mater (Latin for tough mother) is a collagenous membrane which covers the brain and spinal cord. It consists of meningeal and endosteal layers (Figure 8.52). At certain locations, the gap between these two layers results in the formation of the dural sinuses. This layer is the most sensitive of all meninges to painful stimuli. Dural innervation within the anterior cranial fossa is maintained by the meningeal branches of the ophthalmic and maxillary nerves. The meningeal branches of the maxillary nerve also contribute to the innervation of the dura mater in the middle cranial fossa. Within the posterior cranial fossa, the dura is supplied by the meningeal branches of the upper cervical spinal nerves. A recurrent branch of the ophthalmic nerve supplies the tentorium cerebelli. In the same manner the blood supply of the dura mater is secured by branches of the ophthalmic and middle meningeal arteries in the anterior cranial fossa. The dura of the middle cranial fossa is supplied by the middle meningeal, accessory meningeal, ascending pharyngeal, and lacrimal arteries, whereas the

> Vulnerability to ischemia in hypertensive individuals is greater in watershed areas, which represent the cerebral zones supplied by the terminal branches of the anterior, posterior, and middle cerebral arteries that lack adequate blood supply and efficient perfusion pressure.

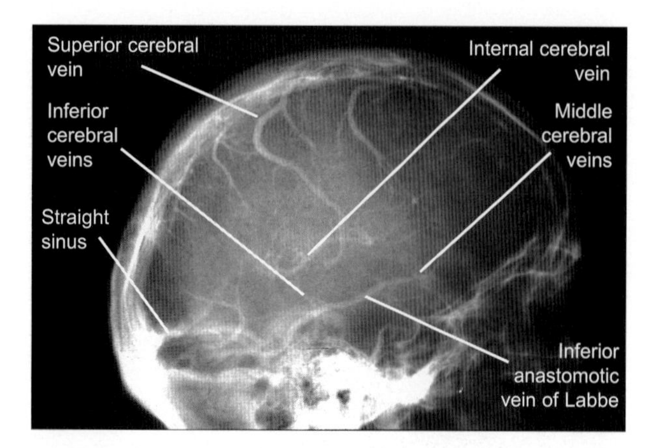

Figure 8.43 Angiogram of the superficial and cerebral deep veins and associated sinuses. Note the bridging veins and their drainage sites into the superior sagittal sinus

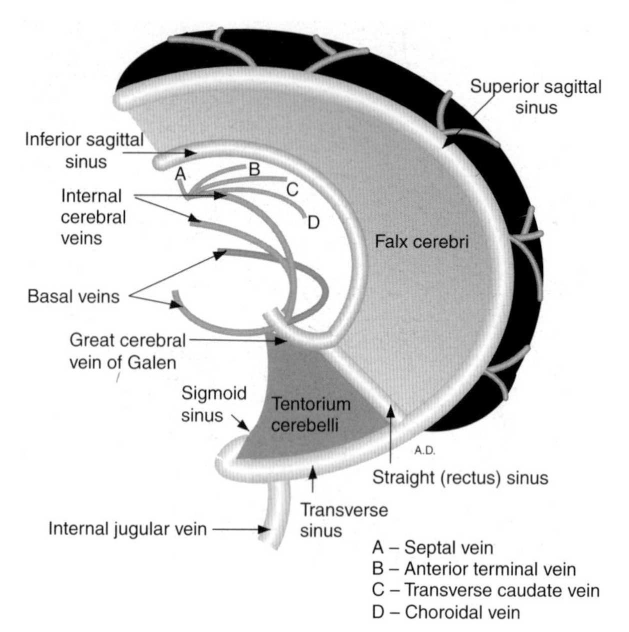

Figure 8.45 Tributaries of the internal cerebral vein and great cerebral vein of Galen are shown in this figure. The straight sinus, which receives the deep cerebral veins, is illustrated in relation to the tentorium cerebelli and falx cerebri

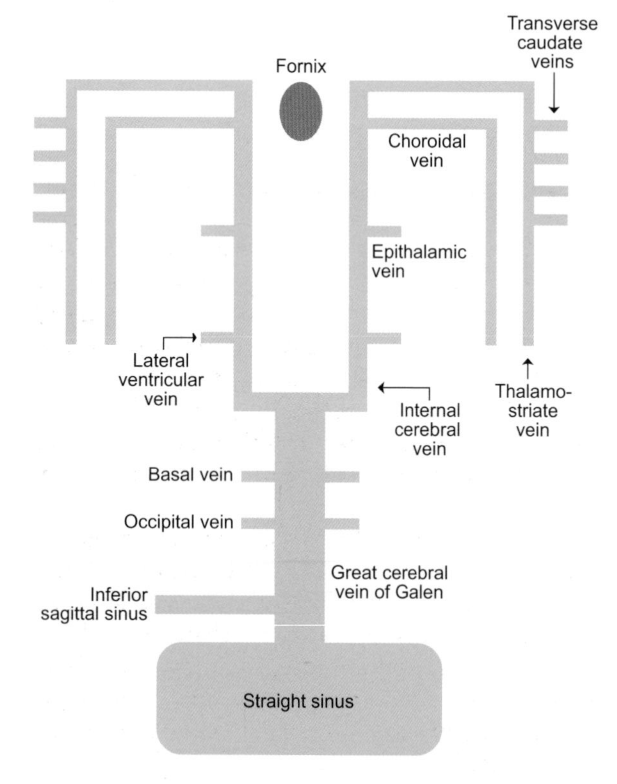

Figure 8.44 Schematic diagram of the deep cerebral vein. The great cerebral vein of Galen joins the venous blood from the inferior sagittal sinus to drain into the straight sinus

meningeal branches of the occipital, vertebral, and the ascending pharyngeal arteries provide blood supply to the dura in the posterior cranial fossa. Dural sinuses and eventually the internal jugular vein receive the venous blood of the dura mater.

Epidural and subdural spaces, which lie superficial and deep to the dura mater, respectively, may be the sites of hematoma that produce serious manifestations. The clinical aspects of epidural and subdural hematomas are discussed below in detail.

Extensions of the dura mater divide the cerebral hemispheres, and separate the cerebellar hemispheres from each other and from the brain. It also forms the diaphragma sella, a covering of the pituitary gland superior to the hypophysial fossa. These dural partition include the falx cerebri, tentorium cerebelli, and falx cerebelli.

- The falx cerebri (Figure 8.45) is a dural partition which arches over the corpus callosum, extending between the two cerebral hemispheres. It stretches from the crista galli of the ethmoid bone rostrally to the tentorium cerebelli caudally. Attached to the superior sagittal sulcus lies the superior sagittal sinus. This partition has a free inferior border that contains the inferior sagittal sinus.
- The tentorium cerebelli (Figure 8.45 & 8.47) occupies the space between the occipital lobes of the brain and the cerebellum. It contains the transverse sinus in its attached border, and the straight sinus at its junction with the falx cerebri (Figure 8.45). Both the great cerebral vein of Galen and the inferior sagittal sinus drain into the straight (rectus) sinus. Rostrally the tentorium forms the tentorial notch, which allows the brainstem to pass through and join the diencephalon. This dural extension produces

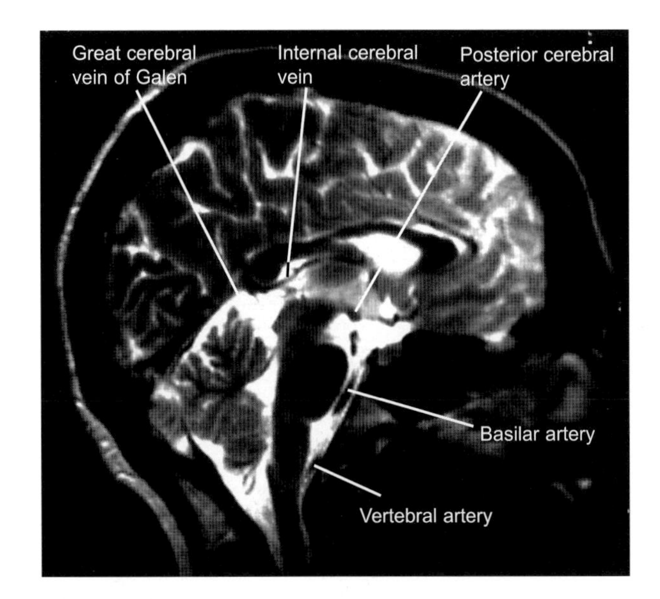

Figure 8.46 In this MRI scan the internal cerebral vein and the great cerebral vein of Galen, as well as the straight sinus are shown. Branches of the vertebrobasilar system are also illustrated

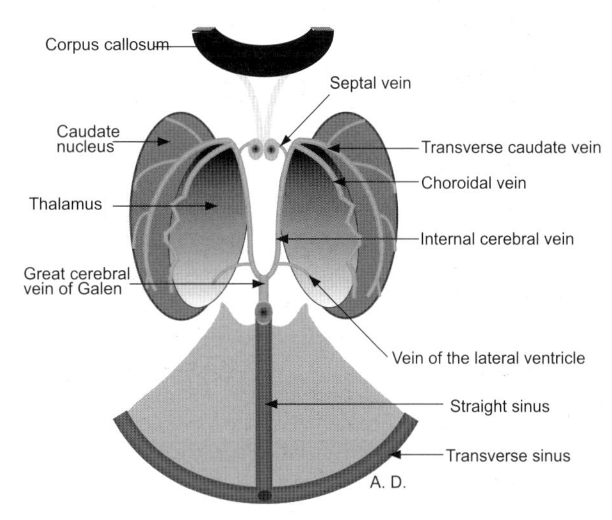

Figure 8.47 A detailed drawing of the various tributaries of the deep cerebral vein

Thinness of the wall of the great cerebral vein of Galen may predispose the vessel to rupture. Additionally, aneurysm of this vessel as a result of congenital arteriovenous malformation may produce hydrocephalus. The latter condition may be associated with congestive heart failure.

supratentorial and infratentorial compartments, respectively. In the supratentorial compartment lies the occipital lobes and diencephalon, whereas in the infratentorial compartment the cerebellum and part of the brainstem are lodged.

• The falx cerebelli is a sickle-shaped dural partition, located in the posterior cerebellar notch, separating the cerebellar hemispheres. It attaches posteriorly to the occipital crest, extending from the internal occipital protuberance to the foramen magnum, containing the occipital sinus.

Dural sinuses

Dural sinuses (Figures 8.43, 8.45, 8.47, 8.55, 8.56 & 8.57) are valveless venous channels, which are devoid of muscular tissue and are commonly located between the meningeal and endosteal layers of the dura mater.

A lesion of the parietal lobe may cause widening of the posterior part of the falx cerebri and a 'square shifting' of the anterior cerebral vein.

Thrombosis of the cerebral veins may occur as a result of mycotic or pyogenic infections, or due to non-infectious conditions (e.g. malnutrition, hypercoagulable states). It may also be associated with trauma, administration of contraceptives, otitis media, and sinusitis. Thrombosis of the cerebral veins occurs as a result of primary or secondary lesions elsewhere in the body. Headache, vomiting or convulsions due to cerebral edema may be seen with or without a focal motor deficit.

Superficial cerebral veins in particular are prone to rupture at the points of their entry into the superior sagittal sinus (bridging veins), resulting in subdural hematoma. This is characterized by sudden and severe headache, vomiting, and dilatation of the facial veins, focal seizures, hemiplegia, dyskinesia, and possibly papilledema (discussed in detail later in this chapter).

These sinuses, which drain intracranial structures, are classified into posterosuperior and anteroinferior groups. The posterosuperior group includes the sup-erior and inferior sagittal sinuses, straight sinuses, confluence of sinuses, transverse (lateral) sinuses, sigmoid sinuses, and the occipital sinus. Other sinuses, such as the sphenoparietal, cavernous, intercavernous, and the superior and inferior petrosal sinuses, form the anteroinferior group.

Arteriovenous malformations (Figure 8.48) are congenital connections between arteries and veins that develop between the fourth and fifth weeks of embryonic life and are commonly seen in the cerebral vessels. The most frequently afflicted vessel with this malformation is the middle cerebral artery. AV malformations are responsible for hemorrhage, focal or generalized seizures. They may mimic signs and symptoms of multiple sclerosis or tumors. Headache and intellectual deterioration are some of the manifestations of these anomalies.

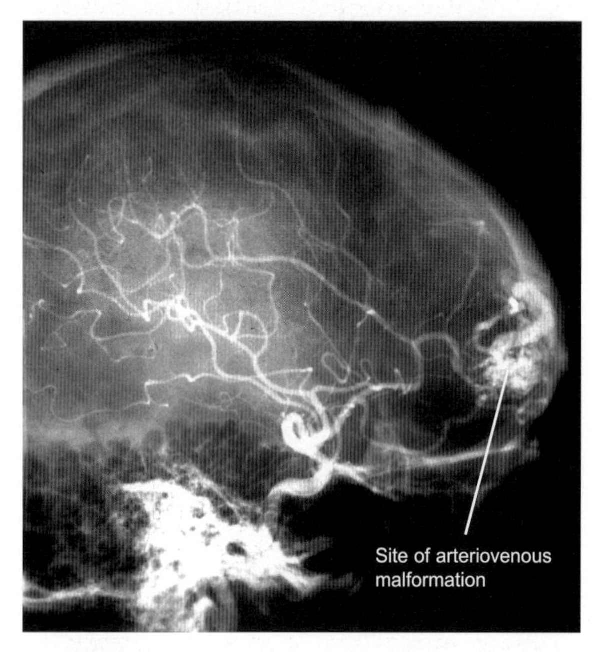

Figure 8.48 An internal carotid artery angiogram showing the general pattern of arteriovenous malformation involving the frontopolar branch of the middle cerebral artery

Posterosuperior group of dural sinuses

• The superior sagittal sinus (Figures 8.43, 8.44, 8.45, 8.54 & 8.55) runs in the inner surface of the frontal bone from the crista gali, where it is connected to the emissary veins and the veins of the nasal cavity, to the internal occipital protuberance, where it continues with the right transverse sinus. The connection to the nasal cavity is usually maintained via the foramen 'cecum' if it is patent. The dilated posterior extremity of this sinus continues with the confluence of the sinuses. The superior sagittal sinus occupies the upper border of the falx cerebri and receives the diploic veins, arachnoid villi, and the superior cerebral veins. The superior sagittal sinus receives venous lacunae, parietal emissary veins, and superior cerebral veins.

Irritation of the epineurium and perineurium may account for the distinctive features of Kernig's sign which is characterized by nuchal rigidity, pain and strong passive resistance upon attempt to extend the knee while the thigh is in flexed position. It may also account for automatic flexion of the hip and knee joints (Brudzinski's sign) upon abrupt flexion of the neck in meningitic patients (Figure 8.49).

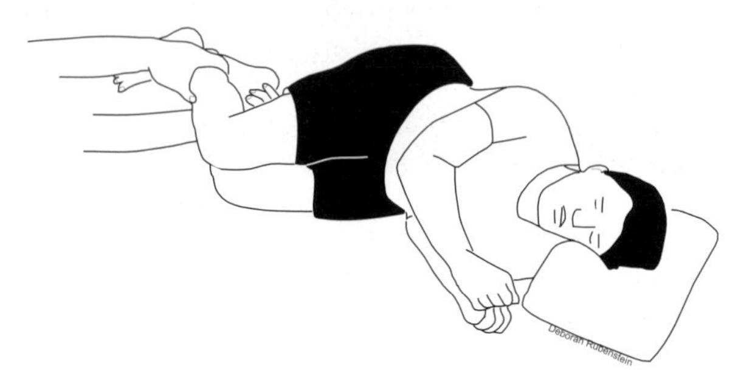

Figure 8.49 This diagram illustrates Brudzinski's sign

Venous lacunae are irregular venous pockets located on each side of the superior sagittal sinus, which receive the meningeal veins and the arachnoid granulations. They also communicate with each other as well as with the superior sagittal sinus.

Emissary veins are small veins, which pierce the skull, establishing a connection between dural sinuses (e.g. superior sagittal sinus) and the extra-cranial veins. They serve as another route by which infection may spread to the cranial cavity from areas outside the skull. Emissary veins are classified into frontal, parietal, temporal, and mastoid veins.

Diploic veins, which are contained in the cancellous tissue between the compact layers of the calvaria, are connected to the emissary veins. These veins are devoid of valves and are absent in the newborn, but well developed in adults.

• The inferior sagittal sinus (Figures 8.44 & 8.45) occupies the inferior margin of the falx cerebri dorsal to the corpus callosum. It continues with the straight sinus at the junction of the falx cerebri and tentorium cerebelli.

The dura mater may be associated with tumors (e.g. meningioma), epidural or subdural hemorrhages.

Epidural (extradural) hematoma refers to a localized accumulation of blood between the bony skull and the endosteal layer of the dura (Figure 8.50). It occurs most commonly as a result of rupture of the anterior division of the middle meningeal artery or vein. Epidural hematoma may be due to a minor trauma to the side of the head that produces fracture of the antero-inferior portion of the parietal bone. Epidural hematoma may lead to a rise in intracranial pressure, a condition which must be relieved before irreversible brain damage and herniation occur. Unsuccessful treatment of this condition may lead to death. It is a rapidly progressive condition, which may involve the posterior cranial fossa, resulting in ipsilateral cerebellar dysfunctions, headache, and upper motor neuron palsy. This is generally a unilateral condition that results in the majority of cases from temporal bone fractures associated with mild head trauma. This condition is usually associated with transient loss of consciousness and contralateral spastic palsy, requiring immediate surgical intervention. Due to the increased arterial blood pressure in epidural hematoma, MRI appears more rounded and isolated.

A subdural hematoma refers to the accumulation of venous blood in the subdural space (Figure 8.51). It results from rupture of the bridging superficial cerebral veins as they drain into the dural sinuses. Most often this condition occurs as a result of severe trauma to the front or back of the head, producing excessive anteroposterior displacement of the brain within the skull. Subdural hematoma is much more common than epidural hematoma and tends to have a poorer prognosis. The progression of this condition is relatively slower than that of an epidural hematoma. Often symptoms do not appear until months after the initial head injury. Displacement of the pineal gland to the contralateral side as a radiographic finding may be an initial sign of subdural hematoma. It is classified as an acute condition when it occurs within 3 days, subacute or chronic condition when it shows manifestations between 3 days and 3 weeks. This condition is characterized by loss or altered state of consciousness, unilateral pupillary dilatation (anisocoria, unequal pupils), and contralateral spastic palsy. However, spastic palsy may be seen ipsilaterally if the subdural hematoma is associated with compression of the midbrain (e.g. against the tentorial or Kernohan's notch).

Progression of this condition is more insidious with symptoms appearing weeks or sometimes months after head injury. A sustained headache and mental obtundation are the most common symptoms. Imaging techniques that elucidate a brain shift may be useful in the diagnosis and assessment of asymptomatic cases. CT scans must be interpreted cautiously, since at certain stages (within weeks following injury) the hematoma itself becomes isodense with the surrounding brain tissue. MRI, which is a less cost-effective method of scanning, overcomes this diagnostic problem. Subdural bleeding, that follows the contour of the lateral surface of the brain, appears flat.

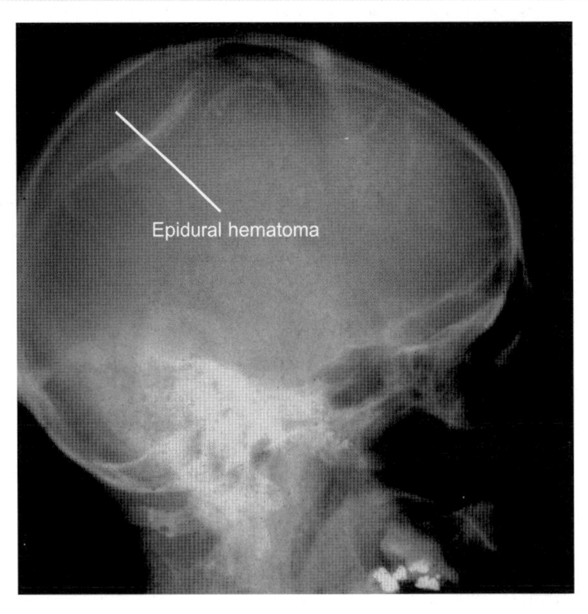

Figure 8.50 This radiographic image clearly illustrates sites of epidural hematoma

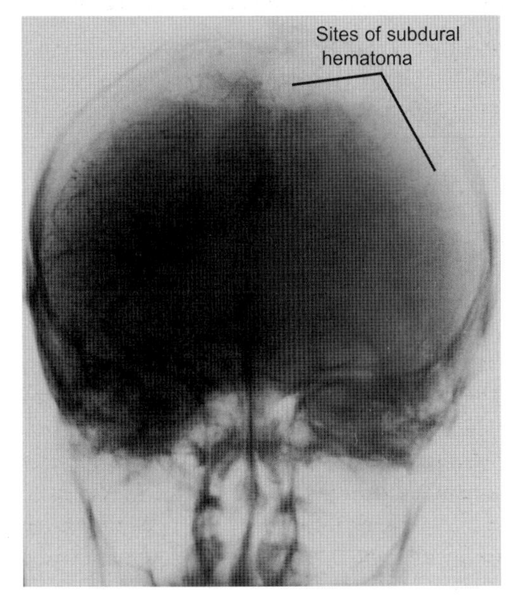

Figure 8.51 Radiographic image of the skull showing the extent of subdural hematoma

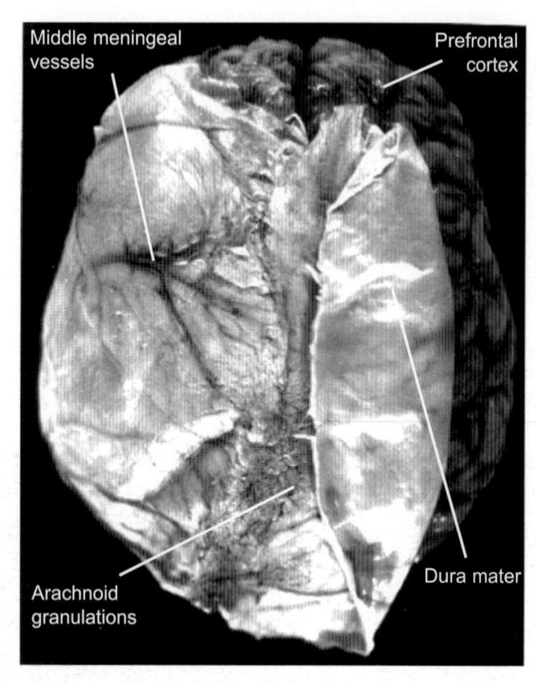

Figure 8.52 Photograph of the dura mater covering the brain. Notice the branches of the meningeal vessels

• The straight sinus (Figure 8.44, 8.45, 8.47 & 8.56) extends from the site of junction of the tentorium cerebelli and falx cerebri to the left transverse sinus. It receives the inferior sagittal sinus, superior cerebellar veins, and the great cerebral vein of Galen.

• The confluence of sinuses (Figures 8.45. 8.47 & 8.56) is formed at the site of union of the superior sagittal, straight, occipital and transverse sinuses near the internal occipital protuberance.

• The transverse sinus (Figures 8.45, 8.47 & 8.56) occupies the transverse sulci and the lateral and posterior borders of the tentorium cerebelli. It appears that the right transverse sinus is continuous with the superior sagittal sinus, while the left sinus continues with the straight sinus. As the transverse sinus drains into the sigmoid sinus, it receives the superficial middle cerebral, inferior cerebral, inferior cerebellar, and some of the diploic veins. Anastomotic veins of Labbe (inferior anastomotic vein) and Trolard (superior anastomotic vein) connect the superficial middle cerebral veins to the transverse and superior sagittal sinuses, respectively. Occasionally, the transverse sinus also receives arachnoid villi.

• The sigmoid sinuses (Figure 8.56) form a curve on the medial surface of the mastoid part of the temporal bone and are separated from the mastoid antrum of the middle ear by a thin plate of bone. They are connected on both sides to the occipital and posterior auricular veins, as well as to the veins of the suboccipital triangle via the emissary veins.

Increased pressure in a tentorial compartment, due to a developing mass, may eventually lead to herniation of part of the brain into the compartment with the lower pressure. Bilateral supratentorial masses (Figure 8.53), which are produced by rostrocaudal displacement and compression of the midbrain, pons, the tectum, reticular formation, oculomotor nerve, and the diencephalon, may result in transtentorial or central herniation. In this condition, the undue pull of the displaced posterior cerebral artery upon the anterior choroidal artery, paramedian and pontine branches of the basilar artery, produces stretching and shearing of these vessels. Stupor (due to compression of the fibers of the ascending reticular activating system), and irregular and/or Cheyne–Stokes respiration (due to compression of the descending pathways to the respiratory center) are the main features of this condition.

Due to its entrapment between the superior cerebellar and posterior cerebral arteries, the oculomotor nerve is most likely to be affected at the early stage of this disease. The supratentorial structures may also enlarge as a result of obstruction of the cerebral aqueduct. As a result of compression of the cerebral peduncle against the edge of the tentorium on the opposite side of herniation, the upper motor neuron dysfunctions will be seen on the side of herniation. Signs of decortication and decerebration will follow, leading to stiff posture and later rigidity. Additionally, paralysis of both vertical and horizontal (doll's eyes) eye movements, and conjugate gaze to the contralateral side (due to destruction of the tectum, pontine tegmentum, and the corticofugal fibers) may also be seen. Uncal herniation occurs as a result of supratentorial mass and is characterized by displacement of the uncus into and inferior to the tentorial notch. A more detailed account on this condition is documented with the temporal lobe and uncus.

• The occipital sinus is contained in the posterior border of the falx cerebelli and lies anterior to the occipital crest, joining the confluence of sinuses near the internal occipital protuberance. It is connected to the occipital vein and the internal vertebral venous plexus, which serve as a collateral venous route upon blockage of the internal jugular vein. As the smallest of all dural sinuses, the occipital sinus begins near the margin of the foramen magnum and terminates in the confluence of sinuses.

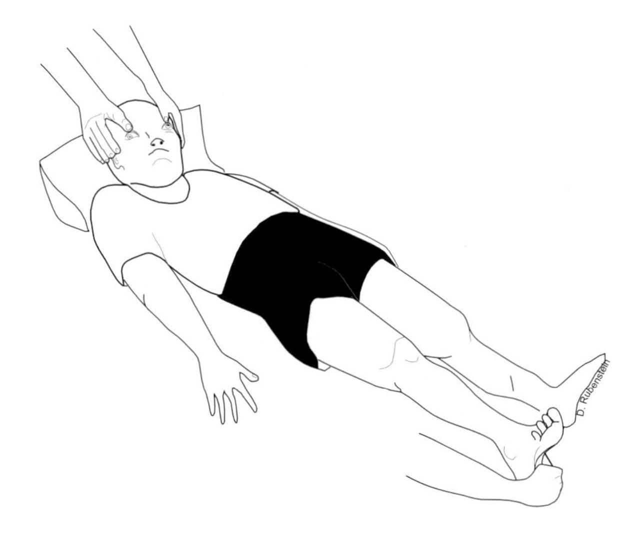

Figure 8.53 This is a depiction of an individual with transtentorial herniation

The venous communications between the superior sagittal sinus and the extracranial veins may serve as a route for spread of infection from the nose and scalp to the dural sinuses and eventually the systemic circulation.

Obstruction of the superior sagittal sinus (Figure 8.54) as a result of thrombosis may impede absorption of CSF and cause increased intracranial pressure.

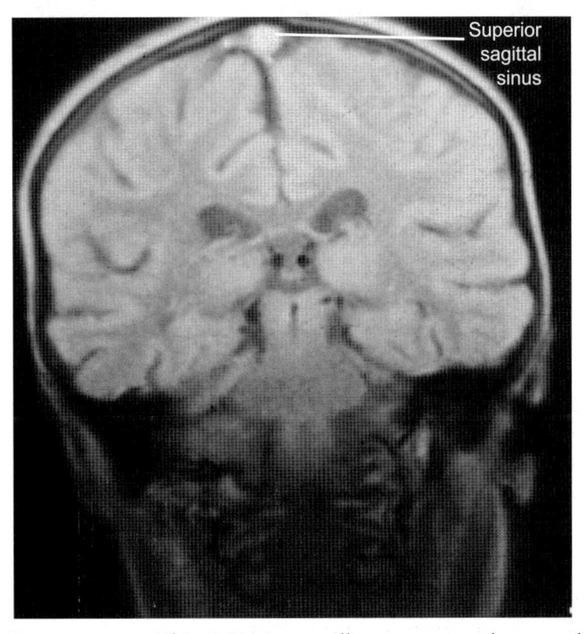

Figure 8.54 This MR image illustrates an obstructed superior sagittal sinus

Anteroinferior group of dural sinuses

• The sphenoparietal sinus (Figures 8.56) runs along the inferior surface of the lesser wing of the sphenoid bone, joining the cavernous sinus on both sides. It receives the anterior temporal diploic veins and a branch of the middle meningeal vein.

• The cavernous sinuses (Figures 8.55 & 8.56) are located on both sides of the sella turcica, sphenoidal body, and the pituitary gland. Each sinus extends from the superior orbital fissure to the apex of the petrous temporal bone, and is connected to the pterygoid venous plexus via the emissary veins of the foramen ovale. It is important to note that this sinus contains the abducens, oculomotor, trochlear, ophthalmic, and maxillary nerves, as well as the internal carotid artery and associated sympathetic plexus.

• The superior petrosal sinus (Figure 8.56) follows the corresponding sulci, connecting the cavernous sinus to the transverse sinus. It receives the inferior cerebral as well as the superior cerebellar veins.

• The inferior petrosal sinus occupies the inferior petrosal sulcus and connects the cavernous sinus to the internal jugular veins. It receives the superior cerebellar and labyrinthine veins and venous tributaries from the pons and medulla. It courses between the petrous temporal bone and the basilar part of the occipital bone.

• The arachnoid (spidery) mater (Figure 8.58) is a delicate trabecular layer that covers the exterior of the brain without following the sulci, and lies between the dura and pia mater. Leptomeninges is a common term that refers to the combined pia–arachnoid layer. The arachnoid mater is separated from the pia and dura mater via the subarachnoid and subdural spaces, respectively. Dilatations of the subarachnoid space around the brain and brainstem that contains the cerebral vessels are termed as cisterns. These cisterns occupy strategic locations around the brain and brainstem, which include the cerebellomedullary cistern, cisterna ambiens, pontine and interpeduncular cistern, the cistern of the lamina terminalis, and the supracallosal cistern.

• The cerebellomedullary cistern (cisterna magna) establishes communication with the fourth ventricle via the foramina of Luschka and Magendie, containing the terminal branches of the posterior inferior cerebellar artery.

• The cisterna ambiens is located posterior to the pineal gland and contains the great cerebral vein of Galen.

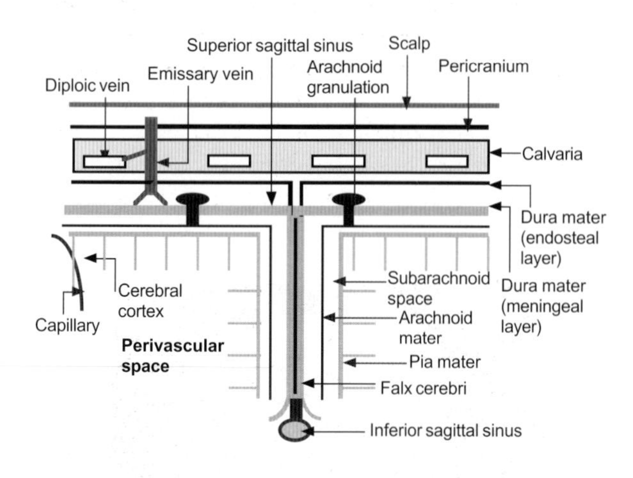

Figure 8.55 Schematic drawing of the scalp, calvaria, and superior sagittal sinus. The emissary veins and arachnoid granulations are shown to indicate the connection of the cerebrospinal fluid and extracranial veins to the superior sagittal sinus

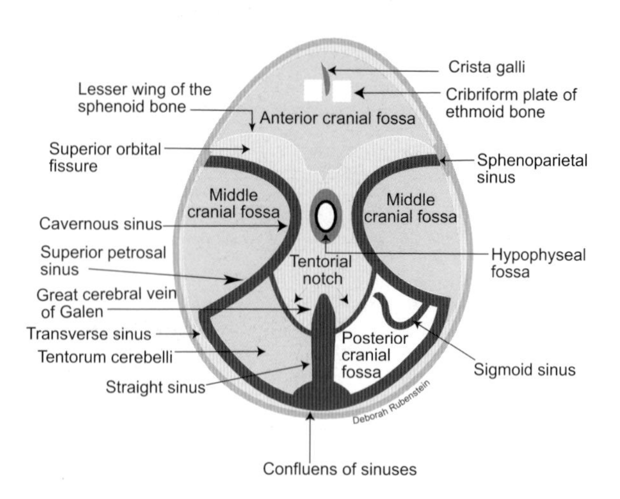

Figure 8.56 Diagram of the dural sinuses in the base of the cranium. Note the S-shaped configuration of certain dural sinuses on both sides of the cranial base

• The pontine cistern is located on the ventral surface of the pons, containing the basilar artery.
• The interpeduncular cistern encircles the arterial circle of Willis, mamillary bodies, and tuber cinereum.
• The cistern of the lamina terminalis and supracallosal cistern bridge the anterior cerebral artery.

Numerous minute arachnoid projections or villi, guarded by one way valves, enter the superior sagittal and/or transverse sinuses. These one-way valves allow CSF to pass from the subarachnoid space to the systemic circulation, while preventing fluid movement in the opposite direction. It is important to note that the low pressure of the dural sinuses relative to the intracranial pressure also dictates the flow of CSF. Arachnoid villi, which are present in the fetus and newborn, form the macroscopic arachnoid granulations (Pacchionian bodies) around 18 months of age that exert pressure on the inner surfaces of the calvarium, producing pressure atrophy and visible depressions. Extensive contacts with the ependymal lining of the ventricles, pia mater, adjacent glial cells, and the capillary endothelium of the choroid plexus may be secured by the long path followed by the CSF.

The pia (faithful) mater, the innermost of the meninges, is a delicate layer that follows the architecture of the sulci and contributes to the formation of the blood–brain barrier. When the pia mater (containing small blood vessels) and ependyma join, they form the tela choroidea, which gives attachment to the choroid plexus. The pia mater consists of an outer epipial layer and an inner intima pia layer. As the cerebral arteries course within the subarachnoid space, they become isolated from the pia–arachnoid layer by a space which is continuous with the intrapial periarterial space, a gap between the smooth muscle of arterial capillary and pia mater. At the point of

The cavernous sinus communicates with the facial (angular) veins and pterygoid venous plexuses via the superior and inferior ophthalmic veins, respectively. Thus, thrombosis of the cavernous sinus may occur as a result of spread of infection from the face (e.g. infected acne near the medial canthus), scalp, or infratemporal fossa.

A fistula between the internal carotid artery and cavernous sinus may occur spontaneously or as a result of trauma, presenting with retrobulbar pain, pulsating exophthalmos, double vision, visual deficits, audible orbital or cranial bruit, and dilatation of the conjunctival vessels. Cavernous sinus thrombosis is associated with swelling of the orbital soft tissue, which eventually leads to proptosis (forward displacement or bulging of the eyeball), disturbances of consciousness, hypersomnia, and often headache, chills, diplopia, visual impairment, and eye pain. Exophthalmos is the most frequent presenting sign, while papilledema, facial edema, and subdural hematoma are less frequent findings. Retinal hemorrhage may also occur. This condition may prove to be fatal if it is left untreated.

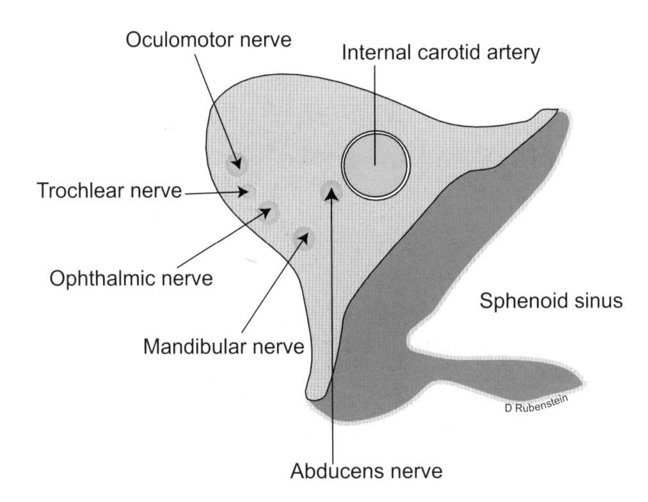

Figure 8.57 Content of the cavernous sinus. This dural sinus is unique in that it contains the internal carotid artery. The abducens nerve is the closest cranial nerve to the internal carotid artery

entrance of the arterial capillaries into the pia mater, one pial layer forms the pia-adventitial layer around the capillaries, which is separated from the external glial membrane by the periarterial subpial space. This layer follows the vessels into the brain forming a perivascular sheath which becomes discontinuous and eventually disappears as the vessels become capillaries. In this manner, leptomeningeal cells separate and form a regulatory interface between the arteries and the brain tissue, preventing the neurotransmitters released from nerves that supply the cerebral vessels from affecting the brain.

Brain barriers that maintain the ionic composition of brain tissue by selectively allowing or excluding certain substances from entering the brain or nerve tissue are formed by contribution from the pia mater. They (Figure 8.61) are classified into blood–brain barrier, blood–CSF barrier, and blood–nerve barrier. These barriers are essential for the optimum functions of the central and peripheral nervous systems. However, small lipophilic molecules can gain access to the brain, utilizing specific transport systems and are not affected by the barrier system. There are independent transporters for D-glucose, and acidic, basic and neutral amino acids.

• Blood–brain barrier is formed by the tight junctions of endothelial cells of the cerebral capillaries that prevent transcapillary movements and selectively impede the diffusion of large molecules, while allowing substances with high lipid solubility (Figure 8.61). It has been suggested that this barrier system depends upon the close interrelationship of the

Occlusion or atresia of the foramina Lushka or Magendie may produce a non-communicating hydrocephalus, which is characterized by enlargement of the ventricles, compression of the brain, and the subsequent thinning of the cerebral cortex (cortical atrophy). Varying degrees of mental retardation and skull enlargement may also occur.

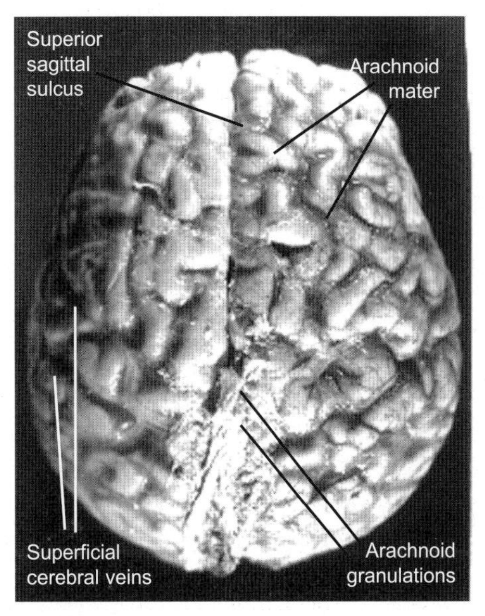

Figure 8.58 This photograph illustrates the arachnoid mater and arachnoid granulations

astrocytic end-feet with the basement membrane of the capillary endothelial cells. These endothelial cells contain large number of mitochondria, and harbor enzymes that govern transport of ions to and from the CSF. They are non-contractile and do not contain actomysin filaments that respond to histamine. This fact may account for the unresponsiveness of brain capillaries to allergic disorders associated with histamine release, although local release of norepinephrine may result in reduction of blood flow into the brain capillaries via its action on the pericytes around these capillaries. In the newborn, the endothelium of the cerebral capillaries of the blood–brain barrier allows large molecules like albumin to be carried by their pinocytotic vesicles. This explains the high level of albumin in the CSF in neonates compared to infants. The blood–brain barrier may be disrupted as a result of brain tumors, or as a sequel to bacterial meningitis. Stroke-induced cerebral edema may also impair the function of the blood–brain barrier.

In general, the blood–brain barrier prevents substances circulating in the blood stream from gaining access to the brain and also plays a role in

Meningiomas (Figure 8.59), which are the most common benign primary intracranial tumors, are believed to arise from the arachnoid villi along the course of the superior sagittal and sphenoparietal sinuses. These tumors compress the brain and produce focal seizures often as an early sign. Also occlusion of the arachnoid villi due to thrombosis, infection or tumors may produce a communicating hydrocephalus.

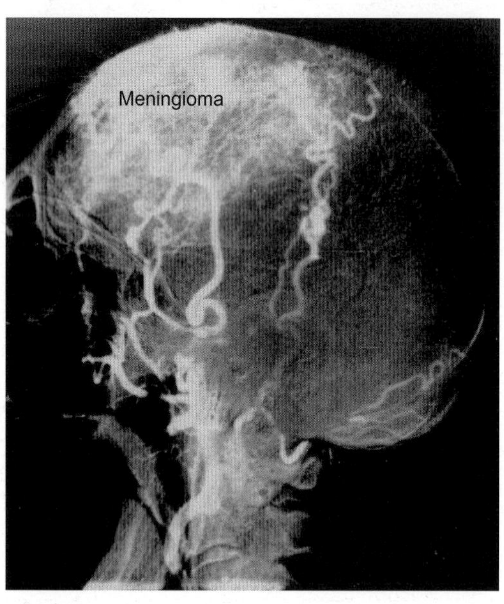

Figure 8.59 An angiogram of the internal carotid artery showing meningioma in the parietofrontal area

their modification and metabolism. Diffusion of large concentrations of D-glucose (2–3 times more than normally metabolized by the brain) through the blood–brain barrier is facilitated by an insulin-dependent GLUT-1 glucose transporter. Reduced GLUT-1 transporter may be associated with seizures, impaired brain development, and mental retardation. Passage of structurally related essential amino acids (precursors of catecholamines and indolamines) through the cerebral capillaries to the brain is mediated by a single transporter. This allows an intense competition among neutral L-amino acids; thus elevation in the plasma concentration of a rival amino acid may account for the inhibition of uptake of others. Therefore, high plasma levels of phenylalanine, as in phenylketonuria, may remarkably reduce the uptake of the competing essential amino acids. However, amino acids that are synthesized in the brain (such as amino acid neurotransmitter) are actively transported in a reverse direction outside the brain. CO_2, O_2, N_2O and volatile anesthetics diffuse rapidly into the brain.

Subarachnoid hemorrhage and/or hematomas are commonly caused by rupture of the congenital berry aneurysm of the arterial circle of Willis. These saccular or oval-shaped, berry-like dilatations frequently occur near the proximal branches of the cerebral arteries (e.g. middle cerebral and posterior communicating arteries) where the muscular and elastic layers are deficient. They are associated with aortic coarctation and polycystic kidney. Compression of the optic, oculomotor, and trigeminal nerves by these aneurysms may occur prior to their rupture. Bleeding into the subarachnoid space produces symptoms that are frequently sudden in onset and include nuchal rigidity, severe ipsilateral headache, nausea, vomiting, transient vertigo, and occasional syncope (transient loss of consciousness that is preceded by lightheadedness). Hearing and visual impairment, as well as seizures (which occur in less than 20% of patients) may also be detected. Consciousness usually remains unaffected; however, disorders of concentration and attention, amnesia, visual and hearing impairment may be observed. Signs of oculomotor nerve palsy and possible hemiplegia are also seen in patients with subarachnoid hemorrhage. CT scan and lumbar puncture (LP) may diagnose this condition. Neurologic complications may also include hydrocephalus, resulting from a clot that blocks the CSF pathway in the posterior fossa, and vasospasm that produces cerebral infarctions.

Arachnoidal (leptomeningeal) cysts (Figure 8.60) are congenital lesions that develop from splitting of the arachnoid mater. Intrasellar cysts occupy extradural positions. They are divisible into simple and complex arachnoid cysts. Simple cysts maintain the ability to actively secret CSF, and are commonly found in the middle cranial fossa. Complex cysts may contain neuroglia and ependyma. Presentations of arachnoid cysts show variation according to their location. Increased intracranial pressure (ICP), visual deficits, developmental retardation, precocious puberty, hydrocephalus, and signs of 'bobble-head doll' syndrome are seen in suprasellar arachnoid cysts. Cysts in the middle cranial fossa may produce hemiparesis, convulsions, and headache. Diffuse supra- or infratentorial cysts may also produce growth retardation, craniomegaly, and increased intracranial pressure.

• Blood–CSF barrier is formed by the epithelial cells of the choroid plexus, which are connected by tight junctions, and the overlying pia–glial and ependyma–glial membranes. Thus, ventricular CSF

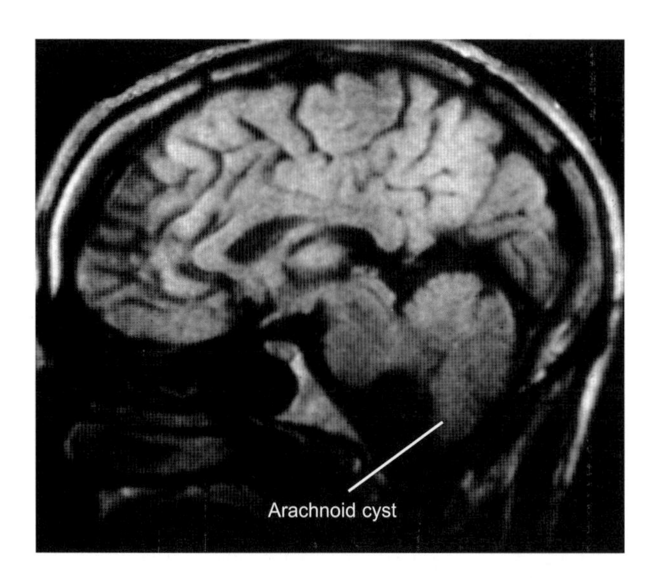

Figure 8.60 This MRI image illustrates an arachnoid cyst in the posterior cranial fossa, inferior and ventral to the cerebellum

diffuses into brain extracellular fluid and eventually into the subarachnoid space. This barrier system is well illustrated in patients with jaundice where the bile is selectively excluded from entering the CSF or brain. The presence of yellowish bile stain only in the stroma of the choroid capillaries substantiates the activity of this barrier. Na^+–K^+–ATPase system, which provides a pump allowing sodium into CSF and potassium into the plasma, is contained in the choroid epithelium. This ionic movement assists the passage of a large amount of plasma water from the capillary bed.

- Blood–nerve barrier comprises the perineurium and the capillaries of the endoneurium. The wall of these capillaries is non-fenestrated and the endothelial cells establish tight junctions. This barrier is functionally much more effective in the dorsal root ganglia and autonomic ganglion.

It must also be remembered that the optimal function of cerebral neurons depends upon the stability of the glucose level in the extracellular space, brain potassium concentration, as well as transmitters' levels in the CSF and the brain, which are maintained by the barrier system. Normal CSF and extracellular fluid levels of potassium are maintained via active transport systems that operate in the choroid plexus and brain capillaries. A rise in the level of blood potassium renders the transport systems non-operational, whereas excess of potassium in the CSF is removed via active transport into the blood.

Not all areas of the brain harbor the barrier systems, in fact structures which occupy positions near the ventricles surrounding the brainstem and the diencephalon known as the circumventricular organs (Figure 8.62) lack the barrier system and operate as points of contact between certain brain receptors and blood. These specialized areas, which may utilize the neurohumoral mechanism to exert their influences, include the subfornical organ, area postrema, and organ vasculosum of the lamina terminalis, neurohypophysis and intermediate lobe of the pituitary gland, medial eminence, pineal gland, and the subcommissural organ.

- The subfornical organ is located near the anterior pole of thalamus and the interventricular foramen of Monro, maintaining massive projections to the supraoptic and paraventricular nuclei and the lateral hypothalamic area. Through these connections, it plays an important role in homeostasis, osmoregulation and the circulation of blood in the choroid plexus. It binds angiotensin II and receives input from the hypothalamus, regulating water intake and inducing vasopressin secretion.
- The area postrema (AP), an emetic chemoreceptor center, is located near the obex at the point of junction of the lateral walls of the caudal half of the fourth ventricle, receiving projection from the spinal cord and solitary nucleus. AP is intimately interconnected with the solitary tract and nucleus. It is sensitive to apomorphine and digitalis glycosides, regulating food and water intake and cardiovascular functions.
- The vascular organ of the lamina terminalis is a highly vascular structure, which lies superior and rostral to the optic chiasma and carries hypothalamic peptides (such as somatostatin, angiotensin II, and atrial natiruretics). Structurally, it is similar to the medial eminence, maintaining functional relationship with the preoptic area and preserving fluid balance.
- The neurohypophysis receives terminals of the hypothalamic neurons that convey oxytocin, neurophysin, and vasopressin via the hypothalamo-hypophyseal tract.
- The medial eminence serves as a link and transducer between the hypothalamic neurons that secrete the hormone regulating factors, and the portal-hypophysial system. It also deals with the transduction of hypothalamic neuronal impulses.
- The pineal gland is discussed in detail with the epithalamus.
- The subcommissural organ is located ventral to the posterior commissure and at the site of junction of the third ventricle and cerebral aqueduct. It secretes proteinaceous materials into the CSF; however, its role in salt and water balance is not yet confirmed.

Since both the ventricular system and the subarachnoid space contain the CSF, a brief review of the

Clinical significance of barriers can be illustrated in: (1) kernicterus, (2) cerebral edema, (3) brain scans, (4) the therapy of Parkinsonism, (5) epinephrine surge, and (6) hyperglycemia and/or hypoglycemia. Other conditions that exhibit the role of the brain barrier are ischemia, and certain viral or autoimmune diseases.

(1) In infants, the immature liver can not conjugate large amounts of bilirubin to serum albumin, leading to an increase of the unconjugated bilirubin. Inability of the blood–brain barrier to block the unconjugated bilirubin from entering the CNS may be followed by its deposition in the basal and brainstem nuclei. Deposition of unconjugated bilirubin in the CNS is commonly detected in kernicterus or erythroblastosis (neonatorum) fetalis, a condition that develops as a consequence of Rh incompatibility between the mother and the fetus. It may also be associated with hemolytic diseases (low albumin reserve), reduced amount of glucuronyl transferase in the preterm infant, biliary atresia, low pH and hepatitis. Infants with this predicament initially exhibit hypotonia, followed by rigidity and gaze palsies. They are likely to die and autopsy may reveal discoloration of the brain tissue particularly of the basal nuclei. Survivors show signs of mental retardation and some forms of choreiform movements. Bilirubin toxicity may also occur as a result of lack or ineffective CNS bilirubin oxidase system due to birth trauma or incomplete development.

(2) Cerebral edema may be caused by bradykinin, which stimulates the influx of protein into the brain and subsequent enlargement of the extracellular space of the white matter.

(3) Brain scans also utilize the blood–brain barrier concept, and contrast medium injected into the blood breaks down the barrier, allowing visualization and localization of lesions (gray matter contains large number of capillaries that greatly exceeds the white matter) via various imaging techniques.

(4) Administration of L-dopa, a precursor of dopamine, which is able to cross the blood–brain barrier by a neutral amino acid carrier, may be used in the treatment of Parkinson's disease.

(5) Neuronal function is maintained and released from the disastrous consequence of epinephrine surge by the blood–brain barrier.

(6) Neuronal activities may be inhibited and coma may result in individuals with hyperglycemia that eventually leads to accumulation of ketone bodies in the brain. Overactivity of the CNS and mental confusion may occur in hypoglycemia. Lack of glucose in this condition may lead to insulin coma.

ventricles with major emphasis on CSF, its pathway and associated clinical conditions will be needed.

The ventricular system (Figures 8.62, 8.63 & 8.65), a derivative of the neural canal, is comprised of the lateral, third, and fourth ventricles, cerebral aqueduct, and the central canal. This system communicates with the venous blood via the subarachnoid space.

Secretion of the CSF is maintained at a rate of 0.3–0.4 ml/min by the choroid plexus (Figures 8.63 & 8.65) and to a much lesser degree by the ependyma. The choroid plexus is formed by the vascular epithelium of the pia mater and the ependyma that invaginates through the choroidal fissure, containing mostly simple cuboidal ependymal cells with microvilli of the brush border type. This plexus is innervated by the postsynaptic sympathetic fibers that emanate from the superior cervical ganglion of the sympathetic trunk. Selective prevention of the free entrance of protein and electrolytes from the blood to brain tissue is maintained by the ependymal cells that form tight junctions. Secretion and transport of certain hormones such as transthyretin, a carrier of thyroxin and retinol and insulin-like growth factor-II into CSF may also be accomplished by the choroid plexus. Hardened bodies called psammoma (sand-like), which are composed of concentric rings of calcium carbonate, calcium and magnesium phosphate, occur normally in the adult choroid plexus. CSF is circulated about three times in 24 hours, and the absorption rate can be 4–6 times the normal rate of formation. Secretion of CSF may occur through a variety of processes that include filtration across the endothelial wall, hydrostatic pressure dependent activity, and an enzymatically controlled active process by the choroidal epithelium. The latter process is activated by ATPase and carbonic anhydrase. Cardiac glycosides produce inhibition of CSF secretion by uncoupling the mitochondrial oxidative phosphorylation. The steady secretion and absorption of the CSF is essential for maintaining a uniform pressure within the ventricles and the cranial vault. Vast numbers of 5-hydroxytryptamine receptors may influence the blood flow in the choroid plexus.

Cerebrospinal fluid regulates ionic transport to and from the extracellular space (sink action), maintaining low concentrations of certain substances in the CSF and brain relative to plasma concentrations. This allows the creation of an efficient milieu for the conduction of nerve impulses. It also acts as a buffer to lessen the impact of head trauma (buoyancy effect), provides nutrients to the leptomeninges, and dramatically reduces the weight of the brain. Increased intracranial pressure and brain volume, as a result of vasodilatation of cerebral vessels or

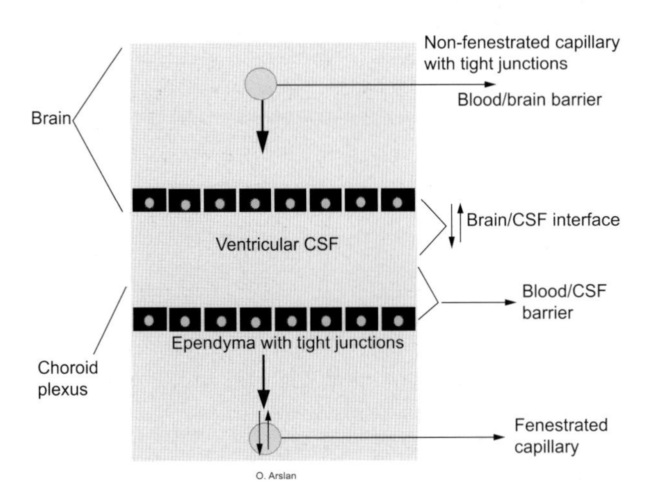

Figure 8.61 Diagram of the neural and vascular elements involved in the formation of the blood–brain and blood–cerebrospinal fluid (CSF) barriers

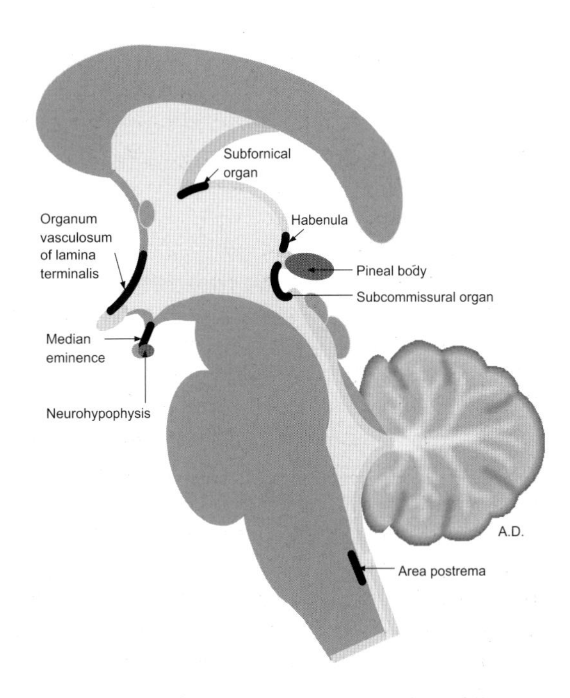

Figure 8.62 Drawing of the midsagittal brain illustrating the sites where blood–brain barriers are absent

parenchymal swelling, may be counteracted by displacement of the CSF. The role of CSF in signal transduction, transport of hormones, and immune reaction has also been suggested.

Respiratory movements influence circulation of the CSF, cardiac systole associated changes in the intracranial blood volume, current created by ependymal cilia, and by arterial pulsation of the choroid plexus. The normal pressure of CSF ranges between 6–14 cmH_2O, which could be monitored by a manometer attached to the lumbar puncture needle. This pressure remains constant at 5–20 cmH_2O unless an increase or decrease in brain size or blood volume occur (e.g. compression of the internal jugular veins increases the venous return of dural sinuses and subsequently produces increased intracranial pressure).

In general, the circulation of the CSF involves the following paths:

Lateral ventricle → foramen of Monro → third ventricle → cerebral aqueduct → fourth ventricle → foramen of Luschka and foramen of Magendie → cerebellomedullary cistern (cisterna magna) → superior sagittal. The CSF is also absorbed at the levels of spinal nerve root sheaths.

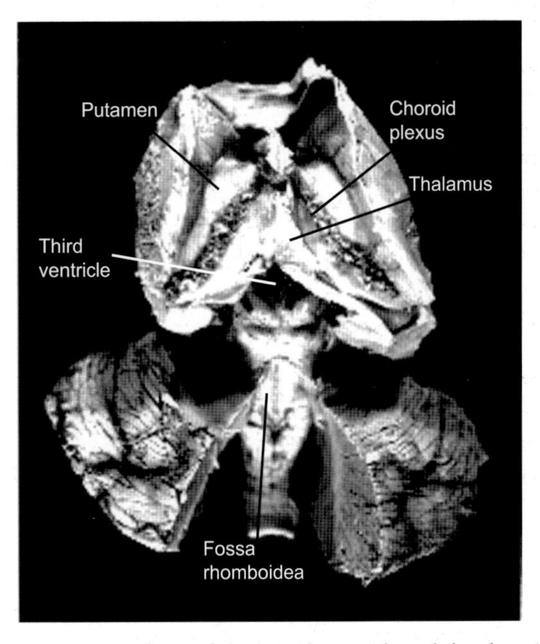

Figure 8.63 The floor of the lateral ventricle and the choroid plexus. The third ventricle and the floor of the fourth ventricle are also illustrated. The cerebral hemispheres, corpus callosum and part of the cerebellum are removed

The CSF is a clear, colorless fluid that maintains a normal pH of 7.3. Turbidity or pinkish discoloration may be associated with presence of fresh blood from subarachnoid hemorrhage; yellowish discoloration of CSF (xanthochromia) may result from disintegrated blood or increased amount of protein and/or bilirubin. It contains trace amount of protein, mainly immunoglobulin and to a lesser extent albumin. It also contains a few leukocytes and a small amount of glucose (80–120 mg/dl) and potassium (2.9 mEq/l), with greater concentration of sodium and chloride. This is in contrast to serum, which contains lower sodium concentration and higher potassium and calcium. The arachnoid mater, which forms part of the blood–CSF barrier, may also play a role in changing the composition of the CSF. Numerous peptides are also found in the CSF, which include luteinizing hormone releasing factor, cholecystokinin, angiotensin II, substance P, somatostatin, thyroid-releasing hormone, oxytocin, vasopressin, etc. Alternations in the CSF concentrations of peptides may be used for the diagnosis of certain neurological diseases.
The changes in the composition of the CSF may serve as a tool in the diagnosis of certain diseases. A low pH value may occur in conditions associated with acidosis, hypercapnia, and in certain pulmonary diseases. Increased protein concentration is observed in spinal shock (complete transection of spinal cord) and in cases of extra- and intramedullary spinal cord tumors. Immunoglobulin G (IgG) is generally elevated in multiple sclerosis. Glucose levels decrease in bacterial, fungal, and viral meningitis. The CSF may contain neoplastic cells in primary and secondary neoplasms.

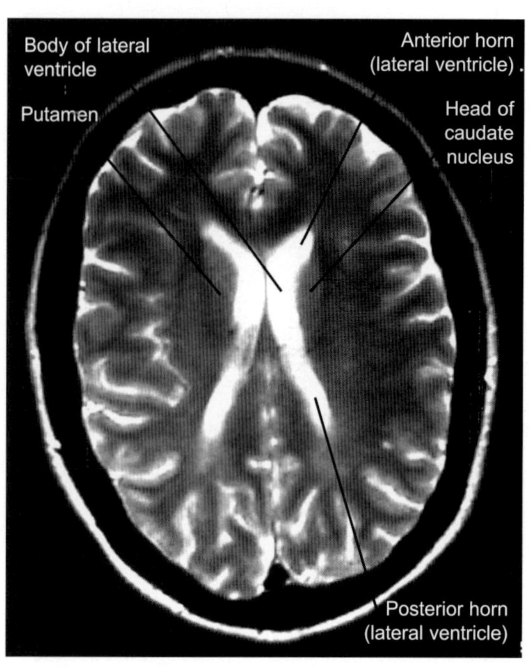

Figure 8.64 MRI scan of the brain showing some of the components of the lateral ventricles

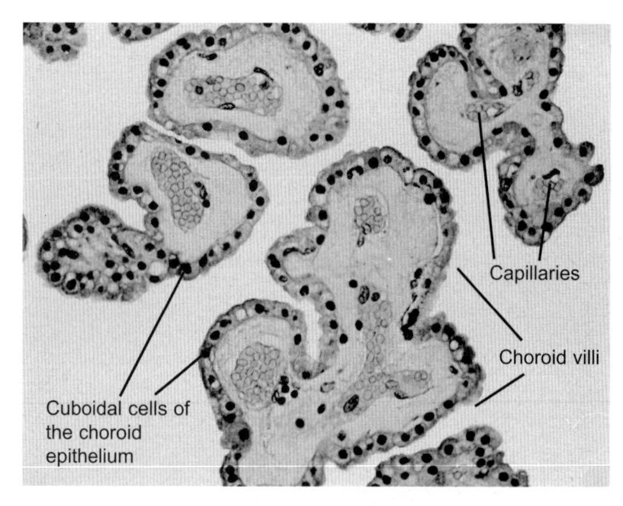

Figure 8.65 Photograph of the choroid plexus. Note the cuboidal cells of the choroid epithelium and associated capillary network

Accumulation of CSF in the ventricles produces hydrocephalus (Figures 8.67 & 8.68). This condition commonly results from obstruction of CSF pathways as in ependymoma; over-secretion of the CSF as in a papilloma of the choroid plexus; venous insufficiency as in dural sinus thrombosis; or defective absorption due to occlusion of the arachnoid villi as in arachnoiditis. Hydrocephalus may also be produced as a result of atrophy or reduction in the total volume of the brain (hydrocephalus *ex vacuo*) as is seen in Alzheimer's disease.

Intracranial pressure (ICP) increase may be associated with an abnormal increase in the CSF volume, cerebral edema, intracranial hemorrhage, or impairment of venous drainage. ICP may produce papilledema, nausea and vomiting (due to activation of the vomiting center in the upper medulla), bradycardia (slowing of the heart due to vagal activation), and coma.

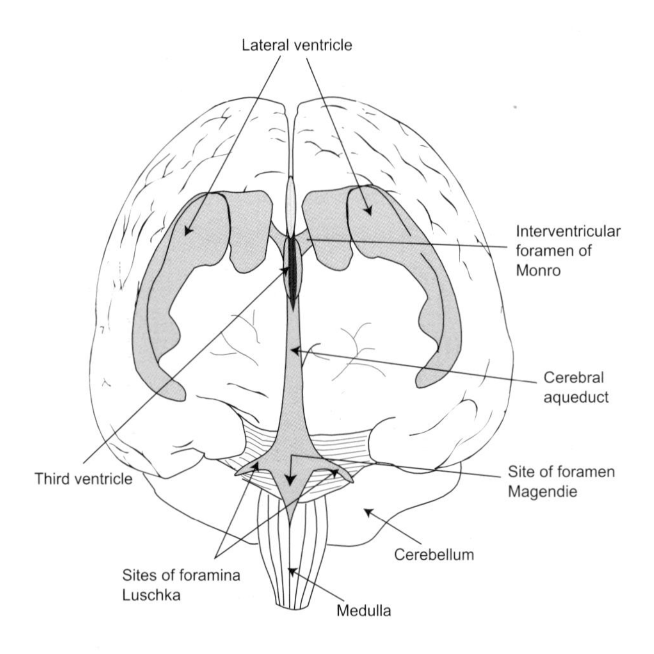

Figure 8.66 The ventricular system (posterior view)

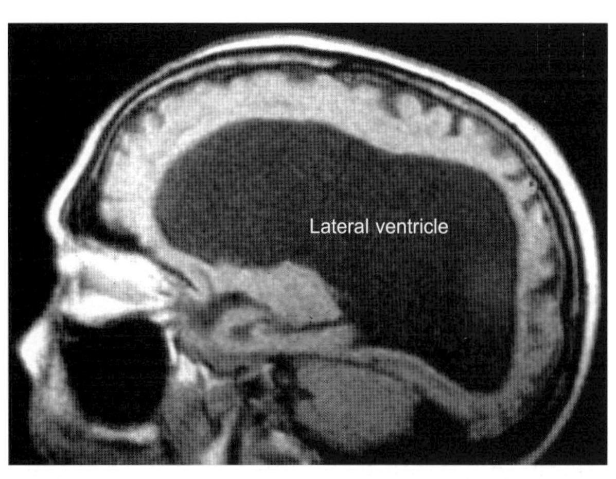

Figure 8.67 This MRI image shows the massive dilatation of the lateral ventricle in an individual with obstructive hydrocephalus. Note the remarkable thinning of the cerebral cortex

Hydrocephalus may be congenital or acquired:

Congenital (overt-infantile) hydrocephalus occurs usually in the first few months of life or inside the uterus. It rarely extends to the fifth decade of life. It is commonly associated with stenosis of the cerebral aqueduct; obstruction of the foramina Luschka and Magendie (Dandy–Walker syndrome) or obliteration of foramen Magendie and cisterna magna (as in Arnold–Chiari Type II malformation). It may also result from asymmetrical fusion of cervical vertebrae, imperfect fusion or non-union of the vertebral arches, meningomyelocele and closure of the foramina Magendie and Luschka (Klippel–Feil deformity). Congenital hydrocephalus commonly presents with meningomyelocele, increased intracranial pressure, separation of the sutures (diastasis), and protrusion of the fontanelles. It is also associated with enlargement of the head, thinning of the scalp, visible superficial vessels, and downward (setting sun sign) position of the eyes. Parinaud's syndrome, which exhibits vertical gaze palsy, may also be seen as a result of the pressure exerted on the pretectal area. Disorders of respiration (apneutic episodes), abducens nerve palsy, and 'cracked pot sound' upon percussion of the enlarged lateral ventricles (Macewen's sign) are also observed in hydrocephalic young children. It

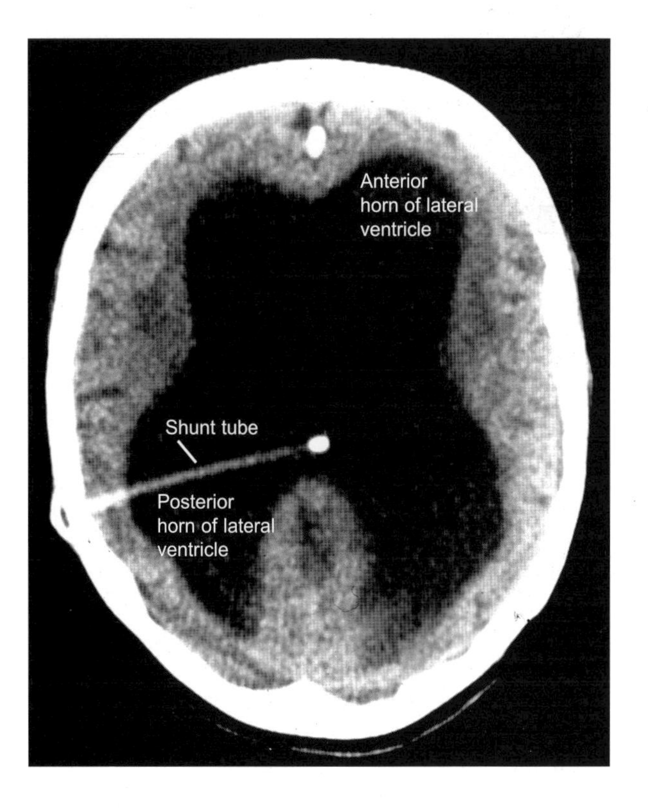

Figure 8.68 This image demonstrates a massive dilatation of the lateral ventricle in an individual with hydrocephalus. A surgical shunt was placed to drain the excess cerebrospinal fluid

is interesting to note that the cerebral cortex is less affected than the white matter and many patients survive this condition because of the capacity of the skull to expand. Signs of upper motor neuron palsy may also be seen. Although impairment of vision and paleness of the optic disc may be observed upon examination of the fundus, papilledema will not be seen in these individuals.

Acquired (occult) hydrocephalus may be caused by subarachnoid hemorrhage, post-meningitis, cysticercosis, vascular malformation, as well as tumors of the third ventricle, thalamus, and cerebral hemispheres. It is characterized by visible papilledema, and less prominent frontal and occipital headache. It exhibits signs of frontal lobe dysfunctions such as inattentiveness, marked slowness in response, and easy distractibility. Memory impairment and reduction in mental and physical capacity are also noted. In older children and adults with rigid calvaria, gait apraxia and amnesia followed by dementia and slowing of thought processes and later urinary incontinence may also occur. Imaging of the brain may reveal herniation of the third ventricle, erosion of the sella turcica, and atrophy of the corpus callosum. The onset of this condition is subacute and can cause intellectual deterioration followed by a restriction of movement. Manifestations of upper motor neuron palsy, which include deep tendon hyperreflexia of the lower extremities, and Babinski's sign, may be detected along with gait disturbances. Surgical treatment involves ventriculo-peritoneal shunting, which is the most commonly used technique in adults.

It may also involve subcutaneous shunting, and draining the CSF into a systemic vein through the internal jugular vein. While improvement of gait is less likely, urinary incontinence is the most likely symptom to improve with shunting, while dementia is the least likely to undergo any changes. Hydrocephalus may also be classified into obstructive and non-obstructive forms:

Obstructive hydrocephalus, the most common type of hydrocephalus, results from obstruction of the flow of CSF inside the ventricular system or at its drainage sites into the dural sinuses. It may occur as a result of stenosis of the cerebral aqueduct, atresia of foramina Magendie and Luschka, obstruction of the fourth ventricle (as in Arnold–Chiari syndrome), or obstruction of the interventricular foramen of Monro. Obstructive hydrocephalus may be divided into communicating and non-communicating types.

Communicating hydrocephalus occurs as a result of obstruction of the CSF pathway outside the ventricular system (e.g. in the cisterns and arachnoid villi), while normal connection between the ventricular system and the subarachnoid space via the foramina Magendie and Luschka is maintained. It is a slowly developing condition, which is characterized by ventricular enlargement, deterioration of mental faculties, and bilateral upper motor neuron palsy. It is seen in Arnold–Chiari malformation and leptomeningitis.

Normal pressure hydrocephalus (NPH) is one form of the communicating hydrocephalus which develops over weeks or months as a result of a subarachnoid hemorrhage, post-traumatic events, post-meningitic conditions, or posterior cranial fossa tumors. NPH is a reversible condition, which exhibits no signs of increased intracranial pressure, papilledema or cranial nerves dysfunctions. It is presumed to be due to partial obliteration of the subarachnoid space around the cerebral hemispheres, combined with defective reabsorption of the CSF through the arachnoid villi. It is seen in 15% of Alzheimer's patients and may be associated with basilar artery ectasia and fluctuations in cerebrospinal pressure. Radiographic imaging (CT and MRI scans) may show expansion of the temporal lobe and ventricles with minimal or no cortical atrophy. The prominent clinical manifestations of this condition includes gait apraxia, dementia, and urinary incontinence. With treatment, gait apraxia is the first sign to improve. Urinary incontinence may involve urgency and frequency of urination, but in severe cases, patients are totally incontinent. Dementia, which includes psychomotor and cognitive impairment, is unusual.

Non-communicating hydrocephalus is caused by an obstructive process within the ventricular system, producing loss of communication between the ventricular system and the subarachnoid space. Obstruction sites may include foramina of Monro, Magendie, or Luschka or the cerebral aqueduct. This form of hydrocephalus is often fatal and progresses very rapidly.

Non-obstructive hydrocephalus is a sequel to papilloma of the choroid plexus that results in oversecretion of CSF. It may also be caused by defective absorption at the arachnoid villi subsequent to hemorrhage, infections, dural fibrosis, or thrombosis of the superior sagittal sinus.

Suggested reading

1. Annett M. Laterality and cerebral dominance. *J Child Psychol Psychiatry* 1991;32:219–32
2. Bertoni JM, Brown P, Goldfarb LG, *et al.* Familial Creutzfeldt-Jakob disease (codon 200 mutation) with supranuclear palsy. *JAMA* 1992;268:2413–15
3. Bouchaud V, Bosler O. The circumventricular organs of the mammalian brain with special reference to monoaminergic innervation. *Int Rev Cytol* 1986;105: 283–327
4. David AS. The split-brain syndrome. *Br J Psychiatry* 1989;154:422–5
5. Davis JM, Davis KR, Crowell RM. Subarachnoid hemorrhage secondary to ruptured intracranial aneurysm: prognostic significance of cranial CT. *Am J Roentgenol* 1980;134:711–15
6. Davson H, Segal MB. *Physiology of the CSF and Blood-Brain Barriers*. Boca Raton, FL: CRC Press, 1996
7. Dudley AW. Cerebrospinal blood vessels: Normal and diseased. In Haymaker W, Adams RD, eds. *Histology and Histopathology of the Nervous System*. Springfield, IL: Thomas, 1982
8. Endo M, Kawano N, Miyasaka Y, *et al.* Cranial buff hole for revascularization in Moyamoya disease. *J Neurosurg* 1989;71:180–5
9. Farah MJ. Agnosia. *Curr Opin Neurobiol* 1992;2:162–4
10. Foster JM. Frontal lobes. *Curr Opin Neurobiol* 1993;3:160–5
11. Galaburda AM. Anatomic basis of cerebral dominance. In Davidson RE, Hugdahl K, eds. *Cerebral Asymmetry.* Boston: MIT Press, 1994:51–73
12. Halliday AL, Chapman PH, Heros RC. Leptomeningeal cyst resulting from adulthood trauma: case report. *Neurosurgery* 1990;26:150–3
13. Howard MA, Gross AS, Dacey RG, *et al.* Acute subdural hematomas: an age-dependent clinical entity. *J Neurosurg* 1989;71:858–63
14. Joynt RJ, Honch GW, Rubin AJ, *et al.* Occipital lobe syndromes. In Frederiks JAM, ed. *Handbook of Clinical Neurology*, volume I. Amsterdam: Elsevier Science Publishers, 1985:49–62
15. Lechtenberg R. *Seizure Recognition and Treatment.* New York: Churchill Livingstone, 1990
16. Masliah E, Terry RD. Role of synaptic pathology in the mechanisms of dementia in Alzheimer's disease. *Clin Neurosci* 1993;1:192–8
17. Miller SL, Delaney TV, Tallal P. Speech and other central auditory processes: Insights from cognitive neuro-science. *Curr Opin Neurobiol* 1995;5:198–204
18. Ondra SL, Troupp H, George ED, *et al.* The natural history of symptomatic arteriovenous malformations of the brain: A 24-year follow-up assessment. *J Neurosurg* 1990;73:387–91
19. Thadani V, Penar PL, Partington J, *et al.* Creutzfeld-Jakob disease probably acquired from a cadaveric dura mater graft. *J Neurosurg* 1988;69:766–9
20. Vassilouthis J. The syndrome of normal-pressure hydro-cephalus. *J Neurosurg* 1984;61:501–9

Section III

Peripheral nervous system

The peripheral nervous system (PNS) consists of somatic, visceral (autonomic), sensory, and motor nerve fibers. These fibers are contained within the spinal and cranial nerves. Virtually all spinal and cranial nerves, with the exception of the olfactory and optic nerves, are integral parts of the PNS. However, the neuronal cell bodies which form the PNS are located within the central nervous system (CNS). The PNS is further classified into autonomic (visceral) and somatic nervous systems. The somatic nervous system (SNS) utilizes acetylcholine as a neurotransmitter. It forms cervical, brachial, and lumbosacral plexuses. Excision of somatic nerve fibers may result in atrophy and paralysis of the denervated structure. The response of the SNS to a stimulus is immediate, fast, and of short duration. This is due to the fact that the synaptic gap between the presynaptic and postsynaptic membranes of the innervated structure is very narrow, allowing a lesser diffusion distance. Additionally, the narrow synaptic gap renders the degradation of the neurotransmitter more efficient. The somatic afferents transmit pain, temperature, touch, and movement sensations from the skin, muscles, and tendons. The neurons of the somatic afferents are located in the dorsal root ganglia, trigeminal and geniculate ganglia, and the superior ganglia of the glossopharyngeal and vagus nerves.

Autonomic nervous system 9

The autonomic nervous system (ANS) consists of neurons that extend in the peripheral and central nervous systems, regulating visceral motor and reflex activities, as well as emotional behavior. It is not a fully autonomous entity, as the name may imply, but rather an interdependent system that functions under massive input from the cerebral cortex. The ANS maintains, through its diverse connections with the somatic nervous system, a stable internal environment or milieu (homeostasis), which is essential for normal physiological functions. It consists of efferent and afferent fibers and central neurons in the spinal cord, brainstem, diencephalon, and the brain. The afferent component transmits visceral pain and organic visceral sensations (e.g. hunger, malaise, nausea, libido, bladder and rectal fullness). The efferent component innervates the smooth muscles, glandular tissue, and sweat glands.

Autonomic neurons and ganglia

The autonomic nervous system (ANS) (Figures 9.1 & 9.2) regulates visceral motor activity, contraction of the blood vessels, glandular secretions and transmission of visceral sensations. This system maintains homeostasis, an essential element for the physiological functions of the visceral organs. As an interdependent entity, the ANS functions under the influence of the cerebral cortex and hypothalamus. In contrast to the somatic nervous system, the innervation of the ANS occurs through two sets of neurons: preganglionic and postganglionic neurons. Preganglionic neuronal cell bodies are located within the CNS (spinal cord and the brainstem), whereas the postganglionic neurons lie within the peripheral nervous system within the paravertebral, prevertebral or the intramural ganglia.

The multipolar neurons of the autonomic ganglia (Figure 9.3) establish facilitatory or inhibitory synapses with the adjacent neurons, interneurons, or afferent cholinergic fibers. Multiplicity and pattern of the synaptic connections between the preganglionic and postganglionic neurons are aimed at intensifying autonomic response.

The axons of the preganglionic neurons consist of thinly myelinated (group B) fibers, whereas axons of the postganglionic neurons are of the unmyelinated type (C-group), rendering the conduction of generated response much slower. In contrast, axons of somatic nerves are principally myelinated.

Terminals of the autonomic fibers in the smooth muscles contain varicosities and synaptic vesicles that resemble that of the somatic nerves. However, these terminals show diffuse and extensive branching in which a terminal nerve fiber may innervate several smooth muscle fibers. This branching is limited and well localized in the fast acting muscles such as the dilator and constrictor pupillae. The synaptic gaps in the ANS are much wider than that of the skeletal (somatic) muscles, which allows far greater diffusion capacity and accounts for the delayed response associated with this system relative to the somatic nervous system. The greater the synaptic gap the slower the degradation processes of the neurotransmitter and the resultant prolonged action.

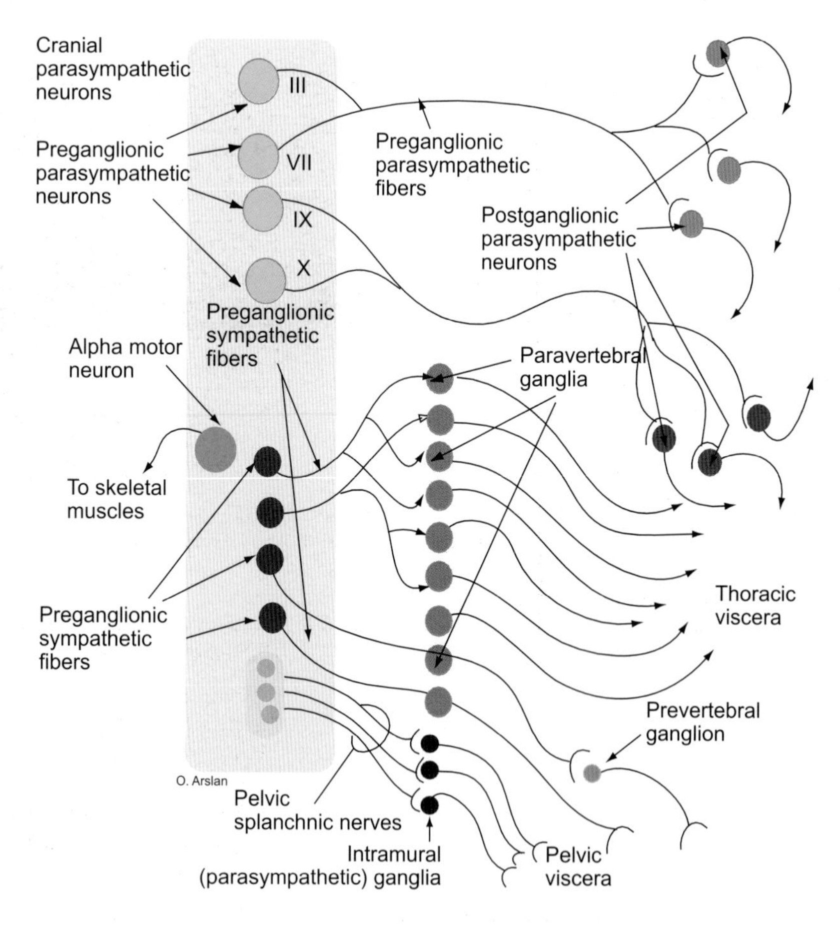

Figure 9.1 Schematic representation of the preganglionic and postganglionic neurons of the sympathetic and parasympathetic systems

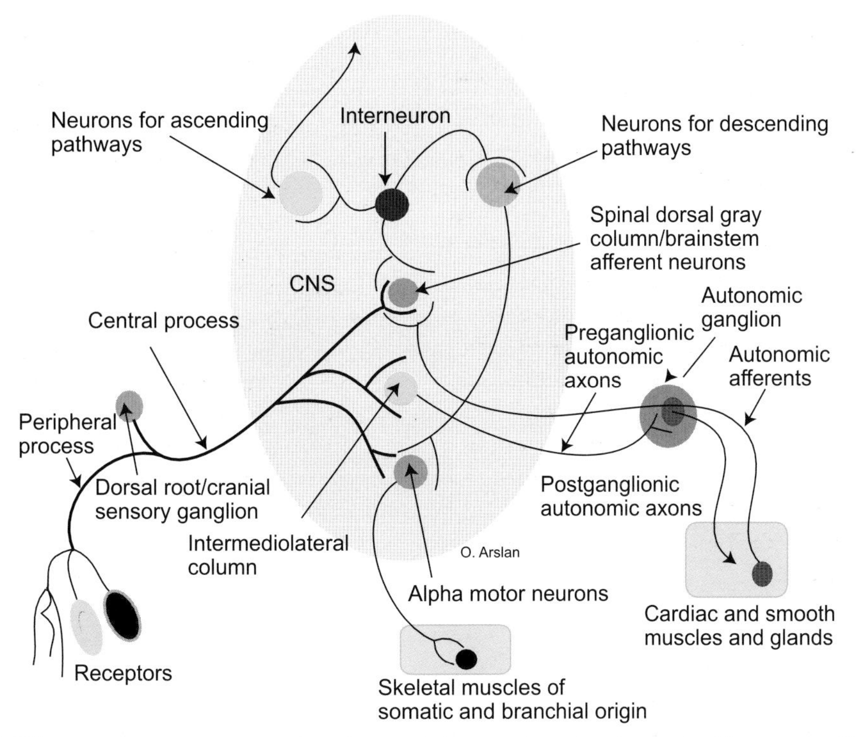

Figure 9.2 General differences between autonomic and somatic innervation, showing both efferent and afferent connections

Due to the direct and uninterrupted connections between the somatic neuronal axons and their targets (e.g. skeletal muscles), the generated response is immediate and rapid.

Excision of the autonomic fibers may not produce atrophy of the denervated organ, as is the case with the somatic denervation.

Autonomic afferents carry a variety of sensations such as visceral pain, thirst, hunger, cramps, and sensations of well-being or malaise, whereas somatic afferents primarily transmit signals associated with tactile, thermal, painful, joint and vibratory sensations.

Additional differences between these two main components of the peripheral nervous system also exist in regard to receptors and type of stimuli.

Visceral receptors are mainly free nerve endings,

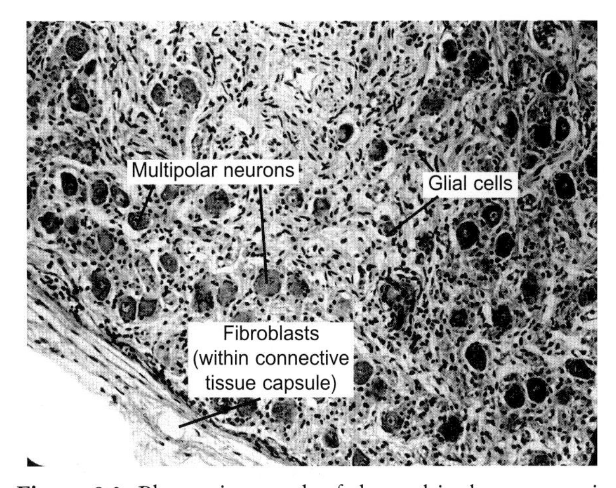

Figure 9.3 Photomicrograph of the multipolar autonomic neurons and associated glial cells

Table 9.1 Ganglia of the peripheral nervous system

General characteristics	*Sensory ganglia*	*Autonomic ganglia*
Location	1. Cranial nerves V, VII & IX 2. Dorsal root ganglia	1. Sympathetic a. Paravertebral ganglia b. Prevertebral ganglia 2. Parasympathetic a. Submandibular, otic, ciliary & pterygopalatine ganglia b. intramural ganglia
Type of neurons	Pseudounipolar or bipolar neurons	Multipolar neurons
Synaptic connections	None	Paravertebral, prevertebral & intramural ganglia

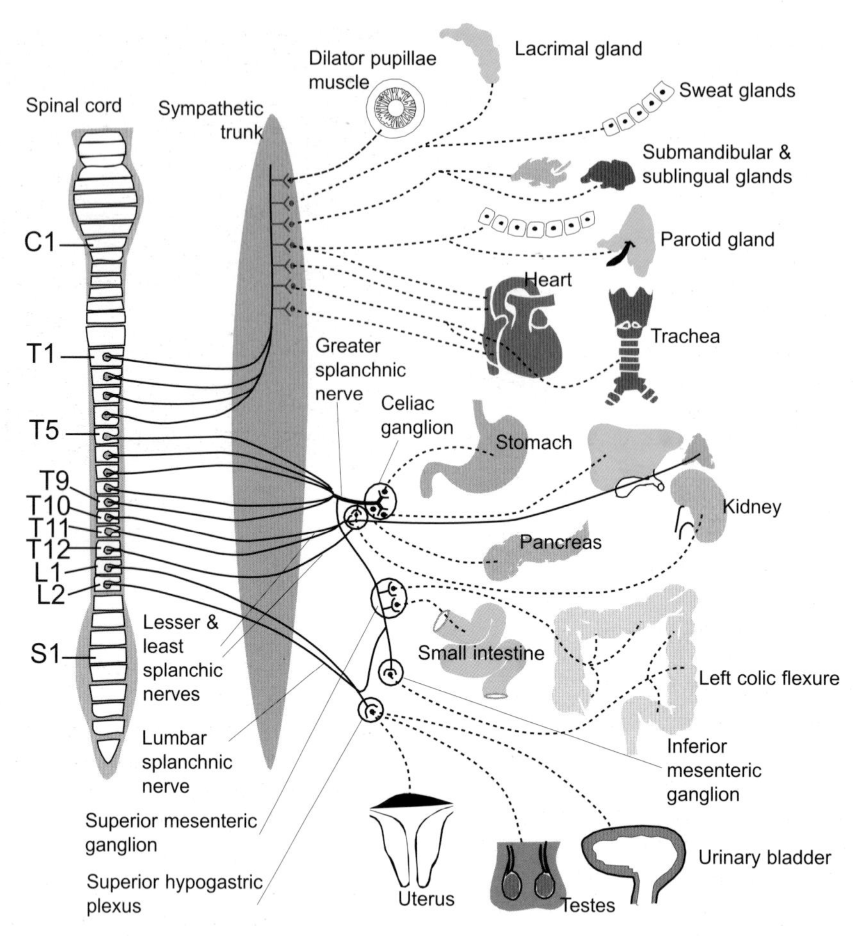

Figure 9.4 The preganglionic and postganglionic neurons of the sympathetic nervous system are illustrated. The splanchnic nerves and their associated ganglia are clearly marked

whereas the somatic system comprises a variety of capsulated and uncapsulated receptors.

Conventional somatic stimuli may not elicit any visceral response; however, ischemia and tension within the wall of a viscus may activate the visceral receptors, eliciting pain.

Despite these differences both visceral and somatic afferents maintain neurons in the dorsal root ganglia. Specifically, cranial somatic sensory neurons constitute the trigeminal, geniculate, and superior ganglia of the glossopharyngeal and vagus nerves, whereas the neurons for visceral afferents are located in the geniculate, and the inferior ganglia of glossopharyngeal and vagus nerves. Additionally, both somatic and autonomic nerves form plexuses. Autonomic plexuses are scattered in the thoracic, abdominal, and pelvic cavities, and surround the corresponding arteries to form the pulmonary, cardiac, celiac, aortic plexuses, etc. Somatic nerve plexuses such as the cervical, brachial and lumbosacral plexuses are formed in the neck, axilla, posterior abdominal wall and pelvic wall.

The ANS utilizes norepinephrine, dopamine, and acetylcholine as primary neurotransmitters, producing either an excitatory or inhibitory response. Modulators, such as hormones or tissue metabolites, may exert influences upon the transmitter release or their action.

There are two functionally diverse and antagonistic systems within the ANS which operate in conjunction with the enteric nervous system. These are the sympathetic and parasympathetic systems.

Sympathetic system

The sympathetic is a system brought into action during emergency and under stressful situations. It releases energy by increasing the level of blood sugar and blood pressure, and intensifying the rate of cardiac contractility and output (positive ionotropic and chronotropic effects). This system directs blood flow to the voluntary musculature at the expense of viscera and skin. Excitation of a disproportionately large number of postganglionic neurons (great degree of divergence) and high concentration of the circulating epinephrine in the blood stream are responsible for the mass-response of the sympathetic system. It utilizes acetylcholine at the preganglionic level and norepinephrine at the postganglionic level. However, there are exceptions to this rule; acetylcholine may be utilized at the postganglionic level in the innervation of the blood vessels of skeletal muscle and the sweat glands. This may not be true in the case of sweat glands of the palm, which receive adrenergic fibers.

Other cotransmitters also play important roles in this system such as ATP and neuropeptide Y (NPY). Sympathetic (adrenergic) receptors are integral membrane glycoproteins, which are classified into presynaptic and postsynaptic groups.

Presynaptic receptors are categorized into α_2 and β_2 groups. Additional groups and subtypes are discussed in detail in Chapter 12.

Group α_2 receptors are found on both cholinergic and adrenergic nerve terminals. They act on pancreatic islets (β) cells and on platelets, resulting in reduction of insulin secretion and platelets aggregation, respectively. The latter effect is due to inhibition of adenylate cyclase and activation of K^+ channels via Gi protein. On the adrenergic endings, α_2 (autoreceptors) receptors inhibit the release of norepinephrine by inhibiting neuronal Ca^{2+} channels.

Group β_2 (hormonal) receptors are found on the vascular, pupillary, ciliary, bronchial, gastrointestinal, and genitourinary tract smooth muscles, initiating relaxation of coronary vessels, bronchi and skeletal arterioles, as well as the ciliary and constrictor pupilla muscles. These series of actions are produced by activation of adenyl cyclase, mediated by Gs protein. They also produce glycogenolysis in the skeletal muscles and liver. β_2 agonists relax the bronchial smooth muscles and therefore can be used in the treatment of asthma. β_2 antagonists are not essential.

Postsynaptic receptors are categorized into α_1 and β_1 groups. α agonists cause contraction of the smooth muscles of the arterioles and sphincters, whereas α antagonists may counteract this effect in individuals with hypertension and peripheral vascular disease.

α_1 group produces vasoconstriction and enhances glandular secretion (odoriferous apocrine sweat glands) via stimulation of phospholipase with formation of IP_3 (inositol-1,4,5-triphosphate) and diacylglycerol, and increased cytosolic Ca^{2+}.

β_1 receptors act particularly on the cardiac muscles, producing increased rates and force of contraction and atrioventricular nodal velocity via activation of adenylyl cyclase and Ca^{2+} channels. The role of β_1 agonists in the stimulation of the nodal and ventricular muscles of the heart may be utilized in the treatment of heart failure. In the same manner β_1 antagonists' (β blockers) role in reducing cardiac rate and force of contractility may be utilized in the treatment of angina pectoris.

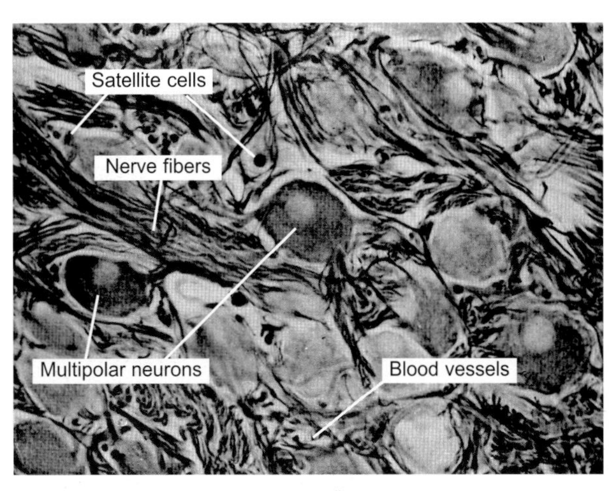

Figure 9.5 Photomicrograph of the multipolar neurons of the sympathetic ganglia and associate satellite cells

Sympathetic innervation of the skin encompasses adrenergic and cholinergic fibers. Cholinergic fibers act upon the muscarinic receptors, enhancing the secretion of the eccrine sweat glands, while adrenergic fibers produce vasoconstriction of the cutaneous arterioles and activate the secretion of the odoriferous apocrine sweat glands.

Activation of the sympathetic nervous system, which may be mimicked by sympathomimetics, produces:

- Mydriasis (dilation of the pupils) by inducing contraction of the dilator pupillae muscles.
- Relaxation of the ciliary muscle.
- Increased contractility and rate of heart beat (positive ionotropic and chronotropic effects).
- Vasodilatation of the skeletal and coronary arteries.
- Vasoconstriction of the bronchial arteries, as well as arteries of the digestive system and skin.
- Dilatation of bronchi and inhibition of bronchial secretion.
- Inhibition of gastrointestinal motility and contraction of the sphincters.
- Increased secretion of the sweat glands.
- Contraction of the erector pilorum muscles.
- Vasoconstriction of the genital arteries and contraction of the vas deferens, seminal vesicle, and prostate.
- Inhibition of the detrusor muscle of the bladder and contraction of the urethral sphincters.

The preganglionic sympathetic neurons (Figures 9.1 & 9.4) of the intermediolateral column of the thoracic and upper two or three lumbar spinal segments are connected to the postganglionic sympathetic neurons via the myelinated fibers of the white communicating rami. Postganglionic neurons form two sets of ganglia: paravertebral and prevertebral.

Figure 9.6 This is a depiction of manifestations of Horner's syndrome

Ipsilateral disruption of the sympathetic fibers to the head produces manifestations of Horner's syndrome (Figure 9.6) which comprises ptosis (drooping of the upper eyelid), miosis (constriction of the pupil), enophthalmos (sunken eyeball), and anhydrasis (lack of sweating).

Activation of the preganglionic sympathetic neurons at T1–T2 spinal segments requires descending autonomic input from the hypothalamus that travels in the lateral medulla and the lateral funiculus of the cervical spinal cord. Therefore, destruction of the lateral part of the medulla may also produce signs of Horner's syndrome, which are seen as a component of lateral medullary (Wallenberg's) syndrome.

Paravertebral ganglia

The paravertebral ganglia (Figures 9.1, 9.4, 9.5, 9.8 & 9.9) form the sympathetic trunk, consisting of two symmetrical chains parallel to the vertebral column that unite anterior to the coccyx at the ganglion impar. Multipolar neurons of the paravertebral ganglia receive presynaptic fibers from the intermediolateral columns of the thoracic and upper two or three lumbar segments via myelinated white communicating rami, and provide postsynaptic fibers that join the spinal nerves via the unmyelinated gray communicating rami. The paravertebral ganglia are divided into cervical, thoracic, lumbar, and sacral parts.

The cervical part of the sympathetic trunk (Figure 9.8) consists of the superior, middle, and inferior ganglia. Since these ganglia do not receive white communicating rami, the presynaptic fibers, which emerge from the upper thoracic spinal nerves, have to travel through the corresponding thoracic ganglia to reach their destination in the cervical ganglia. However, gray communicating rami do arise from these ganglia, supplying the upper four cervical spinal nerves.

The superior cervical ganglion, the largest cervical ganglion, is formed by fusion of the upper four cervical ganglia. It lies anterior to the longus capitis and posterior to the carotid sheath. This ganglion supply fibers to the carotid body and the pharyngeal plexus and cardiac branch (superior cardiac nerve) which contains efferent but not afferent nociceptive fibers. Most of the emerging postganglionic fibers form the external and internal carotid plexus around the corresponding arteries. Fibers of the internal carotid plexus enter the cranial cavity supplying the dura mater and dilator pupilla, superior tarsal, and orbital muscles. Fibers within the external carotid plexus supply vasoconstrictor and sudomotor fibers to the face and neck, as well as secretomotor fibers to the salivary glands via the otic and submandibular ganglia. It also provides gray communicating rami to the upper four cervical spinal nerves. Since the presynaptic fibers from the T1–T2 spinal segments project to the superior cervical ganglion, which innervates the structures in the head, removal of this ganglion may deprive the ipsilateral side of the head of sympathetic innervation.

The middle cervical ganglion (Figure 9.7) is an inconstant ganglion, which is located anterior to the inferior thyroid artery at the level of the sixth thoracic vertebra. It is formed by the fusion of the fifth and sixth cervical ganglia. It provides innervation to the heart, and furnishes gray communicating rami to the fifth and sixth cervical ventral rami. It is connected to the cervicothoracic ganglion via the ansa subclavia (see below). It provides thyroid branches and the middle cardiac nerve, which is the largest sympathetic contribution to the deep cardiac plexus.

Pancoast tumor is a tumor of the apex of the lung, which may result in compression or destruction of the stellate ganglion and the inferior trunk of the brachial plexus, producing pain and numbness in C8–T1 dermatomes. The phrenic nerve may also be affected in this condition, producing diphragmatic palsy. Additional manifestations such as cardiac arrythymias, obstruction of the superior vena cava, and hoarseness due to left recurrent laryngeal nerve palsy may also been. Signs of spinal cord compression may occasionally be seen due to erosion of the vertebral laminae by extension of the tumor.

The inferior cervical ganglion joins the first thoracic ganglion to form the stellate ganglion.

The cervicothoracic (stellate) ganglion (Figure 9.7) lies posterior to the initial part of the vertebral artery, apex of the lung and cervical pleura, occupying the area between the transverse process of C7 and the neck of the first rib. It contributes gray communicating rami to the seventh and eighth cervical, and to the first thoracic spinal nerves. It also supplies postganglionic branches to the subclavian artery and its branches, and to the vertebral plexus, which extends into the cranial cavity. The preganglionic fibers that pass through the stellate ganglion, for the most part, project to the head and neck. However, vasomotor and sudomotor fibers are not contained in the white communicating ramus to the cervicothoracic ganglion. Postganglionic fibers from the stellate ganglion also travel within the inferior trunk of the brachial plexus and then within the ulnar, radial and median nerves. In the hand, the postganglionic fibers leave these nerves and travel with the corresponding arteries. Occasionally a vertebral ganglion may be present near the origin of the vertebral artery, which provides gray communicating rami to the fourth and fifth cervical spinal nerves.

The stellate and middle cervical ganglia are connected via the ansa subclavia, a nerve loop that encircles the subclavian artery on both sides and courses medial to the origins of the internal thoracic and vertebral arteries.

The thoracic part of the sympathetic trunk consists of 11 or 12 ganglia arranged anterior to the costal heads, and covered by the costal pleura. These ganglia are connected to the thoracic spinal nerves via the white and gray communicating rami. As mentioned earlier, the first thoracic and the inferior cervical ganglia frequently join to form the cervicothoracic (stellate) ganglion. The second through the fifth thoracic ganglia provide sympathetic fibers to the posterior pulmonary and deep cardiac plexuses, while the upper five thoracic ganglia provide sympathetic fibers to the aortic plexus. Presynaptic fibers, mainly from the T1–T6 (T7), which are destined to the cervical ganglia en route to the head, neck, and upper extremity, also travel within the cervical ganglia. Since vasoconstrictors to the upper extremity primarily emerge from the second and third thoracic spinal segments, excision of the corresponding thoracic ganglia (second and third) may denervate the vessels of the upper extremity. Presynaptic sympathetic fibers from the fifth through the ninth ganglia form the greater splanchnic nerve (Figure 9.4), presynaptic fibers from the tenth and eleventh ganglia form the lesser splanchnic nerve, whereas fibers that emanate from the twelfth thoracic ganglion form the least splanchnic nerve.

Pain relief from the upper extremity, alleviation of vascular spasm in the hands (seen in Raynaud's disease) or hyperhidrosis (excessive sweating) may be achieved by injection of anesthetic solution into the stellate ganglion (stellate block). Success of this procedure may be ascertained by the appearance of signs of Horner's syndrome and increased temperature of the ipsilateral upper extremity.

The lumbar part of the sympathetic trunk is connected to the thoracic part via the gap posterior to the medial arcuate ligament. The lumbar sympathetic ganglia receive white communicating rami from the upper four lumbar spinal nerves. These ganglia, which are located medial to the psoas major muscle and anterior to the lumbar vertebrae, give rise to the lumbar splanchnic nerves (preganglionic sympathetic fibers) that join the celiac, intermesenteric and the superior hypogastric plexuses.

The sacral part of the sympathetic trunk lies anterior to the sacrum and medial to the pelvic sacral foramina. It joins the sacral ganglia of the opposite side via the ganglion of impar. These ganglia receive preganglionic fibers from the lower thoracic and upper two lumbar spinal segments, giving rise to gray rami that join the sacral and coccygeal plexuses.

Prevertebral ganglia

The prevertebral ganglia (Figures 9.1 & 9.4) lie anterior to the lumbar part of the vertebral column, comprising the celiac, aorticorenal, superior mesenteric, and the inferior mesenteric ganglia. The celiac ganglion, the largest prevertebral ganglia, is located around the celiac trunk and medial to the suprarenal gland. The caudal (lower) part of each celiac ganglion is known as the aorticorenal ganglion. Much smaller ganglia, such as the superior and inferior mesenteric, are lodged within the corresponding plexuses.

Postganglionic fibers from the prevertebral plexuses travel in the femoral and obturator nerves, supplying vasoconstrictor fibers to the femoral and obturator arteries and their branches. Therefore, surgical removal of the upper three or four lumbar ganglia or their preganglionic neurons may completely denervate the lower extremity vessels.

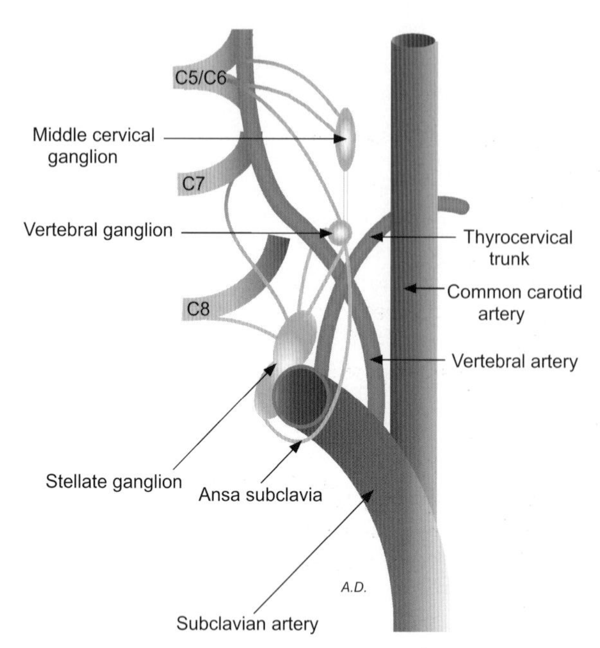

Figure 9.7 Drawing of the middle cervical and stellate ganglia and the connecting ansa subclavia

Course of the sympathetic fibers

Axons of the sympathetic preganglionic neurons that project to the paravertebral ganglia usually follow an orderly course through the ventral root, spinal nerve and the white communicating rami. This is not true in the cervical and sacral segments which lack the corresponding white communicating rami that connect the spinal nerves to the paravertebral ganglia. Postganglionic axons of the sympathetic ganglia may follow a diverse course which depends upon the site of termination (Figures 9.4 & 9.8). Most fibers return to the spinal nerves via the gray communicating rami. This route is followed by fibers that innervate the sweat glands, erector pilorum muscles, and the vessels of the extremities, thoracic and abdominal walls. Other fibers, which ascend to synapse in the superior cervical ganglion, form plexuses around major blood vessels destined to the head and neck region (e.g. sweat glands of the face and the dilator pupillae muscles). The sympathetic postsynaptic fibers to the lower face arise from the external carotid plexus, a network of sympathetic fibers that encircle and follow the course of the corresponding artery. Sympathetic fibers to the sweat glands of the supraorbital region are contained within the supraorbital and supratrochlear branches of the frontal nerve. The latter, a branch of the ophthalmic nerve, receives its sympathetic fibers by communicating with the nasociliary nerve. Innervation of the dilator pupillae muscle is maintained by the postsynaptic sympathetic fibers that travel within the long ciliary branch of the nasociliary nerve. Sympathetic postsynaptic fibers to the superior tarsal muscle of the upper eyelid originate from the internal carotid plexus, as it travels within the cavernous sinus, and are contained within the oculomotor nerve. An interesting point to bear in mind is the fact that both the sympathetic postsynaptic fibers to the superior tarsal and the somatic fibers to the levator palpebrae muscles course within the oculomotor nerve. Sympathetic postganglionic fibers to the thoracic viscera originate from the cervical and upper five thoracic paravertebral ganglia. Some fibers that are destined to the abdominal viscera bypass the sympathetic trunk to terminate in the prevertebral ganglia as the splanchnic nerves (Figures 9.1 & 9.4). The greater splanchnic nerve consists of preganglionic efferent and visceral afferent fibers that penetrate the crus of the diaphragm to enter the abdomen, establishing synaptic connections primarily with the celiac ganglion and partially with the aorticorenal ganglion. The lesser splanchnic nerve synapses in the aorticorenal ganglion, whereas the least splanchnic nerve (renal nerve) contributes to the renal plexus. Fibers that bypass both the paravertebral and prevertebral ganglia remain preganglionic and terminate in the chromaffin tissue of the adrenal medulla.

Parasympathetic (craniosacral) system

The parasympathetic is a local-response system, consisting of pre- and postganglionic neurons that act on the smooth muscles and viscera. The preganglionic parasympathetic fibers are contained in the pelvic splanchnic nerves and certain cranial nerves (Figures 9.1, 9.9 & 9.10). These preganglionic fibers establish connections with the postsynaptic parasympathetic neurons of the intramural ganglia (on pelvic and abdominal viscera) or with parasympathetic ganglia in the head. Parasympathetic responses are manifested in miosis (constriction of the pupil), contraction of the ciliary muscle, decreased contractility and cardiac output (negative ionotropic and chronotropic effect), and constriction of the bronchi and bronchioles. Other manifestations include increased gastrointestinal tract motility, constriction of the coronary arteries and vasodilatation of the vessels of the external genitalia and gastrointestinal tract, and contraction of the muscular wall of the urinary bladder (dilation of cerebral vessels are primarily due to change in CO_2 concentration).

Acetylcholine, the main neurotransmitter at the parasympathetic terminals, is contained in the clear-spherical vesicles, acting primarily in conjunction with cotransmitters such as VIP (vasoactive intestinal peptide) and to a lesser degree ATP. Due to the rapid

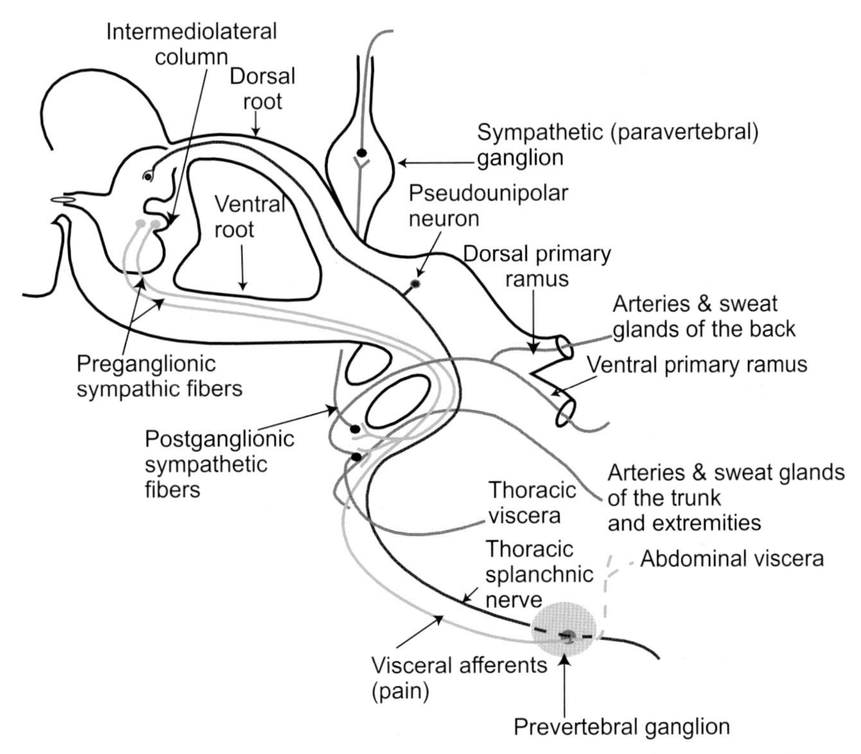

Figure 9.8 The functional components of a spinal nerve including general somatic afferent, general visceral afferent, general somatic efferent and general visceral efferent are shown

degradation of acetylcholine and the lesser degree of divergence (low ratio of preganglionic to postganglionic neurons), the action of the parasympathetic system remains localized and of short duration. Cholinergic receptors, which are activated by acetylcholine, are classified into nicotinic and muscarinic types.

Nicotinic receptors are further subdivided into nicotinic muscle receptor and nicotinic neuronal receptor.

Nicotinic muscle receptors (C-10 receptor) are pentameric proteins, activation of which produces rapid increase in permeability of cells to sodium and calcium ions and subsequent depolarization and contraction of the skeletal muscle. Phosphorylation by cAMP protein kinase, protein kinase C, or trypsin kinase increases the desensitization of these receptors. Muscle receptor contains α, β, γ, and δ or α, β, δ, and ε subunits in a pantameric complex. The reason for the difference is because the ε subunit replaces the γ in the adult. Subunit γ is particularly detected in the embryo or denervated muscle.

Neuronal nicotinic receptors are categorized into two subunits, α and β, with the α occurring in at least seven different forms and β in three forms. They exist in the autonomic ganglia, adrenal medulla, and CNS. Neuronal nicotinic receptors are classified into bungarotoxin-insensitive (C-6) and bungarotoxin-sensitive nicotinic receptors. The former exist in the autonomic ganglia and produce depolarization and firing of the postganglionic neurons in the autonomic ganglia via opening of the cation channel. There are numerous agonists for these receptors such as nicotine, phenyltrimethyl-ammonium, methylisoarecolone, cytosine, and dimethylphenylpiperazinium, as well as a plethora of antagonists such as tubocurarine, lophotoxin, and dihydro-β-erythroidine.

Muscarinic receptors are coupled to G-proteins, and either act directly or indirectly on ion channels or linked to second messenger systems. They are classified on pharmacological basis into M_1–M_3, and on the basis of molecular cloning into M_4–M_5 subtypes. All five subtypes exist in the CNS. M_1 receptors show great affinity to pirenzepine and are found in the autonomic ganglia and glands. AFDX-116 shows high affinity to M_2 receptors in the myocardium and smooth muscles, whereas 4-DAMP (4-diphenylacetoxy-N(2-chloroethyl)-piperdine hydrochloride) displays high affinity to M_3 receptors in the smooth muscles and secretory glands. M_1, M_3, and M_5 are coupled to phosphoinositol (PI) hydrolysis and M_2 and M_4 are coupled to cAMP.

Activation of the muscarinic receptors produces depolarization or hyperpolarization by opening or closing the potassium, calcium or chloride channels. Activation of M_1 receptors produces depolarization in the neurons of the autonomic ganglia. Stimulation

Cholinergic agents like carbachol (stimulates the bladder and bowel) and pilocarpine (produces constriction of the pupil) have similar effects to acetylcholine. Some of these agents act by inhibiting the enzyme cholinesterase and subsequently increasing the concentration of acetylcholine in the synaptic clefts. These cholinesterase inhibitors include physostigmine and di-isopropylfluorophosphate (DFP). Others, such as tubocurarine, act as antagonists by competing with natural mediators at the synaptic site.

Anticholinergic medications may be used clinically to: (1) induce dryness of the bronchi during surgery; (2) maintain dilatation of the pupil for in-depth ophthalmologic examination; (3) block the vagal inhibition in case of cardiac arrest; (4) prevent vomiting (antiemetic); (5) counteract the spastic effect of morphine on the gastrointestinal tract; (6) treat poisoning by overdose of cholinergic drugs; and (7) cause relaxation of the urinary bladder in individuals with cystitis.

of the M_2 receptors elicits hyperpolarization in the SA node, and a decrease in the atrial contractile force and conduction velocity in the atrioventricular node, and a slight decrease in the ventricular contractile force. Activation of the M_3 receptors produces contraction of the smooth muscles and increased glandular secretion.

Acetylcholine acts upon the muscarinic receptors on the exocrine glands, heart, and smooth muscles. Cholinergic receptors are nicotinic in the autonomic ganglia and muscarinic at the postganglionic parasympathetic nerve endings. In the CNS both muscarinic and nicotinic receptors exist. The combined effect of the muscarinic autoreceptors on the nerve endings (comparable to the α_2 autoreceptors of the sympathetic system) and acteylcholinesterase may prevent accumulation of acetylcholine in the synaptic cleft.

The parasympathetic system consists of cranial and sacral parts. The cranial part (Figures 9.9 & 9.10) consists of preganglionic neurons that course within the oculomotor, facial, glossopharyngeal, and vagus nerves.

The oculomotor nerve (III) contains preganglionic parasympathetic fibers, which are derived from the Edinger–Westphal nucleus of the oculomotor nuclear complex. These fibers synapse in the ciliary ganglion, giving rise to postsynaptic fibers that eventually innervate the constrictor pupillae and the ciliary muscles. CN III (inferior branch) → ciliary ganglion → constrictor pupillae & ciliary muscle.

The facial nerve (VII) contains preganglionic parasympathetic fibers that emanate from the neurons of the lacrimal and superior salivatory nuclei, establishing synapses in the pterygopalatine (sphenopalatine) and the submandibular ganglia, respectively. The postsynaptic parasympathetic fibers from the pterygopalatine ganglion supply the lacrimal gland, mucus glands of the palate, nasal cavity, and pharynx. On the other hand, the postsynaptic parasympathetic fibers from the submandibular ganglion innervate the submandibular and sublingual glands. CN VII → greater petrosal nerve → pterygopalatine ganglion → lacrimal gland, mucus glands of the palate, and nasal cavity and pharynx. CN VII → chorda tympani → submandibular ganglion → sublingual & submandibular glands.

The glossopharyngeal nerve (IX) contains preganglionic parasympathetic fibers from the medullary inferior salivatory nucleus that synapse in the otic ganglion. This ganglion sends postsynaptic secretomotor fibers to the parotid gland via the auriculotemporal nerve. CN IX → lesser petrosal nerve → otic ganglion → parotid gland.

The vagus nerve (X) contains preganglionic parasympathetic fibers, which are derived from the medullary dorsal motor nucleus of the vagus. These fibers synapse in the terminal (intramural) ganglia scattered along the thoracic and abdominal viscera (e.g. the pulmonary, myenteric, and submucosal plexuses). Intramural ganglia contain in abundance neurons that are non-adrenergic and non-cholinergic. They may also contain excitatory transmitters such as serotonin and substance P or inhibitory transmitters such as VIP, ATP or enkephalin. The vagal parasympathetic contributions to the abdominal viscera terminate at the junction of the right two-thirds and left one-third of the transverse colon.

Sacral part

The sacral part (Figure 9.1) of the parasympathetic system includes the parasympathetic preganglionic axons, emanating from the intermediolateral column of the second, third, and fourth sacral spinal segments. The axons of these neurons leave through the ventral roots of the corresponding segments and form the pelvic splanchnic nerves, which supply the pelvic viscera and part of the abdominal viscera. The pelvic splanchnic nerves are excitatory to the muscular wall of the descending colon, sigmoid colon, rectum and anal canal, as well as to a portion of the transverse colon. These splanchnic nerves are inhibitory to the urethral sphincters and vasodilatory to the erectile tissue of the external genitalia.

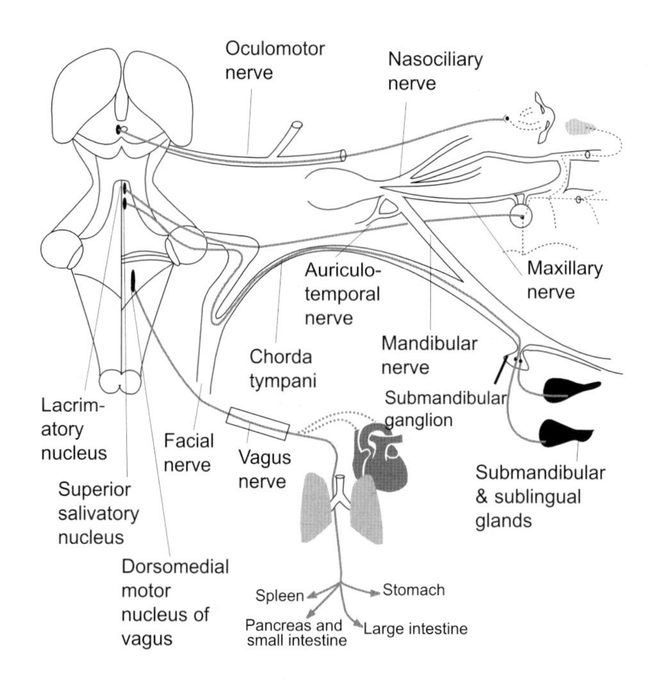

Figure 9.9 Cranial part of the parasympathetic nervous system. Note the associated nuclei, preganglionic fibers, and related ganglia

Autonomic centers

Higher autonomic centers represent specific areas in the cerebral cortex, diencephalon, and the brainstem that closely regulate the ANS. This is based on the fact that stimulation or inhibition of these centers produces a variety of visceral changes. Autonomic centers in the cerebral cortex are scattered in the cingulate gyrus and hippocampal formation. In the diencephalon parasympathetic (anteromedial) and sympathetic (posterolateral) centers are located in the hypothalamus.

Hypothalamic control of the brainstem and spinal autonomic neurons is achieved via the dorsal longitudinal fasciculus (DLF), the mamilllotegmental tract, and the medial forebrain bundle (MFB). The DLF connects the medial hypothalamus to the dorsal motor nucleus of vagus, nucleus ambiguus, salivatory and Edinger–Westphal nuclei, as well as the intermediolateral columns of the spinal cord. Medial hypothalamic neurons also send fibers to the dorsal motor nucleus of vagus, locus ceruleus, and raphe nuclei via the MFB. Brainstem raphe nuclei project to the prefrontal cortex, septal area, and cingulate gyrus also via the MFB.

The mamilllotegmental tract is formed by the axons of the mamilllary neurons that project to the raphe nuclei and other nuclei of the mesencephalic and pontine reticular formation.

Transection of the lower pons disrupts the descending fibers from the pneumotaxic center, resulting in a deep respiratory cycle (apneustic breathing).

In the brainstem, the pontine autonomic is comprised of a cardiovascular center in the caudal pons between the superior olivary nucleus and the root of the facial nerve. The medullary autonomic (respiratory) is comprised of the inspiratory center around the solitary nucleus that contains opiate receptors upon which morphine acts as a depressant, and an expiratory center around the ambiguus nucleus. The latter projects to the motor neurons of the thoracic spinal segments, innervating the internal and innermost intercostals. The pneumotaxic center is comprised of the parabrachial nuclei of the pons, which influences the rate of breathing by shortening the respiratory cycle.

Autonomic reflexes

A reflex is an innate and automatic response to a stimulus that occurs as a basic defense mechanism. It may be inherited or primitive, present at birth, and is common to all human beings. It may also be conditioned, which is acquired as a result of experience. Intactness of the receptor, sensory (afferent) neuron and a motor (efferent) neuron are essential for a typical reflex to occur. Afferent fibers deliver the generated impulses from a receptor to the CNS where it may be inhibited, facilitated, or modified, while the efferent fibers transmit the processed information to the effector organ. An interneuron may exist between the afferent and efferent neurons. Reflexes do not always operate independently; in fact descending supraspinal pathways (somatic and visceral) modulate and regulate the neurons, which form the reflex arc. Conditions that affect the neural elements of a reflex arc or their supraspinal input may produce a variety of deficits. A lesion which disrupts the reflex arc may result in hyporeflexia or areflexia depending on the number of the involved segments. In peripheral neuropathy and poliomyelitis, the receptor, afferent or the efferent neurons of a reflex arc may be damaged, producing hyporeflexia or areflexia.

Damage to the supraspinal pathways may produce hyperreflexia, hyporeflexia, or areflexia (e.g. upper motor neuron palsy manifests both deep tendon hypereflexia, and areflexia or hyporeflexia in the superficial abdominal reflexes).

Reflexes are categorized into superficial reflexes associated with the skin and mucus membrane and deep reflexes pertaining to the muscles and tendons.

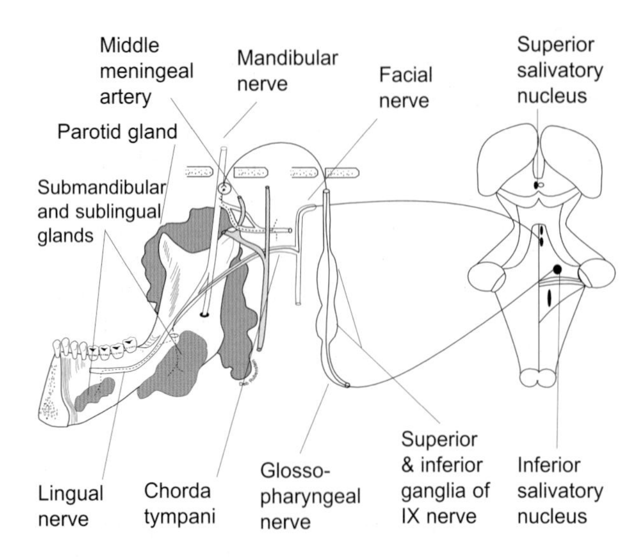

Figure 9.10 The sympathetic neurons associated with innervation of the thoracic viscera (heart, lungs and bronchi)

Reflexes may be mediated by cranial nerves (cranial reflexes) or spinal nerves (spinal reflexes). Additional classifications into visceral and somatic reflexes are based upon the nature of the innervated structure.

Visceral reflexes include viscero-visceral and viscero-somatic reflexes, while somatic reflexes comprise somato-somatic and somato-visceral reflexes. Visceral reflexes facilitate automatic adjustments of the entire organism to the internal and external environments. In order to promote digestion, some of these reflexes produce an increase in blood flow to the digestive tract following food ingestion, and decrease in absorption. Other reflexes may increase the rate and depth of respiration to meet the body's demand for oxygen in response to physical activity. Visceral reflexes are classified into viscero-visceral, viscero-somatic, and somato-visceral reflexes.

Viscero-visceral reflexes include the carotid sinus, Bainbridge, and carotid body reflexes.

• Carotid sinus reflex is mediated by the carotid sinus (receptor), the carotid sinus branch of the glossopharyngeal (afferent limb), reticular formation, and the vagus nerve (efferent limb). An increase in blood pressure will stimulate the carotid sinus and activates the neural mechanism that adjusts the blood pressure to normal level.

• Bainbridge reflex monitors the central venous pressure through afferent nerve endings in the right atrium. These endings are represented by the peripheral processes of the neurons of the inferior ganglion of the vagus nerve. Distention of the right atrium produces reflex tachycardia due to vagal inhibition and sympathetic stimulation.

• Carotid body reflex is initiated by an increase in

Pupillary light reflex is lost in the Argyll Robertson pupil, which results from destruction of the area medial to the lateral geniculate nucleus and is associated with neurosyphilis. In this condition the pupil remains unresponsive to atropine. Loss of light reflex may also be seen in diabetes mellitus, epidemic encephalitis, and alcoholism.

carbon dioxide and a decrease in oxygen tensions of the blood. These changes stimulate the carotid body (chemoreceptor) and eventually the respiratory center through the vagus nerve.

Viscero-somatic reflexes comprise Hering–Breuer and vomiting reflexes.

• Hering–Breuer reflex initiates expiration upon excitation of the terminals of the bronchial tree of the inflated lung. These excited terminals stimulate the expiratory center and the solitary nucleus. The expiratory center inhibits the inspiratory center, eliciting passive and elastic recoil of the lung.

• Vomiting reflex is mediated by receptors that are located in the mucosa of the stomach, gall bladder, and the duodenum. Activation of these receptors results in transmission of the generated impulses via the vagus nerve (afferent limb) to the solitary nucleus, medullary vomiting center, reticulospinal tracts, and neurons of the anterior horn and the intermediolateral columns of cervical and thoracic spinal segments.

Somato-visceral reflexes are comprised of pupillary light, pupillary-skin (ciliospinal), accommodation, bladder and rectal reflexes, and mass reflex of Riddoch.

• The pupillary light reflex produces constriction of the pupil of the stimulated eye (direct light reflex) and the contralateral eye (consensual light reflex) in response to direct light applied to one eye. This reflex is mediated by the optic and the oculomotor nerves. The optic nerve, the optic tract, and the brachium of the superior colliculus form the afferent limb of this reflex. The pretectal nucleus, Edinger–Westphal nucleus, oculomotor nerve, ciliary ganglion, and the short ciliary nerves comprise the efferent limb.

• The pupillary-skin (ciliospinal) reflex is characterized by pupillary dilatation in response to painful stimuli. It may be elicited by a simple scratch, pinch, or a cutaneous wound, especially involving the facial skin. The afferent limb (depending upon the site of the stimulus) may include neurons of the dorsal root ganglia or the trigeminal nerve and ganglion, as well as neurons of the posterior horn or the spinal trigeminal nucleus. The efferent limb includes the reticular formation, reticulospinal tracts, intermediolateral columns of the first thoracic spinal segment, and the sympathetic pathway to the dilator pupilla muscle of the eye.

• The accommodation reflex adjusts both eyes to near vision, involving convergence (adduction) of the eyes, constriction of the pupils, and increased curvature of the lens in both eyes. The input for this reflex is carried from the retina by the optic nerve, optic tract, lateral geniculate body, and optic radiation to the visual cortex. The efferent impulses travel eventually via the short ciliary nerves to the ciliary body and the constrictor pupilla muscle.

The bladder and rectal reflexes regulate the sphincteric control of micturition and defecation via the pelvic splanchnic nerves. Incontinence may occur as a result of disruption of this reflex arc. The urge to urinate or defecate may be lost upon interruption of the afferent fibers.

The mass reflex of Riddoch is characterized by sudden evacuation of the bladder and bowel, flexion of the lower extremity and sweating in response to emotional stimulus.

The mass reflex of Riddoch may be elicited in an individual with spinal shock by stimulating the skin below the level of the spinal lesion.

Autonomic plexuses

Autonomic plexuses represent a network of visceral nerve fibers which innervate structures in the thoracic, abdominal, and pelvic cavities. They are formed by sympathetic and parasympathetic nerve fibers and associated ganglia. Their names are derived from the corresponding arteries that they are associated with. These comprise the cardiac, celiac, suprarenal, renal, ureteric, superior mesenteric, aortic, inferior mesenteric, and superior and inferior hypogastric plexuses.

The cardiac plexus (Figures 9.4 & 9.10) provides innervation to the heart and the coronary arteries, consisting of deep and superficial parts. The superficial part of the cardiac plexus lies below the aortic arch and is formed by the cardiac branch of the left superior cervical ganglion, and by the parasympathetic fibers of the vagus nerve. The deep part of this plexus lies anterior to the bifurcation of the trachea and is formed by the cardiac branches of the cervical (with the exception of the left superior cardiac branch) and upper four or five thoracic ganglia. In contrast branches of the vagus and the recurrent laryngeal nerves form the parasympathetic component. Postganglionic fibers from the right vagus nerve establish synaptic connection with the sinuatrial node and with both atria. On the other hand, the

Cardiac pain (e.g. due to myocardial ischemia) is transmitted by the C fibers of pseudounipolar neurons of the upper four or five thoracic spinal nerves that initially run in the middle and inferior cardiac branches of the sympathetic trunk. These fibers enter the dorsal horns of the corresponding spinal segments and synapse in certain laminae that form the anterolateral system. These connections may explain the referred pain to dermatomes of T1–T5, which is experienced by individuals with acute myocardial infarction.

postsynaptic fibers from the left vagus nerve act on the ventricular myocardium and the atrioventricular (His) bundle. Reduction of the contractile force of the heart and rate of contraction are achieved by stimulation of the vagus nerves that act upon the muscarinic receptors of the cardiac nodal tissue and atria.

The presynaptic muscarinic receptors on the sympathetic fibers are also inhibited by stimulation of the vagus nerves. Sympathetic postganglionic fibers act upon the β_1 and to a lesser degree α_1 receptors in the sinuatrial and atrioventricular nodes, atrioventricular bundle and ventricular myocardium. Activation of the β_2 receptors in the coronary arteries, by circulating epinephrine, produces relaxation of the vessels. Cholinergic presynaptic α_2 receptors on branches of the vagus nerve may also be inhibited by the sympathetic postganglionic fibers.

Subsidiaries of the cardiac plexus are the coronary plexuses that surround the coronary arteries.

The left coronary plexus is an extension of the deep cardiac plexus, supplying the left atrium and left ventricle. The right coronary plexus innervates the right chambers of the heart, and is formed by the fibers of the deep and superficial parts of the cardiac plexus. The sympathetic fibers of this plexus, upon activation, produce coronary vasodilatation, while the parasympathetic fibers elicit vasoconstriction. The cardiac plexus continues with the pulmonary plexuses (Figure 9.11) around the corresponding arteries.

The pulmonary plexus, which lies partly anterior and partly posterior to the pulmonary hilus, receives parasympathetic fibers from the vagus nerve and sympathetic fibers from the second through the fifth thoracic spinal segments. This plexus innervates the pulmonary arteries, bronchi, and bronchial arteries.

The celiac plexus (Figures 9.4 & 9.12) surrounds the celiac trunk, and lies anterior to the diaphragmatic crura and medial to the suprarenal glands. It receives sympathetic fibers via the greater splanchnic (T5–T9 spinal segments) and the lesser splanchnic (T9–T11 spinal segments) nerve. The parasympathetic fibers

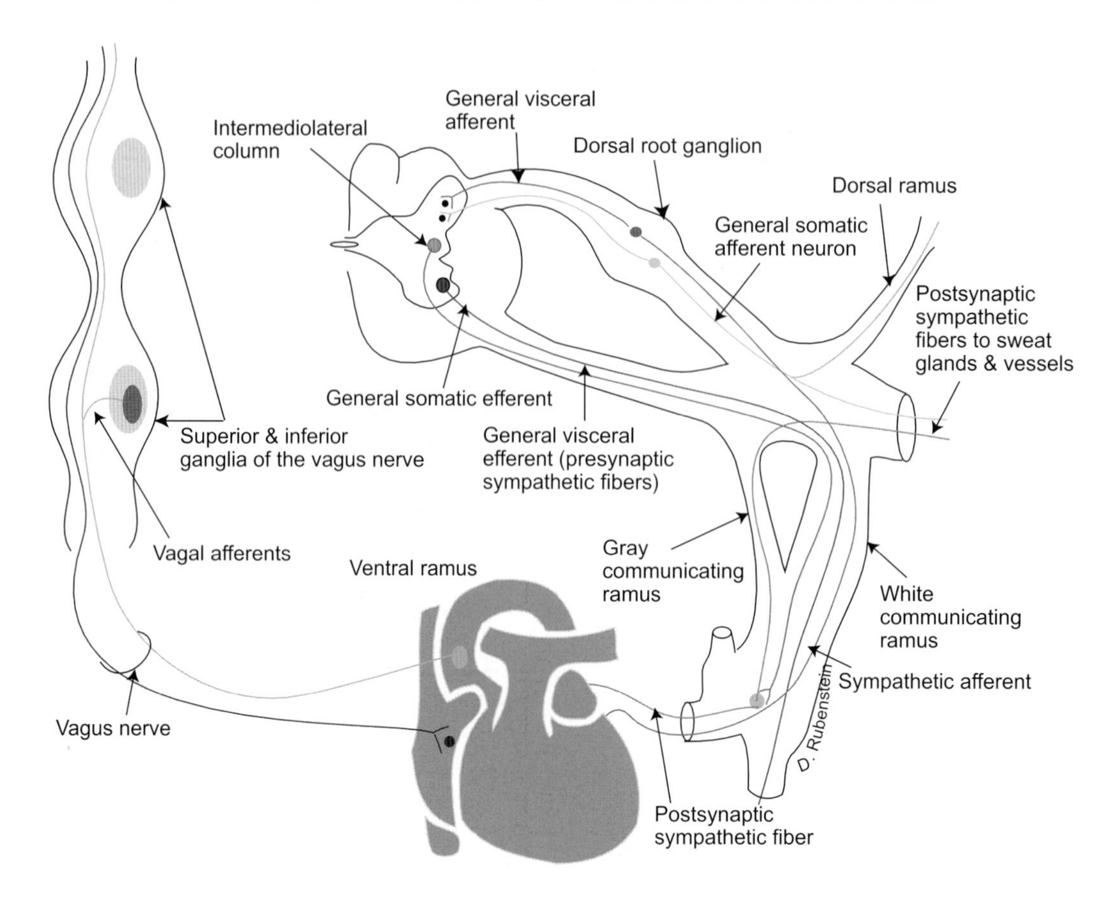

Figure 9.11 Diagram of the sympathetic neurons associated with innervation of the thoracic viscera (heart, lungs, and bronchi)

are derived from the vagus nerve. This plexus, which also receives somatic fibers via the phrenic nerves, contains the celiac ganglia (visceral brain) where the greater and lesser splanchnic nerve establish synaptic connections. Due to the proximity of the lower part of the celiac ganglion (aorticorenal ganglion) to the renal artery, it contributes postsynaptic parasympathetic fibers to the renal plexus. The celiac, as the mother of all abdominal plexuses, has subsidiary plexuses which innervate the liver, gallbladder, diaphragm, stomach, duodenum, spleen, adrenal glands, kidneys, and testes or ovaries (Figure 9.9).

The hepatic plexus, a continuation of the celiac plexus, surrounds the common hepatic artery and its branches and supplies the liver and gallbladder. Activation of the vagal parasympathetic fibers produces contraction of the gallbladder, bile duct, and relaxation of the sphincter of Oddi. This plexus receives sympathetic contribution from the seventh through the ninth spinal segments.

The gastric plexus consists of right and left plexuses; the right plexus, an extension of the hepatic plexus, innervates the pylorus. The sympathetic fibers produce contraction of the pyloric sphincter and inhibition of the gastric muscles, while the parasympathetic fibers maintain an opposing effect. The left gastric plexus, another extension of the celiac plexus, surrounds the left gastric artery. It exerts similar effects upon the stomach and pylorus. The sympathetic fibers, which supply the stomach, are derived from the T6 to the T10 thoracic spinal segments.

The suprarenal plexus is formed largely by the preganglionic sympathetic fibers of T8–L1 spinal segments that synapse in the chromaffin cells of the adrenal medulla.

The renal plexus surrounds the renal arteries and is formed by the sympathetic fibers of the lesser splanchnic nerve (T10–T11), least splanchnic nerve (T12), and the first lumbar splanchnic nerve (L1), primarily exerting a vasomotor action. The vagal parasympathetic fibers serve as afferents, terminating in the wall of the kidney. This plexus also contributes to the ureteric and the gonadal plexuses.

The ureteric plexus receives sympathetic fibers from (T11–L1) spinal segments and parasympathetic fibers from the vagus and pelvic splanchnic nerves. Due to its close relationships to the abdominal and pelvic viscera, the ureteric plexus receives fibers from

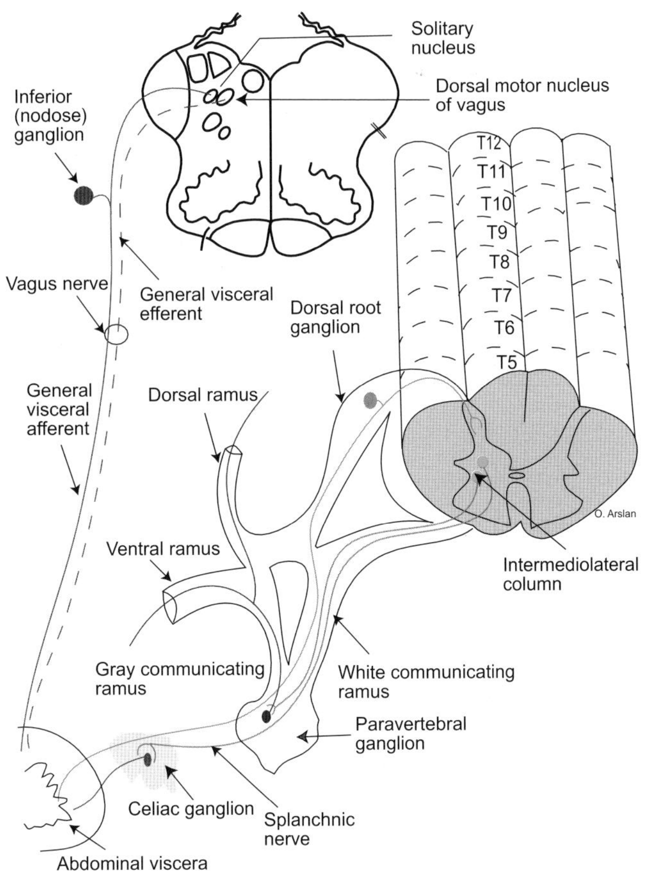

Figure 9.12 Schematic drawing of the sympathetic innervation of the abdominal viscera. Note the role of the celiac ganglia in mediating this innervation

diverse sources including the aortic, renal, vesical and hypogastric plexuses.

The superior mesenteric plexus (Figures 9.4 & 9.12) is a continuation of the celiac plexus which is formed by sympathetic fibers of the ninth, tenth, eleventh, and twelfth thoracic and the first lumbar (T10–L1) spinal segments, and by the parasympathetic fibers of the vagus nerve. This plexus supplies part of the duodenum, jejunum, ileum, and approximately the right 2/3 of the large intestine.

The aortic (intermesenteric) plexus (Figures 9.4 & 9.12) encircles the aorta between the superior and inferior mesenteric arteries. It contributes to the testicular, inferior mesenteric, and the hypogastric plexuses.

The inferior mesenteric plexus (Figures 9.4 & 9.12) surrounds the inferior mesenteric artery, containing sympathetic and parasympathetic fibers. The sympathetic fibers, which are inhibitory to the muscular walls of the descending colon, sigmoid colon, and upper part of the rectum, are derived from the first and second lumbar (L1 & L2) spinal segments. The parasympathetic fibers of the pelvic splanchnic nerves are excitatory, originating from the second, third, and fourth (S2–S4) sacral spinal segments.

The superior hypogastric (presacral nerve) plexus (Figure 9.4) is a continuation of the inferior mesenteric plexus. It receives sympathetic fibers from the eleventh thoracic through the second lumbar (T11–L2) spinal segments, and parasympathetic fibers from the pelvic splanchnic nerves (S2–S4 spinal segments). It runs anterior to the sacrum, sacral promontory and sacral plexus. It then divides into the

Division of the sympathetic fibers of the superior hypogastric plexus (presacral neurectomy) may be performed in attempt to relieve pain associated with diseased pelvic viscera. However, dual transmission of pelvic pain via sympathetic and parasympathetic fibers may render complete analgesia an impossible task to achieve. In the male, removal of the superior hypogastric plexus may lead to loss of contraction of the seminal vesicles, prostate and vas deferens, and eventual sterility.

right and left inferior hypogastric (pelvic) plexuses, supplying the pelvic structures.

The inferior hypogastric (pelvic) plexus (Figure 9.4) runs on both sides of the rectum, the uterus, and bladder, and gives rise to the vesical, middle rectal, prostatic, and the uterovaginal plexuses. It contains parasympathetic fibers from the pelvic splanchnic nerves and sympathetic fibers from the lower thoracic and the upper lumbar (T12–L1) spinal segments. The uterovaginal part of the pelvic plexus supplies the serosa, myometrium, and the endometrium, as well as the vagina. The sympathetic fibers derived from the T12–L1 spinal segments produce uterine contraction and vasoconstriction, while the parasympathetic fibers produce relaxation of the myometrium and vasodilatation.

The vesical plexus, a subsidiary of the inferior hypogastric plexus, causes contraction of the detrusor muscles via the pelvic splanchnic nerves (parasympathetic), mediating micturition. The sympathetic component is derived from T11–L2 spinal segments, which also supply motor fibers to the vas deferens and the seminal vesicle.

The prostatic plexus, another subsidiary of the inferior hypogastric plexus, supplies the urethra, bulbourethral glands, corpora cavernosa, and corpus spongiosum via the lesser and greater cavernous nerves. The sympathetic part of the prostatic plexus controls ejaculation, inhibits detrusor musculature of the bladder, and induces vasoconstriction. The parasympathetic part is formed by the pelvic splanchnic nerves, producing vasodilatation and erection.

Enteric nervous system

The enteric nervous system, an integral part of the autonomic nervous system, consists of a group of neurons within the myenteric plexus of Auerbach and the submucosal plexus of Meissner's and Henle's, as well as the pancreatic and cystic plexuses, which are derived from the neural crest cells. The myenteric plexus is located between the circular and longitudinal muscle layers, extending from the esophagus to the level of the internal anal sphincter. On the other hand, the submucosal plexus lies between the circular muscle and muscularis mucosa, stretching from the stomach to the anal canal. This system exerts a local reflex activity independent from the control of the brain and spinal cord. It is important to note that enteric nerves have more common features with the CNS than with the peripheral nerves. In fact, enteric nerves do not have collagenous coats, as is the case in the PNS. Further more, they lack the endoneurium and are supported by glial cells that resembles the astrocytes that contain glial fibrillary acidic protein (GFAP). Upon

Pain from the fundus and body of the uterus is received by the lower three thoracic spinal segments, while nociceptive stimuli from the cervix are received by the second, third, and fourth sacral spinal segments.

this network of neurons the motility and the secretory functions of the gastrointestinal tract from the middle third of the esophagus to the anorectal junction remain dependent. The number of neurons associated with this system may be equivalent or exceeds the entire population of spinal neurons. This system of ganglia and plexuses is responsible for the induction of reflex peristalsis, independent of the direct commands of the brain. These ganglia maintain a blood–ganglion barrier and are not pierced by vessels or connective tissue septa. Some neurons of this system may subserve sensory function, and respond to changes in the morphology of bowel shape. Others are simply interneurons that receive input from sensory neurons and project to the parasympathetic postganglionic neurons.

Numerous neuropeptides have been identified within this system of neuronal network. These peptides may act to enhance or suppress the effects of transmitters or maintain a trophic role. Intrinsic motor neurons within this system may assume an excitatory role, utilizing acetylcholine and substance P as cotransmitters; others may project an inhibitory effect using ATP (co-transmitter in the large and small intestine), vasoactive intestinal polypeptide (VIP), and NO (nitric oxide) as cotransmitters.

Somatostatin is widely distributed in the gastrointestinal tract and the δ cells of the pancreas where it inhibits the secretion of glucagon and insulin, a fact that may prove significant in diabetic patients. It is present in the dorsal root ganglia and autonomic plexuses. In the CNS, it is concentrated in the hypothalamus, amygdala, and neocortex, where it facilitates responsiveness to acetylcholine. In Alzheimer's disease, formation of somatostatin neuritic processes and depletion of its somatostatin-28 from the cortex are detected. Somatostatin-14, another form of this peptide may show reduction upon the administration of cysteamine as a treatment for the metabolic disease known as cystinosis.

VIP is distributed in the pancreas, autonomic plexuses, and CNS. It is also contained in the parasympathetic cholinergic neurons of the salivary glands. Secretion of this peptide increases the glandular secretion (enhances the secretory function of acetylcholine) and blood flow to the intestine (as a result of vasodilatation). VIP-related peptides include the growth hormone-releasing hormone (GHRH)

Lack of the parasympathetic ganglia in the myenteric and submucosal plexuses, as a result of failure of migration of the neural crest cells, is responsible for congenital megacolon of Hirschsprung's disease, which is characterized by dilatation of the affected segment and constipation. Since constipation in patients with Parkinson's disease may also exhibit deficiency in dopaminergic neurons of the enteric system, a possible correlation between the migration of neural crest cells and dopamine may require further investigation. It may also be possible to use these enteric dopaminergic neurons as donor grafts. Other diseases that affect the enteric nervous system include herpes simplex, diabetes mellitus, amyloidosis and Chagas disease.

and pituitary adenylate cyclase activating peptide (PACAP). GHRH is isolated from the intestine, and PACAP, as the name indicates, from the pituitary gland.

Afferent components of the autonomic nervous system

The peripheral processes of the dorsal root ganglia or certain cranial nerve ganglia are usually accompanied by the afferent autonomic (visceral) fibers. These afferent fibers, which are myelinated and unmyelinated, follow the course of the pre-and postganglionic neurons, terminating in the capsulated receptors of the visceral and vascular walls. In addition to visceral pain, they also mediate visceral reflexes, and transmit organic visceral sensations, libido, distention, hunger, and nausea. Stimuli for visceral pain do not encompass cutting, burning, or crushing, but rather obstruction, and/or ischemia and distention of the visceral wall. Visceral afferents utilize mechanoreceptors, chemoreceptors, thermoreceptors, and osmoreceptors.

Pain from viscera is predominantly carried by the sympathetic fibers and will be felt in the cutaneous areas of the spinal segments that originally provided the presynaptic neurons to the diseased viscus. Pain fibers from the bladder and anterior urethra is conducted by the pelvic splanchnic nerves and the superior and inferior hypogastric plexuses, as well as the lumbar splanchnic nerve. The hypogastric plexuses and the lumbar splanchnic nerves convey uterine pain, with the exception of the cervix, to the lower thoracic and upper lumbar spinal segments. Dysmenorrhea (intractable pain associated with menses) may be alleviated by excision of the superior hypogastric plexus. Nociceptive impulses from the uterine cervix is transmitted via the pelvic splanchnic nerves to the second, third, and fourth spinal segments. Afferents from the testis and ovary run through the gonadal plexuses that terminate in the 10th and 11th spinal segments. General visceral afferents are found in the glossopharyngeal and vagus nerves.

Disorders of the autonomic nervous system

Autonomic disorders occur in a variety of diseases and conditions which affect the autonomic centers in the CNS, descending autonomic pathways, or the preganglionic neurons in the spinal cord. They are also seen as a result of disruption of the postganglionic neurons in the paravertebral and prevertebral ganglia. These dysfunctions are comprised of the Hirschsprung's disease, hyperhidrosis, Raynaud's disease or phenomenon, spinal cord lesions, Horner's syndrome, stellate ganglion syndrome, Shy–Drager syndrome, botulism, Riley–Day syndrome, reflex sympathetic dystrophy, achalasia, and Chagas disease.

Hirschsprung's disease (congenital megacolon) as previously described is a condition which results from absence of the parasympathetic ganglia in the myenteric plexus of Auerbach. Loss of the peristaltic movement and subsequent constriction of the affected segment and retention of feces above the aganglionic segment characterize this disorder. This condition, which frequently involves the sigmoid colon and the rectum, is more common in males (see also developmental aspects).

Hyperhidrosis (excessive sweating) is characterized by increased sweating due to over-stimulation of the sympathetic postganglionic neurons that innervate the sweat glands. It may be associated with peripheral neuropathy and reflex sympathetic dystrophy. Palm sweating, due to social or situational nervousness, may be effectively relieved by the application of a 20% solution of aluminium chloride hexahydrate in absolute ethyl alcohol at night to the palm and covering firmly with a thin film of polyethylene.

Raynaud's disease is a primary idiopathic vascular disorder while Raynaud's phenomenon occurs secondary to other conditions. It is characterized by spasmodic vasoconstriction of the digital arteries of the extremities in response to cold or emotional stress. This phenomenon may occur secondary to a cervical rib, scleroderma, thoracic outlet syndrome, atherosclerosis of the brachial artery and connective tissue disease. It may be attributed to a lack of histamine induced vasodilatation subsequent to a lack of the neural mechanism for histamine release in individuals with an intact hypothalamic sympathetic center. Emotional stimuli and cold may activate the sympathetic system, lowering the threshold for vasospastic response. It is characterized by intermittent pallor due to depletion of the blood in the capillary beds of the digits and cyanosis as a result of deoxygenation of the stagnant blood in the capillary beds. Color changes may involve redness of the affected digits (reactive hyperemia) as a result of dilation of the digital arteries, and engorgement of the capillary beds with oxygenated blood may also be observed. This will confer a ruddy complexion to the skin of the digits. Small painful ulcers may appear on the tips of the digits. Drug treatment should be reserved for severe cases. Oral administration of reserpine in doses of 0.25–0.5 mg once a day have been shown to increase blood flow to the fingers. Infusion of the brachial or radial artery with a single dose of reserpine has been reported to reduce pain and promote healing of ulceration. Mild cases may be controlled by protecting the body and extremities from cold and by using mild sedatives. Prazosin, the calcium antagonist nifedipine, phenoxybenzamine and prostaglandins (thromboxane) are also effective medications for this condition.

Spinal cord lesions produce autonomic disturbances, which vary with the level of injury. Lesions of the cervical and upper thoracic spinal segments are most likely to produce combined sympathetic and parasympathetic dysfunctions, whereas damage to the lower thoracic segments are only associated with parasympathetic dysfunctions. Transection of the cervical part of the spinal cord may result in loss of all sensory and motor activities below the level of affected segment(s), as well as autonomic dysfunctions including loss of sweating, piloerection, loss of micturition, impotence, and hypotension (spinal shock). Recovery of autonomic functions may occur as a result of the release from cortical and hypothalamic control. Since changes in blood pressure in individuals with this condition may no longer be mediated by autonomic centers in the brainstem, cutaneous stimulation below the level of the lesion may produce a rise in blood pressure, mydriasis and sweating. Bladder function becomes automatic and urination may occur when it is full. Following these changes, patients may manifest a triple or mass reflex in which a mild cutaneous stimulus may produce flexion in all joints of the lower extremity (triple reflex) which disappears approximately four months following transection of the spinal cord.

Horner's syndrome is characterized by miosis (constriction of the pupil), ptosis (drooping of the upper eyelid due to paralysis of the superior tarsal muscle), anhydrasis (lack of sweating) and apparent enophthalmos (sinking of the eyeball due to paralysis of the orbital muscle). Heterochromia, which refers to the diversity of colors in part or parts that should normally be one color, is a characteristic of the congenital form of Horner's. In infants, Horner's syndrome may be associated with unpigmented iris that assumes a bluish or mixed gray and blue appearance. It may be caused by a lesion of the intermediolateral column of the first thoracic spinal segment or emerging ventral root, degeneration of the lateral medulla, lesion of the descending autonomic pathways from the hypothalamus, superior cervical gangliotomy, or syringomyelia. It may also be caused by percutaneous carotid puncture for cerebral angiography, intracavernous lesions, birth trauma, enlargement of the cervical lymph nodes, thoracic tumors, destruction of the internal carotid plexus, or hypothalamic lesion.

Stellate ganglion syndrome is produced by compression of the stellate ganglion (as seen in Pancoast tumor of the apical lobe of the lung), exhibiting signs of Horner's syndrome and reflex sympathetic dystrophy. The latter manifests dryness of the skin of the upper extremity and vasodilatation.

Achalasia refers to failure or incomplete relaxation of the lower esophageal sphincter, which is more common in males. In this condition the normal peristalsis of the esophagus is replaced by abnormal contractions. It is classified into vigorous and classic achalasia. Vigorous achalasia resembles diffuse esophageal spasm, exhibiting simultaneous and repetitive contractions with large amplitude, whereas classic achalasia shows contractions of small amplitude. Secondary achalasia may result from infiltrating gastric carcinoma. Dysphagia, chest pain, regurgitation and pulmonary aspiration, and projectile vomiting characterize it. Emotional disorders and hurried eating may predispose the individual to this condition. Although esophageal myenteric plexus lack ganglia, the pathogenesis of this dysfunction is not well understood. Treatment may include administration of anticholinergics and calcium channel antagonists, or balloon dilatation. Surgical intervention in which the lower esophageal sphincter is incised may prove to be effective.

Chagas disease is an infectious and zoonotic disease caused by *Trypanosoma cruzi* and is transmitted from infected animals to humans by Reduviid bugs. Chagoma, an inflammatory lesion, is often seen at the site of entry of the parasite. When the parasite enters through the conjunctiva, edema of the palpebrae and periocular tissue is a characteristic feature (Romana's sign). The heart is the most commonly affected organ, exhibiting cardiomyopathy, ventricular enlargement and thinning of the walls, mural thrombi and apical aneurysm. The right branch of His bundle is frequently damaged, producing atrioventricular block. Patients show signs of malaise, fever, and anorexia, which are associated with swelling of the face and lower extremities. This infectious parasitic agent may also cause destruction of the myenteric plexus in the esophageal, duodenal, colonic, and ureteric wall, producing megacolon, megaduodenum and megaureter. Lymphadenopathy, meningo-encephalitis and increased incidence of esophageal varicosities are characteristics of this disease. This condition may be treated by nifurtimox, an effective drug against *Trypanosoma cruzi* during the acute phase of the disease.

Shy–Drager syndrome (idiopathic orthostatic hypotension) is a multisystem disorder which includes autonomic dysfunctions, ataxia, and upper motor neuron palsy. Autonomic dysfunctions comprise anhydrasis (lack of sweating), impotence, postural hypotension, mydriasis and pupillary asymmetry, bowel and bladder dysfunctions. The hallmark of this disease is postural hypotension, which is greater than 30/20 mmHg on standing from a supine position. Patients also exhibit Parkinsonian manifestations in which rigidity and bradykinesia are very conspicuous. Neuronal loss has been shown in the intermediolateral column of the thoracic spinal segments, peripheral autonomic ganglia, substantia nigra, locus ceruleus, olivary nuclei, caudate nucleus, and the dorsal motor nucleus of vagus. These cellular losses are accompanied by gliosis and in some cases with Lewy bodies, which are typical of Parkinson's disease. Men are more frequently affected than women are and the disease exhibits an insidious onset. Postural hypotension may be treated by medications that increase blood volume and by pressure (antigravity) stockings. Parkinsonian symptoms may be treated by the administration of sinemet or bromocriptine as well as α agonists.

Botulism is caused by ingestion of food contaminated with *Clostridium botulinum* (anaerobic Gram-positive organism), ingestion of spores and production of toxin, or as a result of wound infection with the same bacteria. It is a paralytic disease, which initially affects the cranial nerves, and expands to involve the limbs.

Symptoms of botulism include autonomic disturbances such as nausea, vomiting, dysphagia, extremely dry throat, blurred vision, loss or diminished light reflex and ptosis, in addition to skeletal muscle paralysis. Descending paralysis which is symmetric involving the head, neck, arm and thorax is characteristic of this disease. Deep tendon reflexes are not generally affected, although the gag reflex may be depressed. Patients may die from respiratory failure. Patients may be given antitoxin (equine antitoxin) as well as cathartics and enemas to eliminate the toxin, supplemented with antibiotics.

Riley–Day syndrome (familial dysautonomia) is a familial recessive disorder of infants, which is characterized by a constellation of sensory and motor deficits. These deficits include hypopathia, hearing deficits and loss of taste. The autonomic disturbances in this syndrome include loss of lacrimation and loss of the mechanisms that regulate blood pressure and temperature.

Reflex sympathetic dystrophy exhibits pain and autonomic changes, occurring as a result of bone fracture, trauma to soft tissue, or myocardial infarction. The autonomic changes include increased sweating and vasoconstriction. Causalgia, a burning pain that is often accompanied by trophic cutaneous changes, is a form of reflex sympathetic dystrophy, occurring in partial lesions of a peripheral nerve such as the median or sciatic nerve.

Suggested reading

1. Karczmar AG, Koketsu K, Nishi S. *Autonomic and Enteric Ganglia: Transmission and Its Pharmacology*. Plenum Publishing, 1986
2. Dohn DF, Sava GM. Sympathectomy for vascular syndromes and hyperhidrosis of the upper extremities. *Clin Neurosurg* 1978;25:637–50
3. Elfvin LG, Lindh B, Hökfelt T. The chemical neuroanatomy of sympathetic ganglia. *Annu Rev Neurosci* 1993;16:471–504
4. Fugére F, Lewis G. Celiac plexus block for chronic pain syndromes. *Can J Anaesth* 1993;40:954–63
5. Garrett JR, Howard ER, Nixon HH. Autonomic nerves in rectum and colon in Hirschsprung's disease. A cholinesterase and catecholamine histochemical study. *Arch Dis Child* 1969;44:406–17
6. Higgins CB, Vatner SF, Braunwald E. Parasympathetic control of the heart. *Pharmacol Rev* 1973;25:119–55
7. Hoyle CHV, Burnstock G. Neuronal populations in the submucous plexus of the human colon. *J Anat* 1989;166: 7–22
8. Jit I, Mukerjee RN. Observations on the anatomy of the human thoracic sympathetic chain and its branches, with an anatomical assessment of operations for hypertension. *J Anal Soc Ind* 1960;9:55–82
9. Koltzenburg M, McMahon SB. The enigmatic role of the sympathetic nervous system in chronic pain. *Trends Pharm Sci* 1991;12:399–402
10. Laskey W, Polosa C. Characteristics of the sympathetic preganglionic neuron and its synaptic input. *Prog Neurobiol* 1998;31:47–94
11. Lepor H, Gregerman M, Crosby R, *et al.* Precise localization of the autonomic nerves from the pelvic plexus to the corpora cavernosa: a detailed anatomical study of the adult male pelvis. *J Urol* 1985;133:207–12
12. Loewry AD, Spyer KM. Vagal preganglionic neurons. In: Loewry A D, Spyer K M, eds. *Central Regulations of Autonomic Functions.* New York: Oxford University Press, 1990:68–87
13. MacDonald IA. The sympathic nervous system and its influence on metabolic function. In Bannister R, Mathias CJ, eds. *Autonomic Failure: a Textbook of Clinical Disorders of the Autonomic Nervous System.* Oxford: Oxford Medical Publishers, 1992:197–211
14. Miolan JP, Niel JP. The mammalian sympathetic prevertebral ganglia: Integratine properties and role in the nervous control of digestive tract motility. *J Auton Nerv Syst* 1996;58:125–38
15. Mizeres N. The cardiac plexus in man. *Am J Anat* 1963;112:141–51
16. Nathan PW, Smith MC. The location of descending fibers to sympathetic neurons supplying the eye and sudomotor neurons supplying the head and neck. *J Neurol Neurosurg Psychiatry* 1986;49:187–94
17. Norvell JE. The aorticorenal ganglion and its role in renal innervation. *J Comp Neurol* 1968;133:101–12
18. Schott GD. Visceral afferents: Their contribution to "sympathetic dependent" pain. *Brain* 1994;117:397–413
19. Sinnreich Z, Nathan H. The ciliary ganglion in man. *Anat Anz* 1981;150:287–97
20. Szurszewski IH, King BF. Physiology of prevertebral ganglia in mammals with special reference to inferior mesenteric ganglion. In Wood JD, ed. *Handbook of Physiology – the Gastrointestinal System I.* Bethesda, MD: American Physiological Society: 1989: 519–92

Spinal nerves 10

Spinal nerves are formed by the union of the dorsal and ventral roots which later divide into dorsal and ventral rami. The dorsal rami supply the skin and muscles of the back, while the ventral rami contribute to the formation of the cervical, brachial, and lumbosacral plexuses. Each plexus consists of ventral rami from a series of spinal segments, giving rise to branches that supply muscles and cutaneous areas. Some of these branches are motor, others are sensory, and most carry both sensory and motor fibers. Damage to these branches may occur in conditions that frequently produce a constellation of disorders that involve muscles and/or dermatomes. These conditions and associated deficits are discussed in this chapter.

Formation, distribution and components of the spinal nerves

Cervical spinal nerves
- Cervical plexus

Thoracic spinal nerves

Brachial plexus
- Branches of the roots
- Branches of the superior trunk
- Branches of the lateral cord
- Branches of the medial cord
- Branches of the posterior cord

Lumbar spinal nerves
- Lumbar plexus

Sacral spinal nerves
- Sacral plexus

Spinal reflexes
- Superficial reflexes
- Deep reflexes

Formation, distribution and components of the spinal nerves

The spinal nerves are formed by the union of the dorsal and ventral roots. Both of these roots run in the subarachnoid space to reach their points of exit at the intervertebral foramina. The central processes of the unipolar neurons of the spinal ganglia (Figures 10.1 & 10.3) form the dorsal roots.

Dorsal root fibers enter the posterolateral sulcus of the spinal cord as medial and lateral bundles, receiving coverings from the pia mater, arachnoid and the dural sheath. These roots consist of thickly and thinly myelinated, as well as unmyelinated fibers. The thickly myelinated fibers comprise group Ia (annulospiral) and group II (flower spray endings) fibers that convey information from muscle spindles, as well as group Ib fibers of the Golgi tendon organs. The thickly myelinated fibers of the dorsal roots can selectively be blocked by the application of pressure on the dorsal roots. These thick fibers also show selective degeneration in combined system disease (associated with pernicious anemia), tabes dorsalis, and arsenic poisoning. Smaller, thinly myelinated (Aδ) fibers, and the unmyelinated C fibers that carry nociceptive impulses may be blocked more effectively by local anesthetics and can selectively be affected in beriberi disease (associated with vitamin B_1 deficiency). The skin area supplied by one dorsal root is known as a dermatome. Dermatomes (Figure 10.2) of successive dorsal roots show extensive overlap, which may be limited along the axial line.

The ventral roots are comprised of axons of the α and δ motor neurons (GSE) of the ventral horn of the spinal cord which supply the extrafusal and intrafusal muscle fibers, respectively. They also contain general visceral efferents (GVE), emanating from the intermediolateral columns of the thoracic and upper lumbar segments (sympathetic fibers) or arising from the second through the fourth sacral segments (parasympathetic fibers). In the intervertebral foramina, the dorsal and ventral roots unite to form the spinal nerves, which are accompanied by the meningeal and spinal branches of the segmental arteries. There are 31 pairs of spinal nerves; eight cervical, 12 thoracic, five lumbar, five sacral, and one coccygeal. All spinal nerves emerge via the intervertebral foramina (bounded anteriorly by the intervertebral discs and vertebral bodies, posteriorly by zygapophyseal joints, superiorly and inferiorly by the vertebral notches), with the exception of the first cervical (suboccipital) and the fifth sacral spinal nerves. The first cervical spinal (suboccipital) nerve leaves the vertebral column between the occiput and the atlas, whereas the fifth sacral spinal nerve exits through the sacral hiatus. The eight cervical spinal nerve emerges inferior to the first thoracic vertebra. The sympathetic ganglia are connected to all spinal nerves via the gray communicating rami. The thoracic and upper two or three lumbar spinal nerves have additional connection to the sympathetic ganglia via the white communicating rami (Figure 10.3).

Spinal nerves also give rise to recurrent meningeal branches to the spinal and cerebral dura mater, periosteum, blood vessels, posterior longitudinal ligament, as well as the intervertebral discs. Recurrence of pain in diseases associated with the vertebral column or spinal nerves may be attributed to irritation of these meningeal branches. There is considerable overlap in the innervation of the spinal dura mater, which accounts for the perception of pain over several dermatomes upon irritation of a small area of the dura mater that receives innervation via a single spinal nerve.

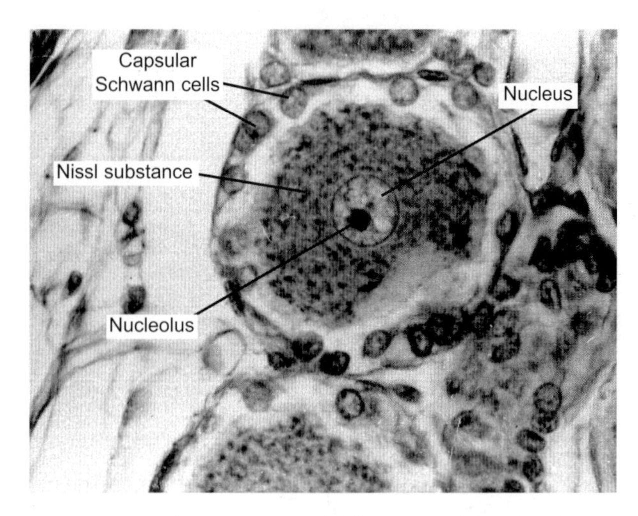

Figure 10.1 Photomicrograph of section of the dorsal root ganglion showing its main components

> Proximity of the dorsal roots to the intervertebral discs may render them more prone to compression by a herniated intervertebral disc in which the nucleus pulposus extrudes posterolaterally into the vertebral canal.

> Recurrence of pain in diseases associated with the vertebral column or spinal nerves may be attributed to irritation of the meningeal branches. There is considerable overlap in the innervation of the spinal dura mater, which accounts for the perception of pain over several dermatomes upon irritation of a small area of the dura mater that receives innervation via a single spinal nerve.

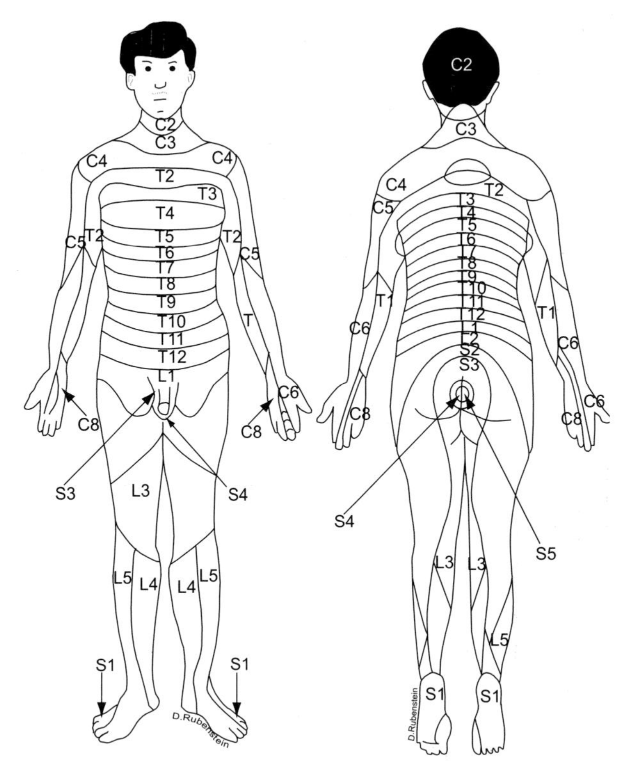

Figure 10.2 Dermatomes are mapped according to the band of skin innervated by dorsal roots of a single spinal segment

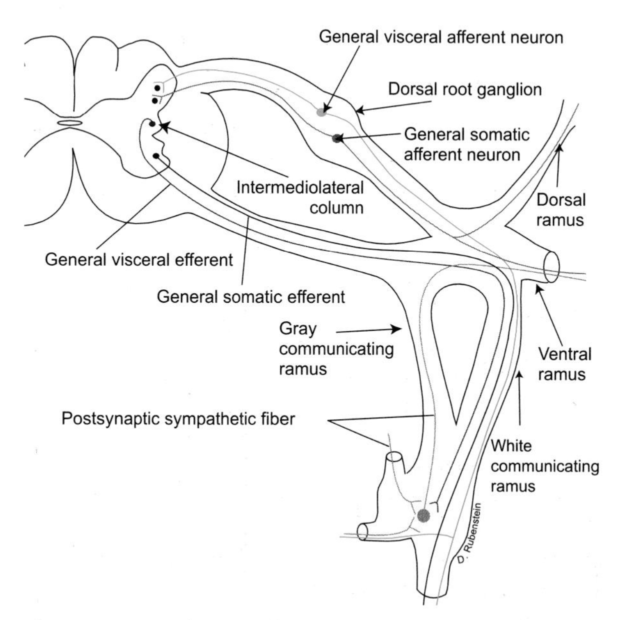

Figure 10.3 The spinal nerve, associated rami, and its functional components

In general, the peripheral nerves are arranged in bundles or fasciculi that join together to form nerve trunks. The epineurium is a collagenous layer with variable amount of fat that covers and enwraps the nerve trunk. The fat content of the epineurium plays a protective role against injuries, and loss of this fatty layer may produce pressure palsies in bed-ridden chronic patients. Each fasciculus within a nerve trunk is encircled by the perineurium, a relatively thicker connective tissue sheath that exhibits epitheloid and myoid characteristics. The perineurium consists of collagen and cells derived from fibroblasts that continue with the coverings of the encapsulated receptors. Individual fibers are surrounded by the endoneurium, a loose connective tissue layer that is derived from the mesoderm. It consists of collagenous fibers, fibroblasts, Schwann cells, and endothelial cells that are immersed in a fluid, maintaining a higher pressure than the surrounding.

The pressure difference in the endoneurium, which is maintained by the perineurium, may be critical in preventing toxic contamination of the endoneurium. In addition, host neurons may send sprouts into the tubes formed by the endoneurium within a donor skin graft, allowing the re-innervation of the graft tissue to proceed.

The nerve fibers and the connective tissue coverings are vascularized by intraneural capillaries called vasa nervorum. These vessels are categorized into extrinsic epineurial vessels and intrinsic longitudinal endoneurial microvessels. The unique course of these vessels may account for the relative resistance of peripheral nerves to ischemia. In addition to these coverings, nerve fibers may be ensheathed by myelin or remain unmyelinated. Thickness of the coverings also varies with the relative site of the nerve fiber and speed of conduction. Generally speaking, thickly myelinated fibers exist in high pressure sites, innervating targets that require fast conduction. Unmyelinated nerve fibers are relatively thin, and display a slower rate of conduction. They remain enfolded by the cytoplasm of the Schwann cells, in which each nerve fiber is enfolded by a single Schwann cell to form the Remak bundle. The unmyelinated fibers form the postganglionic autonomic fibers, olfactory nerve and the nociceptive C fibers. Following their formation, the spinal nerves divide into a large ventral and smaller dorsal rami (Figure 10.3).

The ventral rami are mostly larger than the dorsal rami and supply the extremities, thoracic and abdominal walls. They form elaborate connections within the cervical, brachial and lumbosacral plexuses. These connections enable one spinal segment to contribute to the formation of more than one spinal nerve and also make it possible for one muscle to be innervated

Pathological conditions involving the α motor neurons in the spinal cord, e.g. poliomyelitis, or brainstem, or their axons (polyneuropathy) may produce spinal nerve dysfunctions via demyelination, followed by muscle denervation, and paresis or paralysis.

Paresis or paralysis may be preceded by visible involuntary contractions of the muscle fibers (fasciculation) involving a motor unit. Fibrillation involves a single muscle fiber, may not be visible through skin, and may only be detected through electromyography.

The extent and severity of neuronal damage may determine the degree of dysfunction. On that basis trauma or neurological diseases may produce disorders that are classified into neuropraxia, axonotmesis and neurotmesis.

Neuropraxia is an incomplete, transient and reversible loss of conduction without the loss of anatomic integrity. It results from transient ischemia or paranodal demyelination subsequent to severe compression. Neuropraxia may produce loss of deep tendon reflexes, and sensory dissociation with preservation of pain and thermal sensations, but with no detectable autonomic dysfunctions.

Axonotmesis refers to a complete interruption of an axon and its myelin sheath with preservation of the connective tissue stroma. It is characterized by immediate and complete loss of all sensory, motor, and autonomic functions. The degree of recovery of injured nerve fibers (commonly from a closed crushed injury) is dependent upon the length of the damaged segment and its distance from the innervated structure. The part of the axon distal to the site of injury undergoes Wallerian degeneration. After a latent period of approximately 1 month downward-directed nerve regeneration may occur.

Neurotmesis refers to the complete anatomic disruption of neural and connective tissue elements of an axon. It is caused by injuries that penetrate nerves such as stab or gunshot wounds. Fibrosis and loss of continuity of endoneurial tubules render spontaneous recovery and regeneration almost impossible, and neurosurgical repair a necessity.

by more than one spinal segment. The thoracic spinal nerves, for the most part, retain their segmental arrangement and form the intercostal nerves.

The dorsal rami supply the skin and intrinsic muscles of the back and neck. They usually divide into medial and lateral branches (with the exception of the dorsal rami, the C1 , C4, S5, and CC1 spinal nerves). In general, the medial branches of the dorsal rami of the cervical and upper thoracic spinal nerves provide primarily sensory while the lateral branches provide motor innervation. This pattern is reversed in the lower thoracic and lumbar spinal nerves. The dorsal rami of the sacral spinal nerves exit through the dorsal sacral foramina with the exception of the fifth sacral nerve, which divides into the medial and lateral branches. The dorsal ramus of the coccygeal nerve joins the lower two sacral dorsal rami to supply the coccygeal skin.

Spinal nerves may be affected in entrapment neuropathies, as a result of a localized injury or inflammation caused by mechanical irritation from impinging anatomical structure. Burning pain felt at rest associated with altered sensation is characteristic of these types of nerve injury. Injury to the spinal nerves or roots may also occur as a result of herniated intervertebral discs, tumors, osteoarthritis, spina bifida cystica, or cauda equina syndrome. These clinical conditions are generally dependent upon the extent of damage and the number of affected roots or nerves. Nerve root compression, e.g. as a result of disc prolapse, commonly occurs at sites where the vertebral column is most mobile. The lower cervical and lower lumbar vertebrae are the frequent sites of root compression. Paresthesia or pain may result from compression of the dorsal roots.

Cervical spinal nerves

There are eight cervical spinal nerves that divide into dorsal and ventral rami. The first cervical spinal nerve forms the suboccipital nerve, which runs in the suboccipital triangle and provides innervation to the rectus capitis posterior major and minor and the inferior and superior capitis oblique muscles. The medial branch of the dorsal ramus of the second cervical spinal nerve forms the greater occipital nerve, which encircles the inferior oblique muscle and ascends to supply the skin of the posterior scalp as far as the vertex. The ventral rami of the cervical spinal nerves form the cervical plexus and contribute partly to the brachial plexuses.

Cervical plexus

The cervical plexus (Figures 10.4 & 10.5) is formed by the ventral rami of the upper four cervical nerves, with a small contribution from the fifth cervical segment. It lies deep to the internal jugular vein and anterior to the middle scalene muscle. This plexus gives rise to sensory and motor nerves. It also provides segmental motor innervation to the geniohyoid,

Cervical disc herniation commonly occurs between the sixth and seventh cervical vertebrae. This is due to the fact that C6 is the fulcrum for cervical movements. Degeneration of the intervertebral discs, e.g. in cervical spondylosis, results in motor deficits and pain in the arm or the neck. Prolapse of the intervertebral disc between the fifth and sixth vertebrae is most likely to compress the sixth cervical root. Cervical disc prolapse may protrude centrally to compress the spinal cord and produce combined signs of upper motor neuron palsy and sensory deficits in the lower extremity.

Thoracic spinal nerve roots are rarely affected due to the restricted rotatory movement between the thoracic vertebrae. However, direct trauma or cancer metastasis may cause collapse of the thoracic vertebrae and subsequent compression of the thoracic spinal nerve roots. Violent drawing of the entire body along the ground with one hand may specifically injure the dorsal root of the first thoracic spinal nerve, producing signs of Horner's syndrome and atrophy of the intrinsic muscles of the hand.

Prolapse of the lumbar intervertebral discs commonly occurs between the fourth and fifth lumbar vertebrae or between the fifth lumbar and the first sacral vertebrae. Prolapse of the fourth intervertebral disc (between L4–L5) is most likely to compress the fifth lumbar spinal root. In general disc herniation may be precipitated by flexion injuries and is often seen in middle-aged people who exhibit degenerative changes in the intervertebral discs and the posterior longitudinal ligament. It produces back pain that projects to the leg (sciatica), and movement disorders such as weakness of dorsiflexion of the foot and toes and sensory loss in lateral leg and dorsal surface of the foot (L5). Prolapse of the fourth intervertebral disc may also exhibit weakness of plantar flexion and eversion, pain or loss of sensation in the posterior leg and lateral plantar surface of the foot (S1). The hamstring muscles may show spasm when attempt is made to flex the thigh at the hip joint while the leg is extended (Lasegue's sign). Central protrusion of the herniated disc between L4 and L5 is commonly accompanied by urinary or bowel dysfunctions.

Involvement of sympathetic fibers adjacent to the affected spinal nerve may occur as a result of trauma to soft tissue or bony fracture. This may result in burning pain in a wider territory than the area of distribution of the affected spinal nerve (causalgia) accompanied by autonomic disturbances such as sweating and vasoconstriction (reflex sympathetic dystrophy).

Pain associated with compression of spinal nerves is generally confined to the area of distribution of the affected nerves and may or may not be accompanied by motor dysfunctions. Certain movements such as flexion, extension, or rotation aggravate root pain associated with a lesion or prolapsed disc of one or more spinal roots. Since the ventral rami are the primary contributors to the cervical, brachial, lumbar, and sacral plexuses, a detailed discussion of these plexuses will be appropriate at this point.

rectus capitis anterior and lateralis, longus capitis, and longus colli muscles. This plexus gives rise to ansa cervicalis, phrenic, lesser occipital, great auricular, transverse (colli) cervical, and supraclavicular nerves.

The ansa cervicalis (C1, 2, 3) is a nerve loop formed by the union of the ventral ramus of the first cervical spinal nerve (descendens hypoglossi or superior root) and the ventral rami of the second and third cervical spinal nerves (descendens cervicalis or inferior root). The ansa cervicalis pierces the carotid sheath and runs superficial to the internal jugular vein, innervating the infrahyoid (strap) muscles (omohyoid, sternohyoid and sternothyroid), with the exception of the thyrohyoid muscle which is innervated by the ventral ramus of the first cervical spinal nerve.

The phrenic nerve (C3, 4, 5) is formed by the ventral rami of the third, fourth, and fifth cervical spinal nerves, with the largest contribution coming from the fourth cervical spinal segment. This nerve runs on the anterior surface of the anterior scalene muscle and posterior to the prevertebral fascia. It courses within the superior and middle mediastina, between the mediastinal pleura and fibrous pericardium. This nerve runs anterior to the pulmonary root, separating it from the vagus nerve. It supplies sensory fibers to the central part of diaphragmatic pleura and diaphragmatic peritoneum, pericardium, mediastinal pleura, and to the hepatic plexus. It also provides motor fibers to the muscular diaphragm.

The accessory phrenic nerve is frequently derived from the fifth cervical spinal nerve. The superficial branches of the cervical plexus (listed below) exit at the midpoint of the posterior border of the sternocleidomastoid muscle (SCM) accompanied by the spinal accessory nerve.

The lesser occipital nerve (C2) curves around the sternocleidomastoid muscle and supplies the upper part of the medial surface of the ear and the area of the posterior scalp.

The greater auricular nerve (C2–3) ascends toward the parotid gland, accompanied by the external jugular vein, carrying sensation from the facial skin

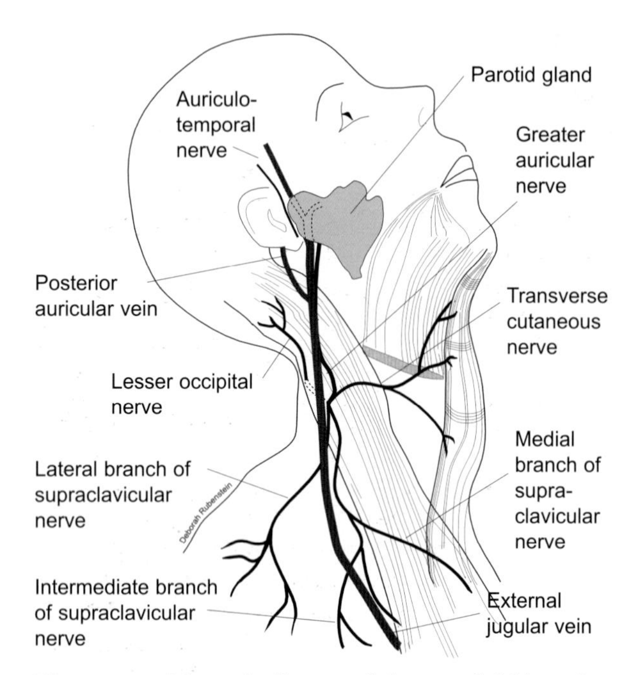

Figure 10.4 Schematic diagram of the superficial branches of the cervical plexus

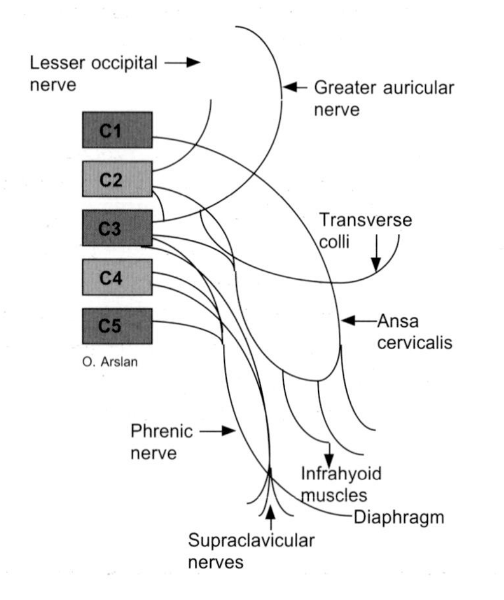

Figure 10.5 Components of the cervical plexus and its area of distribution

Pathological conditions involving the mediastinal and/or diaphragmatic pleura and diaphragmatic peritoneum or the gall bladder may result in pain radiating to the dermatomes of the third, fourth and fifth cervical spinal nerves which correspond to the back and upper part of the shoulder. Paralysis of the hemidiaphragm may result from excision of the phrenic nerve in the neck unless an accessory phrenic nerve exists.

Anesthetics may be injected into the midpoint of the posterior border of SCM to achieve complete cervical nerve block in radical neck dissection.

that covers the parotid gland, mastoid process, and the ear lobule. It is the only cutaneous nerve to the face which is not derived from the trigeminal nerve.

The transverse (colli) cervical nerve (C2–3) arises from the ventral rami of the second and third cervical spinal segments, supplying cutaneous fibers to the anterior and lateral neck.

The supraclavicular nerves (C3–4) are derived from the ventral rami of the third and fourth cervical spinal nerves and descend deep to the platysma. They divide into lateral, intermediate, and medial branches, supplying the lower neck and the upper part of the anterior thorax.

Thoracic spinal nerves

The thoracic spinal nerves emerge from the intervertebral foramina distal to the corresponding vertebrae. Due to the difference between the length of the vertebral canal and the length of the spinal cord, the lower thoracic spinal nerves pursue a longer course in order to exit through the corresponding foramina. The medial branches of the dorsal rami of the upper six thoracic spinal nerves are primarily cutaneous to the back, while the lateral branches of these rami are principally muscular to the iliocostalis and levator costarum muscles. On the other hand, the medial branches of the lower six thoracic spinal nerves innervate the multifidi and the longissimus muscles, and the lateral branches remain sensory. Continuation of the ventral rami of the thoracic spinal nerves form the intercostal nerves that run in the costal sulci and innervate the upper extremity, thoracic wall, anterior abdominal, and the gluteal region. The ventral ramus of the first thoracic nerve contributes to the brachial plexus and to several other nerves such as the ulnar

The origin, course, and the area of distribution of the intercostal nerves may explain the mechanism of projected pain to the thoracic wall associated with inflamed costal or diaphragmatic pleura. It may also account for the pain sensation in the anterior abdominal wall as a result of subluxation of the interchondral joints and compression of the lower intercostal nerves (clicking rib syndrome). Tuberculosis of the thoracic vertebrae may produce pain in the anterior abdominal wall as a result of compression of the intercostal nerves.

Herpes zoster (shingles) commonly affects the dorsal root ganglia of the thoracic spinal nerves producing pain in the thoracic wall. In addition, involvement of the first thoracic spinal segment in the transmission of cardiac pain is responsible for the referred pain felt in the medial arm, medial forearm and medial fifth digit. The first thoracic spinal segment provides both sympathetic fibers to the cardiac plexus and cutaneous fibers to the medial arm and forearm via the medial antebrachial and brachial cutaneous branches (T8–T1) of the brachial plexus. It has been suggested that painful impulses from the heart are conveyed to T1 spinal segments via the sympathetic fibers, lowering the threshold of the cutaneous neurons within that particular segment.

Contraction of the abdominal muscles in response to cutaneous stimulation of the abdomen confirms the fact that the intercostal nerves subserve dual functions of cutaneous and muscular innervation of the anterior abdominal wall.

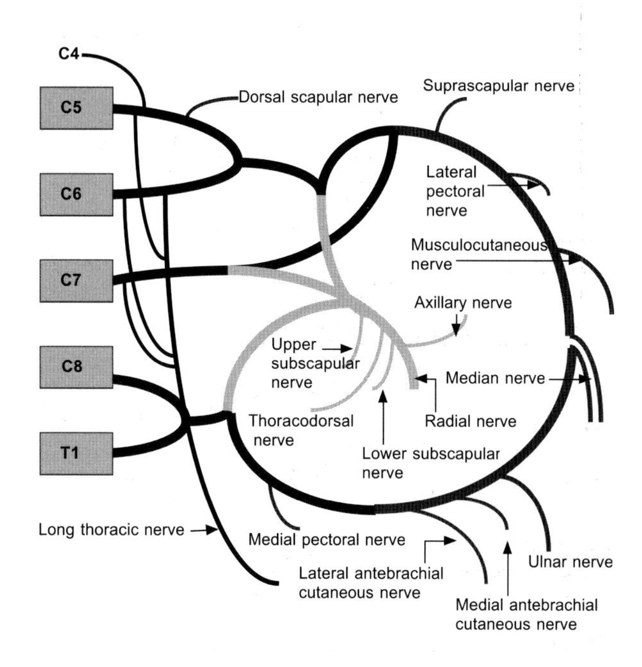

Figure 10.6 Formation of the brachial plexus. Observe the segmental contribution of the spinal cord to the trunks, divisions, cords, and peripheral branches

and median nerves. A particular branch, the intercostobrachial nerve, arises from the second intercostal nerve (sometimes the third intercostal nerve) that joins the brachial plexus and supplies the skin of the upper part of the medial arm. The upper six intercostal nerves supply the thoracic wall, costal pleura, the diaphragm, and the diaphragmatic pleura and peritoneum, while the lower five intercostal (thoracolumbar) nerves course between the internal oblique and transverse abdominis muscles, innervating the skin and muscles of the anterior abdomen, as well as the peritoneum. In particular, the tenth intercostal nerve supplies the skin of the umbilicus, whereas the seventh, eighth, and ninth intercostal nerves supply the supra-umbilical region. The lower two intercostal nerves supply the infra-umbilical region. The subcostal nerve courses posterior to the lateral arcuate ligament, anterior to the quadratus lumborum, and distal to the twelfth rib, accompanied by the subcostal vessels. It pierces the abdominal wall posterior to the anterior iliac spine and innervates the skin of the anterolateral gluteal region.

Table 10.1 Summary

5	Roots:	Ventral rami of C5–T1, give rise to the dorsal scapular and long thoracic nerves	
3	Trunks:	Superior trunk (C5 & C6) give rise to the suprascapular nerve and nerve to the subclavius, middle trunk (C7), inferior trunk (C8 & T1)	
6	Divisions:	Three anterior & three posterior	
3	Cords:	Lateral, posterior, medial	
16:	Branches	Posterior cord:	axillary, radial, thoracodorsal, upper & lower subscapular nerves
		Lateral cord:	musculocutaneous and lateral pectoral nerves, and the lateral root of the median nerve
		Medial cord:	medial pectoral, medial brachial and antebrachial cutaneous, and ulnar nerves, and medial root of the median nerve.

The roots (ventral rami), trunks, divisions, cords, and branches of the brachial plexus may be simplified by the following mnemonic: **R**obert **T**aylor **D**rinks **C**old **B**eer.

Injury to the superior trunk (Erb–Duchenne's paralysis) may be caused by hyperextension of the neck which increases the angle between the shoulder and neck, or undue pull on the suprascapular nerve that anchors to the margins of the suprascapular foramen. A fall from a motorcycle, a careless forceps delivery in breech position, or traction on the head may also precipitate this type of injury.

Patients present with an adducted (deltoid and supraspinatus are inoperative), extended and a medially rotated arm hanging limply at the side (coracobrachialis, infraspinatus and teres minor are no longer functioning). The forearm is extended and pronated (biceps brachii and brachialis are non-functional) and the wrist is slightly flexed, forming the typical configuration of waiter's tip hand (Figure 10.7). Due to the extensive overlap between the cutaneous fibers of the contiguous nerves, sensory loss associated with this type of injury will be confined to a small region of the shoulder.

Injury to the inferior trunk may be caused by excessive abduction of the arm which may occur in an individual who clutches himself to an object while falling from a height (Klumpke's palsy). It may also occur during a difficult breech delivery (birth palsy or obstetric paralysis), or upon a sharp angulation of the inferior trunk over the cervical rib (cervical rib syndrome). It may be seen in individuals with abnormal insertion or spasm of the anterior and middle scalene muscles (scalene anterior syndrome), or as a result of an anomalous fibrous band that extends to the first rib. This condition is characterized by paralysis of the intrinsic muscles of the hand, especially the thenar muscles, as well as the long flexors of the hand and the digits. Pain and numbness is felt along the medial border of the forearm, hand, and medial two digits. Horner's syndrome (ptosis, miosis, and anhydrasis) may also be seen in this condition due to involvement of the first thoracic spinal segment. Klumpke's palsy also exhibits 'claw-hand' configuration.

When it is caused by a cervical rib or a fibrous band that extends to the first rib the subclavian artery may also be compressed in conjunction with the inferior trunk, producing combined neuronal and vascular disorders (thoracic outlet syndrome). In this syndrome the inferior trunk is damaged and the blood flow to the upper extremity within the subclavian artery is substantially decreased, producing coldness, cyanosis and pain in the arm. Patients manifest a positive Adson's test, a clinical finding in which the radial pulse becomes weaker on deep inspiration and also upon turning the head to the affected side.

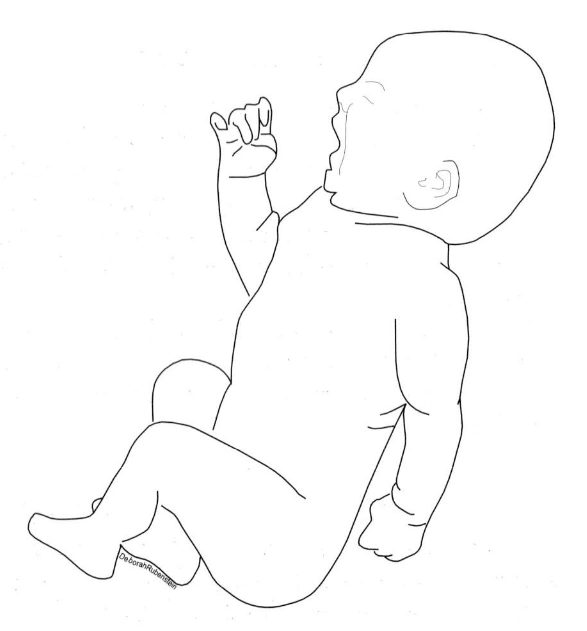

Figure 10.7 This diagram illustrates the manifestations of Erb–Duchenne's paralysis

Brachial plexus

The brachial plexus (Figure 10.6) is formed by the union of the ventral primary rami of the C5–T1 spinal nerves (with a small contribution from the ventral ramus of C4 spinal nerve). This plexus lies in the posterior triangle of the neck, posterior to the clavicle, and between the anterior and middle scalene muscles. It may be prefixed brachial plexus when it is formed by the ventral rami of the C4–C8 spinal nerves or postfixed brachial plexus when the ventral rami of C6–T2 spinal segments form it. The ventral rami of C5 and C6 join to form the superior trunk. The ventral ramus of C7 continues as the middle trunk, whereas the ventral rami of C8 and T1 spinal nerves form the inferior trunk. These individual trunks run in the axilla and then divide into anterior and posterior divisions. Union of the posterior divisions forms the posterior cord. The anterior divisions of the upper and middle trunks form the lateral cord while the anterior division of the inferior trunk continues as the medial cord. The roots of the brachial plexus give rise to the dorsal scapular and long thoracic nerves. The superior trunk gives origin to the suprascapular nerve; the lateral cord provides

Overactivity and fibrosis of the middle scalene muscle as a result of ischemic hypertrophy may damage the dorsal scapular nerve. Entrapment of this nerve within the middle scalene muscle hinders its ability to accommodate changes in position during movement of the head and arm. These movements exacerbate the pre-existing weakness, atrophy, and pain in the rhomboids and levator scapula muscles.

Proximity of the long thoracic nerve to the axillary lymph nodes, and its location superficial to the serratus anterior muscle and lateral to the mammary gland may account for the vulnerability of this nerve to injury in radical mastectomy. Carrying heavy objects on the shoulder or entrapment within the middle scalene may also damage the nerve. Unlike the dorsal scapular nerve, entrapment of this nerve within the middle scalene muscle does not produce pain in the upper extremity. Protrusion of the inferior angle of the scapula (winged scapula), which becomes evident upon protraction, is the main characteristic of long thoracic nerve dysfunction. This is due to paralysis of the serratus anterior and the inability of the muscle to hold the scapula against the thoracic wall. Weakened protraction and lateral rotation of the scapula are also common manifestations of this condition.

the musculocutaneous and lateral pectoral nerves, as well as the lateral root of the median nerve. Numerous branches arise from the medial cord which include the medial pectoral, ulnar, and medial brachial and antebrachial cutaneous nerves, the nerve to subclavius, and the medial root of the median nerve. Branches of the posterior cord are the axillary, radial, upper and lower subscapular, and thoracodorsal nerves.

Branches of the roots

The dorsal scapular nerve (C5) (Figure 10.6) arises from the ventral ramus of the fifth cervical spinal nerve, pierces the middle scalene, and supplies the levator scapula, as well as the major and minor rhomboids.

The long thoracic nerve (C5, C6, C7) (Figure 10.6) arises from the ventral rami of the C5, C6, and C7 spinal nerves and supplies the serratus anterior muscle, accompanied by the lateral thoracic vessels. Initially, the upper two roots (C5, C6) pierce the

The suprascapular nerve may be injured as a result of fibrosis and subsequent narrowing of the suprascapular foramen. Rupture of the rotator cuff and shoulder dislocation may also contribute to suprascapular nerve dysfunction. Fixation within the suprascapular foramen in the immobilized upper extremity of an individual with frozen shoulder and repeated compensatory motion of the scapula may endanger this nerve. Traction exerted on this nerve may eventually produce a pull on the upper trunk of the brachial plexus leading to Erb's palsy. Compression of the suprascapular nerve may result in atrophy of the supraspinatus and infraspinatus muscles and associated weakness in lateral rotation and abduction of the arm, as well as pain sensation confined to the posterior shoulder. The suprascapular nerve may also be damaged in Colle's fracture as a result of downward movement of the thorax when the scapula is fixed and the arm is extended.

middle scalene muscle and later unite with the lower root from seventh (C7) cervical spinal segment.

Branches of the superior trunk

The suprascapular nerve (C5, C6) (Figure 10.6) runs deep to the omohyoid muscle, accompanied by the suprascapular vessels. It enters the suprascapular foramen, inferior to the suprascapular ligament. Then, it courses in the supraspinous and infraspinous fossae, innervating the supraspinatus and the infraspinatus muscles.

The nerve to the subclavius (C5, C6) as the name implies supplies the subclavius muscle, which acts as a cushion that prevents rupture of the subclavian artery in clavicular fracture.

Injury to the musculocutaneous nerve, although rare, may result from a fracture of the humerus, shoulder dislocation, positioning of the arm during surgery, or entrapment inside a hypertrophied coracobrachialis muscle. Common manifestations of this injury are weakened flexion of the arm, markedly weakened flexion of the forearm, weakened supination and instability of the shoulder joint. Impairment of the cutaneous sensation in the lateral half of the forearm will also be observed in this injury. Heavy objects placed on the forearm and supported by the elbow may particularly compress the lateral antebrachial cutaneous branch of this nerve.

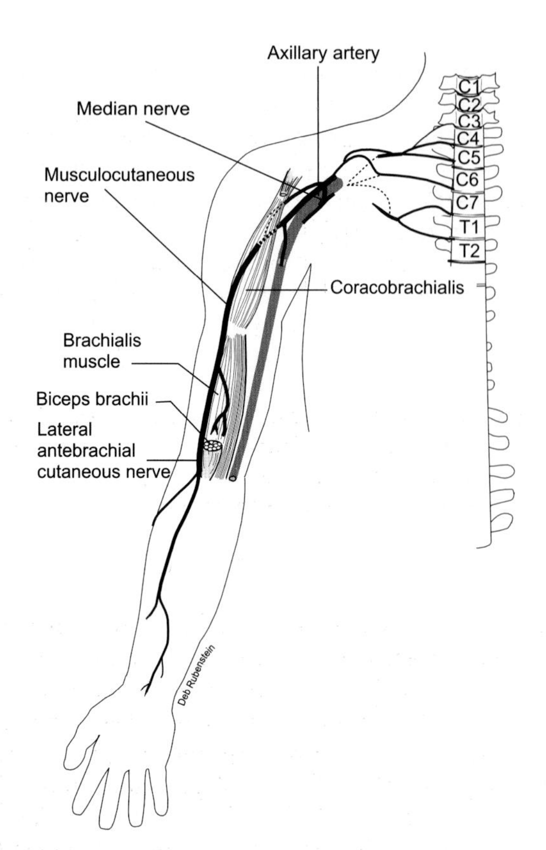

Figure 10.8 The musculocutaneous nerve and its branches

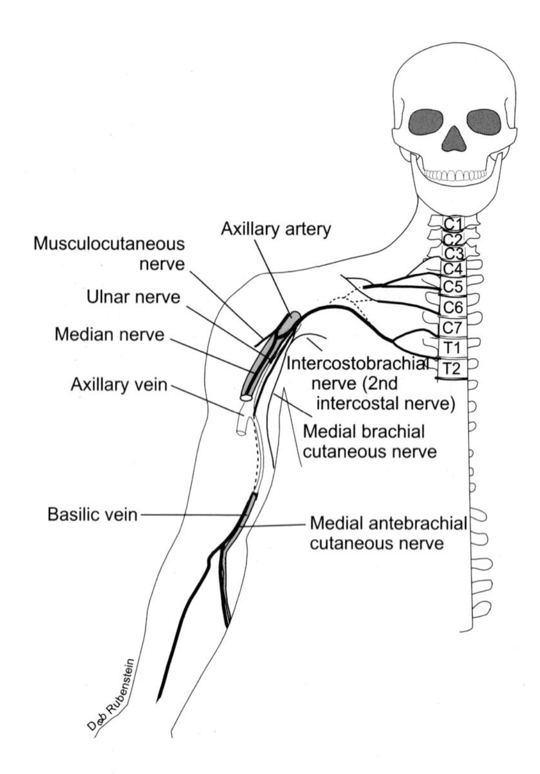

Figure 10.9 The course and areas of distribution of the medial brachial and antebrachial cutaneous nerves

Branches of the lateral cord

The musculocutaneous nerve (C5, C6, C7) (Figures 10.6 & 10.8) is formed by the ventral rami of the fifth, sixth, and seventh spinal nerves. It pierces the coracobrachialis muscle and continues to the forearm as the lateral antebrachial cutaneous nerve. This nerve supplies the flexors of the elbow such as coracobrachialis, brachialis and biceps brachii. It also provides cutaneous innervation to the lateral side of the forearm.

The lateral pectoral nerve (C5, C6, C7) (Figure 10.5) supplies the pectoralis major muscle. The lateral cord also gives rise to the lateral root (C5, C6, C7) and contributes to the formation of the median nerve.

Branches of the medial cord

The medial cord gives rise to the medial pectoral, medial brachial, medial antebrachial, and ulnar nerves, as well as to the medial root to the median nerve.

The medial pectoral nerve (C8, T1) (Figure 10.6) pierces the pectoralis minor muscle and innervates both the pectoralis major and minor muscles.

The medial brachial cutaneous nerve (C8, T1) is the smallest branch of the brachial plexus that supplies the distal one third of the medial surface of the arm (Figure 10.9). It may join the intercostobrachial nerve, or it may be replaced by the combination of the intercostobrachial nerve and a branch from the third intercostal nerve.

The medial antebrachial cutaneous nerve (C8, T1) (Figure 10.9) supplies the anterior and posterior surfaces of the medial side of the forearm as far down as the wrist.

The ulnar nerve (C8, T1) (Figure 10.10) runs medial to the axillary and brachial arteries, coursing in the corresponding sulcus on the medial epicondyle of the humerus. Later, it pierces the flexor carpi ulnaris near its origin and runs toward the wrist, deep to muscle. At the wrist, it crosses the flexor retinaculum between the pisiform bone and the hook of the hamlet (canal of Guyon), a common site of entrapment of the ulnar nerve. Then, the ulnar nerve divides into motor and sensory branches in the hand.

No muscle in the arm receives innervation from the ulnar nerve. In the forearm, it innervates the flexor

Fibers of the ulnar nerve which supply the intrinsic muscles of the hand may run within the median nerve in about 20% of individuals and leave the nerve distal to the elbow to join the ulnar nerve again (Martin–Gruber anastomosis).

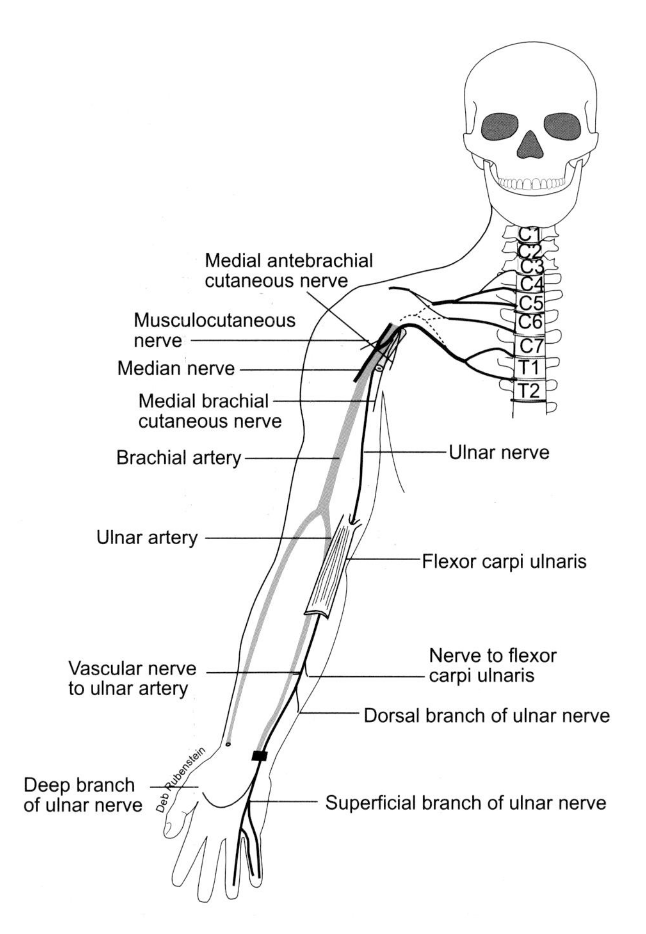

Figure 10.10 The ulnar nerve, its course and areas of distribution are illustrated in this diagram

Ulnar claw hand (Figures 10.11 & 10.12), a characteristic configuration of ulnar nerve damage at the wrist, is due to hyperextension at the metacarpophalangeal joints particularly of the ring and fifth digits and hyperflexion at the interphalangeal joints. Weakened extension and flexion at the interphalangeal joints (IP) of the fourth and fifth digits, hollowing of the palm (empty purse) and guttering of the grooves between the metacarpal bones may also be seen in this type of injury. Impaired sensation, paresthesia and nocturnal pain in the medial one third of the palm and dorsum of the hand may also be experienced by the patient.

Injury to the ulnar nerve at the axilla is rare, however its predisposition to injury is more common at the elbow due to its superficial position in the condylar sulcus of the medial epicondyle (ulnar nerve sulcus).

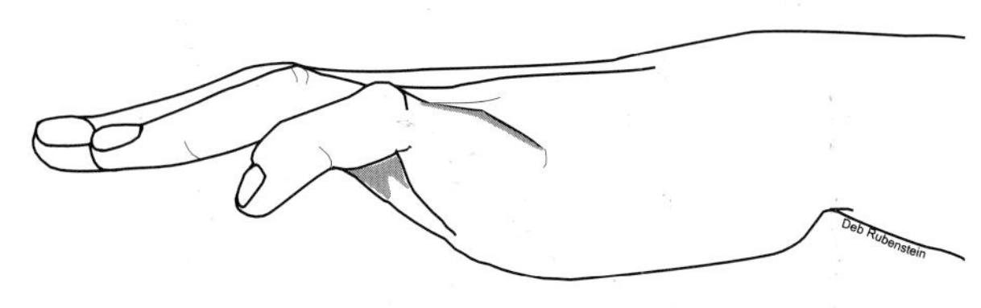

Figure 10.11 The hand and digit disorders associated with ulnar nerve damage at the elbow. Note abnormal posture of the fourth and fifth digits and flattening of the dorsal interossei with normal hypothenar muscle

Ulnar nerve injuries occur in fractures involving the medial epicondyle, dislocation of the elbow, entrapment within the Guyon canal. It may also be compressed as a result of entrapment between the two heads of the flexor carpi ulnaris, prolonged leaning on the elbow (Vegas neuropathy), sustained flexion of the elbow, cubitus valgus deformity (tardy ulnar palsy), or entrapment within the cubital fossa (cubital tunnel syndrome). Injury to the ulnar nerve at the wrist results in atrophy of the hypothenar muscles and subsequent loss of thumb adduction which makes scraping the thumb across the palm and formation of the letter 'O' by the second digit and the thumb an impossible task. Loss of abduction and adduction of the digits is exemplified by the inability of a patient to hold a piece of paper between the digits.

carpi ulnaris and medial half of the flexor digitorum profundus.

After exiting the Guyon canal, it innervates all the hypothenar muscles, the abductor, flexor and opponens digiti minimi, the palmar and dorsal interossei, the adductor pollicis (in about 55% of individuals), and the two medial lumbricals. The ulnar nerve also provides cutaneous innervation to one and a half of the medial portion of the palm and dorsum of the hand via the palmar and dorsal cutaneous branches. However, these cutaneous branches to the hand may leave the ulnar nerve proximal to the canal and flexor retinaculum, bearing significant clinical importance in anesthetic block associated with hand injuries.

The median nerve (C6, C7, C8, T1) (Figures 10.13 & 10.14) is formed by the union of the corresponding roots from the lateral and medial cords, anterior to the axillary artery. Sometimes, the musculocutaneous

Damage to the ulnar nerve at the elbow may occur as a result of recurrent trauma or subluxation and subsequent displacement of the nerve anterior to the epicondyle. It may also occur in gouty tophus or as a result of entrapment in the aponeurosis between the flexor digitorum profundus and the superficialis, producing cubital tunnel syndrome. This syndrome is characterized by numbness or paraesthesia (abnormal sensations such as tingling, prickling, burning, and itching) in the area of distribution of the ulnar nerve, extending to the forearm with possible involvement of the precondylar or intracapsular region. Impaired hand adduction, lateral deviation of the hand upon flexion, loss of flexion at the distal interphalangeal joints of the ring and fifth digits and relatively mild form claw hand may also be observed.

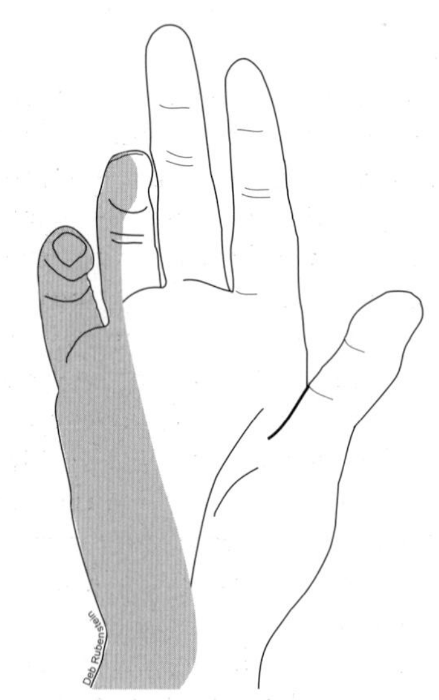

Figure 10.12 Manifestations of ulnar claw hand due to an injury to the ulnar nerve at the wrist. The shaded zone indicates the area of sensory loss

may also join the medial root when the lateral root is smaller than usual. This nerve runs in the middle of the arm with the brachial artery, anterior to the brachialis tendon, and within the cubital fossa. It courses between the humeral and ulnar heads of the pronator teres muscle, descending with the anterior interosseous vessels, on the posterior surface of the flexor digitorum superficialis. At this point, it gives rise to the anterior interosseous branch, which supplies the pronator quadratus, the lateral half of the flexor digitorum profundus, and the flexor pollicis longus. It enters the hand, deep to the flexor retinaculum, within the carpal tunnel. This tunnel is formed by the flexor retinaculum (transverse carpal ligament)

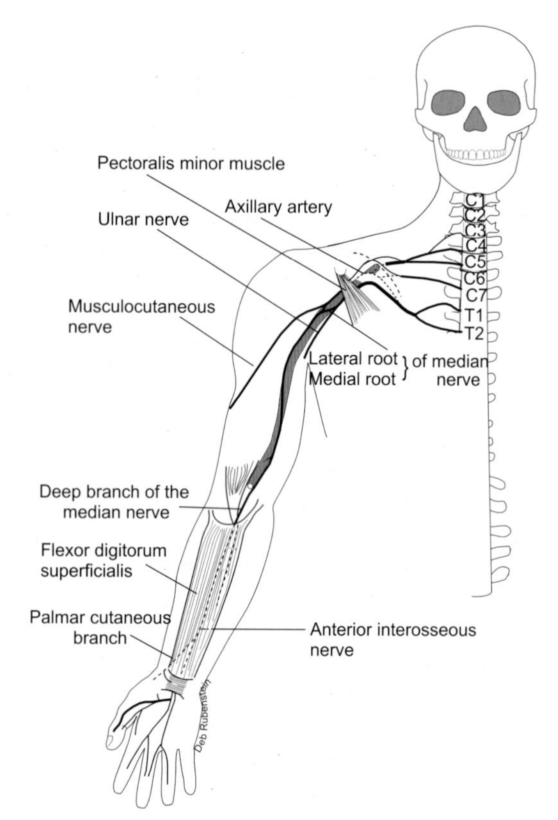

Figure 10.13 Course of the median and its terminal branches

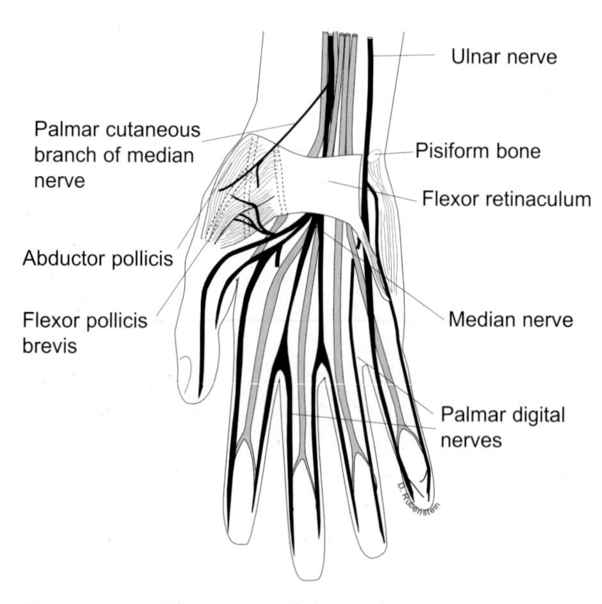

Figure 10.14 The course of the median nerve in the carpal tunnel, its branches in the hand. Notice the origin and course of the palmar branch of the median nerve

between the scaphoid and trapezium on the radial side and the pisiform and hamate on the ulnar side. It gives passage to the median nerve and tendons of the flexor digitorum superficialis, flexor digitorum profundus and the flexor pollicis longus. The ulnar artery and

Damage to the median nerve may occur as a result of fractures of the humerus, antecubital catheterization, elbow dislocation, or low grade pressure to the upper arm. It may also be compressed as a result of sustained muscular force while the forearm is pronated. It may also occur during its course between the ulnar and humeral heads of the pronator teres muscle. Volkmann's ischemic contracture may also lead, if untreated, to compression of the median nerve by the pressure of the swollen muscles of the forearm or pressure buildup by the accumulated fluid and blood under the flexor digitorum superficialis. Cuts across the wrist, anterior dislocation of the lunate bone, or compression in the carpal tunnel may also damage the nerve. Occasionally, both the median nerve and the brachial artery may be entrapped unilaterally or bilaterally by the ligament of Struthers, which extends from the medial epicondyle of the humerus to a bony spur at the distal humerus (seen in 2% of people). It is equally important to note that compression of the brachial artery may simultaneously be accompanied by compression of the median nerve, producing vascular and neuropathic changes.

Damage of the median nerve at the wrist produces loss of thumb opposition as in thumb pinching, atrophy of most thenar muscles and loss of cutaneous sensation from the lateral two thirds of the palm. There may also be anesthesia in the palmar surfaces of the thumb, index and middle fingers, as well as the lateral one half of the ring finger and the dorsal surfaces of the medial four digits as far as the middle phalanges. Weakened flexion at the metacarpophalangeal joints, weakened extension at the interphalangeal joints of the index and middle and impaired thumb abduction are also observed. Weakened thumb abduction and inadequate thumb rotation may produce 'bottle sign', a deficit in which the subject is unable to maintain a grip of a round object. Flattening and atrophy of the thenar eminence and the counter pull of the extensor and the abductor pollicis longus muscles on the thumb result in pulling the thumb to the same level with the other digits and producing the ape hand configuration.

Compression of the median nerve at the wrist may also occur inside the carpal tunnel producing signs of carpal tunnel syndrome. This syndrome is a unilateral and sometimes bilateral condition that often involves the dominant hand, and is common during the second and third trimester of pregnancy. It may be caused by fluid retention, Colle's fracture, acromegaly, hypothyroidism, congestive heart failure, tenosynovitis of the flexor tendons, mucopolysaccharidosis, or tuberculosis of the synovial sheaths. Since repetitive movements at the wrist displace the flexor tendons against the palmar side of the carpal tunnel, continuous and sustained flexion and extension of the wrist (common in industrial occupations) may be a predisposing factor for this condition. Transducers inserted into the canal may measure the increase in intracarpal pressure.

Carpal tunnel syndrome (Figure 10.15), the most common neuropathy of the hand, is characterized by acroparethesia (tingling, numbness), pain in the radial three digits that increases at night or early morning and is frequently relieved by firm grasp of the hand. Demyelination of the median nerve, subsequent atrophy of the thenar muscles, and associated autonomic disturbances such as swelling and alteration in the texture of the skin, are also common manifestations of this disease. Opposition, and to a lesser degree abduction of the thumb may eventually be affected. Carpal tunnel syndrome is confirmed by the application of Tinel's sign and Phalen's maneuver.

Tapping the wrist produces Tinel's sign, which is characterized by tingling and electrical sensation in the area of the sensory distribution of the median nerve. Forced flexion of the wrist (Phalen's maneuver) or forced extension of the wrist (reverse Phalen's maneuver) produces pain and tingling in the cutaneous distribution of the median nerve (Figure 10.16). Anomalous innervation, as in Martin–Gruber anastomosis, must be ruled out in order to confirm the diagnosis of this condition.

In addition to 'ape hand' configuration and the dysfunctions listed with damage at the wrist, injury to the median nerve at the elbow results in greater sensory loss in the palm. Loss of forearm pronation and loss of flexion of the thumb are also experienced by the affected individuals. Loss of flexion at the interphalangeal joints of the index and middle fingers and the subsequent inability to make a fist are also seen in this condition. Inability to make a fist, which allows the patient to flex the digits supplied by the ulnar nerve, produces preacher's hand position. Weakened flexion at the elbow and at the wrist and radial abduction of the hand are additional manifestations of this lesion.

Compression of the median nerve as it courses between the ulnar and humoral heads of the pronator teres or as it runs posterior to the fibrous arch of the flexor digitorum profundus produces pronator teres syndrome. This syndrome, which may also be precipitated by repeated pronation and supination, manifests similar dysfunctions to those seen with median nerve lesion proximal to the elbow. Pronation may be weakened, but not totally lost. Pain in the palmar surface of the hand aggravated by either pronation or elbow flexion, and paresthesia in the proximal forearm, which is elicited by forced supination of the forearm and extension of the hand at the wrist, are characteristics of this syndrome.

The median nerve may also be entrapped in the bicipital aponeurosis, producing signs of lacertus fibrosus syndrome. This type of entrapment produces pain upon forced pronation of the flexed and supinated hand. Damage to the median nerve at the midpoint of the forearm produces partial paralysis of the flexor digitorum superficialis muscle, which results in pointing of the index finger (Figure 10.17). This is due to the uncountered action of the extensors of the index finger.

In general, injury to the median nerve is commonly associated with a burning and tearing pain sensation in the digits and palm of the hand, which is accompanied by vasomotor and sudomotor changes on cutaneous areas wider than the area of distribution of the median nerve (causalgia). The affected part of the hand and digits become extremely sensitive to touch, including contact with clothes or air. Causalgia is attributed to overstimulation of sensory fibers at their point of interruption by the sympathetic fibers. Sympathectomy or blockade of the corresponding sympathetic ganglia may relieve this condition. Causalgia may be also associated with ulnar and sciatic nerves injury.

Compression of the anterior interosseous branch of the median nerve (anterior interosseous nerve syndrome) may result from humeral fracture, percutaneous puncture of the median cubital vein, or fibrous bands extending from the flexor digitorum superficialis across the median nerve. It is characterized by weakened flexion at the distal interphalangeal joints of the thumb and index finger accompanied by an impaired pinch maneuver. Weakened pronation, and more importantly loss of flexion at the distal interphalangeal joints of the index and middle finger, are the primary deficits of this syndrome.

Muscles that are innervated by the ulnar nerve may also be affected if the motor fibers to these muscles run within the anterior interosseous nerve before joining the ulnar nerve (Martin–Gruber anastomosis).

Combined damage to the median and ulnar nerves at the wrist results in true claw hand, which is characterized by hyperextension of the metacarpophalangeal joints and hyperflexion of the distal interphalangeal joints of the digits, unopposed by the action of interossei and lumbricals. Thumb opposition and adduction as well as atrophy of the thenar and hypothenar muscles are additional deficits of this injury.

Combined ulnar and median nerves damage at the elbow or at a more proximal site produces 'ape hand' configuration (but no apparent clawing) and hyperextension of the hand at the wrist (unopposed by the flexors). It also results in forearm supination (due to paralysis of the pronators), and hyperextension of the digits at the metacarpophalangeal joints (due to paralysis of the interossei and lumbricals).

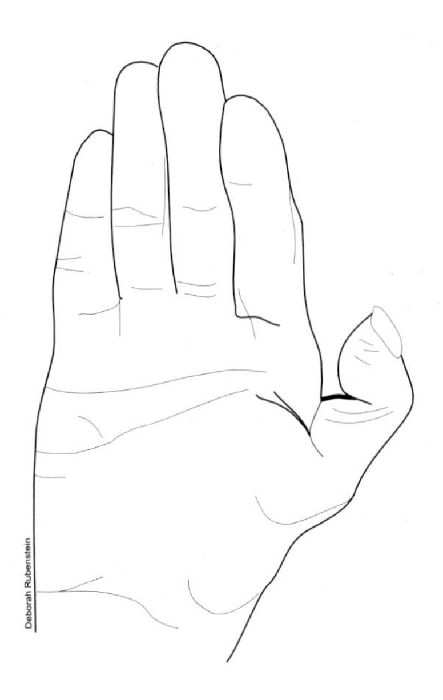

Figure 10.15 Manifestations of carpal tunnel syndrome. Note the wasting of the thenar muscles and ape hand configuration

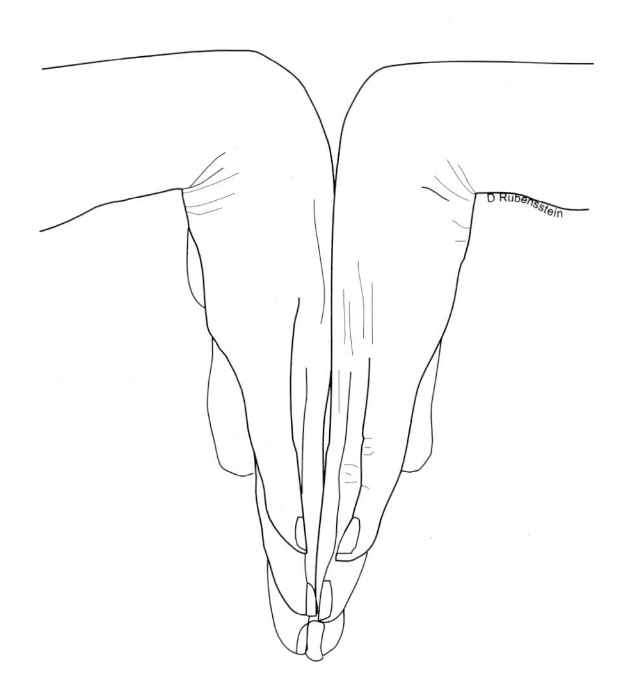

Figure 10.16 This diagram depicts the Phalen's test. Observe the acute flexion at the wrist

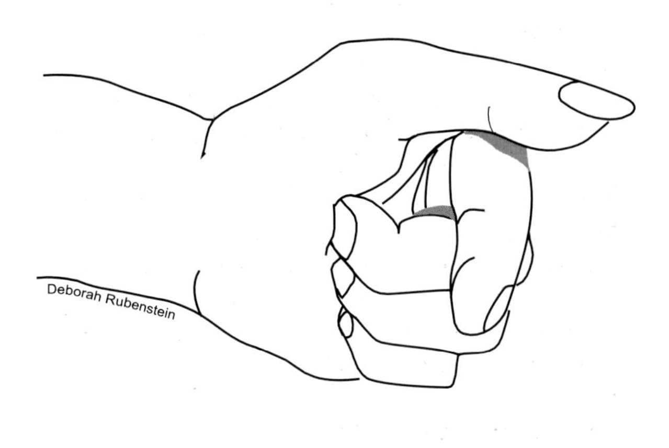

Figure 10.17 This is a drawing depicting the major manifestations of anterior interosseous nerve syndrome

nerve, radial artery and nerve, palmaris longus, flexor carpi ulnaris and the palmar cutaneous branch of the median nerve lie outside of the carpal tunnel, anterior to the flexor retinaculum. It supplies the thenar muscles and provides cutaneous innervation to the hand and digits. Proximal to the flexor retinaculum, the median nerve gives rise to the palmar cutaneous branches to the skin of the thenar eminence. It also innervates the skin of the palmar surfaces of the lateral three and half digits and the dorsal surfaces of the distal phalanges of these digits. It also supplies all muscles of the anterior forearm with the exception of the flexor carpi ulnaris and the medial half of the flexor digitorum profundus. Additionally, it innervates the thenar muscles and the lateral two lumbricals, with the exception of the adductor pollicis.

This innervation can be abbreviated by the mnemonic 'loaf' denoting 1/2 of lumbricals, opponens pollicis, abductor pollicis, and flexor pollicis brevis are supplied by the ulnar nerve.

Branches of the posterior cord

The axillary nerve (C5, C6) runs in the quadrangular space accompanied by the posterior humeral circumflex artery and vein (Figure 10.6). The anterior branch

Damage to the axillary nerve may occur as a result of inferior dislocation of the humeral head or fracture of the humeral neck, manipulation to reduce dislocation of the humerus, intramuscular injections, or pressure from the use of crutches. Axillary nerve palsy is characterized by loss of shoulder contour and severe weakness of arm abduction (due to intactness of the supraspinatus, upper fibers of the trapezius, and serratus anterior muscles). Weakened lateral rotation (due to paralysis of the teres minor) and limited loss of cutaneous sensation from the shoulder are additional deficits of this condition.

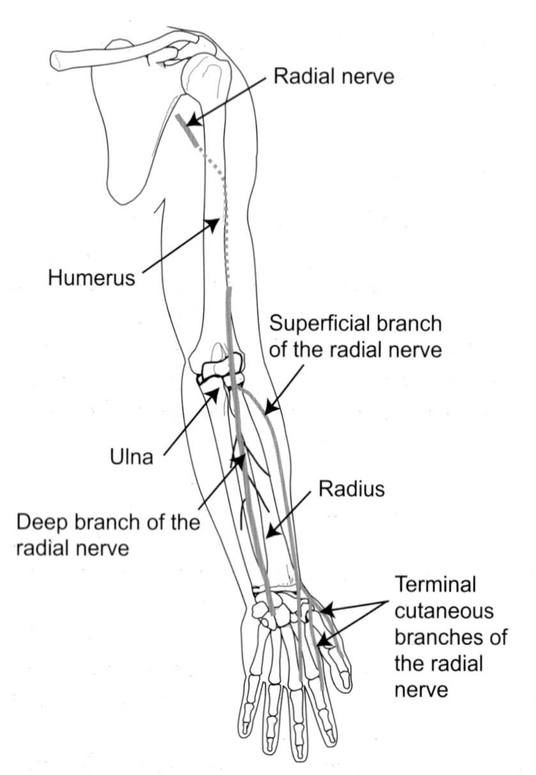

Figure 10.18 The course, branches, and distribution of branches of the radial nerve

Radial nerve damage commonly occurs in fractures that involve the midshaft of the humerus. Sleeping while inebriated with the arm hanging over the edge of a chair (sleep, or Saturday night palsy), crutch misuse (crutch palsy), or misplaced pacemaker catheter may compress and injure the radial nerve. Arcade of Frohse, a fibrous band associated with flexor digitorum superficialis muscle may entrap the radial nerve. In Parkinson's disease and Guillain–Barré syndrome, the radial nerve may be compressed as a result of fibrosis of the triceps brachii muscle. Damage to the radial nerve in the axilla may result in loss of extension of the elbow (due to paralysis of the triceps brachii and anconeus). Loss of extension of the hand at the wrist (wrist drop) and extension of the thumb and the metacarpophalangeal joints, and loss of sensation in the cutaneous area of distribution of the radial nerve (Figure 10.19) may also occur. Weakened flexion of the elbow, abduction of the thumb, radial and ulnar deviation of the hand, and weakened extension of the interphalangeal joints are additional manifestations of this type of injury. Since the triceps muscle receives innervation proximal to the midpoint of the humerus, damage to the radial nerve immediately distal to this point will spare extension of the elbow.

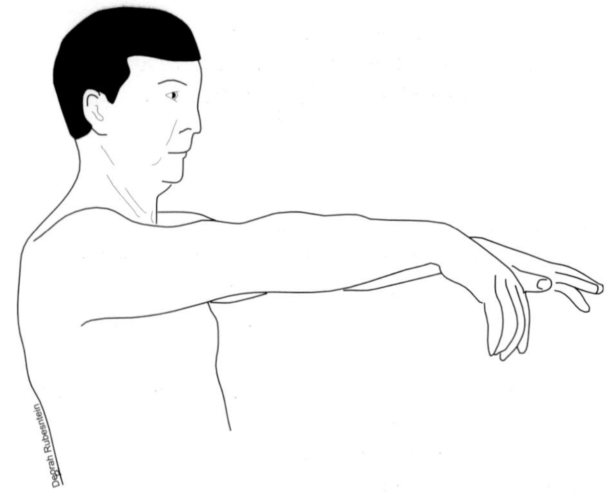

Figure 10.19 This drawing illustrates right-hand wrist drop as a result of radial nerve damage

Compression or injury to the posterior interosseous nerve (PIN) distal to the supinator muscle spares the extensor carpi radialis, resulting in radial deviation of the hand upon extension, wrist drop, and weakened extension at the interphalangeal joints. Sensory changes are not expected to occur since the PIN is purely motor, however some radial paresthesia may be observed. Compression of the superficial branch of the radial nerve by tight watchbands or handcuffs may produce isolated numbness, loss of pain, and more importantly numbness with no motor deficits (cheiralgia paresthetica or Wartenberg's disease). Prolonged orchestral drumming has been implicated in the entrapment of the posterior cutaneous nerve of the arm.

of this nerve supplies the deltoid muscle and the skin that covers the muscle, whereas the posterior branch supplies the teres minor and the posterior part of the deltoid, carrying sensations from the lower lateral part of the arm.

The radial nerve (C5, C6, C7, C8 & T1) is formed by the ventral rami of the fifth, sixth and seventh cervical, and first thoracic, spinal nerves (Figure 10.18). This nerve runs in the radial sulcus of the humerus accompanied by the deep brachial vessels. It innervates the triceps brachii by a branch that arise within this sulcus, and proximal to the mid-diaphysis of the humerus. The site of innervation of this muscle accounts for intactness of triceps brachii muscle in a fracture involving the distal third of the humerus. The radial nerve also innervates the anconeus, a weak extensor of the elbow. Anterior to the lateral epicondyle of the humerus, it divides into superficial

Damage to the subscapular nerve may produce weakened medial rotation and adduction of the arm (due to paralysis of the subscapularis and teres major muscles).

Iliohypogastric nerve damage may occur by a surgical incision in the lower anterior abdominal wall, which weakens the anterior abdominal wall, predisposing the patient to direct inguinal hernia.

and deep branches. The superficial branch supplies sensory fibers to the radial side of the thumb and adjacent area of the thenar eminence, as well as the three and half digits of the dorsum of the hand as far as the mid-portions of the middle phalanges of the index, middle and ring fingers. The deep branch pierces the supinator muscle and continues as the posterior interosseous nerve to supply the brachioradialis, supinator, common extensor digitorum, extensor carpi radialis longus and brevis. It also innervates the abductor pollicis longus, all extensors of the thumb (extensor pollicis brevis and longus) and extensors of the digits at the interphalangeal joints. This branch ends at the dorsal surfaces of the carpal bones as a pseudoganglion, a swelling which provides articular branches to the wrist joint. The cutaneous fibers of the radial nerve also innervate the skin of the lower lateral and posterior surfaces of the arm and forearm via the posterior cutaneous branches.

The upper subscapular nerve (C5, C6) supplies the subscapular muscle. The lower subscapular nerve (C5, C6) supplies the teres major and distal part of the subscapular muscles.

The thoracodorsal nerve (C6, C7, C8) (Figure 10.6) descends between the upper and lower branches of the subscapular nerve accompanied by the corresponding branch of the subscapular vessels (thoracodorsal artery and vein), supplying the latissimus dorsi muscle.

Lumbar spinal nerves

As is the case with other spinal nerves, the dorsal rami of the lumbar spinal nerves divide into medial branches that supply the multifidi and lateral branches innervating the erector spinae. Lateral branches of the upper three dorsal rami form the superior clunial nerves and supply the skin of the gluteal region. The ventral rami accompany the lumbar arteries, receiving the gray communicating rami from the sympathetic ganglia. The upper two or three ventral rami may also receive white communicating rami, conveying presynaptic sympathetic fibers. They form the lumbar plexus and innervate the muscles in the posterior abdominal wall and the lower extremity.

Lumbar plexus

The ventral rami of all lumbar spinal nerves form the lumbar plexus, running posterior to the psoas major muscle. The ventral ramus of the fourth lumbar spinal segment contributes to both the lumbar and sacral plexuses (nervus furcalis). It is considered a prefixed plexus when the ventral rami of the third and fourth lumbar spinal nerves contribute to both the lumbar and sacral plexus. However, when the fifth lumbar ventral ramus splits between the lumbar and sacral plexuses, the lumbar plexus is considered a postfixed plexus. This plexus (Figure 10.20) is commonly formed by the ventral rami of the upper lumbar spinal nerves which run anterior to the transverse processes of the lumbar vertebrae, giving rise to the iliohypogastric, ilioinguinal, genitofemoral, lateral femoral cutaneous, femoral, obturator and possibly the accessory obturator nerves.

The iliohypogastric nerve (L1) arises from the entire ventral ramus of the first lumbar spinal nerve with a smaller contribution from the subcostal nerve (Figures 10.20). It courses initially between the kidney and the quadratus lumborum, then it pierces the transverse abdominis, running between the transverse and internal oblique muscles. It divides into branches that supply the skin of the anterolateral gluteal region and the skin and muscles of the anterior abdominal wall, proximal to the superficial inguinal ring.

The ilioinguinal nerve (L1) maintains similar origin to the iliohypogastric nerve, running between the internal oblique and the transverse abdominis muscles, and through the inguinal canal with the spermatic cord or the round ligament (Figure 10.20). It

Ilioinguinal nerve damage may occur in surgical repair of a direct inguinal hernia, or as a result of a low incision in the anterior abdominal wall while performing an appendectomy operation. It may be compressed by constant and violent contraction of the muscles of the anterior abdominal wall as a result of a fall from a height, abnormalities in the hip joints, or ligamentous disorders of the vertebral column. Weakness in the abdominal muscles innervated by the ilioinguinal nerve may precipitate a direct inguinal hernia. Pain associated with entrapment of the ilioinguinal nerve may mimic urinary tract or gastrointestinal disorders.

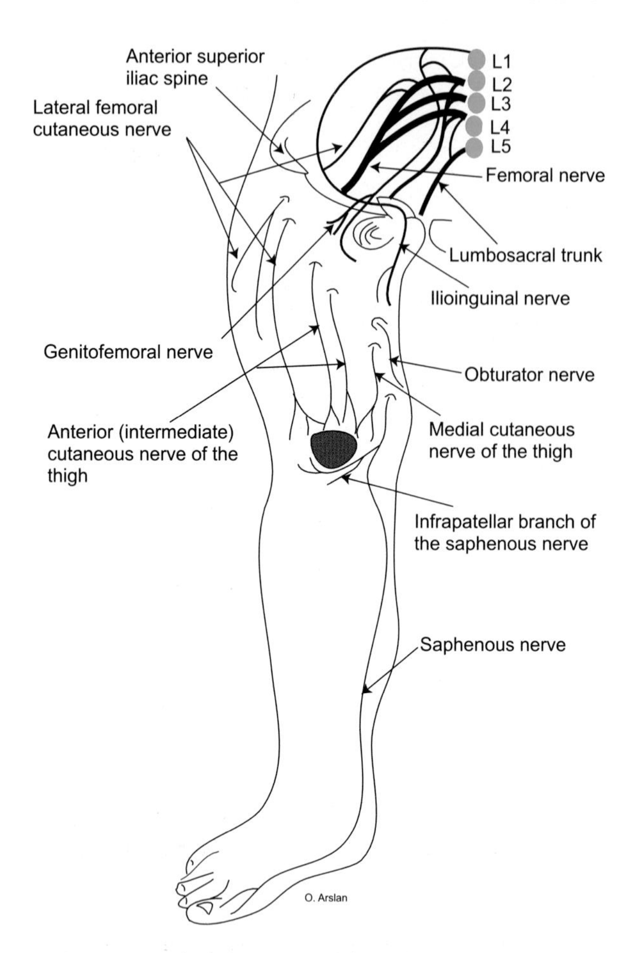

Figure 10.20 Lumbar plexus showing segmental contributions to the individual branches

emerges from the superficial inguinal ring to supply the lower portions of the internal oblique and transverse abdominis muscles, the skin of the upper medial thigh, and the anterior part of the external genitalia.

The genitofemoral nerve (L1, L2) is derived from the anterior branches of the ventral rami of the first and second lumbar spinal nerves (Figure 10.20). It pierces the psoas major muscle, passes posterior to the ureter, and then divides into genital and femoral branches. The genital branch enters the deep inguinal ring, supplying the cremasteric muscle, skin of the scrotum, mons pubis, and the labia majora. The femoral branch pierces the femoral sheath and supplies the skin anterior to the upper part of the femoral triangle.

The lateral femoral cutaneous nerve (L2, L3) originates from the posterior branches of the ventral rami of the second and third lumbar spinal nerves (Figures 10.20 & 10.21). This nerve crosses the iliacus muscle, providing sensory fibers to the parietal peritoneum. It then pierces the skin near the anterior superior iliac spine to distribute to the lateral thigh.

Attempts to extend the leg while the thigh is abducted, particularly following parturition, may produce excessive angulation of the lateral femoral cutaneous nerve and subsequent compression, resulting in meralgia paresthetica. Individuals with this condition may exhibit numbness and tingling or burning sensation in the lateral thigh and knee. This condition is more common in obese individuals following substantial weight loss. It may occasionally be seen after abdominal operations. However, despite its course behind the cecum on the right side and sigmoid colon on the left, no apparent relationship to abdominal disorders have been shown.

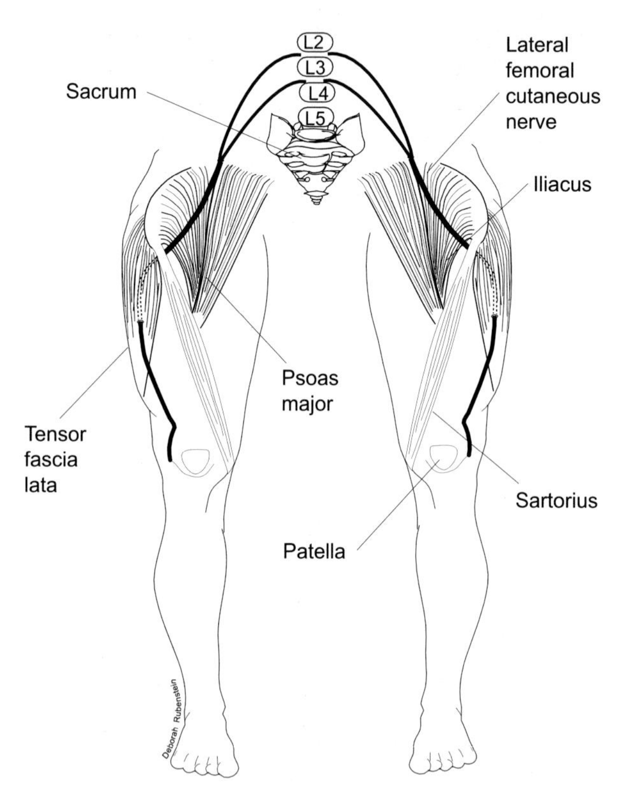

Figure 10.21 Lateral femoral cutaneous nerve. Observe its close relationship to the inguinal ligament

The lateral femoral cutaneous nerve may be damaged in individuals with lumbar lordosis, as a result of entrapment within the inguinal ligament, or compression against the anterior superior iliac spine. Constant adduction (e.g. sitting with crossed legs for prolonged period of time), compensatory stretching of the fascia and muscles around the nerve, and disorders in the ligaments that stabilize the vertebral column may all contribute to damage to the lateral femoral cutaneous nerve.

Femoral nerve damage can occur in dislocation of the hip joint, as a result of stab or gun shot wound, or as a sequel to fractures of the coxa or proximal femur. Retroperitoneal abscesses or tumors, and complication of femoral angiography may also injure the femoral nerve, producing paralysis of the quadriceps femoris and subsequent loss of the patellar reflex and impairment of knee extension. Extension of the knee may still be possible via the iliotibial tract. The patient may be able to stand and walk, experiencing difficulty in going up and down stairs. Patients cannot climb stairs and are unable to swing the lower extremity forward during walking. Complete paralysis of the sartorius, rectus femoris and partial paralysis of the pectineus muscle may also occur, leading to weakened thigh flexion. However, the iliopsoas, which is the main flexor of the thigh, remains intact. Sensory loss on the anterior and lower medial thigh and the medial surface of the leg and foot are also observed in femoral nerve damage.

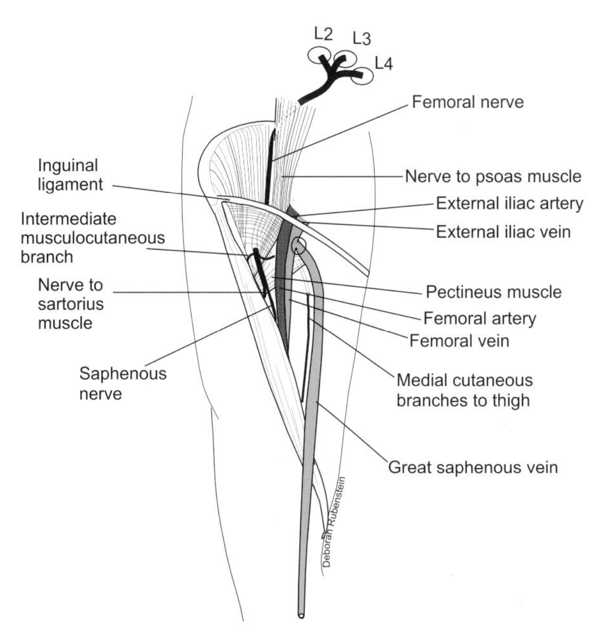

Figure 10.22 In this drawing the origin of the femoral nerve and its innervation in the thigh are illustrated

The femoral nerve (L2, L3, L4) is formed by the posterior branches of the ventral rami of the second, third and fourth lumbar spinal nerves (Figures 10.20 & 10.22). It runs deep to the psoas major, anterior to the iliacus, and then descends posterior to the inguinal ligament. It enters the femoral triangle where it innervates the sartorius, pectineus and quadriceps femoris. It provides sensory fibers to the hip joint,

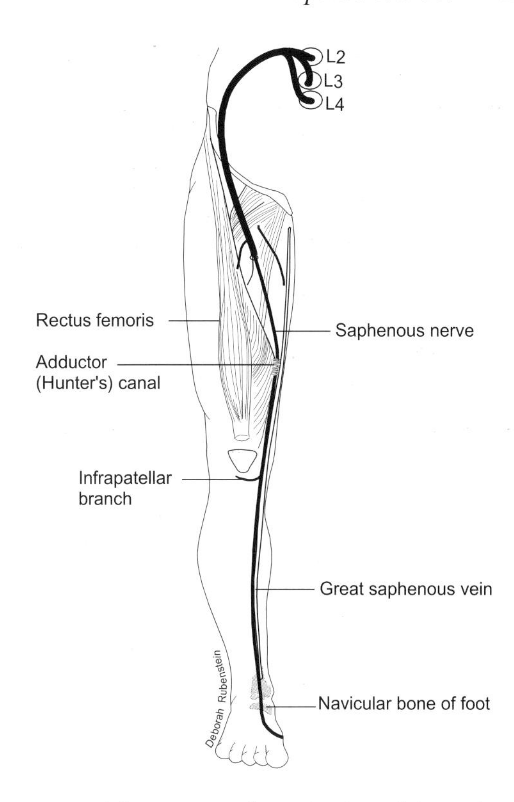

Figure 10.23 The course and termination of the saphenous branch of the femoral nerve, and its infrapatellar branch

The saphenous nerve may be entrapped, as it exits the adductor canal, producing numbness or anesthesia in the medial surface of the leg and medial border of the foot. Accidental excision of the infrapatellar branch of the saphenous nerve during arthroscopic knee surgery may result in the formation of neuroma, eliciting excruciating pain in the area of its distribution.

anterior thigh via the intermediate femoral cutaneous nerve, and to the skin of the lower medial thigh via the medial femoral cutaneous nerve.

The saphenous nerve (Figure 10.23), considered the longest branch of the femoral nerve, runs in the femoral triangle and adductor canal. It leaves the canal by piercing the deep fascia medial to the knee joint, giving rise to the infrapatellar branch. It supplies the medial surface of the leg and the medial border of the foot.

The obturator nerve (L2, L3, L4) arises from the anterior branches of the ventral rami of the second, third, and fourth lumbar spinal nerves (Figure 10.24). It descends posterior to the psoas major muscle, crosses the pelvic brim, and enters the obturator canal. It exits the obturator canal, dividing into anterior and posterior branches, which are separated by

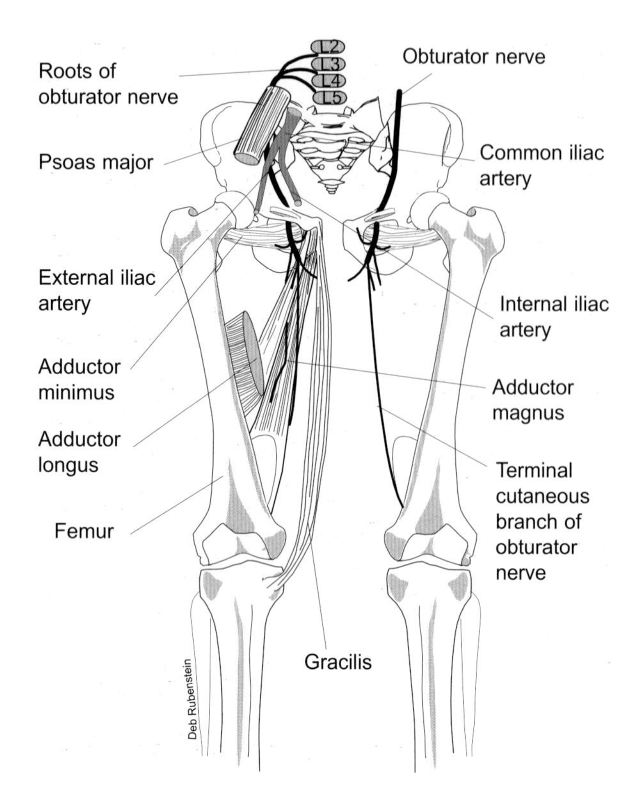

Figure 10.24 Course and distribution of the obturator nerve

The obturator nerve may be compressed by the obturator hernial sac, during hernial repair, or when entrapped within the obturator membrane. It may also be injured in pregnancy as a result of the pressure exerted by the head of the fetus, as a consequence of metastatic pelvic disease, hip replacement, or complicated labor. Surgically, the obturator nerve may be excised in paraplegic patients to relieve spasticity in the adductor muscles. The deficits associated with damaged obturator nerve include weakened or impaired ability to adduct the thigh or cross the legs, weakened medial rotation of the thigh and tendency to move the legs outward during walking. The affected individual may also experience numbness or pain that radiates to the middle portion of the medial thigh. Diseases of the hip and knee joints produce pain that radiates to the cutaneous areas of the obturator nerve.

the adductor brevis muscle. It supplies the adductor longus and brevis, part of the adductor magnus, and pectineus muscles. It also provides sensory fibers to the medial side of the thigh and hip joint.

The accessory obturator nerve (L3, L4) is a variable nerve, which gains origin from the anterior branches of the ventral rami of the third and fourth lumbar spinal nerves. It travels posterior to the pectineus and supplies the obturator muscle and the hip joint.

Sacral spinal nerves

The dorsal rami of the sacral spinal nerves emerge from the dorsal sacral foramina with the exception of the fifth sacral nerve. Each dorsal ramus gives off medial branches that terminate in the multifidi and lateral branches that join the dorsal ramus of the fifth lumbar spinal nerve to form the middle clunial nerves that supply the skin of the posterior gluteal region. Dorsal rami of the fourth and fifth sacral nerves also innervate cutaneous fibers to the skin overlying the coccyx. Visceral efferent fibers of the sacral plexus form the pelvic splanchnic nerves. These parasympathetic efferents arise from the intermediolateral columns of the second, third and fourth sacral spinal segments, and run in the ventral rami. Due to the vasodilator effect upon the penile arteries and the important role they play in erection, these fibers are also known as nervi erigentes. They also control micturition by inducing contraction of the detrusor muscles and relaxation of the urethral sphincters. Additionally, these nerves provide parasympathetic fibers to the left one-third of the transverse colon, descending colon, sigmoid colon and the rectum.

Sacral plexus

The sacral plexus is formed by the union of the lumbosacral with the ventral rami of the first, second, third, and part of the fourth sacral spinal nerves (Figures 10.25 & 10.26). The lumbosacral trunk results from union of part of the ventral ramus of the fourth lumbar spinal nerve and the entire ventral ramus of the fifth lumbar spinal nerve. It is embedded in the digitations of the piriformis muscle on the posterolateral wall of the pelvis, anterior to the sacrum and posterior to the rectum. Branches of the sacral plexus leave through the greater sciatic foramen, proximal and/or distal to the piriformis muscle. A pregnant uterus, malignancies involving the

Damage to the superior gluteal nerve, although rare, causes loss of abduction of the thigh and subsequent tilting of the pelvis toward the unsupported side when the foot is off the ground, producing lurching gait (Trendelenberg sign).

Damage to the inferior gluteal nerve results in atrophy and wasting of the gluteus maximus and impairment of extension of the thigh (although some extension is possible by the hamstrings). Patients are unable to jump, climb stairs, or rise from a seated position.

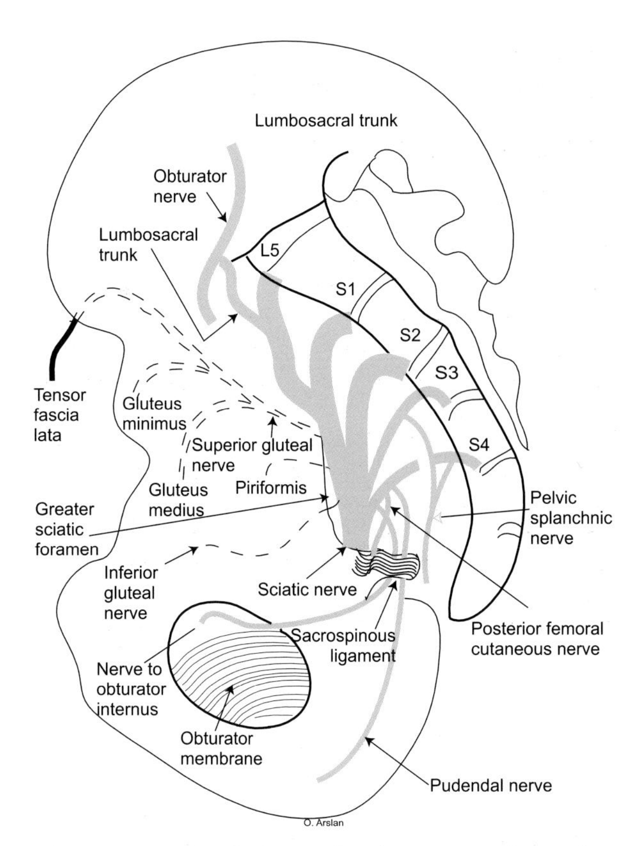

Figure 10.25 Sacral plexus, showing segmental contributions, divisions, and individual branches

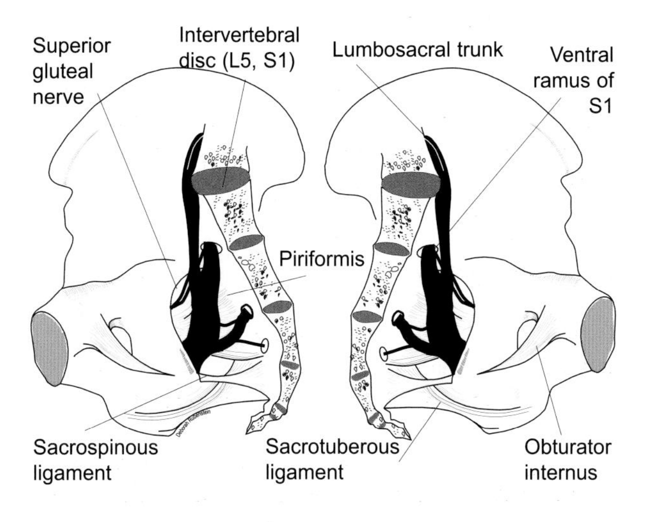

Figure 10.26 The roots of the sacral plexus in relation to the piriformis muscle showing the lumbosacral trunk

rectum and other pelvic structures may compress the sacral plexus. Compression of this plexus produces pain that radiates to the posterior thigh and leg. Aneurysm of the superior gluteal artery may particularly affect the lumbosacral trunk.

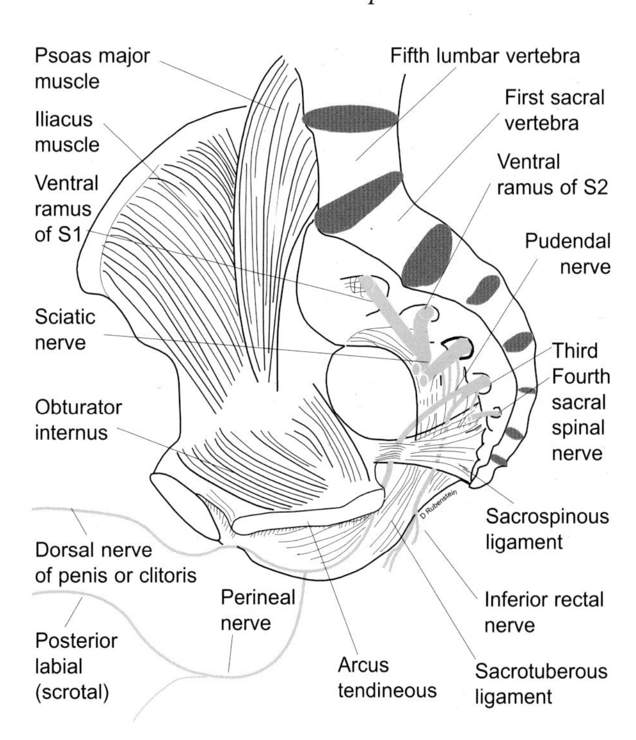

Figure 10.27 The course and branching of the pudendal nerve

> Damage to the posterior femoral cutaneous nerve produces anesthesia primarily in the posterior thigh with no motor deficits. Recurrence of pain after successful pudendal nerve block may be attributed to intactness (not affected by the anesthetic) of the perineal branch of the posterior femoral cutaneous nerve which also supplies the external genitalia.

The sacral plexus gives rise to branches, which supply the gluteal region, posterior leg, and foot. These branches are the superior gluteal, inferior gluteal, posterior femoral cutaneous, pudendal, sciatic nerve, and the pelvic splanchnic nerves. The latter has already been described earlier.

The superior gluteal nerve (L4, L5, S1) is formed by the posterior branches of the fourth and fifth lumbar and the first sacral ventral rami (Figures 10.25 & 10.26). It leaves the pelvis through the greater sciatic foramen and proximal to the piriformis muscle. It supplies the abductors and the medial rotators of the thigh (gluteus medius, gluteus minimus, and tensor fascia lata).

The inferior gluteal nerve (L5, S1, S2) arises from the posterior branches of the fifth lumbar and first and second sacral ventral rami (Figures 10.25 & 10.26). This nerve exits the pelvis via the greater sciatic foramen, proximal to the piriformis muscle, innervating the gluteus maximus muscle.

The pudendal nerve may be injured or compressed within the pudendal canal during horseback riding, as a result of a pressure from a mass or exudate in the ischiorectal fossa, pressure of a pregnant uterus, or fracture of the ischial spine. Damage to the pudendal nerve produces loss of sensation from the posterior part of the external genitalia and the ectodermal anal canal. It may also result in paralysis of the perineal muscles including the external urethral sphincter and the external anal sphincter.

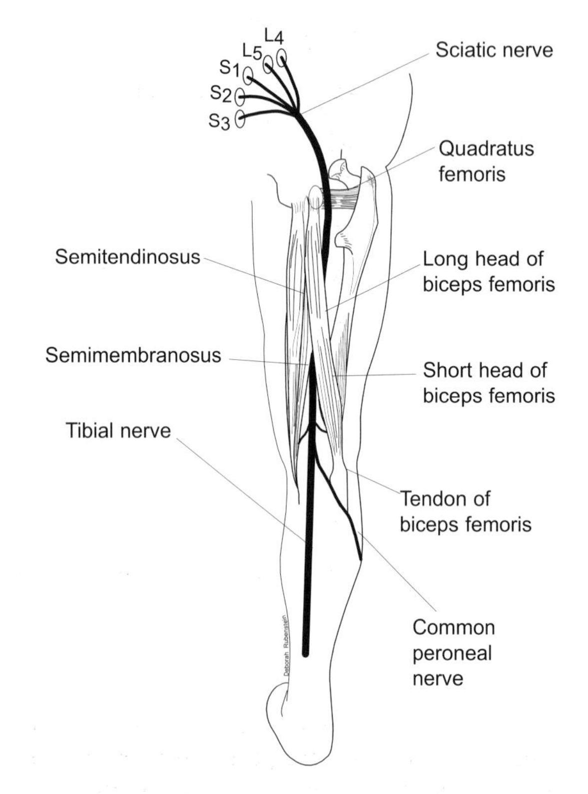

Figure 10.28 Segmental origin of the sciatic nerve, its course, and division into the tibial and common peroneal nerves

The posterior femoral cutaneous nerve (S1, S2, S3) arises from the posterior branches of the first and second sacral and the ventral branches of the second and third sacral ventral rami (Figures 10.25). This nerve leaves the pelvis via the greater sciatic foramen, distal to the piriformis muscle. It supplies sensory fibers to the posterior thigh, as far down as the popliteal fossa. It gives rise to the inferior clunial branches, which innervate the skin of the lower gluteal region. It also supplies the posterior part of the external genitalia via the perineal branch.

The pudendal nerve (S2, S3, S4) (Figures 10.25 & 10.27) originates from the anterior divisions of the ventral rami of the second, third and fourth sacral

The sciatic nerve may be entrapped within the greater sciatic foramen or compressed by an anomalous ligament or fibrous tissue within the greater sciatic foramen. Posterior dislocation of the hip joint, fractures of the femur, and aneurysm of the inferior gluteal artery may also impair the function of this nerve. Signs and symptoms associated with sciatic nerve damage are equivalent to a combination of damage to the tibial and common peroneal nerves. When the sciatic nerve is affected near its origin, or as it crosses within the greater sciatic foramen, or at any site proximal to the midposterior thigh, weakened knee flexion and loss of plantar flexion of the foot will ensue. The patient is able to stand and walk, but exhibits foot drop and toe drop, and inability to move the foot. Trophic and vasomotor changes may also be seen. Sensations from the posterior thigh, posterior leg and dorsum and plantar surfaces of the foot are lost, while sensations from the medial surface of the leg and medial border of the foot remain unaffected. Projection of pain (usually of episodic nature) to the posterior thigh and posterior leg, as a result of over-stretching of the irritated or inflamed sciatic nerve, is commonly referred to as sciatica.

The patient may present with low back or sciatic pain or both. Backache may be acute, severe, and incapacitating. It may also be gradual in onset and diffuse in nature. Lumbar spasm and abnormalities of posture and restriction of spinal movement are usually seen. Complete recovery may be possible; however, the tendency for recurrence of symptoms always exists. Lying down or standing may relieve pain, but it is aggravated by coughing, sneezing or stooping. This condition is diagnosed by either tapping the sciatic nerve or flexing the thigh at the hip while the leg is extended and the patient is in a supine position (Lasegue's sign). Patients attempt to relieve the pain by flexing the leg at the knee (Kernig's sign). It is considered a sign of meningitis if the patient can easily extend the leg in dorsal decubitus, but cannot fully extend it in sitting position or lying with the thigh flexed. However, this sign may be encountered in the absence of nuchal rigidity. Trauma to the sciatic nerve may result in a severe persistent burning pain (causalgia) which may be accompanied by vasoconstriction and sweating in an area larger than the area of distribution of the nerve itself (reflex sympathetic dystrophy). This may be due to possible involvement of the sympathetic nerves that accompany the neighboring arteries.

The common peroneal nerve is prone to damage in a spiral fracture of the neck of the fibula, or from the pressure exerted by a cyst on the lateral side of the popliteal fossa. It may also be affected in individuals with an improperly fitting cast or as a result of squatting. These conditions produce 'foot drop', a common deficit, due to paralysis and atrophy of the dorsiflexors (extensors) and evertors of the foot. It is also associated with limited loss of sensation from the dorsum of the foot and the upper lateral leg. This limited sensory loss is due to the overlap of the cutaneous innervation of the affected areas.

ventral rami. It leaves the pelvis via the greater sciatic foramen and enters the gluteal region, accompanied by the internal pudendal vessels. It then crosses the ischial spine, the sacrospinous ligament, and enters the ischiorectal fossa through the lesser sciatic foramen. It travels in the pudendal canal on the lateral wall of the ischiorectal fossa, giving rise to the inferior rectal branch. The inferior rectal nerve supplies motor fibers to the external anal sphincter and sensory fibers to the lining of the ectodermal part of the anal canal. The pudendal nerve divides into the perineal branch and dorsal nerve of the penis or clitoris. The perineal nerve gives rise to the posterior scrotal (labial) branches, which are sensory to the scrotum or labia majora, and to muscular branches to the urogenital muscles. The dorsal nerve of the penis or clitoris runs in the urogenital diaphragm, then within the suspensory ligament of the penis or the clitoris, accompanied by the corresponding artery and vein. This nerve provides sensory fibers to the penis and clitoris.

The sciatic nerve (L4, L5, S1, S2, S3) is the largest nerve in the body, emerging from the greater sciatic foramen distal to the piriformis (Figures 10.25, 10.28 & 10.29). This nerve is derived from the ventral rami of the fourth and fifth lumbar and the first, second

The superficial peroneal nerve may be compressed in lateral compartment syndrome, resulting in numbness and burning sensation on the dorsum of the foot. Weakened eversion but not total loss due to intactness of the extensor digitorum longus and peroneus tertius may also be observed. Plantar flexion is also affected due to paralysis of the peroneus longus and brevis. Injury to the superficial peroneal nerve may also occur as it pierces the deep fascia of the distal leg to innervate the dorsum of the foot. In this instance the dysfunction is limited to a burning sensation in the area of distribution of the nerve.

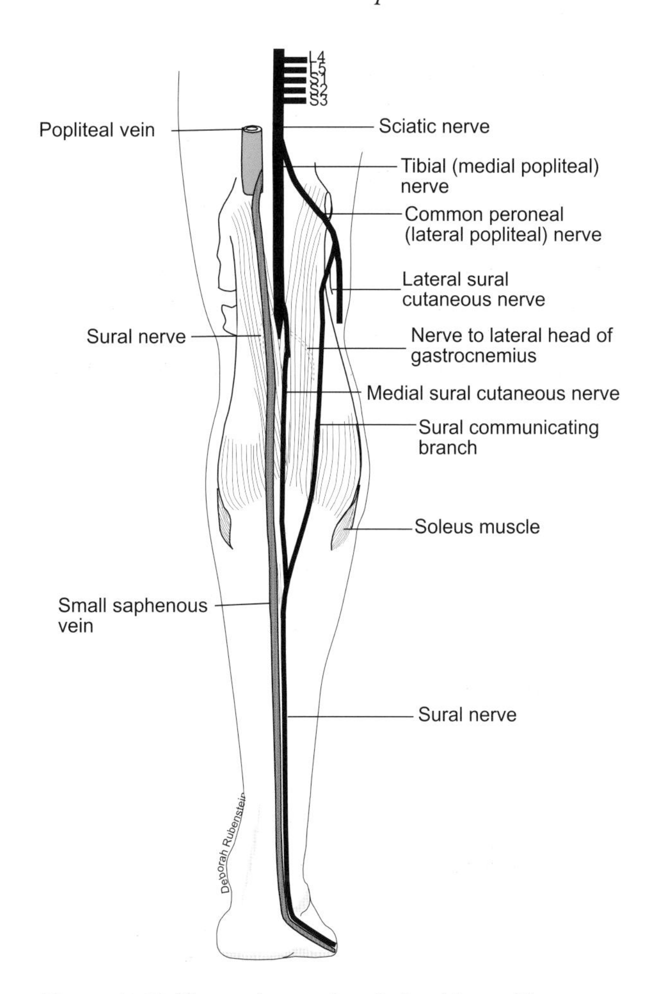

Figure 10.29 The sural nerve, its relationships and its course

Trauma to the dorsum of the foot, poorly fitting casts or shoes, or violent plantar flexion or eversion of the foot may easily damage the deep peroneal nerve. Since, the anterior compartment is a confined space sealed by a bony wall and connective tissue septum which allows no expansion, a leg cast (shin splint) may result in compression of the associated vessels and nerves with resultant edema. The pressure from the developed edema may be sufficient to produce ischemic necrosis of the structures and signs of anterior compartment syndrome. An intense pain, redness and swelling anterior to the tibia characterize this syndrome. Dorsiflexion of foot and toes becomes very painful. Paralysis of the tibialis anterior and extensor digitorum longus may also occur, producing foot drop (Figure 10.30). In addition to the above deficits weakened eversion and inversion of the foot may also be seen.

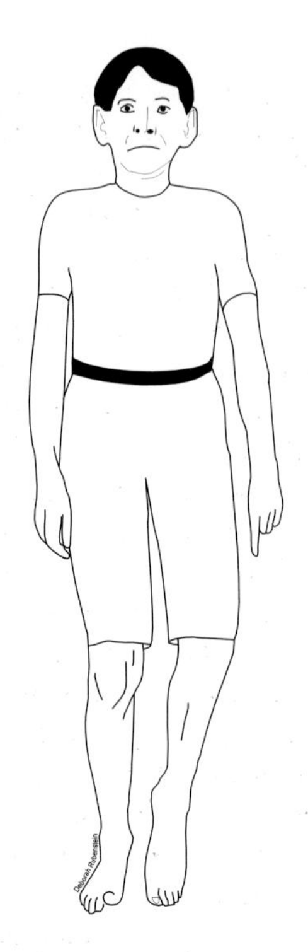

Figure 10.30 This drawing is a depiction of left foot drop in an individual with damage to the common peroneal nerve

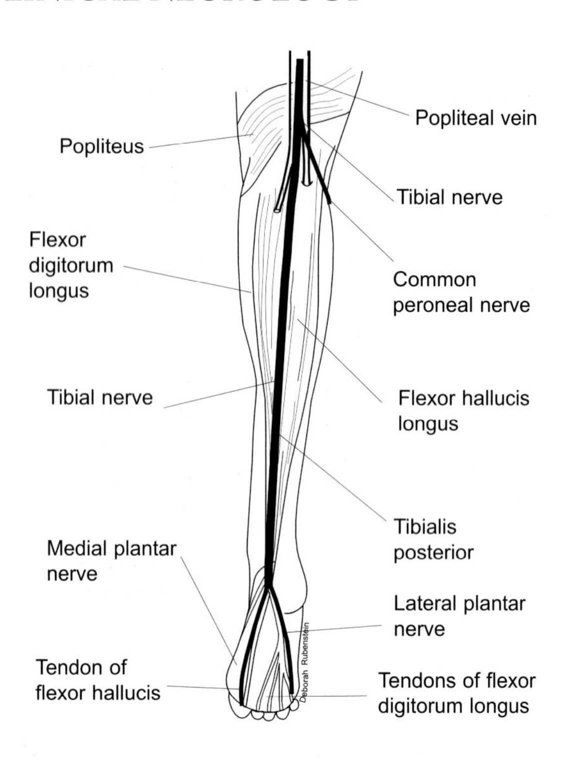

Figure 10.31 The tibial nerve and its terminal branches, the medial and lateral plantar nerves are illustrated

and third lumbar ventral rami. It descends posterior to the gemelli and the quadratus femoris, between the greater trochanter and the ischial tuberosity, accompanied by the posterior cutaneous nerve of the thigh and branches of the inferior gluteal nerve.

The tibial component of the sciatic nerve innervates the hamstring muscles (semimembranosus, semitendinosus, and long head of biceps femoris) and part of the adductor magnus. At the lower third of the posterior thigh, the sciatic nerve, accompanied by the popliteal vessels and sympathetic fibers, divides into a lateral branch, the common peroneal nerve, and a medial branch, which is the tibial nerve.

The common peroneal (fibular) nerve (L4, L5, S1, S2) (Figure 10.25, 10.28 & 10.29) originates from the dorsal branches of the ventral rami of the fourth and fifth lumbar, and the first two sacral spinal nerves. It descends posterior and lateral to the popliteal fossa and medial and posterior to the biceps femoris tendon. It encircles the neck of the fibula and then divides into the superficial and the deep peroneal nerves.

The superficial peroneal (fibular) nerve is a component of the lateral compartment of the leg. It innervates the peroneus longus and brevis and provides cutaneous fibers to the dorsum of the foot and toes (with the exception of the skin web between the great toe and the second toe).

Damage to the tibial nerve may occur in a fracture of the distal end of the femur, as a result of trauma to the popliteal fossa, or entrapment within the tarsal tunnel. Demyelination of fibers of this nerve may also be caused by thiamine deficiency as in beriberi disease. Loss of flexion in all joints of the toes (due to paralysis and atrophy of the intrinsic plantar muscles of the foot) with the resultant pes cavus, an exaggerated plantar arch, may occur when the tibial nerve is damaged. Loss of abduction and adduction of all toes, weakened flexion of the leg at the knee, weakened inversion and impaired plantar flexion may also occur. Due to the extensive overlap of the cutaneous innervation of the foot, sensory deficits are not prominent, although numbness and burning pain may be felt in the sole of the foot, especially upon standing. In general, compression of the tibial nerve may be suspected in individuals exhibiting a burning pain and paresthesia in the foot.

The deep peroneal (fibular) nerve descends in the anterior compartment of the leg, posterior to the tibialis anterior muscle and the extensor retinacula, accompanied by the anterior tibial vessels. It innervates the tibialis anterior, extensor digitorum longus,

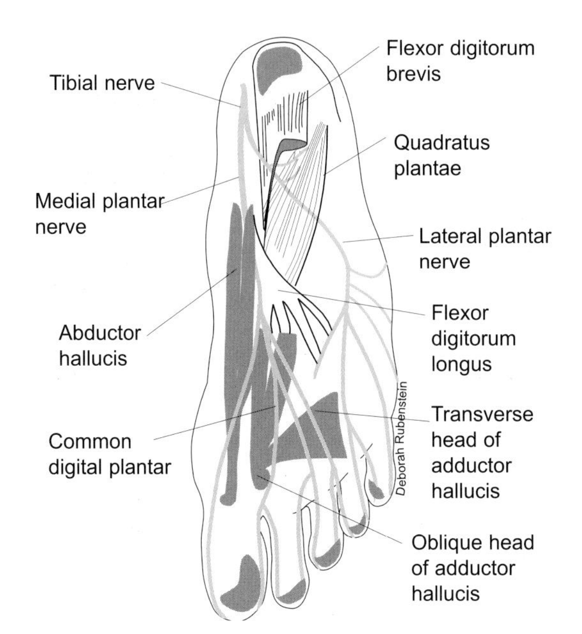

Figure 10.32 The medial and lateral plantar nerves

extensor hallucis longus, and the peroneus tertius muscle. It carries sensation from the dorsal surface of the skin web between the great toe and the second toe.

The tibial nerve (L4 to S3) is formed by the anterior branches of the ventral rami of the fourth, fifth lumbar and upper three sacral spinal nerves (Figures 10.25, 10.28 & 10.29). It descends posterior to the popliteal fossa, superficial to the popliteal vein, where it gives rise to the sural nerve. It descends anterior to the triceps surae muscle (gastrocnemius and soleus), accompanied by the posterior tibial vessels. It innervates the popliteus muscle and other muscles of the posterior compartment of the leg and the sole of the foot. It provides sensory branches to the knee joint, and also to the lower lateral surface of the leg and the lateral border of the foot via the sural nerve (Figure 10.29). The sural nerve joins a communicating branch from the common peroneal nerve, and later courses between the two heads of the gastrocnemius muscle, accompanied by the short saphenous vein. In the lower third of the leg, the tibial nerve becomes superficial and courses posterior to the medial malleolus and deep to the flexor retinaculum, where it bifurcates into the medial and lateral plantar nerves (Figures 10.31 & 10.32). The medial plantar nerve innervates the abductor hallucis, flexor digitorum brevis, flexor hallucis brevis, and the first lumbrical muscles. It carries sensation from the medial two-thirds of the plantar surface of the foot. The lateral plantar nerve innervates the rest of the plantar muscles of the foot and receives cutaneous sensation from the lateral one-third of the plantar surface of the foot.

Prolonged paralysis of the plantar flexors may cause shortening of the calcaneal (achilles) tendon, producing equinovarus deformity, which is characterized by plantar hyperflexion, inversion of the foot and medial rotation of the tibia. Patients attempt to walk on the lateral border of the foot and may develop slapping-gait. This condition may also arise from intrauterine compression of the spinal segments that contribute to the tibial nerve.

Damage to the tibial nerve distal to the middle third of the leg may occur as a result of fractures of the medial malleolus, calcaneous, or talus. It may also occur in dislocation of the ankle joint and compression within the flexor retinaculum (tarsal tunnel syndrome). This syndrome is associated with post-traumatic deformities of the knee, tight shoes and Pott's (Dupuytren's) fracture which involves the distal end of the fibula and the medial malleolus. Deficits may include, depending upon the extent of nerve damage, paresis of the plantar muscles of the foot with no detectable dysfunctions in the muscles of the leg. The metacarpophalangeal joints of the lateral four toes may exhibit hyperextension (extensors are not counteracted by the lumbricals and interossei), whereas the interphalangeal joints show hyperflexion (flexors are not opposed by the lumbricals and interossei). Sensory disturbances are restricted to a burning sensation in the sole of the foot, which may be aggravated by walking. Neuromas of the digital branches of the medial and lateral plantar nerves cause a condition known as Morton metatarsalgia, in which pain is felt in the anterior part of the sole of the foot. Damage to the sural branch of the tibial nerve may result from a Baker's cyst (a synovial cyst of the popliteal fossa) or fracture of the base of the fifth metatarsal bone. Impairment of sensation in the lower lateral leg and lateral border of the foot characterizes this injury.

Spinal reflexes

Spinal reflexes are locally mediated neuronal events, which are constantly modulated by the facilitatory and inhibitory influences of the descending supraspinal pathways. However, the ascending influences from the lower spinal segments are also exerted upon higher spinal levels. The dramatic intensity in the extensor rigidity of the forelimb muscles in a spinal animal, whose spinal cord has been transected at the level of the sixth thoracic spinal cord segment, is thought to be dependent upon the ascending inhibitory pathways (Shiff–Sherrington reflex).

Unilateral absence of the superficial abdominal reflex may be seen in both upper and lower motor neuron disorders. Upper motor neuron lesions that produce this reflex disorder usually involve the cerebral cortex and the descending autonomic pathways, whereas lower motor neuron lesions affect the lower three thoracic spinal segments in order to produce the loss of this reflex.

No significance is attached to the bilateral absence of this reflex, however unilateral absence may indicate upper motor neuron palsy.

The cremasteric reflex is brisk in young adults and is usually absent in conus medullaris syndrome, varicocele, upper motor neuron palsy, and in damages involving the upper lumbar roots.

Spinal reflexes may be classified into superficial and deep reflexes.

Superficial reflexes

The superficial reflexes are comprised of the interscapsular, superficial abdominal, cremasteric, gluteal superficial, plantar, anal, and bulbocavernous reflexes.

- Interscapular reflex refers to the reflex contraction of the rhomboideus muscles and bilateral retraction of the scapula, upon stroking the skin of the interscapular (T2–T4) area.
- Superficial abdominal reflex is elicited by stroking the skin of the abdomen from the periphery toward the umbilicus, stimulating the seventh through the twelfth thoracic (T7–T12) spinal segments. It results in contraction of the oblique abdominal muscles and movement of the umbilicus toward the side of the stimulus. However, obese and pregnant individuals usually do not exhibit this response.
- Cremasteric reflex, on the other hand, is characterized by contraction of the cremasteric muscle, followed by retraction of the ipsilateral testicle, upon a light stroke in a downward direction to the upper medial thigh. This reflex is mediated by the ilioinguinal nerve (L1) as the afferent limb and the genitofemoral nerve (L1, L2) as the efferent limb.
- Gluteal superficial reflex (L4–S1) is characterized by contraction of the gluteus maximus in response to examiner's stroke of the skin of the buttock.
- Plantar reflex (L5–S2) is produced by stroking the lateral aspect of the foot, eliciting either plantar flexion of all the toes or no response at all.
- Anal reflex (S4, S5, CC1) is elicited by stroking the perianal region with a pinwheel, producing puckering of the anal orifice. It is abolished in tabes dorsalis, cauda equina and conus medullaris syndromes.
- The bulbocavernous reflex (S2, S3, S4) may be utilized to reveal the intactness of the bladder, and is particularly important in upper motor neuron diseases. This reflex is characterized by contraction of the bulbospongiosus muscle upon compressing the glans of the penis or clitoris or pinching the prepuce. Interruption of the efferent motor fibers of this reflex produces incontinence, while disruption of the afferent limb abolishes the urge to urinate and defecate.

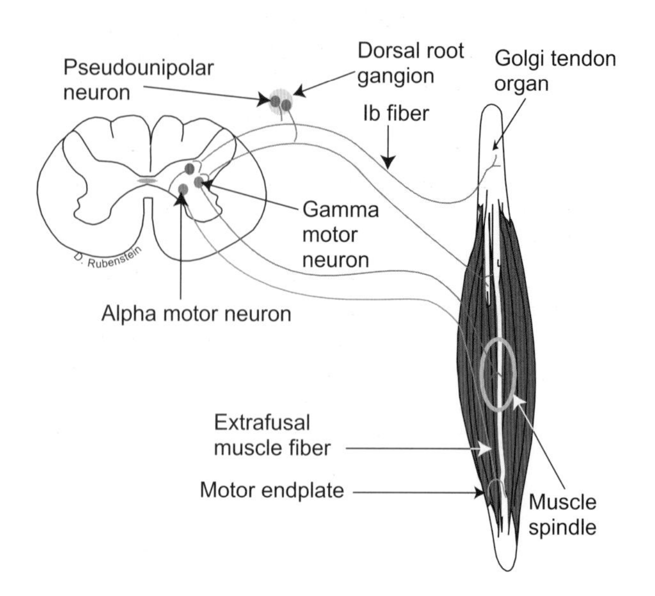

Figure 10.33 This diagram illustrates the components of myotactic stretch reflex. Observe the γ neurons and afferents of the muscle spindle

Deep reflexes

Deep reflexes include the stretch (myotactic), inverse myotactic (clasp knife), flexor, and crossed extension reflexes.

Stretch (myotactic) reflex (Figure 10.33) is elicited by tapping the tendon of a muscle which produces increased length of the muscle fibers and subsequent activation of the muscle spindle. Contraction of the muscle spindle activates the annulospiral (Ia) fibers, which in turn monosynaptically stimulate the ipsilateral α motor neurons, producing contraction of the stretched muscle. Annulospiral afferents establish excitatory monosynaptic connections with α motor neurons of the synergists and disynaptic inhibitory connections to the motor neurons of the antagonistic muscle (reciprocal inhibition). Myotatic stretch reflexes are produced by tapping the patellar ligament in patellar reflex (L2, L3 & L4) and biceps brachii tendon in biceps reflex (C5, C6). It is also elicited upon tapping the tendon of the triceps in triceps

Diminished or absence of myotactic reflex may occur in peripheral neuropathies, tabes dorsalis, poliomyelitis, diabetes mellitus, Holmes–Adie syndrome, sciatica, syringomyelia and cervical spondylosis. Spinal shock, coma, and certain types of hydrocephalus may also diminish or abolish the myotactic reflexes. Upper motor neuron disorders such as stroke, multiple sclerosis, and spinal cord tumors or strychnine poisoning and anxiety disorders may render this reflex hyperactive (intensified). Westphal's sign, failure to produce the patellar (myotactic) reflex, may be reversed by the reinforcement method of Jendrassik which requires the patient to clench his hands while the patellar tendon is tapped.

reflex (C7, C8), tendon of the brachioradialis in radial reflex (C7, C8), gastrocnemius (Achilles) tendon in gastrocnemius or Achilles reflex (S1, S2), etc. Periosteoradial reflex produces flexion and supination of the forearm upon tapping the radial styloid process, while periosteoulnar reflex produces extension and ulnar abduction upon striking the ulnar styloid.

• Contraction of a muscle can also be elicited by activation of the muscle spindle via the γ loop, without stretching the muscle. In the γ loop, the contraction of the muscle spindle activates the primary Ia (annulospiral) endings, which in turn monosynaptically activates the α motor neurons. The firing of these α neurons results in contraction of the extrafusal muscle fibers. Thus, under normal conditions, the cerebral cortex can trigger muscle contraction and initiate postural changes and movement via two mechanisms: (a) activating the α motor neuron directly; and (b) indirectly via the γ loop. The role of the γ loop can be illustrated when assuming an erect posture. Standing stretches the quadriceps muscle, which causes activation of the stretch receptor and subsequent contraction of the quadriceps femoris. However, the muscle begins to relax as soon as the tension in the muscle spindle is reduced, and the rate of discharge from α motor neurons is diminished. In order to maintain erect posture, the γ loop comes into action and activates the muscle spindle. Voluntary and

The inverse myotactic reflex is thought to underlie the mechanism of 'clasp knife' phenomenon, which is observed in upper motor neuron palsy. In this reflex passive stretching of the spastic muscle is met initially with great resistance to an extent, after which the muscle suddenly gives away. Sherrington named this phenomenon because of its similarity to the action of a jack or a switchblade knife.

precise movements are executed by the simultaneous activation of both systems, which are complementary. In general, activation of the α motor system predominates when a quick response is desired, whereas activation of the γ system predominates when a smooth and precise movement is desired.

• Inverse myotactic reflex comes into action upon stimulation of the Golgi tendon organ and the Ib fibers as a result of the tension developed in the contracted muscle. The Ib fibers establish disynaptic inhibitory (autogenic inhibition) contacts with the agonist neurons, and excitatory synapses with the antagonistic neurons. The sum of these actions produces relaxation of the agonistic muscles.

• Flexor (withdrawal) reflex enables an individual to avoid harm by withdrawing from nociceptive or injurious stimuli. This reflex is mediated by the free nerve endings, to a lesser extent by the tactile receptors, as well as group III nerve fibers, conveying the impulses to the spinal cord. These afferent fibers establish polysynaptic excitatory and inhibitory connections with the motor neurons. The net effect of this circuitry is facilitation of ipsilateral flexor (agonists) motor neurons and inhibition of ipsilateral extensor (antagonist) motor neurons.

• Crossed extension reflex is characterized by flexion of the ipsilateral limb and extension of the contralateral limb in response to a strong nociceptive stimulus. This reflex is a byproduct of the flexion reflex, whereby the afferent fibers establish multisynapses at many levels of the spinal cord with the ipsilateral flexor neurons, and with the contralateral extensor neurons via the anterior white commissure.

Suggested reading

1. Callahan JD, Scully TB, Shapiro SA, *et al.* Suprascapular nerve entrapment: a series of 27 cases. *J Neurosurg* 1991;74:893–6
2. Dyck PJ, Thomas PK. *Peripheral Neuropathy*, 3rd edn. Philadelphia: W. B. Saunders, 1993
3. Futagi Y, Tagawa T, Otani K. Primitive reflex profiles in infants: differences based on categories of neurological abnormality. *Brain Dev* 1992;14:294–8
4. Kars HZ, Topaktas S, Dogan K. Aneurysmal peroneal nerve compression. *Neurosurgery* 1992;30:930–1

5. Katirji B, Hardy RW Jr. Classic neurogenic thoracic outlet syndrome in a competitive swimmer: a true scalenus anticus syndrome. *Muscle Nerve* 1995;18:229–33
6. Laha RK, Lunsford LD, Dujovny M. Lacertus fibrosus compression of the median nerve. *J Neurosurg* 1978;48: 838–41
7. Lee C-S, Tsai T-L. The relation of the sciatic nerve to the piriformis muscle. *J Formosan Med Assoc* 1974;73:75–80
8. Liguori R, Krarup C, Trojborg W, *et al.* Determination of the segmental sensory and motor innervation of the lumbosacral spinal nerves. An electrophysiological study. *Brain* 1992;115:915–34
9. Lundborg G. *Nerve Injury and Repair*. Philadelphia: Churchill Livingstone, 1988
10. McKowen HC, Voorhies RM. Axillary nerve entrapment in the quadrilateral space: case report. *J Neurosurg* 1987;66:932–4
11. Nakanishi T. Studies on the pudendal nerve. 1. Macroscopic observations on the pudendal nerve in humans. *Acta Anat Nippon* 1967;42:223–39
12. Nakano KK, Lundergan C, Okihiro M. Anterior interosseous nerve syndromes: diagnostic methods and alternative treatments. *Arch Neurol* 1977;34:477–80
13. Pagnanelli DM, Barrer SJ. Bilateral carpal tunnel release at one operation: report of 228 patients. *Neurosurgery* 1992;31:1030–4
14. Roles NC, Maudsley RH. Radial tunnel syndrome: resistant tennis elbow as a nerve entrapment. *J Bone Joint Surg* 1972;54B:499–508
15. Shea JD, McClain EJ. Ulnar-nerve compression syndromes at and below the wrist. *J Bone Joint Surg* 1969;51A:1095–103
16. Spangfort EV. The lumbar disc herniation. A computer-aided analysis of 2,504 operations. *Acta Orthoped Scand* 1972;142(Suppl):1–93
17. Szabo R, Steinbert D. Nerve entrapment syndromes in the wrist. *J Am Acad Orthopaed Surg* 1994;2:115–23
18. Thornton MW, Schweisthal MR. The phrenic nerve: its terminal divisions and supply to the crura of the diaphragm. *Anat Rec* 1969;164:283–90
19. Tsairis P, Dyck PJ, Mulder DW. Natural history of brachial plexus neuropathy: report on 99 patients. *Arch Neurol* 1972;27:109–17
20. Williams PH, Trzil KP. Management of meralgia paresthetica. *J Neurosurg* 1991;74:76–80

Cranial nerves 11

Topographically, cranial nerves occupy the cranial cavity, though some may extend to the neck, even to the thorax and abdomen. They are classified according to their functions and connections to various parts of the central nervous system (CNS). Some of these nerves are sensory (olfactory, optic), others subserve motor functions (oculomotor, trochlear, abducens, accessory, and hypoglossal), while others carry both sensory and motor components (e.g. trigeminal, facial, glossopharyngeal, and vagus nerves). The facial, glossopharyngeal, and vagus nerves carry taste sensations, and innervate structures both in the head and neck. The nerve fibers that subserve sensory functions are the central extensions of the unipolar neurons (e.g. facial, glossopharyngeal, and vagus nerves), represent the axons of bipolar neurons (e.g. olfactory and vestibulocochlear nerve), or are formed by the axons of multipolar retinal neurons (e.g. optic nerve). It should be noted that the olfactory and optic nerves are considered extensions of the CNS and are therefore not true cranial nerves. Certain cranial nerves, such as oculomotor, facial, and glossopharyngeal, contain presynaptic parasympathetic fibers while the trigeminal nerve contains postsynaptic parasympathetic fibers. Others such as the trigeminal, facial, glossopharyngeal, and vagus nerves innervate muscles of branchial origin.

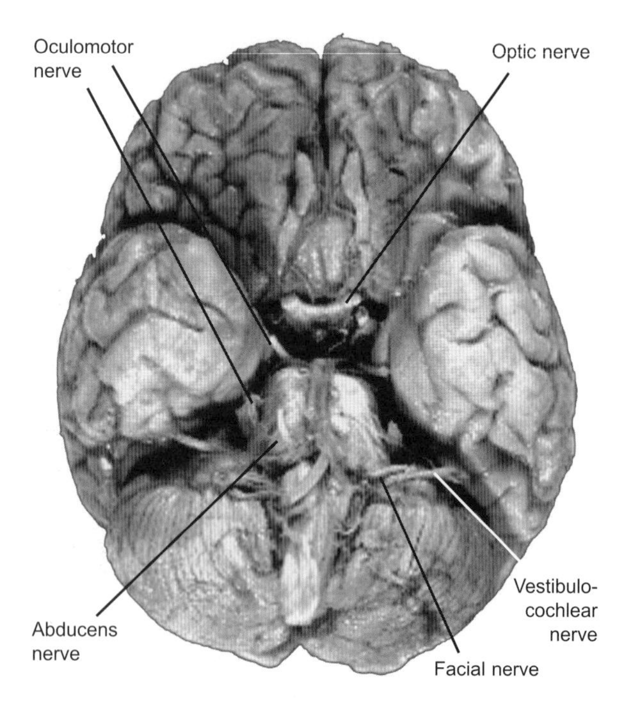

Figure 11.1 Inferior surface of the brain illustrating some of the associated cranial nerves

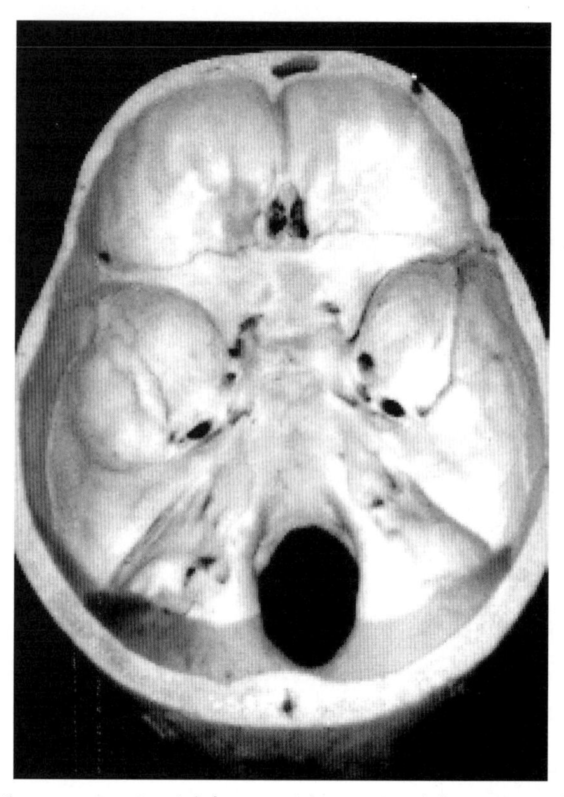

Figure 11.2 Cranial fossae with associated foramina and openings

I Olfactory nerve

The olfactory nerve (SVA) represents the axons of the bipolar neurons of the olfactory mucosa (Figure 11.3). The filaments of this nerve (fila olfactoria) pass through the cribriform plate of the ethmoid bone (Figures 11.2) and enter the olfactory bulb, establishing synapses with the mitral cells. The olfactory bulb (Figures 11.1 & 11.2), located inferior to the frontal lobe, contains the second order neurons that extend in the olfactory tract. The olfactory bulb is an allocortex and consists of three layers, containing mitral and tufted cells, as well as inhibitory granular and periglomerular cells. Periglomerular cells, which are GABAergic, receive excitatory fibers from the bipolar neurons and establish inhibitory connections with the surrounding mitral cells. Inhibition is also accomplished by the dendrodendritic synaptic connections between the dopaminergic and granule cells. Centrifugal fibers from the contralateral anterior olfactory nucleus may activate the inhibitory internuncial neurons. Mitral cells also provide collateral fibers to the anterior olfactory nucleus. Impulses in the olfactory tract pass through the medial, intermediate and lateral olfactory striae.

The lateral olfactory stria terminates in the uncus (Figure 11.1), which constitutes the primary olfactory cortex. The medial and intermediate olfactory striae terminate in the septal area and the anterior perforated substance, respectively.

The olfactory nerve is tested by the application of a mild, non-irritating scent (coffee or herb) to one nostril at a time while the eyes are closed. Smell sensation may be reduced in Paget's disease, diabetes mellitus, and post laryngotomy. Head trauma may lead to contusion of the olfactory nerves, producing anosmia (complete loss of the sense of smell). Post-traumatic anosmia may be detected weeks or months after the insult and may last as long as the post-traumatic amnesia. Unilateral anosmia of non-rhinogenic origin may be a sign of a subfrontal lobe tumor, meningiomas of the sphenoidal ridge, and hypophysial tumors affecting the sella turcica. It may also occur as a sequel to a fracture of the anterior cranial fossa, which is frequently accompanied by leakage of cerebrospinal fluid (CSF) through the nostrils (rhinorrhea). Bilateral anosmia may occur as a result of nasal infection (e.g. rhinitis sicca), common cold, excessive smoking and cocaine use, and is commonly associated with loss of taste. A smaller proportion of individuals with viral influenza-induced anosmia may completely recover the sense of smell. Damage or irritation of the uncus by a developing mass may produce olfactory hallucination with phantom smells.

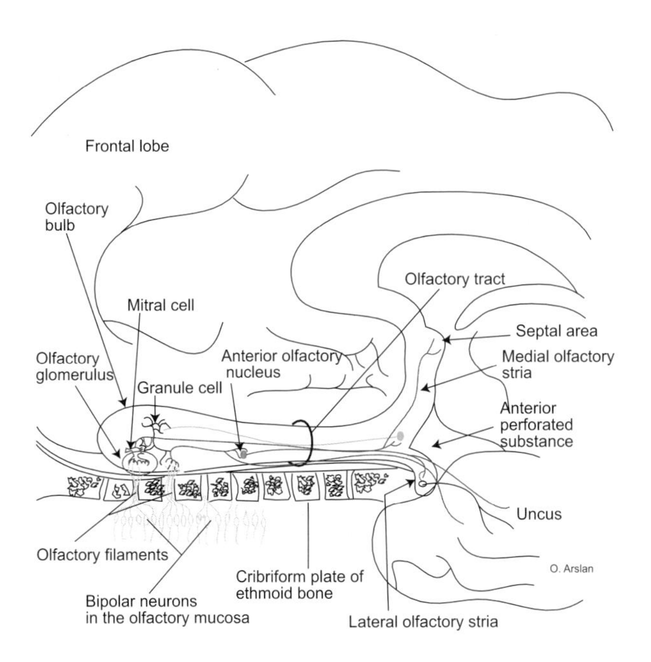

Figure 11.3 Olfactory nerve filaments, synaptic connection, and olfactory striae are followed in this diagram to their terminations in the septal area and uncus

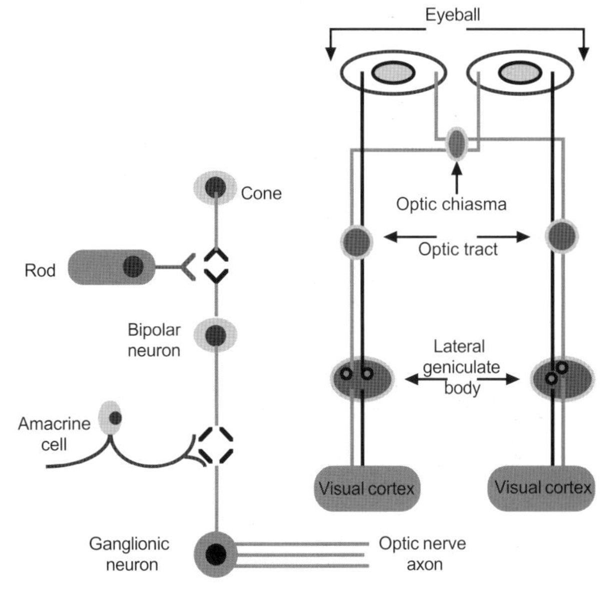

Figure 11.4 Impulses carried by the optic nerve are followed through the optic chiasma and optic tract to their final destination in the visual cortex

In Kallmann's syndrome, which is characterized by secondary hypogonadism and associated dwarfism and occasional color blindness, anosmia may be caused by aplasia (lack of development) of the olfactory bulb and agenesis of the olfactory lobes. Anosmia is often accompanied by ageusia (impaired sense of taste). This combined disorder may be seen in an individual following head trauma, and in patients with scleroderma who underwent treatment with histidine.

Presence of the central retinal artery and vein inside the optic nerve may account for the reduction of arterial and retardation of venous blood flow in these vessels upon compression of this nerve by a growing tumor or by increased intracranial pressure. Additionally, since meningeal coverings of the brain, associated subarachnoid space, and cerebrospinal fluid continue around the optic nerve, conditions that affect the circulation of the CSF may eventually translate into pressure build-up on this nerve.

II Optic nerve

The optic nerve (SSA) is formed by the axons of retinal ganglion cells (Figures 11.1 & 11.4). It is considered an extension of the brain for two main reasons: first, it is surrounded by myelin from the oligodendrocytes; second, it is an embryological derivative of the forebrain diverticulum. Due to these reasons, patches of demyelination along the course of the optic nerve are seen in multiple sclerosis. The optic nerve acquires myelin in the orbit, otherwise myelinated axons in the retina my cause light reflection and blurred vision. Visual information from the temporal and nasal halves of the corresponding retina, as well as impulses concerned with pupillary light and accommodation reflexes are carried by the optic nerve. Fibers of the optic nerve leave the retina medial to the fovea centralis and converge on the optic disc, piercing the choroid layer, sclera, and entering the orbit. In the orbit, it is crossed by the ophthalmic artery. Then, it leaves the orbit and gains access to the cranial cavity through the optic canal (Figure 11.2). Posterior to the optic canal, the nasal fibers decussate to form the optic chiasma (Figures 11.1 & 11.4). In the cranial cavity, the internal carotid artery lies lateral to the optic nerve, and ventral to the anterior cerebral artery.

Integrity of the optic nerve is determined by examination of the visual fields and visual acuity. This is accomplished by confrontational visual field testing which involves closure of the examiner's right eye and patient's left eye while standing at eye level opposite each other. This is followed by the examiner's simultaneous show of one or two fingers in each hand and his request that the patient ascertains the fingers that he has seen. The other eye will be tested the same way from upper to lower quadrants. In normal individuals the fingers will be seen at the same time by the examiner and patient.

Scotoma (focal blindness) which occurs in glaucoma and tumors of the CNS may be detected upon widening of the visual field of a patient by pulling the examiner's hand away from the patient. A flashing light beam or a pencil may also be used and the patient is asked to state the timing of its appearance and direction. Visual acuity may be assessed by using the Snellen eye chart, positioned approximately 20 feet from the patient. Each eye is tested separately and the first number in the standard ratio 20/20 denotes the actual distance of the patient from the chart, while the second number represents the distance at which a person with normal vision can read the chart. Visual acuity of each eye, which reflects the macular function, should be tested independently with and without glasses. For this purpose the examiner may use a newspaper article, or attempt to present a picture or small objects to be identified by the patient. Visual acuity may vary by environmental factors such as illumination and degree of contrast.

To detect the differences between both eyes in response to afferent stimuli, the swinging light test is employed. In this test the patient is asked to look at a distant object while the examiner rapidly swings a light beam from one eye to the other. When directing the light into the blind eye, neither eye will show constriction. However, upon moving the light back to the intact eye, the blind eye shows apparent pupillary dilatation due to the lack of afferents to the retina and optic nerve (Marcus Gunn pupil).

Complete destruction of the optic nerve results in total blindness on the affected side. Due to the bilateral connections of the visual fibers of the optic nerve, pupillary light and accommodation reflexes are lost on both sides when the affected eye is stimulated.

Prolonged elevation of intracranial pressure produces uniform compression of the optic nerve leading to blindness. Papilledema (choked disc) is a condition in which the optic disc protrudes anteriorly as a result of dilatation of the subarachnoid space around the optic nerve, subsequent to increased intracranial pressure. This condition may also be associated with compression of the central retinal artery and vein, and may be observed in hypertensive individuals.

III Oculomotor nerve

The oculomotor nerve (GSE, GVE) emerges from the ventral surface of the midbrain within the interpeduncular fossa, in close relationship to the crus cerebri (Figures 11.1, 11.5, 11.6 & 11.7). This nerve runs between the superior cerebellar and the posterior cerebral arteries, and inferior to the posterior communicating artery. It courses within the cavernous sinus and leaves the middle cranial fossa via the superior orbital fissure. It enters the orbit as a component of the tendinous annulus of Zinn, where it divides into superior and inferior rami. The superior ramus (GSE) supplies the superior rectus and levator palpebrae, whereas the inferior ramus provides innervation (GSE) to the medial and inferior recti, and the inferior oblique. There are preganglionic parasympathetic fibers (GVE) also that run within the inferior ramus. These autonomic fibers constitute the efferent limb for both pupillary light and accommodation reflexes. The oculomotor nerve is formed by neuronal axons of the oculomotor nucleus, a V-shaped nucleus, which lies medial to the medial longitudinal fasciculus (MLF) at the level of the superior colliculus. It consists of somatic and visceral nuclear columns. In the somatic component of this nucleus (Figures 11.5 & 11.7) dorsal, intermediate, ventral, caudal, central, and medial cellular columns exist. The dorsal column innervates the ipsilateral inferior rectus muscle; the intermediate column supplies the ipsilateral inferior oblique muscle; and the ventral column provides innervation to the ipsilateral medial rectus. The latter neuronal column is controlled by the internuclear neurons of the abducens nucleus in lateral gaze. The caudal central column provides bilateral innervation to the levator palpebrae muscle, whereas the medial column innervates the contralateral superior rectus muscle. Preganglionic parasympathetic fibers to the ciliary ganglion, which innervate the constrictor pupilla and ciliary muscles, arise from the Edinger–Westphal nucleus, the visceral nuclear column of the oculomotor nucleus (Figure 11.7).

Descending cortical fibers to the oculomotor nucleus form crossed and uncrossed corticobulbar

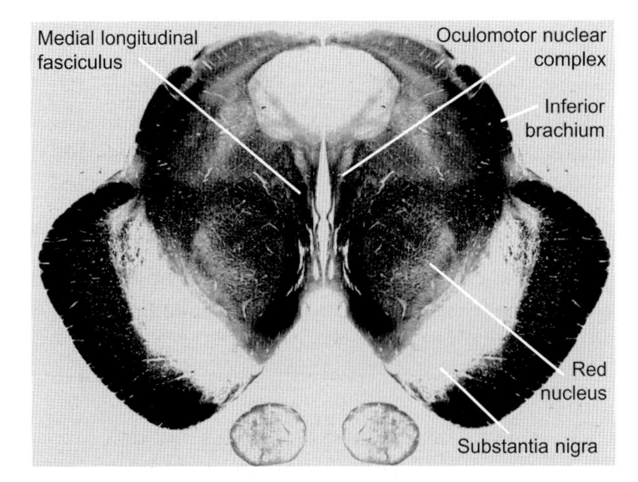

Figure 11.5 The oculomotor nuclear complex is illustrated and its location is identified in reference to the medial longitudinal fasciculus and the periaqueductal gray matter

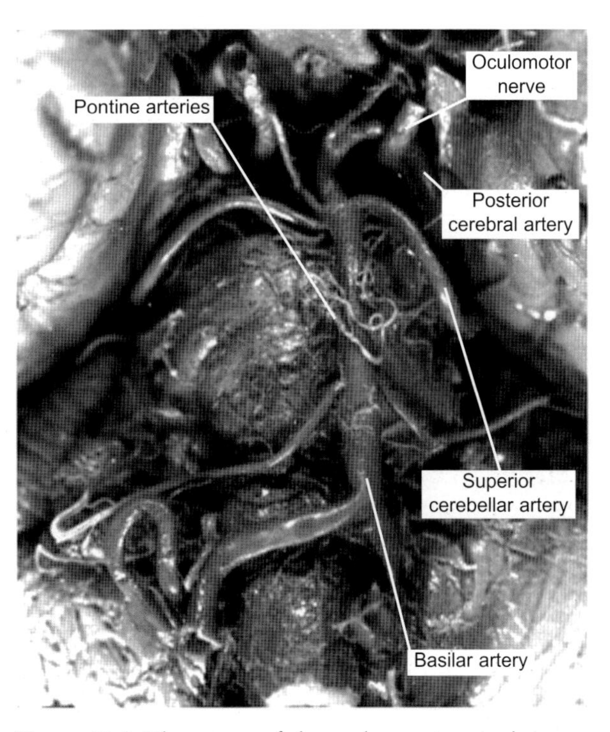

Figure 11.6 The course of the oculomotor nerve between the superior cerebellar and posterior cerebral arteries

Destruction of the oculomotor nerve may occur in combination with the corticospinal tract in Weber's syndrome or in conjunction with the red nucleus, spinothalamic tracts, medial lemniscus, and superior cerebellar peduncle in Benedikt's syndrome. It may also be damaged together with the red nucleus in Claude's (lower red nucleus) syndrome, exhibiting contralateral hemiataxia, but with no apparent hyperkinesia. Oculomotor nerve palsy with contralateral cerebellar ataxia, tremor, and signs of spastic palsy are seen in Nothnangel syndrome. Since the oculomotor nerve courses immediately rostral to the superior cerebellar artery, caudal to the posterior cerebral artery and inferior to the posterior communicating artery, an aneurysm of any of these individual vessels occasionally may produce signs of oculomotor palsy. Transtentorial herniation may also pose undue stretch on these vessels, resulting in oculomotor nerve dysfunction that almost always involve the pupil. Thrombosis of the cavernous sinus, fractures of the middle cranial fossa or superior orbital fissure, parasellar neoplasm, and demyelinating diseases may also produce deficits associated with the oculomotor nerve. In amyotrophic lateral sclerosis and poliomyelitis, the oculomotor nerve remains intact despite involvement of the motor neurons. Some of the above syndromes are discussed with the motor system or under the heading of combined motor and sensory lesions.

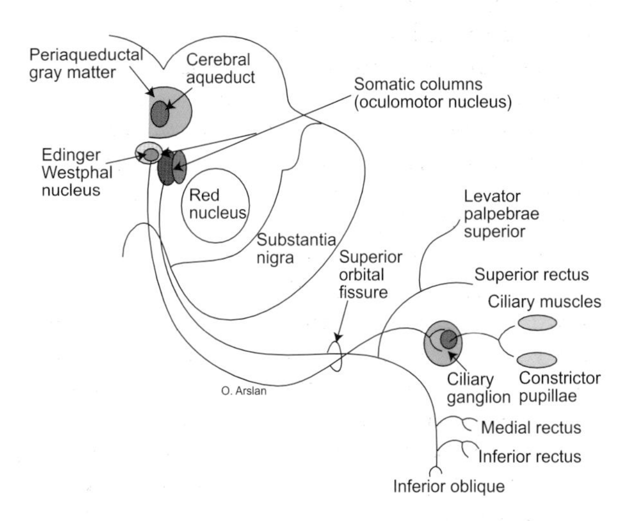

Figure 11.7 The functional components of the oculomotor nerve from their origin in the oculomotor nuclear complex to their site of innervation are shown in this drawing

fibers that establish contact with the nucleus via the reticular formation. Additionally, vestibular fibers also project to the oculomotor nucleus through the MLF, regulating movements of the head and fixation of gaze. Fibers from the accessory oculomotor nuclei

Injury to the oculomotor nerve may be partial or complete and the recovery may occur unevenly. Diabetes-induced oculomotor nerve damage is painful and the parasympathetic fibers within the nerve may be spared. However, the same fibers may be damaged in brainstem lesions. Fibers which mediate the accommodation reflex may be damaged in diphtheria. An acute isolated painful oculomotor nerve palsy is most commonly associated with aneurysm of the branches of the internal carotid artery. Damage to the oculomotor nucleus may occur in multiple sclerosis and produces identical deficits to oculomotor nerve palsy, and if complete it also results in contralateral superior rectus palsy. Unilateral lesion of the lateral tegmentum of the midbrain may produce signs of oculomotor palsy on the same side and trochlear nerve palsy on the opposite side.

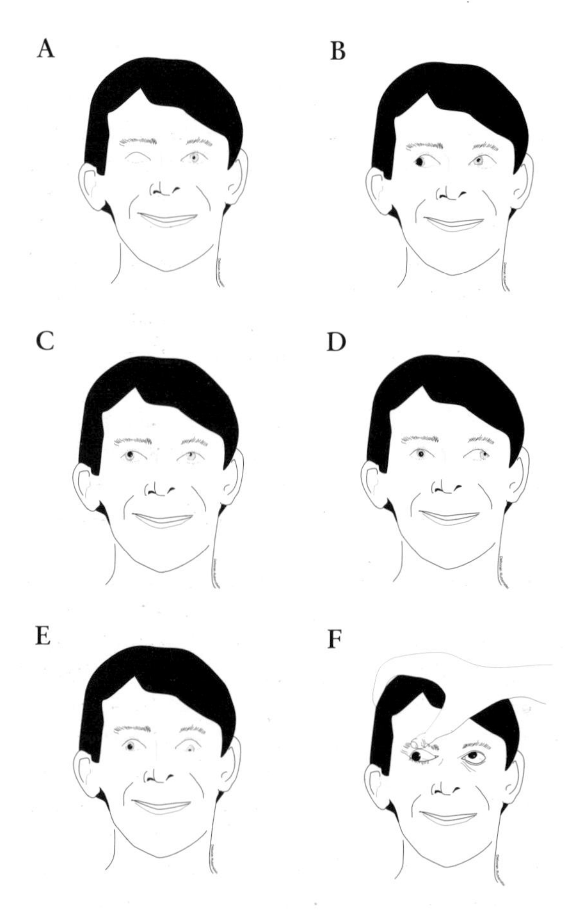

Figure 11.8 Manifestations of right oculomotor nerve palsy are depicted in diagrams A to G. (A) Ptosis; (B) intact lateral gaze to the right side; (C) inability to gaze in an upward direction; (D) inability to adduct the right eye upon left lateral gaze; (E) inability to look downward on the right eye; and (F) paralyzed eyelid is raised, exhibiting lateral strabismus

Oculomotor palsy (Figure 11.8) is characterized by ipsilateral:

- Ptosis (drooping of the upper eyelid due to paralysis of the levator palpebrae superior). This deficit should be distinguished from the ptosis observed in Horner's syndrome. In the latter condition, ptosis is less pronounced and occurs as a result of paralysis of the superior tarsal muscle, subsequent to disruption of the postsynaptic sympathetic fibers.
- Mydriasis (dilatation of the pupil), due to paralysis of the constrictor pupillae muscle and the unopposed action of the dilator pupillae muscle.
- Diplopia (double vision), seen in all directions except in lateral gaze and the distance between true and false images is maximal in the direction of the gaze. The false image will always be peripheral to the true image.
- Lateral strabismus (lateral deviation) and downward deviation of the eye due to activation of unopposed lateral rectus and the superior oblique muscles.
- Enophthalmos (inward displacement of the eyeball), possibly an illusionary feature due to drooping of the upper eyelid.
- Loss of pupillary and accommodation reflexes.

Integrity of the oculomotor nerve can be checked by observing the light reflex and eye movement. To test the light reflex, the patient is asked to look at a distance, while the examiner shines a bright light into one eye of the patient; the examiner then observes the pupils in both eyes (direct and consensual light reflexes). In normal individuals, both pupils will constrict in response to the light applied to one eye. In lesions of the oculomotor nerve, the affected eye remains unreactive regardless of which eye is stimulated. Since both the optic and oculomotor nerves mediate the accommodation reflex, evaluation of near vision may reveal the condition of these nerves. In order to accomplish this task, the patient is asked to look alternately at distant and close objects and observe convergence of both eyes. The functional integrity of the extraocular muscles which receive innervation from the oculomotor nerve may be tested by asking the patient to follow the examiner's finger as it traces an 'H' configuration, first, moving to patient's right, then up, then down, then back to the midline.

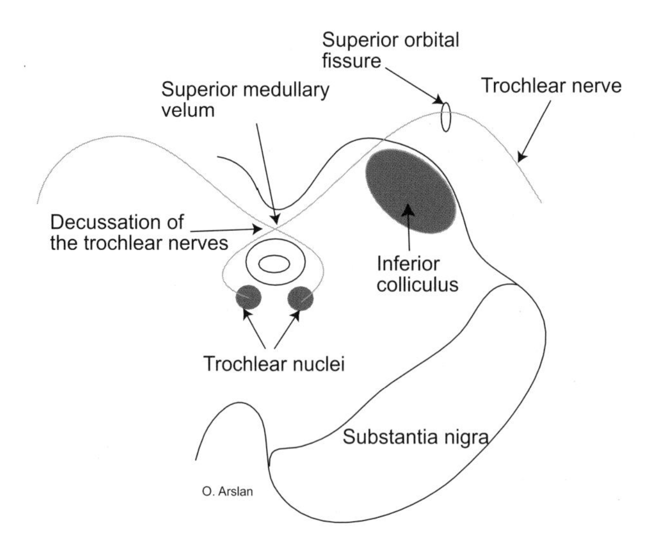

Figure 11.9 The course of the trochlear nerves and their decussation in the superior medullary velum is shown in this drawing

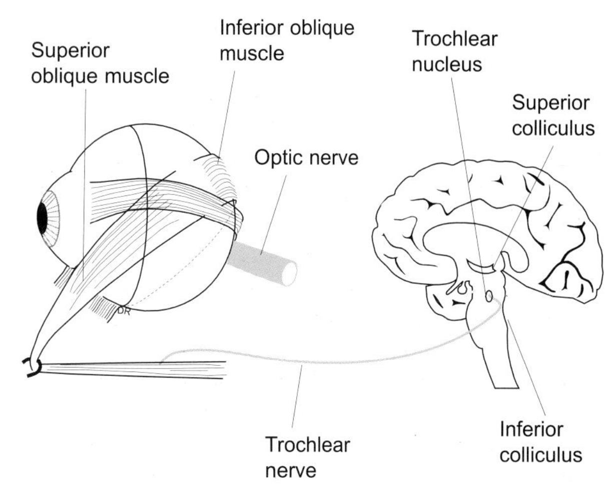

Figure 11.10 The origin, course and final destination of the trochlear nerve

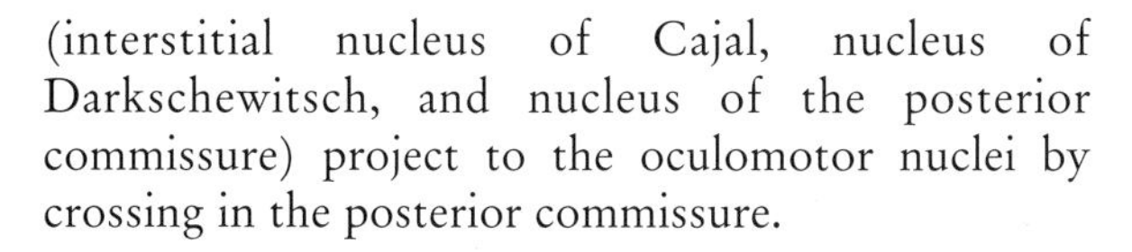

(interstitial nucleus of Cajal, nucleus of Darkschewitsch, and nucleus of the posterior commissure) project to the oculomotor nuclei by crossing in the posterior commissure.

IV Trochlear nerve

The trochlear nerve (GSE) is the only cranial nerve which exits from the dorsal surface of the pons, caudal to the level of the inferior colliculus (Figures 11.9 & 11.10). This nerve, which represents the axons of the trochlear nucleus, completely decussates with the corresponding nerve of the opposite side within the superior medullary velum. Following its exit from the cavernous sinus, the trochlear nerve runs between the superior cerebellar and posterior cerebral arteries, making it vulnerable to damage by an aneurysm of these arteries. It enters the orbit via the superior orbital fissure, innervating the superior oblique muscle. The trochlear nucleus receives ipsilateral vestibulo-ocular fibers and bilateral fibers from the accessory oculomotor nuclei.

V Trigeminal nerve

The trigeminal nerve (GSA, SVE), the largest of all cranial nerves, exits the pons through the middle cerebellar peduncle (Figures 11.1 & 11.18). It has a sensory (trigeminal, Gasserian, semilunar) ganglion which is formed by the unipolar neurons of the sensory fibers. This ganglion lies anterior to the apex of the petrous temporal bone and resides in Meckel's cave. The trigeminal nerve gives off the ophthalmic

Lesions of the trochlear nerve may occur in multiple sclerosis, cavernous sinus thrombosis, and superior orbital fissure syndrome that also involves the oculomotor (CN III) nerve. It may also occur in orbital apex syndrome, affecting the optic (CN II), oculomotor (CN III), and abducens (CN VI) nerves in addition to the trochlear nerve. Trochlear nerve palsy is characterized by impairment of downward gaze of the adducted eye (Figures 11.11 & 11.12).

The eye on the affected side remains elevated and assumes a higher position in the adducted position than when the eye is abducted, and decreases with abduction. Elevation of the eye on the affected side assumes maximal position when the neck is bent toward the affected side, maintaining normal position upon bending of the neck toward the intact side (Bielschowsky's head-tilt test). Vertical diplopia will be more evident as the patient looks down and inward. Patients adopt a characteristic posture, tilting the head toward the opposite side so that the face will be directed toward the affected side (Bielschowsky sign). Maintenance of such a posture for an extended period will lead to torticollis (Wry neck), which refers to the spasmodic contracture of the neck. Bilateral trochlear nerve palsy may occur in superior medullary velum syndrome as a result of a lesion that disrupts the decussating fibers of this nerve within the superior medullary.

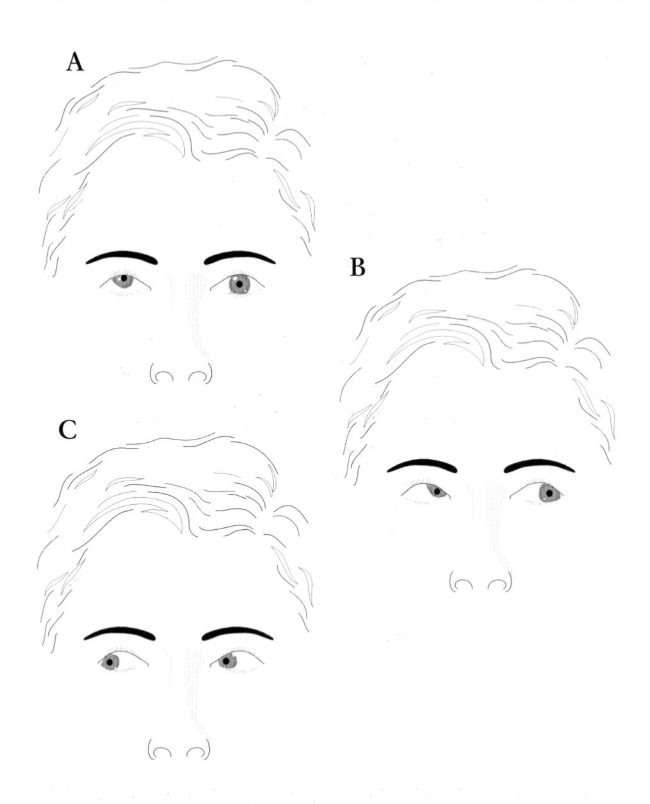

Figure 11.11 The deficits associated with trochlear nerve palsy. Note in (A) the right eye is elevated upon forward gaze, in (B) elevation of the eye is increased with adduction, and in (C) decreased with abduction

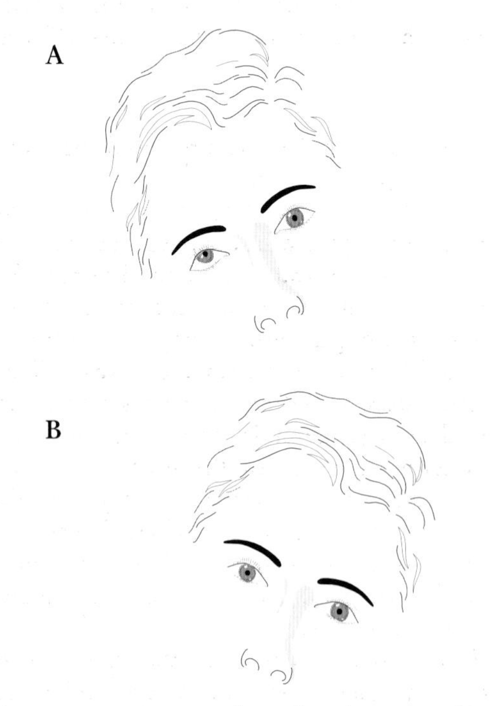

Figure 11.12 Diagrams of manifestations of trochlear nerve palsy. In (A) maximal elevation upon tilting the head to the affected side, while in (B) eye elevation disappears upon head tilting in the opposite direction

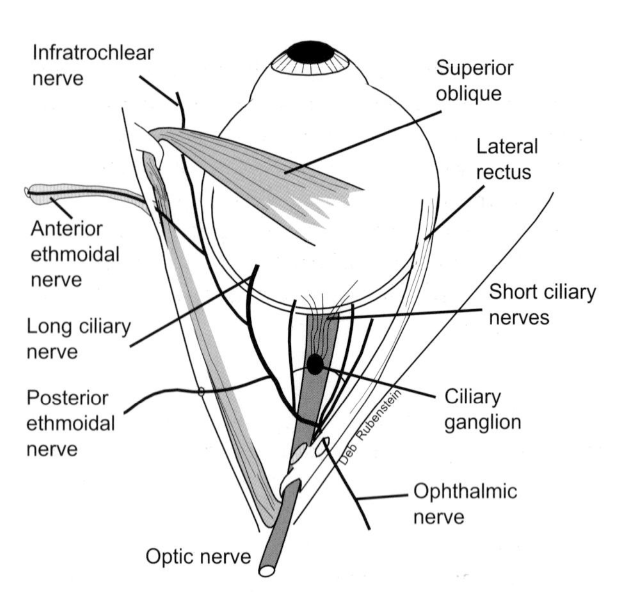

Figure 11.13 The course and branches of the nasociliary nerve

(V1), maxillary (V2), and the mandibular (V3) divisions. Functionally, the ophthalmic and maxillary divisions are sensory while the mandibular division remains mixed. There are postsynaptic sympathetic and postsynaptic parasympathetic fibers within the sensory fibers.

The ophthalmic nerve (V1) runs in the cavernous sinus and reaches the orbit through the superior orbital fissure (Figure 11.15). It supplies the frontal and ethmoidal sinuses, eyeball, the dura of the anterior cranial fossa, nasal cavity, upper eyelid, and the skin of the forehead region and scalp as far as the lambdoid suture. This division has frontal, nasociliary, and lacrimal branches. The frontal nerve provides sensory fibers to the forehead, upper eyelid, and scalp via its supratrochlear and supraorbital branches (Figure 11.14). The lacrimal nerve also transmits postsynaptic parasympathetic fibers to the lacrimal gland (Figure 11.18). The nasociliary branch of the ophthalmic nerve (Figure 11.13) runs with the ophthalmic artery, giving rise to sensory fibers to the lateral nose and the eyeball. It also carries presynaptic parasympathetic fibers that eventually run through the short ciliary branches to the ciliary body and the constrictor pupillae muscle.

The ophthalmic nerve mediates both corneal and lacrimal reflexes. The corneal reflex is a somatic reflex which is elicited by a light touch of the cornea or conjunctiva with a wisp of cotton, as the patient looks to the opposite side. As a result, contraction of the orbicularis oculi muscles and subsequently blinking in both eyes is produced. This reflex is mediated by the

Nasociliary neuralgia is an episodic or prolonged pain sensation in the medial canthus of the eye, eyeball, and external nose. Redness of the forehead, congestion of the nasal mucous membrane, lacrimation, and conjunctivitis usually accompany this condition. This may be triggered by stimulation of the medial canthus. Antibiotics, cortisone, and local anesthetic (5% solution of cocaine) may be used to treat this condition.

indirect bilateral connections of the ophthalmic nerve (afferent limb) to the facial motor neurons (efferent limb) via interneurons of the pontine reticular formation.

Lacrimation (tearing) reflex is also mediated by the ophthalmic nerve (afferent limb), which conveys the signals to the superior salivatory nucleus in the reticular formation of the pons. The latter projects via the intermediate and the greater petrosal nerves to the pterygopalatine ganglion that provides postsynaptic parasympathetic fibers to the lacrimal gland via the zygomaticotemporal branch of the maxillary and the lacrimal branches of the ophthalmic nerve.

The maxillary nerve (V2) runs in the cavernous sinus, leaves the middle cranial fossa through foramen rotundum (Figure 11.15), and enters the pterygopalatine fossa, where it is attached to the pterygopalatine ganglion (Figure 11.16). It provides sensory fibers to the following areas:

- Skin overlying the maxilla, upper lip, lower eyelid and side of the nose, as well as upper canine teeth via the infraorbital branch.
- Molar and premolar maxillary teeth and maxillary sinuses via the superior alveolar branches.
- Mucosa of the palate via the greater and lesser palatine branches.
- Meningeal branches to the dura of the middle cranial fossa.
- Nasal branches to the nasal cavity and nasopharynx.
- Temporal region via the zygomatic branch, which further divides into zygomaticofacial and zygomaticotemporal branches. The latter communicates with the lacrimal nerve, conveying postganglionic parasympathetic fibers to the lacrimal gland. It also gives rise to ganglionic branches to the pterygopalatine ganglion. The nasal branches form the afferent limb of the nasal (sneeze) reflex, which is characterized by contraction of the muscles of the soft palate, pharynx and larynx, diaphragm and intercostal muscles in response to irritation of the nasal mucosa. The afferent limb of this reflex conveys these impulses to the spinal trigeminal nucleus then are transmitted in turn to the trigeminal motor and

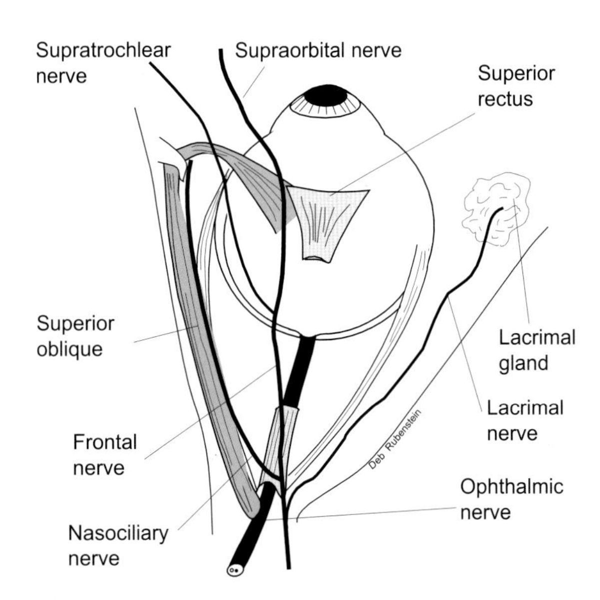

Figure 11.14 Frontal and lacrimal branches of the ophthalmic nerve

Damage to the ophthalmic or facial nerves, or their central connections in the pons may produce loss of corneal reflex. Loss of the corneal reflex due to an ophthalmic nerve lesion may be observed in both eyes when the affected side is stimulated. However, this reflex may still be elicited in both eyes when the unaffected side is simulated (Figure 11.19).

ambiguus nuclei, as well as the intercostal and motor neurons of the phrenic nerves.

The mandibular nerve (V3) is sensory to the mandibular teeth, floor of the mouth, anterior two-thirds of the tongue, and skin of the jaw and chin (Figure 11.17). It innervates the muscles of mastication, and the tensor palatini and tympani. It leaves the middle cranial fossa via the foramen ovale (Figure 11.15) and enters the infratemporal fossa to course lateral to the otic ganglion. It divides into a primarily motor anterior trunk (with the exception of the buccal branch) and a principally sensory posterior trunk, with the exception of the mylohyoid nerve.

The posterior trunk of the mandibular nerve gives rise to the lingual, inferior alveolar, auriculotemporal, and mylohyoid branches.

The lingual nerve is sensory to the anterior two-thirds of the tongue, floor of the mouth, and lingual gingiva. In the infratemporal fossa, it joins the chorda tympani branch of the facial nerve, which carries taste sensation from the anterior two-thirds of the tongue and proceeds to attach to the submandibular ganglion.

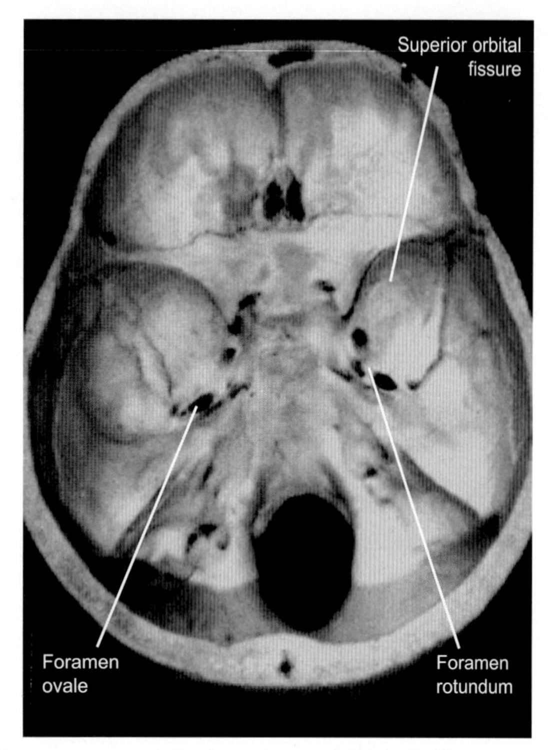

Figure 11.15 Main foramina and fissures that transmit branches of the trigeminal nerve

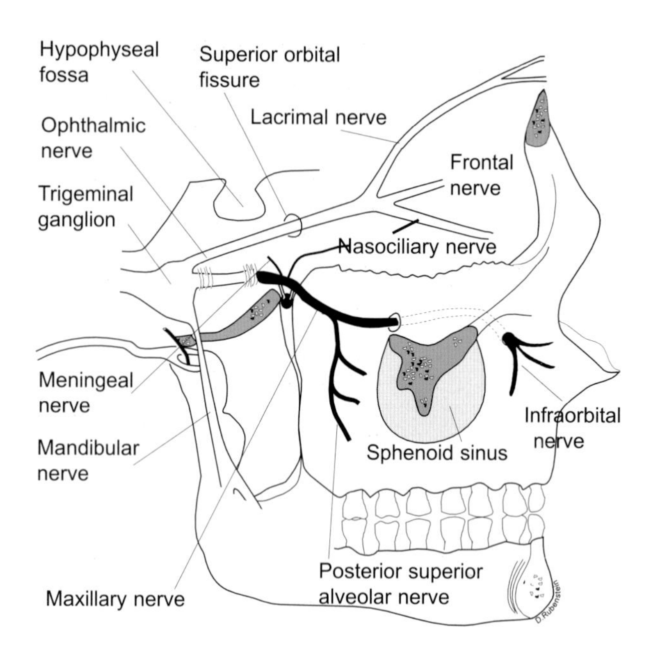

Figure 11.16 The origin, general course and branches of the maxillary nerve

Sluder's neuralgia is a condition which is associated with a lesion of the pterygopalatine ganglion. This disorder is characterized by pain in the areas of distribution of the nasal, pharyngeal, and palatine branches of the maxillary nerve that pierce the pterygopalatine ganglion. Frequent sneezing is common in many cases. Paranasal infection may also induce this condition.

The inferior alveolar nerve (Figure 11.17) runs in the mandibular canal, supplies the mandibular teeth, and gives rise to the mental and incisive nerves. The mental nerve supplies the skin of the chin and is involved in the jaw-jerk reflex. This reflex is monosynaptic, characterized by sudden closure of the mouth (as a result of bilateral contraction of the masseter and temporalis muscles) following a downward tap on a finger placed on the jaw when the mouth is slightly open. It is mediated by the mandibular nerve, through the mesencephalic and the motor trigeminal nuclei.

The auriculotemporal nerve (Figure 11.17) is the only branch that courses posteriorly, arising by two roots that encircle the middle meningeal artery. It carries postganglionic parasympathetic fibers to the parotid gland, and sensory fibers to the temporomandibular joint, anterior temporal region, auricle, and the external acoustic meatus. The mylohyoid nerve innervates the mylohyoid muscle and the anterior belly of the digastric muscle.

The anterior trunk of the mandibular nerve gives rise to nerves that supply the temporalis, lateral and medial pterygoid, masseter, tensor tympani, and tensor palatini muscles. The buccal nerve, the only sensory branch of the anterior trunk, supplies the skin and mucosa of the cheek.

The trigeminal nerve has three sensory and a single motor nucleus (Figures 11.18 & 11.20). The sensory nuclei of the trigeminal nerve are the spinal, principal sensory, and mesencephalic trigeminal nuclei.

The spinal trigeminal nucleus (GSA) extends from the midpons to the upper segments of the spinal cord, representing the rostral extension of the substantia gelatinosa. It receives thermal, painful, and tactile sensations from the head region within branches of the trigeminal, facial, glossopharyngeal, and vagus nerves. As they terminate in this nucleus, they form the spinal trigeminal tract, which is positioned lateral to the nucleus. Within the spinal trigeminal nucleus and tract, the ophthalmic fibers occupy a caudal position to the more rostral mandibular fibers, while the

Division of the lingual nerve distal to the site of union with the chorda tympani may produce loss of general and taste sensations from the anterior two-thirds of the tongue and impairment of salivary secretion from the submandibular and sublingual glands.

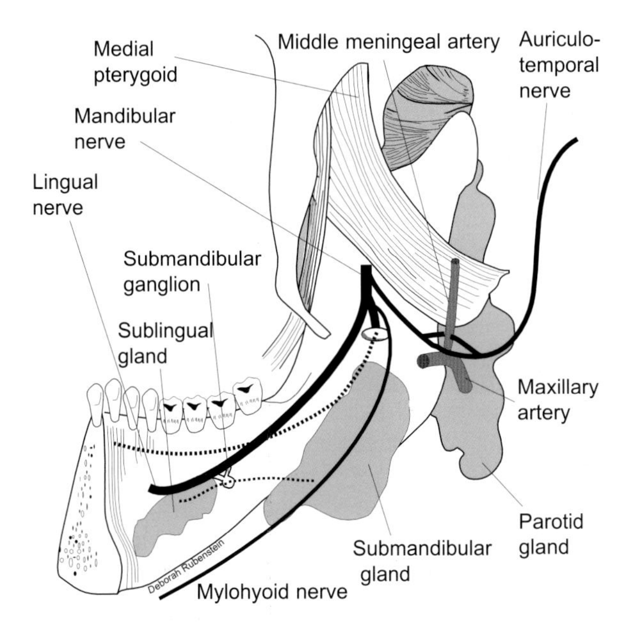

Figure 11.17 The principal branches of the mandibular nerve are shown

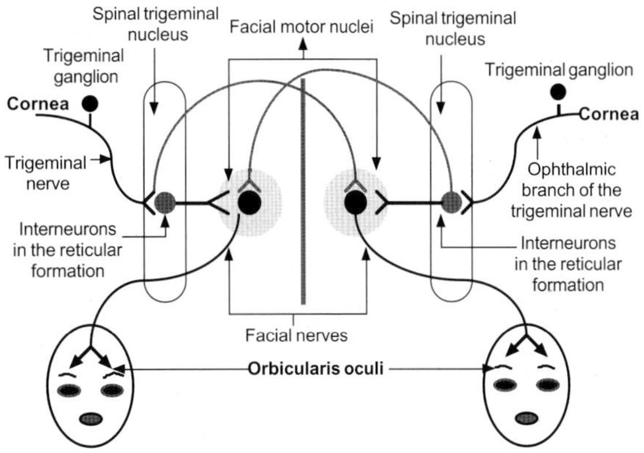

Figure 11.19 Corneal reflex. Both the trigeminal and facial nerves, as illustrated in this diagram, mediate this reflex

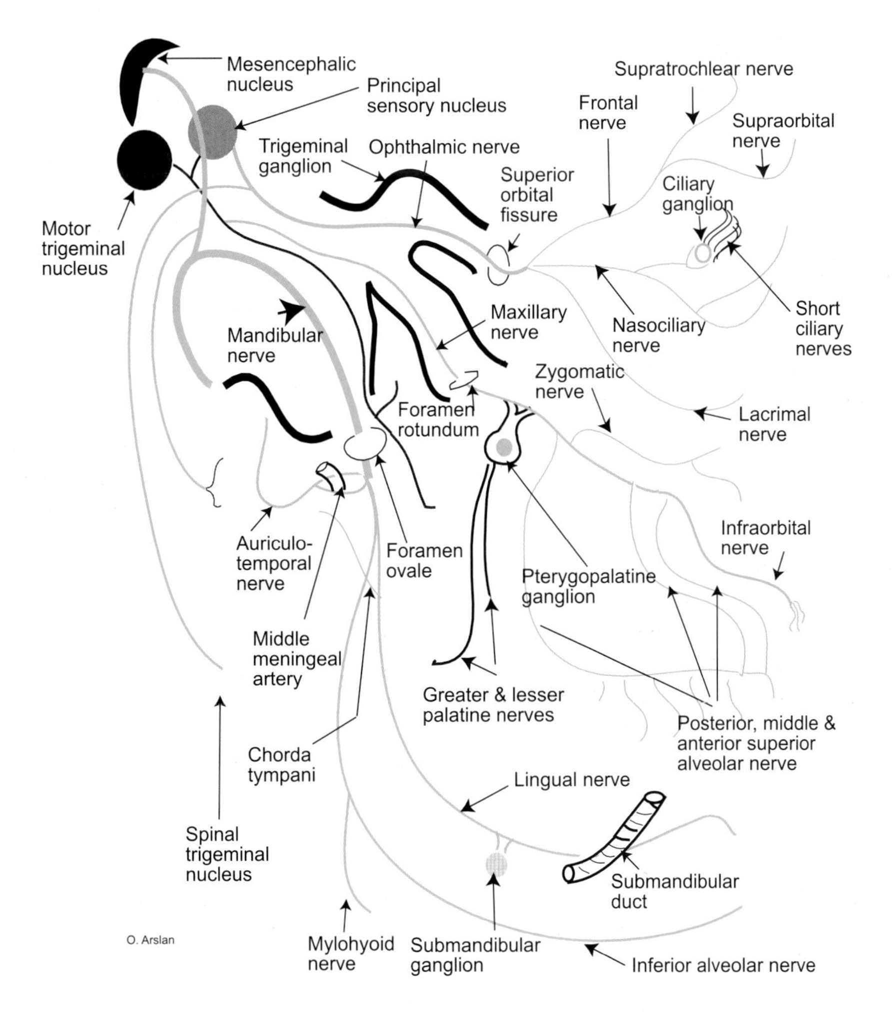

Figure 11.18 Functional components, branches, and associated nuclei of the trigeminal nerves. The course of these branches and their areas of distribution are documented to facilitate the understanding of the central and peripheral courses of this nerve

Failure to elicit jaw-jerk reflex may indicate a pontine lesion involving the trigeminal nerve, mesencephalic, or motor trigeminal nuclei. Hyperactive jaw reflex is a manifestation of cortico-bulbar tract damage.

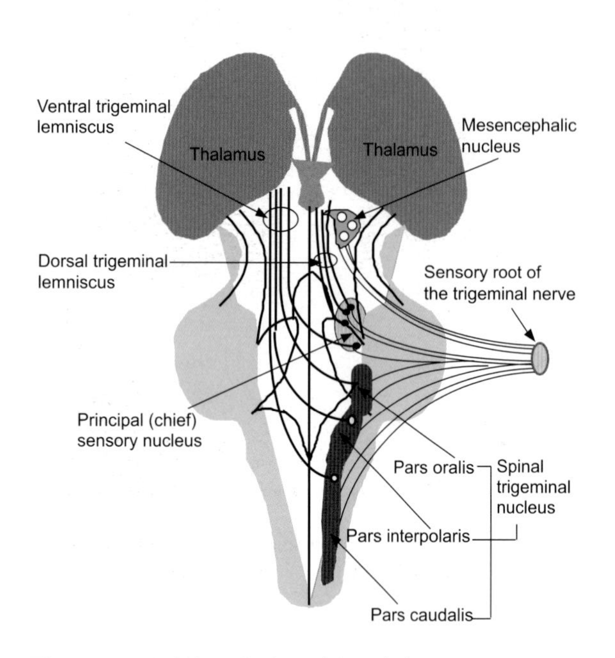

Figure 11.20 Trigeminal nuclei and their projections via the trigeminal lemnisci

maxillary fibers maintain an intermediate position. Since pain and thermal fibers terminate in the most caudal portions of this nucleus and tract, within the caudal medulla, excision of the spinal trigeminal tract (tracteotomy), if it is performed at this level, may alleviate the intractable pain associated with trigeminal neuralgia.

In the midpons, the principal (pontine or chief) sensory nucleus (GSA) lies lateral to the trigeminal nerve fibers. It transmits tactile and pressure sensation, showing identical somatotopic distribution to that of the spinal trigeminal nucleus.

Neurons of the spinal trigeminal nucleus and the ventral part of the principal sensory nucleus give rise to axons regarded as secondary trigeminal fibers. These fibers cross the midline and run through the reticular formation, forming the ventral trigeminal lemniscus or tract on the opposite side. Fibers from the dorsal part of the principal sensory nucleus are virtually derived from the mandibular nerve and ascend ipsilaterally as the dorsal trigeminal lemniscus or tract. Both of the trigeminal lemnisci (tracts) are associated topographically with the medial lemniscus and project to the VPM nucleus of the thalamus. The secondary trigeminal fibers have collaterals that establish contact with specific nuclei in the reticular formation, mediating various trigeminal reflexes.

The mesencephalic nucleus (GSA) receives fibers that transmit proprioceptive input from the temporomandibular joint, muscles of mastication, hard palate, mandibular and maxillary alveoli, and possibly from the extraocular muscles. It projects to the cerebellum and superior colliculi, and mediates the jaw-jerk reflex.

Injury to the trigeminal nerve or its branches may occur as a result of a tumor of the pontocerebellar angle, otitis media, cavernous sinus thrombosis, fractures involving the middle cranial fossa, or metastatic carcinomas. Multiple sclerosis may produce trigeminal neuralgia and transient facial anesthesia in young adults. In particular, the ophthalmic nerve may be damaged as it courses within the superior orbital fissure in conjunction with the oculomotor, trochlear, and abducens nerves. It may also be injured in orbital apex syndrome together with the optic, oculomotor, trochlear, and abducens nerves. A fracture confined to the ramus of the mandible may put the mandibular nerve out of function.

Destruction of the trigeminal nerve produces combined sensory and motor, as well as reflex disorders. These dysfunctions include unilateral anesthesia in the area of distribution of the trigeminal nerve, loss of corneal reflex on both sides when the affected eye is stimulated, and also atrophy of the muscles of mastication, tensor tympani, and tensor palatini muscles. Additional deficits include loss of sensation from the facial region, oral and nasal cavities, and the anterior two thirds of the tongue. Sensations from the temporomandibular joint, paranasal sinuses, and anterior part of the external acoustic meatus may also be lost. The jaw-jerk, oculocardiac (a reflex that mediates slowing of heart rate upon compression of the eyeball), sneezing, and lacrimation reflexes are impaired. Impairment of the post-synaptic parasympathetic innervation to the head region may also be noticed. Projected pain to the area of distribution of branches of the trigeminal nerve is common. For example pain from carious tooth and ulcer of the tongue may produce pain that is felt in the ear and temporal region that corresponds to the area of distribution of the auriculotemporal nerve.

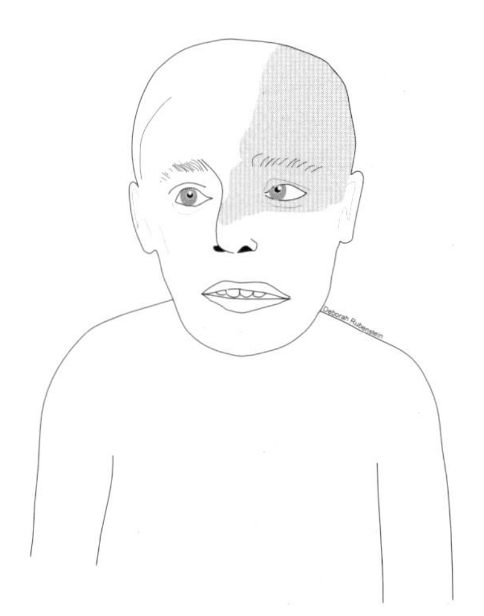

Figure 11.21 Port-wine discoloration (cutaneous vascular nevus) of the territory of the ophthalmic nerve. The patient is retarded and suffers from epileptic seizures

The motor trigeminal nucleus (SVE) is located medial to the entering trigeminal nerve fibers in the rostral pons. It is the source of motor fibers to the muscles of mastication, tensor tympani, and tensor palatini muscles. It receives bilateral corticobulbar fibers, and forms the efferent limb of the jaw-jerk reflex.

Anomalous connections between the postsynaptic parasympathetic fibers of the auriculotemporal nerve which are destined to the parotid gland and the sympathetic postsynaptic fibers that supply the sweat glands of the face may occur following infection, trauma, and, although rarely, after surgical operation involving the parotid gland. These aberrant connections produce signs of Frey syndrome (gustatory sweating) which is characterized by sweating induced by salivatory stimuli. Patients with Frey syndrome exhibit flushing and sweating on the face along the distribution of the auriculotemporal nerve in response to tasting or eating. Auriculotemporal neuralgia may accompany this condition. Although a rare form of neuralgia, this disorder may exhibit burning pain in the preauricular and temporal areas, triggered by chewing or tasting spicy food.

Sturge–Weber syndrome (Figures 11.21 & 11.22), a rare congenital sporadic disease, is associated with cerebral cortical (railroad track) calcification and atrophy. Patients present with seizure, neuro-ocular disorders, mental retardation, and unilateral or bilateral port-wine discoloration (facial nevus) in the areas of distribution of the ophthalmic nerve alone, or in combination with the maxillary and mandibular nerves. An ipsilateral enlarged choroid plexus may be an early anatomic manifestation.

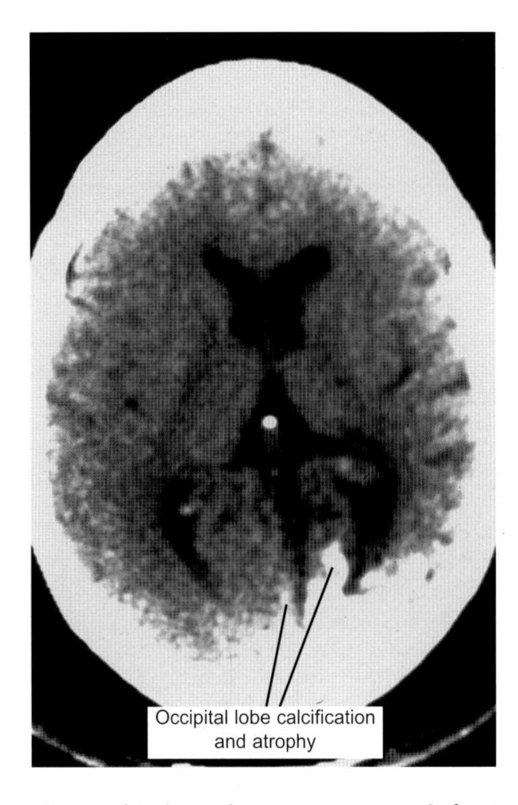

Figure 11.22 This image shows a prominent calcification in an individual with Sturge–Weber syndrome

A lesion confined to the motor trigeminal nucleus may only result in atrophy of the muscles of mastication and deviation of the mandible toward the lesion side.

Trigeminal neuralgia (tic douloureux), a common idiopathic condition, exhibits a lightening or lancinating facial pain on the affected side which lasts for a few seconds. This lifelong recurring disorder affects women twice as often as men and develops relatively late in life between the age of 50 and 60. The paroxysmal pain associated with this condition is frequently elicited by mild stimuli (e.g. touch, application of cold, etc.) in one or more area of distribution of the trigeminal nerve. Patients may also exhibit signs of autonomic disorders (lacrimation, salivation, and flushing of the face), reflex facial muscle spasm, and sensory loss. Aberrant superior cerebellar or cerebral arteries may also produce this condition by compressing the root of the trigeminal nerve. Multiple sclerosis accounts for most cases of trigeminal neuralgia in young adults. This condition may be treated by injection of glycerol into the root of the trigeminal nerve or by relieving the compression by placing a barrier between the trigeminal nerve and the anomalous vessel.

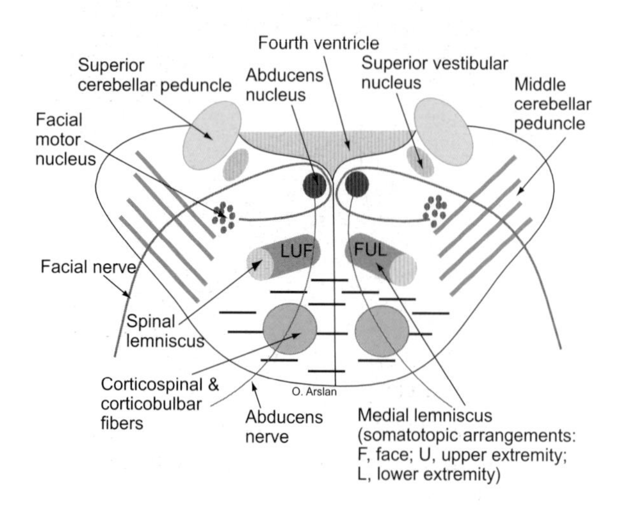

Figure 11.23 The abducens nucleus and nerve are shown in this section of the caudal pons. The relationship of the facial nerve to the abducens nucleus is also illustrated

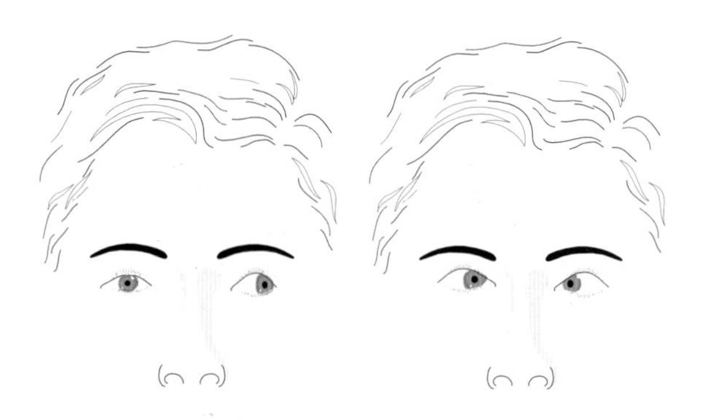

Figure 11.24 Deficits associated with right abducens nerve palsy. Note the eye on the affected side is adducted at rest (left diagram), and cannot abduct when attempt to look to the right (right diagram)

VI Abducens nerve

The abducens nerve (GSE) is formed by the axons of the motor neurons of the abducens nucleus, which lies deep to the ependyma of pontine part of the fourth ventricle (Figures 11.1 & 11.23). The abducens nucleus is surrounded by the motor fibers of the facial nerve, forming the facial colliculus (Figure 11.22). This nucleus is unique among all other cranial nerve motor nuclei in that it contains two populations of neurons: the α motor and internuclear neurons. The α motor neurons give rise to the abducens nerve, whereas the smaller population of interneurons send axons through the contralateral MLF to the motor neurons of the oculomotor nucleus. These axons control the oculomotor neurons that innervate the medial rectus muscle. The duality of neuronal function of the abducens nucleus may account for the distinct deficits produced by a lesion of the abducens nerve versus the manifestations associated with abducens nucleus damage. During its course in the tegmentum and basilar pons, the abducens nerve runs adjacent to the corticospinal tract. Then it leaves the brainstem at the pontobulbar sulcus, maintaining a long intracranial course. Before gaining access into the cavernous sinus, it forms a bend over the apex of the petrous temporal bone. Within the cavernous sinus, it lies adjacent to the oculomotor, trochlear, ophthalmic and maxillary nerves, maintaining close proximity to the internal carotid artery. Then it leaves the cavernous sinus to enter the orbit via superior orbital fissure, innervating the lateral rectus muscle.

Proximity of the abducens nerve to the internal carotid artery may account for the initial signs of abducens nerve palsy in individuals with aneurysm of the internal carotid artery. Also, due to its long intracranial course and sharp bend over the petrous temporal bone, this nerve is more prone to injury than any other cranial nerve. Damage to this nerve may also occur in cavernous sinus thrombosis, fracture of the superior orbital fissure, and aneurysm of the internal carotid artery. Medial strabismus (convergent squint), in which the patient is unable to direct both eyes toward the same object, characterizes abducens nerve palsy (Figure 11.24). This occurs due to paralysis of the lateral rectus muscle. Patients also experience horizontal diplopia (in acute stage) on attempted gaze to the affected side. Chronic abducens nerve palsy may not exhibit diplopia as the image from the affected eye is suppressed psychologically.

Destruction of the abducens nerve and the adjacent fibers of the corticospinal tract on one side may produce signs of middle alternating hemiplegia in which hemiplegia is manifested on the contralateral side while signs of abducens nerve palsy remain ipsilateral. A lesion of the abducens nucleus results in disruption of the abducens nerve and the internuclear neurons that project to the contralateral medial rectus, producing lateral gaze which is characterized by adductor paresis on the contralateral side and abductor palsy on the side of the lesion.

VII Facial nerve

The facial nerve (GSA, GVE, SVA, SVE) leaves the brainstem at the cerebellopontine angle and enters the internal acoustic meatus, accompanied by the

Table 11.1 Cranial nerves I–VI

Cranial nerve	*Component*	*Location of cell bodies*	*Course*	*Distribution*	*Function*
I Olfactory	SVA	Neuroepithelial cells in nasal cavity	Cribriform plate of ethmoid	Olfactory mucus membrane	Olfaction
II Optic	SSA	Ganglion cells in the retina	Optic canal	Retina	Vision & visual reflexes
III Oculomotor	GSE	Somatic column of oculomotor nucleus	Superior orbital fissure	Levator palpebrae & all extraocular muscles, except lateral rectus and superior oblique	Elevates the upper eyelid; adducts, extorts, elevate or depresses the eyeball
	GVE	Edinger–Westphal nucleus	Same as above	Ciliary & constrictor pupillae muscles	Mediates light and accommodation reflexes
IV Trochlear	GSE	Trochlear nucleus	Superior orbital fissure	Superior oblique	Abducts, intorts & depresses the eyeball
V Trigeminal	GSA	Trigeminal (Gasserian) ganglion	V1 – superior orbital fissure V2 – foramen rotundum V3 – foramen ovale	Skin of face, scalp, gingiva, anterior 2/3 of tongue, oral and nasal cavities, eye, paranasal sinuses & meninges	Cutaneous and proprioceptive information
	SVE	Trigeminal motor nucleus	Foramen ovale	Muscles of mastication, anterior belly of digastric, tensor tympani & mylohyoid muscle	Mastication, mandibular movements & deglutition
VI Abducens	GSE	Abducens nucleus	Superior orbital fissure	Lateral rectus	Lateral deviation of the eye

vestibulocochlear and labyrinthine artery (Figures 11.1, 11.25, 11.26 & 11.28). It runs dorsal to the cochlea within the facial canal, emerging through the stylomastoid foramen. It supplies the occipitalis, posterior belly of the digastric, and the mylohyoid muscles. It pierces the parotid gland and then divides into temporal, zygomatic, buccal, marginal mandibular, and cervical branches. The facial nerve has motor and intermediate roots.

The motor root (SVE) is derived from neurons of the facial motor nucleus that provide innervation to the muscles of facial expression. These fibers ascend toward the midline and encircle the abducens nucleus, forming the facial colliculus. These motor fibers mediate both glabellar and corneal reflexes. The corneal reflex (also discussed with the trigeminal nerve, (Figure 11.19) is characterized by contraction of the orbicularis oculi and the resultant blinking upon stimulation of the cornea by a wisp of cotton. Damage to the facial nerve also results in loss of this reflex on the side of the lesion.

The intermediate (root) nerve contains general and special sensory as well as preganglionic parasympathetic fibers. The sensory neurons of the facial nerve are located in the geniculate ganglion, a collection of unipolar neurons at the junction of the vertical and horizontal parts of the facial nerve.

The somatic sensory fibers of this root terminate in the spinal trigeminal nucleus, while the special visceral afferent (taste) fibers establish synaptic connections with neurons of the solitary nucleus. This root also contains preganglionic parasympathetic fibers running within the greater petrosal nerve and chorda tympani, providing secretomotor fibers to the submandibular, sublingual, and lacrimal glands. The facial nerve gives rise to the greater petrosal nerve, chorda tympani, stapedius nerve, posterior auricular nerve, and muscular branches.

The greater petrosal nerve (GVE, GVA, SVA) arises from the facial nerve at the level of the geniculate ganglion, carrying preganglionic parasympathetic fibers (GVE) from the superior salivatory nucleus to

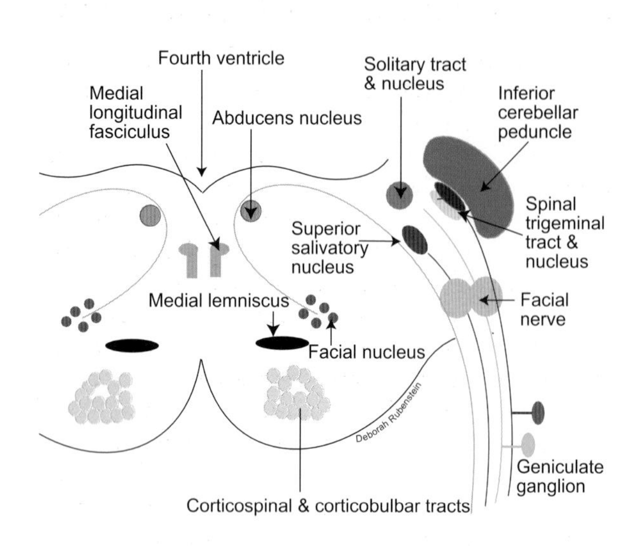

Figure 11.25 The functional components of the facial nerve, course, and associated nuclei are shown in this diagrammatic section of the pons

Ramsay Hunt syndrome is a condition that results from herpetic viral infection of the geniculate ganglion resulting in auricular pain followed by vesicular eruptions in the area of distribution of the auricular branch of the facial nerve in the concha of the ear.

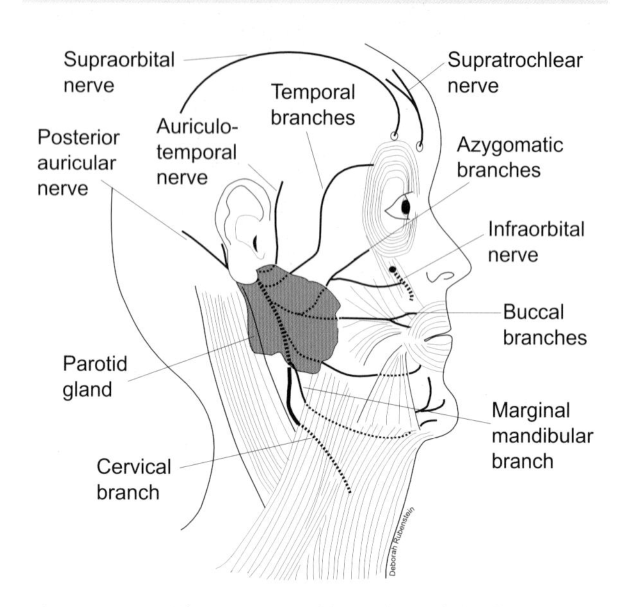

Figure 11.26 The course and branches of the facial nerve, outside the facial canal, are illustrated

The glabellar (McCarthy's) reflex or Myerson sign exhibits forceful, persistent, involuntary and repeated contraction of the orbicularis oculi muscle, which is elicited by repetitive finger taps on the forehead and supraorbital margin in a downward direction to the glabella. Damage to the facial nerve or its nucleus produces glabellar hyporeflexia, a disorder seen in Parkinson's disease and in patients with bilateral frontal lobe lesion, and occasionally in tense and overly stressed individuals.

Anomalous connection between the preganglionic parasympathetic fibers of the chorda tympani and those of the greater petrosal nerve may result in salivation induced by lacrimation (crocodile-tear syndrome).

the pterygopalatine ganglion (Figure 11.28). It also conveys taste (SVA) and general sensory (GVA) fibers from the soft palate to the solitary nucleus. This nerve should not be confused with the lesser petrosal branch of the glossopharyngeal nerve.

The chorda tympani (SVE, GVE) runs in the upper quadrant of the tympanic membrane medial to the malleus and incus, leave the skull via the petro-tympanic fissure and enters the infrartemporal fossa (Figure 11.28). It carries preganglionic parasympathetic (GVE) fibers from the superior salivatory nucleus and conveys taste sensation (SVA) from the anterior two-thirds of the tongue to the solitary nucleus.

The stapedius nerve (SVE) is formed by fibers of the facial motor nucleus, supplying the stapedius muscle. This muscle contracts in response to loud noises.

The posterior auricular branch (SVE) innervates the occipitalis muscle, and derives its axons from the facial motor nucleus (Figure 11.25). Other muscular branches of the facial nerve supply the stylohyoid (SVE) posterior belly of the digastric muscles. Five muscular branches arise from the trunk of the facial nerve within the parotid gland, which include the temporal, zygomatic, buccal, marginal mandibular, and cervical branches (Figure 11.26). It is important to distinguish between the buccal branch of the facial nerve and that of the mandibular nerve. The buccal branch of the facial nerve is motor to the muscles around the mouth and runs with the parotid duct, supplying the masseter muscle, while the buccal branch of the mandibular nerve is sensory to the skin and mucosa of the cheek. The auricular branch (GSA) transmits general sensory impulses from the concha of the ear to the spinal trigeminal nucleus.

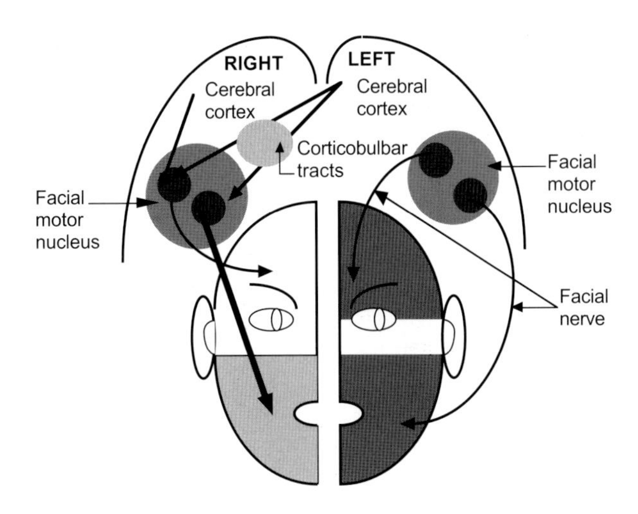

Figure 11.27 Schematic drawing of the preferential distribution of the corticobulbar tract in the facial motor nuclei. Note the bilateral distribution of the corticobulbar fibers to the neurons of the upper face and contralateral projection to the neurons of the lower face

Nuclei associated with the facial nerve

These nuclei are comprised of the facial motor, superior salivatory, solitary, and the spinal trigeminal (Figures 11.25 & 11.28).

The facial motor nucleus (SVE) is located in the tegmentum of the caudal pons, giving rise to the special visceral efferent fibers that encircle the abducens nucleus, forming the facial colliculus. Uniqueness of this nucleus is illustrated in its highly distinctive corticobulbar connections. The part of the facial motor nucleus that provides innervation to the muscles around the mouth and lower face receives only crossed corticobulbar fibers, whereas the part of the nucleus that supplies the muscles around the orbit and the forehead region receives bilateral corticobulbar projections. This diverse cortical input accounts for the selective paralysis of the contralateral muscles around the mouth in individuals with unilateral corticobulbar damage, while sparing the muscles around the eye and orbit (Figure 11.27).

The spinal trigeminal nucleus (GSA) receives general somatic sensations from the concha, to be delivered to the ventral posteromedial nucleus of the thalamus.

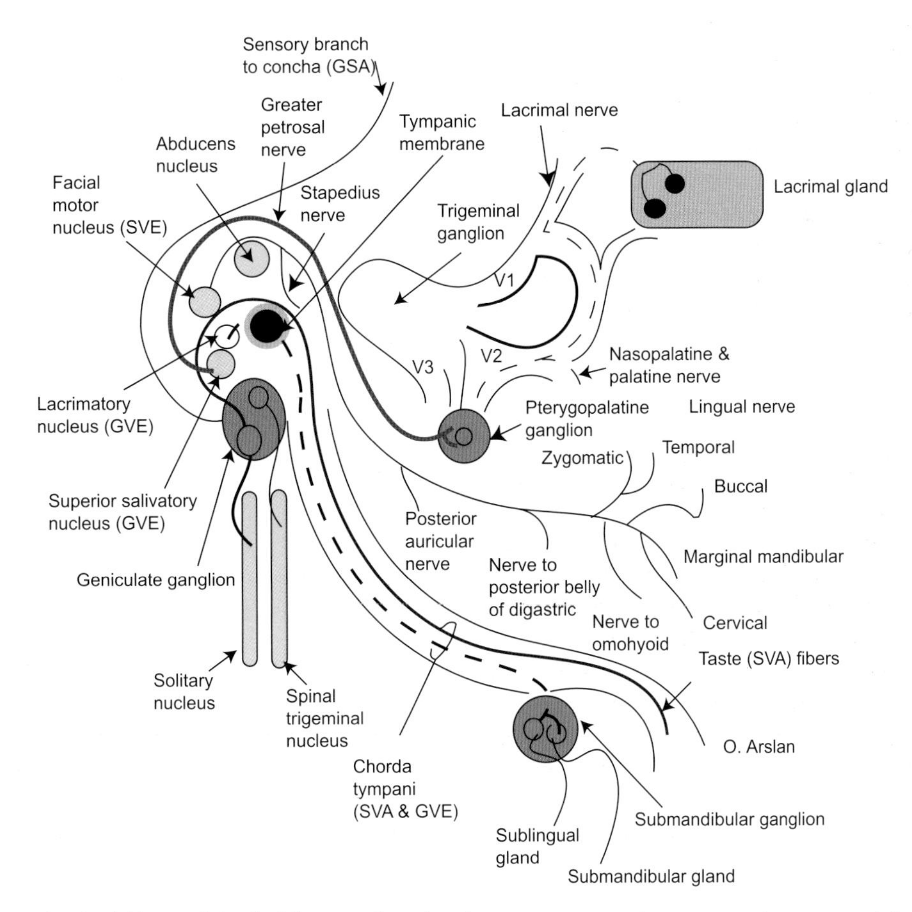

Figure 11.28 The individual fibers of the facial nerve, functional components, and sensory and motor nuclei of the facial nerve are illustrated

Facial palsy occurs as a result of lesions involving the corticobulbar fibers, facial motor nucleus, or the facial nerve. These lesions, as documented below, are categorized into supranuclear and infranuclear lesions.

Supranuclear lesions (Figure 11.29) involve the corticobulbar fibers that emanate from the cerebral cortex and project to the facial motor nucleus. These lesions, as seen in multiple sclerosis, produce upper motor neuron signs in the muscles around the mouth on the contralateral side. Patients cannot voluntarily move the affected muscles (voluntary facial palsy); however, they remain responsive to emotional stimuli. Supranuclear lesions that affect the projections of the limbic system to the facial motor nucleus may result in mimetic facial unresponsive to emotional stimuli, while preserving the ability to produce contraction of the affected muscles upon command.

Infranuclear lesions (Figure 11.30) are caused by damage to the facial nerve and/or the facial motor nucleus. These lesions may be the result of mumps, acoustic neuroma, parotid tumors, geniculate herpes (herpes zoster oticus), leprosy, leukemia, sarcoidosis, or remain idiopathic. In the newborn, absence of the mastoid processes render the facial nerves on both sides unusually vulnerable and exposed to damage by careless use of obstetrical forceps. Facial nerve damage produces ipsilateral symptoms, which vary dependent upon the site of the lesion. A lesion at or above the level of the geniculate ganglion (as in tumors of the cerebello-pontine angle or in fractures of the internal acoustic meatus) results in the ipsilateral loss of lacrimation and loss of general and special visceral (taste) sensations from the palate. These deficits are due to disruption of the greater petrosal nerve. Taste sensation from the anterior two-thirds of the tongue and secretion of the submandibular and sublingual are also lost, due to destruction of the chorda tympani. Hyperacusis (lower hearing threshold) as a result of the destruction of the stapedius nerve and paralysis of all facial muscles of expression, the posterior belly of the digastric, and the stylohyoid muscles may also be visible. Paralysis of muscles of facial expression results in asymmetry of the face, widening of the palpebral fissure, inability to close the eye, loss of the corneal reflex, sagging of the angle of the mouth, and smoothing of facial sulci.

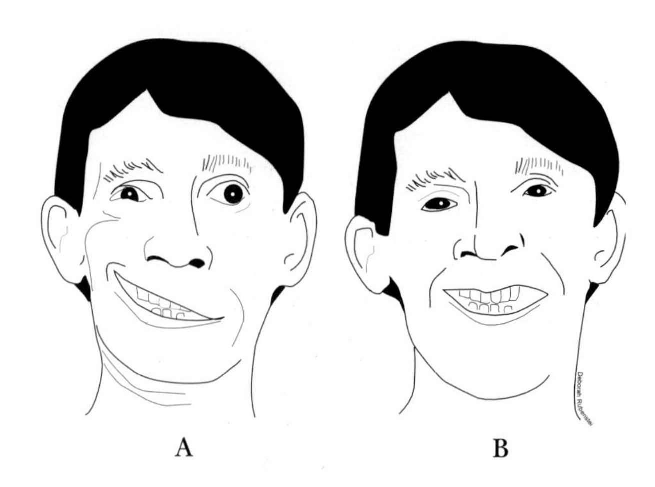

Figure 11.29 Manifestations of supranuclear facial palsy. Note the prominent weakness in the muscles around the mouth when the patient is asked to open the mouth (A) and the retention of normal function of these muscles in response to emotional condition (B)

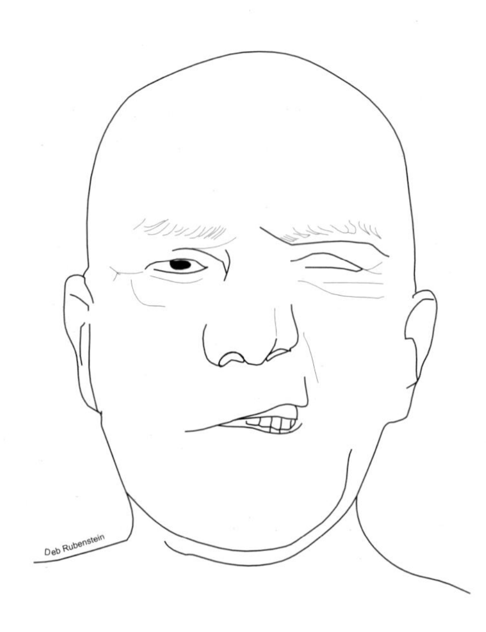

Figure 11.30 Right infranuclear facial palsy. Note the smoothing of the sulci, including the nasolabial sulcus, sagging of the labial commissure on the affected side

The superior salivatory nucleus (GVE) lies adjacent to the caudal end of the facial motor nucleus in the caudal pons and provides preganglionic parasympathetic fibers to the pterygopalatine and the submandibular ganglia via the greater petrosal and chorda tympani branches, respectively. Postsynaptic fibers from these ganglia control the secretion of the lacrimal, submandibular, sublingual glands, and the mucus glands of the palate and pharynx.

Unilateral destruction of the facial motor nucleus produces similar deficits to the lesion that disrupts the facial nerve at or proximal to the geniculate ganglion.

A lesion immediately distal to the geniculate ganglion produces all of the above mentioned deficits while preserving lacrimation and taste sensation from the palate. Paralysis of the facial muscles of expression is the only deficit seen in individuals with a damaged facial nerve at the stylomastoid foramen. Loss of corneal reflex due to facial nerve damage is observed only on the side of the lesion, regardless of which side is stimulated. Inability to close the eye by the unopposed action of the levator palpebrae superioris, and loss of blinking may increase the potential of corneal irritation or ulceration and may lead to keratitis and possible blindness. This complication may be avoided by placing a patch over the affected eye. The oculo-auricular reflex, which is characterized by posterior movement of the ear when the patient directs his or her gaze as far laterally as possible, is lost in Bell's palsy. Prognosis of this condition depends on the severity of ear pain and the degree of paralysis, and it is poorer in individuals with ear pain and extensive facial palsy.

Bell's palsy is an inflammatory disease of unknown etiology, which results from compression or inflammation of the facial nerve. It may occur following exposure to cold or a viral infection. This condition, which accounts for about 80% of all cases of facial palsy, may accompany otitis media, mastoiditis and petrositis. It may occur in diabetic patients and pregnant woman. Patients exhibit abrupt or progressive unilateral paralysis of the facial muscles, which may be preceded or accompanied by earache. Glandular secretion, stapedius muscle function, and taste sensation often remain unaffected.

The solitary nucleus (SVA) receives taste fibers via the central processes of the geniculate neurons that primarily originate from the anterior two-thirds of the tongue. Postsynaptic fibers from this nucleus terminate in the ventral posteromedial nucleus of thalamus.

VIII Vestibulocochlear nerve

The vestibulocochlear nerve (GSA) has cochlear (auditory) and vestibular components (Figure 11.1). The vestibular nerve (SSA) represents the axons of the bipolar neurons of the superior and inferior

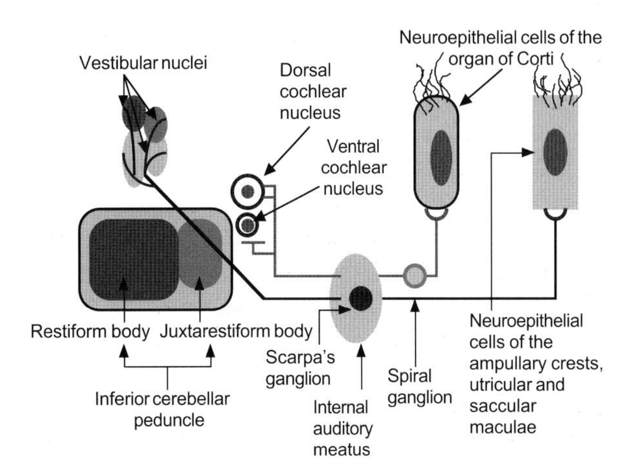

Figure 11.31 In this diagram the vestibular and auditory receptors, associated ganglia, and course of individual nerves are illustrated. The connection of the vestibular nerve to the vestibular nuclei is also shown

Intactness of the vestibular nerve and its connections may be evaluated by instilling cold or warm water (caloric test) or by rotating in the Barany chair. Dysfunction of the vestibular nerve produces ataxia, vertigo, and nystagmus, which are briefly explained below (see also the vestibular system, Chapter 15).

Ataxia (incoordination of motor activity) is due to vestibular nerve dysfunction and is gravity dependent, severe, and often an intermittent incoordination of limb movement. It becomes apparent during walking and standing.

Vertigo is a severe sense of rotation of the environment which is frequently intermittent and may be accompanied by oscillopsia (a back and forth movements of the visual objects with downbeat nystagmus), nausea, and vomiting. Vertigo may result in pallor, depression and falling.

Nystagmus, an abnormal rhythmic oscillation of the eyeball, is produced visually by watching stationary targets from a moving vehicle (optokinetic nystagmus), or by extreme gaze to one side. It may also be produced iatrogenically by instilling cold or warm water into the ear (caloric test), or rotating in the Barany chair. Clinically, it may result from peripheral or central vestibular lesions (see the vestibular system).

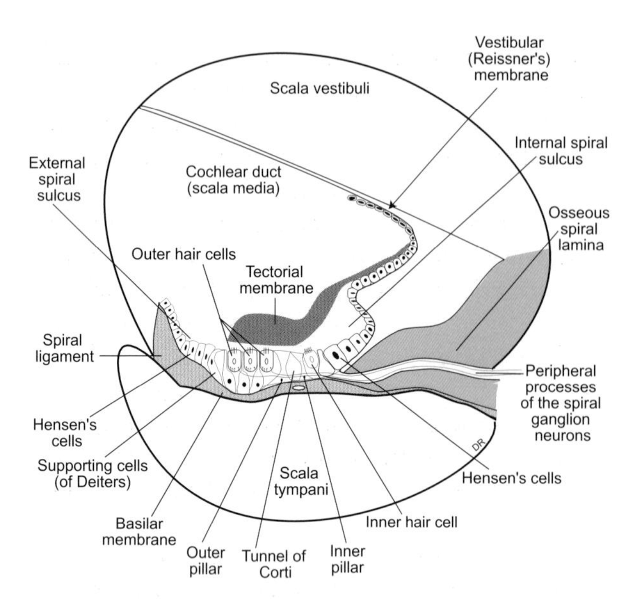

Figure 11.32 In this drawing the neuroepithelial cells within the organ of Corti, spiral ganglia, and course of the cochlear nerve are illustrated

Damage to the cochlear nerve, as a result of acoustic neuroma, Meniere's disease, or fracture involving the petrous temporal bone, produces nerve deafness on the side of the lesion. This type of deafness may be accompanied by tinnitus (ringing of the ear), and is distinguished from conduction deafness by Rinne and Weber tests (see the auditory system, Chapter 14). Tinnitus, a unilateral or bilateral continuous or intermittent hissing sound, is an important manifestation of cochlear nerve damage. It is often severe enough to interfere with the normal daily activities. Fractures of the posterior cranial fossa, involving the internal acoustic meatus and tumors of the cerebellopontine angle (acoustic neuroma) may result in combined vestibular, cochlear, and facial nerve dysfunctions. This type of injury is characterized by deafness on the side of the lesion, vertigo (sense of rotation of the environment or self), nystagmus, ataxia, and signs of ipsilateral infranuclear facial palsy.

vestibular (Scarpa's) ganglia (Figure 11.31). The dendrites of these neurons pass through foramina in the internal acoustic meatus and eventually distribute to the vestibular receptors in a selective manner. For example, the neurons of the superior vestibular ganglion receive information from the ampullary crests of the anterior and lateral semicircular ducts and the macula of the utricle, whereas the inferior vestibular ganglion receives input from the ampullary crests of the posterior semicircular duct and the macula of the saccule. This nerve leaves the brainstem at the cerebellopontine angle, accompanied by the facial, cochlear nerves, and the labyrinthine artery. Therefore, combined vestibular and auditory deficits, as well as facial palsy may be seen in pathological conditions involving the internal acoustic meatus or the cerebellopontine angle (i.e. acoustic neuroma).

The cochlear nerve (SSA) is formed by the central processes of the bipolar neurons of the spiral ganglion, which are located in the modiolus of the cochlea (Figure 11.32). The site of entry of the cochlear nerve into the pons and its course within the internal acoustic meatus are identical to that of the vestibular nerve. The dendrites of the bipolar neurons of the spiral ganglia receive auditory impulses from the organ of Corti (site of auditory neuroepithelial cells) within the cochlea and convey these impulses to the cochlear nuclei in the brainstem.

IX Glossopharyngeal nerve

The glossopharyngeal nerve (GVA, GVE, GSA, SVA, SVE) exits the medulla from the rostral portion of the postolivary sulcus and leaves the skull through the jugular foramen. It has a superior somatic sensory and an inferior visceral sensory ganglion (Figures 11.1, 11.33 & 11.34). This nerve gives off tympanic, carotid sinus, lingual, pharyngeal, tonsillar, muscular, and auricular branches.

The tympanic nerve (GVA), also known as Jacobson's nerve, arises from the glossopharyngeal nerve in the jugular foramen and forms the tympanic plexus. This nerve conveys sensations from the mucosa of the middle ear, auditory tube, and mastoid air cells to the solitary nucleus.

The lesser petrosal nerve (GVE) carries presynaptic parasympathetic fibers from the inferior salivatory nucleus of the medulla to the otic ganglion.

The carotid sinus nerve (GVA) carries baroreceptor information from the common carotid artery to the solitary nucleus, mediating the carotid sinus reflex in response to a rise in the blood pressure, or external massage of the carotid sinus in the neck. These stimuli activate the visceral afferent fibers of the glossopharyngeal nerve that establish contacts via interneurons in the reticular formation of the medulla with the dorsal motor nucleus of the vagus and simultaneously, with the neurons of the reticulospinal tracts. Then the dorsal motor nucleus of the vagus sends preganglionic parasympathetic fibers to the cardiac plexus, reducing cardiac contractility (negative chronotropic effect). The medullary reticulospinal tract inhibits or reduces

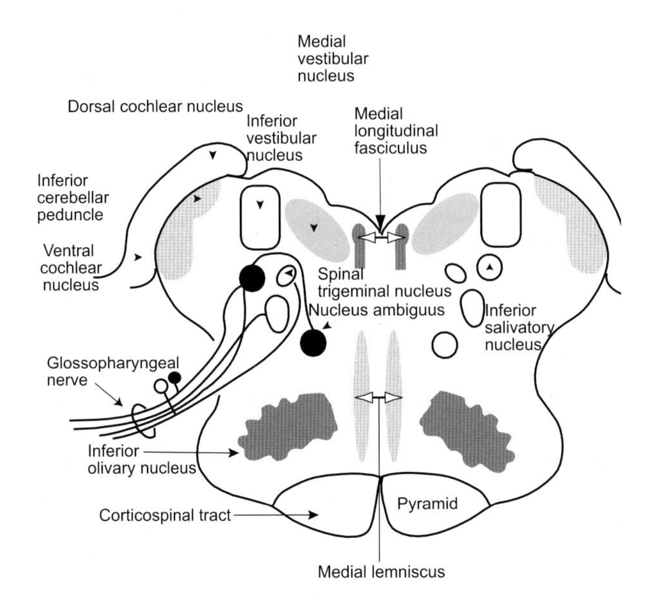

Figure 11.33 The functional components of the glossopharyngeal nerve and related nuclei are shown

the firing rate of the preganglionic sympathetic neurons that supply the cardiac plexus and the cutaneous arterioles. The decrease in the sympathetic output combined with the vagal inhibition results in a decrease of cardiac rate and output. The decrease in the peripheral vascular resistance leads to a decrease in blood pressure. This reflex is hyperactive in some individuals with vasomotor instability, and in response to a mild stimulus (carotid sinus syncope). Lesions of the glossopharyngeal or the vagus nerves abolish this reflex.

The lingual branch carries special visceral afferents (SVA) and general sensory fibers (GVA) from the posterior one-third of the tongue to the solitary nucleus.

The pharyngeal (GVA) branch joins the corresponding branches of the vagus nerve to form the pharyngeal plexus that supplies the oropharyngeal mucosa, mediating the pharyngeal (gag) reflex through its connections to the solitary nucleus and nucleus ambiguus, and the vagus nerve. The gag reflex is characterized by elevation of the stimulated part of the soft palate upon light stimulation of the oropharynx and soft palate. Unilateral loss of gag reflex may indicate damage to the glossopharyngeal nerve or the vagus nerve and may be observed in lateral medullary syndrome. In the elderly, this reflex may be reduced on both sides or may be absent.

The tonsillar branch (GVA) carries sensations from the palatine tonsils, fauces, and soft palate to the

The glossopharyngeal nerve may be damaged in fractures of the posterior cranial fossa and stenosis of the jugular foramen. Demyelination caused by multiple sclerosis, tumors of the posterior cranial fossa, aneurysm of the internal carotid artery, and injuries involving the retroparotid space may also damage this nerve. Occlusion of the posterior inferior cerebellar artery may affect the associated nuclei in the medulla, producing deficits that may also be shared by lesions of the vagus and accessory nerves. Although isolated injury to the glossopharyngeal nerve is rare, the following are some of the conditions associated with irritation, compression, or damage to this nerve.

A lesion that destroys the glossopharyngeal, vestibulocochlear, and vagus nerves, as well as the corticospinal tract may produce signs of Bonnier's syndrome, which is characterized by vertigo, nystagmus, hearing deficits, dysphonia, hoarseness of voice, contralateral hemiplegia, and tachycardia.

Vernet's syndrome may occur as a sequel to fractures of the base of the skull, involving the jugular foramen and its contents, which includes the glossopharyngeal, vagus, and the spinal accessory nerves. Disorders of this condition are analogous to the combined deficits associated with these individual nerves.

Villaret's syndrome is a condition that results from injury to the retroparotid space, involving the glossopharyngeal, vagus, accessory, and hypoglossal nerves. Sympathetic postganglionic fibers may also be disrupted in this condition, producing Horner's syndrome.

Glossopharyngeal neuralgia is a rare condition, which is characterized by spontaneous attacks of excruciating pain in the tonsillar area, posterior third of the tongue, and the external acoustic meatus, radiating to the throat, side of the neck, and the back of the lower jaw. It is provoked by yawning, laughing, chewing, or by swallowing of particularly cold liquid, and may be associated with peritonsillar abscess, oropharyngeal carcinoma, and ossified stylohyoid ligament. It is rarely bilateral and it may accompany trigeminal neuralgia. When ear pain is felt without signs of middle ear disease, oropharyngeal cancer must also be considered. This condition may be associated with episodes of fainting, syncope, and reflex bradycardia as a result of involvement of the carotid sinus nerve.

Table 11.2 Cranial nerves VII–IX

Cranial nerve	*Component*	*Location of cell bodies*	*Course*	*Distribution*	*Function*
VII Facial	SVE	Facial motor nucleus	Internal auditory meatus, facial canal & stylomastoid foramen	Facial muscles of expression, stylohyoid, stapedius & posterior belly of digastric muscle	Facial expression & increase of hearing threshold & elevation of hyoid bone
	GVE	Superior salivatory nucleus	Within the intermediate nerve via the internal auditory meatus; then via greater petrosal & pterygoid canal nerves to the pterygopalatine ganglion	Lacrimal gland, mucus glands of palate, pharynx & nasal cavity	Parasympathetic
	SVA	Geniculate ganglion	Same as above	Via chorda tympani to anterior 2/3 of the tongue	Taste
	GSA	Geniculate ganglion	Via auricular branch	Concha of ear	General somatic sensation
VIII Vestibulo-cochlear					
Cochlear	SSA	Spiral ganglion	Internal acoustic meatus	Organ of Corti	Audition
Vestibular	SSA	Scarpa's ganglion	Same as above	Receptors in the semicircular canals, utricle & saccule	Balance, orientation in three dimensions & fixation of gaze
IX Glosso-pharyngeal	GVE	Inferior salivatory nucleus	Jugular foramen	Parotid gland	Parasympathetic
	SVE	Nucleus ambiguus	Jugular foramen	Stylopharyngeus	Swallowing
	GVA	Inferior ganglion	Same as above	Carotid sinus & body; Posterior 1/3 of the tongue, oropharynx, palatine tonsils, and tympanic membrane	Baroreceptor & chemoreceptor; pain & temperature Sensations from the mucosa of these areas
	SVA	Inferior ganglion	Same as above	Posterior 1/3 of the tongue	Taste
	GSA	Superior ganglion	Same as above	Retro-auricular	General sensations

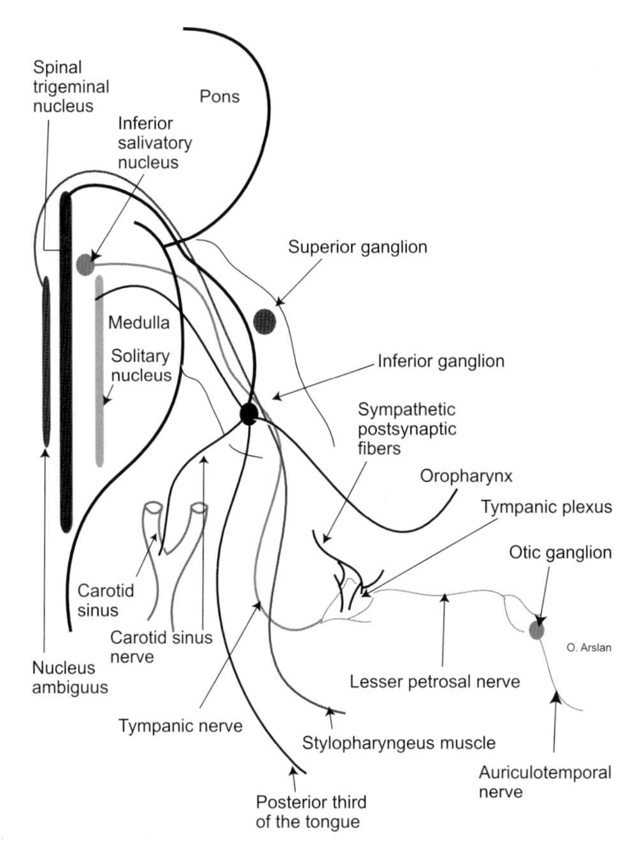

Figure 11.34 Functional components, branches, and associated nuclei of the glossopharyngeal nerve

solitary nucleus. This branch may be overactivated in glossopharyngeal neuralgia. The muscular branches (SVE) arise from the ambiguus nucleus and supply the stylopharyngeus muscle, while the auricular branch (GSA) carries general sensation from the retro-auricular area, terminating in the spinal trigeminal nucleus.

Nuclei of the glossopharyngeal nerve

The nuclei associated with the glossopharyngeal nerve (Figures 11.33 & 11.34) are comprised of the nucleus ambiguus, solitary, spinal trigeminal, and inferior salivatory nuclei.

The nucleus ambiguus (SVE) is located in the medulla and provides special visceral efferent fibers that supply the stylopharyngeus muscle. It also provides fibers to the cranial part of the accessory nerve and to the vagus nerve, innervating the laryngeal, pharyngeal, and palatine muscles. Damage to this nucleus results in deviation of the uvula to the intact side, a unique feature of this lesion.

The solitary nucleus receives taste (SVA) sensations from the posterior one-third of the tongue and general visceral sensations (GVA) from the carotid sinus, and pain and temperature sensations from the middle ear, posterior one-third of the tongue, oropharynx, and the tonsils.

The spinal trigeminal nucleus (GSA) receives cutaneous sensation from the retro-auricular area.

The inferior salivatory nucleus (GVE) parasympathetic fibers are conveyed to the otic ganglion, which eventually innervate the parotid gland.

X Vagus nerve

The vagus (Figures 11.1, 11.35 & 11.36), as in the case of the glossopharyngeal nerve, is a composite nerve with diverse functional entities. It travels ventrolaterally in the caudal medulla through the spinal trigeminal tract and nucleus and adjacent to the nucleus ambiguus and spinal lemniscus. Therefore, the vagus nerve and associated nuclei may be damaged by a single lesion in the lateral medulla, producing signs and symptoms of Wallenberg's or lateral medullar syndrome, which exhibits dysphagia, dysphonia, and alternating hemianesthesia.

This nerve leaves the medulla via the postolivary sulcus as a series of rootlets and exits the skull through the jugular foramen, accompanied by the glossopharyngeal and accessory nerves. Then it runs through the neck as the posterior component of the carotid sheath. During its course in the superior and posterior mediastina, it gives rise to branches to the cardiac and pulmonary plexuses. Later, it contributes to the formation of the anterior and posterior vagal trunks around the abdominal part of the esophagus, entering the abdomen through the esophageal hiatus of the diaphragm. In the abdomen, it contributes presynaptic parasympathetic fibers to the celiac, superior mesenteric, and aortic plexuses. The vagal parasympathetic contribution to the abdominal viscera terminates at the junction of the right two-thirds and left one-third of the transverse colon. The vagus nerve has superior (jugular) and inferior (nodose) ganglia, containing neurons for somatic and visceral sensations, respectively. Within this nerve, the motor fibers belong to the cranial part of the accessory nerve, which distribute to the laryngeal, pharyngeal, and palatal muscles.

Through its course, the vagus nerve gives rise to meningeal, auricular, pharyngeal, carotid body, superior and inferior laryngeal, cardiac, pulmonary, esophageal, celiac, and superior mesenteric branches.

The meningeal branch (GSA) conveys sensations from the dura mater of the posterior cranial fossa to the spinal trigeminal nucleus.

The auricular branch (GSA) carries somatic sensations from the external acoustic meatus to be conveyed to the spinal trigeminal nucleus.

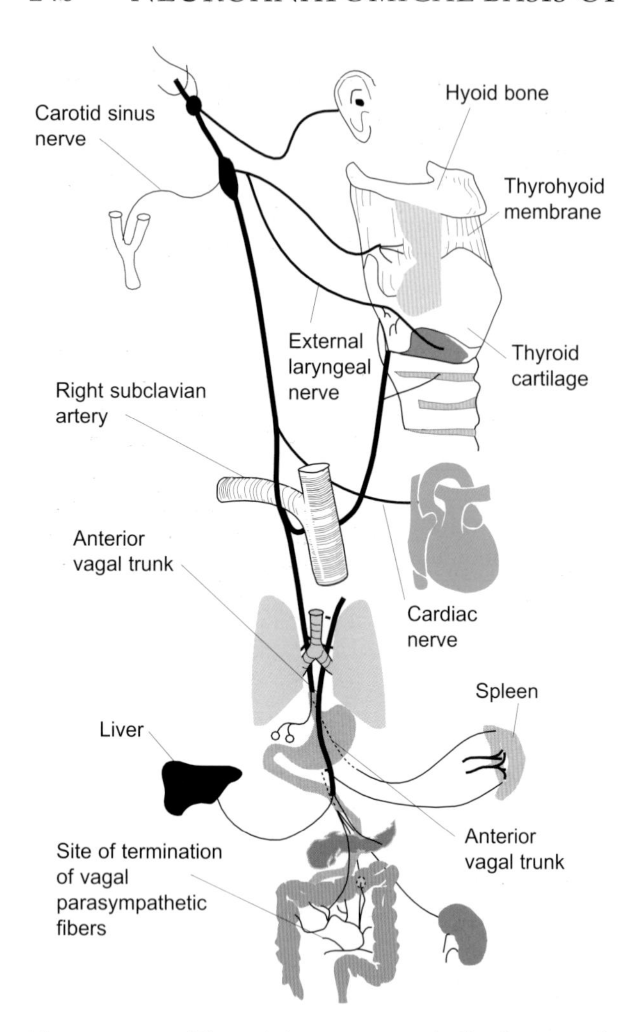

Figure 11.35 The origin, course, and distribution of branches of the vagus nerve

The pharyngeal branch (SVE) contains the cranial part of the accessory nerve that contributes to the formation of the pharyngeal plexus. Through this plexus the vagus nerve innervates most of the pharyngeal muscles (with the exception of the stylopharyngeus) and palatine muscles with the exception of the tensor palatini. Branches that distribute to the epiglottic vallecula (SVA) convey taste sensations to the solitary nucleus.

Branches to the carotid body (GVE) arise from the nodose ganglion and course as a component of the pharyngeal branch and very rarely run within the superior laryngeal branch, transmitting signals about the changes in the level of carbon dioxide and oxygen tension. These fibers join the pharyngeal branches of the glossopharyngeal nerve and the cervical part of the sympathetic trunk to form the pharyngeal plexus.

The superior laryngeal nerve (GVA, SVE) divides into internal (sensory) and external (motor) laryngeal branches. The internal laryngeal nerve (GVA, SVA) accompanies the superior laryngeal vessels in its course in the medial wall of the piriform recess, and distribute branches to the laryngeal mucosa of the vestibule, laryngeal sinus, and the epiglottic vallecula. This branch carries general sensations (GVA) from the laryngopharynx, piriform recess, and most of the laryngeal mucosa to the solitary nucleus. Taste sensation from the extreme posterior part of the tongue and the epiglottic vallecula is also conveyed by this nerve to the solitary nucleus. The external laryngeal nerve (SVE) emanates from the nucleus ambiguus and provides motor fibers to the cricothyroid muscle.

Vagal trunks may selectively be severed in vagotomy, a surgical procedure used in the treatment of chronic gastric or duodenal ulcers. This surgical approach is intended to greatly reduce hydrochloric acid secretion and thus enhance healing of the affected part of the gastrointestinal tract. It may be classified into truncal, selective, and high selective vagotomy. Truncal vagotomy may not be desirable due to the accompanied gastric stasis (dumping), atonia of the gallbladder, and impaired pancreatic secretion. In selective vagotomy, although the gastric branches of the vagus nerve, including nerves of Latarget to the antrum, are specifically cut, gastric dumping occurs, requiring surgical bypass. In high selective vagotomy only the branches to the fundus and body of the stomach (acid secreting areas) are cut and gastric dumping is thus avoided. This procedure may also induce atrophy of the oxyntic cells, rendering it unresponsive to the circulating gastrin.

The inferior (recurrent) laryngeal nerve (GVA, SVE) encircles the subclavian artery on the right side and the aortic arch on the left side and courses medial to the ligamentum arteriosum. This nerve carries general visceral sensation (GVA) from the infraglottic part of the larynx. During its course within the tracheoesophageal sulcus and medial to the thyroid gland, this nerve runs in close proximity to the inferior thyroid artery, a relationship that bears important clinical significance in thyroidectomies.

The cardiac branches contribute parasympathetic fibers (GVE) to the deep and superficial cardiac plexuses, with the superficial part of the plexus receiving innervation from the left vagus, and the deep part receiving branches from the right vagus and the recurrent laryngeal nerves. These fibers slow the heart rate and produce constriction of the coronary arteries. They also carry general visceral sensation (GVA) from the heart.

The vagus nerve is prone to damage in fractures of the posterior cranial fossa involving the jugular foramen, or by an aneurysm of the common or internal carotid artery. A unilateral lesion of the vagus nerve results in slight difficulty in swallowing (dysphagia) and breathing (dyspnea), accompanied by regurgitation of food through the nasal cavity. It also produces hoarseness and a voice of nasal quality, transient tachycardia, loss of gag reflex, and deviation of the uvula toward the intact side on phonation. In unilateral vagal dysfunction, no consistent deficits are associated with the heart, lungs, or bowel functions. Bilateral disruption of the vagus nerves results in a serious condition which manifests cardiac arrhythmia, severe difficulty in breathing (respiratory stridor), dysphagia, dysphonia, abdominal pain, and stomach distention.

Emotional stress, crowded environment, and warmth often precipitate a common condition known as vasovagal syncope. It is characterized by sweating, an aura of nausea, and loss of consciousness. It is similar to the reflex vasovagal syncope in which a fainting spell is caused by venipuncture.

To test intactness of the vagus nerve, the uvular (palatal) reflex is activated and elevation of the soft palate is produced by asking the patient to say 'ah'. The uvular (palatal) reflex is characterized by equal and symmetrical movement of the soft palate and elevation of the uvula, upon stimulation of the mucosa of the soft palate or during phonation. It is mediated by the glossopharyngeal and vagus nerves.

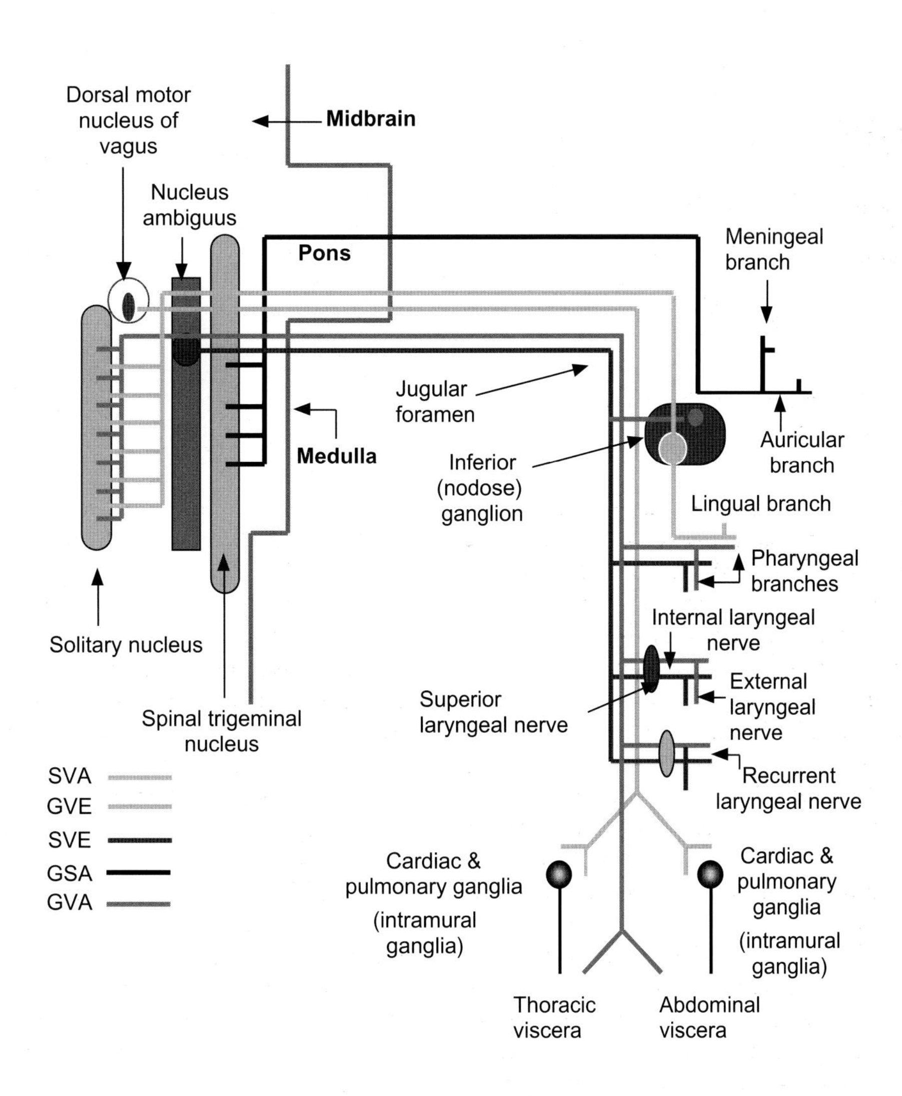

Figure 11.36 Complete descriptive diagram of the vagus nerve, nuclei, and areas of distribution

A specific lesion of the superior laryngeal nerve produces anesthesia in the mucus membrane of the vestibule and the laryngeal sinus. Tension in the affected vocal folds is lost, resulting in a monotonous voice. On the other hand, the recurrent laryngeal nerves are prone to damage in thyroidectomy upon ligation of the inferior thyroid arteries that maintain close relationships to these nerves. The left recurrent laryngeal nerve may also be damaged in an aneurysm of the aortic arch, as a result of bronchial and esophageal carcinoma, or in conditions which produce enlargement of the mediastinal lymph nodes. According to Semon's Law, progressive lesions of the recurrent laryngeal nerve produce dysfunction in the abductors of the vocal folds before any significant deficits in the adductors. In contrast, the recovery involves the adductor muscles first followed by the abductors of the vocal cords. Unilateral damage to the recurrent laryngeal nerve results in paralysis of the intrinsic laryngeal muscles with the exception of the cricothyroid, as well as ipsilateral loss of sensation from the infraglottic part of the larynx. Initially, the voice is weak and altered (like a whisper), but movement of the opposite vocal fold toward the midline may compensate for this deficit, rendering the voice fairly normal. The external laryngeal nerve may be destroyed in ligation of the superior thyroid artery, producing paralysis of the cricothyroid muscle and a monotonous voice.

The pulmonary branches that act as bronchoconstrictors and secretomotor (GVE) to the bronchial mucus glands run within the pulmonary plexus. These branches also carry information from stretch receptors (GVA) from the pulmonary bronchi.

The esophageal branches carry parasympathetic preganglionic (GVE) fibers that facilitate esophageal motility, as well as general visceral sensation (GVA) from the esophagus. These branches form the esophageal plexus and continue to the abdomen around the esophagus as the anterior and posterior vagal trunks. The anterior vagal trunk is formed primarily by the left vagus, while the posterior vagal trunk is principally derived from the right vagus.

Branches to the celiac and superior mesenteric plexuses provide parasympathetic (GVE) fibers to the stomach, small intestine, cecum, ascending colon, and the right two-thirds of the large intestine. It also provides general visceral afferent (GVA) fibers, conveying sensory modalities of hunger, nausea, thirst, and bowel fullness to the solitary nucleus.

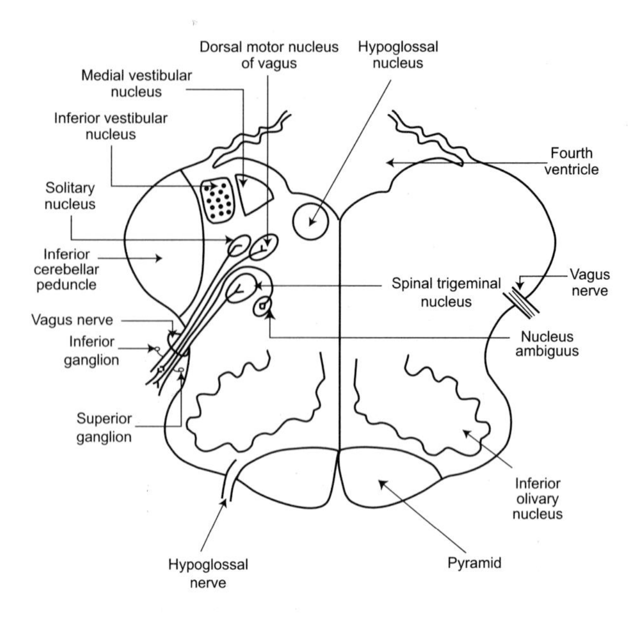

Figure 11.37 Drawing of the medulla at the level of the vagus nerve showing the associated nuclei and ganglia

Nuclei associated with the vagus nerve

The vagal nerve nuclei include the dorsal motor nucleus of vagus, nucleus ambiguus, solitary nucleus, and the spinal trigeminal nucleus (Figures 11.36 & 11.37).

The dorsal motor nucleus of the vagus (GVE) occupies the area dorsolateral to the hypoglossal nucleus, forming the vagal trigone in the floor of the fourth ventricle. This nuclear column maintains similar dimensions to the hypoglossal nucleus both caudally and rostrally, providing parasympathetic preganglionic (GVE) fibers to the thoracic and abdominal viscera.

The nucleus ambiguus (SVE) provides special visceral motor fibers that innervate the muscles of the pharynx, larynx, and soft palate.

The solitary nucleus receives general visceral (GVA), which constitute nearly 80% of the entire vagus nerve fibers, from the bronchi, gastrointestinal tract, and the carotid body. It also receives special visceral afferents (SVA – taste) from the root of the tongue and epiglottic vallecula. These afferents are conveyed to the ventral posteromedial nucleus of thalamus en route to the sensory cortex.

The spinal trigeminal nucleus (GSA) receives general somatic sensations from the external acoustic meatus, concha of the ear, and the dura mater of the posterior cranial fossa.

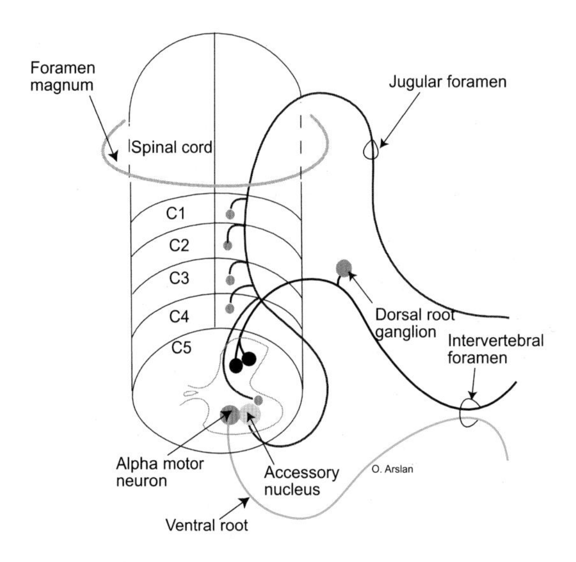

Figure 11.38 Drawing of the spinal accessory nerve; its origin, course, and areas of distribution are visible

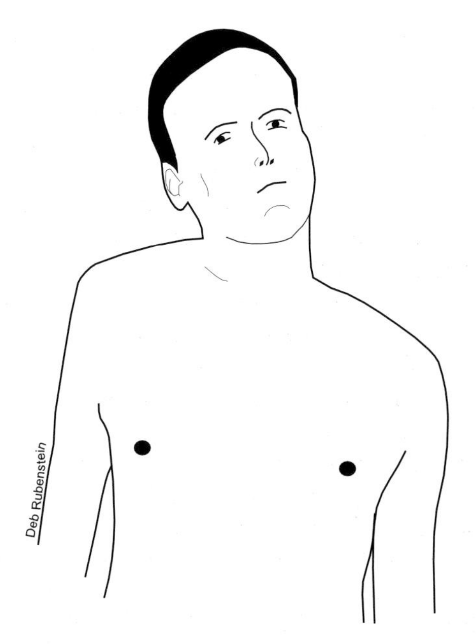

Figure 11.39 This is a depiction of a patient with right side torticollis due to spasmodic contracture of the sternocleidomastoid muscle

The accessory nerve is vulnerable to damage in radical neck dissection, a fairly common surgical procedure employed in metastatic carcinomas of the neck. The pressure exerted by calcified tuberculous cervical lymph nodes, or surgical attempts to remove these nodes may also damage the spinal accessory nerve. A stab wound in the neck, fractures of the foramen magnum or jugular foramen may also injure the accessory nerve. In addition, this nerve may also be damaged when the facemask of a football player is suddenly pulled laterally. Irritation of the spinal accessory nerve may lead to torticollis (Figure 11.39), a spasmodic contracture of the sternocleidomastoid muscle. Damage to the spinal accessory nerve may lead to paralysis of the trapezius and sternocleidomastoid muscles. Paralysis of the trapezius muscle results in winging of the scapula, which becomes more prominent upon attempt to abduct the arm on the affected side. This fact distinguishes winging of the scapula observed in long thoracic nerve damage versus accessory nerve damage. Paralysis of the sternocleidomastoid produces inability to turn the face to the opposite side.

XI Accessory nerve

The accessory nerve (Figure 11.37) has a cranial part (SVE), which is derived from the ambiguus nucleus (Figures 11.41 & 11.43), and a spinal part that originates from the upper five spinal segments (Figure 11.38). The cranial part joins the vagus nerve, distributing through its branches to the muscles of the larynx and pharynx. The spinal part forms the spinal accessory nerve, enters the cranial cavity through the foramen magnum (Figure 11.15), and leaves the skull through the jugular foramen. It courses through the posterior triangle and innervates the sternocleidomastoid and the trapezius muscles.

To test the integrity of the spinal accessory nerve, the trapezius and sternocleidomastoid functions are tested. To check intactness of the trapezius muscle, the patient is asked to shrug his shoulders against resistance. To test the sternocleidomastoid muscle, the patient is asked to turn his head to one side against resistance by the examiner.

XII Hypoglossal nerve

The hypoglossal nerve (Figures 11.40), which is derived from neuronal axons of the hypoglossal nucleus (GSE) of the medulla, descends within the reticular formation, lateral to the medial lemniscus, and between the inferior olivary and the pyramidal tract. The hypoglossal nucleus (Figures 11.40, 11.42) extends from the level of the stria medullaris to an area near the upper pole of the inferior olivary nucleus. Due to its proximity to the medial lemniscus and the pyramids, a single lesion may producing signs of medial medullary syndrome or inferior alternating

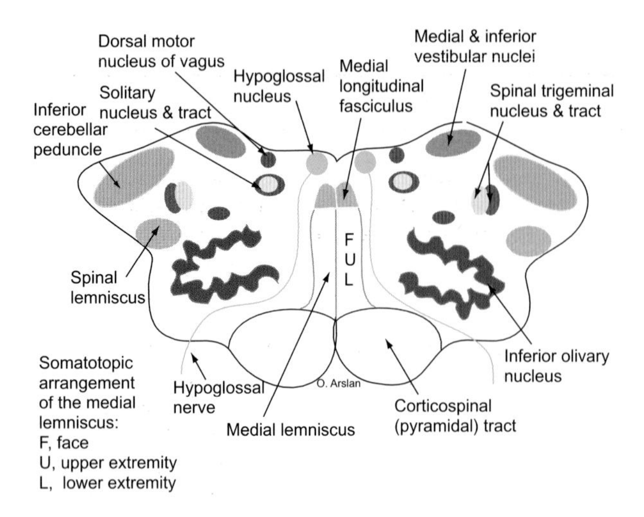

Figure 11.40 In this section of the medulla at the mid-olivary level, the hypoglossal nuclei and nerves are shown. Observe the course of the hypoglossal nerve lateral to the medial lemniscus and between the pyramid and the inferior olivary nucleus

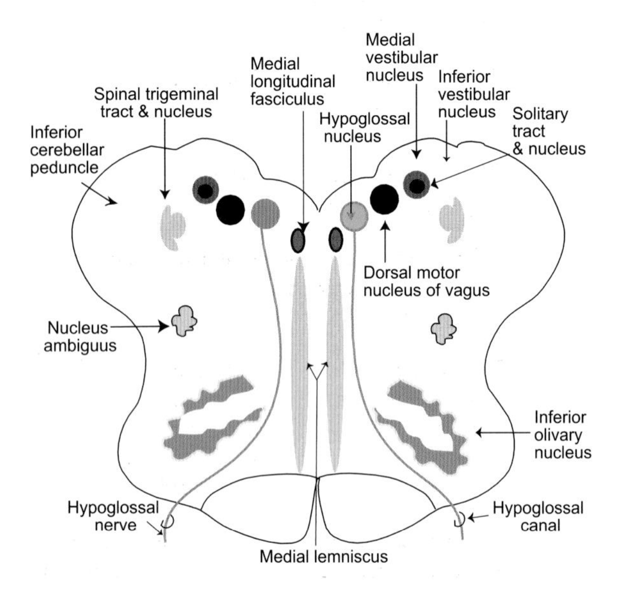

Figure 11.41 This drawing emphasizes some of the structures shown in Figure 11.40

hemiplegia, dependent upon the extent of the damage. It emerges from the medulla at the preolivary sulcus, leaves the skull through the hypoglossal canal and runs in the occipital and carotid triangles, encircling the occipital branch of the external carotid artery. In the neck, a branch of the first cervical spinal segment joins the hypoglossal to later leave as the superior root of the ansa cervicalis. All intrinsic and extrinsic lingual muscles, with the exception of the palatoglossus (innervated by the pharyngeal plexus), receive motor innervation from the hypoglossal nerve.

A unilateral lesion of the hypoglossal nerve (Figure 11.42) may produce ipsilateral atrophy of the lingual muscles and deviation of the tongue toward the lesion side upon protrusion. On retraction, the atrophied part of the tongue rises up higher than the other parts. Bilateral lesion of the hypoglossal nerves results in defective speech and difficulty in chewing. The tongue lies motionless and thus swallowing becomes very difficult, forcing the patient to extend his head back and push the bolus of food into the pharynx with the aid of his or her fingers. The hypoglossal and ambiguus nuclei may selectively be damaged in individuals with Tapia syndrome, which is characterized by paralysis of the muscles of the soft palate, posterior pharyngeal wall, vocal cords, and lingual muscles.

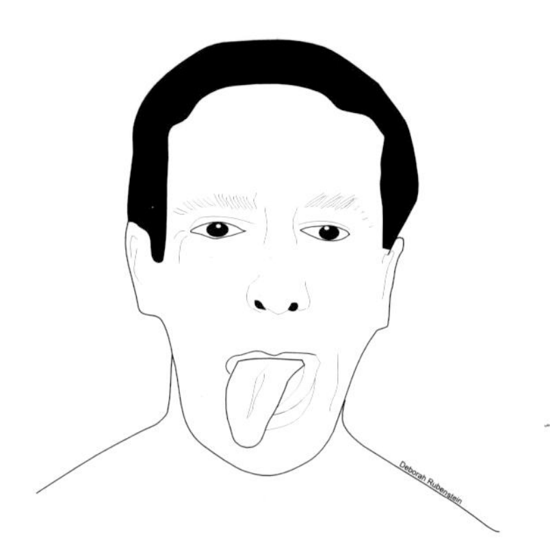

Figure 11.42 This is a drawing of an individual with right hypoglossal nerve palsy

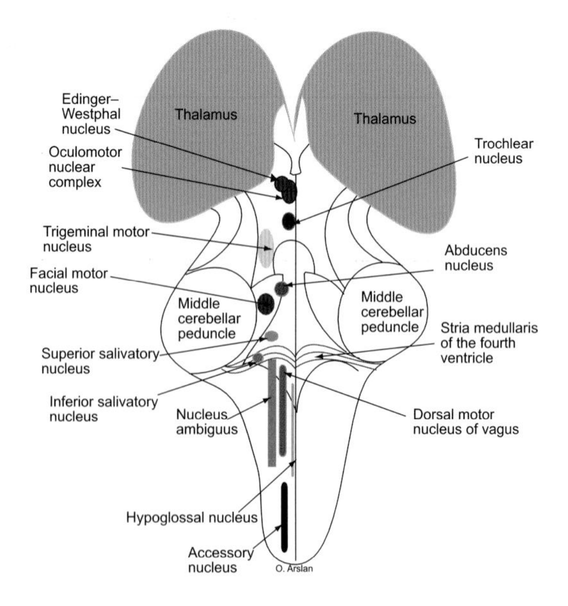

Figure 11.43 Diagram of the somatic and autonomic efferent nuclei of the cranial nerves

Table 11.3 Cranial nerves X–XII

Cranial nerve	*Component*	*Location of cell bodies*	*Course*	*Distribution*	*Function*
X Vagus	GVE	Dorsal motor nucleus of vagus	Jugular foramen	Esophageal, bronchial & muscles of the small intestine & right 2/3 of the large intestine	Secretomotor to the glandular tissue and motor to the intestinal wall
	SVE	Nucleus ambiguus	Same as above	Most muscles of the soft palate, pharynx & larynx	Movements of muscles of the soft palate, pharynx & larynx during deglutition, respiration & phonation
	GVA	Inferior (nodose) ganglion	Same as above	Visceral sensations from the pharynx, larynx, bronchi, aortic arch & body, most of the digestive tract including the right 2/3 of large intestine	Visceral sensations; chemo-receptor & pressure changes
	GSA	Superior (jugular) ganglion	Same as above	External auditory meatus & auricle	Cutaneous sensibility
	SVA	Inferior (nodose) ganglion	Same as above	Taste sensation from epiglottic vallecula	Taste
XI Accessory	SVE	Nucleus ambiguus	Jugular foramen	Within branches of the vagus to pharyngeal & laryngeal muscles	Swallowing & phonation
	GSE	Upper five spinal segments	Foramen magnum rostrally; exit via jugular foramen	Trapezius & sternocleidomastoid muscles	Elevation of scapula & shoulder point; turning the face upward & to the opposite side
XII Hypoglossal	GSE	Hypoglossal nucleus	Hypoglossal canal	Intrinsic & extrinsic muscles of the tongue	Change the shape of the tongue & maintain its movement

Table 11.4 Cranial nerves: somatic and autonomic components

				Efferent		
Nerve		*Afferent*		*Parasympathetic ganglia*	*Somatic*	*Branchial*
I	Olfactory	+	Bipolar neurons in the olfactory mucosa	None	None	None
II	Optic	+	Extension of CNS	None	None	None
III	Oculomotor		None	Ciliary	+	None
IV	Trochlear		None	None	+	None
V	Trigeminal		Trigeminal, (semilunar or Gasserian) ganglion	None	None	+
VI	Abducens		None	None	+	None
VII	Facial	+	geniculate	Pterygopalatine & submandibular ganglia	None	+
VIII	Vestibulo-cochlear	+	Scarpa's & spiral ganglia	None	None	+
IX	Glosso-pharyngeal	+	Superior & inferior ganglia	Otic ganglion	None	+
X	Vagus	+	Superior & inferior ganglia	Intramural ganglia	None	+
XI	Accessory		None	None	None	+
XII	Hypoglossal	XII	None	None	+	None

Suggested reading

1. Battrier-Ennever JA, Akert K. Medial rectus subgroups of the oculomotor nucleus and their abducens internuclear input in the monkey. *J Comp Neurol* 1981;197:17–27
2. Breen LA, Hopf HC, Farris BK, Guttmann L. Pupil-sparing oculomotor nerve palsy due to midbrain infarction. *Arch Neurol* 1991;48:105–6
3. Caliot P, Bousquet V, Midy D, Cabanié P. A contribution to the study of the accessory nerve: surgical implications. *Surg Radiol Anat* 1989;11:11–15
4. Devriese PP, Schumacher T, Scheide A, *et al.* Incidence, prognosis and recovery of Bell's palsy: a survey of about 1000 patients (1974-1983). *Clin Otolaryngol* 1990;15:15–27
5. Dulguerov P, Quinodoz D, Cosendai G, *et al.* Frey syndrome treatment with botulinum toxin. *Otolaryngol Head Neck Surg* 2000;122:821–7
6. Humphrey T. The central relations of the trigeminal nerve. In Kahn EA, ed. *The Surgery of Pain*, 2nd edn. Springfield, IL: Thomas, 1969
7. Katz AD, Catalano P. The clinical significance of the various anastomotic branches of the facial nerve.

Otolaryngol Head Neck Surg 1987;113:959–62
8. Keane JR. Isolated brainstem third nerve palsy. *Arch Neurol* 1988;45:813–14
9. Keller JT, van Loveren H. Pathophysiology of the pain of trigeminal neuralgia and atypical facial pain: a neuroanatomical perspective. *Clin Neurosurg* 1985;32: 275–93
10. Krammer EB, Rath T, Lischka MF. Somatotopic organization of the hypoglossal nucleus. A HRP study in the rat. *Brain Res* 1979;170:533–7
11. Loewry AD, Spyer KM. Vagal preganglionic neurons. In Loewry A D, Spyer KM, eds. *Central Regulations of Autonomic Functions*. New York: Oxford University Press, 1990:68–87
12. Laine FJ, Smoker WR. Anatomy of the cranial nerves. *Neuroimaging Clin North Am* 1998;8:69–100
13. Shima F, Fukui M, Kitamura K, *et al.* Diagnosis and surgical treatment of spasmodic torticollis of 11th nerve origin. *Neurosurgery* 1988;22:358–63
14. Spoendlin H, Schrott A. Analysis of the human auditory nerve. *Heart Res* 1989;43:25–38
15. Tarlov EC. Microsurgical vestibular nerve section for intractable Meniere's disease. *Clin Neurosurg* 1985;33:667–84
16. Tschabitscher M, Hocker K. The variations in the origin of cranial nerves III, IV and VI. *Anat Anz* 1991;173:45–9
17. Uemura M, Matsuda K, Kume M, *et al.* Topographical arrangement of hypoglossal motor neurons. An HRP study in the cat. *Neurosci Lett* 1979;13:99–104
18. Ullah M, Salman SS. Localization of the spinal nucleus of the accessory nerve in the rabbit. *J Anat* 1986;145:97–107
19. Wiegand DA, Fickel V. Acoustic neuromas. The patient's perspective. Subjective assessment of symptoms, diagnosis, therapy, and outcome in 541 patients. *Laryngoscope* 1989;99:179–87
20. Wilson-Pauwels L, Akesson E, Steward P. Cranial nerves: anatomy and clinical comments. Toronto: BC Decker, 1988

Neurotransmitters 12

Classical neurotransmitters are small molecules of neuroactive agents actively involved in synaptic transmission and modulation. They are synthesized in the neurons and are released at the presynaptic terminals in sufficient amounts to affect the membrane potential or conductance of the postsynaptic neurons and produce inhibition or excitation. Their effect is commonly associated with the selective opening of specific ion channels in the postsynaptic membrane and/or phosphorylation of intracellular protein. They may bind directly to a receptor and cause second messenger-mediated changes in the neurotransmission. Exogenous administration of neurotransmitters may mimic the actions of the endogenous transmitters. Certain neurotransmitters may be released after a more sustained activation. Inactivation of these agents may occur locally at the terminals by enzymatic uptake and degradation, or diffusion and release. Some neurotransmitters do not act upon the postsynaptic membrane but project its response to other neuromediators by enhancing or inhibiting their activities. Classical neurotransmitters are comprised of amino acid neurotransmitters, acetylcholine and biogenic amines.

Peptidergic neurotransmitters are scattered in the peripheral, central, and enteric nervous systems, and their synthesis is regulated by mRNA and ribosomes at the soma or dendrites. They are derived from inactive prohormone which is cleaved by certain proteolytic enzymes. Peptidergic neurotransmitters are organized structurally into families and many are neurohormones which are synthesized in neurons and released into the blood circulation, cerebrospinal fluid (CSF), or into the intercellular space by exocytosis.

Amino acid neurotransmitters
- GABA
- Glycine
- Glutamic acid

Acetylcholine

Monoamines
- Catecholamines
- Indolamines

Neuropeptides

Amino acid neurotransmitters

Amino acid neurotransmitters comprise γ-aminobutyric acid (GABA) and glycine as proven inhibitory neurotransmitters, glutamate and aspartate as putative stimulatory neurotransmitters. Others such as proline, serine, and taurine await further study to be considered as neurotransmitters. For our limited purposes we will be dealing with the first group only.

GABA

GABA is the major inhibitory neurotransmitter in the brainstem, spinal cord and Purkinje neurons of the cerebellum. It induces depolarization predominantly in the spinal cord and hyperpolarization in the cortical cells. It acts by increasing the permeability of postsynaptic membrane to chloride. GABA is produced via an irreversible reaction of L-glutamic acid and glutamic acid decarboxylase (GAD), utilizing pyridoxal phosphate (a form of vitamin B6) as a cofactor. Substantial increase in the postmortem level of GABA may be attributed to the transient activation of glutamic acid decarboxylase.

GABA is metabolized principally by GABA transaminase (GABA-T), an extensively distributed enzyme, which binds to pyridoxal phosphate, and may be inhibited by gabaculine. Transamination of GABA produces succinic semialdehyde, which is later reduced to γ-hydroxybutyrate (GHB).

GABAergic neurons are scattered in high concentrations in many brain areas such as the lateral geniculate nucleus, Purkinje cell axons that project to the lateral vestibular and the cerebellar nuclei, and also in the striatal neurons that convey impulses to the substantia nigra and cortical neurons. Most interneurons are GABAergic such as the amacrine and horizontal cells of the retina. GABAergic neurons are lacking or found in trace amounts in the peripheral nervous system. The highest concentrations of GABA are found in the diencephalon, whereas lower concentrations are localized in the cerebral hemispheres and brainstem.

Depolarization of the presynaptic neurons stimulates the release of GABA at the synaptic clefts. Reuptake into both presynaptic terminals and surrounding neuroglial cells terminates the action of GABA. Temperature and ion-dependent transport systems maintain this reuptake. In contrast to nerve terminals, GABA taken up into glial cells can not be utilized, but instead it may be metabolized to succinic semialdehyde by GABA-T. The semialdehyde is oxidized to succinate via succinic semialdehyde dehydrogenase.

Substances that decrease the amount of pyridoxine (e.g. hydrazides) or inhibits its action (e.g. sulfhydryl reagents, chloride, etc.) may repress the action of GABA and subsequently induce reversible epileptic seizures. Experimentally, localization of glutamic acid decarboxylase may be of value in determining the concentration of GABA. Development of autoantibodies against glutamic acid decarboxylase may be associated with Stiffman syndrome, a rare chronic neurological condition that exhibits progressive and fluctuating muscle spasm and atrophy.

Increased levels of γ-hydroxybutyrate subsequent to a deficiency of succinic semialdehyde dehydrogenase may occur as a manifestation of congenital disorder of GABA metabolism, producing dementia and ataxia.

Deficiency of succinic semialdehyde dehydrogenase may produce mental retardation, cerebellar disorders including hypotonia. These patients excrete copious amounts of both succinic semialdehyde and 4-hydroxybutyric acid. Deficiency of GABA-T produces deep tendon hyperreflexia, psychomotor retardation and increased height. The latter effect may be attributed to the ability of GABA to enhance release of growth hormones.

Blockage of GABAergic cortical neurons may be responsible for inducing convulsions and maintaining myoclonus. It has been suggested that lack of GABA in the substantia nigra, putamen, and caudate nucleus, subsequent to degeneration of the GABAergic neurons, may be associated with the involuntary choreiform movements observed in Huntington's disease. Additionally, overactivity of GABAergic neurons, which exert inhibitory effect on dopaminergic nigral neurons, is thought to have a role in producing some of the signs and symptoms of Parkinson's disease. Also, the toxin of Clostridium tetani may bind to the presynaptic GABAergic cells of the α motor neurons of the spinal cord and brainstem, blocking the release of GABA. Inactivation of GABA blocks the inhibitory influences on the motor neurons, resulting in muscle spasm, rigidity, locked-jaw, dysphagia, and opisthotonos.

$GABA_A$ receptors show binding sites for benzodiazepines, which comprise a group of anxiolytic drugs that act particularly on $GABA_A$-γ2 receptors. Barbiturates (e.g. phenobarbitol) and anti-epileptic drugs act on $GABA_A$-α and $GABA_A$-β receptors. These drugs increase chloride current and the duration of channel opening induced by GABA. $GABA_A$ receptors are also the major molecular target for the volatile anesthetics and possibly ethanol. Neuroactive steroids (analogs of the progesterone and corticosterone derivatives) may exert anti-anxiety, sedative and hypnotic effects via their potent modulatory effects of $GABA_A$ receptors.

Glial GABA may be recovered via the Krebs cycle where it is converted to glutamine. Glutamine is transferred to neurons where it is converted by glutaminase to glutamate, which re-enters the GABA shunt.

Overactivity of GABA produces an inhibitory effect on dopaminergic neurons by producing hyperpolarization at the postsynaptic level in the cerebral cortex and hyperpolarization at the presynaptic level in the spinal cord.

GABA receptors are localized in the neuronal cell membranes as well as astrocytes and are classified into $GABA_A$ and $GABA_B$. The $GABA_A$ receptors are more common than $GABA_B$ and belong to the same superfamily of ligand-activated receptors as the nicotinic-acteylcholine receptors. They are G-protein coupled ionotropic receptors that comprise α, β, γ, δ, and ρ subunits with additional subtypes. The ρ subunit in particular is abundant in the retina.

$GABA_B$ receptors encompass two principal types of receptors that differ in regard to location. They are coupled indirectly to calcium and potassium channels via second messenger systems (G-proteins). Inhibitory response of these receptors is produced at both presynaptic and postsynaptic levels by increased potassium or decreased calcium conductance and inhibition of cAMP production. Receptor antagonists such as picrotoxin block $GABA_B$ receptors, which mediate postsynaptic inhibitory potentials (IPSPs). They have selective affinity to baclofen, a GABA-analog (β-(4-chloro-phenyl-γ-aminobutyric acid), which releases intracellular GABA, but not to bicuculline, and are not affected by benzodiazepines or barbiturates. Both groups of drugs increase GABA-induced chloride current either by intensifying the frequency of channels opening or prolonging its duration.

Some investigators suggest that improved cognitive ability may be achieved by blocking the $GABA_B$ receptors that increase $GABA_A$ and neuronal excitability of hippocampal neurons and thus improve memory encoding.

Glycine

Glycine, the simplest of all amino acids in structure, is formed from serine in a reaction catalyzed by serine trans-hydroxy-methylase. It may also be formed by transaminase reaction with glutamate. Glycine is an essential component in the metabolism of peptides, proteins, nucleic acids, and porphyrins. Transmitter glycine potentiates synaptic activity by hyperpolarizing the membrane and increasing chloride permeability. Its inhibitory action is similar to that of GABA, but mainly restricted to the spinal motor neurons and interneurons. It can be blocked by strychnine and not by bicuculline or picrotoxin. Glycine is found in high concentrations within the interneurons of the ventral horn of the spinal cord. It is released from the presynaptic terminals of primary afferent fibers of the retina, pons, and medulla. It has been shown that glycine has a role in both increasing the frequency of NMDA (N-methyl-D-aspartate) receptor channel opening and also preventing desensitization of these receptors, without involving the glycine receptors. β-alanine, taurine, L-alanine, L-serine and proline activate glycine receptors.

The delay in muscle relaxation following voluntary contraction or percussion is seen in myotonia, a condition that exhibits abnormally slow relaxation of the skeletal muscles following active contraction. These changes are ascribed to the decrease in chloride conductance caused by reduction of glycine at synaptic clefts and its excretion in urine. Spasticity in antigravity muscles and hyperreflexia observed in upper motor neuron palsies are also thought to be mediated by glycine.

Glutamic acid

Glutamic acid, the most abundant amino acid in the central nervous system (CNS), is a precursor for GABA, and a short-acting excitatory neurotransmitter. This neurotransmitter is involved in the formation of peptides and proteins as well as detoxification of ammonia in the cerebral cortex. The L-glutamate form of this neurotransmitter is synthesized in the nerve terminals via the Krebs cycle and transamination of α-oxyglutarate and also from glutamine in the glial cells. Glial cells and nerve endings release glutamic acid via a calcium-dependent exocytotic process. It has been found to have a powerful depolarizing effect on the neurons in all areas of the CNS. The sensitivity of glutamate receptors to glutamate agonists N-methyl-D-aspartate (NMDA) is utilized as a basis in the of ionotropic glutamate receptors into NMDA and non-NMDA receptors. There are also metabotropic glutamate receptors, which exert effects via G-protein. Neuronal dysfunctions associated with anoxia, seizures or hypoglycemia may be due to the disproportionate inflow of calcium ions through the NMDA receptor channels and dramatic sustained increase in the level of glutamate (excitotoxicity). Anoxia impairs the sodium/potassium pump by reducing ATP, which is followed by excessive increase in the level of potassium concentration in the extracellular space, promoting depolarization of neurons and inhibiting glutamate uptake and its release by reversing the glutamate transporter. This positive feedback system leads to a dramatic increase in the extracellular glutamate concentration. During this process vast influx of sodium ions via both NMDA and non-NMDA receptor channels may enhance cellular necrosis by increasing the water content of the neuron (cytotoxic edema). Calcium ions gain access through NMDA receptors and voltage-dependent calcium channels and possibly via AMPA (α-amino-3-hydroxy-5-methyl-4-isoxazole propionic acid). Therefore, glutamate receptor antagonists, calcium channel blockers, and antioxidants can achieve suppression of calcium influx. Glutamate produces neuronal necrosis by utilizing calcium and free radicals as mediators. Calcium ions' role in neuronal death occurs initially upon activation of proteases and endonucleases, which lead to proteolysis of the microfilaments and eventual destruction of DNA. These ions also induce neuronal damage by enhancing the release of toxic hydroxyl free radicals via stimulation of NO synthetase (NOS) and phospholipase A_2.

AMPA and NMDA receptor agonists may become excitotoxic in the presence of bicarbonate. This observation is based on individuals with Guam disease, manifesting clinical signs of amyotrophic lateral sclerosis, Parkinsonism and senile dementia complex. This disease occurs in the inhabitants of island of Guam (Chamorros) who consume seeds of cycad-cycas circinalis (false sago) as a food source. The latter contains the neutral amino acid β-N-methyl-amino-L-alanine (BMAA) which exhibits an affinity for NMDA receptors, but does not display *in vitro* toxicity or direct excitatory effect.

Some scientific views indicate that the biochemical changes observed in Huntington's chorea may be directly or indirectly related to the glutamate actions as a neurotransmitter. This observation is based upon the experimental injection of NMDA agonists into the striatum and the subsequent neuronal loss that seems similar to that of Huntington's chorea. This is supported by the fact that NMDA receptor-mediated neurotoxicity may result from presynaptic and postsynaptic abnormalities that affect glutamate synapses and render some cells prone to toxicity by the normal glutamate levels. Lower concentrations of glutamate transporters may lead to toxic extracellular glutamate levels. Inhibition of glutamate release and blockage of NMDA and kainate-mediated processes are the mechanisms utilized by medications for amyotrophic lateral sclerosis such as riluzole.

Substance that act like an AMPA agonist and stimulate its receptors can selectively produce destruction of the upper and lower motor neurons. This is evident in individuals with neurolathyrism, an upper motor neuron disorder that occurs in the inhabitants of certain parts of Africa and Asia subsequent to a dietary reliance on chickpea (*Lathyrus sativus*). The toxin that produces this deficit is β-N-oxalylamino-L-alanine (BOAA).

In global ischemia, which occurs in individuals with cardiac arrest, neuronal death (e.g. in the CA1 pyramidal neurons of the hippocampal gyrus) may also be significantly reduced when AMPA receptor antagonists are administered within the first 24 hours of ischemic episodes. Combination of therapeutic hypothermia and the administration of an NMDA antagonist may achieve neuroprotection in instances of global ischemia. Neuronal protection of the penumbra (area in the immediate surrounding of the site of focal ischemia) can be accomplished to a remarkable degree by the administration of NMDA antagonists in the first few hours of the insult. Experimental evidence points to the significant role that AMPA antagonists can play in neuronal protection following focal ischemic insult.

Since NMDA receptor antagonists appear to prevent the induction of epileptic seizures, the role of NMDA receptors in induction of epilepsy has been strongly suggested.

Reproduction of the positive and negative symptoms of schizophrenia by phencyclidine and other NMDA receptor antagonists and altered postmortem levels of glutamate in schizophrenic brains may support the role of glutamate in this psychiatric disorder.

Glutamate is also released at synaptic sites where long-term potentiation (LTP), a sustained increase in amplitude of excitatory potentials, occurs. LTP, which is thought to be associated with long-term memory, may be facilitated by the depolarization, calcium influx in the hippocampal neurons. It may also be associated with calcium-calmodulin dependent protein kinase and protein kinase C.

Acetylcholine

Acetylcholine is the first neurotransmitter to have been identified in the CNS. Extensive distribution of this neurotransmitter in the limbic system, interneurons of the neostriatum, α motor neurons of the spinal cord and basal nucleus of Meynert has been subsequently confirmed. In the brainstem, the parabrachial nuclear complex, which is located dorsolateral to the superior cerebellar peduncle, contains high concentration of the cholinergic neurons. A prominent subnucleus of this complex, the pedunculopontine nucleus, is involved in the generation of rhythmic movements by projecting to the spinal cord. This neurotransmitter is also found in the retina, cornea, motor nuclei of the cranial nerves, and autonomic ganglia. In general the function of acetylcholine varies with the site of its activity. It maintains an excitatory function in the central and the peripheral nervous systems (with the exception of its inhibitory effect upon the cardiac muscles). Acetylcholine is formed in the cytosol by the reversible reaction of acetyl CoA and choline and conversion of acetyl CoA to CoA which is catalyzed by choline acetyltransferase (ChAT). Choline is derived from the degradation of acetylcholine by acetylcholinesterase within the synaptic cleft and also from the breakdown of phosphatidyl choline from other membrane sources. Lecithin produces a greater and longer lasting rise in plasma choline levels. In general, the amount of acetylcholine at any given moment is dependent upon the amount of calcium influx and the duration of the action potential. Acetylcholinesterase, which is widely distributed in neuronal and non-neuronal tissues, exists in several molecular forms and catalyzes the hydrolysis of acetylcholine.

Cholinergic receptors, found in the brain and spinal cord, are classified into antagonistic muscarinic and nicotinic receptors.

The muscarinic receptors have several transmembrane sparing regions that are linked to GTP-binding protein (G-protein coupled receptors) which respond slowly, stimulated by muscarine and are blocked by atropine or scopolamine. They cause inhibition of adenyl cyclase, stimulation of phospholipase C and regulation of ion channels. There are five muscarinic receptors, which display relatively slow response times to acetylcholine binding. They are categorized into M_1–M_5. The M_1 receptors are postsynaptic, excitatory, and remain particularly unaffected in Alzheimer's disease. On the other hand, M_2 receptors are presynaptically inhibitory, regulating the release of acetylcholine. This group of receptors is reduced in Alzheimer's disease and their binding sites in the hippocampal gyrus, cerebral cortex, and striatum show appreciable age-related decrease. However, major impairments were detected with M_2 control of DA release.

Nicotinic receptors respond quickly, are excitatory, and are activated by nicotine. These receptors are blocked by curare drugs or excess of nicotine or

Cholinergic neurons are thought to have crucial role in learning, memory, and cognitive ability. The role of cholinergic neurons in short-term memory is also assumed on the fact that centrally acting muscarinic blockers such as atropine and scopolamine may produce loss of memory and inability to execute learned tasks. On the contrary, chemicals that inhibit acetylcholinesterase (e.g. physostigmine) may elicit the opposite responses.

Cholinergic over-activation, which may be responsible for the clinical signs of Parkinson's disease, occurs as a result of either decreased dopaminergic activity or increased amount of acetylcholine. Administration of physostigmine increases the striatal acetylcholine concentration and often contributes to the exacerbation of Parkinson's disease.

Treatment of pre-eclampsia (hypertension-induced nervous disorders, e.g. seizure coma, etc.) is based upon the fact that magnesium sulfate, the drug of choice, acts by inhibiting acetylcholine release.

Inhibition of the release of acetylcholine by the toxin of *Clostridium botulinum*, a calcium-dependent substance, occurs peripherally by binding to the external receptors at the synapse sites. Botulinum toxin cannot penetrate the central nervous system or exert any direct influence upon the central cholinergic neurons. Excessive inhibition of acetylcholinesterase produces a surplus of acetylcholine that binds to receptors, leading to exhaustion.

acetylcholine, and are found in the neuromuscular junction, autonomic ganglia, cortex and thalamus. They are desensitized by continued exposure to agonists. Nicotinic receptors have distinct subunits which include two α, β, γ, and δ containing four membrane-spanning α helices. It has been suggested that these subunits are arranged around an ion channel that remains closed at rest, but opens when the acetylcholine binds to a subunit of these receptors. Neuronal nicotinic receptors that contain α2–α6 subunits are distinct from non-neuronal receptors, showing resistance to α-bungarotoxin and related α-neurotoxin.

Cholinergic neurons within the nucleus basalis of Meynert, which project to wide areas of cerebral cortex, receive afferents from the limbic system and the hypothalamus and are involved in the ascending reticular activating system.

Monoamines

Monoamines (biogenic amines) are comprised of the catecholamines and the indolamines. Monoamine transmitters are stored in synaptic vesicles, their release is a calcium-dependent processes, and they may be metabolized by monoamine oxidase (MAO). Plasma membrane transporter terminates synaptic action of the monoamine transmitters.

Catecholamines

Catecholamines are derivatives of β-phenyl ethylalanine, with hydroxy groups on the third and fourth positions. They are synthesized by tyrosine, a common precursor for norepinephrine, epinephrine, and dopamine. In the central and peripheral nervous systems, transformation of tyrosine via a series of chemical changes may lead to the formation of norepinephrine, dopamine, or epinephrine, a process which is dependent upon tyrosine hydroxylase and dopamine-β hydroxylase. Release of catecholamines is stimulated by the influx of calcium.

Catecholamines are formed by L-tyrosine, which is converted to L-dopa (levodopa), via hydroxylation by tyrosine hydroxylase (rate-limiting enzyme). L-dopa is converted to dopamine following decarboxylation by an aromatic amino acid decarboxylase. Dopamine is either stored in the vesicles or hydroxylated to L-norepinephrine by dopamine-β-hydroxylase. L-epinephrine is formed in the adrenal medulla from norepinephrine by the enzyme phenyl-ethanolamine-N-methyltransferase. Norepinephrine and dopamine are metabolized (inactivated) by MAO and catechol-O-methyl transferase (COMT). Inactivation of norepinephrine occurs by the re-uptake mechanism into the presynaptic nerve terminals.

Enzymes which are involved in metabolic degradation of the catecholamines such as monoamine oxidase (MAO) and catechol-O-methyl transferase (COMT) maintain an important role in the expression of emotion. Lower levels of catecholamines may produce depression, while higher levels produce euphoria.

Catecholamines are found in the brain, chromaffin tissue of the adrenal medulla, and the sympathetic nervous system maintaining massive projections throughout the brain. Amine transmitters have slow modulatory influences and most of their receptors are part of the G-protein coupled family. They may play a role in the regulation of visceral activities, emotion, and attention. Norepinephrine is the primary neurotransmitter in the sympathetic (peripheral nervous) system, while dopamine, serotonin, and norepinephrine act primarily in the CNS.

Adrenergic receptors are classified into alpha (α) and beta (β) receptors. Although activation of both receptors produces inhibition of gastrointestinal tract motility, their classification in general is based on their respective excitatory and inhibitory effects on smooth muscles. These receptors are further subdivided into α_1 and α_2 and β_1, β_2 and β_3. Activation of α_1 noradrenergic postsynaptic receptors produces vasoconstriction, while activation of the presynaptic α_2 receptors may inhibit the release of norepinephrine, a process unaffected by pertussis toxin, which inhibits G-protein. α_1 receptors are inhibited by

Norepinephrine regulates the degree of arousal, mood, memory and learning, and also modulates sound transmission (sharpening effect). An excess of norepinephrine has been shown to elicit euphoria, while its depletion may produce depression. The effects of mood elevating drugs (antidepressants) such as MAO inhibitors, re-uptake blockers such as reserpine (antidepressant) and amphetamines (sympathomimetics) is based upon the role of these agents in either increasing the concentration or depleting the endogenous norepinephrine. MAO action on norepinephrine results in the formation of vanillylmandelic acid (VMA), a readily detectable product in the urine. Measurement of VMA levels may bear diagnostic value in conditions such as pheochromocytoma and neuroblastoma.

Checking the levels of 3-methoxy-4-hydroxy-phenethyleneglycol (MHPG), a major CNS metabolite of norepinephrine that is found in the urine, blood, and cerebrospinal fluid (CSF), may assess the functional activity of the central adrenergic neurons.

prazosin, an α_1 blocking substance. β receptors are closely linked to adenyl cyclase activation via G-protein. Stimulation of the β receptors results in changes which include vasodilatation of the coronary and abdominal arteries, and relaxation of the ciliary, gastrointestinal, and detrusor muscles. They also produce activation of glycogenolysis, as well as dilatation of the bronchi. Activation of the β_1 receptors produces increased rate and strength of cardiac contractility as well as increased secretion of renin, whereas activation of β_3 enhances lipolysis. Low concentrations of epinephrine activate presynaptic β_2 receptors.

Norepinephrine

Norepinephrine in the CNS is concentrated in neurons of the locus ceruleus of the rostral pons and caudal medulla and the lateral tegmental nuclei. These noradrenergic neurons project to the entire cerebral cortex, hippocampus, cerebellum, and spinal cord as well as basilar pons and ventral medulla. It is released in small amounts from the adrenal gland during circulatory collapse. In dystonia, the level of norepinephrine drops in the red nucleus, hypothalamus, mamillary body, locus ceruleus, and subthalamic nucleus.

Norepinephrine has potent excitatory but weak inhibitory effects on smooth muscles. It has also stronger affinity than epinephrine to β_3 receptors. Noradrenergic neurons in the brainstem reticular formation are classified, according to Dahlström and Fuxe, into A1 through A8 groups. The A1 group projects to the spinal cord, solitary nucleus and hypothalamus and is located in the caudal lateral medulla. A5 lies in the caudal pons near the superior olivary nucleus, and projects to the intermediolateral column of the spinal cord. The locus ceruleus, which is designated as group A6, occupies the rostral pons and caudal midbrain, projecting via the central tegmental tract, medial forebrain bundle, superior cerebellar peduncle, and the tectospinal tract.

The extensive and global projection of the locus ceruleus to the intralaminar thalamic nuclei via the ascending reticular activating system may account for its role in paradoxical (REM) sleep.

Norepinephrine is thought to have a role in stiff man syndrome, which is characterized by uncontrollable muscular spasm and stiffness. This suggestion is based upon the fact that an increased level of 3-methoxy-4-hydroxyphenoglycol, a metabolite of norepinephrine, is detected in the urine of the affected individuals.

Agonists and antagonists of a receptor exert influences on the firing rate of locus ceruleus neurons. α_2 agonists such as clonidine inhibit the firing of locus neurons in contrast to yohimbine and iadazoxane (α_2 antagonists) which facilitate this neuronal activity. The A7 group is located in the lateral pontine tegmentum of the isthmus sending projections to the spinal cord. In the peripheral nervous system, norepinephrine stimulates presynaptic and postsynaptic receptors.

Epinephrine

Epinephrine is released into the blood stream from chromaffin cells of the adrenal medulla. This neurotransmitter stimulates the vascular smooth muscles, and produces a rise in blood glucose concentration. Thus, arousal from insulin coma may be achieved by activation of glycogenolysis. Epinephrine has a stronger affinity for β adrenergic receptors in the smooth muscles of the vessels, bronchi, gastrointestinal tract and urogenital system. Similarly, it also has stronger affinity to α_1 and α_2 receptors than norepinephrine. Epinephrine-containing neurons are classified into group C1 in the lateral tegmentum, C2 in the dorsal medulla, and C3 in the medial longitudinal fasciculus. Certain noradrenergic neurons of the C1 and C2 groups project to the hypothalamus via the central tegmental tract and periventricular gray.

Dopamine

Dopamine, representing approximately 50% of the total catecholamines, is formed by conversion of tyrosine to L-dopa tyrosine hydroxylase and then to dopamine via L-aromatic amino acid decarboxylase. Orally administered tyrosine does not increase dopamine levels. However, orally administered levodopa (L-dopa-L-dihydroxyl phenylalanine) is

Monoamine oxidase-B (MAO-B) oxidizes and selectively increases the level of dopamine. Inhibitors of this enzyme such as amphetamines have a profound mood-elevating effect, and may even produce agitation.

Degeneration of the nigrostriatal fibers and depletion of dopamine accounts for the manifestations of Parkinson's disease. In Parkinson's disease, the mechanism of degradation of dopamine is maintained but its synthetic machinery is impaired. Dopaminergic neurons in the ventral tegmental nuclei of the midbrain, which are localized medial to the substantia nigra, form the mesolimbic system, a group of neurons that maintain diffuse projections to the septal area, amygdala, entorhinal area, nucleus accumbens septi, olfactory tubercle, and the pyriform cortex. The role of these projections in controlling mood and emotion accounts for the psychiatric disorders that accompany L-dopa therapy.

It has been suggested that overactivity of the mesolimbic dopaminergic neurons may be associated with schizophrenia. Antipsychotic drugs may increase dopaminergic neuronal activity and dopamine synthesis by blocking the postsynaptic dopaminergic receptors. Prolonged usage of these medications may lead to tolerance development. On the same basis, the inhibitory action of neuroleptic drugs upon these neurons may explain the improvement seen in patients with these disorders, as well as the unwanted side motor disorders.

absorbed from the small intestine by active transport and later converted in the dopaminergic nigral neurons into dopamine. Dopaminergic neurons are concentrated in the tuberal nuclei of the hypothalamus, nucleus accumbens, olfactory tubercle, midbrain, carotid body, and the superior cervical ganglion. The latter also contains both cholinergic and noradrenergic neurons. In general, the distribution of dopamine parallels that of norepinephrine, with dopaminergic neurons outnumbering the noradrenergic neurons at the ratio of 3 to 1. High concentrations of dopamine and low concentrations

Dopaminergic neurons located in the tuberal nuclei of the hypothalamus project to the median eminence via the tuberoinfundibular tract and then to the adenohypophysis via the portal-hypophysial system. These projections are postulated to regulate the secretion of prolactin and melanocyte-stimulating hormone (MSH). Dopaminergic neurons in the retina and olfactory bulb may play a role in the phenomenon of lateral inhibition, which sharpens the visual and olfactory impulses and prevents neuronal cross talk.

Dopamine transporter (DAT), a sodium/calcium dependent plasma protein, may bind to drugs like cocaine and amphetamine, producing behavioral and psychomotor changes. Thus, cocaine overdose may be treated by antagonists that prevent this binding. Substances that show high affinity for DAT such as 1-methyl, 4-phenyl 1,2,3,4,6-tetrahydropyridine (MPTP) may prove to be toxic to dopaminergic neurons.

D_1 receptors mediate the dopamine-stimulated increase of adenylate cyclase and subsequently intracellular cAMP. The role of D_2 receptors in motor activity, independent of adenylate cyclase, is clearly illustrated in the reduction of both motor and vocal tics upon administering drugs that act upon these receptors.

Additionally, the newer generation of antipsychotics such as clozapine (Clozaril™) are thought to act upon D_2 receptors, resulting in selective inactivation of the dopaminergic neurons in the ventral tegmentum, but not in the neurons of the substantia nigra. These drugs may also have fewer side-effects than the older generation of antipsychotics, which are known to produce tardive dyskinesia in some patients. D_3 and D_4 receptors maintain primary functional relationships to the limbic system and the telencephalon and secondary connections to the basal nuclei. The anti-psychotic drug clozapine's strong affinity to D_3 and D_4 receptors may account for the suppression of unwanted subcortical motor side-effects.

of norepinephrine exist in the caudate nucleus and putamen, whereas the reverse occurs in the hypothalamus.

Dopaminergic neurons in the pars compacta of the substantia nigra exert inhibitory influence. The pars compacta synthesizes dopamine and delivers it to the neostriatum (caudate and putamen) via the nigrostriatal fibers.

Dopaminergic neurons in the ventral tegmentum and substantia nigra form the mesocortical pathway that projects to the prefrontal cortex (involved in motivation, attention and social behavior) and entorhinal area. In contrast to the mesolimbic system, these neurons do not develop tolerance to continued usage of antipsychotic medications. Midbrain dopaminergic neurons, which are components of the mesocortical system, lack autoreceptors that regulate impulse traffic. They are generally presynaptic and respond to the same transmitter utilized by the neuron that contains them. They have faster firing rate than the mesolimbic dopaminergic neurons and

Dopamine's possible role in Gilles de la Tourette syndrome (hereditary multiple tic disorder), a childhood neurological condition which exhibits multiple motor and vocal tics and compulsive utterance, is based upon the improvement seen in patients with this disease following administration of dopaminergic antagonists.

In psychiatric patients using certain neuroleptic drugs, a reduction of dopamine concentration in the substantia nigra may occur as a result of manganese intoxication or ingestion of levodopa, which produces hydrogen peroxide and hydroxyl free radicals, leading to involuntary motor activities.

are less affected by the dopamine receptor blocking agents such as haloperidol.

Dopaminergic receptors are classified on an anatomical and functional basis into D_1 and D_2 receptors; within the D_1 group of receptors D_3, D_4, and D_5 subtypes exist. D_5 and D_1 have similar concentrations in the hypothalamus and temporal lobes. D_4 subtype is particularly identified in the brains of schizophrenics. D_1 and D_2 receptors in the caudate and putamen showed detectable increase in density in Parkinsonism and decrease in Huntington's chorea.

Indolamines

Indolamines comprise serotonin and histamine.

Serotonin

Serotonin (5-hydroxytryptamine) is synthesized by hydroxylation of tryptophan and carboxylation of the product of the reaction by tryptophan hydroxylase. The latter enzyme also catalyzes the carboxylation process that forms serotonin and converts dopa to dopamine. The level of tissue oxygen, pteridine, and tryptophan (cofactors or substrate) may also influence the rate of 5-HT formation. Although presence of this neurotransmitter in the central nervous system represents a small fraction of its total concentrations in the whole body, serotonin is found in high concentration in the raphe nuclei, spinal cord, and in the hypophysis cerebri, where it is converted into melatonin by acetylation and methylation. Serotonin is actively transported from the cytoplasm to the storage vesicles. The vesicular transport involves vesicular transporter-1 and -2 (VMAT1 and VMAT2) which also function as antiporters to eliminate cytoplasmic toxic materials. For this very reason, vesicular transporters are known as toxin extruding antiporters (TEXANs). Serotonin is stored in vesicles which do not contain ATP, but instead contain specific protein that binds to 5-HT (serotonin binding protein) with high affinity in the presence of Fe^{2+}. Release of serotonin is thought to occur via exocytosis and its rate is determined by the firing rate of serotoninergic soma in the raphe nuclei. Synaptic actions of 5-HT are terminated by binding of these molecules to specific transporter proteins on the serotoninergic neurons.

Changes in 5-HT function have been implicated in the affective disorders schizophrenia, migraine, sleep and anxiety disorders. Neurotransmission at 5-HT receptors may be blocked by antidepressants (e.g. fluoxetine), hallucinogens (e.g. LSD), anxiolytics (e.g. buspirone), antiemetics (e.g. ondansetron), and antimigraine drugs (e.g. sumatriptan).

Serotonin receptors generally operate via a G-protein and are classified according to their coupling to second messengers and their amino acid sequence homology. The 5-HT_1 group of receptors are negatively coupled to adenylate cyclase via the G_i family of proteins, exhibiting high binding affinity to [3H] 5-HT and mediating inhibition.

The 5-HT_1 family of receptors is further classified into 5-HT_{1A}, 5-HT_{1B}, 5-HT_{1D}, 5-HT_{1E}, and 5-HT_{1F}. The 5-HT_{1A} receptors are localized mainly in the hippocampus (CA1 sector), amygdala, neocortex, hypothalamus, and raphe nuclei, mediating emotion. Neuronal hyperpolarization of this group of receptors is achieved by inhibiting adenylate cyclase activity and/or opening of the potassium channels via G_i protein or by inhibition of calcium channels via G_O protein. 5-HT_{1A} autoreceptor agonists such as R(1)8-OH-DPAT are effective in inhibiting the neuronal activities of the midbrain raphe nuclei and the CA1 pyramidal cells of the hippocampal gyrus. Other members of 5-HT_{1A} are also negatively coupled to adenylate cyclase. 5-HT_{1B} also utilizes G_i protein and mediates inhibition of adenylate cyclase. The same principles apply to 5-HT_{1C}, 5-HT_{1D}, and 5-HT_{1E}. Binding sites for [3H] 5-HT in the choroid plexus termed 5-HT_{1C} subtype, whereas the binding sites for [3H] 5-HT in the bovine brain designated as 5-HT_{1D} subtype. Inositol phosphate is liberated by the activation of 5-HT_{1C}, which leads to the opening of the calcium-dependent chloride channel.

The 5-HT_2 group contains receptors that maintain amino acid homology and are coupled to phospholipase C possibly via G_q protein. They can selectively be blocked by ketanserin and ritanserin and may produce excitatory effects, displaying high affinity to H_3-spiperone. 5-HT_2 antagonists may block the excitatory effects of glutamate and 5-HT receptors in the

5-HT_6 receptors exhibit high affinity for LSD, antipsychotic and tricyclic antidepressants such as clozapine, clomipramine, mianserin, and ritanserin, and positively couple to adenylate cyclase.

Uptake of 5-HT is accomplished by an active process that utilizes Cl^- and Na^+ and remains temperature-dependent. Therefore, inhibitors of Na^+–K^+–ATPase may impair the uptake process of serotonin. Tricyclic antidepressants, such as imipramine and amitriptyline, inhibit reuptake of both serotonin and norepinephrine. Therefore, selective serotonin reuptake inhibitors (SSRIs) may also be used for the treatment of clinical depression. It has been suggested that serotonin reuptake inhibitors may alleviate symptoms of obsessive–compulsive disorder (OCD). Atypical antipsychotic drugs such as clozapine may inhibit central dopaminergic neurons but primarily maintain more powerful antagonistic action on 5-HT_{2A} and 5-HT_{2C} receptors.

Agonists at 5-HT_{2A} and 5-HT_{2C} receptors may be responsible for the hallucinogenic activity of certain drugs such as LSD. Sumatriptan, an effective medication in the treatment of migraine, is thought to derive its therapeutic role from the agonist action at 5-HT_{1D} and 5-HT_{1F}. Depletion of 5-HT content of serotonin is correlated with the tranquilizing action of reserpine, a hypotensive drug, which also depletes norepinephrine and dopamine contents.

Chemotherapeutic agents, such as cisplatin and dacarbazine, induce severe forms of nausea and vomiting. This results from a series of events that involve release of 5-HT from the chromaffin tissue of the gastrointestinal tract and the enteric nervous system. Released 5-HT specifically activates 5-HT_3 receptors, which are ligand-gated ion channel receptors, producing depolarization of the afferent nerves and increasing their firing rates. This eventually leads to activation of the chemoreceptor (emetic) trigger area. Thus, antagonistic agents that act on 5-HT_3 receptors in the GI tract such as ondansteron and granisteron, and not on the emetic center, break these series of events and produce relief from nausea and vomiting.

facial motor nucleus. This group of receptors is comprised of 5-HT_{2A}, 5-HT_{2B}, and 5-HT_{2C}. Prolonged stimulation of the 5-HT_{2B} and 5-HT_{2C} receptors may produce reduction in receptor density or sensitivity. Sustained administration of 5-HT antagonists may result in down-regulation of both 5-HT_{2A} and 5-HT_{2C}.

5-HT_3 is a member of ligand-gated ion channels that function independent of G-protein. They are densely populated at the nerve terminals in the entorhinal and frontal cortices, hippocampus, and area postrema, as well as the peripheral nervous system. These receptors are excitatory in the peripheral, enteric and autonomic nervous systems, facilitating membrane depolarization to serotonin. They resemble the nicotinic cholinergic receptors, mediating fast synaptic transmission.

Additional receptors within the 5-HT family that are positively coupled to adenylate cyclase include 5-HT_4, 5-HT_6, and 5-HT_7. 5-HT_4 receptor-binding sites are identified in the striatum, substantia nigra, olfactory tubercle, and atrium. They may mediate striatal dopamine release by 5-hydroxytryptamine. However, 5-HT_5 receptors do not couple to adenylate cyclase and consist of 5-HT_{5A} and 5-HT_{5B}. The 5-HT_{5A} mRNA transcripts are localized in the hippocampal gyrus, granule cells of the cerebellum, medial habenular nucleus, amygdala, thalamus, and olfactory bulb, whereas 5-HT_{5B} mRNA are found in the dorsal raphe nucleus, habenula and hippocampus. 5-HT_6 receptor mRNA has been detected in the striatum, olfactory tubercle, and nucleus accumbens.

5-HT_7 is identified as a receptor in the vascular smooth muscle cells and astrocytes of the frontal cortex.

Dahlström and Fuxe described nine groups of serotoninergic neurons that are designated as B1 through B9. B6 and B7 represent the dorsal raphe nucleus; B8 corresponds to the median raphe (superior central) nucleus, whereas B9 forms part of the ventrolateral tegmentum of the pons and midbrain. B1–B4 groups are localized in the midpons through the caudal medulla and they project mainly to the spinal cord. B1, which is observed in the caudal part of the ventral medulla, has no known projections. B3, which corresponds to the nucleus raphe magnus, projects to the spinal laminae I & II and the intermediolateral lateral column. B2 (nucleus raphe obscurus) projects to lamina IX. B6–B9 nuclear groups project to the telencephalon and diencephalon. B8 group appears to project largely to the limbic system, whereas B7 maintains specific

Destruction of the raphe nuclei and subsequent depletion of serotonin may produce reversible insomnia. The use of serotonin blockers such as ergotoxin, tricyclic antidepressants, MAO inhibitors, and sumatriptan in the treatment of migraine headache may support the role of serotonin in this condition.

Neurons of the dorsal raphe nucleus seems to be more prone to neurotoxicity generated by certain amphetamine derivatives, such as D-fenfluramine, 3-4 methylenendioxymethamphetamine (MDMA or 'ecstasy') or parachloramphetamine (PCA). 5-hydroxyindoleacetaldehyde (5-HIAA), a serotonin metabolite, shows reduction in the CSF of MDMA users. Blocking the serotonin transport systems may avert this neurotoxicity. In contrast the median raphe nucleus appears to be unaffected by these neurotoxic effects.

projections to the neostriatum, thalamus, cerebral, and cerebellar cortices. Ascending serotoninergic projections from the rostral midbrain form the dorsal periventricular tract and the ventral tegmental radiation, which join the medial forebrain bundle and the dopaminergic and noradrenergic projections in the hypothalamus. The dorsal raphe nucleus projects to the striatum, whereas the median raphe nucleus sends fibers to the hippocampus, septal area, and hypothalamus. A somatotopic representation of the ascending serotoninergic projections exists in which the rostral and lateral parts of the dorsal raphe nucleus predominantly project to the frontal cortex. It is well established that serotoninergic neurons produce a combination of depolarization and increased membrane resistance of the neurons of the facial motor nucleus, which enhances the response of these neurons to other excitatory input.

Serotonin is implicated in the regulation of the slow phase of sleep, pituitary functions and activities of the limbic system (behavior, thermoregulation, mood and memory). It also plays an important role in the inhibition of pain transmission. Medullary raphe nuclei exert analgesic influence via projections to the spinal cord, whereas pontine and mesencephalic neurons contribute to the ascending reticular activating system via its rostral projections.

Histamine

Histamine acts both as a neurotransmitter and neuromodulator in the central nervous system, mainly occupying the midbrain, tuberal and mamillary nuclei of the hypothalamus and their extensions in the principal mamillary and the tuberoinfundibular tracts, respectively. In the hypothalamus, it coexists with substance P, met-enkephalin, and glutamic acid decarboxylase. Non-neuronal histamine is contained in a substantial amount in the mast cells, where it is depleted by mast-degranulating drugs. Histamine exerts its influences on autonomic activity, temperature regulation, food and water intake (suppressant), vestibular function (may mediate motion sickness), sleep–wake cycles, and neurohumoral mechanism (release of vasopressin, prolactin, adrenocorticotropic hormone, etc.) It maintains the ability to excite CNS neurons and may be involved in locomotor and exploratory behavior, as well as diurnal changes in other CNS functions. Learning and retention of information may be enhanced by histamine. Histamine release is mostly non-synaptic and widely diffuse, and release of neuronal histamine may be increased by stimulation of the D_2 dopaminergic and some serotoninergic and NMDA receptors. Histaminergic neurons project to glial cells, blood vessels, neurons, as well as capillary networks. Histamine receptors are identified as H_1, H_2, and H_3 according to their order of detection. H_1 receptors are shown to be involved in hormonal release, food intake, increased free calcium ion concentration, contraction of the smooth muscles, and increased capillary permeability. In the ventrolateral hypothalamus, H_1 receptor is involved in wakefulness. Both H_1 and H_2 are involved in regulation of pituitary gland function, whereas H_2 receptors may mediate endogenous analgesia.

It has been suggested that histamine may alter the blood–brain barrier, suppress the immune system and produce certain vascular changes, contributing via these neurotoxic effects to the development of certain neurodegenerative diseases, such as multiple sclerosis, Alzheimer's disease and Wernicke's encephalopathy. The H_1 receptor-mediated effect of histamine reduces seizure activity. Therefore, H_1 antagonists increase seizure onset and/or duration. H_1 antagonists that induce sedation include diphenhydramine and mepyramine, as well as meclizine (anti-motion sickness medication). H_2 receptors are involved in inhibition of gastric secretion, positive ionotropic and chronotropic effects upon the cardiac muscles, and inhibition of contraction of the smooth muscles. Therefore, H_2 antagonists (cimetidine and ranitidine) reduce gastric secretion and thus can be used for the treatment of gastric and duodenal ulcers. On the other hand, H_3 receptors may be involved in the regulation of histamine release and inhibition of acetylcholine, dopamine, norepinephrine and other peptides.

Neuropeptides

Neuropeptides are substances that arise from inactive precursors. They are synthesized on ribosomes in the perikaryon or dendrites of a neuron that is regulated by mRNA and packaged for release in the endoplasmic reticulum. Their eventual release is by a calcium-dependent process. Neuropeptides include the enkephalins, endorphins, substance P, cholecystokinin (CCK) and hypothalamic peptides. Many peptides such as bradykinin, somatostatin, gastrin, secretin, and vasoactive polypeptide (VIP) are also shown to act upon the autonomic intestinal neurons and enteric nervous system, which will be discussed later with the autonomic nervous system. In view of the vast amount of information available in this area, the discussion will only be restricted to certain peptides.

Enkephalins are pentapeptides that are present in the interneurons of the substantia gelatinosa, nucleus raphe magnus and the small intestine. Enkephalinergic neurons modulate pain via presynaptic inhibition upon afferents in the brainstem and spinal cord. They are classified into methionine enkephalin and leucine enkephalin.

Endorphins are peptides (naturally occurring opiates) consisting of C-terminally extended forms of Leu-enkephalin that bind to opiate receptors in the brain, and induce analgesia similar to morphine. These receptors have the ability to bind to opiate agonists (e.g. morphine) or to antagonists (e.g. naloxone). Endorphins may be implicated in states of depression and generalized convulsions. They are derived from different genes and they are classified into dynorphin-A, dynorphin-B, and α and β-endorphins. β-endorphins are contained in neurons within the diencephalon and pons, and may be involved in the acquired intellectual deterioration in adults.

Substance P is an 11 amino acid oligopeptide that is present in the nerve endings of the unmyelinated class C or myelinated Aδ fibers which carry nociceptive (painful) stimuli to the substantia gelatinosa. Therefore, axotomy may reduce the level of this peptide. Substance P is found in the dorsal root ganglia, gastrointestinal tract, and sensory ganglia of cranial nerves, spinal trigeminal nucleus, basal nuclei, nucleus raphe magnus, periaqueductal gray, and the hypothalamus. Other peptides that relate closely to substance P are neurokinin A and neurokinin B with their specific receptors NK1 and NK2. The neurokinin A gene is located on chromosome 7, whereas the gene for neurokinin B is positioned on chromosome 12. Substance P exerts a more powerful effect than both neurokinins at NK1, however it remains less potent at NK2.

CCK is a neuromediator concentrated in the amygdala, hypothalamus, cerebral cortex, periaqueductal gray matter, and the spinal dorsal gray columns. It coexists with other peptides in the substantia nigra, ventrotegmental area, and the medulla. CCKa and CCKb are known receptors for cholecystokinin, with CCKb being the predominant receptor in the brain. CCKa is found in the nucleus accumbens septi, posterior hypothalamus, and area postrema of the medulla.

Hypothalamic peptides are 3–14 amino acid peptides that include thyrotropin-releasing hormone, somatostatin, corticotropin-releasing hormone, melanocyte-stimulating hormone release-inhibiting factor, and luteinizing hormone-releasing hormone.

Suggested reading

1. Amara SG, Kuhar MJ. Neurotransmitter transporters: recent progress. *Annu Rev Neurosci* 1993;16:73–93
2. Chessell JP, Francis PT, Pangalos MN, *et al.* Localization of muscarinic (m1) and other neurotransmitter receptors on corticofugal-projecting pyramidal neuons. *Brain Res* 1993;632:86–94
3. Cimino M, Marini P, Fomasafi D, *et al.* Distribution of nicotinic receptors in cynomolgus monkey brain and ganglia: localization of a3 sub-unit mRNA, a-bungarotoxin and nicotine binding sites. *Neuroscience* 1992;51: 77–86
4. Di Figlia M, Aronin N. Amino acid transmitters. In Paxinos G, ed. *The Human Nervous System*. New York: Academic Press, 1990:1115–34
5. Dougherty PM, Paleck J, Zorn S. Combined application of excitatory amino acids and substance P produces long lasting changes in responses of primate spinothalamic tract neurons. *Brain Res Rev* 1993;18:227–46
6. Everitt BJ, Hökfelt T. The coexistence of neuropeptide-Y with other peptides and amines in the central nervous system. In Mutt YV, Hökfelt T, Lundberg J, eds. *Neuropeptide*. New York: Raven Press, 1989:61–77
7. Finle PR. Selective serotonin reuptake inhibitors. *Ann Pharmacother* 1994;28:1359–69
8. Hills JM, Jessen KR. Transmission: gamma-amino butyric acid (GABA), 5-hydroxytryptamine (5-HT) and dopamine. In Burnstock G, Hoyle CHV, eds. *Autonomic Neuroeffector Mechanisms*. Switzerland: Harwood Academic Publishers, 1992:465–507
9. Homung J-P, De Tribolet N. Distribution of GABA-containing neurons in human frontal cortex: a quantitative immunocytochemical study. *Anat Embryol* 1994;189:139–45
10. Kelly RB. Storage and release of neurotransmitters. *Neuron* 1993;10(Suppl):43–53
11. Kosaka T, Tauchi M, Dahl JL. Cholinergic neurons containing GABA-like and/or glutamic acid decarboxylase-like immunoreactivities in various brain regions of the rat. *Exp Brain Res* 1988;70:605–17

12. Matthews G. Neurotransmitter release. *Annu Rev Neurosci* 1996;19:219–33
13. Onodera K, Yamadotani A, Watanabe T, Wada H. Neuropharmacology of the histaminergic neuron system in the brain and its relationship with behavioral disorders. *Prog Neurobiol* 1994;42:685–702
14. Ottersen OP, Mathisen JS. *Handbook of Chemical Neuroanatomy. Volume 18: Glutamate.* Amsterdam: Elsevier Science, 2000
15. Palermo-Neto J. Dopaminergic systems: dopamine receptors. *Psychiatr Clin North Am* 1997;20:691–703
16. Pearson J, Halliday G, Sakamoto N, Michael JP. Catecholaminergic neurons. In Paxinos G, ed. *The Human Nervous System*. New York: Academic Press, 1990:1023–50
17. Piero EP, Mai JK, Cuello C. Distribution of substance P- and enkephalin-immunoreactive fibers. In Paxinos G, ed. *The Human Nervous System*. New York: Academic Press, 1990
18. Pin JP, Duvoisin R. Neurotransmitter receptors I. The metabotropic glutamine receptors: structure and function. *Neuropharmacology* 1995;34:1121
19. Tecott LH, Julius D. A new wave of serotonin receptor. *Curr Opin Neurobiol* 1994;3:310–15
20. van Ree JM, Matthysse S. Psychiatric disorders: neurotransmitters and neuropeptides. *Prog Brain Res* 1986;65:1–228

Section IV

Sensory systems

The sensory systems transmit special somatic sensations (vision, auditory and vestibular), general somatic sensations (tactile, thermal, painful, and proprioceptive), general visceral sensations (visceral pain, changes in blood pressure, hunger, libido, etc.), and special visceral sensations (olfactory and taste). Most of these sensations project to the cerebral cortex while others are perceived at subcortical levels (e.g. cerebellum).

Visual system 13

The visual system is a special somatic afferent system which receives, processes, and recognizes visual impulses. It forms binocular images and regulates reflexes associated with vision. It is the only sensory system which is totally dependent upon the cerebral cortex. In order for visual images and associated memory to be constructed, visual impulses must pass through a chain of structures and neurons which are located in the eye and along the visual pathway, encompassing the cornea, iris, anterior and posterior chambers of the eye, vitreous body, retina, optic nerve, optic tract, visual radiation, and visual cortex.

Peripheral visual apparatus

The eyeball is the peripheral visual organ which is situated in the bony orbital cavity. It is surrounded by the orbital fat, separated by a thin fibrous (Tenon's) capsule.

Eyeball

The eyeball (Figure 13.1) consists of an outer fibrous, an intermediate vascular, and an inner neuronal layer.

Tunica fibrosa

The tunica fibrosa, the outermost layer of the eyeball, consists of the cornea and sclera.

The cornea (Figure 13.1) is an avascular structure that forms one sixth of the fibrous tunic and represents the main refractive medium of the eyeball. It has no lymphatics, receives a rich nerve supply by the long ciliary nerves, and is highly resistant to infection. It forms the anterior wall of the anterior chamber of the eye and joins the sclera at the corneoscleral junction and the iris at the irideocorneal angle.

The Kayser–Fleischer ring, a greenish gray pigmentation around the corneoscleral junction, is formed by the deposition of copper in the Descemet's membrane and is seen in Wilson's disease (hepatolenticular degeneration). See also the subcortical motor system.

At the corneoscleral junction the Schlemm's canal (sinus venosus sclera), a venous channel, exists and receives the aqueous humor (fluid within the anterior and posterior chambers). Occlusion at this site may lead to accumulation of aqueous humor, increased intraocular pressure, and resultant glaucoma.

The sclera (Figure 13.1) is a fibrous structure which preserves the shape of the eyeball, resists intraocular pressure, and provides smooth surface for eye movements, giving attachment to the extraocular muscles. It is continuous anteriorly at the limbus with the connective tissue stroma of the cornea, posteriorly with the dural sheath of the optic nerve. At the lamina cribrosa sclera (weakest part of this layer), the sclera is pierced by fibers of the optic nerve as well as the posterior ciliary vessels.

Tunica vasculosa

The tunica vasculosa represents the intermediate layer of the eyeball, consisting of the choroid layer, ciliary body, and the iris (Figure 13.1). Arteries and veins are the main components of the vascular choroid layer.

In some individuals, the lens may be absent as a developmental anomaly (primary aphakia), or as a result of degeneration (secondary aphakia). Corneal opacity and cataracts of the anterior lens are seen in congenital anomaly of Peter's, in which gradual impairment of vision and diplopia are observed. The opacity may be confined to the nucleus of the lens (central cataract), producing myopia and poor vision during the day and better vision in dim light. Peripheral cataracts result in poor vision in dim light and better vision in bright daylight. Congenital cataracts may be seen at birth due to metabolic or chromosomal abnormalities, infection, or maternal diseases. The lens may be affected by a myriad of clinical conditions that include presbyopia and sunflower cataract.

Presbyopia develops as a result of aging and by the conversion of the lens into a less pliable structure, rendering it less reactive to contraction of the ciliary muscles. Presbyopic patients are hyperopic, exhibit difficulty in reading fine print and endure the inconvenience of holding reading materials farther away to achieve optimum vision.

Sunflower cataract (chalcosis lentis) is a condition which is seen in Wilson's disease due to the impregnation of the subcapsular area of the lens with the radiating metallic green grayish opacity.

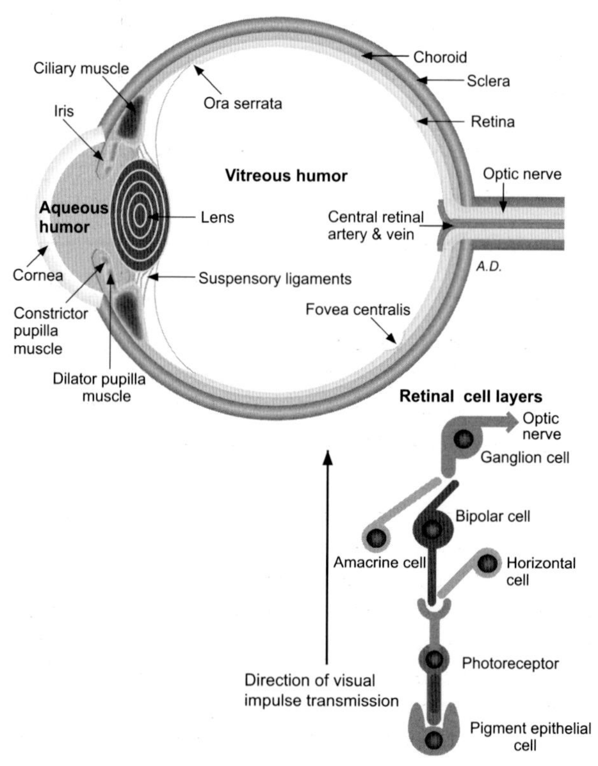

Figure 13.1 Section of the eyeball and associated layers. The neuronal organization of the retina is also illustrated

Miosis refers to constriction of the pupil that is commonly seen in Horner's syndrome. Bilateral miotic pupils may result from metabolic encephalopathies, destructive pontine lesions or opiate use. Miosis of pontine origin may be due to disruption of the descending sympathetic pathways. Disruption of the efferent sympathetic fibers in the carotid sheath, near the apex of the lung or base of the neck, may also produce unilateral miosis.

Mydriasis (dilatation of the pupil) is manifested in oculomotor palsy as a result of the unopposed action of the dilator pupilla muscle. This sign may be seen subsequent to aneurysms of the posterior communicating, superior cerebellar, and posterior cerebral arteries, and also in uncal herniation. Traumatic mydriasis is usually unilateral, occurs in response to direct trauma, and may not be accompanied by ocular muscle dysfunction. Bilateral mydriasis may be seen in trauma patients with poor vascular perfusion subsequent to hypotension or increased intracranial pressure. Prompt restoration of pupillary response may follow adequate vascular perfusion. Patients with Cheyne–Stokes respiration may exhibit mydriasis in hyperventilation phase and miosis in apneustic phase.

Anisocoria refers to the unequal pupils in which one eye may show constriction. Physiological anisocoria occurs in 20% of the population, exhibiting mild difference (up to 2 mm) in pupil size. Determination of whether the smaller or larger pupil is abnormal may require comparing pupil size in the dark and the light. Sympathetic anisocoria will be more marked in dim light due to the subnormal constriction of the affected (small) pupil. This may result from iritis, disruption of the cervical sympathetics or application of miotic (miosis inducing) medications. Parasympathetic anisocoria will be evident in light since the affected (larger) pupil constricts subnormally. This type of aniscoria may be seen in individuals with oculomotor palsy, glaucoma or as a result of application of mydriatic (mydriasis inducing) drugs such as atropine.

Argyll Robertson pupil is characterized by the inability of the eye to constrict in response to light, while remaining responsive (constricts) in accommodation. This condition occurs in tabes dorsalis or tertiary syphilis, diabetes mellitus, and in severe vitamin B deficiency. The lesion is usually located medial to the lateral geniculate nucleus, disrupting the afferent limb of the pupillary light reflex, while preserving the afferent limb of the accommodation reflex. Reverse Argyll Robertson pupil is seen in syphilis and Parkinson's disease.

Hippus is a phenomenon in which the pupil exhibits spontaneous, intermittent rhythmical constriction and dilatation. Although the diagnostic value is questionable, this condition may be associated with hysteria, multiple sclerosis, brain abscess, and Cheyne–Stokes respiration.

Marcus Gunn pupil is characterized by a slow reaction to light or the inability to constrict in response to direct light. It is observed in individuals with a lesion of the optic nerve or retina, and in ipsilateral retrobulbar neuropathy which produces relative afferent pupillary defect (RAPD). In RAPD, the presence of normal consensual reflex when the contralateral eye is stimulated indicates that the oculomotor nerve is intact. When the light is quickly passed from the intact to the affected eye, both eyes show dilatation (positive swinging flashlight test). The consensual reflex is preserved, but the depth of perception of moving and colored objects is lost. The visual deficits are exacerbated by exercise or by any efforts which increase the body temperature (Uhthoff's sign). The latter sign is due to the possible change in the conduction of the affected nerve or variation in the sodium and potassium concentration around the myelin of the optic nerve following physical activity. The Gunn pupil, a congenital anomaly that is characterized by ptosis and retraction of the eyelid on the affected side in response to opening the mouth or deviation of the jaw, is not equivalent to Marcus Gunn.

Within this layer, the veins join together and form 4 or 5 vorticose veins. These vorticose veins drain into the anterior ciliary veins. The arterial part of the choroid layer is formed by the ciliary arteries that extend to the iris, forming the major and minor arterial iridal circles. In addition to the central retinal artery, these vessels provide a supplementary source of blood to the retina. The intermediate part of the tunica vasculosa is known as the ciliary body, which extends from the lateral end of the iris to the ora serrata (site of junction of the light sensitive and nonsensitive parts of the retina). It consists of ciliary muscles and processes, giving attachment to the suspensory ligaments of the lens (zonular fibers). When viewing distant objects, the ciliary muscle is relaxed and the zonular ligaments are stretched and become taut.

On the other hand, when viewing near objects, the ciliary muscle is contracted, accompanied by movement of the ciliary body toward the iris and relaxation

Glaucoma is a condition in which the intraocular pressure is elevated independent from any other diseases of the eye (primary glaucoma) or as a result of ocular diseases (secondary glaucoma). Primary (chronic, or open angle) glaucoma may be congenital or acquired and may result from obstruction at the canal of Schlemm, aqueous veins, or the trabecular meshwork at the irideocorneal angle.

Open angle glaucoma is the most common form of glaucoma, which is produced by a gradual increase in intraocular pressure, accompanied by a gradual loss of peripheral vision, ending in total blindness. It usually begins in the fourth or fifth decades of life in individuals with familial history of a variety of glaucoma. Symptoms are absent at the onset, and its diagnosis may be confirmed by examination of the fundus of the eye and detection of increased intraocular pressure. In later stages of this disease, the optic cup is abnormally deep and permanently put out of function. Lack of clear symptoms in the initial stages of this disease is an important indication that regular ophthalmologic examination is highly recommended for individuals over the age of forty. Pilocarpine, which constricts the pupil and increases the outflow of the aqueous humor through the irideocorneal angle, may be used topically to treat this condition. Timolol may reduce the production of aqueous humor, but its side effects on the cardiovascular system may make it a less desirable medication. Marijuana may also lower intraocular pressure in this type of glaucoma.

Closed angle glaucoma results from increased intraocular pressure subsequent to adhesion of the iris to the cornea and closure of the irideocorneal angle. It may not always be spontaneous, but iatrogenic, resulting from the application of medications that dilate the pupil and block the irideocorneal (filtration) angle by the iris itself. Tricyclic antidepressants with anticholinergic properties may precipitate this condition. Patients are usually older than 40 years with family history of glaucoma. A few patients may complain of seeing halos around lights. This condition may be acute or chronic.
(a) Acute (closed) angle glaucoma is produced by the sudden obstruction of aqueous humor circulation, which produces pain and visual impairment of the affected eye. In this condition, the eye appears red, the cornea seems hazy, and the blood vessels are dilated.
(b) Chronic (closed) angle glaucoma occurs as a result of gradual obstruction of the irideocorneal angle, showing similar signs to the acute closed angle glaucoma. Topical and systemic treatments and even laser iridectomy may have to be promptly applied to reduce production of aqueous humor. Glaucomas may be associated with lesions involving the peripheral retina or the optic nerve, producing peripheral scotoma (focal blindness in the form of dark or colored spot).

of the suspensory ligaments, which results in an increase in lens curvature. The ciliary processes secrete the aqueous humor into the posterior chamber by active transport and diffusion from the capillaries.

The lens (Figure 13.1), a main component of the eye chambers, is a biconvex, colorless, avascular structure, derived from the surface ectoderm, and positioned between the iris and the anterior chamber. It lies posterior to the iris, embedded in the hyaloid fossa, receiving the suspensory ligaments of the lens. It is functionally similar to the cornea with a less refractive (diopteric) power. It has a transparent elastic capsule, a cortical zone, and a nucleus. The anterior and posterior poles represent the most convex parts of the lens. Changes in the lens curvature are regulated by the suspensory ligaments of the lens and the muscles of the ciliary body.

The iris (Figure 13.1) forms the adjustable diaphragm of the eye that encircles the pupil, consisting of circular muscle fibers (constrictor pupilla) and radial muscle fibers (dilator pupilla), pigment cells, and epithelium. Constrictor pupilla, which acts as a pupillary sphincter, is innervated by the postganglionic parasympathetic fibers, whereas the dilator pupilla receives innervation from the postganglionic sympathetic fibers. The iridal pigment is the same in all individuals; however, the amount of pigment and the pattern of its distribution determine the eye color. Malclosure of intraocular fissures results in a defect (coloboma) in the iris, choroid layer, retina, or optic disc. The pupil, an opening in the center of the iris, may exhibit abnormalities such as miosis, mydriasis, aniscoria, Argyll Robertson pupil, hippus, Adie's (tonic) pupil, and Marcus Gunn pupil.

The anterior chamber (Figure 13.1) of the eye is located between the cornea anteriorly, the iris and the lens posteriorly. It contains the aqueous humor, which crosses the trabecular network to gain access to the irideocorneal angle and the canal of Schlemm. The area between the iris anteriorly and the lens and zonular fibers posteriorly represents the posterior chamber. Both eye chambers communicate via the pupil, containing the aqueous humor, which maintains the intraocular pressure. Aqueous humor also serve as a path for metabolites from the cornea and lens, carries nutrients such as glucose, and plays a role in respiratory gaseous exchange. Failure of this communication may result in glaucoma.

The vitreous body (Figure 13.1) is a clear and colorless substance, occupying most of the eyeball posterior to the lens. It primarily consists of water, hyaluronic acid, trace amounts of mucoproteins, and some salts. It contains fibrils, which may be visible as floating objects. The hyaloid membrane surrounds the vitreous body, thickens at the ora serrata of the retina to form the ciliary zonule. The ciliary zonule forms the suspensory ligaments of the lens and the hyaloid membrane. The hyaloid canal pierces the vitreous body and stretches from the optic disc to the posterior pole of the lens.

Refractive disorders

In a normally relaxed eye (emmetropia), no optical defects exist and it is adapted to the far vision. The lens is flat and the suspensory ligaments of the lens (zonular fibers) are tense (Figures 13.1 & 13.3)). However, near vision requires the use of accommodative power that includes an increase in lens curvature, constriction of the pupil, and convergence. The increase in lens curvature shortens the focal distance and allows images from closer objects to fall on the retina. The changes in near vision occur when the eyeball has a normal anteroposterior dimension and functional refractive media within normal range. An abnormally long or short eyeball, an uneven curvature of the refractive media, or a non-functional or overly functional accommodative apparatus may produce a variety of disorders, including myopia, hyperopia, astigmatism, presbyopia, and anisometropia.

Myopia (near-sightedness) is an optical disorder characterized by the inability to see far objects. It may be due to an abnormally long eyeball axis or a refractive power which is too strong (Figure 13.2). Therefore, the image from a distant object falls anterior to the retina. As the object moves closer, the focal point moves back to the same degree until the image is spotted on the retina and clear vision is achieved. This condition may be associated with rhegmatogenous retinal detachment in which the retina is detached and breaks up into pieces. Myopia is corrected by a concave lens.

Hyperopia (far-sightedness) is the most common optical disorder associated with an abnormally short axis of the eyeball or weak refractive power (Figure 13.2). Due to shortness of the optical axis, images from distant objects fall behind the retina in the relaxed eye. However, accommodation focuses the images on the retina, allowing clear vision of distant objects. A convex lens corrects hyperopia.

Astigmatism results from an uneven curvature of the refractive surfaces (egg-shaped), leading to a change in the angle of refraction of the horizontal and perpendicular light rays and subsequent focusing of these rays on different spots on the retina. As a result blurred vision ensues. A cylindrical lens, which is concave or convex on one axis and flat on other, corrects this disorder.

Presbyopia, a condition that develops with aging, is characterized by an inelastic, hard and less pliable lens, which lacks or has limited power of accommodation. Individuals with this disorder are hyperopes (far-sighted).

Anisometropia is a rare disorder in which the refractive powers between both eyes remain different.

Tunica nervosa

The tunica nervosa consists of the pigment epithelial layer and retina (Figure 13.1). The pigment epithelium is loosely bound to the retina that contains the photoreceptor layer. The pigment epithelial layer offers mechanical support, minimizes light reflection, absorbs excess light, and provides nutrients for photoreceptors.

The retina (Figure 13.1 & 13.3) is the inner layer of the tunica nervosa that develops as the optic vesicle from the diencephalon, consisting of the pars optica, pars ciliaris, and pars iridica. The pars optica joins the pars ciliaris at the ora serrata. The pars optica is the only light-sensitive part, which contains the photoreceptors, and it is loosely bound to the pigment epithelium. The retina is incompletely fused to the pigment epithelium and is separated by a potential space.

At the ora serrata, the sensory layers of the retina and the retinal pigment epithelium fuse thus limiting the spread of any pathological subretinal fluid. The tunica nervosa contains photoreceptors, which are divided into cones and rods. Cones vary in number from 6–7 million and are distributed among the rods except in the fovea centralis. They occupy a central position, whereas the rods are localized in the periphery. The optic disc is the site where axons of the ganglionic layer leave the eyeball, the optic nerve is formed, and the photoreceptors are absent. The physiologic cup is the lighter-colored central part of the disc which is penetrated by the retinal vessels. The macula lutea, a yellowish spot on the temporal side of

Detachment of the retina from the pigment epithelium may be complete or focal, occurring as a result of trauma or disease processes. It may be associated with a break up of the retina (rhegmatogenous detachment) subsequent to direct trauma. Retinal detachment, as is seen in diabetic vitreoretinopathy, is associated with intact retina that underwent undue traction by the fibrovascular bundles between the vitreous body and the retina. Retinal detachment may also occur when pathological processes allow exudate derived from the choroid layer to enter the subretinal space (exudative retinal detachment). Retinal detachment may cause blurred vision, light flashes or the appearance of floating bodies. Cherry red spots may be seen on fundoscopic examination of the retina in individuals with Tay–Sachs disease.

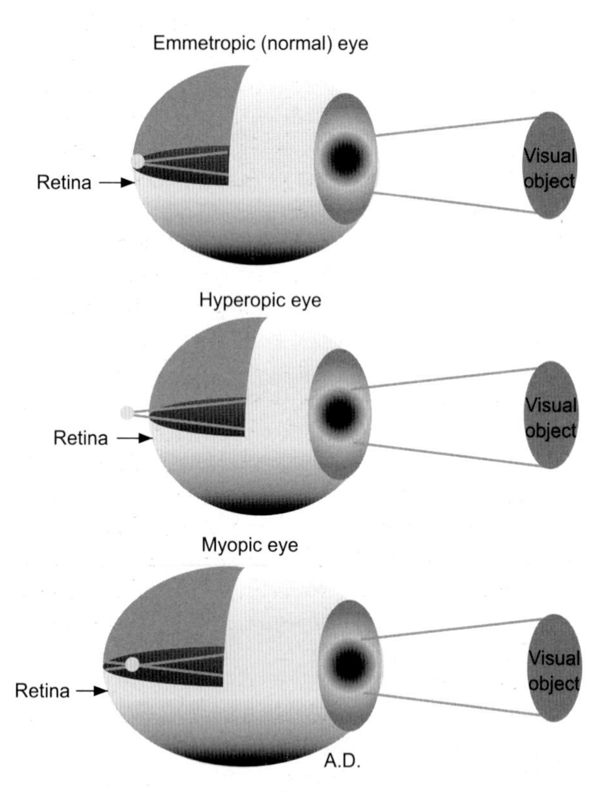

Figure 13.2 Refractive defects associated with vision

the optic disc, contains the fovea centralis. The latter is for acute vision occupied only by cones. The physiologic cup is the lighter-colored central part of the disc, which is penetrated by retinal vessels. The normal cup-to-disc ratio of 1:5 may be lost in glaucoma.

The cones possess a higher threshold of excitability and have a 1:1 synapse ratio with the dendrites of the bipolar neurons. Cones are specialized to day vision (phototopic) and color discrimination.

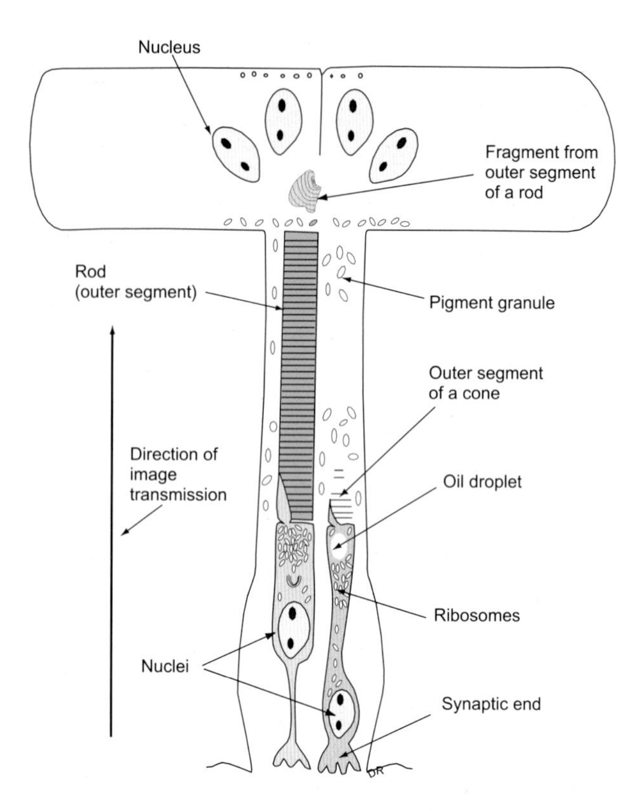

Figure 13.3 Structural organization of the photoreceptors

Rods (Figure 13.3) are the most numerous, averaging between 100–130 million per retina. They are peripherally located and activated by lower illumination. Rods visualize black, white, and gray colors under twilight or scotopic (achromatic) vision. In dim light, rods contract to maximize the surface area exposed to the limited light. The outer segments of rods contain discs, which are sloughed and removed by pigment cells. Dendrites of horizontal cells in the inner nuclear layer interconnect cones and rods. The processes of the large glial (Müller) cells, which hold the retinal layers together, constitute the outer limiting membrane. The photoreceptors form synaptic linkage with the dendrites of the bipolar neurons at the outer plexiform layer.

The bipolar neurons (Figure 13.1) are depolarizing or hyperpolarizing neurons that represent the primary (first order) neurons in the visual pathway. Depolarizing (invaginating) bipolar neurons are inhibited by darkness. They stimulate the 'on' type ganglionic cells and are released from inhibition by illumination. Hyperpolarizing (flat) neurons are inhibited by light, and excite the 'off' type ganglionic neurons, maintaining different receptors. The nuclei of these neurons are located in the inner nuclear layer while the axons are spread in the inner plexiform

A lesion rostral to the optic chiasma involving the foveal part of the retina and the corresponding part of the optic nerve is commonly seen in multiple sclerosis. These lesions may be associated with optic neuritis (inflammation of the optic disc) or retrobulbar neuritis (inflammation of the optic nerve).

Dark or colored spots in the center of the visual field are referred to as central scotoma. The point of exit of axons of the optic nerve from the eyeball marks the optic disc of the retina, commonly known as the blind spot. The focal blind spot attributed to the optic disc is termed physiological scotoma.

Arcuate scotoma is a pathological focal visual deficit, which results from a lesion in the retina or optic nerve fibers. It occurs near the optic disc and arches superiorly or inferiorly toward the nasal field of the retina and in the direction of the axons of the ganglionic multipolar neurons.

Scintillating (flittering) scotoma (teichopsia) is characterized by floating of irregular and lucid spots, sometimes with a zigzag or wall-like outline, which may last for up to 20–25 minutes. It usually occurs secondary to occipital lobe lesion. It may also be associated with migraine (migraine aura).

layer, establishing contacts with the dendrites of the ganglion cells.

The amacrine cells (Figure 13.1), which resemble the granule cells of the olfactory bulb, form inhibitory synapses upon the dendrites of the ganglion cells, and maintain reciprocal connections with the bipolar neurons. These cells contain different transmitters and, together with the ganglionic neurons, are the only excitable (produce action potentials) neurons of the retina. The horizontal cells establish inhibitory dendrodendritic synapses with the bipolar neurons, intensifying contrast by inactivating the bipolar and ganglionic neurons.

Ganglionic (multipolar) neurons (Figures 13.1 & 13.5) form the second order neurons in the visual pathway, which have the capacity to fire at a fairly steady rate even in the absence of visual stimuli. They are either 'on' type or 'off' type, dependent upon their synaptic connections with the bipolar neurons. They are classified into sustained X, transient Y, and intermediate W cells. Sustained X cells subserve constant on or off response, analyzing the field in regard to their shapes and colors of objects in the visual field. Transient Y cells are relatively few in number, giving rise to momentary response to rapidly moving objects. Y cells project to the superior

Color blindness is an inherited (sex-linked) or acquired condition. Patients may exhibit blindness toward all colors (achromatopia) or to one (monochromatopia) or two colors (dichromatopia). Color vision is mediated by the cones, segregated from other visual information in the retina, and eventually processed in a specialized pathway in the visual cortex utilizing the lateral geniculate nucleus and the optic radiation. The inherited variation in the amount of photopigments in the blue cones, green cones, and red cones may account for the sex-linked condition of color blindness. This condition affects 8% of males compared to 2% of females. This is due to the fact that red and green genes exist as a recessive trait on the male X chromosome. One per cent of males lack the red gene (protanopes, lack the long wave mechanism) and 2–3% lack the green gene (deuternopes, lack the medium length mechanism). The gene for the blue color is present on an autosome on the seventh chromosome, and is rarely affected by mutation. It is thought that all three-cone genes maintain a common ancestral red gene. The red gene may have given rise to the blue cone pigment which in turn has given origin to the red and green cone pigments. Trichromats are individuals with normal three-color vision or with one normal color vision and two feeble red vision (protanomaly). Trichromats may also have three feeble green vision (deutranomaly) or four weak blue vision (tritanomaly). These weaknesses are due to the reduction in the amount of cone pigment and are unrelated to neuronal circuitry associated with processing of the color vision. Dichromats, individuals with two-color vision (color blind) and who lack one of the pigments, may not perceive red (protanopes) due to lack of erythrolabe, green (deutranopes) as a result of absence of chlorolabe, or blue (tritanopes, lack the short wave length mechanism) due to absence of cyanolabe. Deutranopia is much more common than protanopia in the ratio of 3:1.

Gene loss or recombination between genes, which produce a hybrid gene on the X chromosome may occur in individuals with red–green color blindness. However, disorders of color vision may also be acquired. Pathological conditions that affect the outer layer of the retina may produce blindness to blue color (tritanopia) as a result of loss of the processing mechanism of short wavelength. In this manner, pathological elements that affect the optic nerve and the inner retinal layer may cause loss of red–green color vision.

Accumulation of pigment cells and floating discs of the outer segments may account for the retinal degeneration and black colored lesions in the fundus seen in retinitis pigmentosa. Disc debris in the outer part of the retina may hinder the diffusion of nutrients from capillaries of the choroid layer to the photoreceptors, thus accounting for the retinal degeneration observed in this blinding disease. A small lesion or petechial hemorrhage in the retina near the optic disc produces a focal blindness or scotoma in which central visual acuity is impaired. Vitamin A plays a significant role in vision. Darkness causes vitamin A to undergo reverse changes into retinin, which bonds with opsin to form rhodopsin. Nyctalopia (night blindness) is associated with vitamin A deficiency.

colliculus and thalamus to detect the movement of objects in the visual field. W cells are small and project to the pretectum to mediate the pupillary light reflex. The dendrites of the ganglionic neurons are connected with the axons of the bipolar neurons in the inner plexiform layer. Some ganglionic neurons in the nasal halves of the retina may fixate the visual image on the fovea centralis, preventing the image from slipping off. This occurs when axons of these specific ganglionic neurons project via the midbrain reticular formation to the inferior olivary nucleus. This is followed by projection of the olivocerebellar fibers to the same Purkinje neurons that receive input from the medial vestibular nucleus.

The axons of the ganglionic neurons leave the eyeball through the lamina cribrosa sclera, forming the optic nerve. The retina is divided by the optic axis into a medial (nasal) half and a lateral (temporal) half; a horizontal plane further divides the retina into upper and lower nasal and upper and lower temporal quadrants. The object seen by the retina represents part of the visual field. The nasal half of the retina of each eye receives visual impulses from the temporal half of the visual field and vice versa. The upper quadrant of the retina of one eye sees the lower quadrant of the contralateral visual field and vice versa.

The retina is dependent upon the arterial supply of the ophthalmic artery (Figure 13.4). This artery arises from the cerebral part of the internal carotid artery medial to the anterior clinoid process. It enters the orbit via the optic canal accompanied by the optic nerve. It supplies the eye, forehead, dura matter, ethmoidal sinuses, and the nasal cavity. In the orbit, it frequently runs superior and then medial to the optic nerve, accompanied by the nasociliary nerve. It gives rise to the central retinal, anterior ciliary, and the long

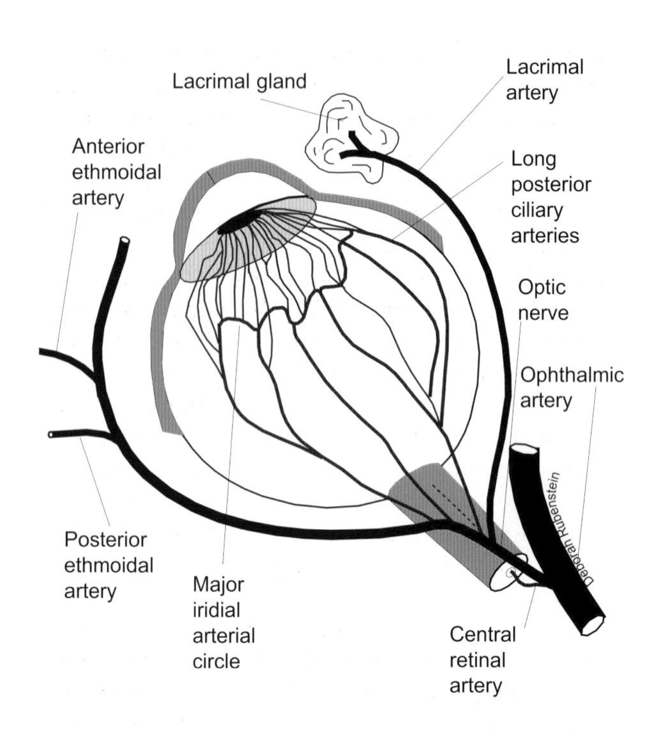

Figure 13.4 The ciliary and central retinal branches of the ophthalmic artery

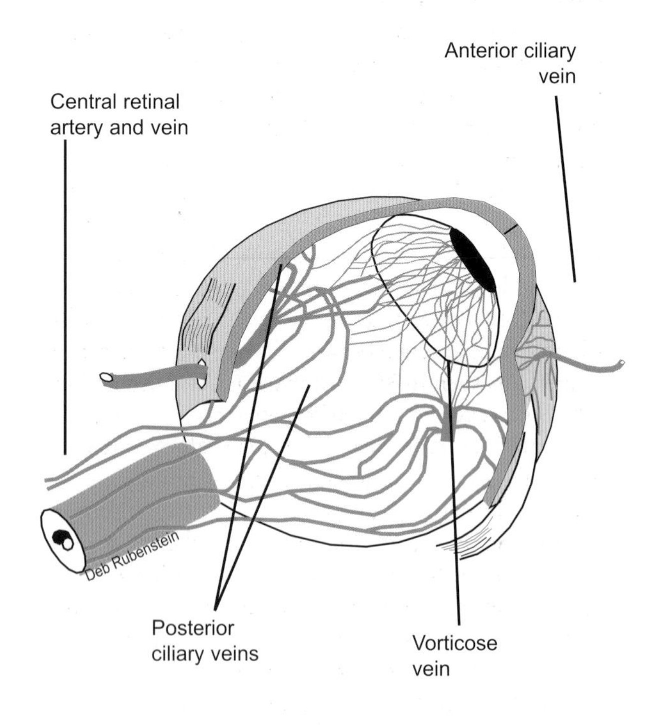

Figure 13.5 The central retinal vein and the anterior and posterior ciliary veins are illustrated

Occlusion of one branch of the central retinal artery may produce quadranopsia. A decrease of blood flow within the central retinal artery may indicate possible occlusion of the internal carotid artery. Incomplete occlusion of the internal carotid artery exhibits sudden transient monocular blindness in the form of a blackout or misty vision appearing as a shade or curtain which covers the visual field from side to side or from above without permanent visual loss (amaurosis fugax). This transient visual attack, which lasts from seconds to minutes, may also result from compression of the ophthalmic artery by the intraocular pressure subsequent to reduction in the pressure of the carotid system. Bilateral reduction in the blood pressure of the ophthalmic artery relative to the pressure of the brachial artery may indicate bilateral carotid disease.

Examination of the fundus of the eye may reveal in a normal person a more sharply defined temporal edge than the nasal edge. The optic disc appears pinkish in light-skinned persons and yellowish-orange in dark-skinned individuals. Pallor of the disc may suggest atrophy of the optic nerve. Retinal vessels radiate from the center of the optic disc and divide into branches that distribute to the retinal quadrants. Hypertension and arteriosclerosis alter the morphology of these vessels.

Essential or malignant hypertension may produce retinal exudate, hemorrhage into the plexiform layer of the retina, cotton wool patches, a lipid star in the macula, and irregular narrowing of the retinal arteries.

The physiologic cup is the lighter-colored central part of the disc, which is penetrated by retinal vessels. The normal cup-to-disc (c/d) ratio of 1:5 is genetically determined. Only 2% of normal eyes have a ratio more than 0.7. Unequal c/d ratios, in which the difference between the two eyes is more than 0.1, is seen in 8% of normal individuals and in 70% of patients with early glaucoma. A changing c/d ratio is significant because glaucomatous expansion of the optic cup is superimposed upon the amount of physiological cupping present before the onset of raised intraocular pressure. During the early stages of glaucoma the increase in size of a small cup may not be detected because its dimensions may still be smaller than the physiological cup. Therefore, estimation of the cup size does not by itself carry diagnostic value, unless the increase is profound. Glaucomatous cups are usually larger than physiological cups, although a large cup may not pathological.

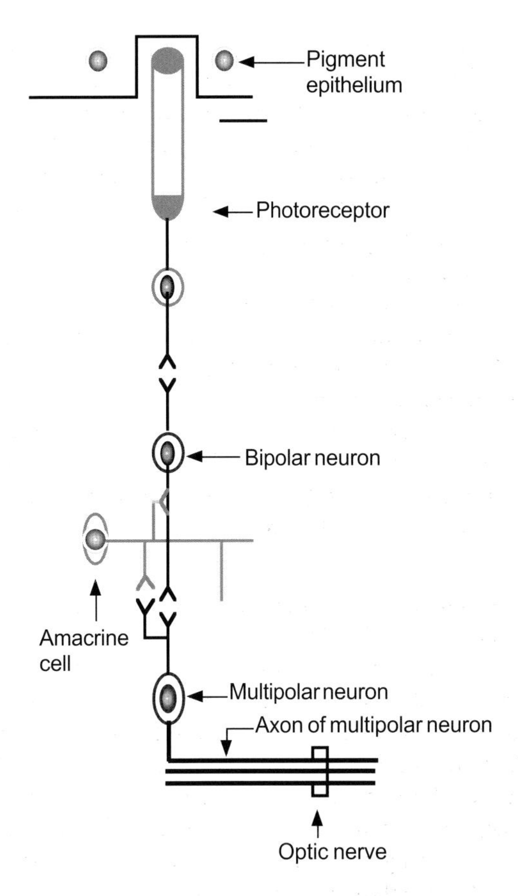

Figure 13.6 Neuronal series associated with transmission of visual image to the optic nerve

and short posterior ciliary arteries. The central retinal artery (Figure 13.4) penetrates the optic nerve near the eyeball and divides into four branches. These branches are end arteries that supply the four quadrants of the retina. The branches appear thinner and brighter red than the corresponding vein with a normal artery-to-vein ratio of 2:3. In hypertensive individuals, the central retinal artery may exhibit narrowing or spasm and become thickened or sclerotic, changing color to orange-metallic.

The central retinal vein (Figure 13.5) may be concealed by the more superficial and widened arterial wall branches that give the appearance of a discontinuous venous column. The long and short posterior ciliary arteries supply the choroid and ciliary processes and establish anastomosis with branches of the central retinal artery.

Visual pathways

The optic nerve (Figures 13.1, 13.6 &13.11) is formed by the axons of the ganglionic layer of the retina, acquiring myelin outside the eyeball.

The arcuate fibers reaching the superotemporal and inferotemporal aspects of the optic nerve are most vulnerable to glaucomatous insult and the fibers of the papillomacular bundle are the most resistant. The nasal fibers of the optic nerve partially decussate in the optic chiasma, while the temporal fibers continue on the same side as part of the optic tract. Fibers from the lower nasal quadrant form a short loop into the medial part of the contralateral optic nerve prior to joining the optic tract (anterior knee fibers of von Willebrand). This accounts for superior temporal quadranopsia in the contralateral eye, which accompanies optic nerve lesion. Fibers from the superior nasal quadrant of the retina also form a short loop (posterior knee fibers of von Willebrand) extending into the ipsilateral optic tract. The optic nerve also forms the common afferent limb for both the pupillary light and accommodation reflexes. It enters the orbit through the optic canal and is enfolded by the meninges. Stenosis, Paget's disease, suprasellar tumors, or fractures involving the optic canal may damage the optic nerve.

The optic nerve may undergo inflammation outside the eyeball (retrobulbar neuritis), or intraocularly (papillitis). Retrobulbar (optic) neuritis produces unilateral visual deficits and may be caused by multiple sclerosis (MS), measles, mumps, or varicella viruses. Optic neuritis in MS patients may produce scotoma and defects in temporal halves of the visual fields. Papillitis produces swelling of the margins of the optic disc as a result of a local inflammatory process

Papilledema (choked disc) is a condition characterized by bilateral passive elevation of the margins of the optic discs as a result of increased intracranial pressure. Since the subarachnoid and subdural spaces of the brain also extend around the optic nerve, increased intracranial pressure can be transmitted along these spaces, producing edema around the nerve and retardation of venous drainage. Tumors which involve the optic nerve sheath, tectum, cerebellum, fourth ventricle (ependymoma), cerebral hemisphere, and the corpus callosum may also produce papilledema. Cerebellar tumors (e.g. medulloblastoma) may protrude into the fourth ventricle and obstruct the pathway of the cerebrospinal fluid, producing increased intracranial pressure and papilledema earlier than any other tumors of the CNS. Cavernous sinus thrombosis as well as sinusitis may contribute to this condition by impeding the venous blood flow.

It is rarely seen in congenital cyanotic conditions of the heart or in Guillain–Barré syndrome (an idiopathic acute febrile inflammatory disease that produces polyneuropathy). Since pontine or medullary tumors do not generally interfere with the circulation of the cerebrospinal fluid, these masses do not usually induce papilledema. Therefore, patients may die from brainstem compression before developing papilledema. Papilledema can be detected by examining the dilated retinal veins in the fundus of the eye. Inflammation of the optic disc is suspected when exudate and hemorrhage with moderate elevation of the disc margin are present. Headache, nausea, vomiting, hemiparesis, hemianopsia, and diplopia due to involvement of the abducens nerve may also be seen in individuals with papilledema.

Embryologically, it develops with the retina as an extension of the telencephalon. Fibers arising from the macula follow a straight course to the optic disc forming a spindle-shaped area termed the papillomacular bundle. Those fibers arising from the nasal retina follow a relatively straight course. Fibers of the temporal retina follow an arcuate path around the papillomacular bundle to reach the optic disc.

The optic chiasma (Figures 13.7, 13.12, 13.14 & 13.16) is formed by the nasal fibers of the optic nerve which decussate rostral to the hypothalamus and tuber cinereum and superior to the pituitary gland.

The optic tract (Figures 13.8, 13.9, 13.10, 13.11, 13.13, 13.14 & 13.15) is formed by the crossed nasal and ipsilateral temporal fibers of the optic nerve. Each optic tract carries impulses to the opposite visual field, and runs adjacent to the internal carotid artery, hypothalamus, cerebral peduncles, and thalamus. The fibers which originate from the macula lutea occupy an intermediate position, fibers from the upper quadrant of the retina reside in an anterior and medial location, while the fibers from the lower quadrant of the retina occupy a more lateral and posterior position in the optic tract. Most of the fibers of the optic tract project to the lateral geniculate body (LGB), a visual relay nucleus of the thalamus, where they establish synaptic links with its neurons. Some fibers of the optic tract bypass the LGB and terminate in the pretectum and tectum, which contains the efferent neurons for the pupillary light reflex. The fibers that bypass the LGB enter the superior colliculus and the pretectum, and activate the Edinger–Westphal nucleus of both sides via the posterior commissure. The Edinger–Westphal nucleus provides preganglionic parasympathetic fibers to the ciliary ganglion that controls the curvature of the lens and the contraction of the sphincter pupillae muscle through the short ciliary nerves.

Due to the close relationship of the optic chiasma to the adenohypophysis, adenomas of the pituitary or craniopharyngioma may compress the optic chiasma and disrupt the nasal fibers of the retina, producing bitemporal heteronymous hemianopsia (tunnel vision). Bitemporal heteronymous hemianopsia (Figure 13.14) may be suspected if a patient can read left-hand letters only with the right eye, and right-hand letters with the left eye. Binasal heteronymous hemianopsia may be produced by aneurysms of the internal carotid arteries. Unilateral nasal hemianopsia may possibly occur if the aneurysm of the internal carotid artery is ipsilateral. This type of aneurysm may mimic a pituitary tumor, producing visual deficits and radiographically detectable sellar enlargement. A chiasmal lesion may also produce junctional scotoma, superior or inferior bitemporal quadranopsia, or monocular temporal hemianopsia.

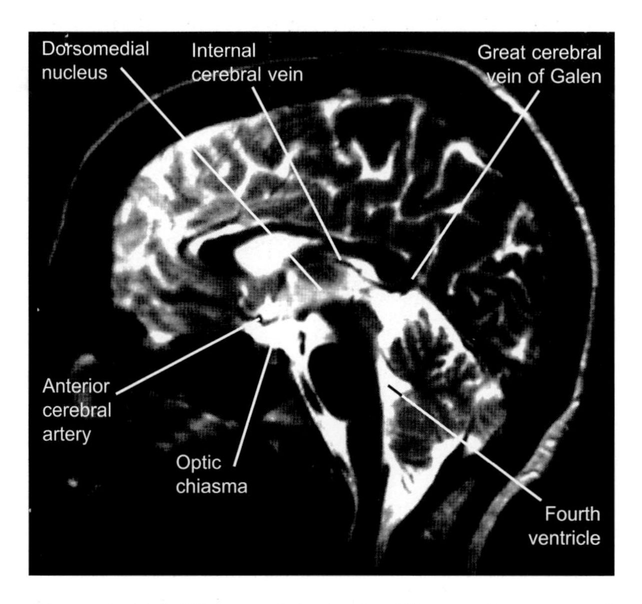

Figure 13.7 MRI scan of the brain. Observe the course of the optic chiasma and its relationship to the anterior cerebral artery

Destruction of the optic tract produces homonymous hemianopsia on the contralateral visual field (Figure 13.12). Due to proximity of the internal carotid artery to the optic tract, aneurysm of this vessel may also compress the ipsilateral (temporal) fibers of the optic tract, resulting in nasal hemianopsia (Figure 13.10).

Tumors or lesions involving the optic tract near the optic chiasma may result in congruous or complete homonymous hemianopsia. A lesion which involves the area medial to the lateral geniculate body, as is seen in tabes dorsalis or tertiary syphilis, may selectively disrupt the fibers that mediate constriction of the pupil in the light reflex, leaving the accommodation reflex arc unaffected. Individuals with this type of lesion may exhibit pupillary response in accommodation but not in response to light (Argyll Robertson pupil).

The lateral geniculate body (LGB) is a visual relay nucleus which lies lateral to the posterior part of the crus cerebri and ventral to the pulvinar and consists of six layers (Figures 13.9, 13.10, 13.11, 13.14, 13.15 & 13.16). Layers 1, 4, and 6 receive fibers from the contralateral retina, while layers 2, 3 and 5 receive fibers from the ipsilateral retina. Layers 1 and 2 form the ventral (magnocellular) subnucleus and layers 3 to 6 comprise the dorsal (parvicellular) subnucleus. Most fibers of the optic tract terminate in the lateral geniculate body. Fibers that bypass the LGB synapse in the pretectal area and the superior colliculus. The synaptic connections between the optic tract and the neurons of the lateral geniculate nucleus are somatotopically arranged. The medial part of the lateral geniculate body receives fibers from the upper retinal quadrant; the lateral part receives fibers from the lower retinal quadrant; and the central part of the lateral geniculate body receives fibers from the macula. The axons of the dorsal (parvicellular) subnucleus of the LGB neurons form the geniculocalcarine tract (optic radiation). The medial part of this projection terminates in the superior bank of the calcarine fissure, whereas the lateral part projects in a similarly precise manner to the inferior bank of the visual cortex.

The optic radiation (Figures 13.9, 13.11, 13.1, 13.14 & 13.15) represents the postsynaptic fibers of neurons of the dorsal (parvicellular) subnucleus of the lateral geniculate body. The optic radiation (geniculocalcarine tract), shaped like a crescent, has ventral and dorsal parts which course within the retrolenticular part of the internal capsule en route to the visual cortex. Each part represents one fourth of the visual field of the contralateral side. The ventral part of the optic radiation known as Meyer's loop forms a curve into the temporal lobe running adjacent to the tip of the inferior horn of the lateral ventricle. It terminates in the lower bank of the calcarine fissure (lingual gyrus). The upper fibers of the optic radiation end in the upper bank of the calcarine fissure, represented by the cuneus. Fibers derived from the upper

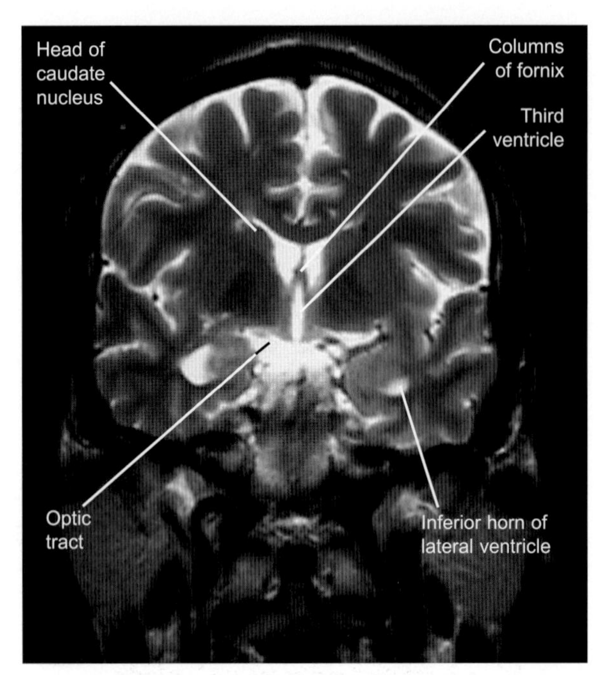

Figure 13.8 MRI scan of the brain (coronal view). Note the course of the optic tract in relation to the third ventricle

Since the ventral portion of the optic radiation (Meyer's loop) follows a separate course within the temporal lobe before joining the bulk of the geniculocalcarine tract, a selective damage to the optic radiation, a frequently occurring lesion, may produce quadranopsia in the opposite visual field. Edema caused by bleeding from the medial striate artery (a branch of the middle cerebral artery) may also compress the optic radiation, resulting in transient homonymous hemianopsia which lasts until the edema subsides. Vascular lesions affecting the optic radiation may also be caused by occlusion of the anterior choroidal and posterior cerebral arteries. An abscess that develops in the temporal lobe, above the level of the auditory meatus, may also compress and disrupt the fibers of Meyer's loop, producing quadranopsia in the contralateral visual field. Homonymous visual field defects due to lesions of Meyer's loop tend to be incongruous. Those due to damage to the optic radiation near the visual cortex are congruous (edges of the visual field defect in each eye is identical in shape).

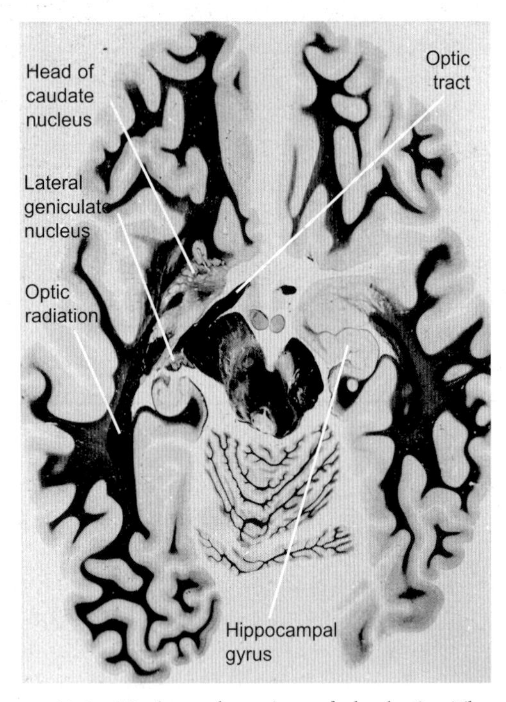

Figure 13.9 Horizontal section of the brain. The optic tract, lateral geniculate body, and the optic radiation are prominently displayed

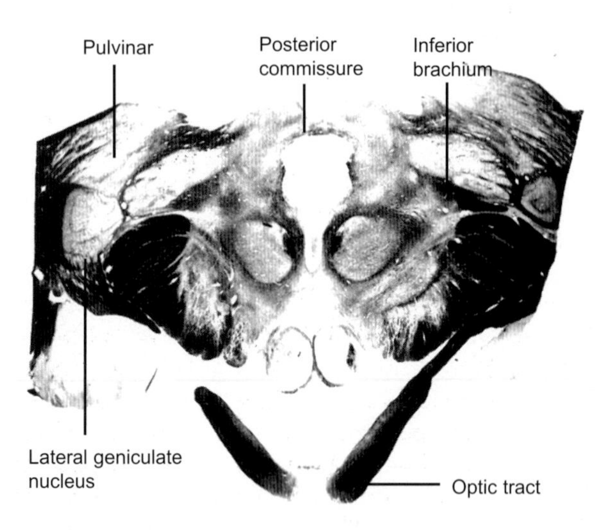

Figure 13.10 In this section the optic tract and the lateral geniculate bodies are also clearly visible

retinal quadrant run in the superior part of the optic radiation, whereas fibers from the lower retinal quadrant fibers shift to the lower part of the optic radiation. The foveal fibers occupy the most lateral position of the optic radiation.

The primary visual or striate cortex (Brodmann's area 17) is the principal area for visual perception, integration, and formation of binocular image. It has a point to point connection with the lateral geniculate body (Figures 13.12, 13.15 & 13.16). Due to this precise connection, a small lesion in the visual cortex may result in scotoma (focal blindness). Area 17 includes portions of the lingual and cuneate gyri, extending to the lateral surface of the occipital lobe. It consists of a very thin granular cortex, in which

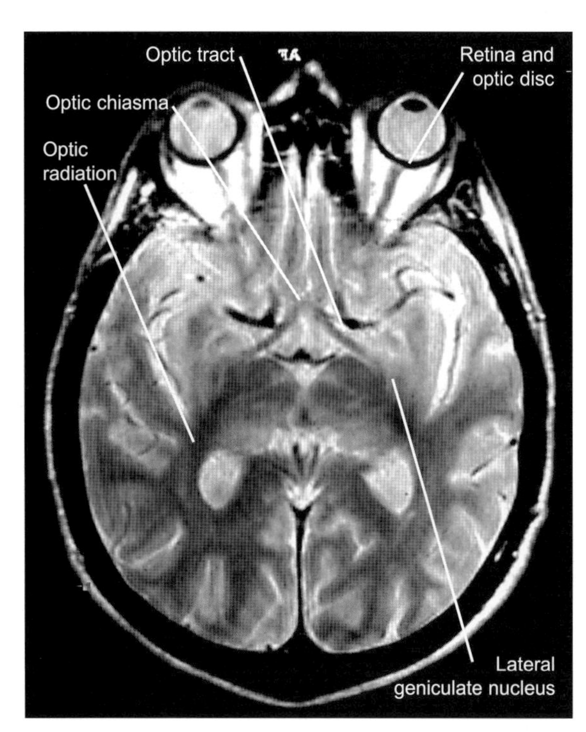

Figure 13.11 This MRI scan illustrates some of the elements associated with the visual system. Note the optic nerve, optic chiasma, optic tract, lateral geniculate nucleus, and the optic radiation

layer IV is divided into densely packed upper and lower sublayers and a lighter middle layer with fewer small cells between the giant stellate cells. The light middle layer has a thickened outer band, which is visible to the naked eye in sections of the fresh brain, is known as the band of Gennari. Area 17 receives information from all neurons of the lateral geniculate body and projects to Brodmann's areas 18 and 19.

The interconnection between areas 17 of both cerebral hemispheres is not well developed. Visual fibers that reach the pulvinar from the contralateral visual field project to layers I, III and IV of the cortical areas 18 and 19 and also to the supragranular layers of Brodmann's area 17. The latter projection constitutes the extrageniculate visual pathway.

Perception of visual images (e.g. an individual's ability to read an article if brought into focus) remains intact even with bilateral damage to the striate cortex as long as the occipital pole is spared. Bilateral destruction of the occipital poles, on the other hand, markedly impairs the ability to clearly and accurately observe visual fields.

Dark bars against a light background, and straight edges separating areas of different degrees of brightness effectively stimulate the visual cortex. The primary visual cortex consists of functional units arranged in columns of cells exhibiting different

Macular sparing is a phenomenon in which a lesion involving the occipital lobe or occipital pole results in no detectable deficit of the central vision. This is thought to be due to the efficient arterial anastomosis and development of collateral circulation between the middle and posterior cerebral arteries. Bilateral representation of the macular area in both cerebral hemispheres is also considered a possible reason for sparing of macular vision. However, cortical blindness and contralateral incongruous homonymous hemianopsia may also be seen in individuals with intact central vision.

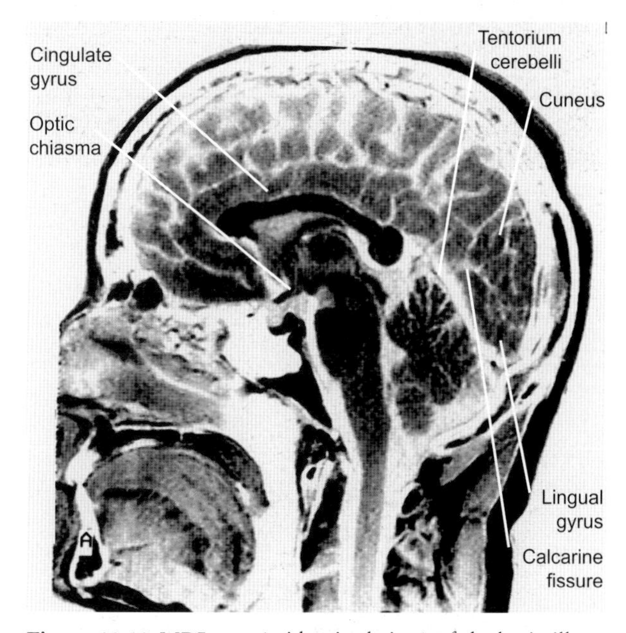

Figure 13.12 MRI scan (midsagittal view) of the brain illustrating the visual cortex within the lingual gyrus and cuneus

receptive fields. These functional units include the ocular dominance and orientation columns that are arranged perpendicular to the cortical surface.

The ocular dominance columns are partially formed at birth, receive visual information from both eyes, and are arranged in such a manner that one eye will be dominant. Segregation of the visual impulses into right and left laminae of the dominance column occurs in layer IV. However, no ocular dominance columns exist in the parts of the striate cortex which receive impulses from the optic disc, and the most peripheral temporal visual field of the ipsilateral eye. The orientation columns are smaller than the dominance columns and extend from the white matter to the pial surface of the cerebral cortex. These orientation columns contain cells which possess the

Unilateral or bilateral occlusion of the posterior cerebral artery (Figure 13.13) is commonly associated with a variety of deficits and syndromes. Infarction of the posterior cerebral artery is the most common etiology of visual deficits of occipital lobe origin (Figure 13.11). Transient occlusion of the vertebral arteries on both sides, which may occur as a result of cervical spondylosis and subsequent narrowing of the transverse foramina, may dramatically reduce the blood flow in the labyrinthine and posterior cerebral arteries. A patient with this condition may experience vertigo and transient blindness, which last for a few seconds, without remembering that these disorders ever have happened.

Anton's syndrome (cortical blindness) is an expression of psychological ramification of cortical blindness which is caused by disruption of the corticothalamic connection between area 17 and the thalamus. It is also observed in individuals with non-dominant hemispheric damage. It commonly results from bilateral occlusion of the posterior cerebral arteries. Patients have normal sized and reactive pupils and may show indifference or pay no attention to half of the visual field of the affected side. They are generally unaware of their blindness. Patients attempt to name objects in the visual field and describe the surrounding objects, though they cannot tell illuminated from non-illuminated areas. They consistently deny their blindness and insist that poor lighting or disinterest is the cause of their visual problems.

Occlusion of the posterior cerebral arteries may result in bilateral degeneration of the parieto-occipital cortex between Brodmann's areas 19 and 7, producing signs and symptoms of Balint syndrome. This syndrome is characterized by the inability to appreciate or scan the peripheral visual field (due to lack of coordination between the visual cortex and the oculomotor system) or use visual cues to grasp an object. Infarction of the posterior cerebral artery may also produce a combination of hemianopsia or quadranopsia, macular sparing, and hemianesthesia with no muscle paralysis. If the infarct involves the dominant hemisphere, Charcot–Wilbrand syndrome may develop, producing visual agnosia.

Gertsmann syndrome, transcortical sensory aphasia, and alexia without agraphia are also seen in posterior cerebral artery infarcts.

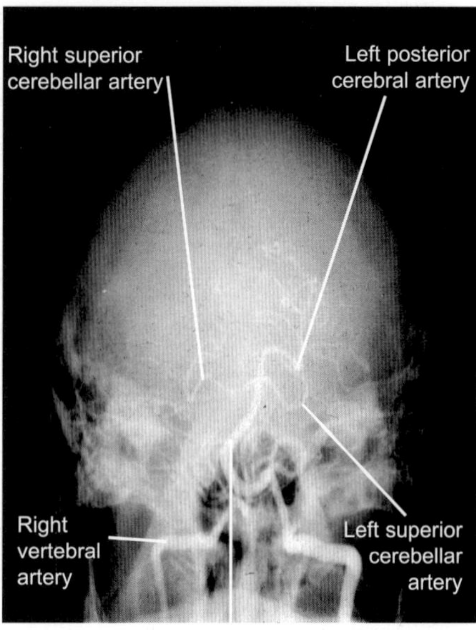

Figure 13.13 In this angiogram of the vertebrobasilar system, the right posterior cerebral artery is obstructed

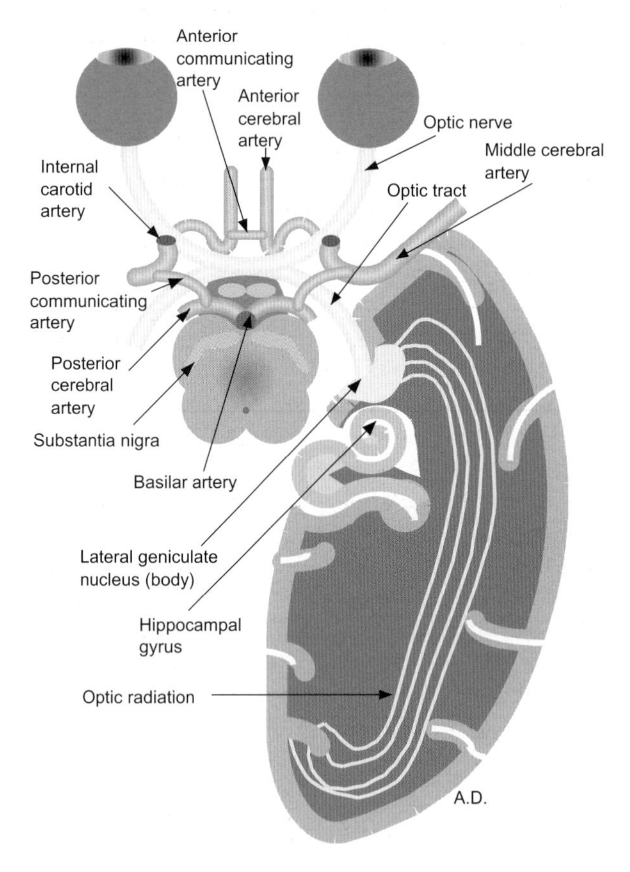

Figure 13.14 Drawing of the visual pathway from the optic nerve to the visual cortex in the occipital lobe. The relationship of the arterial circle of Willis to the optic nerve, optic chiasma, and optic tract is shown

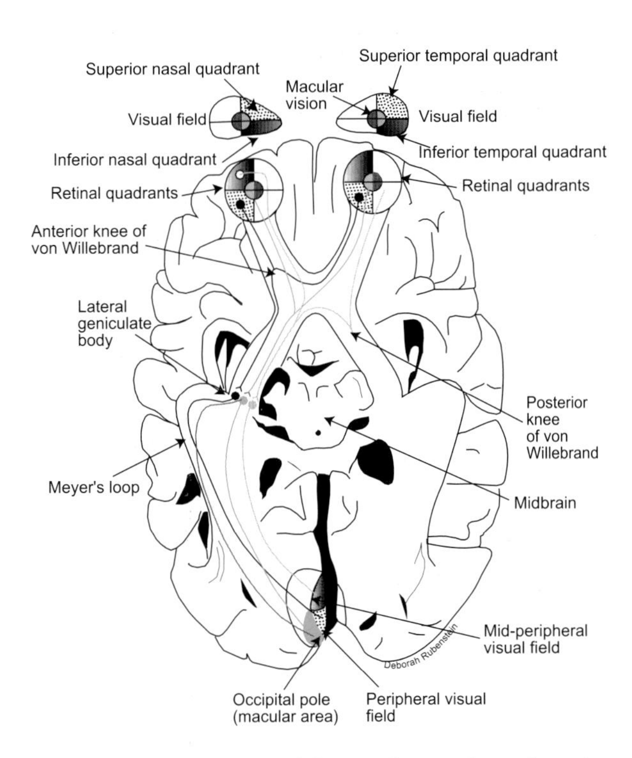

Figure 13.15 The course of the visual image from the retina through the optic nerve, optic tract, lateral geniculate body, optic radiation, to the visual cortex is shown

Amblyopia (lazy eye) is a disorder that develops from a prolonged suppression of an image in one eye between the second and fourth year of life. It may be the result of congenital strabismus and the inadequate stimulation of one eye by visual image. It occurs in children who exhibit diplopia as a sequel to functional imbalance between the extraocular muscles and subsequent attempts to eliminate the image in one eye by constantly utilizing the other eye. As the crossed-eyed child favors one eye over another, the unused eye eventually loses visual acuity and may permanently be blind. In this condition no deficits are recorded in the refractive media or ocular apparatus. Amblyopia may also occur as a result of nutritional deficiency and in alcoholics. This condition may be associated with bilateral scotoma subsequent to damage to the optic nerves. Blurred vision and optic atrophy may also occur in this condition.

same receptive field axis of orientation and have 'on' and 'off' centers. The visual cortex is primarily supplied by the posterior cerebral artery, although the middle cerebral artery also contributes blood supply through its anastomotic connections.

The secondary visual cortex (Brodmann's area 18) adjoins the striate cortex, regulates visual memories, and receives visual impulses from Brodmann's area 17. It is a mirror image representation of the Brodmann's area 17, which consists of a six-layered granular cortex. It interconnects areas 17 and 19 and does not contain the band of Gennari. This cortex, as in the case of area 17, responds best to dark bars and edges. The majority of cells in Brodmann's area 18 are complex cells arranged in columns. Usually, in the dominant hemisphere the upper lateral portion of Brodmann's area 18 deals with memories for inanimate objects while the lower medial portion is concerned with memories for living parts or individuals. In order for the visual object to be recognized, information must project to Brodmann's area 18 of the dominant hemisphere (for fine feature analysis) via the splenium of the corpus callosum.

Disruption of the connection between Brodmann's area 18 of both cerebral hemispheres may occur upon excision of the corpus callosum producing unilateral visual agnosia. Patients with this type of deficit are unable to recognize images received by the right (non-dominant) hemisphere of the brain. Bilateral visual agnosia results from a lesion of Brodmann's area 18 in the dominant cerebral hemisphere. Visually agnostic patients cannot recognize objects without using tactile, auditory, gustatory, or olfactory clues.

Lesions that damage the upper lateral or lower medial parts of the secondary visual cortex in the left dominant hemisphere may result in autotopagnosia, which is characterized by failure of the patient to distinguish living people from objects. Lesions of the dominant hemisphere confined to the upper part of Brodmann's area 17 and the occipital association cortex adjacent to the angular gyrus produce finger agnosia. This condition manifests inability to name objects, identify fingers, write, do arithmetical calculations, or recognize left from right. Achromotopsia, the inability to recognize color in only one-half of the visual field, may occur independently. Stimulation of Brodmann's area 18 results in visual hallucinations in the form of sparkling lights.

The tertiary visual cortex (Brodmann's area 19), a mirror image of Brodmann's area 18, occupies the area lateral to the secondary visual cortex. It is responsible for recalling (revisualizing) formerly seen images. The hypercomplex cells are the primary neurons in Brodmann's area 19 that receive visual input from both eyes. Stimulation of the tertiary cortex produces colorful visual images of moving

Lesions involving the parietal lobe and Brodmann's area 19 of the occipital lobe may cause dysfunction similar to astereognosis. Therefore, the ability to recall objects by using tactile stimuli may be lost.

Visual changes in migraine headaches include blurred vision, flashing lights, wavy lines and scotoma. Hemiparesis, ophthalmoplegia or aphasia may accompany these symptoms. Bilateral disruption of the connections between the association visual cortices and the entorhinal cortex (Brodmann's area 28) may occur as a result of basilar artery insufficiency that extends to involve the posterior cerebral arteries. This disconnection which is associated with degeneration of the occipitotemporal area results in anterograde visual amnesia and the difficulty in visually adapting to new and unfamiliar territory despite intactness of the visual apparatus.

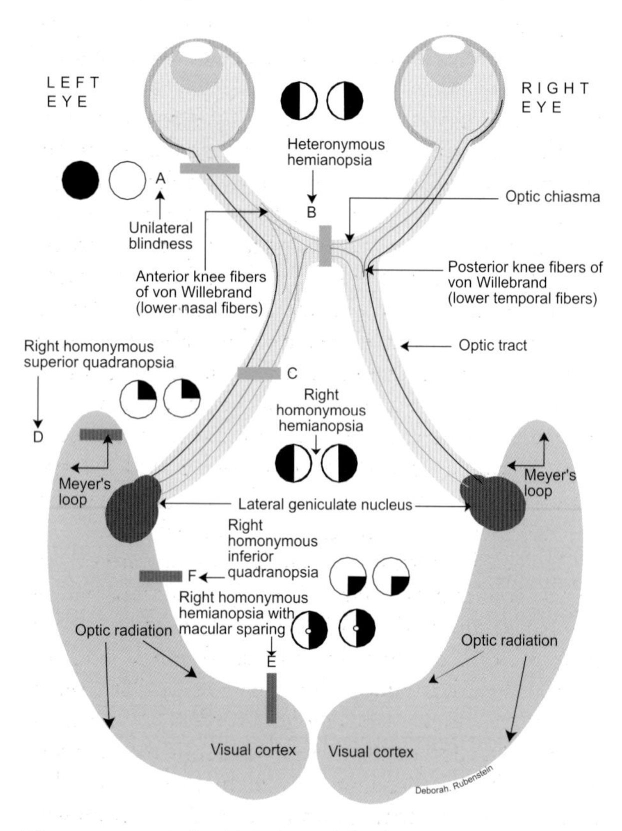

Figure 13.16 A detailed view of the lesions associated with the visual system and pertinent dysfunctions

events and objects. The middle part of area 19 relates to the macula and object sizes, whereas the inferior part of this area responds exclusively to color. Movement activates a small area, anterior to the macular zone of area 19. In order for the images to be

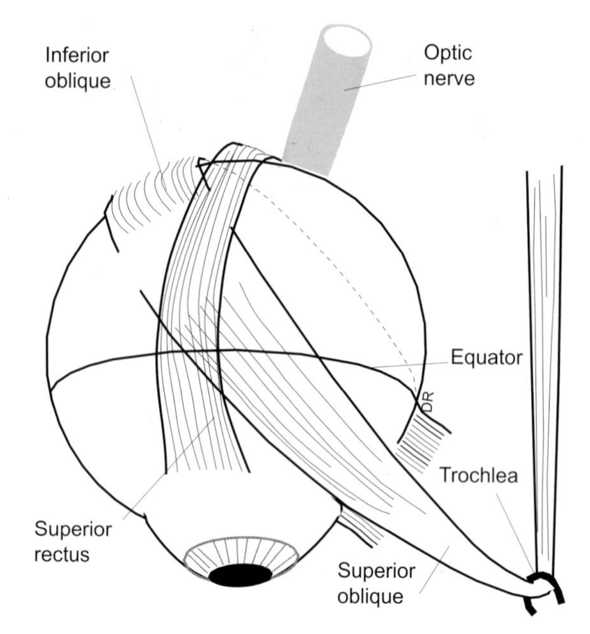

Figure 13.17 Diagram of the eyeball demonstrating the sites of attachment of the oblique muscles in relation to the optic axis

recalled, visual information must project from Brodmann's areas 18 to area 19 where they are activated by various types of stimuli (e.g. auditory, tactile, olfactory, etc.). It is important to note that recalling symbols is a function of the angular gyrus.

It is important to remember that the inferior temporal gyrus serves as a visual association cortex, contains visual memory stores, and receives input from the entorhinal cortex (Brodmann's area 22) and Brodmann's areas 7, 18 and 19. These connections may explain the visual hallucination associated with temporal lobe epilepsy, as well as the vivid scenes experienced by a patient undergoing brain operation. Beginning by the fifth postnatal month, the visual association cortices (Brodmann's areas 18 and 19) become involved in stereopsis, a mechanism that enables the brain to measure the incongruity between the two retinal images, thus constructing a complete three-dimensional image.

Ocular movements

Eye movements provide a significant index of the functional activity of the motor nuclei of the extraocular muscles and the neurons within the brainstem reticular formation. They are classified into conjugate (version) movements, where the visual axes of both eyes remain parallel (for far vision), and disconjugate (vergence) movement, in which the visual axes intersect (for near vision) by the contraction of the medial recti muscles (Figures 13.17, 13.18 & 13.19). Disconjugate (vergence) movements deal with

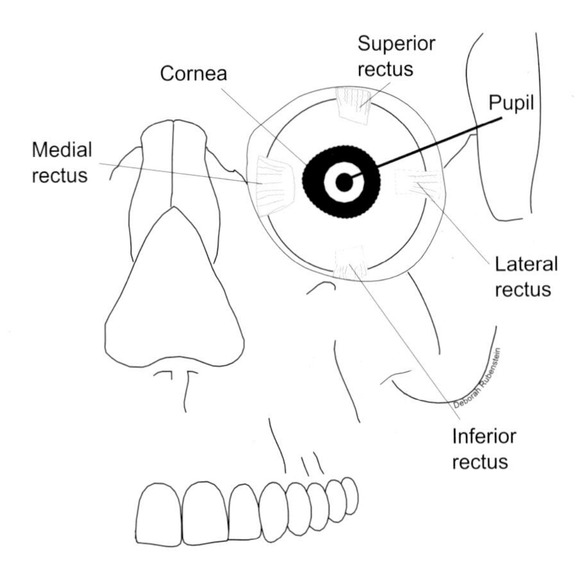

Figure 13.18 Diagram of the recti muscle

tracking of approaching (convergence) or receding (divergence) objects which require slow movements of the eyes in opposite directions. Conjugate movements depend upon the integrity of certain gaze centers and the medial longitudinal fasciculus (MLF). The MLF is the principal internuclear pathway that interconnects the motor nuclei of the extraocular muscles, coordinates eye movements, and ensures binocular vision.

Conjugate eye movements are further categorized into saccadic and smooth-pursuit movements. Saccadic eye movements are involuntary and rapid movements that include successive jumps of the eye from one point of visual fixation to another. These movements are mediated by the contralateral superior colliculus that projects to the ipsilateral parapontine reticular formation (PPRF) and the medial longitudinal fasciculus. The PPRF stimulates the ipsilateral abducens nucleus and subsequently the contralateral medial rectus neurons of the oculomotor nuclear complex. It is the only conjugate movement that could be produced voluntarily.

Saccadic movements are controlled by the contralateral frontal cortex and are not affected by sedatives or analgesics. They are lost in Huntington's chorea and ophthalmoplegia of supranuclear origin. Cerebellar diseases may produce overshooting and undershooting of saccadic movements.

A lesion which destroys the contralateral pontine lateral gaze center, ipsilateral frontal eyefield or the ipsilateral corticomesencephalic tract may result in the inability to look to the opposite side.

Saccades occur virtually in all voluntary eye movements with the exception of smooth pursuit eye movements. In contrast to smooth pursuit movement, visual acuity is diminished during saccades. Saccadic movements are used to improve reading speed by increasing the numbers of words read in a single fixation. Smooth-pursuit movement is a slow conjugate eye movement that becomes active in tracking moving targets. Cerebellar dysfunctions, administration of sedatives and analgesics produce fragmentation of this movement into a series of saccades. Corticotectal fibers that are derived from the occipital lobe (Brodmann's areas 17, 18 & 19) control involuntary smooth-pursuit eye movement.

Disorders of ocular movements

Disorders of ocular movements include nystagmus, conjugate gaze palsy, ocular dysmetria, oculogyric reflex, opsoclonus, ocular flutter, ocular bobbing, and ocular myoclonus.

- Nystagmus is an involuntary, rhythmic oscillation of the eye in response to an imbalance in the vestibular impulses (See also the vestibular system).
- Conjugate gaze palsy includes lateral gaze and vertical gaze palsies. Lateral gaze palsy refers to the inability to look to the side of the lesion that results from destruction of the abducens nucleus. Vertical gaze palsy is characterized by the inability to look upward. This condition results from destruction of the vertical gaze center in the superior colliculus.
- Ocular dysmetria denotes an error in ocular fixation, producing overshooting of the intended target followed by oscillation of the eyeball. This is commonly seen in cerebellar vermian lesion.
- Oculogyric reflex is characterized by upward or side to side rolling movements of the eyes accompanied by abnormal contractions of the facial muscles. It is a manifestation of acute dysgenic condition, a developmental disorder induced by neuroleptics. This reflex may be the result of metabolic disorders of dopamine and may be alleviated with anticholinergic medications.
- Opsoclonus (dancing eyes in infants) is another ocular disorder that exhibits a random, conjugate saccadic movement of the eyes in all directions with unequal amplitudes. It is a manifestation of pretectal lesions or viral encephalitis.
- Ocular flutter, seen in individuals with cerebellar lesions, is characterized by sudden, rapid, and spontaneous to-and-fro oscillations of the eyes. It is associated with blurred vision and may be seen with changes in fixation regardless of the direction of the gaze.

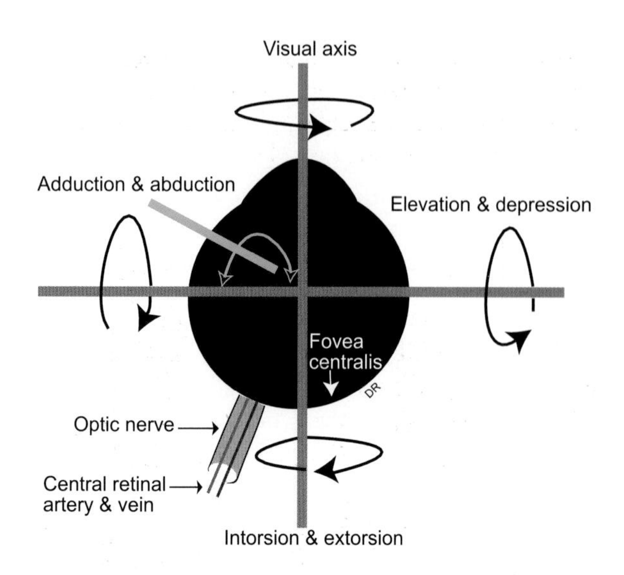

Figure 13.19 Diagram of the ocular movement around the visual axis. These movements encompass adduction, abduction, elevation, depression, intorsion, and extorsion

- Ocular bobbing refers to the fast, spontaneous (non-rhythmic) downward deviation of both eyes, followed by slow synchronous return of the eyes to the original position. This phenomenon may be seen in comatose individuals with lesions of the pons, cerebellum, or cerebral cortex.
- Ocular myoclonus is a term used to describe the rhythmic, rotatory, or pendular movements of the eyes synchronously with similar movements of the palatal, pharyngeal, laryngeal, lingual, and diaphragmatic muscles.
- Congenital ocular motor apraxia (Cogan syndrome) is a disorder of conjugate deviation of the eyes in which voluntary saccades are absent. The eye movements only occur when the head is in motion. The head, and not the eyes, abruptly turns to the side to visualize the object. The eyes move in the opposite direction of the movement.

Ocular reflexes

Ocular reflexes are comprised of the direct pupillary, consensual pupillary, accommodation, ciliospinal, oculocardiac, and oculoauricular reflexes, as well as the blink reflex of Descartes.

The direct pupillary light reflex (Figure 13.20) is produced by shining a beam of light into one eye and observing the pupillary constriction on the stimulated eye. This reflex is mediated by the optic nerve (afferent limb) and the oculomotor nerve (efferent limb). Information which is carried by the optic nerve is delivered to the optic tract, and bilaterally to the oculomotor nuclei. It is lost in Argyll Robertson

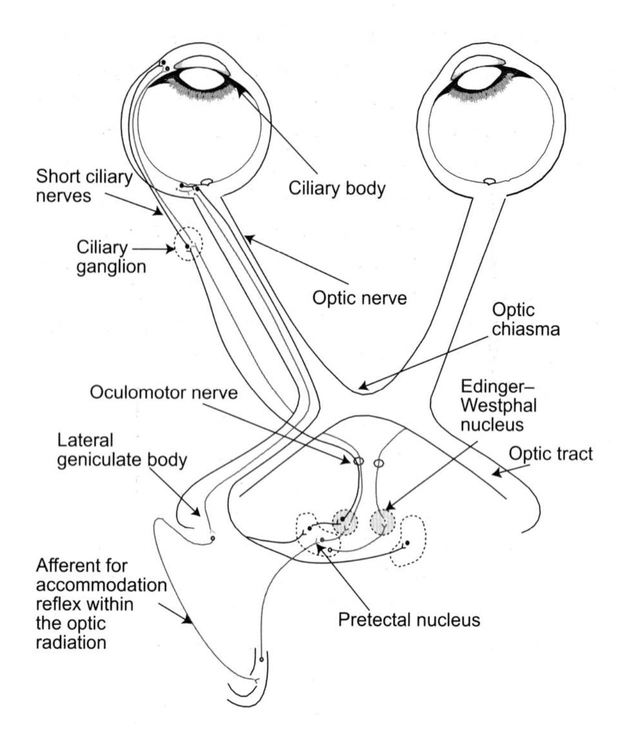

Figure 13.20 The reflex arcs of accommodation and pupillary light reflex

pupil, a pupillary disorder that occurs in neurosyphilis, diabetes mellitus, and individuals with epidemic encephalitis and alcoholism.

The consensual pupillary light reflex (Figure 13.20) is characterized by constriction of both pupils in response to application of light to one eye. It is mediated by the bilateral connection of the optic tract to the oculomotor neurons via the central commissural connections. Disruption of the optic tract fibers which are destined to the oculomotor nuclei, as a result of a lesion medial to the lateral geniculate body, produces manifestations of Argyll Robertson pupil. This exhibits loss of pupillary constriction in the light reflex, while maintaining it in accommodation.

The accommodation reflex ((Figure 13.20) exhibits certain changes in the eye which are associated with near vision. These changes include convergence (adduction of the eyes), miosis, and increased curvature of the lens. It requires the utilization of the visual cortex as well as the optic nerve, optic tract, and oculomotor nuclei.

The ciliospinal reflex exhibits pupillary dilatation in response to painful stimulation of a dermatomal area (e.g. pinching the neck or face). This reflex is dependent upon the integrity of the cervical postsynaptic sympathetic fibers as well as the presynaptic neurons of the first and second thoracic spinal segments.

The oculocardiac reflex is characterized by bradycardia (slowing of heart rate) in response to a pressure applied on the eyeball. It is mediated in the medulla by the ophthalmic nerve's (afferent limb) connections, via interneurons, to the spinal trigeminal nucleus, dorsal motor nucleus of the vagus, and the cardiovascular center in the medulla (efferent limb).

The oculocephalic reflex (doll's eye movement) is produced by rotating the head to one side with the eyes held open. This abrupt head movement results in initial movements of both eyes contralaterally followed by movements to the midline, regardless of the direction of rotation. This movement is dependent upon the integrity of the vestibular, oculomotor, and abducens nerves and nuclei, and the medial longitudinal fasciculus. It is is inhibited in the awake individual by the descending cortical influences. Closure of the eyelids facilitates this reflex by eliminating the cortical input. Patients who have bilateral cortical lesions, as in comatose individuals, with intact brainstem connections between the oculomotor nerve and the vestibulocochlear nuclei, exhibit a brisk doll's eye movement.

Loss of the oculocephalic reflex is an ominous finding of metabolic depression or a sign of a lesion in the brainstem that has disrupted the connection between the oculomotor and vestibular nerves. Suppression of the ascending reticular activating system and loss of consciousness occurs when the lesion is located rostral to the pontine and midbrain gaze centers. Therefore, loss of this reflex in comatose patients may indicate that the trauma has damaged the caudal pons and did not spare the lateral gaze center, requiring urgent intervention. Impaired oculocephalic response may also occur as a result of malpositioning or inadequate head rotation.

The oculoauricular reflex is elicited by asking the patient to look to the extreme temporal side, which induces contraction of the posterior auricular muscles and the subsequent movement of the ear posteriorly, contralateral to the stimulated side. This reflex is absent in Bell's palsy.

The blink reflex of Descartes is produced by an object that abruptly and unexpectedly approaches the eye. This reflex is mediated by the optic and facial nerves and is characterized by contraction of the orbicularis oculi in response to this stimulus.

Gaze centers

Gaze centers are represented by the lateral and vertical gaze centers in the pons and midbrain, respectively.

Lesions of the contralateral pontine lateral gaze center, the ipsilateral frontal eyefield, or the ipsilateral corticomesencephalic tract may produce gaze palsy to the opposite side.

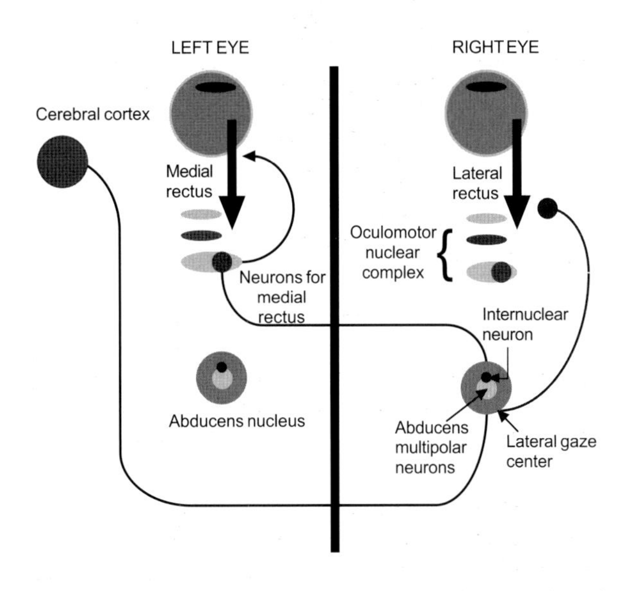

Figure 13.21 The role of the abducens nucleus as a lateral gaze center in adduction of the contralateral eye and abduction of the ipsilateral eye. The role of the cerebral cortex in influencing eye movement is also illustrated

The lateral gaze (horizontal) center (Figure 13.21) is located in the abducens nucleus and the adjacent parapontine reticular formation (PPRF). This region includes a pulse generator for fast eye movements and an integrator that determines the ultimate resting position of the eye. It controls the contralateral medial rectus muscle, and the ipsilateral lateral rectus muscle. The corticotectal tract, which is derived from the frontal eyefield (Brodmann's area 8), carries information that projects to the contralateral gaze center, regulating contralateral voluntary conjugate eye movements. Corticotectal fibers that are derived from the occipital lobe (Brodmann's areas 17,18 and 19) control involuntary smooth pursuit eye movement.

The vertical gaze center is located in the superior colliculus of the midbrain. Parinaud's syndrome (Figure 13.22) is characterized by the inability to gaze upward, manifesting weakness of convergence and sometimes loss of pupillary light reflex and mydriasis. It primarily occurs as a result of a lesion in the superior colliculi or the posterior commissure, subsequent to a pineal gland tumor. The supranuclear mechanism for upward gaze is situated closer to the third ventricle than the center for downward gaze.

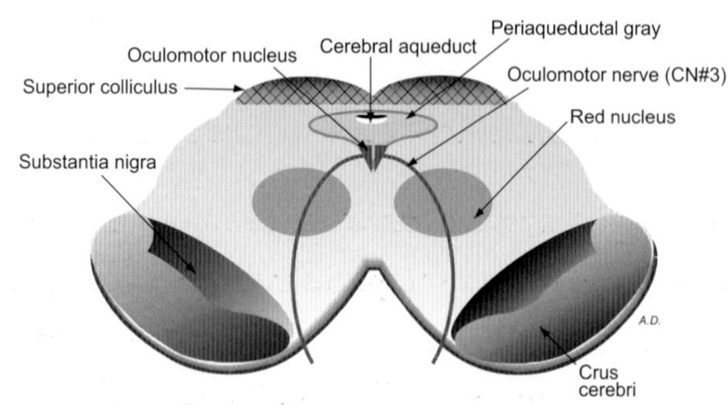

Figure 13.22 Section of the midbrain showing the lesion associated with Parinaud's syndrome. In this syndrome the vertical gaze center which is represented in the superior colliculus is disrupted

Posterior tumors of the third ventricle may result in paralysis of upward gaze (not downward gaze). Upward gaze palsy may also be seen in individuals with subdural hemorrhage or hydrocephalus. Damage to the rostrally situated interstitial nucleus of Cajal, which lies dorsal and medial to the MLF, results selectively in downward gaze palsy. Posterior thalamic hemorrhage may be associated with downward deviation of the eye. Pretectal syndrome results from vascular occlusion or neoplasms that are confined to the pretectum or the tectum and is characterized by bilateral paralysis or paresis of vertical gaze, nystagmus, and lid retraction.

Suggested reading

1. Anderson JC, Martin KAC, Whitteridge D. Form, function, and intracortical projections of neurons in the striate cortex of the monkey Macacus nemestrinus. *Cereb Cortex* 1993;3:412–20
2. Blasdel GG. Differential imaging of ocular dominance and orientation selectivity in monkey striate cortex. *J Neurosci* 1992;12:3115–38
3. Boussaoud D, Barth TM, Wise SP. Effects of gaze on apparent visual responses of frontal cortex neurons. *Exp Brain Res* 1993;93:423–34
4. Breen LA, Hopf HC, Farris BK, Guttmann L. Pupil-sparing oculomotor nerve palsy due to midbrain infarction. *Arch Neurol* 1991;48:105–6
5. Garey LJ. Visual system. In Paxinos G, ed. *The Human Nervous System.* New York: Academic Press, 1990:945–77
6. Gaymard B, Pierrot-Deseilligny C, Rivaud S, Velut S. Smooth pursuit eye movement deficits after pontine nuclei lesions in humans. *J Neurol Neurosurg Psychiatry* 1993;56:799–807
7. Gilbert CD. Circuitry, architecture, and functional dynamics of visual cortex. *Cereb Cortex* 1993;3:373–86
8. Hogan MI. Role of the retinal pigment epithelium in macular disease. *Trans Am Acad Ophthalm Otol* 1972;76:64–80
9. Hogan MJ, Alvarado IA, Weddell JE. *Histology of the Human Eye.* Philadelphia: W. B. Saunders Co., 1971
10. Holmes G. Disturbances of vision by cerebral lesions. In Phillips CG, ed. *Selected Papers of Gordon Holmes.* Oxford: Oxford University Press, 1979:337–67
11. Huerta MD, Harting IK. The mammalian superior colliculus studies of its morphology and connections. In Vanegas H, ed. *Comparative Neurology of the Optic Tectum.* New York: Plenum, 1984:687–773
12. Kaas JH. Theories of visual cortex organization in primates. *Cereb Cortex* 1997;12:91–125
13. Legothetis NK, Sheinberg DL. Visual object recognition. *Annu Rev Neurosci* 1996;19:577–621
14. Leigh RJ, Zee DS. *The Neurology of Eye Movements,* 2nd edn. Philadelphia: F. A. Davis, 1991
15. Mays LE. Neural control of vergence eye movements: convergence and divergence neurons in the midbrain. *J Neurophysiol* 1984;51:1091–108
16. Moschovakis AK, Highstein SM. The anatomy and physiology of primate neurons that control rapid eye movements. *Annu Rev Neurosci* 1994;17:465–88
17. Peters A, Payne BR, Budd J. A numerical analysis of the geniculocortical input to striate cortex in the monkey. *Cereb Cortex* 1994;4:215–29
18. Sengpiel F, Blakemore C. The neural basis of suppression and amblyopia in strabismus, part 2. *Eye* 1996;10:250–8
19. Sparks DL, Mays L. Signal transformations required for the generation of saccadic eye movements. *Annu Rev Neurosci* 1990;13:309–336

Auditory system 14

The auditory system is a special somatic afferent (SSA) system which transmits airborne vibrations via the external acoustic meatus, middle ear cavity and the perilymph to the organ of Corti. Mechanical displacement of the neuroepithelial hair cells is eventually converted into auditory impulses and conveyed via the cochlear nerve and ascending auditory pathways primarily to the contralateral primary and secondary auditory cortices. This occurs through a chain of neurons in the brainstem and diencephalon.

Peripheral auditory apparatus

The peripheral auditory apparatus includes, according to phylogenetic, developmental, structural and functional criteria, the external, middle, and inner ear.

External ear

The auricle (pinna) is a funnel-like structure which collects and directs air vibrations through the external acoustic meatus (Figure 14.1). It is primarily a cartilaginous structure, which consists of numerous irregular curvatures and eminences. These eminences include the helix and antihelix, separated by the scaphoid fossa. The antihelix surrounds the concha, a depression that leads into the external acoustic meatus. The upper part of the antihelix divides into two crura by the triangular fossa. The part of the concha above the anterior end of the helix is known as the cymba concha. The latter overlies a triangle area superior to the external acoustic meatus, which marks the lateral wall of the mastoid antrum. The concha is guarded anteriorly and posteriorly by a cartilaginous projection known as the tragus and antitragus, respectively. These two projections are separated by the inter-tragic notch. The lobule (ear lobe), a non-cartilaginous part of the auricle, consists of fibrous and fatty tissues. The auricle gives attachment to a group of primitive muscles of facial expression, which are innervated by the facial nerve. The auricle receives sensory innervation from the great auricular, lesser occipital, facial, auriculotemporal, and vagus nerves.

The external acoustic meatus (Figure 14.1) is an S-shaped tube which is cartilaginous in its lateral one-third and bony in its medial two-thirds. It extends from the concha to the lateral wall of the tympanic cavity.

In the newborn, the anteroinferior wall of the bony external acoustic meatus contains the foramen of Huschke that persists until approximately 5 years of age. The cartilaginous part continues laterally with the concha. The bony part forms a sulcus medially for the insertion of the tympanic membrane. The external acoustic meatus (Figure 14.1) lies superior to the parotid gland, anterior to the mastoid air cells, and inferior to the middle cranial fossa.

Cerumen (earwax) is a secretion of the subcutaneous glands in the medial part of the external acoustic meatus. These glands receive sensory innervation from the vagus nerve and the auriculotemporal branch of the mandibular nerve. Protection of the ear canal from foreign bodies may also be provided by the adhesive qualities of the cerumen. The auriculotemporal nerve is responsible for referred earaches associated with tooth decay or lingual ulcer.

Middle ear

The middle ear (Figures 14.1, 14.2, 14.3, 14.4 & 14.5) consists of irregular air-filled cavities within the temporal bone, containing the ossicles (malleus,

In the adult, the S-shaped curve of the external acoustic meatus is corrected and proper visualization of the canal during otoscopic examination may be maintained by pulling the auricle upward and posteriorly. In children, shortness of the canal and the equal length of the bony and cartilaginous parts render the tympanic membrane more vulnerable to injuries during examination. Additionally, otoscopic examination and proper visualization of the canal and tympanic membrane may necessitate pulling the auricle inferiorly and posteriorly. The angle of junction between the bony and cartilaginous parts is a common site of entrapment of foreign bodies.

The intimate attachment of skin to the underlying cartilage and bone of the external acoustic meatus accounts for the excruciating pain associated with inflammatory conditions involving the external auditory canal.

Vagal innervation of the external acoustic meatus explains the coughing and sneezing reflexes and bradycardia associated with excessive irrigation of the external acoustic meatus.

Inflammation of the perichondrium of the cartilaginous pinna (perichondritis) may be caused by bacterial infection subsequent to traumatic injury, insect bites, or an incised superficial abscess. It is characterized by accumulation of pus between the cartilage and perichondrium, occasionally leading to avascular necrosis and deformed external ear. Suction drainage and systemic antibiotics are required for treatment.

External otitis (pseudomonas osteomyelitis of the temporal bone) is another condition that commonly occurs in diabetic patients, particularly the elderly. It begins as *Pseudomonas aeruginosa* infection which progresses to become a pseudomonas osteomyelitis. Persistent and severe earache, and development of granulation tissue that blocks the external canal are some of the symptoms of this condition. Conductive hearing loss, and facial palsy may also occur. This condition may spread to the entire temporal bone, if it is not controlled. Surgical intervention or intravenous antibiotic therapy may be required.

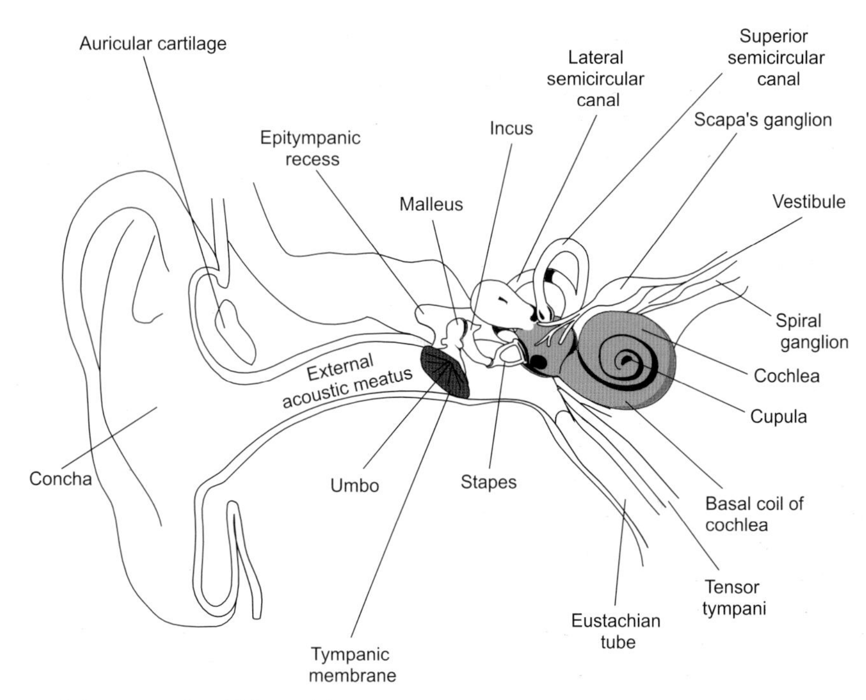

Figure 14.1 Diagram of the external, middle, and inner ear

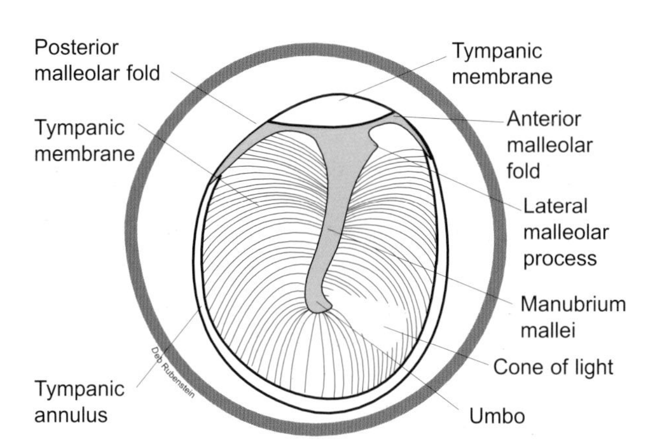

Figure 14.2 The tympanic membrane and associated ligaments

incus, and stapes). This cavity serves a mechanical function in transmitting mechanical energy, in the form of the airborne vibrations of sound waves, from the external environment to the inner ear. This cavity is connected anteriorly to the nasopharynx via the pharyngotympanic (Eustachian) tube and posteriorly to the mastoid antrum. It has lateral, medial, anterior, and posterior walls in addition to the roof and floor. The lateral wall of this cavity is formed by the tympanic membrane within the tympanic sulcus, and the epitympanic recess. The tympanic membrane (Figure 14.1, 14.2, 14.3 & 14.5) in the adult has a fibrocartilaginous periphery lodged in the tympanic sulcus at a 45 degree angle with the horizontal plane. In children this membrane maintains a horizontal plane.

Structurally it consists of a connective tissue layer, covered externally by the skin and lined by mucosa. It has a large, taut distal part (pars tensa) and a smaller, proximal, and triangular part known as the pars flaccida. The flaccid part lies above the malleolar folds, extending between the lateral process of the malleus and the deficient edges of the tympanic sulcus. The handle of the malleus, which attaches to the medial surface of the pars tensa of the tympanic membrane, forms the umbo, a central depression on the lateral surface of the tympanic membrane.

The anterior and inferior quadrant of the tympanic membrane is known as the cone of light, or triangle of Politzer. The chorda tympani, a branch of the facial nerve, runs between the inner and intermediate layers of the tympanic membrane, medial to the handle of the malleus.

The medial wall (Figure 14.4) contains the promontory, a bony prominence formed by the basilar part of the cochlea. The promontory contains the tympanic plexus, which is formed by the tympanic branch of the glossopharyngeal nerve and the carotid-tympanic nerve (sympathetic) fibers. This plexus supplies sensory fibers to the tympanic cavity, auditory tube, and mastoid air cells. The opening that lies posterior

The umbo may disappear in individuals with middle ear infections (otitis media) due to the pressure generated by the accumulated inflammatory fluid in the tympanic cavity.

Drainage of inflammatory exudate from the tympanic cavity is commonly performed through an opening in the posteroinferior quadrant of the tympanic membrane. This quadrant is less vascular and contains no prominent nerves or ossicles.

Barotitis media (aerotitis) is a condition that results from sudden change in the atmospheric pressure relative to the pressure in the tympanic cavity. Descent of an airplane or deep sea diving usually bring about this abrupt ambient pressure change. This is mediated by reflex swallowing and widening of the auditory (Eustachian) tube. Partial or complete occlusion of the auditiory (Eustachian) tube due to allergy, upper respiratory tract infection, or enlarged tubal tonsils may render the pressure in the tympanic cavity lower than the atmospheric pressure. An individual with allergy or respiratory tract infection may be advised not fly or apply nasal vasoconstrictor when flying. This leads to retraction of the tympanic membrane and the transudation of blood from the blood vessels of the lamina propria of the mucous membrane. Bleeding into the tympanic cavity and rupture of the tympanic membrane may occur in severe pressure differentials. Conductive deafness and severe pain usually accompany sudden pressure changes. Sometimes a perilymphatic fistula from the oval or round window may accompany the bleeding, and is generally accompanied by sensorineuronal hearing loss and vertigo.

Bulbous (infectious) myringitis is another condition that affects the tympanic membrane as a result of viral or bacterial infections. It is characterized by the formation of small fluid-filled vesicles on the tympanic membrane. This may progress to produce otitis media with fever and hearing loss. Infectious myringitis persists for two days and is usually caused by *Streptococccus pneumoniae* or mycoplasma infections. Antibiotics, analgesics, and induced rupture of the vesicles are common therapeutic measures of this condition.

and superior to the promontory is known as the oval window (fenestra vestibuli). This opening establishes communication between the tympanic cavity and the scala vestibuli and is covered by the stapes. Immediately above the fenestra vestibuli, the facial canal forms an eminence. The fenestra cochlea (round window) is located posterior and inferior to the promontory, connecting the middle ear cavity to the

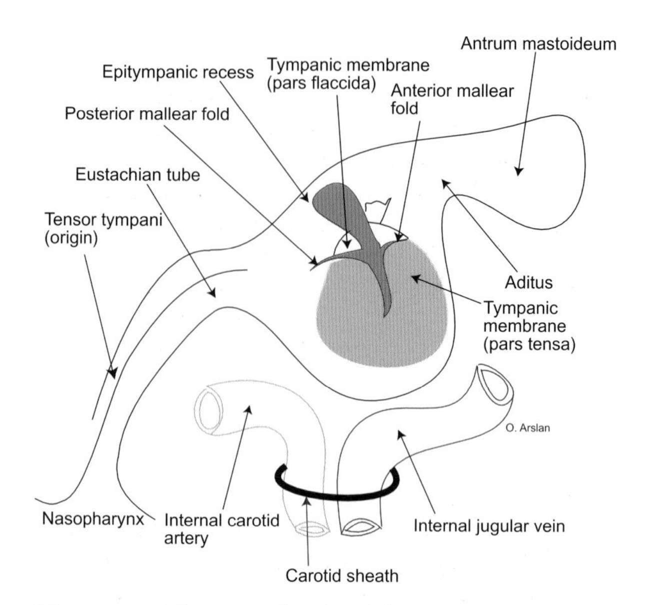

Figure 14.3 The external and middle ear. The relationships of the middle ear are clearly documented

scala tympani. The round window is occupied by the secondary tympanic membrane.

The anterior wall (Figures 14.1, 14.3 & 14.4) is formed by the auditory tube, carotid canal, and the tensor tympani muscle. The auditory (Eustachian) tube connects the tympanic cavity to the nasopharynx, equalizing the pressure between these two cavities. This tube may also serve as a route for spread of infection from the pharynx to the middle ear. The auditory tube consists of lateral bony and medial cartilaginous parts. The medial cartilaginous part forms the tubal torus, a mucosal eminence in the lateral wall of the nasopharynx, which continues inferiorly with the salpingopharyngeal fold. The cartilaginous part gives attachment to the tensor palatini, salpingopharyngeus and levator palatini muscles. The tensor palatini may be responsible for opening of the pharyngeal opening of the tube during swallowing. The carotid canal, which is located within the petrous temporal bone, forms the anterior wall of the middle ear, transmitting the internal carotid artery.

The posterior wall (Figure 14.3) of the tympanic cavity bears several features which include the aditus for the mastoid antrum, pyramidal eminence, and the fossa for the short process of the incus. The aditus is the gate to the mastoid antrum, containing on its medial wall a prominence formed by the lateral semicircular canal. The mastoid antrum is an air sinus which is connected anteriorly to the epitympanic recess, and inferiorly to the mastoid air cells.

The pyramidal eminence contains the stapedius muscle and lies between the facial canal posteriorly and the oval window anteriorly.

The antrum lies immediately medial to the suprameatal triangle, a common site for surgical intervention into the middle ear cavity. It also lies inferior to the tegmen tympani and anterior to the sigmoid sinus, separated from the latter by a bony lamella. The relationship of the antrum to these areas provides possible routes by which a middle ear infection may spread to the temporal lobe of the brain, sigmoid sinus and mastoid air cells.

Acute mastoiditis is an inflammatory condition which involves the soft tissues surrounding the mastoid air cells, usually subsequent to untreated or inadequately treated otitis media (middle ear infection). Surgical mastoiditis, a medical and surgical emergency, encompasses osteitis and periosteitis of the mastoid bone, accompanied by transverse and sigmoid sinus thrombosis. Swelling, pitting edema, erythema, and percussion tenderness are some of the clinical signs of this condition. Displacement of the auricle downward and anteriorly may also occur. An abscess may be formed in the mastoid bone, and may occasionally involve the facial nerve, producing facial nerve palsy. Intravenous administration of antibiotics, and surgical intervention that involves myringotomy, drainage of any abscess, and mastoidectomy may be required. If treatment of mastoiditis is inadequte, sepsis, meningitis, brain abscess, and even death may ensue. Antibiotics may convert temporarily an acute mastoiditis into masked mastoiditis. Similar treatment methodologies are used, and if not successful, cortical mastoidectomy and myringotomy are needed.

The roof (Figure 14.3) of this cavity is formed by the tegmen tympani, which separates it from the meninges and the temporal lobe of the brain. It contains the superior petrosal sinus giving passage to veins that serve as a conduit for the spread of infection from the tympanic cavity to the temporal lobe and the meninges. In infants, unossified areas of the roof may also serve as a route for the spread of infection.

The floor (Figure 14.3 & 14.4) of the middle ear cavity is formed by a thin bony lamina which lodges the internal jugular vein. Spread of infection to the systemic circulation from the middle ear cavity may occur if the separating bony laminas are unossified.

The middle ear cavity also contains bony ossicles (Figures 14.1 & 14.5) which transmit the vibration from the tympanic membrane to the perilymph of the inner ear. These ossicles include the malleus, incus, and the stapes. These ossicles form synovial joints with each other to facilitate their movements.

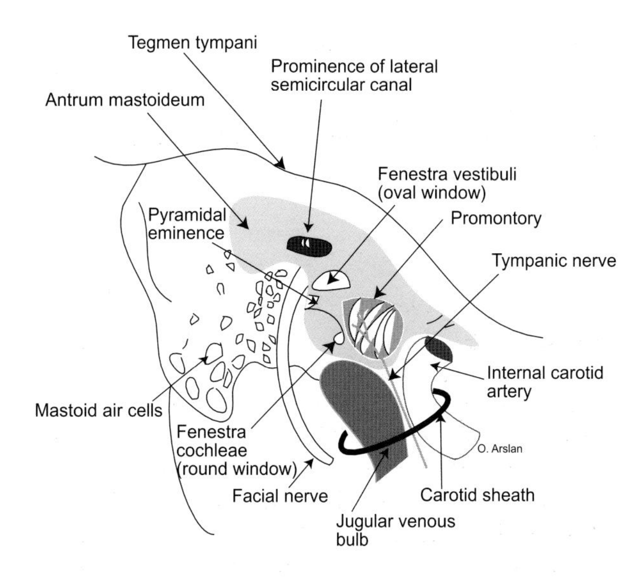

Figure 14.4 This schematic drawing illustrates the roof, floor, medial and anterior walls

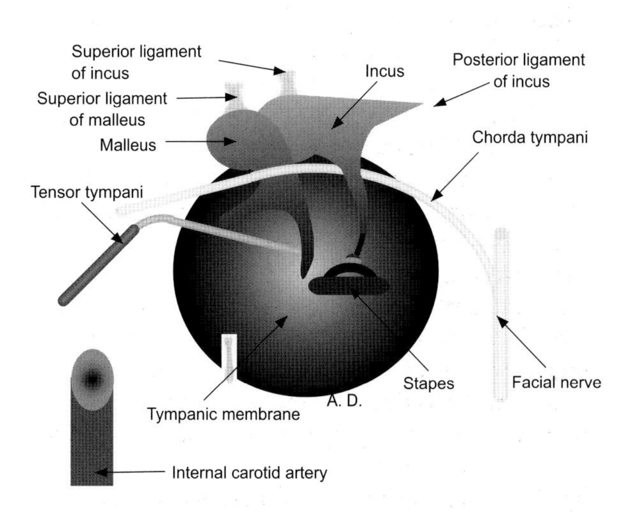

Figure 14.5 The tympanic membrane in association with the ossicles, chorda tympani, and tensor tympani muscle

The malleus has a head, neck, manubrium, and anterior and lateral processes. The head forms a sellar-type joint with the incus in the incudomalleolar articulation. Both the body of the incus and the head of the malleus are located in the epitympanic recess, proximal to the tympanic membrane. The manubrium, which is embedded in the medial surface of the tympanic membrane, forms the umbo. The tendon of the tensor tympani muscle inserts into the upper end of the manubrium. The anterior process is short and is connected to the petrotympanic fissure, while the

Otitis media refers to a bacterial or viral infection of the middle ear, usually secondary to upper respiratory tract infections. This condition, which is more common in children, may be acute or chronic. Acute otitis media may be suppurative or serous. Acute suppurative otits media most commonly develops as a result of bacterial contamination via the Eustachian tube in the presence of preexisting inflammation in the middle ear. Eustachian tube dysfunction results in absorption of oxygen and its replacement by carbon dioxide which initiates an inflammatory response followed by accumulation of transudate. Persistence of this condition produces exudate that may be contaminated by the infected nasopharyngeal content. Parainfluenza, coxsackieviruses, and adenoviruses are the most frequent viruses involved in this condition. Bacterial otitis media is most commonly caused by *Streptococcus pneumoniae*, *Staphylococcus aureus*, *Haemophilus influenzae*, *Mycoplasma pneumoniae*, and group A *Streptococcus pyogenes*. Persistent and severe earache, and temporary hearing loss are some of the initial symptoms. Fever, nausea, and diarrhea may occur in young children. Additional signs include bulging of the tympanic membrane or its severe retraction. Perforation of the tympanic membrane may occur and pulsatile discharge may be seen. Mastoiditis, labyrinthitis, and meningitis are some of the complication of the disease. Therapy may include antibiotics, and myringotomy (a surgical opening in the ear drum) to drain the accumulated pus.

Acute serous otitis media is a disorder in which a middle ear effusion develops subsequent to persistent occlusion of the auditory (Eustachian) tube. Upper respiratory infection, chronic rhinosinusitis, cleft palate, and nasopharyngeal adenoids may lead to persistent occlusion of the Eustachian tube. The serous transudate may become mucoid and glue-like exudate after few days, containing bacterial contaminants. This condition is more common in young children because of the presence of a narrow Eustachian tube, enlarged adenoids, and frequency of nose and throat inflammation. This also occurs when maximum acquisition of speech skills is needed. In this condition the tympanic membrane thickens and frequently is retracted by the negative pressure of the tympanic cavity. The pars flaccida of the tympanic membrane may also be retracted leading to cholesteatoma (keratoma). The latter is a growth of normal stratified squamous epithelium that enlarges and eventually destroys the ossicles and even the inner ear and cranial cavity. Cholesteatoma becomes an excellent site for the growth of bacteria. Tympanometry may be employed to measure the pressure on both sides of the tympanic membrane. Effusion, in general, is a self-limiting process which resolves in a period of two weeks. Antibiotics, nasal decongestants, and antihistaminic medications may be used. A blocked Eustachian tube may be forced open by asking the patient to breathe out through his mouth while his nostrils are pinched shut. Drainage of fluid from the tympanic cavity may be accomplished by myringotomy.

Chronic suppurative otitis media is always associated with central or peripheral perforations of the tympanic membrane with chronic purulent otorrhea. These perforations result in conductive hearing loss. Exacerbations of this condition due to upper respiratory tract infection may lead to the formation of aural polyps and destructive changes in the middle ear cavity. Peripheral perforations usually occur in the posterior-superior quadrant of the pars tensa, destroying large areas of the tympanic membrane, including annulus tympanicus and the mucous membrane. Labyrinthitis, facial palsy, and intracranial suppuration are more likely complications of peripheral than central perforations. Pars flaccida (attic) perforations may also involve the epitympanic recess. Peripheral and pars flaccida perforations are frequently associated with cholesteatoma. In central perforation, chronic otitis media may be exacerbated after upper respiratory infection, resulting in painless discharge from the ear. Persistence of infection may lead to conductive hearing loss.

Topical antibiotics are an initial step in the treatment of chronic suppurative otitis media, although neomycin-containing preparations may be contraindicated in patients with a perforated tympanic membrane because of the possible ototoxicity-induced neuronal deafness. When otorrhea (ear discharge) exists, daily irrigation with Burrow's solution may be advised. Surgical closure of the tympanic membrane may be required.

Paralysis of the stapedius muscle may directly affect the movement of the stapes, thereby lowering the hearing threshold and resulting in hyperacusis.

Otosclerosis is the most common cause of progressive conductive hearing loss in the adult. It should also be remembered that only 10% of the population with this disease develop hearing loss. This hereditary condition is characterized by excessive bony growth and the formation of irregularly arranged immature bones, and ankylosing joints between the ossicles, particularly the stapedial base. When otosclerotic plaques impinge upon the scala media, neuronal deafness occurs. Microsurgery and replacement of the affected stapes with a prosthesis may be required for treatment. Hearing aids may serve the same purpose.

lateral process is longer and gives attachments to the malleolar folds.

The incus resembles the premolar tooth, having long and short processes and a body. The body articulates with the head of the malleus, and the long process articulates in a spheroidal joint with the stapes. The short process attaches to the fossa incudis.

The stapes resembles a stirrup consisting of a head, neck, and oval base that covers the oval window via the annular ligament. The head articulates with the lenticular (long) process of the incus. The neck gives attachment to the tendon of the stapedius muscle. Movement of the stapes is piston-like, producing waves within the perilymph of the scala vestibuli.

Inner ear

The inner ear (Figures 14.4 & 14.6) contains the receptors for the auditory and vestibular systems, consisting of bony and membranous labyrinths. The bony labyrinth encloses the membranous labyrinth. The bony labyrinth is lined by periosteum and is filled with perilymph. It consists of the cochlea, vestibule, and semicircular canals. The vestibule represents the middle portion of the bony labyrinth, connected to the tympanic cavity via the oval window (fenestra vestibuli), and the round window (fenestra cochlea). The membranous labyrinth forms the utricle and saccule. The utricle responds to linear acceleration and deceleration, and gravitational pull, while the saccule is thought to be activated by vibratory stimuli.

The medial wall of the vestibule (Figure 14.6)

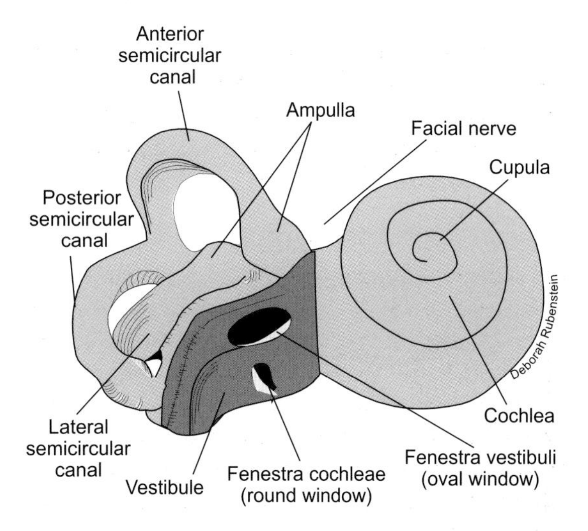

Figure 14.6 In this schematic diagram the main components of the inner ear are illustrated. Observe the close relationship of the facial canal and nerve to the vestibule

Abrupt compression or decompression injuries, head trauma, heavy lifting or straining my result in the formation of round or oval window fistulas. These fistulas may produce fluctuating hearing loss and/or tinnitus, which may improve overnight and worsen during the day.

contains the aqueduct of the vestibule, lodging the endolymphatic duct. The latter duct connects the utricle to the saccule. The posterior part of the bony labyrinth is formed by the semicircular canals which enclose the membranous semicircular ducts. The semicircular canals are organized perpendicular to each other, representing the three dimensions in space. They comprise the anterior (superior), lateral (horizontal), and posterior semicircular canals. The anterior canal is parallel to the posterior canal of the contralateral side. Each canal has a dilated lower part called the ampulla, which contains the ampullary crest and the neuroepithelial receptor cells.

The anterior part of the bony labyrinth is known as the cochlea. The cochlea (Figures 14.1 & 14.6) forms two and one-half turns which end at the apex or cupula. The basilar part of the cochlea forms the promontory on the medial wall of the tympanic cavity. The cochlea is connected to the middle ear cavity by the fenestra vestibuli (oval window) which is covered by the stapes and by the fenestra cochlea (round window). The fenestra cochlea is covered by the secondary tympanic membrane. The bony cochlea encloses the membranous cochlear duct (scala media) containing the auditory receptors. It also contains the organ of Corti, the scala vestibuli, and the scala

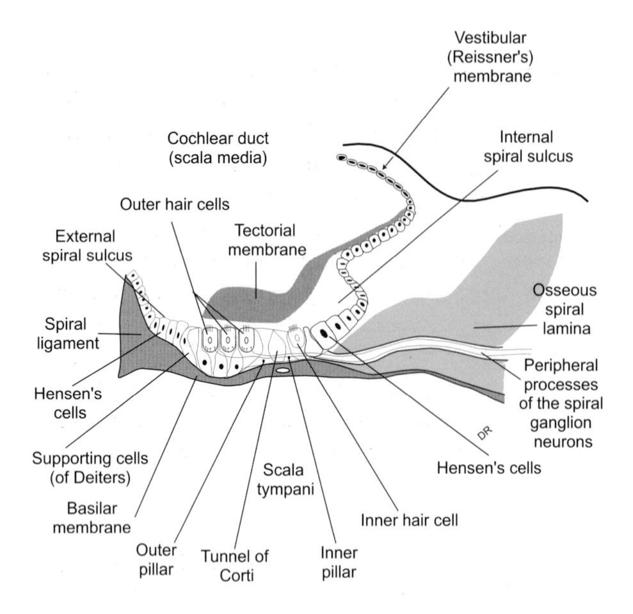

Figure 14.7 The cochlear duct and the organ of Corti

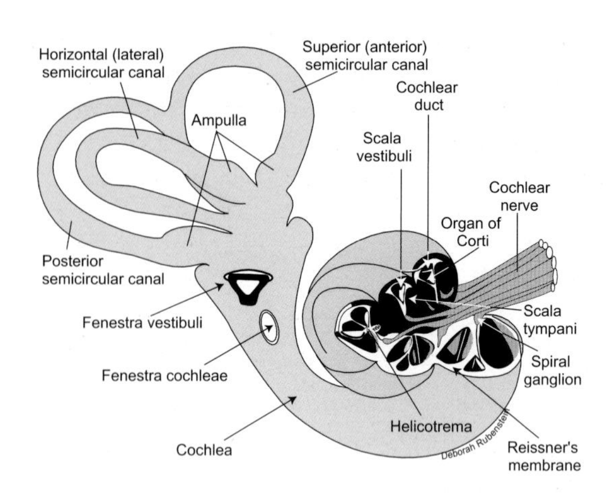

Figure 14.8 Section through the cochlea showing the scalae vestibuli, tympani, and media, as well as the organ of Corti

tympani. The scala tympani is connected to the subarachnoid space by the cochlear canaliculus.

The cochlea consists of a bony shell, a central axis (modiolus), and a spiral lamina that protrudes into the auditory canal. The modiolus, which is comprised of trabecular tissue, forms the pillar around which the cochlea and spiral lamina make two and half turns, containing foramina for the cochlear nerve branches. The spiral bony lamina protrudes inside the cochlear canal, and is connected to the spiral ligament of the cochlear bony wall by the basilar and vestibular membranes. The basilar membrane (Figure 14.7), which contains the outer and inner hair cells and rods, is an integral part of the mechanism that transduces the mechanical energy created by the sound waves to chemo-electric potentials. The tips of the hair cells are embedded in the tectorial membrane. High frequencies activate the neuroepithelial hair cells in the basal turn of the cochlea, whereas low frequency sounds stimulate the corresponding neuroepithelial cells in the apical turn.

The vestibular membrane (Figure 14.7) of Reissner extends from the upper lip of the spiral lamina to the stria vascularis. The latter is a vascular area on the inner wall of the cochlea. The vestibular and the basilar membranes and the cochlear wall constitute the boundaries of the endolymph-filled cochlear duct or scala media. The cochlear duct is separated from the perilymph-filled scala tympani and scala vestibuli via the vestibular and basilar membranes, respectively. The scala vestibuli and tympani are separated from each other throughout the entire length of the cochlear duct except at the cupula, where a connection is formed via the helicotrema. The cochlear canaliculus interconnects the scala tympani to the subarachnoid space.

Tears of the intracochlear (Reissner's) membrane have been identified frequently in patients with cochlear hydrops. This condition may produce potassium poisoning of the sensorineural structures of the cochlea, leading to fluctuating hearing loss.

The perilymph resembles cerebrospinal fluid and extracellular space, and occupies the perilymphatic space between the bony and membranous labyrinths. It is connected to the subarachnoid space via the cochlear canaliculus. It is believed that the perilymph is an ultrafiltrate of blood, or a possible derivative of the cerebrospinal fluid surrounding the vestibulocochlear nerve or contained within the cochlear canaliculus. The perilymphatic space of the semicircular canals and vestibule are connected to the scala vestibuli, and through the helicotrema, to the scala tympani.

The endolymph fills the membranous labyrinth and resembles the intracellular fluid in its ionic composition. In contrast to the perilymph, the endolymph contains a high concentration of potassium ions and a lower concentration of sodium ions. The endolymph may be derived from the stria vascularis, a specialized epithelium with rich blood capillaries, and from the dark epithelial cells of the inner layer of the utricle and semicircular ducts, and also from the platinum semilunatum (a specialized inner epithelium near the ampullary crests and macula).

Airborne vibration is initially collected by the auricle, where binaural audition is enhanced via curves

The organ of Corti may be damaged by a variety of agents and conditions. Degeneration of the hair cells in the basal coil of the cochlea may occur in the elderly, producing deafness to high-tones. Mid-frequency hearing loss may occur as a result of degeneration of the hair cells in the middle cochlea. The latter deficit is seen in industrial workers and rock band performers who are exposed to loud noises. Prolonged treatment with certain antibiotics such as streptomycin, neomycin, kanamycin and dihydrostreptomycin, or aspirin may produce ototoxicity and degeneration of the outer hair cells nearest to the tunnel of Corti. The trend of degeneration will continue to involve the lateral outer hair cells and then the inner hair cells. At first the hearing loss will be confined to higher frequencies and then as the degeneration progresses lower frequencies will also be affected.

and depressions of this cartilaginous structure. Airborne vibration funneled through the external acoustic meatus creates vibrations on the tympanic membrane and subsequent movement of the malleus and incus. Movement of the incus produces a piston-like movement of the stapes on the oval window, and a current in the perilymph of the scala vestibuli. The traveling current, which induces pressure changes, passes through the helicotrema to the scala tympani and then to the compliant basilar membrane. Vibration of the basilar membrane, upon which the neuroepithelial (outer and inner hair) cells rest, results in mechanical displacement of the hair cells within the tectorial membrane accompanied by a shearing force. This results in the propagation of a wave of electrochemical energy, its conversion via transductive process into nerve impulses, and eventual transmission by the peripheral and then the central processes of the spiral ganglion neurons. High frequency sounds generate a current in the basilar membrane near the base of the cochlea, whereas low frequency sounds with peak amplitude of wave in the basilar membrane near cupola. This supports the concept that tontopic localization of sound frequencies may be determined by the mechanical features of the basilar membrane.

The spiral ganglia (Figure 14.7) are located in the modiolus of the cochlea, and consist of bipolar neurons, representing the first order neurons of the auditory system. The cochlear nerve, which is formed by the central process of the spiral ganglion neurons, constitutes the largest component of the vestibulocochlear nerve. This nerve exits at the cerebellopontine angle, and travels within the internal acoustic meatus

Herpes zoster oticus (Ramsay Hunt syndrome) is a viral disease caused by herpes zoster virus that invades the vestibulocochlear nerve and the geniculate ganglion of the facial nerve. Symptoms include neuronal hearing loss, vertigo, and facial palsy. Vesicular eruptions along the course of the cutaneous branches of the facial nerve in the concha can be seen. These symptoms may be transient or complicated by the involvement of other cranial nerves. Corticosteroid and antiviral therapy (acyclovir) may be used in the treatment of this condition. Analgesics for pain and diazepam for the treatment of vertigo may also be used.

accompanied by the vestibular and facial nerves, and the labyrinthine artery.

The internal acoustic meatus lies within the temporal bone, anterior and superior to the jugular foramen. It is divided into upper and lower parts by a transverse crest. The upper part gives passage to the facial nerve, and to vestibular nerve branches to the utricle and the lateral semicircular ducts. The lower part of the meatus transmits the cochlear nerve, and the vestibular nerve branches to the posterior semicircular duct.

The labyrinthine artery also pierces the internal acoustic meatus to supply the structures in the inner ear. This artery, which arises frequently from the anterior inferior cerebellar artery and less often from the lower basilar artery, runs with the facial and vestibulocochlear nerves.

Auditory pathways

Virtually all cochlear nerve fibers (Figures 14.7 & 14.8) distribute to the ventral and dorsal cochlear nuclei, which are located dorsolateral to the inferior cerebellar peduncle. The dorsal cochlear nucleus forms the acoustic tubercle in the rhomboid fossa. Fibers from the basal part of the cochlea that convey high frequencies terminate mainly in the ventral portions of both cochlear nuclei. The cochlear nerve fibers from the apical turn of the cochlea that receive low frequencies terminate in the dorsal parts of the cochlear nuclei. The postsynaptic fibers from the cochlear nuclei form the acoustic striae, contributing to the lateral lemniscus.

The ventral acoustic stria (Figures 14.9 & 14.10), the largest acoustic stria which arises from the ventral portion of the ventral cochlear nucleus, forms the trapezoid body by decussating with the corresponding stria of the opposite side. The trapezoid body lies ventral to the medial lemniscus in the ventral

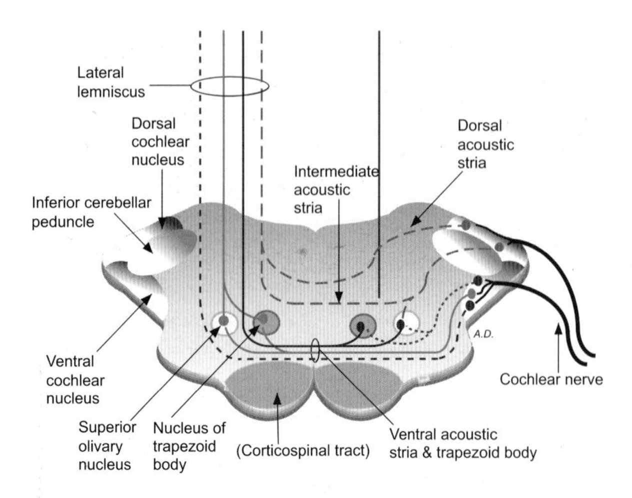

Figure 14.9 The auditory pathway from the cochlear nerve to the lateral lemniscus

tegmentum of the pons, adjacent to the median raphe that contains the raphe nuclei. It continues on the contralateral side dorsal to the superior olivary nucleus as the lateral lemniscus. Some fibers of the ventral acoustic stria terminate bilaterally in the superior olivary nuclei. Postsynaptic fibers from the superior olivary nuclei project predominantly to the lateral lemniscus of the same side.

The intermediate acoustic stria (Figures 14.9 & 14.10) is the smallest stria which originates from the dorsal portion of the ventral acoustic nucleus, crosses the midline, and contributes to the lateral lemniscus of the opposite side. This stria also projects bilaterally to the periolivary and retrolivary nuclei which form the inhibitory olivocochlear bundle. However, there are no fibers from this stria that terminate in the superior olivary nuclei.

The dorsal acoustic stria (Figures 14.9 & 14.10) arises from the dorsal cochlear nucleus, crosses the midline and becomes part of the contralateral lateral lemniscus. Some fibers of this stria project to the superior olivary nuclei, which consist of bipolar neurons, and plays an important role in sound localization. These nuclei contribute fibers only to the ipsilateral lateral lemniscus. All acoustic striae eventually contribute fibers to the contralateral lateral lemniscus.

The lateral lemniscus (Figures 14.9 & 14.10) contains predominantly contralateral and some ipsilateral secondary auditory fibers which are derived from the cochlear nuclei. It also contains tertiary fibers originating from the superior olivary and trapezoid nuclei and the nuclei of the lateral lemniscus. Damage to the lateral lemniscus results in auditory deficits primarily on the opposite side. Complete deafness will not occur in either ear since the lateral lemniscus contains fibers from both sides.

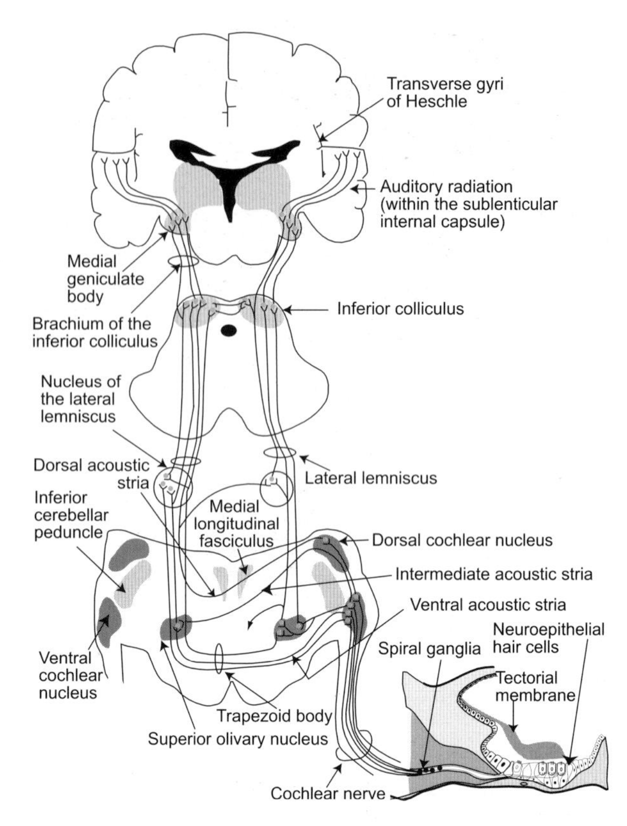

Figure 14.10 Complete neuronal sequence associated with the transmission of the auditory impulses from the external ear to primary auditory cortex in the temporal lobe

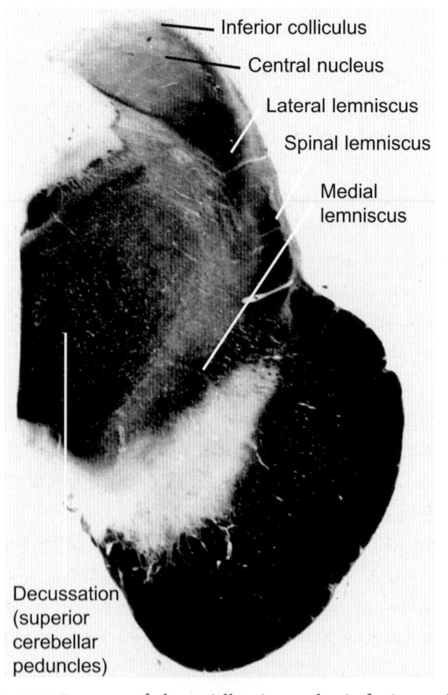

Figure 14.11 Image of the midbrain at the inferior collicular level showing the central nucleus and the termination of the lateral lemniscus

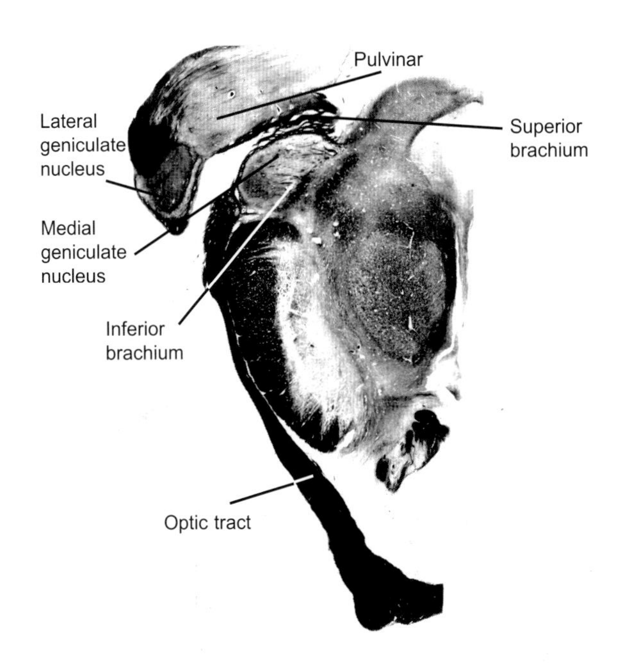

Figure 14.12 Section through the diencephalomesencephalic junction to show the medial and lateral geniculate bodies

Disruption of the connection between the auditory association and entorhinal cortices may produce anterograde auditory amnesia in which patient is unable to retain spoken language. Bilateral damage to the auditory cortex may result in impairment of the ability to distinguish and decipher tones in specific models. Perception of sound and discrimination of tones remain intact in these individuals.

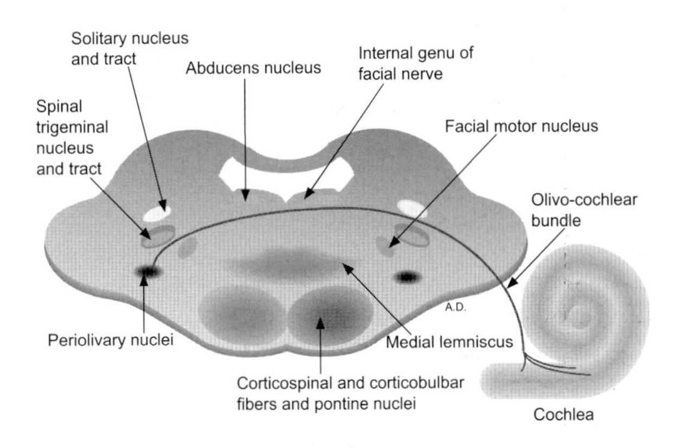

Figure 14.13 The origin of the inhibitory olivochochlear bundle and its course to the outer hair cells of the cochlea

Most of the fibers of the lateral lemniscus terminate in the central nucleus of the inferior colliculus. Some fibers may project to the central nucleus of the opposite side via the inferior collicular commissure, and others may reach the medial geniculate nucleus.

The inferior colliculus (Figure 14.11) is an auditory relay and reflex center. It consists of a central nucleus which receives bilateral auditory impulses, and a pericentral nucleus which is concerned with the auditory impulses from one ear. The dorsal and ventral parts of the inferior colliculus receive high and low frequency auditory impulses, respectively. Neurons of the central nucleus of the inferior colliculus project through the brachium of the inferior colliculus to the ventral laminated part and the magnocellular divisions of the medial geniculate nucleus.

Axons from the medial geniculate nucleus (Figure 14.12) form the auditory radiation which projects to the transverse gyrus of Heschl (Brodmann's areas 41 and 42) via the sublenticular portion of the posterior limb of the internal capsule. It is important to note that the posterior limb of the internal capsule also contains the optic radiation, as well as sensory and motor fibers.

The primary auditory cortex (Brodmann's area 41, 42) is located medial to the superior temporal gyrus, where low frequencies (low-pitched tones) are received laterally and high frequencies (high-pitched tones) are perceived caudally and medially in close proximity to insula. The auditory association cortex (Brodman's area 22) occupies the posterior part of the superior temporal gyrus, processes signals that pertain to spoken language, and maintains connections with the entorhinal cortex.

In summary, neurons associated with the auditory system include the spiral ganglia, the cochlear nuclei, superior olivary nuclei, trapezoid body, lateral lemniscus, inferior colliculus, medial geniculate body, and the transverse gyrus of Heschl. Most of the auditory fibers are contralateral but some remain ipsilateral.

Mechanisms that regulate the transmission of auditory impulses include the descending cortical and subcortical projection, olivocochlear bundle and stapedius reflex. The cortical projection, as the name implies, encompasses cortical input to the medial geniculate body. Descending fibers from the subcortical areas, such as the medial geniculate body, form the geniculotectal tract. Other descending fibers that act upon the lower auditory neurons are also included in this regulating mechanism. The descending fibers have the reverse course of the lateral lemniscus, exerting an inhibitory influence on unwanted background noise. Through this inhibition, the desired auditory signals are sharpened. The olivocochlear bundle (Figure 14.13) contains crossed and uncrossed fibers that originate from the retro-olivary and periolivary nuclei. These inhibitory fibers, that initially run through the vestibular nerve and later project to the outer hair cells via the cochlear nerve fibers, may

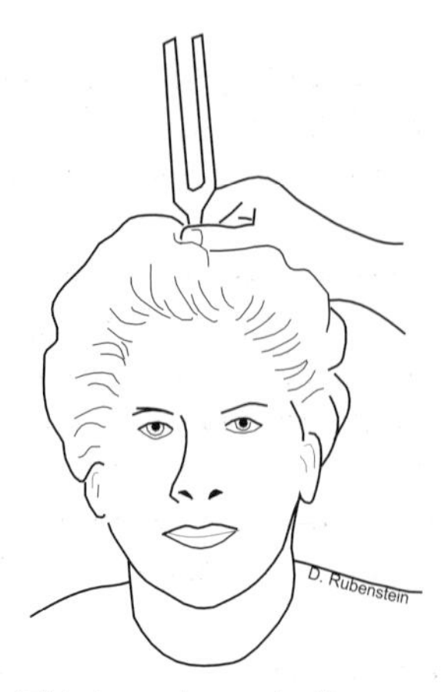

Figure 14.14 This is a schematic diagram of the Weber test. The sound is lateralized to the impaired side

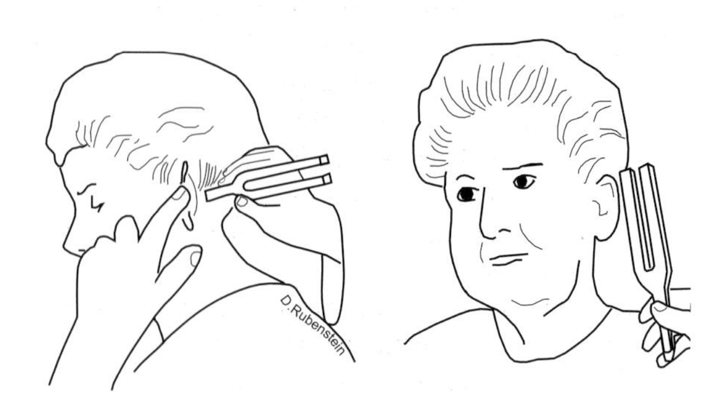

Figure 14.15 This diagram illustrates the Rinne test

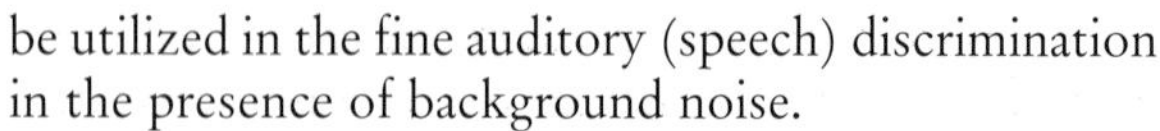

be utilized in the fine auditory (speech) discrimination in the presence of background noise.

The stapedius reflex (Figure 14.13) involves activation of the facial and trigeminal motor nuclei in response to loud noises. It comprises bilateral projection of the auditory impulses to the facial and trigeminal motor nuclei via the superior olivary nuclei. Activation of the facial and trigeminal motor neurons results in contraction of the stapedius and tensor tympani muscles. Paralysis of the stapedius muscle in facial nerve palsy may produce hyperacusis (low hearing threshold). Paralysis of the tensor tympani muscle as a result of mandibular nerve damage produces hypacusis (high hearing threshold).

Sudden hearing loss may develop over a period of a few hours or less usually in only one ear. One of the initial sign of this condition is the loud sound heard in the affected side. Viral disease such as mumps, measles, influenza, chicken pox, or infectious mononucleosis may be the culprit. Strenuous activities such as heavy weight lifting may exert undue pressure on the inner ear, producing sudden deafness. Most patients recover their hearing within two weeks. Hearing impairment, which refers to a defect in the identification of acoustic information, may involve any portion of the transducer mechanism of the ear including defects in the mechanical conduction of sound waves, sensorineuronal coding, sensory transmission to the CNS, or any combination of these defects. Therefore, hearing loss may be viewed in two categories: conductive and sensory neuronal.

Conductive deafness occurs as a result of either mechanical defects or inefficiencies. These may include occlusion of the external acoustic meatus by cerumen, or infection of the soft tissue associated with this canal (exostosis), or nodular overgrowth in the external canal which are seen in the long distance swimmer as a result of exposure to cold water. Congenital atresia or stenosis of the external canal due to fibrosis or cicatrix are some additional causes. Middle ear infection or trauma and build-up of purulent exudate in the tympanic cavity that impair the mobility and the mechanical efficiency of the tympanic membrane may also produce conductive hearing loss. Tympanosclerosis as a result of scar tissue formation and deposition of calcium in the membrane or healed perforation may also impair the elasticity of the tympanic membrane and lead to hearing loss. Bony fusion of the ossicles (otosclerosis), formation of a ball sac within the tympanic cavity (cholesteatoma) following perforation or retraction pocket in the tympanic membrane, and perilymphatic fistula of the round window or the vestibule that extends through the oval window to the middle ear cavity also result in hearing loss.

Sensorineuronal hearing loss (nerve deafness) may occur as a result of lesions of the cochlea, the cochlear nerve or nuclei. Noise trauma as a result of prolonged exposure to excessive levels of noise can produce nerve deafness by damaging the hair cells of the basal turn of the cochlea, which receive high frequency sound waves. However, this type of deafness should not be confused with temporary threshold shift, a temporary hearing loss resulting from sudden exposure to loud noise. Deafness due to this condition improves within 24–48 hours. Presbycusis, which manifests deafness mainly to high frequencies, is age related gradual degeneration of the neuroepithelial cells of the organ of Corti and is considered to be the second most common etiology of neuronal deafness. Destruction of the organ of Corti as is seen in Meniere's disease, congenital atresia of the labyrinth (deaf mutism), acquired atresia of the labyrinth subsequent to meningitis, loud noise, administration of ototoxic medication (i.e. streptomycin, neomycin, or quinone) or acoustic neuroma may also produce nerve deafness.

Viral or bacterial infections of the inner ear may result in rapid, progressive, and total hearing loss. Malarial infections may produce permanent neuronal deafness. Neuronal deafness due to labyrinthitis is usually associated with vertigo and ataxia. Fluctuating hearing loss and ataxia and gradual progression of hearing loss occurs in syphilitic labyrinthitis. Meniere's disease, a condition that develops as a result of disorder of endolymphatic circulation and formation of endolymphatic hydrops, produces fluctuating hearing loss, vertigo, ataxia, and tinnitus.

Ototoxic effects of medications such as salicylates and aminoglycosides may also produce sensorineuronal deafness. Salicylates, in particular, produce reversible symptoms which include tinnitus, deafness to all frequencies, vertigo, and ataxia. Reversible ototoxic effects may result from ethacrynic acid and furosemide. Chemotherapeutic agents such as nitrogen mustard and cis-platinum may also produce profound hearing loss. Streptomycin and gentamicin produce greater adverse effects on the vestibular system and to some degree on the organ of Corti. Kanamycin, tobramycin, and neomycin produce significant damage to the cochlear system. Aminoglycosides damage the inner hair cells, producing neuronal deafness to high frequencies which progresses to involve all frequencies.

Congenital disorders, such as Mondini deformity, in which most of the cochlea fails to develop in both ears, produces neuronal deafness. This may affect acquisition of language skills at an early age. It may sometimes be misdiagnosed as mental retardation. Retrocochlear deafness due to lesions of the cochlear nerve or the central auditory pathways is seen as the result of cerebrovascular diseases, stroke, intracranial hemorrhage, demyelinating diseases, or head trauma. Cerebellopontine angle tumors and acoustic neuroma may also produce nerve deafness.

Acoustic neuroma, the most common tumor of the cochlear nerve, is of unknown etiology that arises from the perineurium of the cochlear nerve accounting for 5–10% of intracranial tumors. It may be associated with neurofibromatosis (von Recklinghausen's disease). Genetic causes are rarely associated with the bilateral form of this disease. It may result in destruction of the vestibulocochlear, facial, and trigeminal nerves. It may expand and destroy the lateral part of the pons. Symptoms have an insidious onset and include hearing impairment which is progressive and sometimes associated with tinnitus, facial numbness, ataxia, vertigo, vomiting, involuntary movements and rarely facial pain. Erosion and widening of the internal acoustic meatus may be observed radiographically in this condition.

Tinnitus is a continuous or intermittent noise heard, unilaterally or bilaterally, by the patient alone in the form of hissing, whistling, or machinery-like noise (subjective tinnitus). It may be severe enough to interfere with normal sleep. Unilateral tinnitus may be a sign of Meniere's disease, acoustic neuroma or otosclerosis. Meniere's disease may produce a low-pitched continuous tinnitus similar to an ocean roar that intensifies before the attack of vertigo. Otosclerosis also produces low-pitched and continuous tinnitus. High pitched tinnitus may result from physical trauma. Chronic exposure to noise, and toxicity with drugs such as streptomycin, salicylates and quinine may also produce bilateral high-pitched tinnitus. However tinnitus induced by drugs such as indomethacin, quinidine, levodopa, propranolol, carbamazepine, salicylates, aminophylline, and caffeine, is not commonly associated with hearing loss.

Objective tinnitus can be heard by the examiner when he places a stethoscope (with bell removed) into the patient's external acoustic meatus, or with the bell around the ear. An abnormally patent (patulous) Eustachian tube may produce a blowing sound that coincides with the inspiration and expiration. Extensive weight loss due to dieting or a consequence to debilitating disease may result in objective tinnitus. It may also be heard in an individual with tetany and as a result of contracture of the palatal muscles and tensor tympani. Objective tinnitus may be audible in individuals with arteriovenous malformation, aneurysms, and vascular tumors. In these cases the tinnitus remains pulsatile and in synchrony with the heart beat. Turbulence within the internal jugular vein produces a venous hum which is heard by the examiner in the form of a 'whistling' or continuous machine-like sound. Deafness may be distinguished by sophisticated audiometers, Weber's or Rinne tests.

Audiometry, a procedure in which an electronic device that produces auditory stimuli of known frequency and intensity is used to measure the hearing ability. In Békésy audiometry, continuous and interrupted tones are presented to the patient by pressing a signal button. Monaural tracings are made to measure the increments by which the patient must increase the volume in order to hear the continuous and interrupted tones just above the threshold. The intensity of the tone decreases as long as the button is depressed and increases when it is released. Four basic configurations have been established based on analysis of tracings. Type I is considered normal, type II indicates a lesion in the cochlea, and types III or IV usually points to a retrocochlear lesion. An objective method of determining auditory acuity may also be achieved by recording and averaging the electrical potentials elicited by the cortex in response to stimulation by pure tones (cortical audiometry).

Electrocochleographic audiometry is another procedure that measures the electrical potentials from the inner ear in response to auditory stimuli. Neonatal deafness may be determined by electrodermal audiometry, in which a patient receives a mild shock that activates sweat glands, a physiological event readily measured as a resistance change between electrodes attached to the hand. Then the patient is conditioned to pure tones paired with harmless electric shock. The anticipation of the shock results in brief electrodermal response (activation of sweat glands) to the sound alone. Hearing threshold is determined by the lowest intensity at which electrodermal response is recorded.

On the other hand, speech audiometry measures (in decibels) the threshold of speech perception (speech reception threshold (SRT)), tolerance for loud speech, degree of hearing impairment, and the ability to understand monosyllabic words (speech discrimination). SRT is performed by presentng a series of standardized words to a patient and then measuring the lowest intensity at which these words are comprehended. Pure tone audiometry, as the name indicates, measures hearing at various tones of selected frequencies with varying intensity. In normal individuals the threshold varies with frequencies, and the ability to hear below 45 Hz is poor.

One of the newest methods that ascertains the integrity of primary and secondary neural auditory pathways is the brainstem auditory evoked response (BAER). In this method, a large number of clicks are delivered to each ear at one time. The series of waves are recorded via scalp electrodes and maximized by computers. Within 10 ms after each stimulus a series of seven waves appear. The first five waves relate to the auditory neurons and pathways. Reduction in amplitude, lower voltage, or delay in the appearance of the waves indicate a structural lesion within the auditory system.

In the Weber test (Figure 14.14), a vibrating tuning fork is placed on the vertex (top) of the patient's head or on the nasal bone and the patient is asked to tell whether the sound is heard equally on both sides (positive Weber test) or whether the sound is greater on one side. In conduction deafness, the sound is lateralized (heard better) on the affected side, while in nerve deafness, the sound is heard louder in the unaffected ear.

In the Rinne test (Figure 14.15) air conduction is compared to bone conduction by using a tuning fork. The examiner first applies a tuning fork of 256–512 Hz frequency to the mastoid process, and when the patient can no longer hear the sound the tuning fork is placed in front of the ear and lateralization of the sound is observed. Because air conduction lasts longer than bone conduction, the sound is heard through the air longer and louder than bone (positive Rinne test). In conduction deafness, sound will be perceived longer and louder through the bone than through the air (negative Rinne test). In neuronal deafness, both air and bone conduction are compromised, however air conduction remains longer and louder than bone conduction.

Suggested reading

1. Bajo VM, Merchan MA, Lopex DE, Rouiller ER. Neuronal morphology and efferent projections of the dorsal nucleus of the lateral lemniscus in the rat. *J Comp Neurol* 1993;334:241–62
2. Comacchio F, D'Eredita R, Marchiori C. MRI evidence of labyrinthine and eighth-nerve bundle involvement in mumps virus sudden deafness and vertigo. *ORL J Otorhinolaryngol Relat Spec* 1996;58:295–7
3. Dallos P, Popper AN, Fay RR, eds. *The Cochlea*. New York: Springer, 1996
4. Edelman GM, Gall EW, Cowan WM, eds. *Functions of the Auditory System*. New York: Wiley, 1988
5. Ehret G, Romand R. *The Central Auditory System*. Oxford: Oxford University Press, 1997
6. Faye-Lund H, Osen KK. Anatomy of the inferior colliculus in rat. *Anal Embryol* 1985;171:1–20
7. Gordon MA, Silverstein H, Willcox TO, Rosenberg SI. A reevaluation of the 512-Hz Rinne tuning fork test as a patient selection criterion for laser stapedotomy. *Am J Otol* 1998;19:712–17

8. Grasby PM, Frith CD, Friston KJ, *et al.* Functional mapping of brain areas implicated in auditory-verbal memory function. *Brain* 1993;116:1–20
9. Masterton RB. Role of the central auditory system in hearing: the new direction. *Trends Neurosci* 1992;15:280–5
10. Maw AR, Parker AJ. A model to refine the selection of children with otitis media with effusion of adenoidectomy. *Clin Otolaryngol* 1993;18:164–70
11. Miller SL, Delaney TV, Tallal P. Speech and other central auditory processes: Insights from cognitive neuroscience. *Curr Opin Neurobiol* 1995;5:198–204,
12. Moore JK. The human auditory brainstem: a comparative view. *Hearing Res* 1987;29:1–32
13. Pandya DN, Rosene DL, Doolittle AM. Corticothalamic connections of auditory-related areas of the temporal lobe in the rhesus monkey. *J Comp Neurol* 1994;345:447–71
14. Robertson D, Anderson CJ. Acute and chronic effects of unilateral elimination o auditory nerve activity on susceptibility to temporary deafness induced by loud sound in the guinea pig. *Brain Res* 1994;646:37–43
15. Shmonga Y, Takada M, Mizuno N. Direct projections from the non-laminated divisions of the medial geniculate nucleus to the temporal polar cortex and amygdala in the cat. *J Comp Neurol* 1994;340:405–26
16. Sichel JY, Eliashar R, Dano I. Explaining the Weber test. *Otolaryngol Head Neck Surg* 2000;122:465–6
17. Spoendlin H, Schrott A. Analysis of the human auditory nerve. *Hearing Res* 1989;43:25–38
18. Starr A, Sininger Y, Winter M, *et al.* Transient deafness due to temperature-sensitive auditory neuropathy. *Ear Hear* 1998;19:169–79
19. Swan IR. The Rinne tuning fork test. *Hosp Pract (Off Ed)* 1989;24:99–102
20. Tanaka Y, Kamo T, Yoshida M, Yamadori A. "So-called" cortical deafness. *Brain* 1991;114:2385–401

Vestibular system 15

The vestibular system is a special somatic afferent (SSA) system which maintains the orientation in the three dimensions, fixation of gaze during head rotation, and control of muscle tone. In order to accomplish these diverse activities, generated impulses at the receptor level are carried by the vestibular nerve to the cerebellum and the vestibular nuclei. Projections from these nuclei eventually reach the extraocular motor nuclei, cerebellum, and the spinal cord.

Peripheral vestibular apparatus and central vestibular pathways

Vestibular receptors are contained in the inner ear within the semicircular canals and the vestibule. These receptors convey vestibular inputs through the dendrites of the bipolar neurons of Scarpa's ganglia to the cerebellum and brainstem. Massive projections from the brainstem to subcortical and spinal neurons allow the generated impulses to exert influences on vast areas of the nervous system.

Semicircular canals (Figure 15.1) are perpendicular to each other and enclose the endolymph-filled membranous semicircular ducts, separated by the perilymphatic space. The ampulla, a dilatation at each end of the semicircular canals and ducts contains the crista ampullaris upon which the neuroepithelial cells rest. These neuroepithelial (hair) cells of the ampullary crest have stereocilia embedded in a gelatinous substance known as the cupula. A single cilium exists at one edge of the stereocilia that is known as the kinocilium. Depolarization occurs when the cilia move toward the kinocilium, and movement in the reverse direction induces hyperpolarization. This movement occurs when the current of the endolymph deflects the cupula to one side or the other. The stereocilia and the kinocilium receive the peripheral processes of the neurons of the vestibular ganglia. Since the cupula and the enclosed ampullary crests of the semicircular ducts are sensitive to angular acceleration and deceleration, these ducts are known as kinetic receptors.

Movement of the endolymph in the semicircular ducts occurs when the head is rotating. However, due to the inertia created by head rotation, the endolymph lags behind, and the cupula is deflected in the direction opposite the head rotation. The movement of the cupula is followed by movement of the stereocilia toward the kinocilium, producing depolarization of the hair cells. As the rotation continues, the endolymph catches up with the head rotation and moves in the same direction. When the head rotation suddenly stops, the endolymph continues to move in the original direction, and the cupula and stereocilia move in the reverse direction. In general, the lateral ampullary crista is excited by movement of the endolymph into the utricle from the lateral semicircular duct, whereas excitation of the superior (anterior) and posterior ampullary cristae occurs upon movement of the endolymph out of the utricle. Deflection of the hair cells in these structures results in excitation of the vestibular nerve endings. The generated nerve impulses in the excited neuroepithelial cells are transmitted to the neurons of the vestibular ganglia.

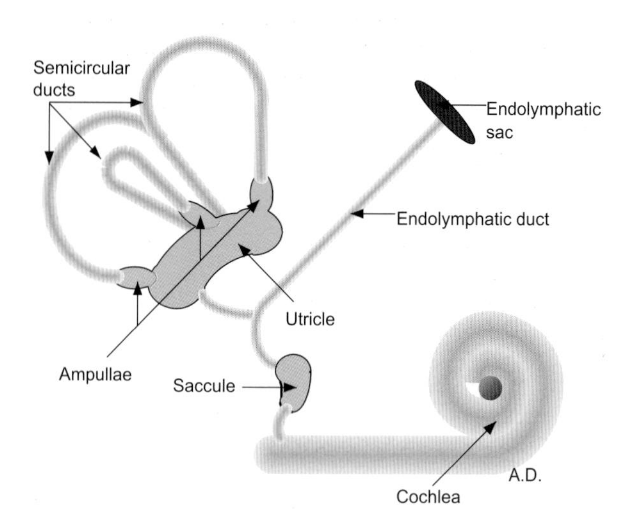

Figure 15.1 Various components of the vestibular receptors

The vestibule (Figure 14.6) is part of the bony labyrinth containing the endolymph filled membranous utricle and saccule. The neuroepithelial cells of the utricle rest on the macula, which is a horizontal structure in the erect posture. This allows changes in its firing pattern when the head is flexed or extended. The saccular macula contains neuroepithelial cells on its medial wall, which assume a vertical position, responding to head tilt to one side. The stereocilia of neuroepithelial cells of both maculae are embedded in a gelatinous material covered with otoliths consisting of crystals of calcium carbonate particles. The fact that kinocilia in the utricular and saccular maculae are located in different positions expands the sensitivity of the receptor neurons to head tilts in different directions. In general, the utricle and saccule are static receptors that respond to gravitational pull, head tilt, and linear acceleration and deceleration. The macula of the saccule is also thought to be sensitive to vibration.

Deflection of the stereocilia of the macula utriculi or sacculi upon head tilting, exerted by the drag on the underlying cilia, is translated into vestibular impulses carried by the dendrites of the bipolar neurons of the Scarpa's ganglia. In this process, the neuroepithelial hair cells are facilitated in one half of the macula and inhibited in the other half.

The maculae project to the lateral vestibular nucleus, which forms the lateral vestibulospinal tract.

Excessive stimulation of the utricle may produce signs of motion sickness.

Vestibular neuronitis is a benign, self-limiting disorder of unknown etiology, although some evidences indicate a viral cause. Prostrating vertigo that usually lasts for 2 or 3 days is the main dramatic symptom of this disease. A unique feature of the condition is the absence of any hearing deficit. Minor episodes of ataxia may be experienced by the patient after the acute episode has subsided. Symptomatic treatment with vestibular sedatives may be essential.

Thus, increase extensor muscle tone via the ipsilateral vestibulospinal tract may occur upon tilting the head to one side. Flexion or extension of the head may activate the maculae on both sides and eventually the vestibulospinal tracts bilaterally. Contraction of the extensor muscles is coordinated by corticocerebellar input or directly via the fastigiovestibular fibers.

The vestibular receptors receive the peripheral processes (dendrites) of the bipolar neurons of the superior and inferior vestibular (Scarpa's) ganglia. The superior vestibular ganglion is comprised of neurons that receive information from the anterior (superior) and lateral semicircular ducts, and the utricle. The inferior vestibular ganglion is formed by neurons which innervate the posterior semicircular ducts and the saccule. Central processes of the bipolar neurons of the vestibular ganglia (primary vestibular fibers) form the vestibular nerve. The vestibular nerve runs in close association with the cochlear and facial nerves and the labyrinthine artery in the pontocerebellar angle and later within the internal acoustic meatus (Figure 15.2). Therefore, tumors of the pontocerebellar angle result in combined vestibular and auditory dysfunctions, as well as facial nerve palsy.

Fibers of the vestibular nerve terminate primarily in the vestibular nuclei. Some of these fibers bypass the vestibular nuclei and terminate in the cerebellum as primary vestibulocerebellar fibers, projecting to the ipsilateral uvula and the flocculonodular lobe. These projections terminate as mossy fibers through the juxtarestiform body, which represents the medial part of the inferior cerebellar peduncle. The lateral, larger part of the inferior cerebellar peduncle is known as the restiform body. Most fibers of the vestibular nerve terminate in the vestibular nuclei (Figure 15.2).

Vestibular nuclei are located in the vestibular area of the rhomboid fossa, consisting of the superior, inferior, medial, and lateral nuclei. They receive afferents from the interstitial nucleus of Cajal, the flocculonodular lobe, the fastigial nuclei and the vermal part of the cerebellum. No known afferents to these nuclei are derived from the cerebral cortex or basal nuclei.

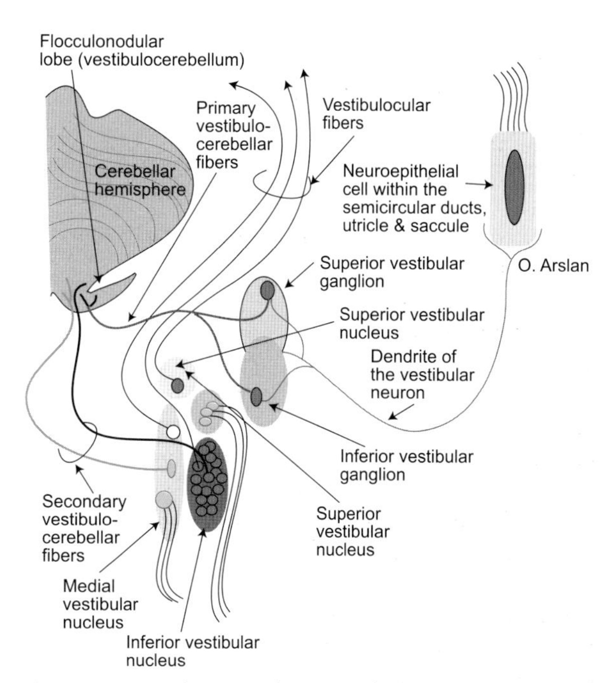

Figure 15.2 Schematic diagram of the vestibulocochlear apparatus, associated nerves and nuclei

The vestibular nuclei (Figure 15.3) project to the spinal cord, cerebellum, other nuclei of the brainstem, and the thalamus.

The superior vestibular nucleus is the most rostral vestibular nucleus which lies ventral and medial to the superior cerebellar peduncle, adjacent to the mesencephalic nucleus and the principal sensory nucleus.

The inferior vestibular nucleus has a speckled appearance as it is crossed by the descending vestibular fibers. It lies lateral to the medial vestibular nucleus, extending from the caudal medulla (at the level of the lateral cuneate nucleus) to the level of the superior vestibular nucleus. It receives input from the vestibular nerve, giving rise to the bilateral ascending and descending projections through the medial longitudinal fasciculus (MLF) to the motor nuclei of the extraocular muscles.

The medial vestibular nucleus is the longest of the vestibular nuclei, extending from the level of the hypoglossal nucleus to the level of the abducens nucleus. It consists of small and medium sized cells which give origin to the medial vestibulospinal tract.

The lateral vestibular (Deiters') nucleus contains giant neurons, which are the source of the main excitatory vestibulospinal tract. It extends from the level of the abducens nucleus rostrally to the site of entrance of the primary vestibular fibers caudally. It receives input from B zone of the anterior vermal cortex, but no input from the fastigial nucleus or labyrinth. Damage to B zone may produce disinhibition

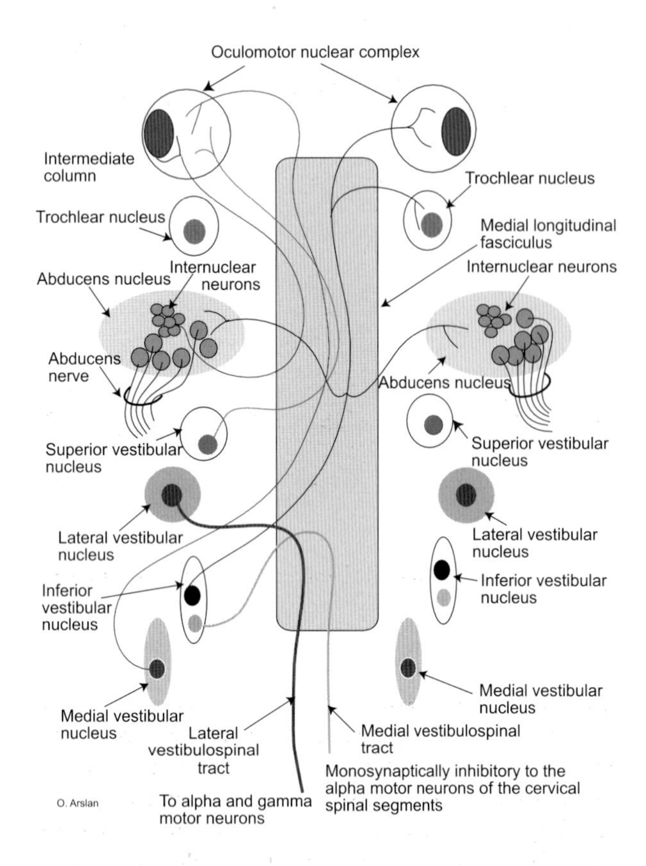

Figure 15.3 Diagram of the medial longitudinal fasciculus showing its principal ascending and descending components

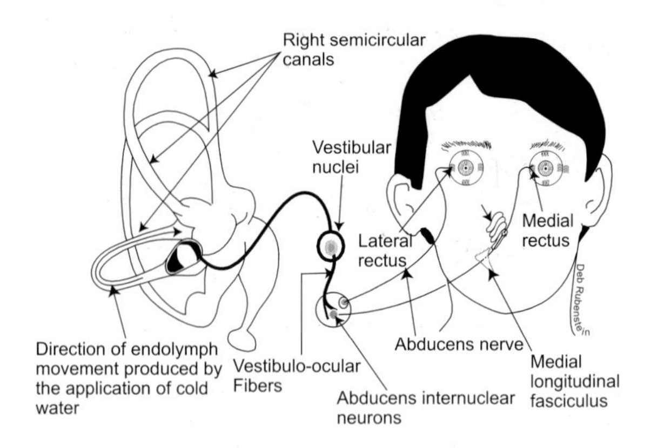

Figure 15.4 This drawing illustrates the mechanism by which vestibular impulses, generated by caloric test (e.g. application of cold water), produce eye movements. Note the vestibular input to the abducens nucleus (lateral gaze center) and ensuing fibers that act upon the medial and lateral recti muscles

of Deiters' nucleus and eventual extensor hypotonia. It mediates cerebellar influences on postural and labyrinthine reflexes.

Central vestibular pathways

Secondary vestibulocerebellar fibers originate from the vestibular nuclei and distribute to the cerebellum, spinal cord, and motor nuclei of the extraocular muscles and thalamus. Those fibers derived from the medial and inferior vestibular nuclei project to the flocculonodular lobe of the cerebellum through the juxtarestiform body.

Secondary vestibular fibers to the spinal cord (Figures 15.3 & 15.4) form the medial and lateral vestibulospinal tracts.

The medial vestibulospinal tract, a monosynaptically inhibitory pathway that originates from the medial vestibular nucleus, extends to the upper cervical spinal segments. It is a bilateral tract with an ipsilateral predominance contained within the medial longitudinal fasciculus.

The lateral vestibulospinal tract is the principal vestibular projection to the spinal cord. It originates from the lateral vestibular nucleus and runs ipsilaterally through the entire length of the spinal cord. It is excitatory to the α motor neurons of the extensor muscles.

Secondary vestibular fibers to the motor nuclei of the extraocular muscles (Figure 15.3) enable the vestibular system to coordinate eye movements via the MLF. The MLF (Figure 15.3) is a composite bundle, containing ascending and descending components. The ascending component includes the vestibulo-ocular fibers and the axons of the internuclear neurons of the contralateral abducens nucleus. The vestibulo-ocular fibers project bilaterally to the motor nuclei of the oculomotor, trochlear, and abducens nerves. These fibers mediate the vestibulo-ocular reflex, coordinate the contraction of the extraocular muscles, and fixate our gaze during head rotation. In the vestibulo-ocular reflex, impulses received by the abducens nucleus, mainly from the contralateral medial vestibular nucleus, project via the internuclear neurons to the neurons of the medial rectus muscle of the contralateral oculomotor nuclear complex, and to the ipsilateral rectus via the α motor neurons.

Activation of the abducens internuclear and abducens motor neurons produces movement of both eyes to the side of original stimulus. As mentioned earlier, abducens internuclear neurons mediate conjugate adduction of one eye with the abduction of the

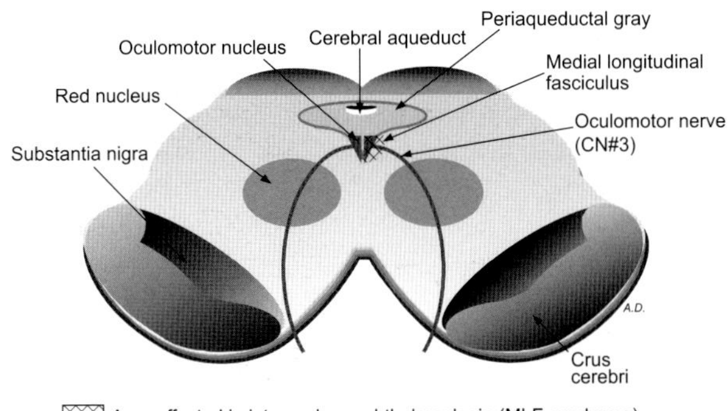

Figure 15.5 Section of the midbrain at the level of the superior colliculus. Observe the lesion that involves the medial longitudinal fasciculus producing internuclear ophthalmoplegia

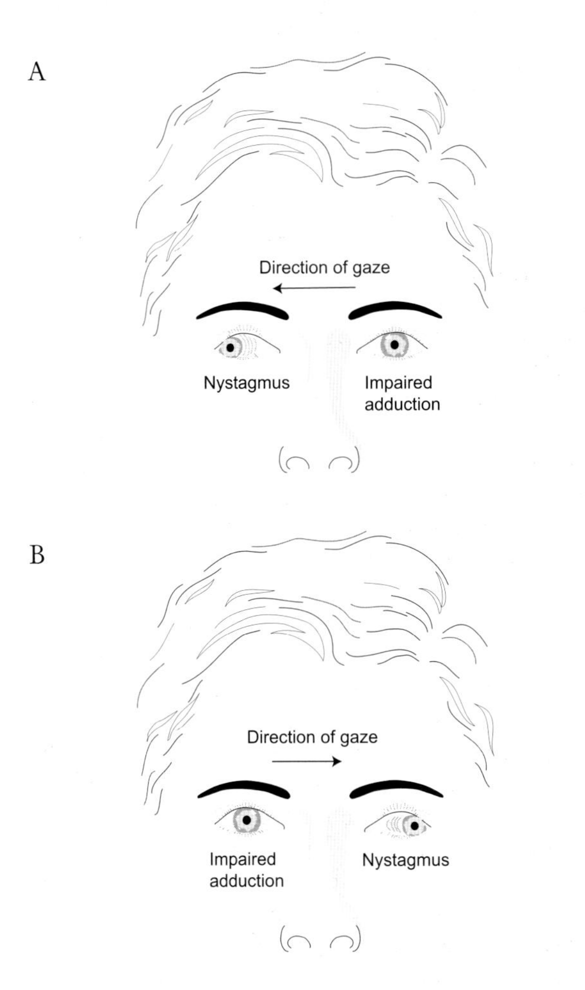

Figure 15.6 Manifestations of internuclear ophthalmoplegia (medial longitudinal fasciculus (MLF) syndrome). Left MLF lesion (A) with adductor paresis of the left eye and monocular nystagmus in the right eye. Right MLF lesion (B) with impaired adduction of the right eye and nystagmus in the left eye

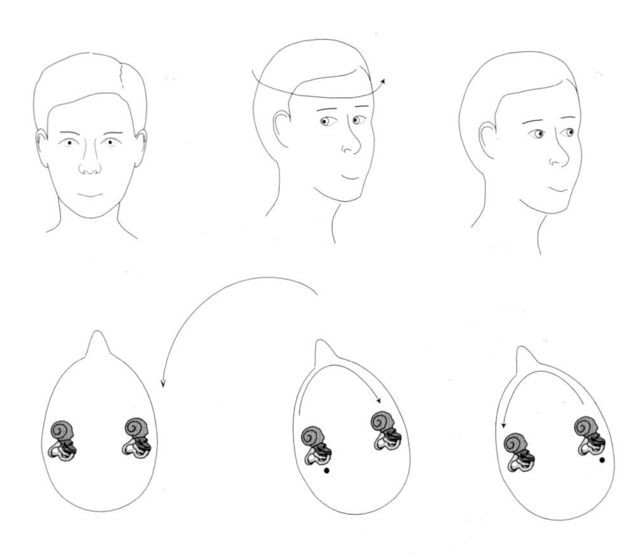

Figure 15.7 Movement of the endolymph in relation to rotation of the head to the left

opposite eye in lateral gaze. The vertical gaze center is also activated by the input generated from the medial vestibular nucleus. This occurs when bilateral movement of endolymph into the superior ampullae (during neck extension) activates the corresponding ampullae, and subsequently the medial vestibular nuclei. Although the superior vestibular nucleus is activated during head movement, its inhibitory role is also conveyed via the MLF. The descending components contain the medial vestibulospinal tract, the pontine reticulospinal tract, the interstitiospinal tract and the tectospinal tract (Figure 15.3).

In summary: vestibular receptors → peripheral processes of the vestibular neurons → vestibular ganglia → primary vestibular fibers → vestibular nuclei → cerebellum. The postsynaptic fibers from the vestibular nuclei form the secondary vestibular fibers → spinal cord, motor nuclei, and the cerebellum.

Vestibular dysfunctions

Vestibular deficits may result from peripheral labyrinthine damage, or central lesions which involve the vestibular nuclei and associated pathways. Below are some of the most important conditions associated with vestibular dysfunctions.

Anterior internuclear ophthalmoplegia (MLF syndrome) is commonly seen in multiple sclerosis as a result of degeneration of the MLF rostral to the abducens nuclei (Figures 15.5 & 15.6). It may also result from occlusion of the paramedian pontine branches of the basilar artery (brainstem stroke). This condition is characterized by paresis of adduction on attempted lateral gaze to the opposite side. Adductor paresis is due to disruption of the axons of the internuclear neurons which project to the oculomotor neurons and innervate the medial rectus. Ataxic nystagmus of the abducting eye (Harris's sign), due to compensatory augmentation of the vestibular output, is seen as a result of damage to the vestibular fibers within the affected MLF. Horizontal diplopia is marked upon gazing to the contralateral side; however, convergence remains unaffected due to intactness of the oculomotor nuclei and associated pathway. In young adults, bilateral internuclear ophthalmoplegia is virtually pathognomonic of multiple sclerosis.

'One-and-a-half' syndrome is characterized by symptoms of anterior internuclear ophthalmoplegia to the contralateral side and a conjugate horizontal gaze palsy to the same side. In other word, one eye remains motionless in the midline while the other eye can only abduct. This syndrome results from a lesion of the caudal MLF and the paramedian reticular formation.

Meniere's disease is a condition that primarily affects middle aged individuals and is commonly unilateral. It does affect both sexes equally with a peak onset between 20 and 50 years of age. It is thought to result from abnormal circulation of the endolymph within the vestibular apparatus, as well as progressive distension of the endolymphatic sac, cochlear duct, and saccule. The etiology of distension has yet to be determined, but theories usually suggest either overproduction or decreased resorption of the endolymph. Some evidence suggests that individuals with Meniere's disease have abnormal temporal bone and skull features along with an abnormal endolymphatic sac from birth. It is associated with vomiting, aural pressure, prostration, and also tinnitus which lasts from several minutes to an hour. Reduced hearing ability to low frequencies that progresses with time may also be observed. Stress, premenstrual fluid retention, and specific food may precipitate these symptoms. The most serious symptom of this condition is recurrent episodes of severe and explosive vertigo, which may last from 20 minutes to several hours. Hearing loss and tinnitus may precede the first episode of vertigo by several months or years in Lermoyez's variant of this disease. In this disease spontaneous nystagmus may be observed during an episode, however caloric-induced nystagmus is lost on the affected side. This disease should be differentiated from basilar artery ischemia which exhibits additional manifestations such as visual deficits, and diplopia.

Meniere's disease is classified into five stages, which determine the appropriate management plan. In stage I only the cochlea is involved and the patient experiences tinnitus, fullness, and low tone hearing loss. This stage responds best to treatment, which include diuretics, vasodilatory drugs (e.g. promethazine) and a low-salt diet. Stage II involves more widespread endolympathic hydrops of the saccule and other parts of the labyrinth. Hearing loss for low-tones, tinnitus, fullness, and dizzy spells are experienced by the patient at this stage. Treatment may include diuretics, a low-salt diet, and dexamethasone. Persistence of symptoms may require insertion of a shunt tube into the endolymphatic duct. If the shunt is not successful streptomycin perfusion may be essential to destroy the end organ. This is achieved by aminoglycoside that replaces the calcium ions in the cell membrane and produces the loss of stereocilia. In stage III the endolymphatic sac becomes smaller and more displaced. The symptoms also include tinnitus, vertigo, and fullness. At this stage vestibular neurectomy, which carries the complication of hearing loss as a side-effect, may be required. In stage IV the obstruction to the flow of the endolymph is complete and the hydrops fills the vestibule. The patient experiences no vertigo since the hydrostatic pressure can not increase. If the disease is bilateral, the patient can not get motion sickness. No known treatment for the disease at this stage. In stage V multiple obstruction and/or rupture of the membranous labyrinth may occur. No treatment is available, although cochlear implants may be performed when the patient's biggest concern is bilateral deafness and not vertigo.

Combined vestibular and auditory deficits are seen ipsilaterally in lesions occurring at the internal acoustic meatus or cerebellopontine angle. Fractures of the temporal bone may result in disruption of the vestibular and cochlear nerves.

Acoustic neuroma (tumor of the pontocerebellar angle) is a benign Schwannoma which arises in the vestibular nerve and may expand to involve the cochlear nerve, compressing the adjacent vestibular nuclei, inferior cerebellar peduncle, trigeminal nerve, glossopharyngeal nerve, and the spinal trigeminal tract. Middle-aged or elderly patients are more prone to this disease than the rest of the population. Despite its vestibular origin, acoustic neuroma initially presents with tinnitus followed by vertigo and gradual deafness that may take a span of months or years. Intactness of the facial nerve may be ascertained by checking the corneal reflex. Patients may also feel pain in the area of distribution of one or more branches of the trigeminal nerve. Radiating pain in the oropharynx, dysphagia, and hoarseness of voice may also occur if the tumor expands further downward.

Deficits associated with the lesions of the vestibular apparatus, nuclei, and pathways include vertigo (sense of rotation of the environment or self), ataxia, past-pointing, nystagmus, and some other deficits depending upon the affected structure. These deficits include:

Vertigo refers to a complete or partial loss of spatial orientation (should be distinguished from dizziness and blackouts). Patients may have a feeling of moving in space or of objects moving around them. This deficit also results from otogenic disorders such as Meniere's disease, neurologic disorders as in multiple sclerosis, visual disorders such as diplopia, or in ischemic attacks of the vertebrobasilar system. Other etiologies of vertigo may include hysteria and blood disorders such as leukemia. Vertigo due to a peripheral lesion produces severe, episodic, and paroxysmal attacks. Central vertigo is a persistent condition and is generally accompanied by nystagmus and ataxia. Damage to one labyrinth results in signs generated by overactivation of the intact labyrinth. The patient will fall to the side of the dysfunction or swerve to that side during walking. Vertigo subsides within hours when the patient assumes a recumbent position. This occurs by the inhibitory action of the flocculonodular lobe upon the intact vestibular nuclei. Sufficient sense of position is maintained, when standing, via the visual cues and utilization of proprioceptive stimuli in the dorsal columns. Darkness may impede this compensatory system. It is not usually seen in lesions of the vestibular nuclei or in cold water caloric testing of the affected side.

Ataxia (incoordination) is characterized by a wide-based, cautious, and unsteady gait, especially during turns and when walking on uneven surfaces. It is severe, gravity dependent, and is generally associated with a lack of coordination of voluntary movements.

Past-pointing refers to overshooting of the patient's finger when attempting to reach a target. This is used to check the integrity of the vestibular system by asking the patient to touch the examiner's finger with his right index finger following rotation in a revolving (Barany) chair. In individuals with an intact vestibular apparatus, the patient's index finger is placed to the right of the examiner finger following rotation to the right. The reverse is true of rotation to the left.

Nystagmus is an alternating, rhythmic oscillation of the eye in response to an imbalance in the vestibular impulses. This condition may be produced artificially by visual (optokinetic nystagmus) or thermal (caloric nystagmus) stimuli. A lesion of the temporal or parietal cortices may abolish visually induced (optokinetic) nystagmus. It may be seen with eyes at rest, or may be accentuated by ocular movement. Congenital nystagmus, which is characterized by pendular movement of the eyes at rest, is triggered by head or eye movements and may be familial in origin. It is generally described as having fast and slow phases (phasic or jerk nystagmus). Occasionally, nystagmus may develop in normal individuals when they attempt to maintain extreme gaze to one side. Pathological nystagmus is generally seen at rest. Clinically, it is described according to the direction of the rapid phase. The slow phase of the jerk nystagmus is due to the tonic vestibular stimuli, while the rapid phase is the result of the action of the frontal cortex.

Nystagmus may occur as a result of peripheral (labyrinthine) disorders, cerebellar diseases, ocular dysfunction (amblyopia), or circulatory disturbances of the endolymph (Meniere's disease). It may also be observed in individuals with central lesions involving the ascending vestibular pathways (e.g. MLF). Impairment of ocular fixation, associated with myasthenia gravis, optic atrophy, and poor illumination (miner's nystagmus) may generate pendular nystagmus with two phases of equal velocity. Pendular nystagmus is also seen in congenital ocular diseases. Nystagmus caused by peripheral lesions is generally horizontal and has a rapid phase toward the intact side, and is commonly associated with diplopia and dysarthria. Diseases of the inner ear may produce positional nystagmus. The latter may also be elicited upon sudden movement of the head 30 degrees backward over the edge of the bed and simultaneously rotating to one side or the other. A left turn results in nystagmus in a clockwise direction, whereas a right turn produces nystagmus in a counter clockwise direction.

Central vestibular lesions produce nystagmus

with the rapid phase toward the opposite side. Vertical nystagmus may be observed in individuals with lesions involving the structures within the posterior cranial fossa as in Arnold–Chiari malformation. Lesions of the ascending component of the MLF (rostral to the abducens nuclei) result in anterior internuclear ophthalmoplegia. Brainstem lesions may cause gaze-dependent, coarse, and unidirectional nystagmus, which may be horizontal or vertical.

Optokinetic testing is performed by observing displacement of the entire visual field (e.g. watching striped colors on a revolving drum). The fast phase of this nystagmus is in the direction opposite to the movement of the object. This test may be of value in the determination of validity of malingerers' claims that they cannot see, and seems of particular value in the examination of hysterical patients.

Caloric testing is based upon initiating a current within the endolymph by cooling or warming the inner ear through irrigation of the external acoustic meatus. The patient will be asked to tilt his head 60° from the vertical plane to bring the horizontal semicircular canals to a vertical position. Rotating the head in the vertical plane, which produces compensatory upward and downward gaze, tests vertical gaze. Instillation of cold water in both ears produces tonic upward movement of the eyes, while bilateral warm water tests produce tonic downward gaze. The effect of angular acceleration on movement of the endolymph is shown in Figures 15.4 & 15.7.

The response to caloric testing is decreased on the affected side in Meniere's disease. In comatose patients with supressed reticular activating system, the caloric test would only elicit tonic deviation of the eye. Supratentorial lesions produce loss of oculocephalic reflex, with the preservation of caloric response. Frontal lesions do not interfere with either the caloric test or oculocephalic reflex. An easy mnemonic to remember the effect of cold and warm water on the direction of nystagmus is 'COWS' (Cold water Opposite, Warm water Same side).

Suggested reading

1. Anoh-Tanon MJ, Bremond-Gignac D, Wiener-Vacher SR. Vertigo is an underestimated symptom of ocular disorders: dizzy children do not always need MRI. *Pediatr Neurol* 2000;23:49–53
2. Beynon GJ, Baguley DM, Moffat DA, Irving RM. Positional vertigo as a first symptom of a cerebellopontine angle cholesteatoma: case report. *Ear Nose Throat J* 2000;79:508–10
3. Bukowska D. Reticular afferents from vestibular nuclei. *Folia Morphol (Warsz)* 1996;55:224–6
4. Compoint C, Buisseret-Delmas C, Diagne M, *et al.* Connections between the cerebellar nucleus interpositus and the vestibular nuclei: an anatomical study in the rat. *Neurosci Lett* 1997;238:91–4
5. Doi K, Seki M, Kuroda Y, *et al.* Direct and indirect thalamic afferents arising from the vestibular nuclear complex of rats: medial and spinal vestibular nuclei. *Okajimas Folia Anat Jpn* 1997;74:9–31
6. Gacek RR, Schoonmaker JE. Morphologic changes in the vestibular nerves and nuclei after labyrinthectomy in the cat: a case for the neurotrophin hypothesis in vestibular compensation. *Acta Otolaryngol (Stockh)* 1997;117:244–9
7. Gdowski GT, Boyle R, McCrea RA. Sensory processing in the vestibular nuclei during active head movements. *Arch Ital Biol* 2000;138:15–28
8. Gdowski GT, McCrea RA. Integration of vestibular and head movement signals in the vestibular nuclei during whole-body rotation. *J Neurophysiol* 1999;82:436–49
9. Gordon CR, Zur O, Furas R, *et al.* Pitfalls in the diagnosis of benign paroxysmal positional vertigo. *Harefuah* 2000;138:1024–7, 1087
10. Karlberg M, Halmagyi GM, Büttner U, Yavor RA. Sudden unilateral hearing loss with simultaneous ipsilateral posterior semicircular canal benign paroxysmal positional vertigo: a variant of vestibulo-cochlear neurolabyrinthitis? *Arch Otolaryngol Head Neck Surg* 2000; 126:1024–9
11. Matesz C, Nagy E, Kulik A, Tönköl A. Projections of the medial and superior vestibular nuclei to the brainstem and spinal cord in the rat. *Neurobiology (Bp)* 1997;5:489–93
12. Pennington CL, Stevens EL. Ultrasonic ablation of the labyrinth in the treatment of endolymphatic hypertension (Meniere's disease). *South Med J* 1967;60:34–43
13. Roy MK. Vertigo: approach to the diagnosis and management. *J Assoc Physicians India* 1999;47:355–6
14. Sato H, Imagawa M, Kushiro K, *et al.* Convergence of posterior semicircular canal and saccular inputs in single vestibular nuclei neurons in cats. *Exp Brain Res* 2000;131:253–61
15. Solomon D. Distinguishing and treating causes of central vertigo. *Otolaryngol Clin North Am* 2000;33:579–601
16. Staecker H, Nadol JB Jr, Ojeman R, *et al.* Hearing preservation in acoustic neuroma surgery: middle fossa versus retrosigmoid approach. *Am J Otol* 2000;21:399–404
17. Tarlov EC. Microsurgical vestibular nerve section for intractable Meniere's disease. *Clin Neurosurg* 1985;33:667–84

18. Watanabe Y, Aso S, Ohi H, *et al.* Vestibular nerve disorder in patients suffering from sudden deafness with vertigo and/or vestibular dysfunction. *Acta Otolaryngol* 1993;504(Suppl):109–11
19. Wiegand DA, Fickel V. Acoustic neuromas. The patient's perspective. Subjective assessment of symptoms, diagnosis, therapy, and outcome in 541 patients. *Laryngoscope* 1989;99:179–87
20. Ylikoski I, Palva T, House WF. Vestibular nerve findings in 150 neurectomized patients. *Acta Otolaryngol (Stockh)* 1981;91:505–10

Olfactory system 16

Olfaction, a special visceral sensation, is involved in the perception, recognition, and fine discrimination of olfactory impulses. This system integrates incoming impulses via its connections to the septal area, entorhinal cortex, and amygdala. The olfactory impulses are received by the olfactory nerves, conveyed to the olfactory bulb, and course through the olfactory tract and striae to the uncus and other parts of the limbic system. Chemicals that dissolve in liquid stimulate both olfactory and taste receptors in a similar way. The perception of food flavor is dependent upon the interaction between the olfactory and gustatory (taste) systems.

Characteristics of the olfactory receptors, nerve and pathways

Characteristics of the olfactory receptors, nerve and pathways

The olfactory apparatus consists of sensory epithelium, containing the chemoreceptor cells that detect and discriminate between different airborne odorants, as well as the supporting cells. Olfactory information is conveyed, in the form of electrical signals, via the olfactory nerves to the olfactory bulb and eventually to the primary olfactory cortex. Olfactory receptors are bipolar cells with apical dendrites extending to the epithelial surface and basally directed axons which stretch to the olfactory nerve filaments to synapse with the second-order neurons of the olfactory bulb. Plasma membrane containing high concentrations of intramembranous particles covers the exposed surfaces of the olfactory dendrites. These particles are thought to be the site of olfactory reception and ion channels related to sensory transduction. Their concentrations accord with their ability to detect low concentrations of odorants. The olfactory receptor cells and their axons contain a number of immunohistochemically demonstrable characteristic proteins, such as 19 kDa olfactory marker protein and carnosine. They also express surface molecules that cross-react with the ABO blood group antigens, and carbohydrates that bind several lectins. Genes that code for olfactory receptor protein molecules have also been isolated via *in situ* hybridization methods.

Olfactory transduction involves stimulus (odorant) dissolution in the thin layer of the olfactory mucosa, diffusion to the exposed receptor membrane, and its activation. This is followed by activation within the cilium of one or more secondary messenger systems, opening of the sodium and calcium channels, and eventual graded depolarization of the receptor cells. An action potential is initiated in the olfactory nerve when the receptor cell is sufficiently depolarized and the voltage sensitive ion channels in the proximal part of its axon are stimulated. The second messenger pathways in olfactory transduction follow two distinct routes: either the guanosine triphosphate (GTP)-dependent adenylate cyclase pathway, or the GTP-dependent phospholipase C (phosphoinositide) pathway. The olfactory receptor G-protein (Golf) is present at the distal segments of the olfactory cilia where maximum exposure to stimuli is maintained.

The supporting cells of the olfactory apparatus play important roles in the regulation of the ionic environment of the receptors, removal of debris and toxic substances, maturation of the receptors, and their insulation. In contrast to the rapid turnover of the receptors cells, the supporting cells remain fairly stable. Presence of high levels of cytochrome P450

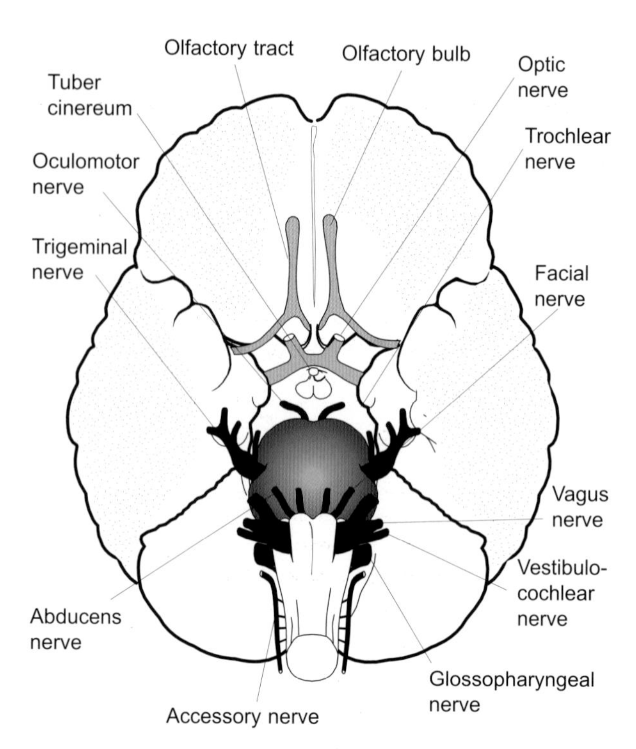

Figure 16.1 General scheme of the cranial nerves. Note the olfactory bulb and tract

and other detoxifying agents in the supporting cells of the olfactory epithelium may suggest the possible role of these cells in ridding the epithelium of unwanted odorants. The phagocytic activities of secondary lysosomal (residual) bodies in the bases of the supporting cells are responsible for the pigmentation of the olfactory area.

The olfactory nerve or filaments represent the central processes of the bipolar neurons located in the olfactory mucosa of the superior nasal concha and the corresponding part of the nasal septum. These filaments, which are among the smallest and slowest in conduction, enter the olfactory bulb in the anterior cranial fossa by traversing the cribriform plate of the ethmoid bone. Extension of the dura around the olfactory nerve fibers continues with the periosteum of the nasal cavity. The arachnoid–pia layer merges with the perineurium.

The olfactory bulb (Figure 16.1) contains synaptic glomeruli, which are formed by the dendrites of the mitral, internal tufted, and periglomerular cells as well as axons of the mitral cells. The olfactory bulb is located ventral to the frontal lobe, and rests on the cribriform plate of the ethmoid bone. The olfactory bulb is comprised of the olfactory nerve layer, molecular layer, mitral cell layer, internal granular layer, and the layer of olfactory tract fibers. The GABAergic granule cells (inhibitory) establish reciprocal synapses

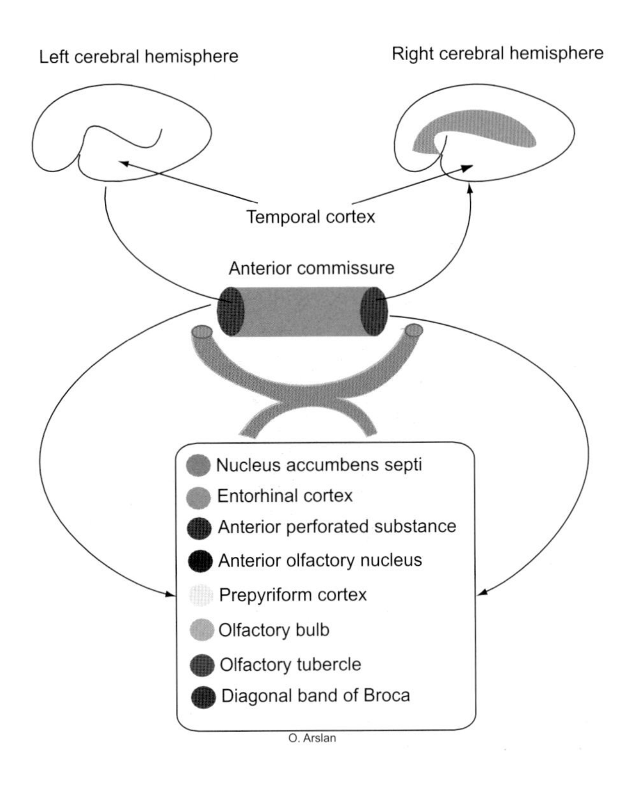

Figure 16.2 Schematic drawing of the anterior commissure. Observe the identical structures in both hemispheres connected by this commissure

with the excitatory mitral cells. The olfactory bulb acts as relay, integration, and feedback centers for complex pathways. It is the synaptic site of the olfactory filaments with the mitral cells, and with the axons that originate from the anterior olfactory nucleus, and the olfactory bulb of the opposite side. The anterior olfactory nucleus consists of pyramidal neurons, which are located in the posterior portion of the olfactory bulb and may extend into the olfactory trigone and the striae. These neurons receive collaterals from axons of mitral and tufted cells and provide recurrent collaterals to synapse with the dendrites of the ipsilateral tufted and granule cells. They also project to the contralateral olfactory bulb and anterior olfactory nucleus through the anterior commissure (Figure 16.2).

The anterior commissure (Figure 16.2) is shaped like the handle of a bicycle, crossing the lamina terminalis and dividing the fornix into pre- and postcommissural columns. It is comprised of an olfactory anterior portion which interconnects the olfactory bulbs, tubercles and tracts and contains the centrifugal axons of the anterior olfactory nucleus which project to the corresponding nucleus on the contralateral side and to the granule cells. The granule and tufted cells activate other parts of the olfactory system through their connections. The centrifugal axons act as a feedback loop, representing the efferent system that regulates the incoming olfactory pathways.

The larger, caudal part of the anterior commissure courses inferior to the lenticular nucleus, running through the external capsule, and terminates in the anterior temporal and parahippocampal gyri. It interconnects the medial, and to a lesser degree the inferior temporal gyri, entorhinal area, diagonal band of Broca, anterior perforated substance, prepyriform cortex, amygdala, and the bed nucleus of the stria terminalis of both sides.

The olfactory tract (Figure 16.2) runs in the olfactory sulcus and expands rostral to the anterior perforated substance to form the olfactory trigone. It consists of the axons of the mitral cells as well as the centrifugal axons of the anterior olfactory nucleus. It divides into lateral, medial, and smaller intermediate olfactory striae. The lateral and medial striae are surrounded by the corresponding gyri.

The lateral olfactory stria forms the principal olfactory pathway, which runs medial to the anterior perforated substance and terminates in the primary olfactory cortex (uncus).

The medial olfactory stria runs anterior to the lamina terminalis accompanied by the diagonal band of Broca. It terminates ipsilaterally in the paraterminal (septal area) and subcallosal (parolfactory) gyri. The subcallosal gyrus lies between the anterior and posterior olfactory sulci, separated from the paraterminal gyrus (pre-commissural septum) by the posterior olfactory sulcus. Olfactory impulses from the septal area travel through the medial forebrain bundle to the lateral hypothalamus, preoptic area, and the tegmentum of the midbrain. This pathway enables the olfactory impulses to influence the autonomic centers in the brainstem.

Additionally, olfactory impulses from the septal area project to the ipsilateral habenula via the stria medullaris thalami. The habenula in turn projects to the interpeduncular and tegmental nuclei via the fasciculus retroflexus, and to the opposite habenula through the habenular commissure. The autonomic (visceral) nuclei of the brainstem and the spinal cord receive projections from the reticular (tegmental) nuclei of the midbrain via the dorsal longitudinal fasciculus of Schütz and the reticulospinal tracts, respectively. These massive connections mediate the emotional reaction as well as the somatic and visceral reflexes, such as tongue, eye and neck movements, salivation, increased gastric secretion, retching and vomiting, in response to odors. The diagonal band of

Broca is located caudal to the anterior perforated substance, continuing with the periamygdaloid area caudally and the paraterminal gyrus rostrally.

The intermediate olfactory stria projects to the anterior perforated substance. The latter is differentiated into an outer plexiform, intermediate pyramidal, and inner polymorphic layers. The anterior perforated substance continues with the prepyriform cortex and the semilunar gyrus laterally and the septal area anteriorly. This area is located between the diverging medial and lateral olfactory striae, medial to the uncus and lateral to the optic chiasma. It is pierced by central branches of the internal carotid and anterior and middle cerebral arteries. The granule cells located in the pyramidal layer of the anterior perforated substance form the islands of Calleja, which contain the nucleus accumbens septi. The olfactory tubercle, a prominence caudal to the olfactory trigone, is separated from the lentiform nucleus by the substantia innominata, fibers of the anterior commissure, and the ansa lenticularis.

The primary olfactory cortex (Brodmann's area 34) includes the periamygdaloid cortex which consists of the corticomedial nucleus of the amygdala and the associated semilunar gyrus and the prepyriform cortex. The latter, which comprises the lateral olfactory and ambiens gyri, projects to the entorhinal cortex, basolateral amygdaloid nucleus, preoptic area, and the dorsomedial nucleus of the thalamus. The pyriform lobe refers to the uncus (intralimbic and uncinate gyri), lateral olfactory gyrus, and the entorhinal cortex (Brodmann's area 28). Through the medial forebrain bundle projections from the prepyriform cortex are carried to various structures of the limbic system. In this manner, the stria terminalis conveys olfactory information from the corticomedial nucleus of the amygdala to the preoptic area and the ventromedial nucleus of the hypothalamus. The basolateral nucleus of the amygdala, septal area, dorsomedial nucleus of the thalamus, and the hypothalamus project back to the primary olfactory cortex. Massive connections of the entorhinal cortex (Brodmann's area 28) to the primary olfactory cortex have already been established.

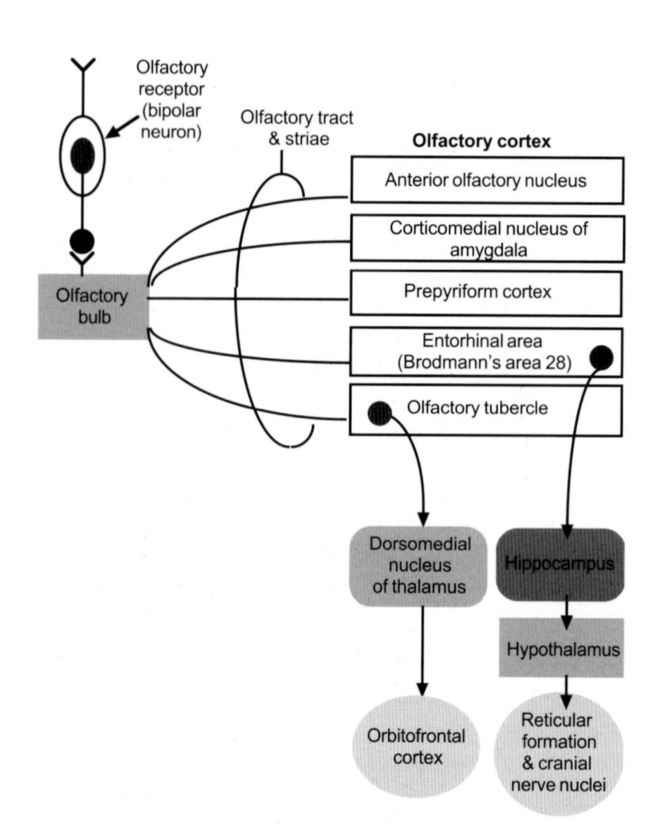

Figure 16.3 Neuronal organization of the olfactory pathway. Olfactory impulses are followed from their receptor to the site of perception and recognition at the olfactory cortex. Projection of the olfactory impulses to the thalamus, hypothalamus, and reticular formation are also shown

The reciprocal connections between the dorsomedial nucleus and the posterior prefrontal and orbitofrontal cortices (tertiary olfactory cortex) may explain the possible role of the prefrontal cortex in odor discrimination. A tumor or lesion involving the uncus, parahippocampal gyrus, interpeduncular fossa, or amygdala may cause uncinate fits. The gustatory cortical area (inferior parts of the post- and precentral gyri) may also be involved in this condition. Patients with this condition may have periods of hallucinatory olfactory perception (olfactory aura) or cacosmia, a perception of foul odors accompanied by minor or major seizures and fear of the unreality of the environment. Chewing or lip smacking may also be observed.

The basal forebrain nucleus of Meynert, which consists predominantly of cholinergic neurons in the substantia innominata, are implicated in Alzheimer's disease. These cholinergic neurons project extensively to the cerebral cortex. Numerous neurofibrillary tangles (NFT) and neuritic plaques are also seen in the CA1 zone of the hippocampal gyrus. The tau protein that cross-links the microtubules in the normal perikaryon becomes excessively phosphorylated and unable to cross-link microtubules, which leads to the formation of the paired helical filaments as an early stage of NFT formation. Hyperphosphorylation of tau proteins occurs as a result of activation of kinases and/or deactivation of phosphatases. As Alzheimer's disease progresses, the NFT and its associated apical dendrites become filled with abnormally phosphorylated tau protein and helical filaments. In the end stages of the disease, the cytoskeleton of the pyramidal neurons disintegrates and the remaining tau protein and paired helical filaments form a 'ghost NFT'. NFTs, often referred to as 'intracellular amyloid' also contain ubiquitin and glycosaminoglycans (GAGs). When aluminum has been demonstrated within NFT in cortical areas of an Alzheimer's disease-affected individual, some investigators have suggested the role of this metal as a cause of this disease. Presence of NFTs is commonly associated with dystrophic neurites (neuropil threads or curly fibers) in the cortical neuropil.

Although dystrophic neurites are not specific to Alzheimer's disease, their abundant presence may correlate with clinical dementia and presence of NFT. MAP-2 is another microtubule-associated protein that is found in healthy cells becomes also abnormally phosphorylated. A-68 protein, which may be an early marker for Alzheimer's disease, is suspected to be a modified form of tau. Neuronal loss in the nucleus basalis of Mynert, characteristic of this disease, is closely related to the severity of dementia. Additional sites of neuronal loss include the association cortices of the frontal, parietal, and temporal lobes.

Cortical neurons most affected in Alzheimer's disease are the pyramidal-shaped neurons, which project to cortical layers I, II, III, and VI. Other limbic structures such as the amygdala, locus ceruleus, median raphe nuclei, hypothalamus, cingulate gyrus and orbitofrontal cortex may also be involved. Some investigators have suggested that Hirano bodies, which are esinophilic rods composed of α-actinin, vinculin, and tropomyosin epitopes, are possible products of the degradative process that the cytoskeletal elements undergo. However, no clear-cut evidence has been documented to support this suggestion.

The role of cholinergic neurons in normal intellectual activity is confirmed by studies in which central anticholinergic medications produce cognitive dysfunctions that mimic Alzheimer's dementia. In Alzheimer's disease, other biochemical imbalances and morphological abnormalities or degeneration are noticed in neurons associated with somatostatin, norepinephrine, substance P, and vasopressin. New evidences have supported the concept that Aβ (diffuse plaques) deposition may occur prior to Alzheimer's type neuronal or glial changes. This assumption that Aβ arises from aberrant (mutant) proteolysis of βAPP gene (on the long arm of chromosome 21) following membrane injury is held because Aβ is derived from an integral membrane sequence and highly insoluble when derived from senile plaques and meningovascular deposits. Molecular evidences suggest that mutations in the βAPP gene may initiate β-amyloidosis prior to any existing pathological changes. Some attribute the familial Alzheimer's disease to the mutations in the βAPP gene. Drugs that inhibit the protease, which produces the C terminus of Aβ (γ-secretase), may be used as therapeutic agents in the treatment of Alzheimer's disease. Detection of soluble Aβ in the CSF and plasma of normal and Alzheimer's disease patients, as well as in a variety of cultured cells, has led to the development of ELISA assays. These assays have established that many of Alzheimer's disease patients show lower CSF concentrations of the soluble Aβ42 peptide than do normal elderly individuals.

In addition to βAPP gene mutations, presenilin 1 & 2 (PS1 & PS2) genes, which are required for embryogenesis, have also been implicated in the familial Alzheimer's disease. The toxic effects of mutant PS1 and PS2 genes are manifested in the disregulation of γ-secretase(s) that selectively intensifies the proteolysis of βAPP at Ab42, resulting in dramatic increase in Aβ42 level. Another major risk factor for the development of Alzheimer's disease is the excess of natural ε4 polymorphism of the ApoE gene in individuals with Alzheimer's disease. Inheritance of ApoE ε3 alleles has been shown to significantly increase the likelihood of developing late appearance of the disease, while inheritance of ApoE ε3 allele is most likely to decrease the possibility of developing the disease. The increase in the likelihood of Alzheimer's disease may also occur upon inheritance ApoE ε4 protein on chromosome 19 that lacks cysteine and thus is unable to undergo disulfide cross-linking.

It has been shown that Alzheimer's disease subjects with two ε4 alleles have higher concentrations of Aβ peptide deposits (particularly the highly aggregation-prone, 42 residue form) in the brain. In fact 30–40% of individuals with ε4 alleles are at risk to develop this disease. However it should also be emphasized that NFTs may also form in a variety of other diseases, indicating that its presence is a mere response to brain insults. The linkage between the development of Alzheimer's disease in individuals with Down's syndrome has been reported. Non-fibrillar, diffuse, and amorphous forms of Aβ deposits have been found in the limbic, striatum, and association cortices of trisomic patients. This early accumulation of diffuse plaques may result from the elevated APP gene and its expression and eventual increase in Aβ concentration. Synaptic losses near the sites of these deposits have also been observed.

Clinically, Alzheimer's disease exhibits progressive intellectual dementia, which includes impairment of judgment and memory. In the initial stages, patients may seem sociable, alert, but show signs of confusion and depression in a new and unaccustomed environment. Patients also exhibit the inability to recall a list of items after a few minutes delay without cues or hints (delayed free recall) while the ability to recall items from the same list with the aid of cues or multiple cues (recognition memory) remains unaffected. Language impairment, depression, and anomia may be seen following the initial stage of this disease. Comprehension of both written and verbal communication (lack of spontaneity in speech) are also affected. Babinski's sign and hyperreflexic jaw-jerk may also be seen at this stage. Patients do not exhibit hemiparesis or visual deficits. The profound deficits in memory and other cognitive therapy are striking in the late stages of this disease. As a late manifestation, patients become unresponsive, incapacitated, and in a fetal posture at this stage. Cholinergic agonist tacrine, which increases the level of acetylcholine by inhibiting degradation of acetylcholinesterase, and selegeline, a monoamine oxidase B inhibitor and powerful antioxidants may be used in the treatment of Alzheimer's disease.

Suggested reading

1. Yilmazer-Hanke DM, Hudson R, Distel H. Morphology of developing olfactory axons in the olfactory bulb of the rabbit (Oryctolagus cuniculus): a Golgi study. *J Comp Neurol* 2000;426:68–80
2. Hummel T, Klimek L, Welge-Lüssen A, *et al.* [Chemosensory evoked potentials for clinical diagnosis of olfactory disorders] Chemosensorisch evozierte Potentiale zur klinischen Diagnostik von Riechstörungen. *HNO* 2000;48:481–5
3. Johnson DM, Illig KR, Behan M, Haberly LB. New features of connectivity in piriform cortex visualized by intracellular injection of pyramidal cells suggest that "primary" olfactory cortex functions like "association" cortex in other sensory systems. *J Neurosci* 2000;20: 6974–82
4. Shoji M, Kanai M, Matsubara E, *et al.* Taps to Alzheimer's patients: a continuous Japanese study of cerebrospinal fluid biomarkers. *Ann Neurol* 2000;48:402
5. Parvizi J, Van Hoesen GW, Damasio A. Selective pathological changes of the periaqueductal gray matter in Alzheimer's disease. *Ann Neurol* 2000;48:344–53
6. Wolstenholme GEW, Knight J. *Taste and Smell in Vertebrates (Ciba Foundation Symposium)*. London: Churchill, 1970:51–67
7. Bojsen-Moller F. Demonstration of terminalis, olfactory, trigeminal and perivascular nerves in the rat nasal septum. *J Comp Neurol* 1975;159:245–56
8. Brennan P, Kaba H, Keveme EB. Olfactory recognition: a simple memory system. *Science* 1990;250:1223–6
9. DeVries SH, Baylor DA. Synaptic circuitry of the retina and olfactory bulb. *Neuron* 1993;10(Suppl):139–49
10. Dionne VE. How do you smell? Principle in question. *Trends Neurosci* 1988;11:188–99
11. Filley CM, Heaton RK, Nelson LM, *et al.* A comparison of dementia in Alzheimer's disease and multiple sclerosis. *Arch Neurol* 1989;40:157–61
12. Finger TE, Silver WL. *Neurobiology of Taste and Smell.* New York: John Wiley & Sons, 1987
13. Katzman R. Clinical and epidemiological aspects of Alzheimer's disease. *Clin Neurosci* 1993;1:165–70
14. Masliah E, Terry RD. Role of synaptic pathology in the mechanisms of dementia in Alzheimer's disease. *Clin Neurosci* 1993;1:192–8
15. Miller R. Schizophrenia as a progressive disorder: Relation to EEG, CT, neuropathological and other evidence. *Prog Neurobiol* 1989;33:17–44
16. Mullan M, Crawford F. Genetic and molecular advances in Alzheimer's disease. *Trends Neurosci* 1993;16:398–403
17. Pearson RCA, Powell TPS. The neuroanatomy of Alzheimer's disease. *Rev Neurosci* 1989;2:101–21
18. Stewart WB, Shepherd GM. The chemical senses: taste and smell. In Swash M, Kennard C, eds. *Scientific Basis of Clinical Neurology.* Edinburgh: Churchill Livingstone, 1985:214–24
19. Yankner BA. Mechanisms of neuronal degeneration in Alzheimer's disease. *Neuron* 1996;16:921–32
20. Zatorre RJ, Jones-Gotman M, Evans AC, Meyer E. Functional localization and lateralization of human olfactory cortex. *Nature* 1992;360:339–40

Limbic system 17

The limbic system (visceral brain) directly or indirectly affects somatic and visceral motor (autonomic) functions by modulating the activities of the brainstem reticular formation, spinal cord, and the hypothalamus. It maintains homeostasis, integrates the olfactory impulses, changes the behavioral pattern and modifies or inhibits reactions to stimuli via rewarding and punishing centers. It also influences the activities of the pituitary gland. This system also plays an important role in encoding and establishing memory patterns. Virtually all sensory systems including olfactory, visual, and auditory impulses maintain connections to the limbic system. It encompasses structures associated with the olfactory system, hippocampal formation, thalamus, hypothalamus, epithalamus, and cerebral cortex. These structures include the septal area, induseum griseum, amygdala, prefrontal cortex, cingulate gyrus, parahippocampal gyrus, hypothalamic nuclei, certain thalamic nuclei, pineal gland, habenula, and stria medullaris thalami.

Features, constituents and connections of the limbic system

The limbic system is an intricate system which regulates emotion, behavior, drive, and learning processes and is associated with all the activities that preserve the individual and species. It comprises the hippocampal formation, septal area, induseum griseum, amygdala, limbic lobe, prefrontal cortex, hypothalamus, certain thalamic nuclei, and epithalamus, as well as all the pathways that interconnect these structures.

The hippocampal formation represents the archipallium and includes the hippocampal gyrus (cornu ammonis), fornix, dentate gyrus, and the subiculum. The hippocampal gyrus (cornu ammonis) develops as an enfolding of the cerebral cortex into the inferior horn of the lateral ventricle (Figures 17.1, 17.2 & 17.3). It is one of the simplest and most primitive structures of the human brain, which stretches from the foramen of Monro to the tip of the inferior horn of the lateral ventricle. It lies superior to the parahippocampal gyrus, expanding anteriorly to form the Pes hippocampi. The cornu ammonis is comprised of the ependyma, the alveus (which is formed primarily by axons of the pyramidal neurons), stratum oriens, stratum pyramidalis, stratum radiatum, stratum lacunosum, and the stratum moleculare. The cornu ammonis is divided into four zones designated as CA1, CA2, CA3, and CA4. In CA1 zone (also known as Sommer's sector) the pyramidal neurons are small and are located near the subiculum, distant from the dentate gyrus. This zone is particularly susceptible to anoxia, which occurs in an epileptic seizure involving the temporal lobe. Axonal branches of the pyramidal neurons known as Schaffer's collaterals extend into CA1 zone, establishing synaptic contacts with the apical dendrites of adjacent pyramidal neurons in CA3. Stimulation of Schaffer's collaterals produces dramatic increase in the number of their synapses on the CA1 pyramidal neurons. However, prolonged stimulation may reduce the formation of new synapses. CA3 and CA2 are sandwiched between CA1 and CA4.

CA4 is the region of the cornu ammonis adjacent to the hilus and the dentate gyrus. The CA4 and CA3 regions contain the largest neurons, which form synapses with the mossy fibers of the granule cell layer of the dentate gyrus.

The dentate gyrus (Figures 17.1 & 17.2) lies lateral to the cornu ammonis and medial to the fimbria. It is separated from the subiculum by the hippocampal sulcus. The induseum griseum and the fasciolar gyrus are considered as posterior extensions of the dentate gyrus. The dentate gyrus lacks pyramidal neurons, consisting of an external molecular layer, which is adjacent to the molecular layer of the cornu ammonis, a dense granule cell layer, and an innermost polymorphic layer. The axons of the granular cell neurons project to the hippocampal gyrus through the polymorphic layer. Dendrites of the granule cells remain confined

Memory traces are believed to be scattered in the stellate cells of association cortices via the entorhinal cortex and fornix.

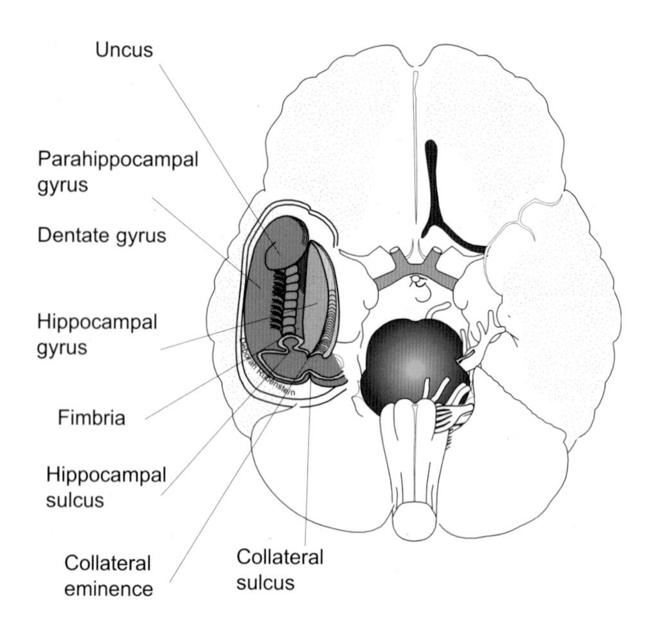

Figure 17.1 The hippocampal formation and its various constituents

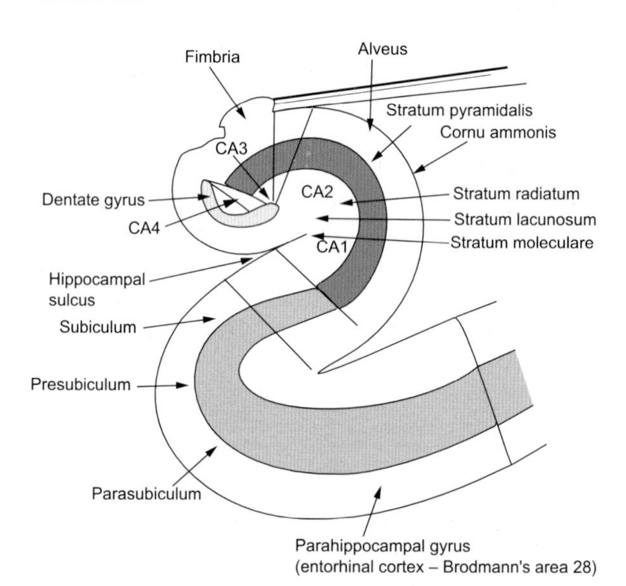

Figure 17.2 The hippocampal formation and its main components are shown. Note also the various sectors of the hippocampal gyrus (cornu ammonis)

It has been suggested that destruction of the hippocampal gyrus may account for serious deficits associated with temporal epilepsy and schizophrenia. Schizophrenia, a term that replaced 'dementia precox' in the psychiatric literature, is a disease that affects as high as 1–1.5% of the adult population and involves a group of disturbances that share a common phenotype. It equally affects men and women with some differences in course and onset. Men have an earlier onset of schizophrenia than women do. Greater than half of all male schizophrenics, but less than one-third of female schizophrenics may show manifestations before the age of 25. Studies have shown that the peak age of onset for males is 15–25 years, whereas for women it ranges between 25–35. In general, the outcome for female schizophrenics may be better than the outcome for male schizophrenics. A particularly interesting finding is that individuals who later develop schizophrenia are more likely to have been born in the winter and early spring months and less likely to have been born in late spring or summer. Monozygotic twins have a 50% concordance rate.

The disturbances associated with schizophrenia include incoherence of thought (hallucinations or false perception), feeling, and behavior. Patients may have difficulty establishing social relationships, experience delusions, mood disturbances, and auditory hallucination, and even motor over-activity and violent or bizzare behavior. These symptoms which are variable in severity may be categorized into positive, negative, and disorganized symptoms. Positive manifestations principally comprise delusions (paranoia, grandiosity, bizarre thoughts, tactile delusions) and hallucinations (mainly auditory). Withdrawal, lethargy, loss of spontaneity and initiative, motivational impairment and indifference to emotional stimuli constitute the negative manifestations of this psychotic disorder. Schizoid individuals present with the negative symptoms. Lack of thought content, illogical ideas and incoherence are the primary disorganizational symptoms of this disease. Some patients are classified as schizotypal when mild positive and disorganizational symptoms are present. Patients are also depressed and exhibit anxiety. When mood-related changes are predominant in this disease, a schizoaffective disorder is produced.

Genetic, environmental, and neurophysiological factors may be responsible for this disease. One model for the integration of biological, social, and environmental factors is the stress-diathesis model. This model states that a person may have a specific vulnerability (diathesis) and when acted upon by particular environmental stress factors, the symptoms of schizophrenia then develop. The stress may be biological, environmental or both. Recent research has implicated a pathophysiological role for the limbic system, prefrontal cortex, and basal nuclei.

According to the ontogenic hypothesis of schizophrenia, the dendrites of the pyramidal cells appear to undergo disorientation subsequent to deranged embryological development. The degree of deviation of the apical shafts of the pyramidal cells may remain proportional to the severity of symptoms and multiple hospitalization. The abnormal deviation varies from 70–180° from normal. Due to this directional deviation, abnormal afferents converge on these dendrites. Additionally, during development primitive neurons migrate from the neuroepithelial zone to the hippocampus via the radial glial cells. These radial glial cells function both as directional guides and as a structural support for the migrating neuroblasts. Without this support and guidance the neuroblasts fail to develop and migrate properly. Other investigators challenge the ontogenic theory on the basis of the fact that schizophrenia occurs in teenagers and that the symptoms wax and wane with progressive deterioration and remission. They claim that infection or injury may result in miswiring and aberrant regeneration, leading to abnormal sprouting and synaptic reorganization of projection sites and increased excitability and abnormal behavioral pattern of the affected neurons. This re-organization may also lead to abnormal functioning of the structures that receive input from the hippocampal gyrus.

During embryogenesis, the close proximity or adhesion of the migrating neuroblasts to the radial glial cells is maintained by the neuronal cell adhesion molecules (NCAM). This mechanism is essential for proper migration, alignment, and lamination resulting in cluster of cells lined up side by side with particular polar orientation. Uncoupling of the NCAM from the neuronal–glial complex allows the migratory neurons to leave the radial glia within the hippocampus. Since neuronal migration occurs in the second trimester of gestation, cellular derangement, subsequent to insult by maternal illness, must occur during this stage of development. This theory is supported by the observation that schizophrenia showed significant increase in the offspring of mothers infected with influenza virus. Since capsular neuraminidase-producing viruses (e.g.

influenza virus) affect the amount of available sialic acid (an N- and O-acyl derivative of neuraminic acid), and subsequently the binding property of the NCAM, the migratory pattern of neurons during embryogenesis will also be adversely affected.

An interesting observation needs to be pointed out in regard to development of schizophrenia is that the myelination of the neuronal pathways of the limbic system occurs during late adolescence under the influence of the neurotropic gonadal hormones. These hormones are involved in the development and lengthening of the dendritic spines in CA1 and CA3. In fact numerous glucocorticoid receptors exist in the hippocampal gyrus and a hormonal role in the growth of dendrites of the granule cells seems obvious. During adolescence the frequency of hippocampal neuronal firing is proportional to the hormonal release. This massive firing predisposes the cells to damage, initiating reorganization and sprouting.

Dopamine is considered the primary neurotransmitter involved in the development of schizophrenia. The simplest form of dopamine hypothesis is based upon the overactivity of the mesolimbic dopaminergic system. This theory relies upon the fact that most antipsychotics are antagonists to the dopamine type (D2) receptors, and the drugs that increase dopaminergic neuronal activity, such as amphetamine, are psychotomimetic. The theory also suggests that dopamine type (D1) receptor may be responsible for the so-called negative symptoms of schizophrenia. Another part of the dopamine theory is the significant increase in plasma levels of dopamine metabolite, homvanillic acid, in schizophrenics. The problems with the dopamine hypothesis are two-fold; the fact that antipsychotic medications are useful in the treatment of virtually all psychotic states suggests that dopaminergic hyperactivity is unique to schizophrenia; and some electrophysiologic data indicate that the firing rate of dopaminergic neurons may actually increase in response to long-term treatment with antipsychotics (see also dopamine in Chapter 12). Excitatory amino acids may also play a role in the pathophysiology of schizophrenia.

Treatment of schizophrenia may be accomplished by dopamine receptor antagonists (D2 receptors that are coupled to adenylate cyclase) such as risperidone, clozapine, melperone, sertindole and ziprasidone. Risperidone is an antipsychotic drug, which is antagonistic to the serotonin type 2 (5-HT2) receptor as well as the dopamine type 2 receptor. Both positive and negative symptoms are improved by this medication. Clozapine is also an effective antipsychotic drug which primarily antagonizes the D4 receptor and to a lesser degree D2 and serotonin receptors. Since clozapine is associated with agranulocytosis in 1–2% of patients, careful monitoring of the blood may be required.

Dopamine receptor antagonists are associated with side-effects such as akathesia (restlessness), rigidity, tremor, tardive dyskinesia (tongue darting), and neuroleptic malignant syndrome.

to the molecular layer. The molecular and the granule cell layers form the dentate fascia. Examining the subiculum (Figure 17.2) reveals a transitional area that shows graded variation between a three-layered cortex of the cornu ammonis, the adjacent entorhinal cortex, and the parahippocampal cortex. It is characterized by a marked thickening of the pyramidal layer. The subiculum is divided into the prosubiculum (which is the closest to and merges with CA1), presubiculum, a region adjacent to the entorhinal cortex, containing large neurons, and the subiculum proper.

Afferents of the hippocampal formation (Figure 17.8) arise from diverse areas of the limbic system (with the exception of the olfactory cortex) and are topographically organized but not modality specific. These afferent fibers arise from the entorhinal area, cingulate gyrus, hippocampal formation of the opposite side, and the medial septal nucleus. The entorhinal area, the secondary olfactory cortex (Brodmann's area 28) forms the main afferents to the hippocampal and dentate gyri. These fibers, which arise from the medial and lateral parts of the entorhinal area, pursue alvear (superficial) or perforant (deep) paths. Alvear fibers run towards the ventricular surface, originate from the medial part of the entorhinal area, and terminate in the subiculum at CA1. Perforant axons originate from the lateral part of the entorhinal area that traverses the subiculum, terminating in the dentate gyrus and virtually all the layers of the cornu ammonis (except CA4). The cingulate gyrus projects indirectly to the hippocampal gyrus via relays in the presubiculum and the entorhinal cortex. These fibers run within the cingulum and some of the cortical association fibers, dorsal to the corpus callosum. The medial septal nucleus projects to the dentate gyrus and to certain parts of the cornu ammonis (CA3, CA4) via the fornix. Cholinergic neurons of the septal area and the diagonal band of Broca also project in a diffuse

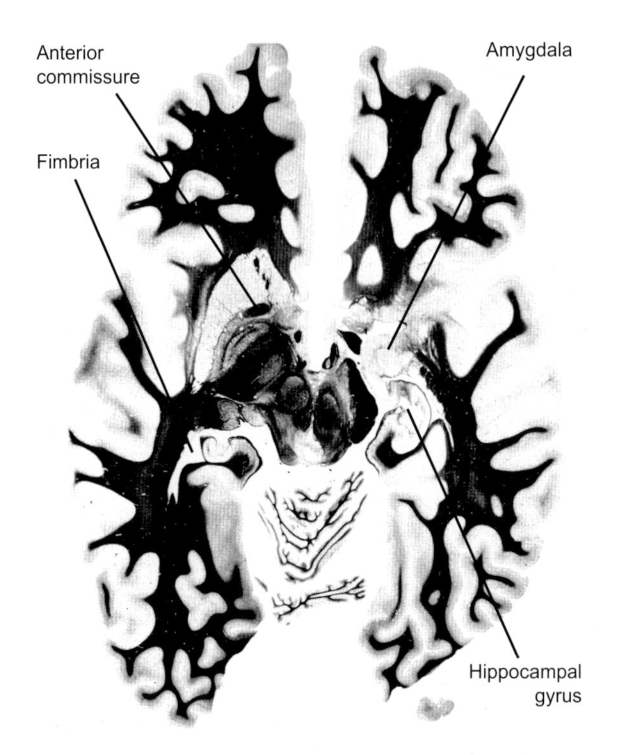

Figure 17.3 In this horizontal section of the brain the hippocampal gyrus, fimbria of fornix, and the amygdala are indicated

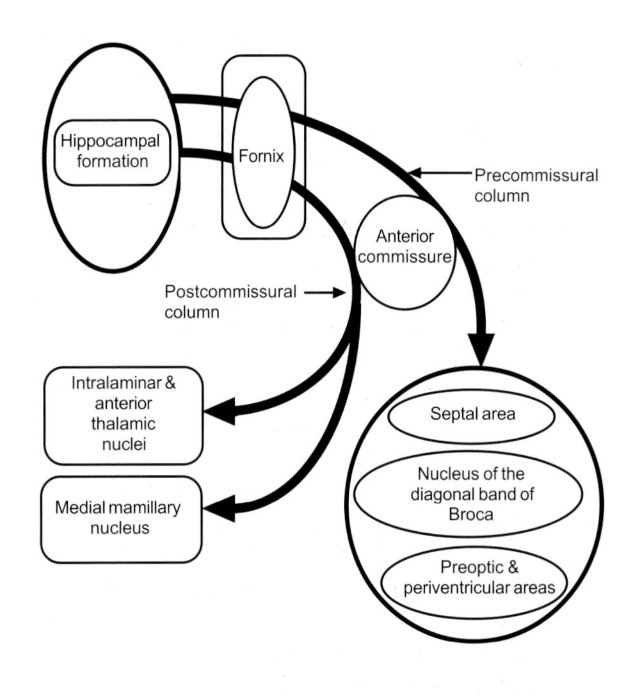

Figure 17.4 This diagram depicts the main efferent pathway from the hippocampal formation which is represented in the fornix. The fornix divides into pre- and post-commissural columns

manner to all layers the hippocampal formation, via the fornix. Noradrenergic input to the hippocampal gyrus is derived from the locus ceruleus. The anterior thalamic nucleus specifically projects to the presubiculum. Pyriform and prefrontal cortices also provide input to the hippocampal formation. Visual, auditory, and somatosensory cortical projections to the hippocampal formation are mediated via the entorhinal cortex.

The hippocampal formation of one hemisphere is connected to the corresponding area of the opposite hemisphere via the hippocampal commissure, which connects the crura of the fornix on both sides.

Efferents of the hippocampal formation

The fornix is the main output of the hippocampal formation, originating from the hippocampal gyrus (Figures 17.4, 17.5 & 17.8). Through this tract, the hippocampal formation influences the activities of the anterior thalamic nucleus and midbrain reticular formation. The fornix is formed primarily by the axons of the pyramidal neurons of the cornu ammonis, and to a lesser degree by the subiculum (with exception of the presubiculum). The subicular fibers of the fornix project to the cingulate gyrus as

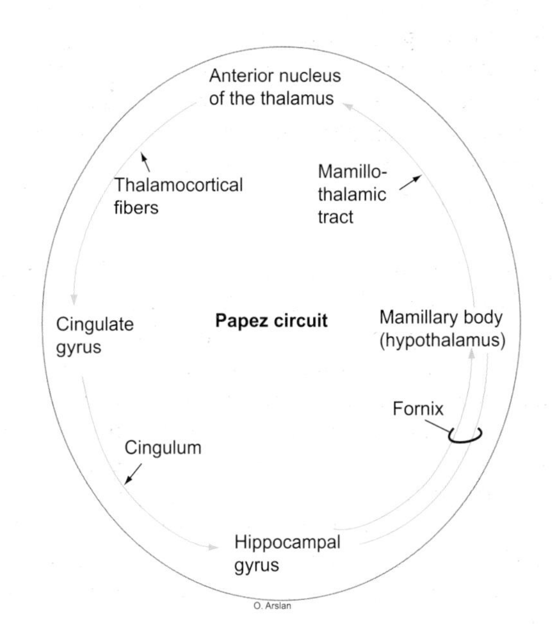

Figure 17.5 This feedback circuit represents the Papez circuit of emotion in which the efferent from the hippocampus gyrus projects to the mamillary body and through the mamillothalamic tract to the anterior nucleus of the thalamus. Corticothalamic fibers enable impulses received by the anterior nucleus of thalamus to project to the cingulate gyrus and via the cingulum back to the hippocampal gyrus where the circuit is completed

well as to the entorhinal area. Through this massive projection, the hippocampal gyrus reaches the septal area, preoptic area, and the hypothalamus. Some efferents of the hippocampal formation are also be derived from the dentate gyrus.

Afferents to the hippocampus are also carried by the fornix. The axons which comprise the fornix converge on the ventricular surface of the hippocampal gyrus as the alveus, and then continue as the fimbria. The fimbria of the fornix and the longitudinal stria of the induseum griseum, which cross the corpus callosum, are collectively known as the dorsal fornix. The latter conveys information from the hippocampal gyrus to the fasciolar, cingulate gyri, and the septum pellucidum. The fimbria extends rostrally to the uncus, constituting the inferior border of the choroidal fissure. It stretches caudally with the crura of the fornix around the thalamus, and inferior to the splenium of the corpus callosum. The crura then interconnect through the hippocampal commissure (psalterium), which also connects CA3 and CA4 zones of both cerebral hemispheres. Between the psalterium and the corpus callosum, an anomalous cyst (cavum vegae) may exist which communicates with the lateral ventricle.

The union of the crura, above the roof of the third ventricle, leads to the formation of the body of the fornix, which is attached to the inferior margin of the septum pellucidum. Above the anterior tubercle of the thalamus, the body of the fornix divides into two bundles and later into two columns. The pre-commissural column arises from the cornu ammonis and the subicular fields (with the exception of the presubiculum) and terminates in the septal area, preoptic nuclei, anterior hypothalamus, and the cingulate gyrus. The post-commissural column arises from the subiculum and terminates in the mamillary body, intralaminar thalamic nuclei, lateral hypothalamus, habenula, and midbrain tegmentum (Figure 17.4). It is fair to conclude from the above data that the hippocampal formation serves as a closed feedback circuit involving the fornix, mamillary body, mamillothalamic tract, anterior nucleus of the thalamus, cingulate gyrus, and cingulum. These connections form the basis for the Papez circuit of emotion (Figure 17.5).

Anterograde amnesia, which results from destruction of the hippocampal gyrus, may explain the role of this gyrus in encoding short-term memory. Seizures of hippocampal origin exhibit unique low threshold activity that remains generally localized with no behavioral change or loss of consciousness, unless other areas of the limbic system are involved. Individuals with these seizures appear to be confused and may show signs of aggressive behavior as well as auditory and gustatory hallucinations. Bilateral removal of the hippocampal gyrus causes short-term memory loss, confusion and compensatory confabulation (tendency to fabricate, to recite imaginary experiences to fill gaps in memory, and to give irrelevant answers to reasonable questions).

Wernicke–Korsakoff's syndrome refers to the combined Wernicke's encephalopathy and Korsakoff's psychosis. However, it must be emphasized that Wernicke's encephalopathy and Korsakoff's psychosis are not separate diseases. In fact Korsakoff's psychosis represents the psychic component of this condition. Wernicke's encephalopathy is an acute, subacute or chronic condition that results from the failure to ingest thiamine and continued ingestion of carbohydrate. However, the most common precipitating condition is alcoholism. Patients develop nystagmus, ophthalmoplegia, confusion, gaze palsy, ataxic gait, stupor and sometimes fatal autonomic insufficiency. If untreated, this condition may lead to Korsakoff's psychosis (amnestic confabulatory syndrome) in which the patient's conversation becomes unintelligible, accompanied by disturbance of orientation, susceptibility to external stimulation and suggestion, amnesia, confabulation, and hallucination. This disorder is due to the lack of ability to encode the semantic component of information at the initial stage of learning. Mental confusion is also obvious, but consciousness and intellect are apparently preserved. Patients are able to learn but do so at a much slower rate, yet they appear to forget retained information over a period of time similarly to healthy individuals. It is commonly associated with bilateral degeneration of the hippocampus, mamillary body, and possibly the dorsomedial nucleus of the thalamus. Patients with this condition exhibit a striking difficulty in remembering events after the onset of the disease and difficulty in retaining newly acquired information and skills. Temporal lobotomy, a rarely performed neurosurgical operation that involves the hippocampus and is used for the treatment of certain types of epilepsy (psychomotor type), may cause similar deficits to Korsakoff's syndrome. It is important to note that transient global amnesia, which results from bilateral temporal lobe ischemia subsequent to atherosclerosis, presents with a sudden impairment of recent memory (retrograde amnesia often taking part) lasting hours, days or weeks.

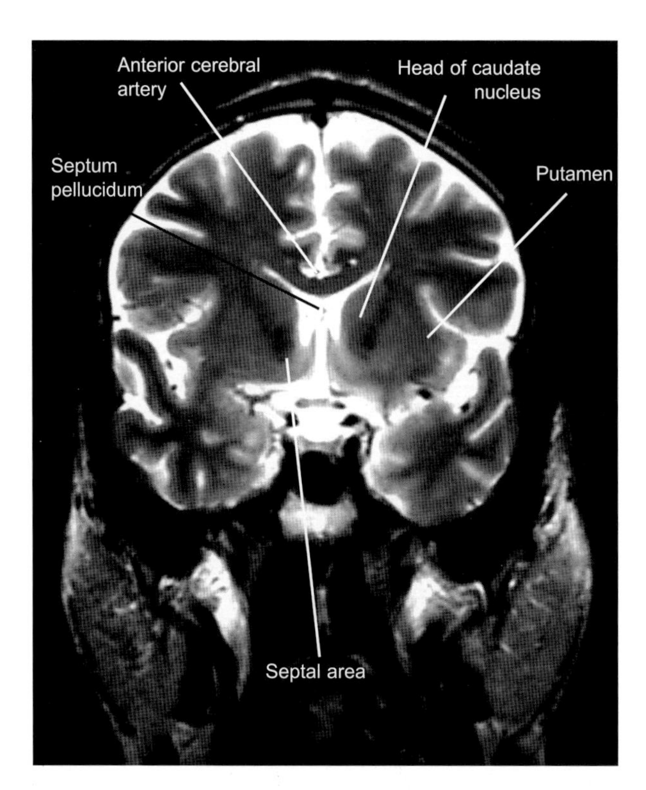

Figure 17.6 This MRI scan shows the septal area adjacent to the striatum and columns of the fornix. The septal area represents the nodal point of the limbic system, allowing impulses to be dispersed to diffuse areas of the brainstem and cerebral cortex

Functional considerations

Hippocampal linkage to the hypothalamus via the septal area may account for the important role that the hippocampal formation plays in the control of behavior. Expression of aggressive behavior is also induced by the effect of the hippocampus gyrus upon the supplementary motor areas of the cerebral cortex. The corticosteroid and estradiol containing hippocampal neurons that act upon the hypothalamus mediate endocrine function. The connection of the hippocampus to the hypothalamus is also responsible for the respiratory and cardiovascular changes observed upon stimulation of the hippocampus.

The septal area lies rostral and superior to the lamina terminalis and anterior commissure, representing the nodal point for integration of impulses associated with the limbic system (Figures 17.6 & 17.8). It is comprised of the precommissural and supracommissural parts. The supracommissural part corresponds to the septum pellucidum. The precommissural part continues with the diagonal band of Broca, medial olfactory stria, and the induseum griseum. It contains the septal nuclei, corresponding to the cortical area between the lamina terminalis and

Dopaminergic receptors in the nucleus accumbens are implicated in the neural mechanism of reward and addiction. The addictive role of psychomotor stimulants in enhancing dopamine release (e.g. amphetamine) or blocking its reuptake (e.g. cocaine) may be attributed to their interaction with the dopaminergic system. Therefore, dopamine receptor antagonists may reduce the effect of cocaine.

The septal area sends inhibitory commands to the hypothalamus, preoptic areas, and the tegmentum of the midbrain via the medial forebrain bundle. This connection may account for the hyperactivity and septal rage associated with septal lesion, although amygdalar fibers may also be involved in this condition.

the paraolfactory sulcus. The anterior extension of the the septal area is known as the prehippocampal rudiment. Septal nuclei are found in the vicinity of the columns of the fornix, which includes the lateral and medial septal nuclei, nucleus of the diagonal band of Broca, and the nucleus accumbens septi. The nucleus accumbens septi is localized between the head of the caudate nucleus and septal area, and collectively with the olfactory tubercle represents the ventral striatum, which receives projections from the basolateral nucleus of the amygdala.

The septal area and the nucleus of the diagonal band of Broca maintain reciprocal connections, via the fornix, with the hippocampus (Figure 17.11). The nucleus of the diagonal band of Broca has connections with the olfactory, cingulate, and prefrontal cortices as well as the amygdala, habenula, and the dorsomedial nucleus of the thalamus.

Information received by the hypothalamus is delivered to the midbrain tegmentum via the mamillotegmental tract and the dorsal longitudinal fasciculus; and to the anterior nucleus of the thalamus via the mamillothalamic tract. The septal area constitutes an integral part of the limbic system, maintaining connections via closed circuit loops. It exerts a significant influence on neuroendocrine, somatic, and visceral motor functions. Reciprocal connections exist between the amygdala, with the septal area, and diagonal band of Broca via the stria terminalis.

Transfer of memory traces from the hippocampus to the prefrontal cortex may involve the septal projection to the dorsomedial nucleus of thalamus.

Due to the ascending projections of the adrenergic and serotoninergic neurons of the brainstem reticular formation to the cortical components of the limbic system and the hypothalamus, mood-elevating drugs have probable antagonistic effects on these pathways.

Through its massive connections, the septal area may be implicated in the regulation of behavior-related autonomic activities. This area may also be involved in θ rhythm detected in EEG during rapid eye movement (REM) sleep and motor activity. Lesions involving the septal area cause heightened reactions including hyperemotionality and an increase in aggressive behavior. Stimulation of this area reduces the aggressive behavior.

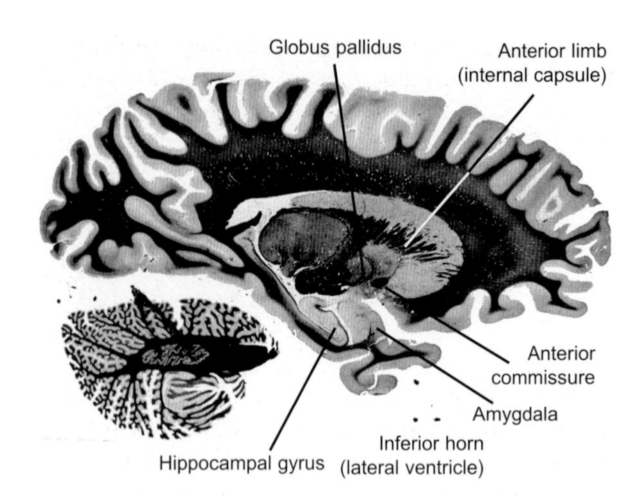

Figure 17.7 Coronal section of the brain through the anterior commissure. The amygdala, a significant structure in the make-up of the limbic system, occupies a prominent area rostral to the inferior horn of the lateral ventricle, adjacent to the uncus

The septal area receives input from and conveys impulses to the habenular nuclei. The habenular projection to the interpeduncular nucleus and the midbrain tegmentum via the fasciculus retroflexus enables the septal area to affect the activities of the reticular formation.

The induseum griseum (supracallosal gyrus), located above the corpus callosum, is continuous rostrally with the septal area, diagonal band of Broca, and the anterior perforated substance. The fasciolar (splenial) gyrus, which is continuous with the dentate gyrus, represents the caudal connections of the induseum griseum. The induseum griseum contains the medial and lateral longitudinal striae. Some fibers of the longitudinal stria intersect the corpus callosum, contributing to the dorsal fornix, while other fibers continue rostrally with the paraterminal gyrus and caudally with the fasciolar gyrus.

The amygdala (Figures 17.3, 17.7 & 17.9) is a nuclear complex which is embedded within the anterior pole of the temporal lobe deep to the uncus and rostral to the tip of the inferior horn of the lateral ventricle. It continues with the claustrum and the external capsule, and joins the tail of the caudate nucleus. The amygdala is adjacent to the pyriform lobe, which consists of the prepyriform (cortical area near the lateral olfactory stria) and periamygdaloid cortex. The amygdaloid nuclear complex encompasses a corticomedial nucleus, which continues with the anterior perforated substance, and a basolateral nucleus merging with the claustrum and parahippocampal gyrus.

The basolateral nucleus consists of smaller lateral, basal, and accessory basal nuclei, whereas the corticomedial nucleus is comprised of a central, medial, and cortical nuclei. Due to its direct, and often reciprocal, connections to the cerebral cortex and thalamus, the basolateral nuclear complex is considered a polycortical structure. Excitatory amino acid neurotransmitters such as aspartate and glutamate are also used by this nuclear subdivision. The lateral nucleus, which is the largest, lies ventrolateral to the basal nucleus. The basal nucleus consists of dorsal magnocellular, intermediate parvicellular, and paralaminar subnuclei. Somatostatin and neuropeptide Y are found in large concentration in the lateral nucleus. Medial to the basal nucleus lies the accessory basal nucleus that consists of similar subdivisions to the basal nucleus. The basolateral nuclear subdivision receives serotoninergic projections from the raphe nuclei that ascend through the medial forebrain bundle.

Dopaminergic projections from the midbrain ventral tegmental area (A10) are mainly received by the lateral and central nuclei, and parvicellular part of the basal nucleus. Dense cholinergic projections from the magnocellular division of the nucleus basalis of Mynert reach the basal and parvicellular accessory basal nuclei. The high concentration of benzodiazepine binding sites, particularly $GABA_A$ binding sites in the lateral and accessory basal nuclei, may explain the anxiolytic action of benzodiazepines on the amygdala and also form a possible neural basis of fear and anxiety. This subdivision contains the highest concentration of opiate receptors in the entire brain, which may be activated under stressful situation.

The corticomedial nucleus consists of several smaller nuclei such as the central, medial, and cortical nuclei and associated subnuclei. These nuclei receive noradrenergic projections from the locus ceruleus and lateral tegmental nucleus via the medial forebrain bundle. The central nucleus lies dorsal and medial to the basal nucleus and, together with the medial nucleus, it merges with the bed nucleus of the stria terminalis and periamygdaloid cortex to form the centromedial amygdaloid nuclear complex. It retains chemical and cellular continuity with the bed nucleus via the stria terminalis and the sublenticular basal forebrain. The paraventricular nucleus of the hypothalamus sends oxytocin and vasopressin-immunoreactive terminals to the central amygdaloid nucleus. Vasopressin-immunoreactive neurons are also found in the medial nucleus, which also receives projections from the suprachiasmatic nucleus of the hypothalamus. The medial and central nuclei contain β-endorphin and enkephalin immunoreactive neurons.

Certain nuclei of the amygdala contain high concentrations of steroid hormones and their binding sites. Estrogen-concentrating neurons are found in the medial, accessory basal, and posterior cortical nuclei, whereas dihydrotesterone-concentrating neurons are abundant in the lateral, parvicellular basal, and accessory basal nuclei. Enzymes such as reductase, which converts testosterone into non-aromatizable 5α-dihydrotestosterone, and aromatase, which converts testosterone and androstenedione to estradiol, are abundant in the amygdala. Thus, changes in hormonal levels that accompany the menstrual cycle may affect the functions of amygdaloid neurons.

Undirectional intrinsic connections exist among amygdaloid nuclei in which the lateral nucleus projects primarily to all divisions of the basal, accessory basal, paralaminar, and anterior cortical nuclei and less densely to the central nucleus. Small nuclei within the basal nucleus project to the accessory basal nucleus, which maintains profuse connections to the central nucleus. The central nucleus projects to the anterior cortical nucleus.

In general, impulses that reach the corticomedial nucleus of the amygdala are derived from the olfactory bulb and the anterior olfactory nucleus of both sides. The basolateral nucleus (Figures 17.10, 17.11 & 17.14) receives input from the pyriform lobe, inferior temporal gyrus, nucleus of the diagonal band of Broca, and the orbitofrontal cortex. The orbitofrontal, cingulate, and temporal neocortex (the largest contributor) maintain reciprocal connections with the amygdala. Amygdalar connections are generally reciprocal with the exception of its connections to the striatum, thalamus, and parts of the cerebral cortex. It is worth noting that massive cholinergic projections from the magnocellular

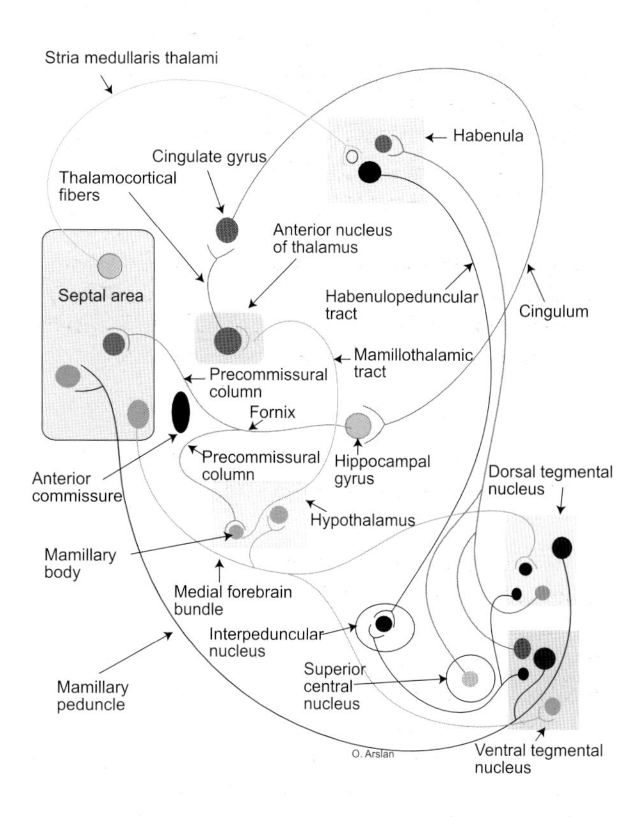

Figure 17.8 This diagram illustrates the major structures associated with the limbic system and their connections. The septal area projects through the habenula to the reticular formation and receives input from the hippocampal gyrus and from the tegmental nuclei via the medial forebrain bundle and the mamillary peduncle

nucleus basalis of Mynert terminate in the magnocellular part of the basal nucleus and accessory basal nucleus. In turn, the central nucleus, parvicellular part of the basal nucleus and the magnocellular accessory basal nucleus project back to the nucleus basalis of Mynert and the nucleus of the diagonal band of Broca. Striatal projections form a substantial component of the amygdaloid output from the basal and accessory basal nuclei to the nucleus accumbens septi, ventromedial parts of the caudate nucleus and putamen. The projections to the nucleus accumbens septi interdigitate in a complex manner with other afferents from the hippocampal gyrus, midline thalamic nuclei, and the prefrontal cortex.

The central nucleus projects to the periaqueductal gray matter, substantia nigra, ventral tegmental area, parabrachial nuclei, dorsal motor nucleus of vagus and the solitary nucleus. Afferents to the central nucleus arise directly from the parabrachial nucleus, which acts as a nodal point in conveying impulses to the

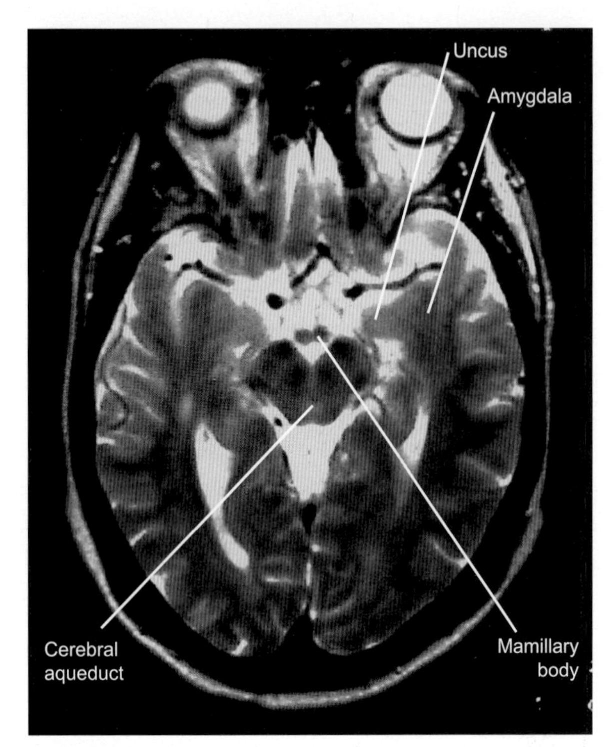

Figure 17.9 In this MRI scan some of the components of the limbic system, e.g. amygdala, hippocampal gyrus, etc. are shown

central nucleus. These connections account for the important role that the amygdala plays in the regulation of cardiovascular, respiratory and gustatory systems. The lateral, basal, and accessory basal nuclei project (not reciprocated) mainly to the magnocellular part of the dorsomedial thalamic nucleus that conveys impulses to the prefrontal cortex, enabling the amygdala to influence the activities of this part of the cerebral cortex. On the other hand the central and medial amygdaloid nuclei maintain reciprocal connections with the midline thalamic nuclei. Some experimental data also suggest projections from the ventral posteromedial nucleus to the lateral nucleus of amygdala.

Amygdaloid efferents are conveyed primarily via the stria terminalis and the ventral amygdalofugal pathways. The stria terminalis runs parallel to the fornix, containing bilateral afferent and efferent fibers to and from the amygdala that cross in the anterior commissure. It initially courses in the roof of the inferior horn of the lateral ventricle medial to the tail of the caudate nucleus. Then it travels in the floor of the central part of the lateral ventricle between the caudate nucleus and the thalamus, adjacent to the thalamostriate vein, dividing into supracommissural, commissural, and subcommissural components. Most of the fibers of the supra- and subcommissural components are amygdalofugal, distributing to the septal and preoptic areas as well as the hypothalamus. Some may terminate in the pyriform lobe, anterior perforated substance, and the dorsomedial nucleus of the thalamus. Fibers of the stria terminalis that joins columns of the fornix and the stria medullaris are primarily subcommissural. Commissural fibers interconnect the amygdala of both sides via the anterior commissure.

The ventral amygdalofugal fibers (Figures 17.10, 17.11 & 17.14), the largest output from the amygdala, originate from the basolateral nucleus, course ventral to the globus pallidus, and distribute to the hypothalamus and preoptic area. These fibers reach the brainstem reticular formation via the medial forebrain bundle. Both nuclear components of the amygdala project to the hypothalamus, dorsomedial nucleus, and the midline nuclei of the thalamus. Stimulation of the amygdala may produce activities associated with food intake, behavioral and visceral changes, and affect. It may also evoke feelings of relief, relaxation, pleasant sensations, and autonomic response such as bradycardia and pupillary dilatation. The amygdala poses antagonistic effects upon behavior and endocrine functions, a characteristic shared by other constituents of the limbic system. Stimulation of the corticomedial and basolateral nuclei inhibits and facilitates aggressive behavior, respectively. Activation of the basolateral nucleus enhances the release of the growth and adrenocorticotropic hormones.

The limbic lobe is the cortical ring which surrounds the corpus callosum and consists of the cingulate, parahippocampal, paraterminal gyri, and the isthmus.

The cingulate gyrus borders the corpus callosum, bounded superiorly by the cingulate sulcus and inferiorly by the callosal sulcus. This gyrus continues with the parahippocampal gyrus through the isthmus. It is connected to the parahippocampal gyrus via the cingulum, and to the dorsomedial nucleus of the thalamus through the thalamocortical and corticothalamic fibers. The dorsomedial nucleus acts as an indirect pathway between the hypothalamus and the cingulate gyrus. The cingulate gyrus establishes

Estrogen-containing neurons of the amygdala (estradiol receptors) may induce ovulation if the corticomedial component is stimulated, whereas transection of the stria terminalis abolishes ovulation. The stria terminalis may exert both excitatory and inhibitory effects on the secretion of the gonadotropic hormones. This hormonal effect may account for the hypersexuality observed in individuals with bilateral amygdalectomy.

Preoptic, anterior hypothalamic nuclei and septal area

↑ Stria terminalis

Amygdaloid nuclear complex

↓ Ventral amygdalofugal fibers

Preoptic nuclei, dorsomedial nucleus, septal area and prefrontal cortex

Figure 17.10 Diagram of the afferent and efferent connections of the amygdala

Experimental ablation of the anterior parts of the temporal lobes, including the amygdala and uncus in monkey, produces the commonly known Klüver–Bucy syndrome. Manifestations of this disease include sexual hyperactivity accompanied by change in appetite and dietary habits and loss of fear. Changes in behavior from aggressive to placid are also seen in this syndrome. Temporary memory loss may also occur. Damage to the visual association areas of the temporal cortex cause visual deficits and hallucinations. Lack of the ability to visually identify objects leads to a persistent tendency and strong compulsion to orally examine edible and non-edible objects (hypermetamorphosis). The human Klüver–Bucy syndrome following traumatic injury to the temporal lobes, Alzheimer's disease, Pick's disease, or herpes simplex encephalitis may exhibit similar manifestations to that of the monkeys, however sexual activity is not prominent. Dementia, aphasia, amnesia, and excessive eating are common manifestations. Signs of this condition may also be observed in post-traumatic apallic syndrome (coma vigil) which is also characterized by reactionless gaze, lack of spontaneous movements, appearance of primitive reflexes, hyperkinesia, and rigidity.

connections with the temporal lobe via the uncinate fasciculus, and with the ipsilateral corpus striatum via the internal capsule. It is also connected to other limbic structures, receiving afferents from the hippocampal gyrus, anterior nucleus of the thalamus, nucleus of the diagonal band of Broca, and dopaminergic input from the midbrain ventral tegmentum. Despite the fact that no output from the

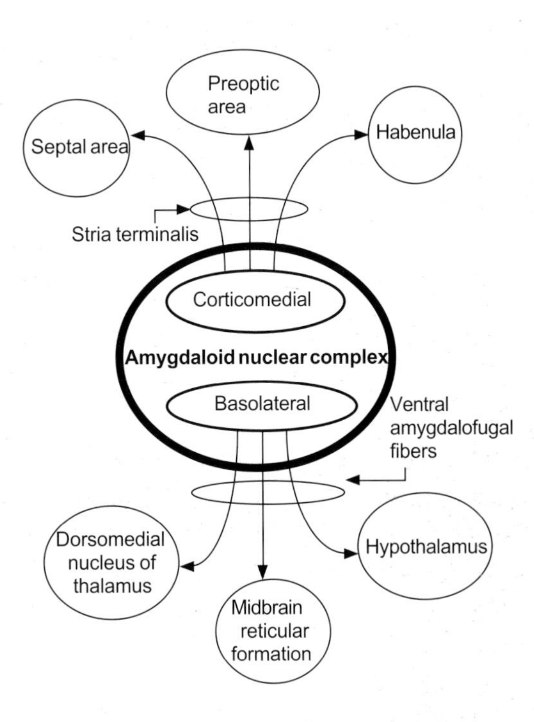

Figure 17.11 Schematic diagram of the outputs of the corticomedial and basolateral nuclei of the amygydaloid nuclear complex

Decortication (excision of the cerebral cortex including the limbic lobe) may produce sham rage, which is characterized by combative, violent reactions to mild stimuli. Dementia, a cardinal manifestation of limbic encephalitis, exhibits profound neuronal loss as a result of gliosis, perivascular cuffing, and infiltration of microglia in the cerebral cortex. These deficits are reversible as long as the treatment is not delayed for a long period of time. Patients with limbic system disease may also exhibit 'florid delirium' and overreaction to a situation or inappropriate behavior.

cingulate gyrus is found to project to the hypothalamus, autonomic activities including changes in blood pressure, cardiac output, respiration, and motility of the digestive tract may be seen upon stimulation of the cingulate gyrus. Tumors of the cingulate gyrus are associated with changes in behavior.

The parahippocampal gyrus (Figure 17.12) represents the inferior portion of the limbic lobe, expanding rostrally into the uncus between the collateral and hippocampal sulci.

The prefrontal cortex (Figure 17.12) develops to a great extent in humans and is linked to both intellectual and emotional processes. It includes the orbital

The role of the prefrontal cortex in the regulation of emotion and behavior is best demonstrated in individuals who have developed frontal lobe tumors and subsequently undergone bilateral prefrontal lobotomy. This surgical procedure was once a commonly employed operation in the treatment of psychotic patients. Bilateral lobotomy alters aggressive behavior, and tends to alleviate the emotional distress associated with chronic intractable pain. In humans who have sustained frontal lobe damage as a result of tumors or wounds, perceptual and intellectual deficits, as well as derangement of behavioral programming and abstract thinking may become evident. Patients lose the ability to identify the goal of their intended action and their reactions change between gloom and elation in a superficial and abrupt manner. Patients also exhibit signs of indifference to their position in society, and become less concerned with monetary matters. Patients may acknowledge pain, but they show little emotional distress.

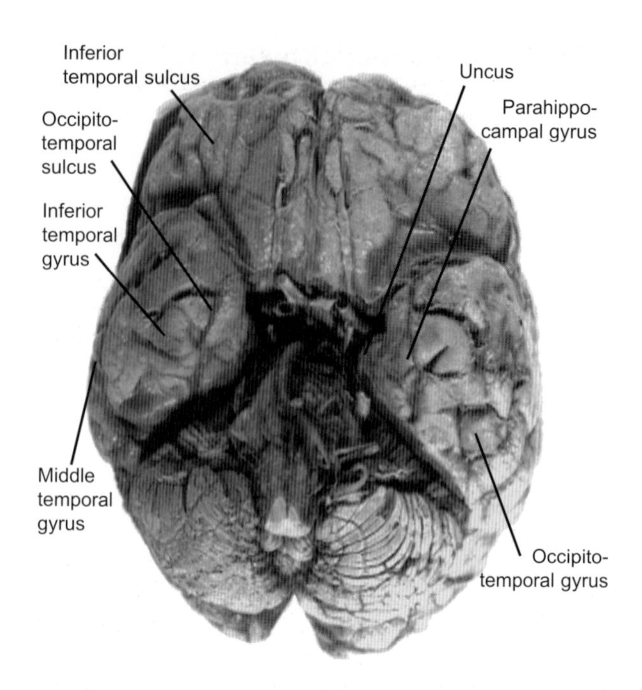

Figure 17.12 Photograph of the inferior surface of the brain illustrating the temporal gyri and the uncus. Note the proximity of the uncus to the oculomotor nerve

gyri and Brodmann's areas 9 and 10, and is connected to the parietal, temporal, and the occipital association areas. It receives information from the dorsomedial nucleus of the thalamus, other parts of cerebral cortex, septal area, monoamine neurons, and the hypothalamus. It conveys this information to the hippocampus, the amygdala, and the subiculum. Stimulation of the prefrontal cortex inhibits aggressive behavior. There is no direct projection from the prefrontal cortex to the hypothalamus.

The hypothalamus contains centers that govern behavior, feeding and drinking habits, expression of emotion, and hormonal control which maintains homeostasis. Certain thalamic nuclei also maintain connections with the limbic lobe as well as the prefrontal cortex. For instance, the dorsomedial and anterior thalamic nuclei maintain bilateral connections with the cingulate and the prefrontal cortices. Lesions of the dorsomedial thalamic nucleus have been associated with anterograde amnesia. The epithalamus, an integral part of the diencephalon, contains neurons that mediate activities of the limbic system via massive connections to the septal area, hypothalamus, and the brainstem reticular formation.

The limbic system utilizes a series of feedback loops to exert its influence on the cerebral cortex and subcortical structures. These circuits involve the hippocampal gyrus, septal area and amygdala. The Papez circuit of emotion (Figure 17.5) is formed by fibers of the fornix that carry impulses from the

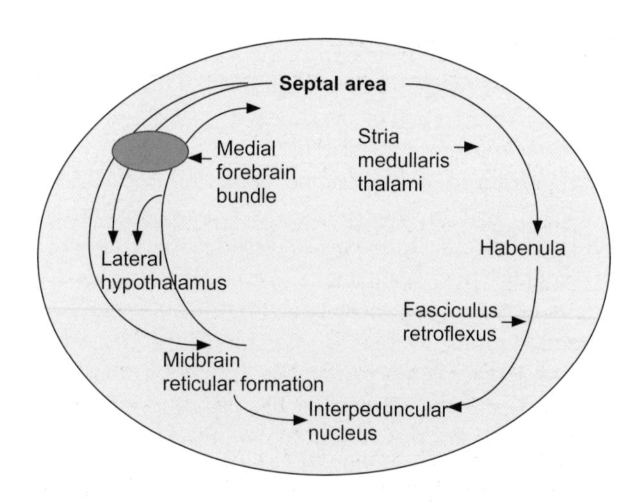

Figure 17.13 This diagram depicts the afferent and efferent fibers of the septal area. This area maintains massive connections to other parts of the central nervous system via the stria medullaris thalami, fasciculus retroflexus, and medial forebrain bundle

hippocampal gyrus to the mamillary body. Information received by the mamillary body is conveyed via the mamillothalamic tract to the anterior nucleus of thalamus and later to the cingulate gyrus within the thalamocortical fibers. The final connection between the cingulate gyrus and the hippocampal gyrus via the cingulum completes this feedback loop.

The second feedback loop involves the septal area (Figures 17.6 & 17.13) and serves as a nodal point for the afferents and efferents that converge on this area, mediating limbic system functions. As can be seen (Figure 17.8), the septal area receives input from the midbrain reticular formation via the medial forebrain bundle and conveys this information via the stria medullaris thalami to the habenula, and via the fasciculus retroflexus to the interpeduncular nucleus. The connections of the interpeduncular nucleus to the habenula and reticular formation complete this circuit.

The amygdaloid nuclear complex (Figure 17.9, 17.10, 17.11 & 17.14) also serves through a feedback circuit to promote the function of the limbic system. As discussed earlier, the amygdala projects to the septal area which through various connections establishes links to the reticular formation. Since the amygdala also projects directly to the reticular formation via the amygdalofugal fibers, the circuit is thus completed. Amygdalofugal fibers also participate in a feedback loop through projections to the prefrontal cortex.

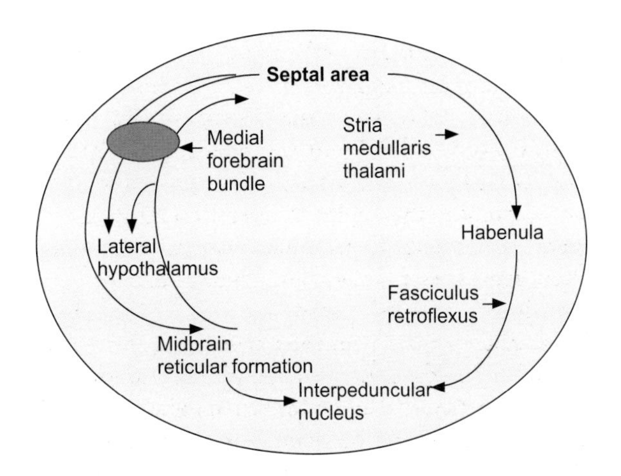

Figure 17.14 Summary of the efferent projections of the amygdala. As indicated the stria terminalis projects to the septal and preoptic areas as well as the habenula. The amygdalofugal fibers enable the amygdala to project to the thalamus, hypothalamus, cerebral cortex, and reticular formation

Suggested reading

1. Amaral DG. Emerging principles of intrinsic hippocampal organization. *Curr Opin Neurobiol* 1993;3:225–9
2. Andreasen NC. Pieces of the schizophrenia puzzle fall into place. *Neuron* 1996;16:697–700
3. Armony JL, LeDoux JE. How the brain processes emotional information. *Ann N Y Acad Sci* 1997;821: 259–70
4. Cohen RZ, Gotowiec A, Seeman MV. Duration of pretreatment phases in schizophrenia: women and men. *Can J Psychiatry* 2000;45:544–7
5. Cohen NJ, Eichenbaum H. *Memory, Amnesia, and the Hippocampal System.* Cambridge, MA: MIT Press, 1993
6. Ellison G. Stimulant-induced psychosis, the dopamine theory of schizophrenia, and the habenula. *Brain Res Brain Res Rev* 1994;19:223–39
7. Gallagher BJ, McFalls JA Jr, Jones BJ, Pisa AM. Prenatal illness and subtypes of schizophrenia: the winter pregnancy phenomenon. *J Clin Psychol* 1999;55:915–22
8. Helgason C, Caplan LR, Goodwin J, *et al.* Anterior choroidal artery-territory infarction: report of cases and review. *Arch Neurol* 1986;43:681–6
9. Jacobson R. Disorders of facial recognition, social behavior and affect after combined bilateral amygdalotomy and subcaudate tractotomy – a clinical and experimental study. *Psychol Med* 1986;16:439–50
10. LeDoux JE, Muller J. Emotional memory and psychopathology. *Trans R Soc Lond* 1997;352:1719–26
11. LeDoux JE. The amygdala: contributions to fear and stress. *Semin Neurosci* 1994;6:231–7
12. McEntee WJ, Mair RG. The Korsakoff syndrome: a neurochemical perspective. *Trends Neurosci* 1990;13:340–4
13. Rosenzweig MR. Aspects of the search for neural mechanisms of memory. *Annu Rev Psychol* 1996;47:1–32
14. Sadikot AF, Parent A. The monoaminergic innervation of the amygdala in the squirrel monkey: An immunohistochemical study. *Neuroscience* 1990;36:431–47
15. Seeman P, Guan HC, Van Tol HH. Dopamine D4 receptors elevated in schizophrenia. *Nature* 1993;365:441–5
16. Strittmatter WJ, Roses AD. Apolipoprotein E and Alzheimer's disease. *Annu Rev Neurosci* 1996;19:53–77
17. Swanson LW, Petrovich GD. What is the amygdala? *Trends Neurosci* 1998;21:323–31
18. White FJ. Synaptic regulation of mesocorticolimbic dopamine neurons. *Annu Rev Neurosci* 1996;19:405–36
19. Winn P. Schizophrenia research moves to the prefrontal cortex. *Trends Neurosci* 1994;17:265–8
20. Zola-Morgan S, Squire LR. Neuroanatomy of memory. *Annu Rev Neurosci* 1993;16:547–64

Gustatory system 18

The gustatory system regulates the perception, transmission, and recognition of taste sensation. It involves the stimulation of the taste buds which are scattered on the vallate, foliate, and fungiform papillae of the dorsum of the tongue, epiglottic vallecula, and the palate. Taste sensations are carried by fibers of the facial, glossopharyngeal, and vagus nerves.

Gustatory receptors and pathways

Gustatory receptors and pathways

Taste, olfactory and thermal sensations, and information that pertains to texture form an integral component of the subjective perception of flavor. Sight, surrounding environment, and even sound influence the gustatory system. Since taste sensation is much simpler and maintains higher threshold than olfaction, appreciation of flavor may be intensified by olfactory impulses (e.g. coffee and chocolate flavors may be appreciated by olfactory neurons). Ability to taste may vary according to gender, age, and fluctuation in metabolic activities. It may also be affected in individuals with fever. Taste also determines our eating habits, weight, and indirectly the electrolyte balance within our system.

Taste receptors (buds) are renewed periodically because of their short life span. Taste buds contain parietal (precursors), sustentacular, basal, and gustatory cells. They begin to decline in number at a rate of 1% a year, increasing after the age of 40. This fact may explain the tendency for elderly people to eat more spicy food than the average population. Gustatory cells are located on the tongue (primarily receive sweet and salty modalities), palate (principally perceives bitter and sour modalities), and also in the laryngopharynx. Lingual papillae that subserve gustatory functions contain groups of neuroepithelial, and supporting sustentacular cells, which provide, on a weekly basis, the taste cells upon their natural death. Taste buds do not show structural variation and the neuroepithelial cells within these buds respond to multiple stimuli. Each neuroepithelial cell may receive terminals from several neurons and several cells may be innervated by one neuron. Therefore the 50 cells of the taste buds receive 50 nerve fibers. Taste buds are contained in the foliate, circumvallate, and the fungiform papillae.

Neurons (Figure 18.1) which receive taste sensation are unipolar contained in the geniculate and the inferior ganglia of the glossopharyngeal and vagus nerves. Peripheral processes of these neurons run within the chorda tympani branch of the facial nerve, lingual branch of the glossopharyngeal nerve, and the internal laryngeal branch of the vagus nerve. Central processes of these neurons project to the medial and

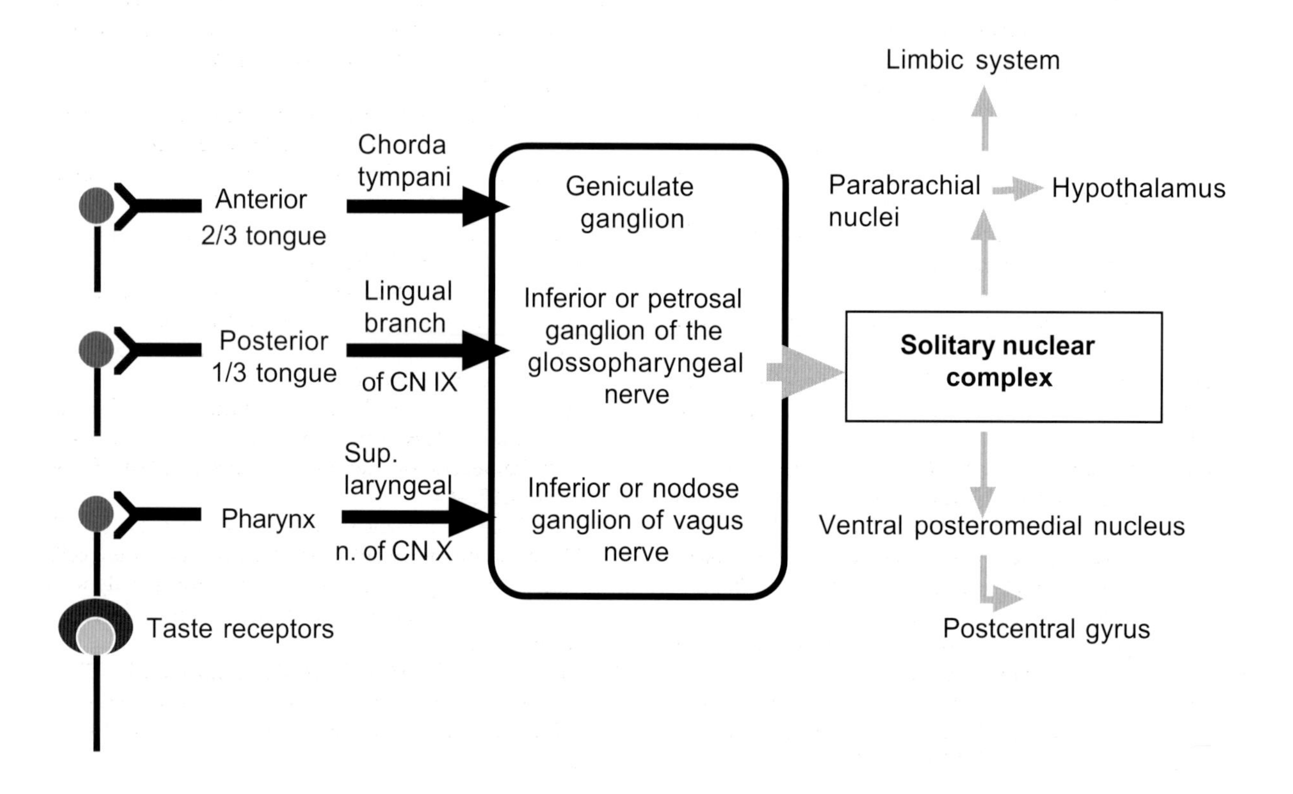

Figure 18.1 This is a schematic representation of the peripheral gustatory apparatus, associated cranial nerves and ganglia, and their central connections

rostral parts of the solitary nucleus (gustatory nucleus), forming the solitariothalamic tract, which ascends ipsilaterally as a component of the ventral tegmental tract. This tract which terminates in the medial part of the ventral posteromedial nucleus of thalamus, ascend in conjunction with the medial lemniscus. From the thalamus the taste fibers project to the primary gustatory cortex located adjacent to the tongue area of the primary sensory cortex, and around the lateral cerebral fissure (parietofrontal operculum). Some taste fibers leave the solitary nucleus to establish synaptic linkages with the ambiguus and hypoglossal nuclei, mediating reflex activities, while others may terminate in the hypothalamus to enhance emotional reactions associated with taste sensation.

Impairment of gustatory sense (aguesia) often accompanies anosmia. Aguesia may result from the toxic effects of certain chemicals (e.g. lead). Unilateral aguesia may be seen in facial nerve dysfunction. Transient aguesia may be a side-effect of oral administration of drugs such as penicillamine, L-dopaphenindione, H_2 receptor antagonist (e.g. ranitidine), and coronary artery dilators. Reduction in gustatory sense (hypoguesia) may normally occur in the elderly or may accompany Sheehan's syndrome, hypothyroidism, sarcoidosis, and possibly diabetes mellitus. Giant cell arteritis in the elderly may present with gustatory disorders.

Suggested reading

1. Ahne G, Erras A, Hummel T, Kobal G. Assessment of gustatory function by means of tasting tablets. *Laryngoscope* 2000;110:1396–401
2. Fujikane M, Itoh M, Nakazawa M, *et al.* Cerebral infarction accompanied by dysgeusia – a clinical study on the gustatory pathway in the CNS. *Rinsho Shinkeigaku* 1999;39:771–4
3. Helcer M, Schnarch A, Benoliel R, Sharav Y. Trigeminal neuralgic-type pain and vascular-type headache due to gustatory stimulus. *Headache* 1998;38:129–31
4. Kaddu S, Smolle J, Komericki P, Kerl H. Auriculotemporal (Frey) syndrome in late childhood: an unusual variant presenting as gustatory flushing mimicking food allergy. *Pediatr Dermatol* 2000;17:126–8
5. King AB, Menon RS, Hachinski V, Cechetto DF. Human forebrain activation by visceral stimuli. *J Comp Neurol* 1999;413:572–82
6. Leonard NL, Renehan WE, Schweitzer L. Structure and function of gustatory neurons in the nucleus of the solitary tract. IV. The morphology and synaptology of GABA-immunoreactive terminals. *Neuroscience* 1999;92:151–62
7. Lundy RF Jr, Contreras RJ. Gustatory neuron types in rat geniculate ganglion. *J Neurophysiol* 1999;82:2970–88
8. Nakashima M, Uemura M, Yasui K, *et al.* An anterograde and retrograde tract-tracing study on the projections from the thalamic gustatory area in the rat: distribution of neurons projecting to the insular cortex and amygdaloid complex. *Neurosci Res* 2000;36:297–309
9. Ohishi Y, Komiyama S, Shiba Y. Predominant role of the chorda tympani nerve in the maintenance of the taste pores: the influence of gustatory denervation in ear surgery. *J Laryngol Otol* 2000;114:576–80
10. Pritchard TC, Macaluso DA, Eslinger PJ. Taste perception in patients with insular cortex lesions. *Behav Neurosci* 1999;113:663–71
11. Saito H. Gustatory otalgia and wet ear syndrome: a possible cross-innervation after ear surgery. *Laryngoscope* 1999;109:569–72
12. Sbarbati A, Crescimanno C, Benati D, Osculati F. Solitary chemosensory cells in the developing chemoreceptorial epithelium of the vallate papilla. *J Neurocytol* 1998;27:631–5
13. Scott TR, Giza BK. Issues of gustatory neural coding: where they stand today. *Physiol Behav* 2000;69:65–76
14. Scott TR, Plata-Salamán CR. Taste in the monkey cortex. *Physiol Behav* 1999;67:489–511
15. Smith DV, Margolis FL. Taste processing: whetting our appetites. *Curr Biol* 1999;9:R453–5
16. Smith DV, St John SJ. Neural coding of gustatory information. *Curr Opin Neurobiol* 1999;9:427–35
17. Spielman AI. Chemosensory function and dysfunction, *Crit Rev Oral Biol Med* 1998;9:267–91
18. Streefland C, Jansen K. Intramedullary projections of the rostral nucleus of the solitary tract in the rat: gustatory influences on autonomic output. *Chem Senses* 1999;24:655–64
19. Whitehead MC, Ganchrow JR, Ganchrow D, Yao B. Organization of geniculate and trigeminal ganglion cells innervating single fungiform taste papillae: a study with tetramethylrhodamine dextran amine labeling. *Neuroscience* 1999;93:931–41
20. Yang R, Crowley HH, Rock ME, Kinnamon JC. Taste cells with synapses in rat circumvallate papillae display SNAP-25-like immunoreactivity. *J Comp Neurol* 2000;424:205–15

General sensations 19

The ascending pathways convey conscious (cortical) and unconscious (subcortical) sensory information to the higher levels of the central nervous system (CNS). These pathways are concerned with the regulation of muscle tone and mediation of intersegmental reflexes, as well as transmission of a variety of sensations. They may exhibit monosynaptic connections or utilize an extensive network of neurons. The modalities of the general somatic sensations include pain, temperature, tactile, joint, vibration, and pressure sensations. These modalities are categorized into epicritic (discriminative) and protopathic sensations. The epicritic (discriminative) modalities include sensations such as fine touch, two-point discrimination (the ability to distinguish two blunt points from one another), joint sensation, and vibratory sense. These sensations are received by the encapsulated receptors and transmitted by the thickly myelinated and fast conducting fibers. The protopathic sensations, which include pain, temperature, and crude touch, are received by the free nerve endings (uncapsulated receptors) and conveyed to the spinal cord by small, thinly myelinated and/or unmyelinated fibers. These fibers represent the peripheral processes of the pseudounipolar neurons of the dorsal root ganglia which are contained in the peripheral nerves. Visceral pain also projects to the cerebral cortex; however, its peripheral transmission is maintained predominantly by the sympathetic fibers. The central transmission of this modality is identical to the somatic sensations. Non-cortical sensations emanate from Golgi tendon organs, muscle spindle, and other exteroceptive receptors, conveying information to the cerebellum, inferior olivary nucleus, and tectum via the spinocerebellar, trigeminocerebellar, spinolivary, and spinotectal tracts.

Peripheral and central transmission of the sensory pathways

Sensory receptors
- Encapsulated receptors
- Non-encapsulated receptors

Cortical ascending pathways
- Cortical body sensations
- Cortical sensations from the head

Ascending subcortical pathways
- Spinocerebellar tracts

Peripheral and central transmission of the sensory pathways

Sensory systems transmit a variety of somatic and visceral modalities, including epicritic (discriminative) sensations such as fine touch, joint sensation, vibratory sense, and two-point discrimination; and propathic sensations including pain, temperature, and crude touch. They are classified into cortical and subcortical sensory subdivisions that utilize the peripheral processes of the pseudounipolar neurons of the dorsal root and sensory ganglia of the cranial nerves in the transmission of impulses from receptors to the CNS. Cortical sensations project to the cerebral sensory cortex, whereas non-cortical sensations are conveyed to the subcortical structures such as the tectum, cerebellum, and inferior olivary nucleus.

Peripheral nerve fibers that transmit sensory impulses are classified by Erlanger and Gasser on the basis of the fiber diameter, rate of conduction, and the degree of myelination into group A, B, and C fibers. Olfactory and optic nerves are not included in this classification. The 'A' fibers, the largest and the fastest in conduction, are classified into Aα, Aβ, Aγ, and Aδ fibers. Each group of fibers, with a few exceptions, is comprised of efferent and afferent fibers. Afferents are confined to groups A and C. Components of group 'A' afferents include Aα, Aβ, and Aγ. Group Aα consists of sensory fibers from the muscle spindles which terminate in the deeper layers of the dorsal horn and also in the ventral horn; group Aβ fibers carry sensations from Golgi tendon organs and Meissner's and Pacinian corpuscles, and end in laminae III–VI. Class Aδ fibers carry nociceptive and visceral stimuli to laminae I, IV, and V. Class C fibers are non-myelinated and considered to be the slowest in conduction, conveying nociceptive and olfactory stimuli. Efferents are classified into class A, B, and C group fibers. Class A includes Aα, Aβ, and Aγ. Class Aα fibers supply fast-twitching extrafusal muscles, Aβ fibers form the plaque endings of the muscle spindle fibers, and collaterals of the Aα axons and Aγ fibers constitute the plate and trail endings of intrafusal muscles. Class B fibers comprise the preganglionic sympathetic and parasympathetic fibers. Class C fibers are non-myelinated and of small-caliber fibers, which convey postganglionic autonomic fibers.

Lloyd also classifies sensory fibers of peripheral nerve into groups I, II, III and IV. Group I fibers consists of the primary sensory Ia fibers associated with the muscle spindle (equivalent to Aα), and Ib fibers that convey impulses from Golgi tendon organs. The Ia group consists of thickly myelinated and fast conducting fibers that receive information from the muscle spindle and establish excitatory monosynaptic connections with the spinal motor neurons of the synergists and inhibitory connections with the antagonists. They are sensitive to stretch stimuli, mediating the monosynaptic myotactic reflex. Tapping a deep tendon (e.g. patellar tendon) with a reflex hammer produces a stretch in the attached muscle, activating the Ia fibers and subsequently the spinal motor neurons. Activation of the spinal motor neurons results in contraction of the innervated muscle. The Ib afferents consist of smaller diameter fibers which mediate the inverse myotactic (clasp-knife) reflex and convey inhibitory impulses to the synergist motor neurons of the spinal cord. It has been shown that interneurons in the reflex pathways of Ib afferents receive short-latency excitation from low threshold cutaneous and joint afferents. When a limb movement is initiated and suddenly meets an obstacle, inputs from cutaneous and joint receptors will trigger the Ib afferent inhibitory system, producing reduction in the muscle tension. Group II fibers include afferents from the pacinian corpuscles, hair follicles, and secondary fibers from muscle spindles. This group of fibers is inhibitory and plays a major role in protecting a muscle when it experiences undue tension. Group III (comparable to Aδ) and group IV (comparable to C fibers) fibers carry nociceptive stimuli, including the free nerve endings in the walls of the blood vessels and in the follicles of the finer hair.

All sensory systems use receptors to transduce various forms of energy into neuronal activity, which, if of sufficient intensity, results in the generation of nerve impulses whose frequency and pattern may determine the sensory experience. Receptors convey the generated stimuli, through a series of neurons, primary, secondary or the tertiary, to the cortical and subcortical areas of the CNS. For the most part, the type of modality of sensation mediated by a particular axon is specific for each axon, and furthermore, not all activities in the sensory receptors are consciously perceived.

Sensory receptors

Receptors, activated by one or more modalities, are classified into mechanoreceptors, chemoreceptors, photoreceptors, and osmoreceptors. The mechanoreceptors convey tactile, pressure, and auditory impulses, while chemoreceptors are concerned with taste, smell, and chemical changes, such as the fluctuation in levels of carbon dioxide. Photoreceptors, on the other hand, react to electromagnetic waves, while osmoreceptors are sensitive to the changes in osmotic pressure. Regulation of blood flow and pressure are accomplished by receptors in the adventitia of the blood vessels. Receptors are also classified according to their areas of distribution into exteroceptors, proprioceptors, and interoceptors. Exteroceptors are the receptors for cutaneous stimuli. The special sensory receptors receive olfactory, visual, acoustic, and taste sensations. Receptors that respond to stimuli associated with the direction and extent of movement such as the Golgi tendon organ, muscle spindle, and pacinian corpuscles, are known as proprioceptors. The interoceptors are primarily free nerve endings located in the walls of the viscera and the blood vessels, which are sensitive to excessive tension and stretch of the visceral wall.

General sensory receptors are also categorized into encapsulated and non-encapsulated receptors. The encapsulated receptors consist of the muscle spindles, Golgi tendon organs, pacinian and Meissner's corpuscles, and Ruffini endings. Less distinct, and not totally recognized by anatomists, are the Krause's end bulb and Golgi–Mazzoni corpuscles.

Encapsulated receptors

Muscle (neuromuscular) spindles are numerous in the intrinsic hand muscles, antigravity axial muscles, and the flexor muscles of the upper extremity and extensors of the lower extremity (Figures 19.1 & 19.2). They are few in numbers in the muscles that contain predominantly fast glycolytic (white) fibers. Each muscle spindle consists of intrafusal muscle fibers surrounded by a connective tissue capsule. The intrafusal muscle fibers are classified into nuclear bag and nuclear chain fibers, acting as a length detector that fires in response to stretch, and mediate the monosynaptic myotactic reflex. This passive stretch may be produced when a reflex hammer strikes the patellar ligament. They receive γ efferents as well as Ia and II afferents. Group Ia and II fibers, that carry impulses generated by passive stretch of a muscle, establish excitatory synaptic contacts upon the spinal α motor neurons. The rate of firing of the Group Ia fibers depends upon the rate of stretch; the more rapid the stretch, the more impulses are generated (phasic response). The plate ending motor neurons intensify the latter response. The magnitude of impulses generated by the secondary (II) fibers depend upon the extent of maintained stretch (tonic response). Tonic response correlates with the intensity of the trail ending motor neurons. Excitation of the α motor neurons produces contraction of the quadriceps muscle and extension of the leg. Ia and II fibers also establish inhibitory synaptic connections with the α motor neurons of the antagonist muscles.

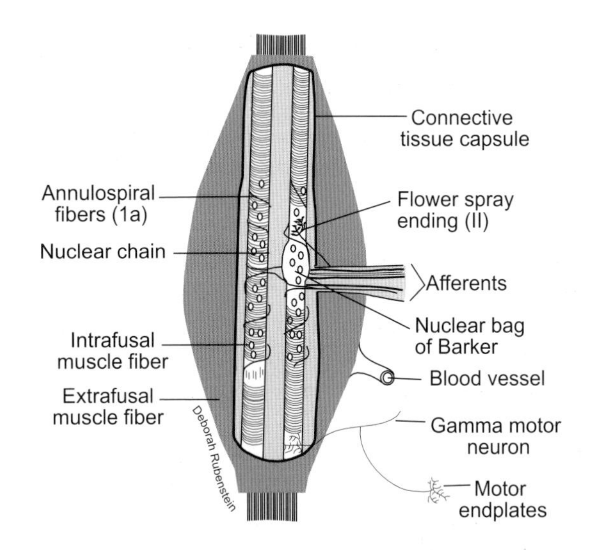

Figure 19.1 This diagram illustrates the two forms of the intrafusal muscle fibers, Ia and II afferents and the terminals of the γ fibers

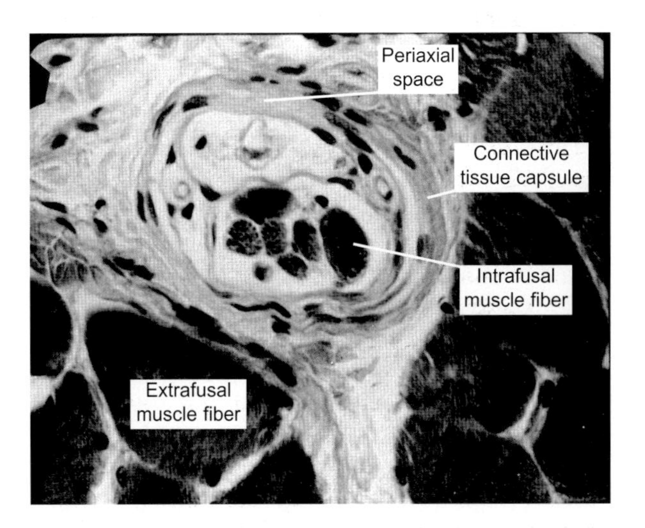

Figure 19.2 Photomicrograph of a section through the muscle spindle. The connective tissue capsule and intrafusal muscle fibers are clearly visible

The Golgi tendon organs (GTO) are found at the junction of a muscle and its tendon, mostly in the aponeurosis. They consist of collagenous fibers that run parallel to the extrafusal muscle fibers. They receive unmyelinated nerve terminals, and are enclosed by a capsule. GTO are activated by excessive tension and provide conscious and unconscious proprioceptive information to the cerebral cortex and cerebellum, respectively. In contrast, a decrease in muscle tension, as a result of fatigue, produces the opposite effect of reducing activation of these

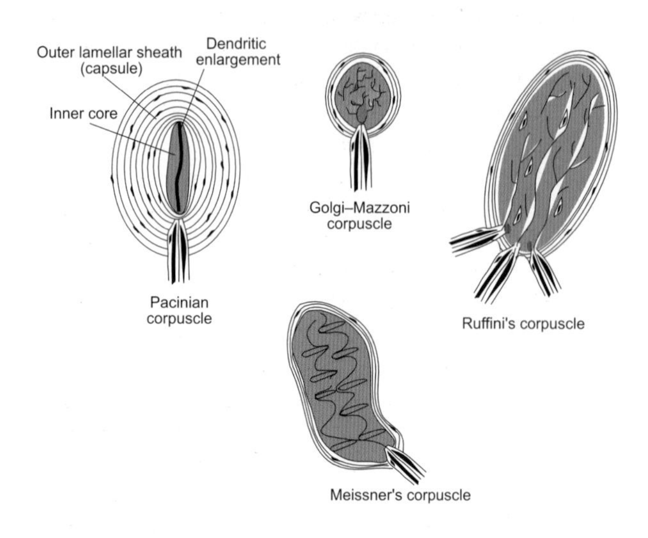

Figure 19.3 Various forms of encapsulated receptors

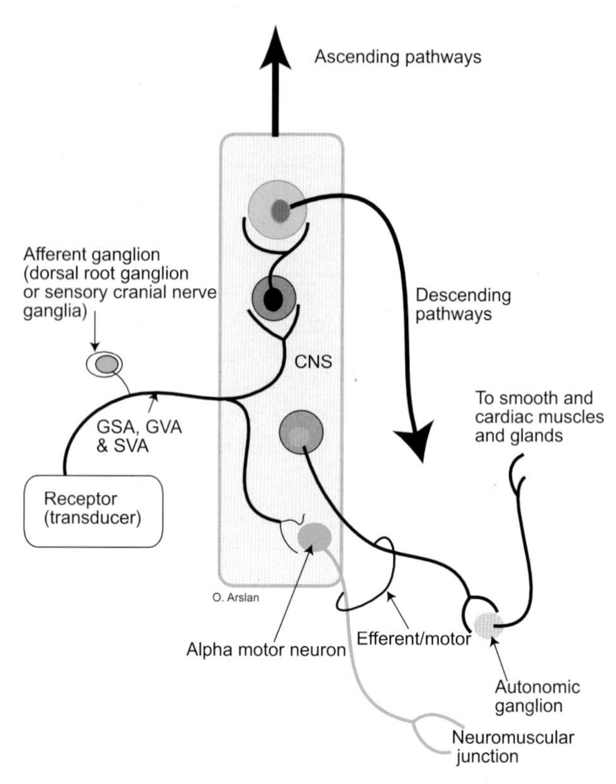

Figure 19.4 In this schematic diagram the peripheral course of the somatic and visceral sensory fibers and their central connections are explored. Relevant reflex connections, and somatic and visceral innervation are also depicted

receptors. The signals from GTO are carried by the Ib fibers to the spinal cord, establishing disynaptic connections with neurons of the agonist and antagonist muscles. Central connections of GTO are studied later in this chapter.

The largest of all receptors are the pacinian corpuscles (Figure 19.3) that consist of a capsule, concentric lamella, and terminals of Aα fibers. These receptors are activated by vibration and tension, and are abundant in the genital organs, nipple, periosteum, ligaments, pancreas, and mesenteries.

The Meissner's corpuscles (Figure 19.3) are a group of receptors that consist of a capsule and a central core, receiving terminals of tactile sensory nerve endings. They are found in the glabrous skin of the hand and foot, lips, and tongue of infants and young adults. These receptors usually disappear with aging. They respond best to movement across a textured surface, as when touching cloth or reading Braille.

Ruffini endings (Figure 19.3) are cylindrical sensory organs that respond to pressure and warmth, as well as to simultaneously applied blunt two-points. They consist of group II fibers embedded in a granular matrix and are surrounded by a thin capsule. These receptors respond upon grasping heavy objects, carrying a brief luggage, and when the nails are used in scratching.

Non-encapsulated receptors

The non-encapsulated receptors are comprised of the free nerve endings, and Merkel's discs.

The free nerve endings are abundant in the cornea, epidermis, mucous membranes, and tooth pulp. These receptors carry primarily pain and thermal sensations. They also respond to mechanical and thermal nociceptors.

The Merkel's discs (tactile menisci) are highly modified basal keratocytes that exist in the glabrous skin and are activated by tactile sensations. They are also considered as sensory receptors with expanded nerve terminals.

As mentioned earlier, sensory systems utilize series of neurons, in an orderly fashion, to convey impulses to higher centers in the CNS. The first order neurons, for all general somatic afferents from the body, are located in the dorsal root ganglia (DRG) of the spinal nerves. The second order neurons may be positioned in the gray matter of the spinal cord or in the brainstem, whereas the third order are formed by the neurons of the ventral posterolateral (VPL) nucleus of the thalamus (Figures 19.4 & 19.5). There are numerous pathways that convey general somatic afferents to the cerebral cortex (ascending cortical pathways). These pathways ascend in the dorsal, lateral, and anterior funiculi of the spinal cord, and

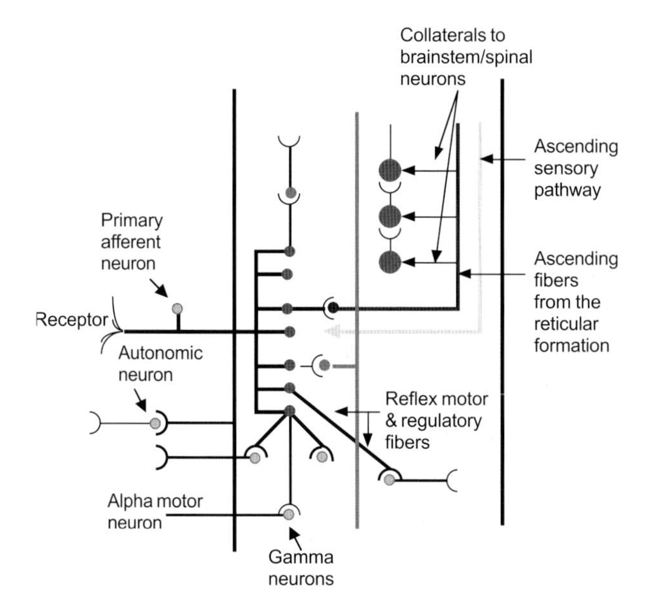

Figure 19.5 This diagram traces the connection of the primary afferent neurons with spinal neurons. The role of these neurons in the mediation of reflex activity on the ipsilateral and contralateral side as well as in the formation of ascending sensory pathways are clearly illustrated

carry joint, tactile, vibratory, nociceptive, and thermal sensations. They are mainly crossed pathways, project to the thalamus, and then to the cerebral cortex. Conscious cortical pathways are comprised of the dorsal column-medial lemniscus, spinothalamic tracts, and the trigeminal lemnisci. Pathways that subserve special cortical sensations such as visual and auditory have already been described.

Cortical ascending pathways

Cortical body sensations

The dorsal white columns (Figures 19.6, 19.7, 19.8 & 19.9) transmit fine tactile sensation, vibratory sense, position and movement sense (kinesthesia), two-point discrimination of simultaneously applied blunt pressure points, and stereognosis (ability to recognize form, and the size, texture and weight of objects). The receptive fields for this pathway are small and are primarily scattered in the fingers and tongue and are less numerous in the skin of the back. Sensations are conveyed by the thickly myelinated, fast conducting fibers, which represent the central processes of the pseudounipolar neurons of the dorsal root ganglia. These fibers occupy the medial portion of the dorsal roots, bifurcating in the spinal cord into ascending and descending branches. The descending branches of the dorsal root fibers form the fasciculus interfascicularis (comma tract of Schultze) in the cervical and thoracic spinal segments, and the fasciculus septomarginalis (located near the posterior median septum) in the lumbar and sacral spinal segments. The long ascending branches travel ipsilaterally in the posterior funiculus and are somatotopically organized. The most medial fibers convey discriminative sensations from the lower extremities and lower half of the trunk, extend the entire length of the spinal cord, forming the gracilis fasciculus. It is important to note that fibers that carry unconscious proprioceptive sensations from stretch receptors of the lower extremity and enter L3–S5 spinal segments are also contained within the gracilis fasciculus. All unconscious and most of the conscious proprioceptive fibers within the gracilis fasciculus terminate in the Clarke's nucleus, giving rise to the ipsilateral dorsal spinocerebellar tract. The latter projects to the cerebellum via the inferior cerebellar peduncle. Before entering the inferior cerebellar peduncle, the conscious proprioceptive fibers from the lower extremity e.g. position and movement senses, leave the dorsal spinocerebellar tract as collaterals to synapse in the nucleus Z of Brodal and Pompieno (Figures 19.6 & 19.7). Since proprioceptive fibers from the lower extremity leave earlier, the gracilis fasciculus in the cervical spinal segments contain, for the most part, cutaneous impulses, terminating in the nucleus gracilis. Fibers that transmit joint and fine tactile sensations from the upper extremities and upper half of the trunk are positioned lateral to the fasciculus gracilis and form the fasciculus cuneatus. Fasciculus cuneatus extends in the cervical and upper six thoracic spinal segments, terminating in the ipsilateral cuneate nucleus. Postsynaptic axons from cuneate and gracilis nuclei and nucleus Z constitute the internal arcuate fibers, which decussate in the caudal medulla, continuing as the medial lemniscus on the contralateral side through the rostral medulla, pons, and midbrain. Here, the medial lemniscus exhibits a somatotopic arrangement that differs from spinal levels (Figure 19.10). In the medulla, the medial lemniscus assumes a vertical position, in which the cervical fibers are dorsal, the sacral fibers are ventral, while trunk fibers maintain an intermediate position. In the pons and midbrain the medial lemniscus shifts to a horizontal configuration, where the cervical fibers are lateral, the sacral fibers are more medial, and the trunk fibers occupy an intermediate region (Figure 9.10). The medial lemniscus terminates in the VPL nucleus of the thalamus. The VPL then relays this information to the cerebral cortex via the posterior limb of the internal capsule and corona radiata. Areas of the cerebral cortex which receive this information include the primary somatosensory and the

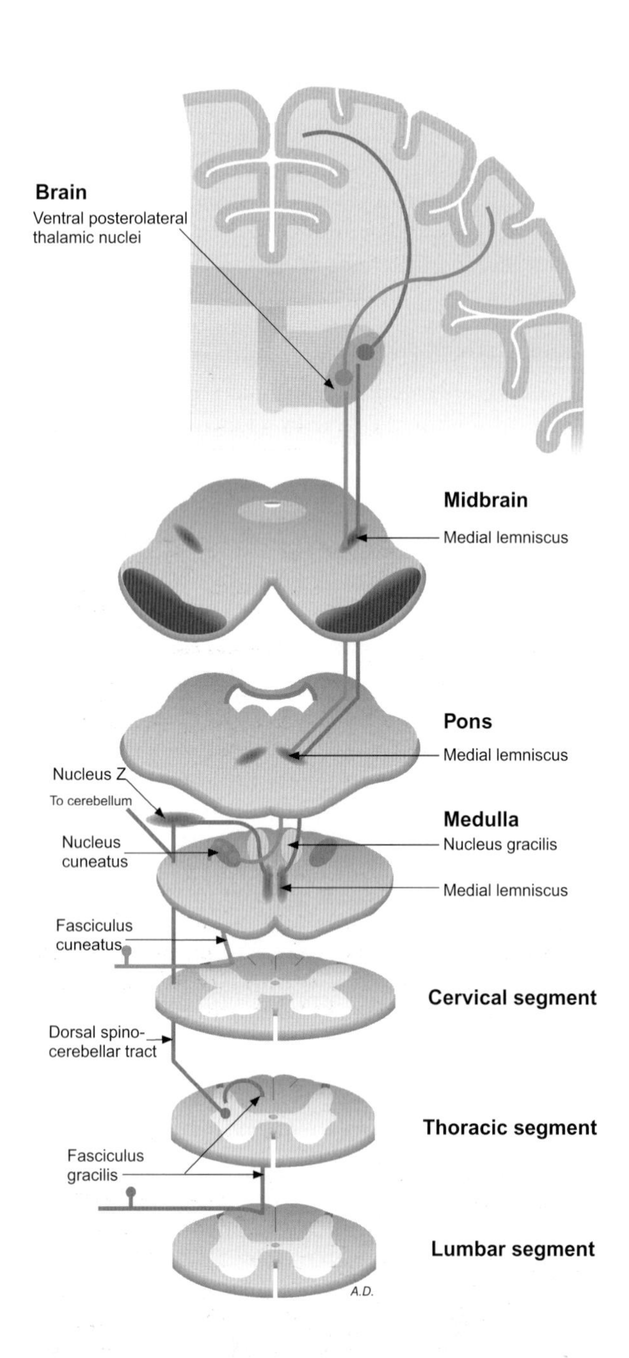

Figure 19.6 This diagram shows the course of most of the fibers that carry conscious proprioception from the lower extremity via the dorsal column-medial lemniscus pathway

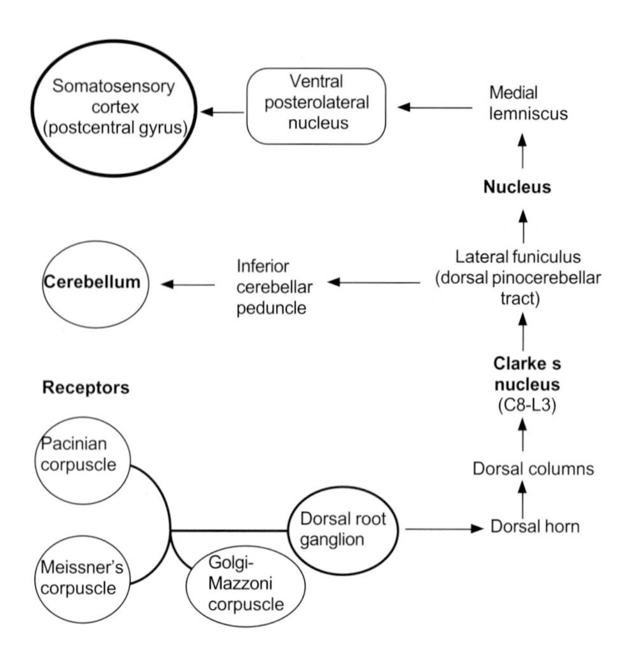

Figure 19.7 This is a more simplified schematic drawing of Figure 19.6. The dorsal column-medial lemniscal pathway and its association with the dorsal spinocerebellar tract in the transmission of proprioceptive fibers from the lower extremity are shown. Most of the fibers that convey proprioceptive impulses from the lower extremity travel in the dorsal spinocerebellar tract before terminating in the nucleus

posterior insular cortices of the contralateral side, and the secondary somatosensory cortices of both sides. The posterior insular cortex receives input from the secondary somatosensory cortex and Brodmann's areas 5 and 7 of the parietal lobe. The dorsal column-medial lemniscus pathway operates on the basis of a 1:1 synapse ratio and lateral inhibition, and is governed by the regulatory influence of the corticospinal tract. Variations in the sensory experiences when the individual is attentive or alert may be attributed to the influences exerted by the corticospinal tract. Rostral to the eight cervical spinal segments, the cuneatus fasciculus transmits unconscious proprioception fibers from the upper extremities that terminate upon the accessory cuneate nucleus. This nucleus forms the ipsilateral cuneocerebellar tract (which will be discussed shortly with the non-cortical sensory pathways).

Some fibers of the dorsal column which pursue a deeper course (phylogenetically older) synapse in the most peripheral neurons of the medullary nuclei of the dorsal column, and travel within the medial lemniscus. These fibers establish synaptic contacts with the reticular formation and eventually become part of the spinoreticulothalamic pathway. This connection may explicate the pain sensation felt upon electrical stimulation of the dorsal column-medial lemniscus pathway

The anterolateral system (Figures 19.12 & 19.13) is a morphological designation for the combined lateral and ventral spinothalamic and the spinotectal tracts. The lateral spinothalamic tract (neospinothalamic-lateral system) is a crossed pathway, which conveys

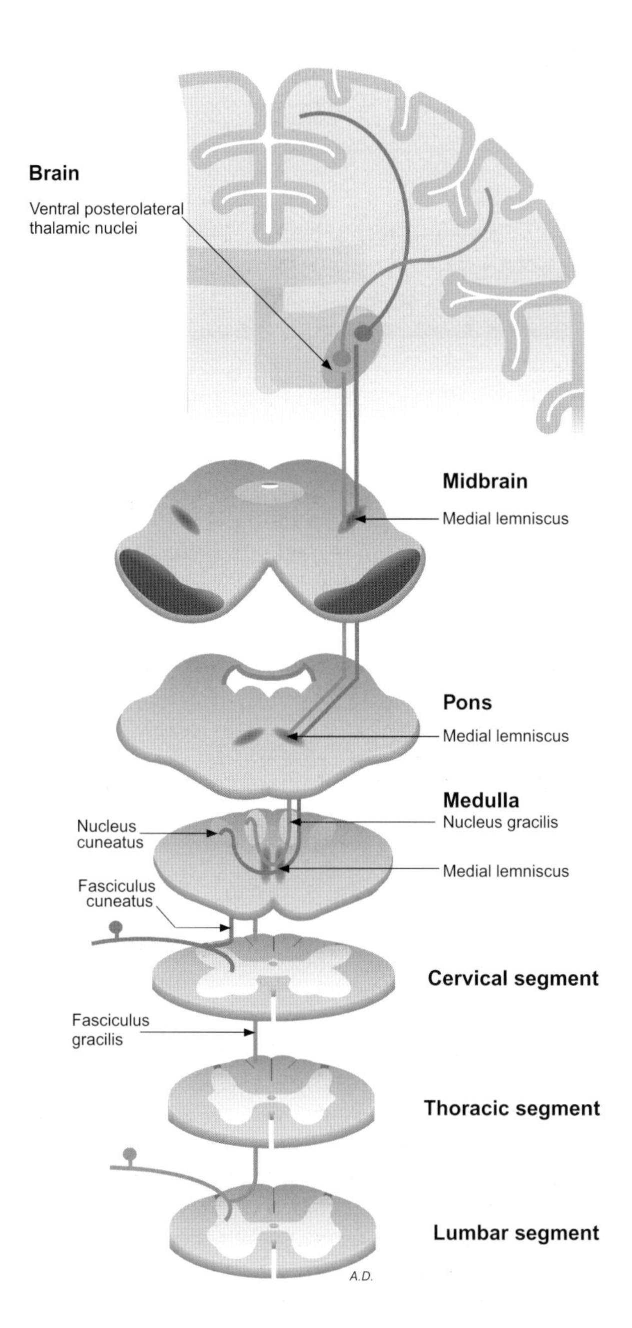

Figure 19.8 Course of all discriminative sensations from the upper extremity and upper half of the trunk and discriminative cutaneous sensations from the lower extremity. Note the ipsilateral route of this pathway in the spinal cord and contralateral course at supraspinal levels

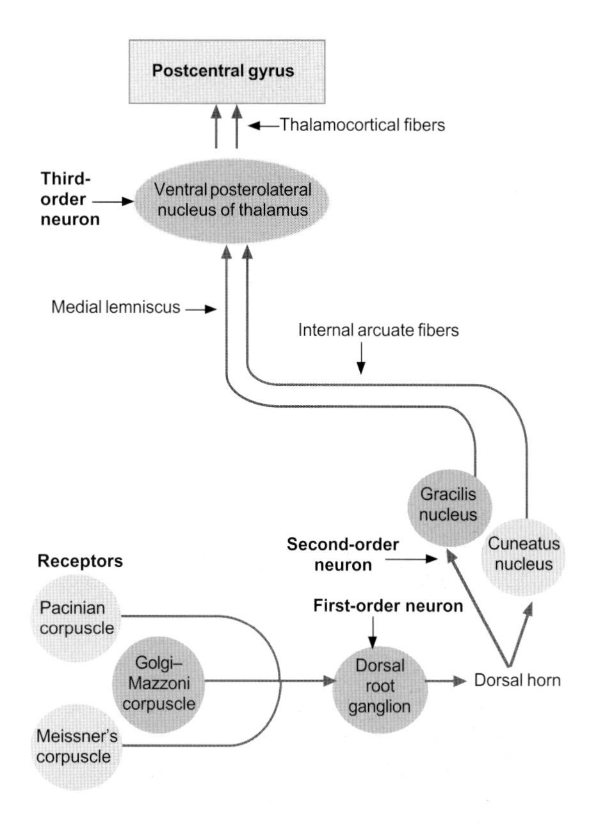

Figure 19.9 Dorsal column-medial lemniscus pathway that conveys conscious proprioception and discriminative cutaneous sensations from the upper extremity and discriminative cutaneous sensations from the lower extremity

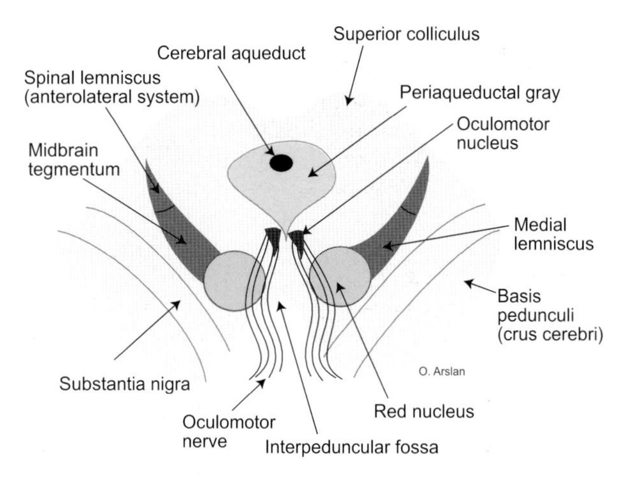

Figure 19.10 Somatotopic arrangement of the fibers within the medial lemniscus. As illustrated, lower extremity fibers occupy lateral and dorsal positions to the upper extremity fibers

Dorsal columns show degeneration in tabes dorsalis (Figure 19.11), a slowly progressive condition which is commonly seen as a late manifestation of neurosyphilis and also in diabetes mellitus. In this disease, the dorsal root fibers of the lumbosacral and sometimes the cervical spinal nerves are affected. Patients may present with early signs of incontinence (atonic bladder) and sexual impotence. They may experience bouts of lightning pain, identical in severity to trigeminal neuralgia. Pain may not be uniform and may be reduced or lost later in certain regions. Argyll Robertson pupil, in which the patient can accommodate to near vision but remains unresponsive to light, is another feature of the disease.

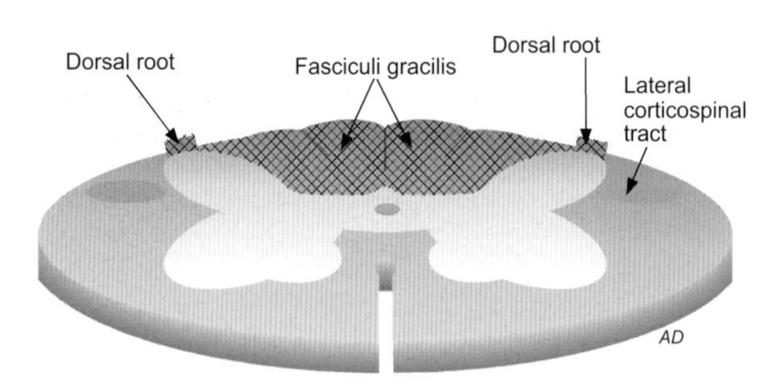

Figure 19.11 Section of the spinal cord showing degeneration in the posterior funiculus associated with tabes dorsalis

thermal and painful sensations from somatic and visceral structures. Painful and thermal stimuli activate the free nerve endings and generate impulses that enter the spinal cord via the C and Aδ fibers of the lateral bundle of the dorsal root. Group C fibers, which carry slow, diffuse, and aching (visceral) pain, as well as thermal sensations, terminate primarily in the substantia gelatinosa (lamina II). The Aδ fibers convey sharp and fast pain in addition to cold stimuli, and establish synapse with neurons of spinal laminae I–V. These two types of fibers may be inhibited by stimulation of the adjacent thickly myelinated and fast conducting (Aβ) fibers.

Most of the Aδ and C fibers ascend one or two segments above the level of their entry in the dorsolateral tract of Lissauer and then synapse in certain gray laminae of the spinal cord; others may establish synaptic connections at the same level of entry into

Due to the loss of position and vibratory senses in the lower limb, patients may exhibit positive Romberg's sign and a high steppage-broad gait, compensating for the loss of proprioception from the feet. Romberg's sign may be utilized to determine the patient's ability to maintain balance while standing in an erect position. The patient's feet must be approximated with his eyes gazing straight ahead and arms by his sides. Some degree of swaying is expected; however, tendency to fall must be reassessed by repeating the test. Anxiety oriented individuals and hysterical patients are naturally prone to test positive. Patients usually fall down in poorly illuminated areas due to the lack of visual cues as well as joint sensation. This test may be repeated with eyes closed to exclude the utilization of visual cues in maintaining equilibrium, and preventing the patient from falling. Degenerative changes may cause swelling and pain in the joints, loss of sensation from the bladder and optic atrophy. Individuals also exhibit loss of two-point discrimination (healthy individuals should be able to distinguish two blunt points applied simultaneously to the finger tips with a range of 4–5 mm. It is important to note that in cerebellar lesions the swaying is very marked, occurring when the eyes are either closed or open.

Note that in some patients with severely damaged dorsal columns, joint and vibratory senses and the two-point discrimination remain preserved. This may be explained on the basis that the posterior spinocerebellar tract transmits joint sensation from the lower extremity. Vibratory sensation may be shared by the spinothalamic tract. Position sense from the upper extremity may be mediated via collaterals of the primary afferents (Ia fibers) which are given to the spinothalamic tract.

the spinal cord. The postsynaptic fibers from laminae I, IV, & V cross the midline in the anterior white commissure to form the lateral spinothalamic tract, a component of the anterolateral system, on the contralateral side. However, not all nociceptive fibers contribute to the lateral spinothalamic tract, some may synapse on interneurons that send processes to inhibit the C and Aδ fibers by secreting enkephalins. Information within the lateral spinothalamic tract is somatotopically arranged. The dorsal portion of this tract transmits temperature, whereas the ventral portion conveys pain sensation. Visceral pain fibers

The gate theory of analgesia and the rationale behind using transcutaneous electrical nerve stimulation (TENS) is based upon the inhibitory role that thickly myelinated fibers exert, when stimulated, upon the unmyelinated C and thinly myelinated Aδ fibers. TENS is employed in the treatment of intractable pain associated with rheumatoid arthritis and postherpetic neuralgia.

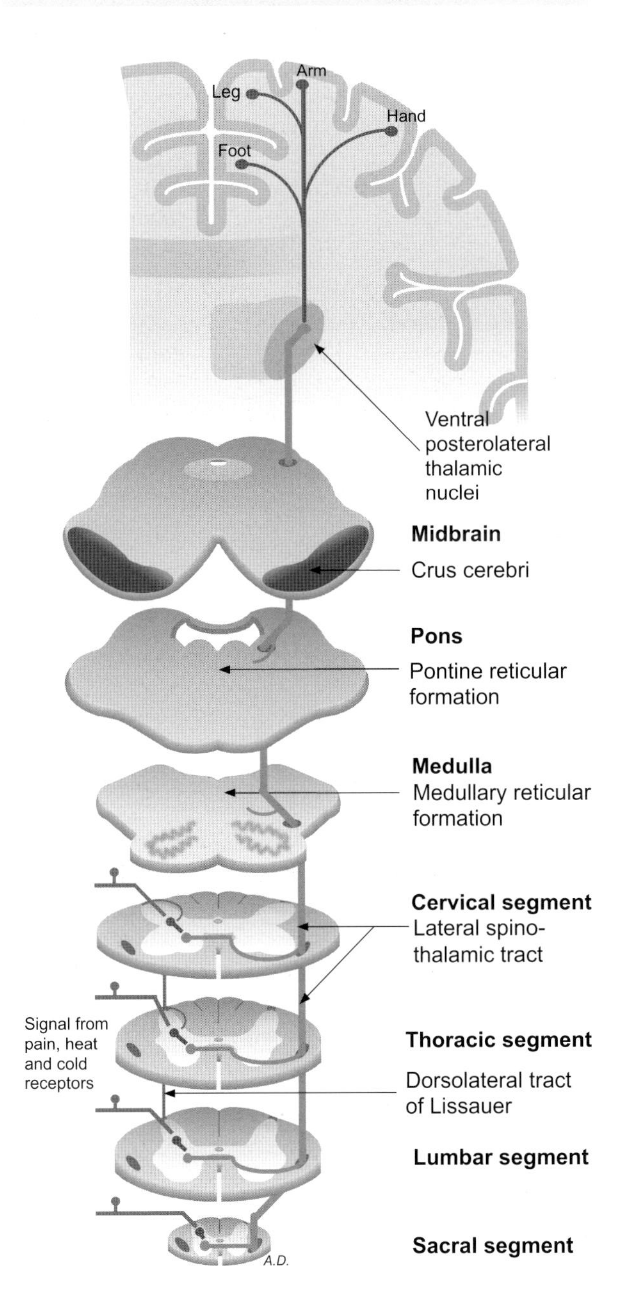

Figure 19.12 This drawing shows the course of pain and temperature impulses through the anterolateral system. This is a contralateral pathway which shows a reverse somatotopic arrangement to that of the dorsal columns

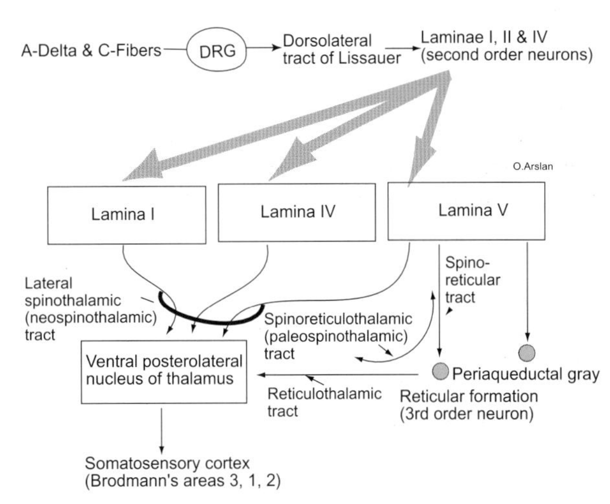

Figure 19.13 This is a simplified diagram of the organization of the anterolateral system. Other pathways associated with transmission of pain such as spinoreticular are also illustrated

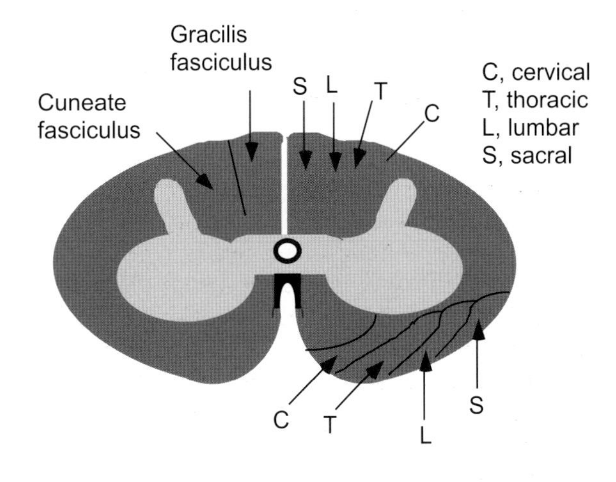

Figure 19.14 In this drawing the somatotopic arrangement of the anterolateral system is compared to that of the dorsal column. Note the sacral fibers are the most lateral whereas the cervical fibers occupy the most medial portion of the pain pathway

Projections of painful stimuli to the orbitofrontal cortex may account for the indifference that patients with prefrontal cortical damage exhibit toward pain despite acknowledging its existence. The secondary sensory cortex plays an important role in the distinction of sharp pain from dull pain.

occupy the medial part of the lateral spinothalamic tract, adjacent to the gray matter. Fibers that originate from the lower half of the body occupy a lateral position to the fibers from the upper half of the body (Figure 19.14).

The lateral spinothalamic tract runs in the lateral part of the brainstem and intermingles with the fibers of the ventral spinothalamic tract, forming the anterolateral system (spinal lemniscus). As it ascends, it provides collaterals to the reticular nuclei of the brainstem. It then projects to the ipsilateral VPL nucleus of the thalamus, intralaminar thalamic nuclei of both sides (with the exception the centromedian nucleus), and to the posterior thalamic zone. Experiments on decorticated animals have confirmed the role of the thalamus in pain recognition. The VPL nucleus projects to the postcentral gyrus, which is believed to deal with the localization of painful stimuli. Information from the posterior thalamic zone and the intralaminar nuclei of the thalamus are transmitted to the secondary sensory and orbitofrontal cortices, respectively. These thalamocortical projections are contained in the internal capsule and the corona radiata. Through these thalamocortical connections, the anterolateral system excites various cortical centers, producing pleasurable or displeasurable (affective) sensations associated with pain. Unlike the dorsal column-medial

Visceral pain is perceived as a dull, aching, slow, diffuse, and deeply seated pain. The site of actual visceral pain sensation is referred to a site distant from the location of the diseased organ (referred pain). Referred pain should be distinguished from projected somatic pain (sciatica) which is induced by irritation of a nerve trunk. Visceral pain may result from over-distention of the walls of the viscera (ureteral or biliary stone) or ischemia due to thrombosis or emboli, impairing the blood supply to the affected organ. In other words, conventional stimuli such as cutting or burning of a viscus do not elicit pain. Visceral pain may be also be produced by palpation of the abdominal wall, in search of the diseased organ, through the parietal peritoneum.

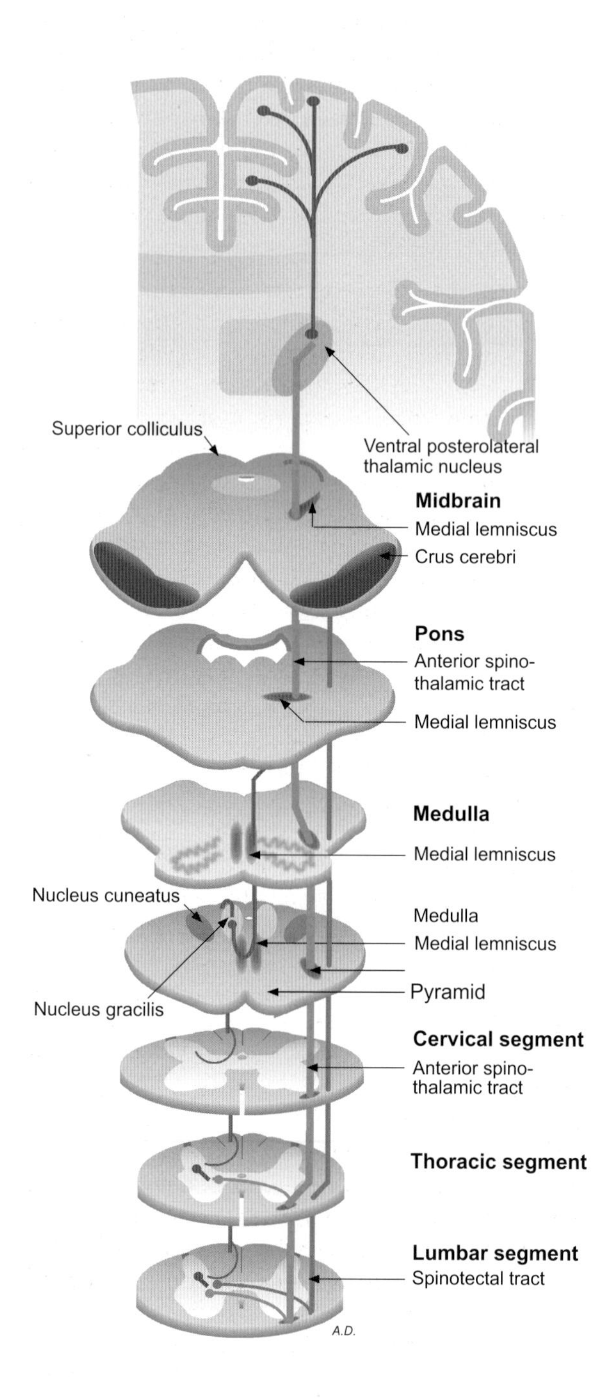

Figure 19.15 Transmission of crude touch is maintained by the ventral spinothalamic tract which is depicted in this picture. This pathway, which forms the anterior component of the anterolateral system, is also thought to transmit itching, tickling, and possibly libidinous sensation. Note that the spinotectal tract, an integral part of the anterolateral system, is also illustrated

The spinoreticulothalamic tract, a phylogenetically old system, carries poorly localized, diffuse, and vague pain from visceral organs. The connection of the spinoreticulothalamic tract to the hypothalamus accounts for the visceral activity and emotional responses associated with pain (e.g. facial expression and GI activity in response to pain).

lemniscus, the lateral spinothalamic tract does not function on the basis of a one to one synapse ratio or lateral inhibition. However, it is regulated by the corticospinal tract, accounting for variation in the sensory experiences and perception by attentive or alert individuals. Organic visceral sensations that signal a bodily need such as hunger, thirst, libido, nausea, and bladder and bowel fullness are generally carried by parasympathetic fibers to the spinal cord.

The spinoreticular tract (Figure 19.13) is an integral part of the ascending reticular activating system, maintaining an important role in the synchronization and desynchronization of the electrocortical activity of the brain, as well as transmission of pain sensation. It establishes a direct link between the spinal cord and the brainstem reticular formation, extending the entire length of the spinal cord. This tract terminates primarily upon cells of the nucleus reticularis gigantocellularis in the ipsilateral medulla. However, some fibers may terminate bilaterally upon cells of the nucleus reticularis pontine oralis and caudalis. This pathway extends from the brainstem reticular formation to the thalamus, constituting the spinoreticulothalamic (paleospinothalamic) tract. The spinoreticulothalamic tract is a multisynaptic, multineuronal pathway, which terminates in the intralaminar nuclei of the thalamus, establishing synaptic connections with the hypothalamus.

The ventral spinothalamic (paleospinothalamic) tract, also known as the anterior system, runs in the ventral funiculus as a component of the anterolateral system, transmitting signals associated with light touch, possibly tickling, itching (via the Merkel's discs and Meissner's corpuscles), as well as libidinous sensations (Figure 19.15). Since fine touch and discriminative tactile sensations are also conveyed in the dorsal columns, this pathway remains clinically insignificant. This tract is formed by axons of the second order neurons located in laminae III and IV which decussate in the anterior white commissure, and ascend contralaterally within the spinal lemniscus. It shows somatotopic arrangement, where fibers carrying information from the caudal parts of the body are located lateral to those originating from the cranial parts. This pathway projects to the VPL nucleus of the thalamus (third order neuron) that

Lesions localized in the spinal cord or brainstem may disrupt the lateral spinothalamic (neospinothalamic) tract or its fibers. In the spinal cord, this pathway may be damaged in the lateral funiculus or at the anterior white commissure. Unilateral destruction of the lateral spinothalamic tract results in analgesia and thermal anesthesia on the contralateral side of the body, extending one or two segments below the level of the lesion. Due to the bilateral representation of the painful stimuli from the perineal region, abdominal and pelvic viscera, destruction of this tract on one side may not be sufficient to produce complete anesthesia in these regions. It is important to remember that the return of some pain sensibility following a lesion may be attributed to the presence of uncrossed intersegmental spinothalamic tracts. Destruction of the posterior part of the lateral spinothalamic tract may extend to involve the posterior spinocerebellar tract, producing incoordination of motor activity (cerebellar ataxia) as well as analgesia.

Disruption of the decussating pain fibers in the anterior white commissure produces bilateral analgesia. Destruction of the anterior white commissure at the cervical and the upper thoracic segments results in a jacket-type anesthesia, a common manifestation of syringomyelia (commissural syndrome), which involves the upper extremities and upper half of the trunk.

Syringomyelia (Figure 19.16) is a condition which results from cavitations around the central canal (hydromyelia). These cyst-like cavities are irregular in shape and contain yellowish fluid. They may communicate with the central canal or remain isolated, and are commonly seen in the cervical segments, extending to the thoracic segments. These cavities tend to extend anteriorly and laterally, destroying the anterior gray columns and the lateral spinothalamic tract. They may even extend to involve the lateral corticospinal tract. Clinical features associated with syringomyelia may appear before the age of thirty and are characterized by dissociated sensory loss, which refers to the bilateral loss of pain (analgesia) and thermal sensation (thermoanesthesia) while preserving other modalities of sensations. Failure to appreciate thermal and painful stimuli may result in cigarette burns and ulceration of the digits. Analgesia may follow signs of upper motor neuron paralysis after years. Wasting of the intrinsic muscles of the hand, as a result of the damage to the a motor neurons of C8–T1 spinal segments, may mimic features of ulnar nerve palsy.

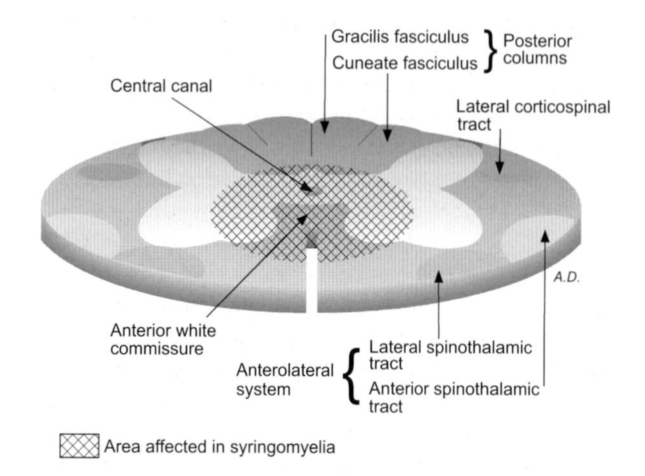

Figure 19.16 In this figure the lesion is caused by cavitation around the central canal disrupting the decussating fibers of the anterolateral system. Pain and temperature sensations are lost bilaterally

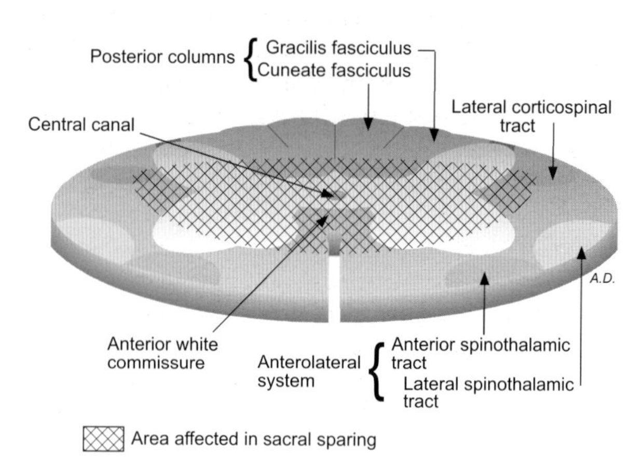

Figure 19.17 In this picture an expanding intramedullary lesion disrupted the pain fibers from the entire body with the exception of the sacral region, resulting in the phenomenon of sacral sparing

Horner's syndrome, the most frequently seen feature of syringomyelia, is the result of destruction of the intermediolateral column. Autonomic dysfunction together with loss of pain sensation may cause mutilation of the fingers.

Syringobulbia is a condition that affects structures in the medulla and pons, producing damage to the nucleus ambiguus, hypoglossal and vestibular nuclei, and the spinal trigeminal tract and nucleus. The symptoms and signs of this condition are characteristically ipsilateral and consist of dysphagia, nystagmus, analgesia and thermoanesthesia of the face, dysarthria, and hoarseness of voice. Patients may exhibit increased curvature of the vertebral column among other signs of poor physical condition.

Due to the somatotopic arrangement of pain fibers, an intramedullary tumor of the spinal cord may expand destroying the cervical, thoracic, and lumbar region, while sparing the sacral segments (sacral sparing). Patients with this type of tumor may exhibit pain and thermal anesthesia in the body without any detectable loss in the dermatomes of the sacral segments (Figure 19.7).

conveys the stimuli to the primary sensory cortex. During its course toward the thalamus it gives off numerous collaterals to the medullary reticular formation.

Pain transmission is not only modified by the corticospinal tract, but also by the descending serotoninergic and noradrenergic neurons and neurotransmitters, such as endorphins, adrenocorticotropic hormone, naloxane, and enkephalin. These polypeptides which are synthesized in the CNS maintain similar analgesic effect to morphine, develop tolerance, and may induce euphoria, mood changes, and possible respiratory depression. They are located in the spinal cord, periaqueductal gray matter, and limbic system. Ability of certain individuals to tolerate pain under extreme physical and emotional conditions as in a professional boxer's tolerance to repeated blows, or a soldier's emotional indifference to the pain produced by a shrapnel wound, is attributed to the dampening effect of the pyramidal tract fibers upon the internuncial neurons of the posterior gray columns and the analgesic role of endogenous opioids (β-endorphin). This endogenous opioid is released by activation of the tuberal (arcuate) nucleus of the hypothalamus and projects to the periaqueductal gray matter. Noxious stimuli applied to one part of the body may produce analgesia in pain stricken area elsewhere (diffuse noxious inhibitory control – DNIC). This phenomenon may have practical application in relief of pain associated with tooth decay by application of ice to the skin of the dorsum of the hand. In the same manner DNIC may explain the analgesic

Phantom limb pain or sensation may be experienced by individuals subsequent to amputation of a limb weeks, months, or years after a traumatic accident. Patients experience the feelings of movements in that particular limb. It commonly develops in patients in their third or fourth decades of life. Patients will be aware of the distal parts of the phantom limb more markedly than the proximal parts. Body parts that maintain larger areas in the sensory cortical homunculus may have more lasting phantom sensation than areas with smaller cortical representation. Phantom perception has some psychogenic basis, although activation of hypersensitive thalamic and cortical neurons may have a role in this phenomenon. The phantom limb may be abolished by excision of the sympathetic nerve fibers to the amputated part. Stump pain, as a result of a surgical procedure, may be attributed to development of neuroma or irritation by the scar tissue. Neuroma may occur as a result of regenerating nerve sprouts within the scar tissue of the amputated limb.

Low back pain ranks second only to common cold as a reason for primary care visits. It is the second leading cause of work absenteeism. Most low back pain is self-limiting and patients recover without specific treatment. Small numbers of cases may exhibit persistent or progressive neurologic dysfunction associated with radiographic findings. Low back pain may be associated with sciatica, in which the pain follows a dermatomal distribution, radiating to the buttock and down the leg often with numbness stretching into the lower leg and foot. Very few patients with low back pain exhibit urinary retention, saddle anesthesia, or bilateral neurological dysfunctions. Serious conditions that produce low back pain, such as abdominal aortic aneurysm, tumors, abscesses, emboli, and cauda equina syndrome, should be excluded before initiating any treatment. Several methods may be employed to confirm the positive findings of low back pain including: (1) reduced Schober index, which measures the distance between the L5 spinous process and a point 10 cm proximal to it in both erect and maximally flexed lumbar vertebrae (normal distance ranges 10–15 cm), (2) forward bending that produces reflex flexion of the knee on the affected side (Neri's sign), (3) well-leg, straight leg raising test. Hematological tests such as erythrocyte sedimentation rate (ESR) may be used to distinguish the medical etiology of back pain from the mechanically induced pain. Malignancies, systemic lupus erythematosus and rheumatoid diseases may show elevation of ESR. C-reactive protein may show an increase in individuals with back pain associated with tuberculosis and rheumatoid arthritis. Hematocrit value, rheumatoid factor, platelet count, level of serum alkaline phosphatase and uric acid may also be used to differentiate the medical causes of this condition. Computed tomography and MRI techniques and myelogram have proved to be useful in the evaluation of abnormalities of the lumbosacral spine. Back pain can occur as a result of herniation of the nucleus pulposus of the intervertebral disc. Most common sites of herniation are L4–L5 and L5–S1, accounting for 95% of the total disorders. One-third of all patients with disc herniation develop sciatica approximately 6–10 years after the onset of pain. Disc herniation at L4–L5 causes pain and numbness which radiate to the posterior thigh, anterolateral leg, medial foot, and great toe. Foot drop and weakness of dorsiflexion may be associated with this herniation.

Disc herniation at L5–S1 presents with pain and numbness of the posterior thigh and leg, posterolateral foot and lateral toes. Plantar flexion of the foot and toes may also be weakened. Crossed Lasegue's sign, which refers to radiation of pain down the back of the affected leg when the contralateral leg is passively elevated may also be used to confirm this condition. Pain may be felt along the course of the sciatic nerve as far distal as the calcaneal tendon (Valleix's pressure points). High lumbar herniation at the L2–L4 level is less frequent and may present with pain and numbness of the posterolateral or anterior thigh radiating to the anteromedial leg. There will also be weakness of knee extension and hyporeflexic knee joint. Osteoarthritis, spondylolisthesis spondylitis, chronic inflammatory disease that involves the sacroiliac and shoulder joints, Reiter's disease (manifests triad of urethritis, arthritis, and conjunctivitis) and vertebral osteomyelitis may also cause back pain.

Rest, back exercise, transcutaneous electrical nerve stimulation (TENS), acupuncture, and injection of corticosteroid may be of value. Acetaminophen is a standard treatment for this condition, although morphine and meperdine may be essential in individuals with osteoporosis or malignancies. Non-steroidal anti-inflammatory drugs (NSAIDS) may be used as analgesics and to counteract the inflammatory process. Surgery may be an option if other therapeutic measures fail.

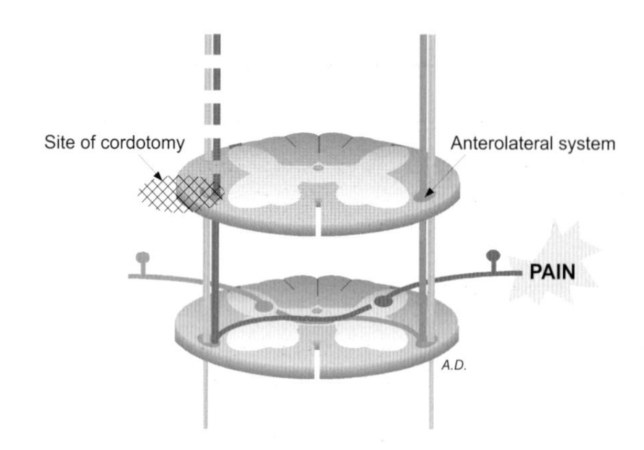

Figure 19.18 Cordotomy, a surgical procedure by which a pain pathway is excised, is depicted in this drawing. Pain and temperature sensations are lost on the contralateral side

effect of acupuncture produced by insertion of a needle into the skin. Transmission of pain may also be suppressed and profound analgesia may be achieved by the descending fibers of the periaqueductal gray matter that act upon the primary afferent neurons. Pain may be reduced or abolished from the thorax and abdomen by the injection of alcohol to a peripheral nerve, although complete blocking of a nerve may produces paralysis, a side-effect that renders this procedure of no clinical value.

Cortical sensations from the head

The trigeminal tracts or lemnisci (see also the trigeminal nerve) convey general sensation (GSA) from the head region to the thalamus. The first order neurons for these pathways are located in the sensory ganglia of the trigeminal, facial, glossopharyngeal, and vagus nerves. Depending upon the modality of sensation, the central processes of the first-order neurons enter the brainstem, establishing synapses with the sensory nuclei of the trigeminal nerve. Fibers carrying pain and temperature sensation synapse in the spinal trigeminal nucleus, fibers conveying pressure and touch sensation terminate in the principal sensory nucleus, whereas proprioceptive impulses are conveyed to the mesencephalic nucleus. The axons of the neurons of the spinal trigeminal nucleus and the principal sensory nuclei form the dorsal and ventral trigeminal tracts. The ventral trigeminal lemniscus (tract) is a crossed tract, which originates from the spinal trigeminal nucleus and the ventral part of the principal sensory nucleus. It ascends with the medial lemniscus and terminates in the ventral posteromedial (VPM) nucleus of the thalamus. The VPM conveys this information through the thalamocortical fibers to the postcentral gyrus (primary sensory cortex). These ascending thalamocortical fibers occupy the posterior limb of the internal capsule. Damage to the ventral trigeminal tract may occur in lesions of the pontine and midbrain tegmentum, resulting in contralateral loss of pain, thermal, and to some degree of tactile

Analgesia may also be attained, although very rarely, by cordotomy (Figure 19.18), a surgical technique that requires excision of the lateral spinothalamic tract in the lateral funiculus of the spinal cord. In this procedure, the dentate ligaments, which are the pial extensions that attach to the spinal cord at sites demarcating the dorsally located lateral corticospinal tract from the ventrally positioned lateral spinothalamic tract, are utilized. Cordotomy may be used in cases of malignant intractable pain, which is unresponsive to conventional analgesics. It is more useful in conditions where the pain is restricted to one side of the body or one limb.

In this procedure a needle is introduced into the anterolateral quadrant of the spinal cord through the intervertebral space between the atlas and axis. To prevent any serious adverse effects, an electrode may be passed through the needle. The path of the needle and the stimulating electrode are monitored via imaging techniques. Any motor response from the ipsilateral upper or lower extremity should be considered a serious sign that the needle has entered the lateral corticospinal tract and that it must be rerouted. Tingling sensations are usually felt as the needle enters the lateral spinothalamic tract, the site of cordotomy.

Bilateral cordotomy, which may be employed for more generalized and diffuse pain, carries the risk of causing incontinence and fatal respiratory impairment. Because of this, and to avoid damage to the phrenic nucleus, a surgical approach at the T1 spinal segment is preferable.

Reduction in the ability to appreciate pain and thermal sensations will be detected upon excision of the lateral spinothalamic tract via the surgical needle. The resultant anesthesia is not permanent and it may disappear after a period of one year (intact uncrossed pain fibers may take over). Since fibers of the lateral spinothalamic tract ascend 2–3 segments in the dorsolateral tract of Lissauer before entering the dorsal gray column, the anesthesia induced by this procedure will start 4–8 segments below the level of cordotomy and therefore the pain from the upper extremity may not be abolished completely.

Table 19.1 General sensory dysfunctions

Conditions	*Affected structures*	*Deficits*
Syringomyelia	1. Anterior commissure 2. Ventral gray column 3. Intermediolateral column	1. Bilateral loss of pain & temperature sensations 2. Bilateral or unilateral flaccid paralysis 3. Signs of Horner's syndrome
Syringobulbia	1. Vestibular nuclei 2. Nucleus ambiguus 3. Spinal trigeminal nucleus 4. Hypoglossal nucleus	1. Vertigo & nystagmus 2. dysphagia & hoarseness of voice 3. Analgesia & thermal anesthesia on ipsilateral face 4. Weakness of lingual muscles & dysarthria
Tabes dorsalis	1. Dorsal columns 2. Dorsal roots	1. Ataxia due to loss of proprioception 2. Bilateral discriminative touch
Friedreich's ataxia (spinocerebellar ataxia)	Clarke's nucleus, dorsal and ventral spinocerebellar tract and the dorsal columns	Gait ataxia involving all extremities due to bilateral impairment of proprioception and cerebellar deficits

sensation from the facial region. Axons which are derived from the dorsal part of the principal sensory nucleus, where the mandibular nerve fibers terminate, form the ipsilateral dorsal trigeminal tract (lemniscus). This pathway projects to the VPM nucleus, conveying this information to the postcentral gyrus via the thalamocortical fibers. Destruction of the dorsal trigeminal tract results in ipsilateral loss of pressure and tactile sensation. However, sensory loss will not be significant since these sensations are also carried by the ventral trigeminal tract.

Ascending subcortical pathways

Spinocerebellar tracts

The spinocerebellar tracts project to the spinocerebellum, carrying proprioceptive, stretch, and tactile sensations. They consist of the dorsal and ventral spinocerebellar, cuneocerebellar, and the rostral spinocerebellar tracts.

The dorsal spinocerebellar tract represents the axons (the largest in the entire CNS) of the Clarke's column which extends in the entire thoracic and upper two or three lumbar spinal segments. It is an ipsilateral tract, which ascends close to the surface of the dorsolateral funiculus and enters the spinocerebellum through the inferior cerebellar peduncle. This tract carries proprioceptive, tactile, and pressure impulses from individual muscles and joints of the lower limb and lower half of the trunk. The upper limb equivalent of this pathway is the cuneocerebellar tract, which originates from the accessory cuneate nucleus and terminates ipsilaterally in the spino- and pontocerebellum via the inferior cerebellar peduncle.

The ventral spinocerebellar tract is derived from the gray columns of the intermediate and the border cells of the ventral gray column in the thoracic, lumbar, and sacral segments. It is a crossed tract, conveying proprioceptive information from the entire lower extremity, particularly from the internuncial neurons that mediate flexor reflexes and provide input to the cerebellum. It also conveys information from collaterals of primary afferents emanating from the muscles and joints of the lower extremity. This pathway reaches the contralateral anterior lobe of the cerebellum through the superior cerebellar peduncle. The upper limb equivalent of this tract is the ipsilateral rostral spinocerebellar tract, which arises from lamina VII of the cervical enlargement and upper thoracic spinal segments. It enters the cerebellum through the inferior and superior cerebellar peduncles to be distributed to the anterior lobe of the cerebellum. The ventral and rostral spinocerebellar tracts also relay

Degeneration of the spinocerebellar tracts, dorsal columns, and Clarke's nucleus occur in Friedreich's (hereditary) ataxia, an autosomal recessive disease which is seen in childhood and before the end of puberty. It is characterized by gait ataxia (high-arched steppage), disturbances of speech, scoliosis (exaggerated lateral curvature of the vertebral column), and paralysis of the muscles of the lower extremities. This disease is often associated with hypertrophic cardiomyopathy. It also manifests areflexia, pes cavus (exaggerated longitudinal arch of the foot), nystagmus, vertigo, hearing impairment, visual deficits, and intention tremor.

information about neuronal activities of the descending motor pathways. Unconscious proprioception from the muscles of mastication, alveoli of the maxillary and mandibular bones, and extraocular muscles is delivered to the mesencephalic nucleus of the trigeminal nerve via the trigeminocerebellar tract, and later conveyed to the ipsilateral spinocerebellum via the superior cerebellar peduncle.

The spino-olivary tract originates from laminae of the spinal cord, ascends in the anterolateral funiculus, terminating in the accessory olivary nuclei. This pathway carries information from Golgi tendon organs, muscle spindles, and some other exteroceptive receptors. A corresponding dorsal spino-olivary tract also exists in the dorsal funiculus, which terminates in the inferior olivary nucleus.

The spinotectal tract (Figures 19.13 & 19.15) is composed of axons that are derived from lamina VII of the spinal gray matter. These axons cross the midline in the anterior white commissure and ascend on the contralateral side as a component of the anterolateral system (spinal lemniscus), terminating in the superior colliculus and the periaqueductal gray matter. The periaqueductal gray, which contains enkephalin and β endorphins, receives input from the frontal lobe and hypothalamus, projecting to the serotinergic tegmental nuclei of the medulla. Although the functional significance of this pathway is not clear, its role in modulating the transmission of pain, thermal, and tactile sensations is currently being studied.

Suggested reading

1. Aghabeigi B. The pathophysiology of pain. *Br Dent J* 1992;173:91–7
2. Apkarian AV, Hodge CJ. Primate spinothalamic pathways. I. A quantitative study of the cells of origin of the spinothalamic pathway. *J Comp Neurol* 1989;288: 447–73
3. Apkarian AV, Hodge CJ. Primate spinothalamic pathways. II. The cells of origin of the dorsolateral and ventral spinothalamic pathways. *J Comp Neurol* 1989;288:474–92
4. Apkarian AV, Hodge CJ. Primate spinothalamic pathways. III. Thalamic termination of the dorsolateral and ventral spinothalamic pathways. *J Comp Neurol* 1989;288:493–511
5. Basbaum AI, Fields HL. Endogenous pain control systems: brainstem spinal pathways and endorphin circuitry. *Annu Rev Neurosci* 1984;7:309–338
6. Cervero F. Visceral pain: mechanisms of peripheral and central sensitization. *Ann Med* 1995;27:235–9
7. Cummings JF, Petras JM. The origin of spinocerebellar pathways. I. The nucleus cervicalis centralis of cranial cervical cord. *J Comp Neurol* 1977;173:655–92
8. Devor M. Pain mechanisms. *Neuroscientist* 1996;2:233–44
9. Frymoyer IW. Back pain and sciatica. *N Engl J Med* 1988;318:291–300
10. Gybels JM, Sweet WH. *Neurosurgical Treatment of Persistent Pain. Physiological and Pathological Mechanisms of Human Pain*. Basel: Karger, 1989
11. Jánig W. Spinal visceral afferents, sympathetic nervous system and referred pain. In Vecchiet L, Albe-Fessard D, eds. *New Trends in Referred Pain and Hyperalgesia.* Elsevier, 1993:93–9
12. Jeanmonod D, Magnin M, Morel A. Thalamus and neurogenic pain: physiological, anatomical and clinical data. *Neuroreport* 1993;4:475–8
13. Logue V, Edwards MR. Syringomyelia and its surgical treatment. *J Neurol Neurosurg Psychiatry* 1981;44: 273–94
14. Moffie D. Spinothalamic fibers' pain conduction and cordotomy. *Clin Neurol Neurosurg* 1975;78:261–8
15. Osborn CE, Poppele RE. Sensory integration by the dorsal spinocerebellar tract circuitry. *Neuroscience* 1993;54:945–56
16. Schott GD. Visceral afferents: Their contribution to "sympathetic dependent" pain. *Brain* 1994;117:397–413
17. Sidall PJ, Cousins MJ. Neurobiology of pain. *Int Anesthesiol Clin* 1997;35:1–26
18. Stevens RT, London SM, Apkarian AV. Spinothalamocortical projections to the secondary somatosensory cortex (SII) in squirrel monkey. *Brain Res* 1993;631:241–6
19. Wall PD. The gate control theory of pain mechanisms. A re-examination and re-statement. *Brain* 1978;101: 1–18
20. Wall PD, Melzaack R. *Textbook of Pain*, 3rd edn. New York: Churchill Livingstone, 1994

Section V

Motor systems

Motor activities are regulated by profuse projections from higher cortical centers to lower neurons that are scattered in the brainstem and spinal cord. These projections reach the subcortical centers via fibers that descend in the corona radiata, internal capsule, and pyramids of the medulla. They exert influences upon the α motor neurons and motor neurons of the cranial nerves. A second subcortical (extrapyramidal) system that operates in conjunction with the cortical motor neurons modulates stereotyped motor activities, adjusts muscle tone, and suppresses cortical motor activities. Neurons of this system are located in the cerebral hemispheres, diencephalon, and brainstem, projecting to the thalamus and brainstem reticular formation. The extrapyramidal system has no direct output of its own; rather, it exerts its influences via the cerebral cortical motor neurons.

Motor neurons 20

Motor neurons are comprised of upper and lower neurons and their axons. Upper motor neurons (UMNs) are represented by the corticospinal and corticobulbar tracts, and also by the descending autonomic pathways which emanate from higher centers in the cerebral cortex and diencephalon. Lower motor neurons (LMNs) are concentrated in the α motor neurons of the ventral horn of the spinal cord and motor nuclei of cranial nerves. UMNs regulate the activities of the α motor neurons and as a result they also control the reflex activities at spinal and brainstem levels.

Upper motor neurons

Upper motor neurons (UMNs) are comprised of neurons of the primary motor, premotor, and supplementary motor cortices (Figure 20.1). The premotor cortex (Brodmann's area 6) occupies the frontal lobe rostral to the motor cortex, while the supplementary motor area lies on the medial side of the superior frontal gyrus, controlling patterning and initiation of motor activities. The primary motor cortex (Brodmann's area 4) is arranged somatotopically (Figure 20.2), in which parts of the contralateral body are represented in a distorted fashion (motor homunculus). In this homunculus, the thumb in particular, and the hand in general occupy larger areas than other parts of the body. This is attributed to the relative density of the neurons associated with movements of the digits. The thumb and index finger occupy an area near the face, while the body lies superior to the head between neurons of the shoulder and hip. The head occupies the lower part of the homunculus near the lateral cerebral (Sylvian) fissure, whereas the lower extremity is confined to the paracentral lobule, and lies rostral to the precuneus on the medial surface of the cerebral hemisphere. These neurons project to the spinal cord and brainstem via the corticospinal and the corticobulbar tracts, respectively. The corticospinal and corticobulbar tracts exert a wide range of influences on motor and reflex activities and sensory transmission. These effects are facilitated by their projections to the spinal and cranial motor neurons, as well as thalamic, striatal, and other brainstem neurons.

The corticospinal tract (CST) is a phylogenetically new pathway that exists in humans and other mammals. It continues to develop throughout the first 2 years of life. This principal motor pathway extends the entire length of the spinal cord, intermingling with fibers of the rubrospinal tract in the lateral funiculus of the spinal cord (Figure 20.4). It regulates the voluntary motor activities, especially the skilled and fine movements of the digits. It exerts facilitatory influences on the flexor neurons and inhibitory effects on neurons of antigravity muscles, and acts as an antagonistic pathway to the vestibulospinal tract. This major pathway is derived from the precentral gyrus (Brodmann's areas 4), premotor area (Brodmann's areas 6, 8) and the postcentral gyrus (Brodmann's areas 3, 1, 2). The largest fibers originate from the giant pyramidal cells of Betz. Primary motor cortex contributes 80% of the total fibers of this tract, premotor supplies 10%, and the remaining fibers originate from Brodmann's areas 3,1,2 and 5. Therefore, the corticospinal tract maintains motor

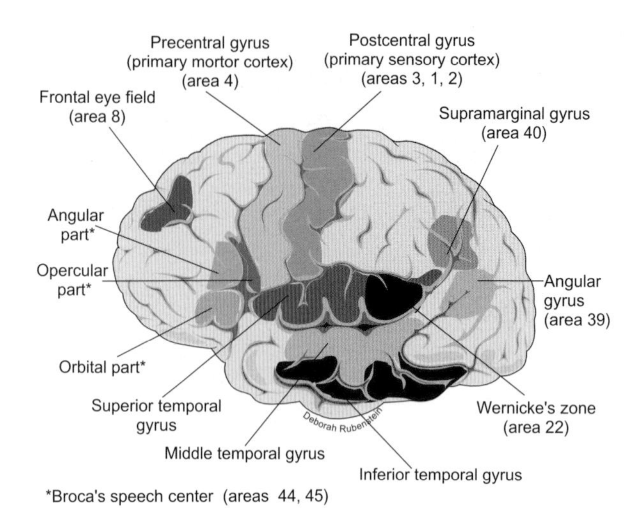

Figure 20.1 The motor and premotor cortices are illustrated

and sensory components. The motor component sends collaterals to the striatum, primary sensory cortex, thalamus, red nucleus, inferior olivary nucleus, pontine nuclei, and the reticular nuclei in the medulla and pons. It also conveys impulses to the motor nuclei of the cranial nerves, and to the α and γ spinal motor neurons. The sensory component, which is derived from the postcentral gyrus and paracentral lobule, sends substantial number of fibers to the striatum, gracilis and cuneatus nuclei, locus ceruleus, and the substantia gelatinosa, exerting inhibitory effects on γ-aminobutyric acid (GABA) neurons. This tract descends through the corona radiata (Figures 20.3, 20.4 & 20.5) and the posterior limb of the internal capsule, continuing through the basis pedunculi of the midbrain, basilar pons, and the pyramids of the ventral medulla (also known as the pyramidal tract). During its descending course it gives collaterals to the striatum.

As the corticospinal tract travels in the brainstem, it sends collateral fibers to the locus ceruleus, and to the red, pontine and inferior olivary nuclei. Other descending fibers may terminate in the dorsal

The corticospinal tract travels in close proximity to the oculomotor nerve in the midbrain, abducens nerve in the pons, and the hypoglossal nerve in the medulla. Therefore, a single lesion may involve both the corticospinal tract and a cranial nerve producing various forms of alternating hemiplegia that combines both of these motor deficits.

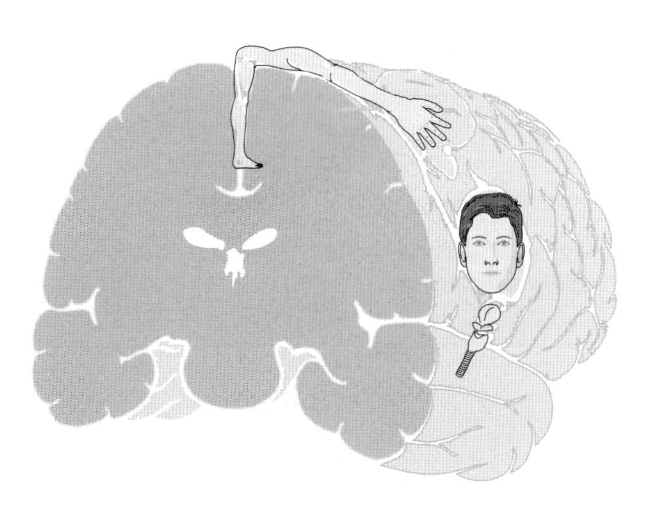

Figure 20.2 Diagram of the motor homunculus in which the entire body is represented in a distorted fashion on the motor cortex. The motor areas for the head and hand are considerably larger than other areas due to neuronal density

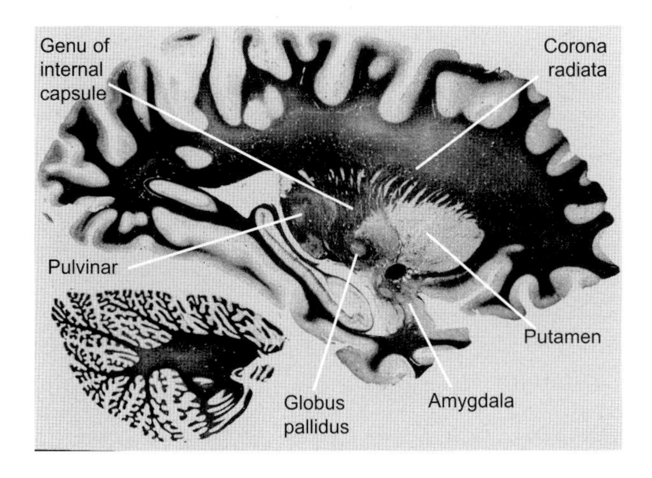

Figure 20.3 Photograph of a sagittal section of the brain showing the corona radiata and basal nuclei

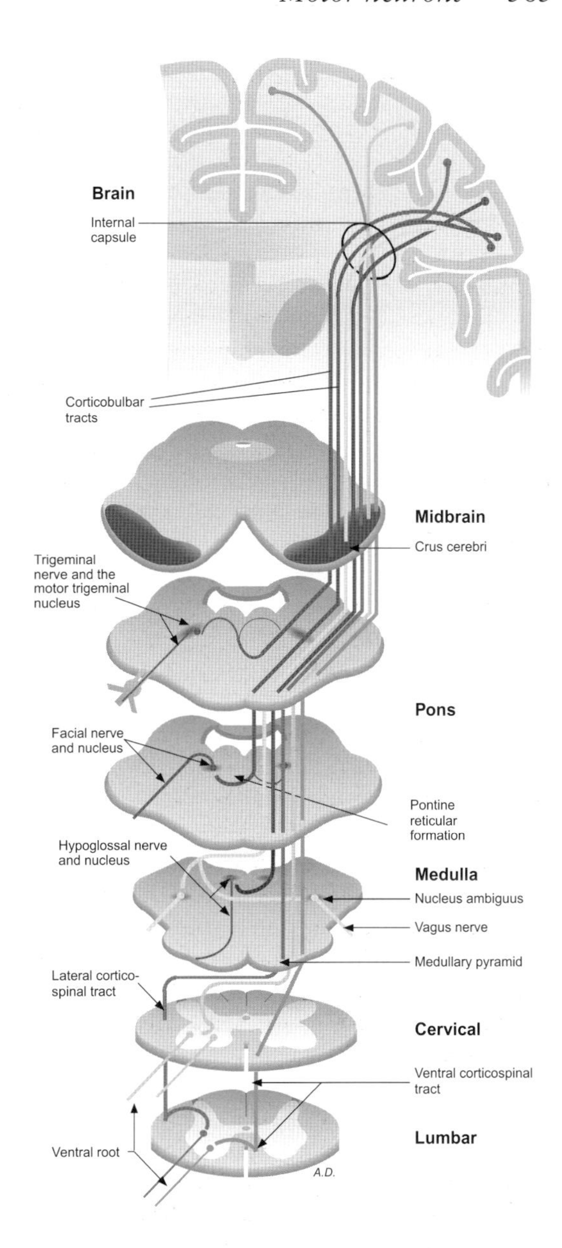

Figure 20.4 The corticospinal and corticobulbar tracts

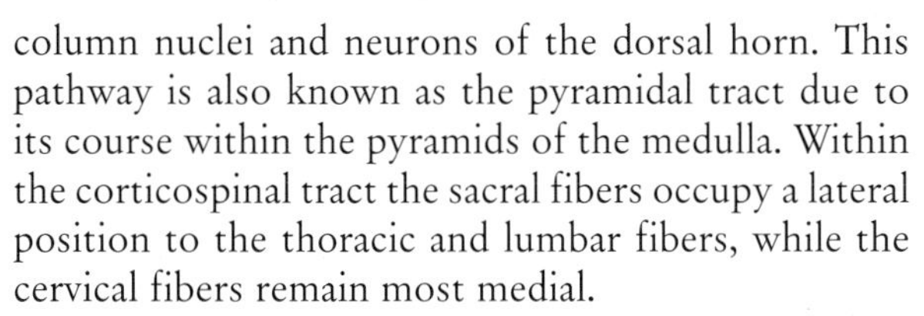

column nuclei and neurons of the dorsal horn. This pathway is also known as the pyramidal tract due to its course within the pyramids of the medulla. Within the corticospinal tract the sacral fibers occupy a lateral position to the thoracic and lumbar fibers, while the cervical fibers remain most medial.

Most of the fibers of the corticospinal tract (80–85%) decussate in the caudal part of the medulla, forming the lateral corticospinal tract (Figure 20.6). The remaining ipsilateral fibers form the anterior and anterolateral corticospinal tracts. The lateral cortico-spinal tract primarily terminates in the cervical segments (55%), other fibers end in the lumbosacral segments (25%), and the remaining terminate in the thoracic segments (20%). The ventral fibers of this tract are derived from the motor cortex, whereas the dorsal fibers originate from the neurons of the primary sensory cortex. Additionally, fibers that have a longer course assume superficial position relative to the shorter fibers. Sacral fibers remain the most lateral; the cervical fibers are medial, whereas the thoracic and lumbar fibers occupy an intermediate position within the lateral corticospinal tract.

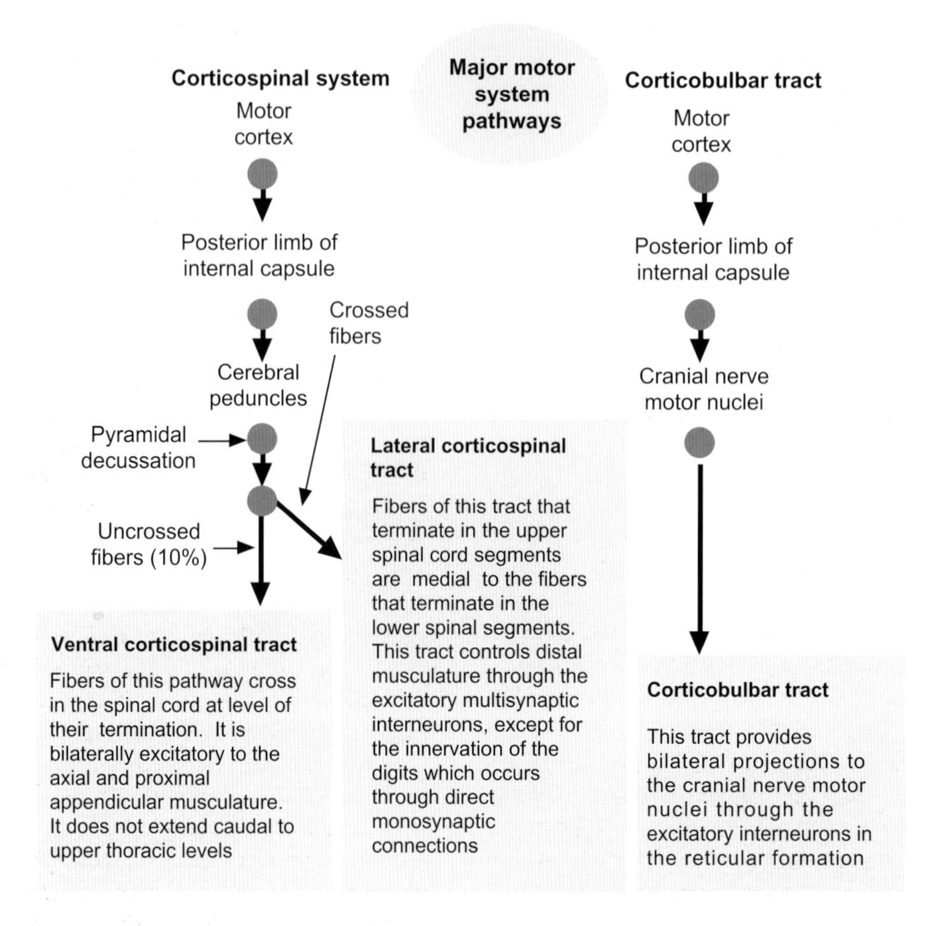

Figure 20.5 The course and termination of the corticospinal and corticobulbar tracts. The cerebral cortex also influences motor activity via the red nucleus and the rubrospinal tract

Some ipsilateral corticospinal fibers may form the anterolateral corticospinal tract that joins the lateral corticospinal tract in the lateral funiculus. Presence of these ipsilateral fibers may hasten the recovery from severe motor weakness that accompanies extensive cordotomy (surgical removal of the lateral spinothalamic tract) damaging the lateral corticospinal tract. Motor recovery may involve early return of flexion and adduction of the arm and extension and adduction of the thigh. Recovery of hand functions remains very poor. Compensatory activation of the ipsilateral fibers of the corticospinal tract and their role in motor recovery may be illustrated in individuals with progressive glioma who have suffered stroke or undergone surgical removal of one cerebral hemisphere. Ipsilateral corticospinal fibers may also play an important role in execution of certain motor activities upon command with one hand only in individuals with excised corpus callosum.

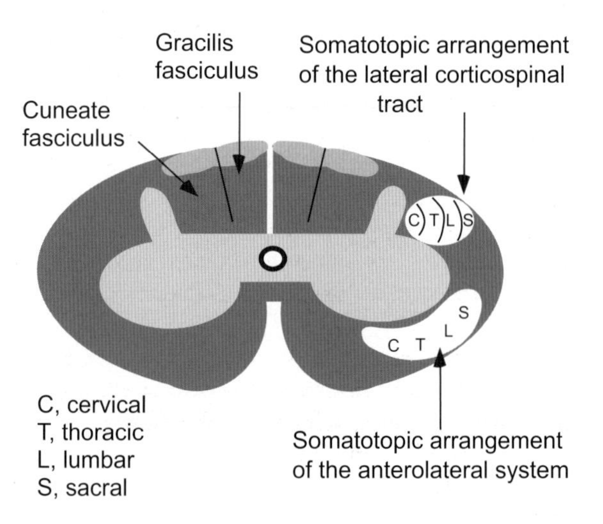

Figure 20.6 In this drawing the somatotopic arrangement of the corticospinal tract is shown. Note the similarity between the arrangement of the fibers of the corticospinal tract and the anterolateral system

Damage to the motor cortex or the corticospinal tract may occur as a result of cerebrovascular or degenerative lesions. These lesions produce a constellation of signs of upper motor neuron palsy (Figures 20.7 & 20.8). When the adjacent cranial nerve is also involved signs of alternating hemiplegia is produced including spastic (UMN) palsy on the contralateral side and cranial nerve dysfunction (LMN) on the same side.

Occlusive lesion of the basilar artery may produce bilateral flaccid palsy followed by spasticity, and possible coma. Bilateral occlusion or stenosis of the carotid arteries may produce quadriplegia and conjugate horizontal deviation of both eyes and eventually coma. Transient episodes of hemiparesis which may lead to hemi-paralysis, aphasia, light-headedness, carotidynia (pain over the occluded artery), mental confusion may also be detected in some individuals, although presence of good collateral circulation may prevent these deficits from appearing in young individuals.

Occlusion of the middle cerebral artery, which supplies the motor areas of the head and upper extremity, may produce signs of UMN palsy that vary considerably with the affected arterial branch. In this manner motor disorders may be restricted to the head and/or upper extremity, sparing the lower extremity. On the other hand, occlusion of the anterior cerebral artery, which provides blood supply to the medial surface of the brain and the paracentral lobule (motor center for the lower extremity) may result in motor dysfunctions restricted to the lower extremity. Irritation of the precentral gyrus by a trauma induced scar is a common culprit that causes an orderly and progressive or marching-type of convulsions (e.g. twitching) of the distal muscle groups that advances proximally (Jacksonian seizure).

Signs of UMN palsy (Figure 20.7 & 20.8) include spasticity, Babinski's sign, Gordon's leg and finger signs, Oppenheim's and Hoffman's signs, Chaddock's response, inverse myotactic reflex, clonus, hyperreflexia, and loss of superficial abdominal reflex.

(1) Spasticity refers to the gradual resistance to a passive movement of a joint that predominate in the antigravity muscles, which are the flexors of the upper extremity and the extensors of the lower extremity. It is observed within a matter of weeks following the insult.

(2) Babinski's sign (extensor plantar response) or upgoing toe reflex (Figure 20.7) is elicited by a gradual stimulation of the lateral plantar surface of the foot with a blunt object. Despite some variation this reflex is characterized by dorsiflexion of the big toe and abduction (fanning) of the other toes. It is part of a generalized nociceptive reflex which also includes flexion of the lower extremity. It is important to note that presence of this reflex may indicate upper motor neuron disease only in adults. Children under two years of age may normally exhibit this reflex due to incomplete development of the corticospinal tract. It may also be seen in unconscious individuals, during seizures and sleep, following anesthesia, in fatigue states, or as a result of intoxication or following an epileptic seizure. It is important to note that despite the appearance of upper motor neuron deficits, Babinski's sign may be absent in chronic paraplegic or hemiplegic patients and in individuals with paralyzed extensor hallucis longus subsequent to common peroneal nerve damage.

Babinski's reflex is absent in 25% of individuals with other UMN disorders. Equivocal Babinski may be confirmed by Gordon's leg, Gordon's finger, Oppenheim's and Hoffman's signs, Chaddock's response, inverse myotactic response, and clonus.

- Gordon's leg sign is a Babinski-like sign produced in response to squeezing of the calf muscle in an anteroposterior direction.
- Gordon's finger sign is characterized by extension of the flexed thumb and index finger in response to application of pressure on the pisiform bone.
- Oppenheim's sign is another Babinski-like response produced by a firm downward stroking of the medial surface of the tibia and tibialis anterior muscle.
- Hoffman's sign is characterized by clawing movement of the fingers, which includes a sharp and sudden flexion and adduction of the thumb and flexion of the fingers upon flicking the distal phalanx of the middle finger of the pronated hand while grasping it between the index and the thumb. Although this reflex may indicate an increase of muscle tone, it should not be used as the only diagnostic sign for upper motor neuron palsy.
- Chaddock response is a Babinski-like response, which is elicited by the application of a stimulus in a downward direction on the lateral malleolus and the lateral side of the foot extending toward the fifth toe. Since frontal lobe lesions generally produce inhibition of the extensor-plantar and excitation of the grasp reflexes, this response may be used to confirm frontal lobe damage.
- Inverse myotactic (clasp-knife) reflex is elicited by passive flexion of the spastic upper or lower extremity. It is mediated by the disynaptic

inhibitory action of the Golgi tendon organs on the agonist muscles and the excitatory influence upon the antagonist muscles.

• Clonus is a repetitive series of myotactic reflexes elicited by passive dorsiflexion of the foot at the ankle joint (ankle clonus) or by pulling the patella downward (patellar clonus). It may also be elicited by an abrupt and sustained application of passive stretch as in tapping the quadriceps tendon. In ankle clonus, forcible and rapid dorsiflexion of the foot stretches the gastrocnemius muscle, activating the muscle spindle and leading to the firing of the annulospiral (1a) fibers. Group 1a fibers convey the generated impulses from the annulospiral fibers to the α motor neurons of the spinal cord. These activated motor neurons produce contraction of the gastrocnemius muscle, resulting in plantar flexion of the foot. Plantar flexion of the foot at the ankle joint stretches the tibialis anterior muscle, stimulating the muscle spindle. This repetitive series of rapid dorsi- and plantar flexion of the foot continues for a prolonged period of time in UMN palsy. Ankle clonus seen in tetanus may also be a quickly exhaustible deficit. Patellar clonus (trepidation sign) exhibits clonic contraction of the quadriceps muscle in response to tapping the quadriceps tendon or upon a downward pull on the patella and stretching of the quadriceps muscle.

(3) Hyperreflexia denotes an increase in reflex activity (as in the patellar reflex) due to lowered threshold of the deep tendon reflexes. Presence of myotactic appendages may produce 'dead beat' rather than pendular termination.

(4) Superficial abdominal reflex is elicited in normal individuals by scratching the abdominal wall and observing the contraction of abdominal muscles and deviation of the umbilicus to the side of the stimulus. In UMN palsy the umbilicus remains stationary and the reflex will be absent.

Damage to the corticospinal tracts may also occur in ventromedian and dorsolateral syndromes, as well as in cerebral palsy.

• In ventromedian syndrome (cruciate hemiplegia) signs of spastic palsy are manifested in the ipsilateral upper extremity and the contralateral lower extremity (Figure 20.9). This syndrome results from a lesion at the site of pyramidal decussation that disrupt the fibers of the lower extremity superior (before) to the site of decussation, and the fibers to the upper extremity neurons inferior (after) to the level of the decussation.

• In dorsolateral syndrome, the lesion is at pyramidal decussation that may also involve the spinal accessory nucleus, producing paralysis of the trapezius and sternocleidomastoid muscles, and spastic palsy in the ipsilateral upper extremity and contralateral lower extremity.

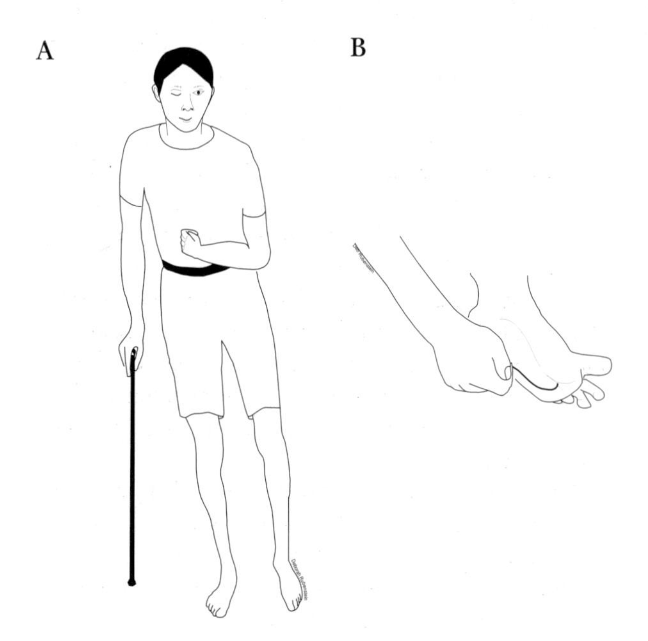

Figure 20.7 (A) depicts a patient with signs of upper motor neuron palsy. Note the characteristic gait and the flexed upper extremity. (B) is an illustration of Babinski's sign

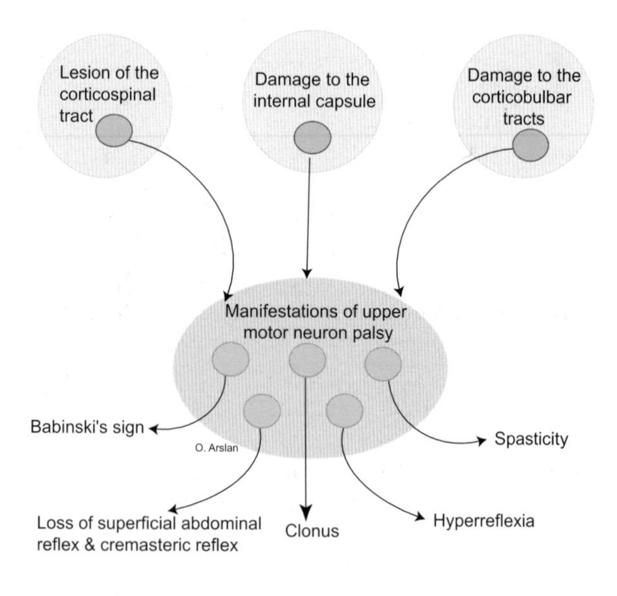

Figure 20.8 Summary of the deficits associated with the lesions of the corticospinal tract (upper motor neuron palsy). Spasticity and hyperreflexia are some of the characteristics of this type of dysfunction

Another important condition that exhibits UMN deficits is cerebral palsy. This refers to a group of prenatal or postnatal heterogeneous neuromotor dysfunctions that occur as a result of periventricular hemorrhage, malformation of the brain or asphyxia. It may be the result of hypoxia or ischemia following infection or trauma. Cerebral palsy may be of a progressive or non-progressive nature. Most infants have normal social and mental development, although one-third of these individuals exhibits varying degrees of mental retardation. Children with minimal cerebral palsy are clumsy and exhibit some visual deficits and signs of upper motor neuron palsy. Others may show dyskinesia, ataxia, or combinations of spastic and dyskinetic forms of cerebral palsy. Patients with the spastic form may develop hemiplegia, diplegia, diparesis, tetraplegia or tetraparesis and it is occasionally associated with epileptic seizures.

In the dyskinetic form of this disease, the least common type of cerebral palsy, abnormal involuntary movements such as athetosis, choreiform movements or a combination of choreoathetoid movements and speech disturbances may occur. The ataxic form presents hypotonia, unsteady gait, and loss of coordination.

In determining the integrity of the cerebral motor cortex in the newborn and infant certain maneuvers may be employed to elicit specific reflexes and signs. These infantile reflexes are primitive, present at birth or appear during the first six to eight months, and disappear later in life. They include Moro, grasp, rooting, incurvation, placing, crossed extension, parachute, Landau, and tonic neck reflexes.

- The Moro (startle) reflex is evoked by sound or by sudden extension of the neck, which may persist up to three years of age. It is characterized by symmetric abduction and extension of the arms, extension of the lower extremity, flexion of the thumb and great toe, abduction and extension of the digits. Asymmetry in these movements may indicate upper motor neuron palsy or brachial plexus injury.
- Forceful grasp of the infant to the examiner's finger upon stroking the palm of the infant's hand is known as grasp reflex.
- The rooting reflex is elicited by stroking the cheek or perioral region of the infant, which leads to opening of the mouth and turning of the head to the stimulated side.
- All normally developed newborns present with the incurvation (Galant's) reflex that exhibits arching of the legs and head of the infant on the same side in response to stroking the back of a prone-positioned infant. The arching in this reflex is produced by unilateral contraction of the muscular columns of the erector spinae muscle.
- In the placing reflex or reaction, passive placement of the dorsum of the infant's foot or hand into contact with the under surface of the edge of a table results in flexion of the hip and knee joints and relocalization of the stimulated foot on the top of the table. Its absence or presence on one side indicates a motor abnormality.
- Crossed extension reflex is characterized by flexion followed by extension and abduction of the contralateral leg upon stimulation of the plantar aspect of the foot.
- The 'parachute reflex' appears at the age of 6–9 months and commonly persists until 2 years of age. It is characterized by abduction and extension of the arms symmetrically upon tilting the head forward from an upright position while holding the infant by the waist. Asymmetry indicates maldevelopment of the motor system.
- The Landau reflex is characterized by extension of the neck and spine and to some degree the lower extremities upon suspending the infant in prone and horizontal positions.

Assessing the degree of brain development can be done via the tonic neck reflex, which is elicited by passively turning the head to one side. It is characterized by extension of the upper and lower extremities on the side to which the head is turned and flexion on the joints of the opposite side.

In the adult patient with upper motor neuron deficit, other reflexes may be elicited such as grasp reflex and mass reflex of Riddoch. Generally, these reflexes are expression of the release of intact structures from the inhibitory and regulatory effects of the supraspinal pathways.

- The adult grasp reflex is characterized by forcible flexion of the toes and the reluctance to release the grasp in response to stroking the center of the plantar surface of the foot by a blunt object. This reflex indicates a lesion of the frontal lobe on the contralateral side.
- The mass reflex of Riddoch, although normal in infants, may occur in adults under stressful conditions. It is characterized by abrupt evacuation of the bladder and sometimes bowel and sweating, accompanied by flexion of the lower extremity.

Fibers of this tract that originate from the post-central gyrus terminate in the proper sensory nucleus, modulating the transmission of sensory impulses. For the most part, the corticospinal tract acts upon the α motor neurons through interneurons of laminae VII and VIII of the spinal cord gray matter. Some fibers exert direct influence upon the α motor neurons in lamina IX, innervating muscles which are involved in fine and skilled movements. The remaining uncrossed fibers (10–15%) of the corticospinal tract descend in the ipsilateral anterior funiculus, forming the anterior corticospinal tract (Figures 20.4 & 20.5).

The fibers of the anterior corticospinal tract decussate in the anterior white commissure before terminating in the neurons of the intermediate gray and central parts of the ventral horn. This tract, which exerts a facilitatory effect on the cervical and axial muscles, may only extend to the upper thoracic segments.

The corticobulbar tract (CBT) conveys impulses from the cerebral cortex to the motor nuclei of the cranial nerves (Figures 20.4 & 20.5). It is primarily a bilateral tract with a contralateral predominance. The fibers of this tract project to the motor nuclei of the cranial nerves either directly via a monosynaptic route or indirectly (corticoreticulobulbar tract) through interneurons in the reticular formation. The motor nuclei of the trigeminal, facial, and hypoglossal nerves receive direct and bilateral corticobulbar fibers. The facial motor nucleus is unique with regard to its connections to the corticobulbar fibers. The part of the facial motor nucleus which provides innervation to the muscles around the mouth receives only contralateral corticobulbar projections, while the part of the nucleus that innervates the forehead and upper facial muscles receives bilateral corticobulbar fibers from both sides.

Unilateral lesion of the CBT is most likely to result in paralysis or paresis of the lower facial muscles on the opposite side, while preserving or minimally affecting all other motor nuclei. Interestingly, ipsilateral lesion of the corticobulbar tract in the pons produces glabellar hyperreflexia.

Bilateral destruction of the corticobulbar tract may result in a group of deficits collectively known as pseudobulbar (supranuclear) palsy. This form of dysfunction may occur in amyotrophic lateral sclerosis, which may also produce bulbar palsy, multiple cerebral infarcts, Alzheimer's disease, bilateral cortical atrophy, cerebral palsy, bilateral lesions of the frontal lobes, and in rostral brainstem damage. It is also seen in occlusive conditions of the posterior cerebral and basilar arteries. Pseudobulbar palsy is characterized by spastic paralysis of the lingual, pharyngeal, laryngeal, masticatory, and palatal muscles. Dysphagia due to spastic palsies of the pharyngeal and palatal muscles may lead to accumulation of food particles in the nasopharynx or trachea. Patients may develop a tendency to choke when swallowing. Therefore, it is advisable that patients be given a diet in the form of paste. Sensory deficits and cerebellar dysfunctions are absent. Stimulation of the posterior pharynx and palate results in strong gag sensation and forceful contraction of the palatal muscles and elevation of the soft palate (gag hyperreflexia). Severe dysarthria (speech difficulty) in the form of explosive speech with nasal sound is also seen as a result of weakness (paresis) of the facial and buccal muscles.

Articulation is not commonly impaired with regards to syntax, grammar or comprehension of the spoken language, but aphasia may be a presenting problem in some patients. The inability to voluntarily control breathing (e.g. coughing and blowing out) may cause death as a sequel to aspiration pneumonia. Due to disruption of the limbic system connections to the motor nuclei of the affected muscles, individuals with this disorder may also exhibit signs of emotional disturbances including expressionless gaze, sudden and unrestrained (pathological) crying or laughing or giggling in response to minimal stimuli. These emotional outbursts may last for a few minutes or until exhaustion. Some can not smile or express sadness by frowning. Patients with pseudobulbar palsy also exhibit jaw-jerk hyperreflexia, dementia and intellectual deterioration.

Bulbar (medullary) palsy is the result of the damage to the motor nuclei of the IX–XII cranial nerves. This condition may be seen in poliomyelitis, amyotrophic lateral sclerosis (Lou Gehrig's disease) and in patients suffering from occlusion of the vertebral artery. Patients may also exhibit dysphagia (difficulty in swallowing) and dysarthria (thickened speech with heavy nasal intonation). However, the gag and jaw-jerk reflexes remain depressed and the uvula deviates towards the intact side. There are no associated emotional or intellectual deficits.

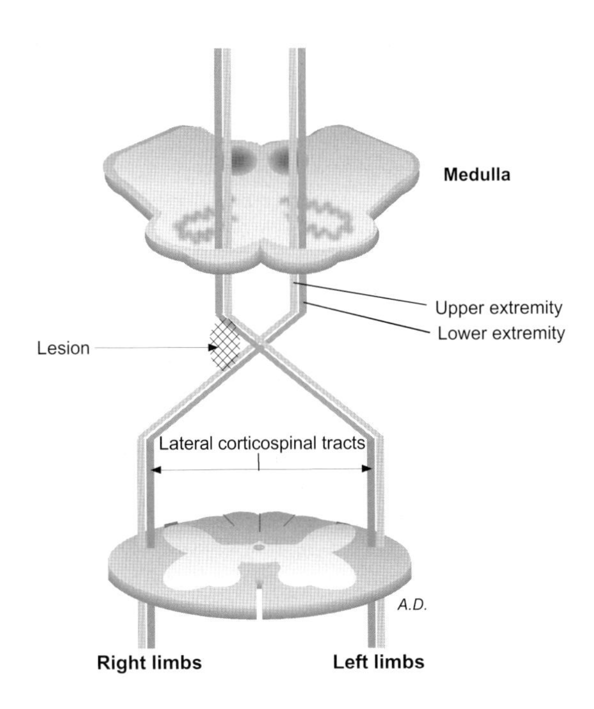

Figure 20.9 Lesion of the corticospinal tracts at the site of their decussation in the caudal medulla. Due to the somatotopic arrangements of these fibers, upper motor dysfunction will be seen in the ipsilateral upper extremity and contralateral lower extremity

Lower motor neurons

Lower motor neurons (LMN) are a group of neurons that lie in the spinal cord and brainstem. They are represented by the α and γ motor neurons of the ventral horn, intermediolateral column neurons of the lateral horn, and the motor nuclei of the cranial (III, IV, VI, & XII) nerves. These neurons provide general somatic, general visceral, and/or special visceral fibers, and they constitute the final common pathway for the motor system (Figure 20.10). The α and γ motor neurons occupy Rexed lamina IX that shows a somatotopic arrangement within the ventral horn of the spinal cord. In this arrangement, neurons that innervate the hand and forearm occupy a lateral position to those that supply the shoulder and trunk. Neurons for the foot and leg are lateral to the neurons of the hip and trunk. Additionally, the flexor neurons are dorsal to the extensor neurons and the adductor neurons are dorsal to the abductor neurons. The axons of α and γ motor neurons, which receive descending motor fibers either directly or indirectly, run through the ventral roots and then the spinal nerves, α and γ neurons. Both types of neurons are involved in voluntary movement via simultaneous contraction of the intrafusal and extrafusal muscles. These activities are mediated via the γ loop, which enables the brain to continuously monitor the state of contraction of the agonists and relaxation of the antagonists.

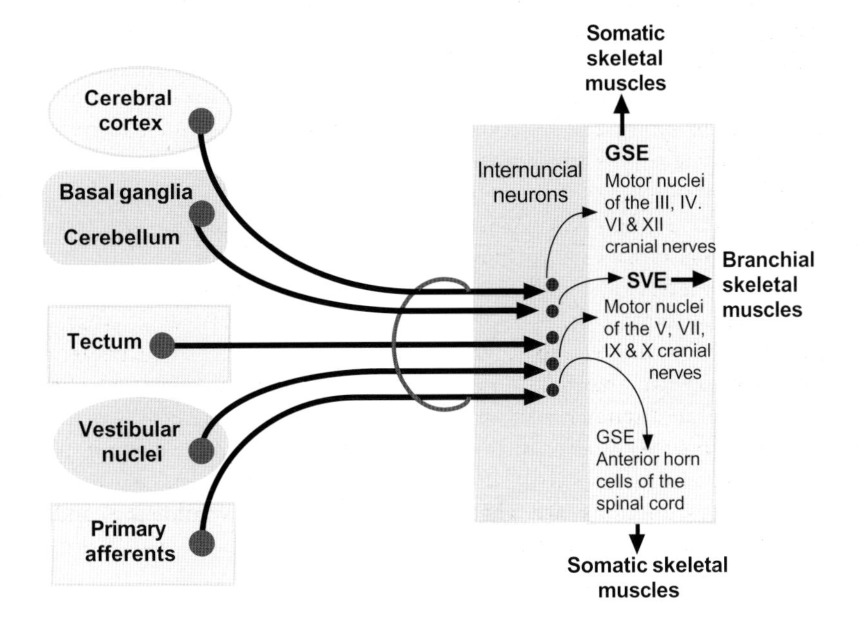

Figure 20.10 The somatic and visceral lower motor neurons within the spinal and cranial nerves. Note the general somatic efferent neurons are represented in the spinal α motor neurons and the motor nuclei of the oculomotor, trochlear, abducens, and hypoglossal nerves, whereas the special visceral motor neurons innervate muscles of branchial origin. As is seen in this diagram these lower motor neurons are influenced by activities of the cerebral cortex, basal nuclei, vestibular system, and tectum

Lesions of the LMNs may involve the anterior horn, ventral root, spinal nerve, or the motor nuclei of cranial nerves, producing signs of lower motor neuron palsy. These deficits consist of ipsilateral flaccid paralysis, atrophy (wasting of the affected muscles), fasciculation and areflexia. Fasciculation results from the action potentials in the terminal branches of the nerve fibers prior to their degeneration. They are visible through the skin and may be elicited by strenuous exercise.

These deficits are exemplified at the spinal level in poliomyelitis, a viral disease that affects the anterior horn neurons of the lumbar and cervical enlargements and, to a lesser degree, the brainstem. The lesions are random and the degeneration is followed by gliosis in the affected areas. Some neurons that escape the disease may increase in size and form relatively giant motor units. Usually the neurological symptoms appear late in the course of the disease including paralysis, areflexia, and atrophy of the affected muscles. Involuntary spasm of the hamstring muscles and pain (Kernig's sign) may be elicited in this disease when the patient's hip is flexed and the knee extended (Figure 20.11). The paralytic stage of this disease is characterized by tenderness and painful contractions of muscles, which may be elicited by movement or exposure to cold. Usually the appendicular muscles are affected, although muscles of mastication, respiration, and swallowing may also be involved. The facial muscles are less frequently involved. Paralysis of these muscles may be severe and generalized and the recovery remains slow and lasts for weeks. Urine retention may be observed, but it disappears in a short period of time. No sensory disturbances or deficits will be detected in this disease. In infants, bone growth will also be adversely affected. Recovery is possible if the respiratory muscles and diaphragm remain unaffected. The cerebrospinal fluid (CSF) shows an increase in protein and cellular components.

Combined UMN and LMN deficits may occur as a result of lesions in the cerebral cortex, brainstem, or spinal cord. Spinal lesions that produce combined motor deficits are manifested in amyotrophic lateral sclerosis (ALS; Lou Gehrig's disease), a relatively uncommon and progressive degenerative disease (Figure 20.12), which is characterized by bilateral degeneration of both the corticospinal and corticobulbar tracts (UMNs) and the anterior gray columns (LMNs). Signs of LMN palsy are seen at the levels of the involved segments, whereas UMN deficits are manifested contralaterally below the levels of the affected segments. ALS affects both the proximal and distal musculature and its distribution is asymmetrical and may remain confined to one side for a period of time. The spasticity may or may not precede the muscular atrophy. No sensory loss will be detected. This disease may be fatal when it involves the anterior horns of the cervical and thoracic segments of the spinal cord due to respiratory depression. Despite the diffuse nature of the destruction of central nervous system (CNS) motor neurons, mental faculties remain unaffected. Surprisingly, deep tendon reflexes show hyperactivity with muscle atrophy. However, motor impairment does not involve bladder and bowel functions or ocular movements. Involvement of the motor nuclei of the trigeminal, facial and hypoglossal nerves in bulbar ALS may result in facial asymmetry, dysarthria, and dysphagia.

Figure 20.11 Kernig's sign. Observe the examiner's attempt to extend the knee

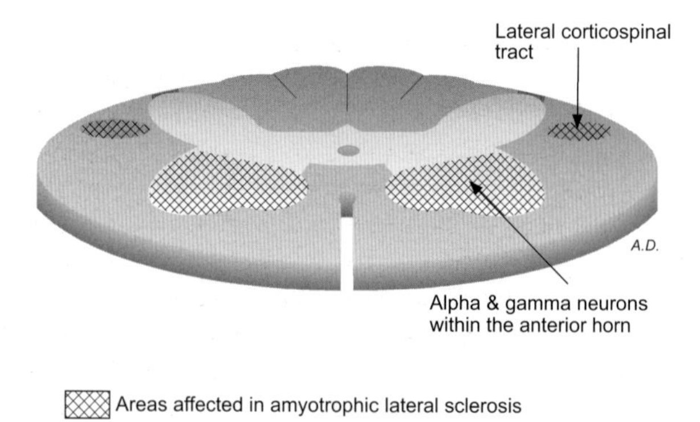

Figure 20.12 Section of the spinal cord shows bilateral degeneration of the lateral corticospinal tracts and anterior horns (hatched) in amyotrophic lateral sclerosis

Table 20.1 Motor dysfunctions

Conditions	*Affected structures*	*Deficits*
Ventromedian syndrome	Corticospinal tracts	Cruciate hemiplegia
Dorsolateral syndrome	Corticospinal tracts and spinal accessory nucleus	Cruciate hemiplegia and paralysis of the trapezius and sternocleidomastoid muscles
Cerebral palsy	Pre- or postnatal brain damage	Upper motor neuron palsy (hemiplegia or diplegia), mental retardation, and visual deficits
Pseudobulbar palsy	Corticobulbar tracts	Spastic palsy of the lingual, pharyngeal, masticatory, and palatal muscles and emotional outbursts
Bulbar palsy	Ambiguus and hypoglossal nuclei	Dysphagia, dysarthria, loss of gag and jaw reflexes and deviation of the uvula to the affected side
Poliomyelitis	α motor neurons of the lumbar and cervical spinal segments. Trigeminal motor nucleus may also be affected	Bilateral flaccid paralysis of the affected muscles, Kernig's sign. Masticatory and respiratory muscles may also be involved

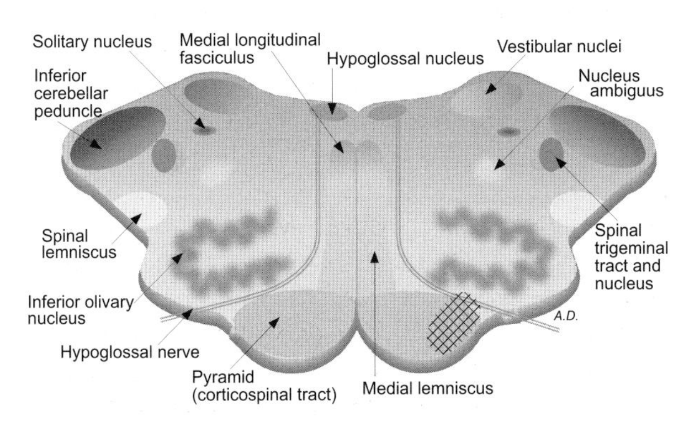

Figure 20.13 Image of the medulla with a lesion (hatched area) involving the corticospinal tract and hypoglossal nerve in inferior (hypoglossal) alternating hemiplegia

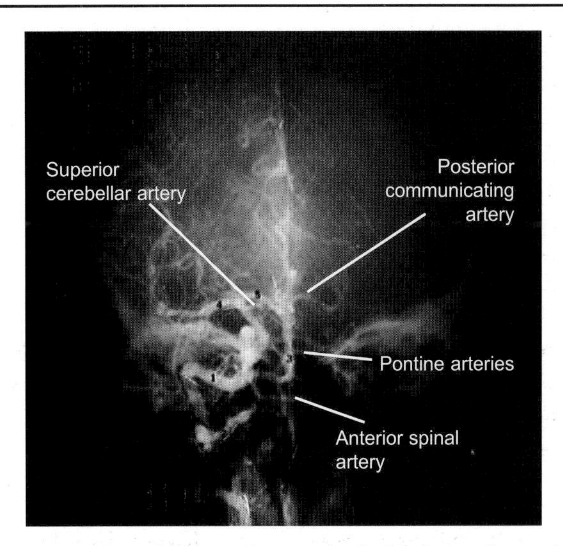

Figure 20.14 In this angiogram both the internal carotid artery and vertebrobasilar system are indicated. The connection between these two arterial systems is maintained by the posterior communicating artery, which is also shown in this angiogram. Note the pontine arteries that supply the corticobulbar and corticospinal tracts are involved in locked-in syndrome

The general somatic motor efferent (GSE) component of the oculomotor nucleus provides innervation to most of the extraocular muscles, whereas the GSE of the trochlear nucleus innervates the superior oblique muscle. GSE fibers in the abducens nucleus innervate the lateral rectus muscle, and also provide fibers to the medial rectus neurons of the oculomotor nucleus. These two sets of GSE fibers from the abducens nucleus become functional in lateral gaze, which produces abduction in the ipsilateral side and adduction in the opposite eye. GSE fibers of the hypoglossal nucleus innervate the lingual muscles.

General visceral motor neurons (GVE) consist of a spinal part contained in the intermediolateral columns of the thoracolumbar (presynaptic sympathetic) and sacral spinal (presynaptic parasympathetic) segments, and a cranial part contained in the motor nuclei of the III, VII, IX, and X cranial nerves (see autonomic nervous system). Presynaptic parasympathetic neurons form the Edinger–Westphal nucleus of the oculomotor nerve, superior salivatory nucleus of the facial nerve, inferior salivatory nucleus of the glossopharyngeal nerve, and dorsal motor nucleus of vagus.

Brainstem lesions may also produce combined upper and lower motor neuron dysfunctions, which are seen in the medulla, pons, and midbrain.

Medullary lesions produce combined upper and lower motor neuron deficits that are manifested in inferior alternating hemiplegia, Jackson's, and Schmidt's syndromes.

Inferior alternating hemiplegia (Figure 20.13) results from a single lesion involving both the pyramidal tract and the hypoglossal nerve on the same side. Flaccid paralysis of the ipsilateral lingual muscles and contralateral spastic paralysis of the upper and lower extremities characterize this condition.

Jackson's syndrome is caused by a lesion that damages the ambiguus and hypoglossal nuclei, as well as the pyramidal tract. It is characterized by all the features of Avellis syndrome with additional finding of ipsilateral paralysis of the lingual muscles.

Schmidt's syndrome is characterized by contralateral spastic hemiplegia, and ipsilateral paralysis of the muscles of the soft palate, vocal cords and the muscles supplied by the spinal accessory nerve (trapezius and sternocleidomastoid).

Pontine lesions also produce combined motor deficits that are seen in Foville's, Millard–Gubler, Raymond's, and locked-in syndromes.

Foville's syndrome results from a combined lesion of the abducens nucleus, facial nerve, and the corticospinal tract. In addition to ipsilateral facial palsy and contralateral hemiplegia, patients also exhibit lateral gaze palsy which is characterized by loss of conjugate horizontal movement to the side of the lesion and deviation of both eyes to the opposite side of the lesion.

Millard–Gubler syndrome (alternating facial hemiplegia) may be caused by intrapontine hemorrhage, thrombosis or tumor, which result in damage to the facial and abducens nerves, as well as the corticospinal tract. This syndrome is characterized by diplopia, facial palsy, and medial strabismus on the ipsilateral side, and hemiplegia on the contralateral side.

Raymond–Cestan syndrome (middle or abducens alternating hemiplegia) is another pontine disorder that results from a lesion of the abducens nerve and the corticospinal tract. Patients exhibit ipsilateral signs of abducens nerve palsy and contralateral hemiplegia (Figure 20.15).

Locked-in syndrome (pseudocoma) is secondary to sustained multiple vascular infarcts caused by occlusion of the pontine branches of the basilar artery (Figure 20.14). This syndrome is associated with bilateral destruction of the corticobulbar and corticospinal tracts in the basilar pons, while preserving the dorsal tegmentum. Patients remain conscious and aware of their surroundings (intact cognitive and affective functions), but suffer from mutism (inability to articulate) and quadriplegia. They are unresponsive to external stimuli and unable to move their trunk or limbs, while sparing movements of the eye and eyelid. This condition may mimic signs and symptoms of severe cases of Guillain–Barré syndrome and myasthenia gravis. Due to intactness of the midbrain and cerebrum, cerebral cortical activity remains normal despite the physical impairment, and partial recovery is possible. One of the important aspects of this condition is the ability of patients to comprehend the events surrounding them, and understanding peoples' comments and reactions.

In a similar fashion, midbrain lesions produce combined motor deficits, which are clearly seen in Weber's and interpeduncular syndromes.

Weber's syndrome (Figures 20.16 & 20.17) is a fairly common condition which results from damage to the oculomotor nerve and the corticospinal tract. It is characterized by signs of ipsilateral oculomotor palsy and contralateral hemiplegia. Hemorrhage due to a ruptured aneurysm of the posterior communicating artery, unilateral thrombosis of paramedian branches of the posterior cerebral artery, or metastatic lesions are common etiologies of this condition.

Interpeduncular syndrome is usually of vascular origin, which results from bilateral damage to the corticospinal tracts and the oculomotor nerves. It is characterized by quadriplegia, Babinski's sign, bilateral loss of convergence and vertical and medial eye movements, ptosis and mydriasis.

Special visceral efferent neurons (SVE) constitute the motor nuclei of the trigeminal, facial, glossopharyngeal, vagus, and accessory nerves that provide innervation to muscles of branchial arch origin. Neurons of the trigeminal motor nucleus innervate the muscles of mastication, tensor tympani, and tensor palatini. In a similar fashion, the facial motor nucleus (SVE) is responsible for the innervation of facial muscles of expression, stapedius, stylohyoid, and posterior belly of the digastric muscle. The nucleus ambiguus (SVE) sends fibers through the glossopharyngeal, vagus, and the accessory nerves to the muscles of pharynx, larynx, and soft palate.

Cerebral lesions may also produce combined upper and lower motor dysfunctions that are evident in decerebrate rigidity.

Decerebrate (γ) rigidity (Figure 20.18) may be attributed to the facilitatory effect of the pontinereticulospinal and the vestibulospinal tracts and to the lack of inhibitory effect of the cortically dependent medullary reticulospinal tract. It is caused by heightened activities of the γ neurons that result from transection of the midbrain at the intercollicular level or from a pathological condition anywhere between the midbrain and the first cervical spinal segment. It may occur spontaneously or in response to mild or noxious stimuli. Decerebrate rigidity may be seen as a result of vascular occlusion, compression, metabolic disorders such as hypoxia or hypoglycemia, or inflammatory process. The antigravity muscles show increased tone as an expression of the facilitation of the γ motor neurons, augmenting the firing rate of the muscle spindles and the α motor neurons. Patients with this condition exhibit a rigid posture in which the jaw is clenched, the extremities are fully extended, the feet are plantar flexed, the hands are extended or clenched, and the wrists are usually flexed and facing forward. Some patients may exhibit tonic neck reflex, which is characterized by flexion of the left elbow and extension of the right elbow upon passive turning of the head to the right side with no detectable movements of the lower extremity. Signs of decerebrate rigidity can be reduced or abolished by destruction of the vestibular nuclei, vestibulospinal tract or inner ear labyrinth. Segmental transection of the dorsal roots may prove beneficial in relieving the patient from signs of rigidity. This condition should not be confused with decorticate rigidity.

Decorticate rigidity (Figure 20.18) exhibits flexion of the elbow, wrist and fingers, extension of the thigh and leg, internal rotation of the leg, and plantar flexion of the foot. The arm and forearm are held tight to the body. It is seen in comatose individuals with a diffuse lesion of the cerebral hemispheres.

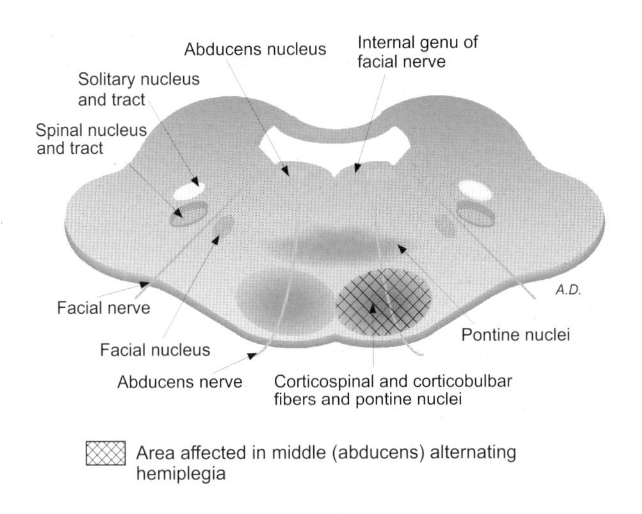

Figure 20.15 The caudal pons at the level of the abducens and facial nerves. Observe the lesion, which disrupts the abducens nerve and the corticospinal tract, producing Raymond's syndrome.

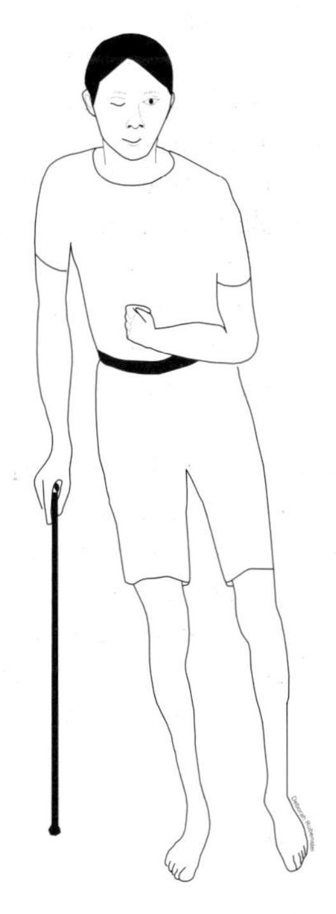

Figure 20.16 A patient with Weber's syndrome. Oculomotor dysfunctions (e.g. ptosis) on the right side and spastic palsy on the left half of the body are evident

Combined sensory and motor dysfunctions occur as a result of lesions involving the spinal cord, brainstem, or cerebral cortex.

Spinal lesions that produce combined system deficits may be seen in the anterior spinal artery, Brown–Sequard (spinal cord hemisection), cauda equina, and conus medullaris syndromes, transection of the spinal cord, ependymoma, multiple sclerosis, spina bifida, pellagra, subacute combined system degeneration, and syringomyelia.

Anterior spinal artery (Beck's) syndrome (Figure 20.19) is characterized by softening of the anterior two-thirds of the spinal cord and the resultant destruction of the spinothalamic and pyramidal tracts. It may be produced by occlusion of the anterior spinal artery as a result of thrombosis secondary to atherosclerosis or developing tumor. There is bilateral loss of pain and temperature sensations, which extend one or two segments below the level of the lesion. Tactile, vibratory, and position senses remain unaffected.

Brown-Séquard syndrome (hemisection of the spinal cord) is produced by disruption of the dorsal root, dorsal horn, ventral gray column, and the ascending and descending pathways on one side of the spinal cord (Figures 20.20 & 20.21). This condition may result from missile wounds, fracture–dislocation of the vertebrae or spinal arteriovenous malformation. It is characterized by ipsilateral loss of motor functions (upper and lower motor neurons) and discriminative tactile as well as joint sensations below the level of the lesion. All sensory and motor activities are impaired at the level of the affected segment(s) on the lesion side. Pain and temperature sensations are lost on the contralateral side extending one or two segments below the level of the lesion. Signs of upper motor neuron palsy (e.g. hyperactive tendon reflexes, spasticity, positive Babinski's sign, Hoffman's sign, clonus etc.) are seen ipsilaterally below the level of the lesion. Symptoms of lower motor neuron palsy, such as flaccidity, areflexia, and muscle atrophy may be observed at the level of the affected segment.

Cauda equina syndrome is a condition produced by compression of the lumbosacral roots as a result of a tumor or prolapse of the intervertebral disc below the first lumbar vertebra. Pain, as a first sign of this condition, is constant and radiates across the dermatomes of the involved roots, followed by flaccid paralysis (wasting of the tibialis anterior muscle), areflexia and saddle-shaped anesthesia in the gluteal region (when the lower sacral roots are involved). Pain may be felt in the lumbar area or in both legs. It may be aggravated by physical activity and coughing and, in contrast to the pain generated by herniated disc, it remains relatively unaffected by bed rest. Disturbance of micturition may appear late in the course of this condition, but no pyramidal signs may be observed. Numbness, tingling or burning sensation may be felt long before any objective findings of sensory loss are detected. Both knee and ankle jerk reflexes may be lost. Sensory loss over the anterior thigh may indicate compression of the upper lumbar roots, whereas compression of the lower lumbar roots produces sensory loss in the anterior and lateral leg.

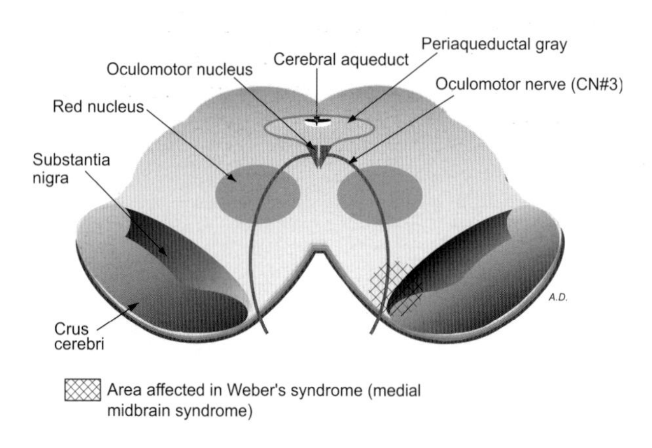

Figure 20.17 Drawing of the midbrain at the level of the superior colliculus. The hatched area represents the lesion site associated with Weber's syndrome, involving the corticospinal tract and the oculomotor nerve

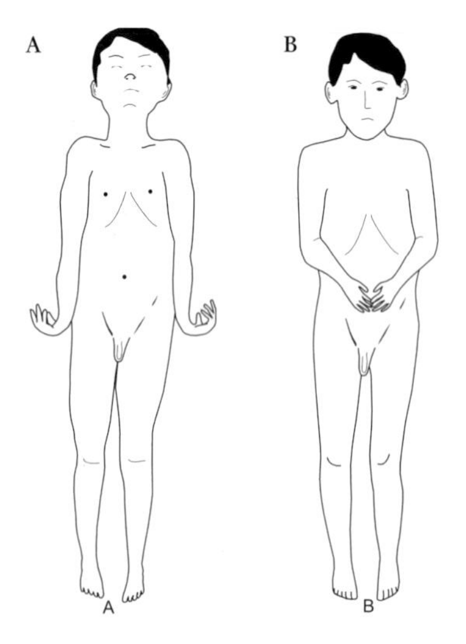

Figure 20.18 Diagrams illustrate decerebrate rigidity (A) and decorticate rigidity (B)

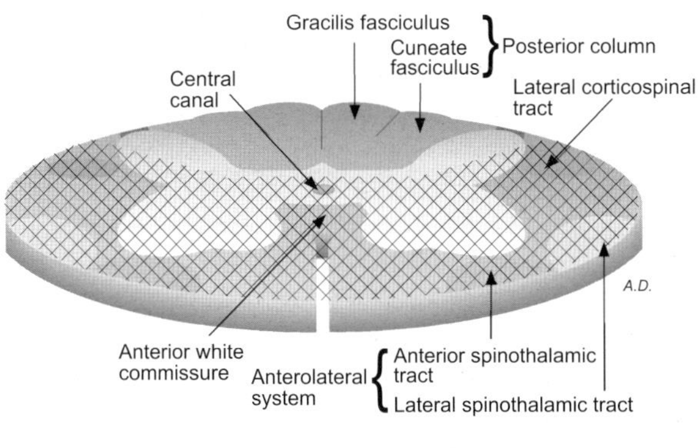

Figure 20.19 Drawing of a section of the spinal cord showing degeneration of its anterior two-thirds associated with occlusion of the anterior spinal artery

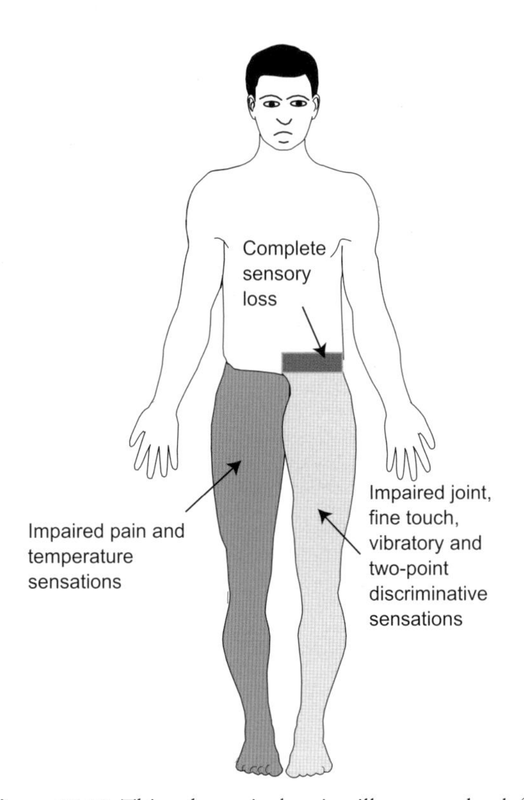

Figure 20.20 This schematic drawing illustrates the deficits associated with hemisection of the spinal cord at the left tenth thoracic spinal segment (Brown-Séquard syndrome)

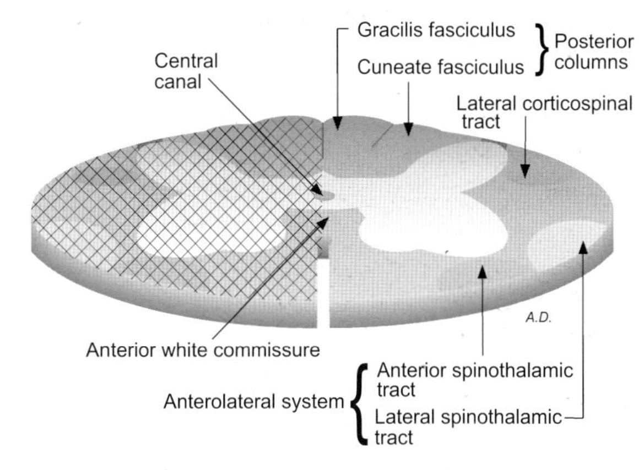

Figure 20.21 This spinal cord section illustrates degeneration in its right half. This type of lesion disrupts the fibers of the dorsal columns, and the lateral corticospinal, reticulospinal, spinothalamic, spinocerebellar and the rubrospinal tracts, as well as the ventral horn and the anterior commissure

Conus medullaris syndrome is a condition which may result from aortic aneurysm, prolapse of the L1/L2 intervertebral disc, nephrectomy and sympathectomy, and neoplasms that damage all or many of the sacral spinal segments. It is characterized by impaired bladder (atonic bladder that has lost the ability to initiate or inhibit urination) and bowel functions, as well as sexual disturbances (impotence). Sensory loss in the gluteal region but not in the posterior thigh or leg (extensive sensory overlap) will be detected with no major motor deficits, although weakness of movement in the feet may occur in an individual where the lesion has destroyed the anterior and posterior gray horns.

Table 20.2 Combined upper and lower motor neuron (LMN) dysfunctions

Conditions	*Affected structures*	*Deficits*
Amyotrophic lateral sclerosis	Ventral horns (LMN), lateral corticospinal, and corticobulbar tracts, as well as trigeminal, facial and hypoglossal motor nuclei	Bilateral flaccid and spastic palsies of the affected muscles, bilateral paralysis of the masticatory, pharyngeal, and laryngeal muscles
Bulbar amyotrophic lateral sclerosis	Trigeminal motor, facial motor, hypoglossal, and ambiguus nuclei	Bilateral atrophy of masticatory, facial, and lingual muscles. Cardiac arrhythmia may also be seen
Cestan–Chenais syndrome	Nucleus ambiguus, corticospinal tract, and descending autonomic fibers	Dysphagia, dysphonia, contralateral hemiplegia, and Horner's syndrome
Inferior alternating hemiplegia	Ipsilateral corticospinal tract and hypoglossal nerve	Flaccid palsy of the lingual muscles and contralateral hemiplegia
Jackson's syndrome	Ambiguus and hypoglossal nuclei, and the corticospinal tract	Signs of Avellis syndrome and ipsilateral paralysis of the lingual muscles
Schmidt's syndrome	Ambiguus nucleus and corticospinal tract	Ipsilateral paralysis of the vocal cord, palate, trapezius, and sternocleidomastoid muscles. Contralateral hemiplegia may also be observed
Foville's syndrome	Abducens nucleus, facial nerve, and corticospinal tract	Ipsilateral facial palsy, contralateral hemiparesis, and lateral gaze palsy
Millard–Gubler syndrome	Facial and abducens nerves, and corticospinal tract	Facial palsy, medial strabismus, diplopia, and contralateral hemiplegia
Locked-in syndrome	Corticospinal and cortico-bulbar tracts on both sides	Quadriplegia and inability to articulate. Eye movements remain intact
Raymond–Cestan syndrome	Abducens nerve and corticospinal tract, and sometimes the trigeminal nerve	Signs of abducens nerve palsy, contralateral hemiparesis, and sometimes sensory loss in the face
Interpeduncular syndrome	Corticospinal tracts and oculomotor nerves on both sides	Oculomotor palsy and quadriplegia
Superior alternating hemi-plegia (Weber's syndrome)	Oculomotor nerve and corticospinal tract	Ipsilateral oculomotor palsy and contralateral hemiplegia
Decerebrate rigidity	Corticospinal and medullary reticulospinal tracts	Rigid posture with extended arms, legs and feet and flexed hand
Decorticate rigidity	Cerebral cortex	Flexion of the elbow, wrist, and fingers, and extension of the hip, knee, and ankle joints

Transection of the spinal cord may result from shrapnel wounds, expanding intra- or extra-medullary tumors, trauma, or occlusion of the artery of lumbar enlargement. This condition produces manifestations of spinal shock, which includes loss of all somatic and visceral sensations, muscle tone, and reflexes below the level of the lesion. There may be a narrow zone of hyper-esthesia at the upper margin of the anesthetic region. Sensory loss may not corresponds to the level of the lesion since a lesion that starts from the periphery inward is most likely to initially affect the outermost fibers that carry pain and temperature sensations from the lower extremity. In contrast, a lesion that expands from the center in an outward direction disrupts these sensations in a reverse way. The recovery stages in humans may last more than six months and terminate by the appearance of Babinski's sign. Recovery phases are orderly, starting with the appearance of minimal reflexes, followed by flexor muscle spasm and alternate flexor–extensor muscle spasm and eventual extensor spasm. Exaggerated flexor reflex activity such as triple flexor reflex (flexion of the hip, knee, and thigh) in response to a mild stimulus may occur during the stage of flexor spasm. Hyperreflexia may be due to hypersensitivity of spinal motor neurons and interneurons and/or release from the inhibitory influences of the upper motor neurons. Bladder and bowel dysfunctions are the most distressing symptoms associated with this condition.

Ependymoma is a rare glioma of the ependymal lining of the ventricular system and the spinal cord. This tumor typically occurs in the vicinity of the fourth ventricle in the first two decades of life and in the sacral spinal segments in middle age. In some patients, the onset of the condition is associated with trauma. Symptoms usually include progressive muscle paresis with insidious onset. Incontinence, hydrocephalus, paraplegia, sensory disorders and priapism are also seen. Patients may be relieved of pain upon standing or walking and show a positive Lassegue test (pain and limitation of movements upon flexion of the thigh at the hip when the knee is extended). The average survival of these patients is 12 years.

Spina bifida (myeloschisis), as was explained earlier in relation to the development of the nervous system, is a congenital abnormality that results from failure of closure of the posterior neuropores. It is commonly seen in the lumbosacral region and in association with Arnold–Chiari malformation. This condition is classified into spina bifida occulta and spina bifida cystica. Spina bifida occulta is usually asymptomatic, unless a mass develops and compresses the spinal cord at the site of anomaly, producing bladder dysfunction and retardation of lower limb development. Back pain and sciatica may accompany this condition. Spina bifida cystica, seen with syringomyelia, is associated with herniation of the meninges (meningocele) or with herniation of spinal cord tissue, nerve roots and meninges (meningomyelocele). Bladder and bowel inconti-nence, sexual dysfunction, sensory and motor loss, or combined manifestations of spinal root and spinal cord lesions may be observed in this condition.

Multiple sclerosis (MS), also discussed earlier, is an autoimmune disease which produces demyelination in the CNS. Demyelination affects certain tracts with a predilection for the optic nerves, spinal cord, periventricular area, brainstem and cerebellum. The most common initial sign is visual deficits (monoc-ular visual loss and diplopia) with retrobulbar pain. Bilateral anterior internuclear ophthalmoplegia (MLF syndrome), resulting from plaque formation in the medial longitudinal fasciculus, is almost pathognomic for this disease. Due to ipsilateral adductor paresis, MS patients with MLF syndrome may exhibit diplopia on lateral gaze. However, it is important to note that unilateral anterior ophthalmo-plegia also occurs in young patients with systemic lupus and in older individuals with basilar artery infarctions. In a third of patients, bed-wetting and the frequent urge to urinate as a result of bladder dysfunction is observed initially. Involvement of the posterior columns of the spinal cord leads to sensory ataxia, resulting in disturbances of gait, equilibrium and a positive Romberg's sign.

Subacute combined system degeneration (Figure 20.22) is commonly associated with pernicious anemia and is due to the lack of intrinsic factor, which facilitates the absorption of vitamin B12. A megaloblastic anemia may be a late sign of this disease and may be detected a few weeks prior to death. This disease initially presents with acroparesthesia, an abnormal sensation (tingling and numbness) in the toes and later in the digits of the hand, progressing to an extent that the patient avoids hand-to-hand contact. Others may exhibit granular foot sensations (as in standing on gravel), or sciatica on the same side, becoming eventually bilateral. Lhermitte's sign, an electrical sensation that travels down the vertebral column to the lower extremity, is seen when the patient flexes his neck. A positive Romberg's sign, which refers to the exaggerated swaying of the trunk and body when the eyes are closed, is also seen in this condition. This sign is negative in cerebellar and labyrinthine diseases. Degeneration of the pyramidal tract, lateral and dorsal columns occur bilaterally, following no particular order. Disruption of the pyramidal tracts produces signs of upper motor neuron palsy, while degeneration of the dorsal columns produces loss of vibratory sense, sense of two-point discrimination, and locomotor ataxia. Sensory changes are attributed to degeneration of the lateral columns and the dorsal root fibers.

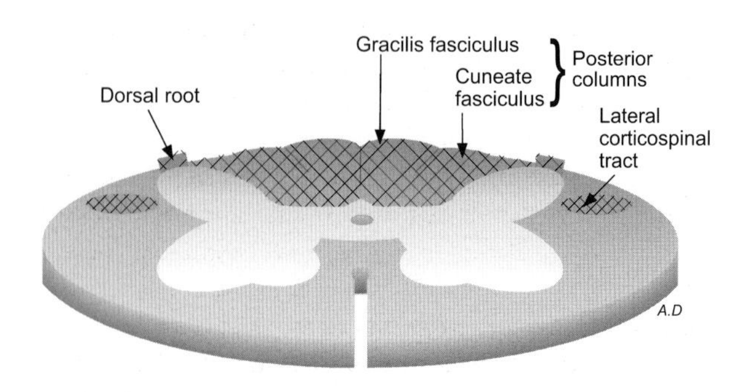

Figure 20.22 A section of the spinal cord demonstrating degeneration in the dorsal column-medial lemniscus pathway and the lateral corticospinal tracts on both sides in subacute combined degeneration. The dorsal root fibers may also be involved. These lesions produce combined sensory and motor deficits commonly associated with deficiency of vitamin B12

Pellagra is caused by a deficiency of nicotinic acid or tryptophan and is seen in individuals who are dependent upon corn as the main component of their diet. Acute encephalopathy, cog-wheel rigidity, locomotor ataxia, confusion, depression, polyneuropathy, subcortical motor signs, and paresis characterize this disease. Degeneration of the neurons of the anterior horn and cerebral cortex, dorsal columns, and spinal nerves is clearly evident in this disease.

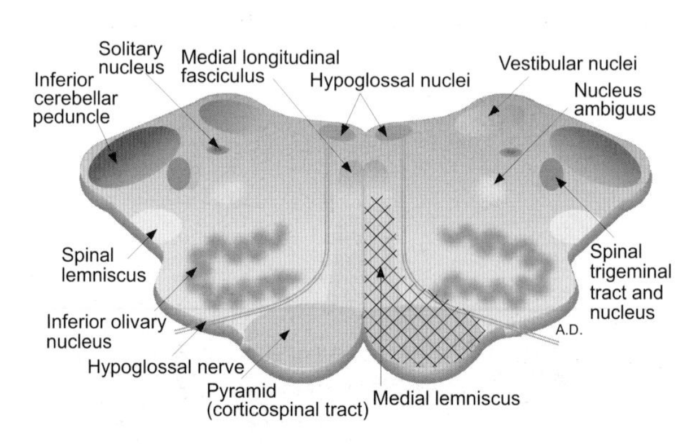

Figure 20.23 Medulla at midolivary level illustrating an ipsilateral lesion that involved the medial lemniscus, hypoglossal nerve, and corticospinal tract

Combined system dysfunctions may be associated with brainstem lesions affecting the medulla, pons or midbrain. Medullary lesions elicit signs of Avellis, medial medullary and lateral medullary syndromes.

Avellis syndrome is caused by a lesion that destroys the nucleus ambiguus, lateral spinothalamic, and corticospinal tracts. Patients present with ipsilateral paralysis of the soft palate and vocal cords, contralateral hemianesthesia, and spastic paralysis of the extremities. If the lesion is extensive enough to involve the sympathetic fibers, this condition is called the Cestan–Chenais syndrome.

Medial medullary syndrome (Figure 20.23) results from occlusion of the paramedian (bulbar) branches of the anterior spinal artery, producing degeneration of the medial lemniscus, hypoglossal nerve and the pyramidal fibers on the ipsilateral side. Signs and symptoms of this condition include contralateral spastic paralysis in the muscles of the extremities and ipsilateral flaccid palsy of the intrinsic lingual muscles. Paralysis of the lingual muscles leads to atrophy of the ipsilateral side of the tongue and its deviation to the affected side upon protrusion. Patients with this disease may also exhibit loss of positional sense, discriminative tactile sensation, vibratory sense, and two-point discrimination.

Lateral medullary (Wallenberg's) syndrome (Figures 20.24 & 20.25) results from occlusion of the posterior inferior cerebellar artery (Figure 20.29), producing degeneration of the spinal trigeminal tract and nucleus, spinal lemniscus, nucleus ambiguus, vestibular and cochlear nuclei, as well as the inferior cerebellar peduncle. It is characterized by dysphagia, dysphonia (hoarseness of voice), Horner's syndrome, and alternating hemianesthesia (anesthesia of the ipsilateral face and contralateral body). Ataxia, vertigo, loss of the gag reflex, and loss of taste sensation from the posterior third of the tongue are additional manifestations of this disease. Its onset is sudden, exhibiting violent rotatory vertigo that mimics labyrinthitis, nausea, vomiting, nystagmus, and hiccups. Consciousness and upper motor neurons remain intact and the prognosis for a complete recovery is good. Death may occur suddenly in some patients with lateral medullary syndrome due to possible overactivation of the dorsal motor nucleus of the vagus. Although recovery from this syndrome is generally good, large ischemia involving the inferior cerebellum may cause compression of the medulla and increased pressure of the posterior cranial fossa, dramatically changing the prognosis. Bilateral lateral medullary ischemia may result in an inability to initiate breathing during sleep and subsequent death (Ondine's curse).

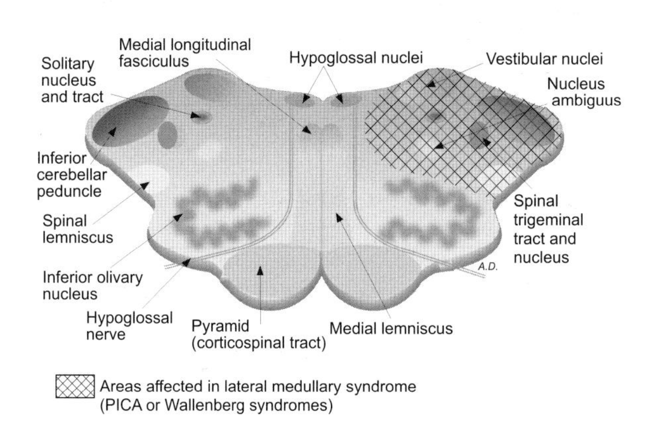

Figure 20.24 Drawing of the site of degeneration in the lateral medulla involving the anterolateral system, spinal trigeminal tract and nucleus, nucleus ambiguus, inferior cerebellar peduncle, and the descending autonomic fibers

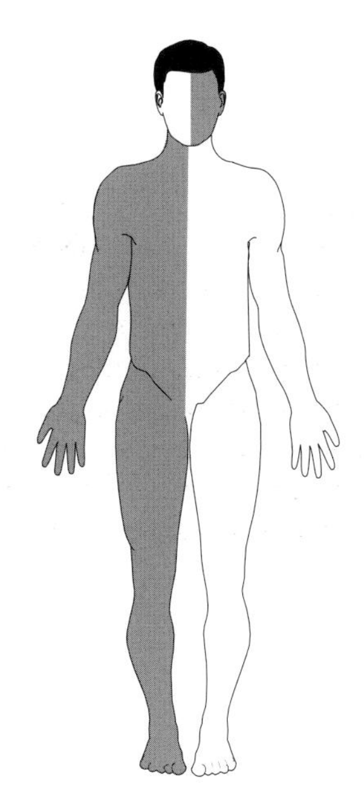

Figure 20.25 Schematic drawing of an individual with lateral medullary syndrome. The shaded areas exhibit loss of pain and temperature sensations

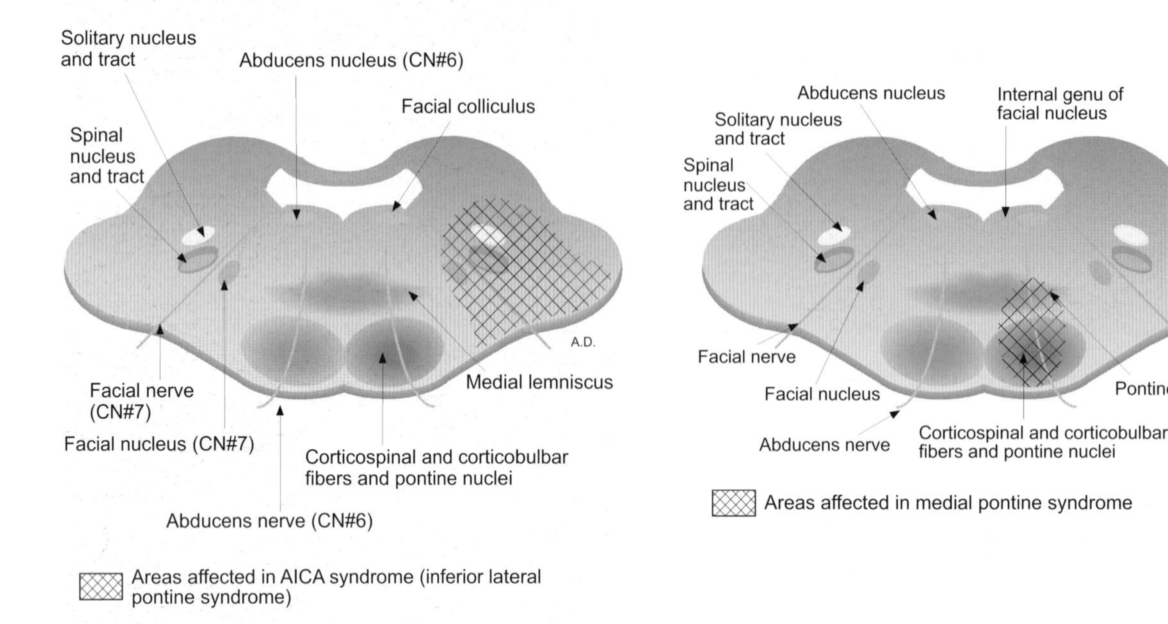

Figure 20.26 Areas of degeneration in the lateral pons associated with occlusion of the anterior inferior cerebellar artery

Figure 20.27 View of the medial pons showing degeneration of the corticospinal tract, abducens and facial nerves, medial lemniscus, and facial motor nucleus. This lesion is associated with occlusion of the paramedian and circumferential branches of the basilar artery

Pontine lesions that produce combined sensory and motor deficits are seen in the anterior inferior cerebellar artery and superior cerebellar artery syndromes, as well as medial pontine syndromes.

Anterior inferior cerebellar artery (inferior lateral pontine) syndrome (Figure 20.26), also known as AICA syndrome, results from occlusion of the anterior inferior cerebellar artery. It is characterized by anesthesia of the ipsilateral face and contralateral body, paralysis of the lower facial muscles, lateral gaze palsy, possible deafness, vertigo, nausea, and cerebellar deficits. Cerebellar deficits include ataxia, hypertonia, nystagmus, and intention tremor.

Superior cerebellar artery syndrome results from occlusion of the corresponding vessel, producing paralysis or paresis of conjugate eye movements, hemianesthesia on the contralateral side and signs of cerebellar dysfunctions. Loss of fine touch, vibratory and position senses are also seen. Impaired optokinetic nystagmus or skew deviation of the eyes may occur.

Medial pontine syndrome (Figure 20.27) is commonly associated with occlusion of the paramedian and short circumferential branches of the basilar artery, resulting in destruction of the abducens and facial nerves, medial lemniscus, corticospinal tract and possibly the medial longitudinal fasciculus (MLF). This syndrome combines signs of abducens alternating hemiplegia, contralateral ataxia and diplopia, which is caused by paralysis of the lateral rectus muscle. Involvement of the MLF may produce manifestations of anterior internuclear ophthalmoplegia, which is characterized by ipsilateral adductor paresis and contralateral monocular nystagmus.

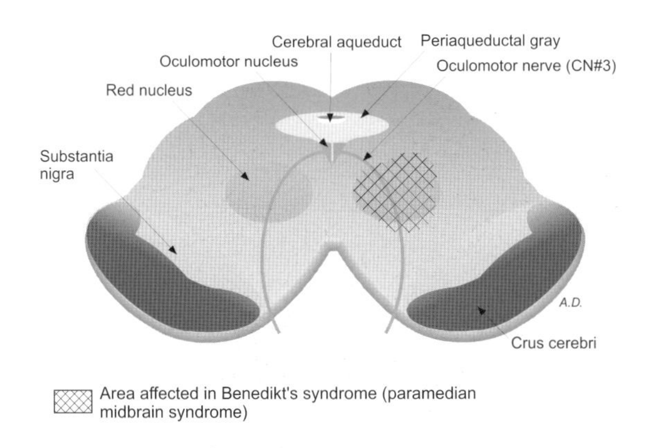

Figure 20.28 This drawing represents sites of degeneration in Benedikt's syndrome. Note that the oculomotor nerve, medial lemniscus, red nucleus, and possibly the spinal lemniscus are affected

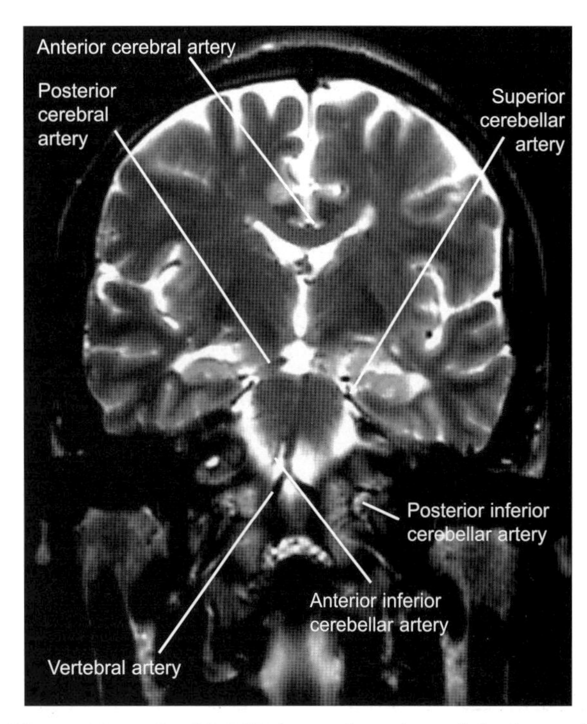

Figure 20.29 In this MR image, branches of the internal carotid artery and vertebrobasilar system are shown on the inferior surface of the brain. Anterior inferior cerebellar, superior cerebellar, and pontine branches of the basilar artery are clearly visible

Midbrain lesions that affect sensory and motor functions are primarily manifested in Benedikt's syndrome (Figure 20.28) which results from occlusion of the circumferential branches of the posterior cerebral artery (Figure 20.29) and subsequent destruction of the red nucleus, superior cerebellar peduncle, medial lemniscus, and spinothalamic tracts. It is characterized by signs of oculomotor palsy (ipsilateral ptosis, mydriasis, lateral strabismus, diplopia, loss of light and accommodation reflexes), intention tremor, ataxia chiefly in the upper extremity, nystagmus, vertigo, loss of pain and temperature, position and vibratory senses on the contralateral side. If the corticospinal tract is involved, hemiparesis without Babinski's sign and sensory loss may be detected in conjunction with upper motor neuron deficits.

An expanding tumor in the frontoparietal area (paracentral lobule) or a traumatic injury to the cerebral cortex may damage the sensory and motor cortical areas, producing deficits similar to that of the vascular lesions. Excessive and constant stimulation of the temporal lobe by a tumor may lead to generalized tonic–clonic seizures, disorder of consciousness, urinary incontinence, tongue biting (clonic phase), aura and fear sensation, followed by fatigue and confusion.

Cerebral lesions that produce combined motor and sensory disturbances may result from occlusion of the internal carotid or its cerebral branches, or capsular hemorrhage involving the lenticulostriate arteries.

Occlusion of the anterior cerebral artery distal to the anterior communicating branch may result in contralateral spastic palsy and sensory disturbances in the lower extremities and possibly mental confusion. Occlusive diseases affecting the callosomarginal branch, in the absence of an efficient anastomosis with the middle cerebral artery, may produce an infarction of the paracentral lobule leading to paralysis of the contralateral leg, cortical sensory loss, and urinary and bowel incontinence. It may also lead to an isolated infarction of the head of the caudate nucleus with no detectable deficits. Unilateral occlusion of the trunk of the anterior cerebral artery is less likely to result in a significant deficit since the anterior communicating artery allows blood to flow to the affected side from the intact side. Obstruction of the terminal portion of the anterior cerebral artery distal to the origin of the callosomarginal artery may not produce any clinically recognizable symptoms. Aneurysm of the pericallosal branch of the anterior cerebral artery (Figure 20.30) may produce headache, mental confusion, and seizure.

Occlusion of the middle cerebral artery, although infrequent, may be secondary to thrombi within the internal carotid artery (Figures 20.29, 20.30 & 20.31). Depending upon the site of occlusion in the dominant or non-dominant hemisphere and the affected arterial branch, a constellation of deficits which range from spastic palsy and supranuclear facial palsy to sensory disturbances, visual deficits, and speech disorders (aphasia) may be seen. These deficits are commonly seen in the morning, subsequently showing progression and fluctuation. In general, middle cerebral artery syndrome exhibits contralateral hemiplegia (especially of the face and arm), contralateral cortical hemianesthesia, contralateral hemianopsia; and aphasia, agraphia and alexia when the lesion is in the dominant hemisphere, apraxia and anosognosia when the non-dominant hemisphere is affected.

Occlusion of the middle cerebral artery proximal to the origin of the lenticulostriate branches, the most serious occlusion, produces extensive damage to the anterior and posterior limbs of the internal capsule. This type of occlusion results in sensory and motor deficits on the contralateral half of the face, hand, arm and leg, homonymous hemianopsia, and global aphasia if the dominant hemisphere is involved. A deep infarction of the middle cerebral artery at or above the described site may produce lacunar infarcts restricted to the internal capsule sparing the cerebral cortex. This selectivity may occur as a result of poor circulation in the lenticulostriate arteries and relatively good collateral circulation on the lateral surface of the brain.

When occlusion of the middle cerebral artery occurs distal to the lenticulostriate artery, it may produce an infarct of the opercular cortex. The cortical deficits may include global dysphasia, right–left disorientation, homonymous hemianopsia and graphic language disturbances when the dominant hemisphere is involved. Involvement of the non-dominant hemisphere may result in dyspraxia, lack of initiative, and a failure to acknowledge the presence of any neurological deficits. Contralateral hemiplegia and hemisensory loss which involves the upper extremity and face may also be observed.

Occlusion of the superior branch of the middle cerebral artery that supplies the precentral and postcentral gyri as well as Broca's center may produce contralateral hemiplegia and hemisensory loss in the upper extremity and face, and expressive (Broca's) aphasia. Infarcts associated with the inferior division of the middle cerebral artery result in contralateral homonymous hemianopsia, disorders of spatial thoughts, and stereognosis. It is also associated with failure to recognize the contralateral extremity, anosognosia (loss of ability to recognize bodily deficits associated with lesions of the non-dominant hemisphere) and receptive aphasia upon involvement of the dominant hemisphere. The rolandic branch of the middle cerebral artery may also be occluded, producing sensory loss and motor paralysis of the contralateral arm and lower face. Infarcts due to occlusion of the ascending frontal branch in the dominant hemisphere may result in Broca's aphasia, supranuclear facial palsy, and tonic deviation of the eyes to the left side of the lesion.

Occlusion of the angular artery produces receptive (Wernicke's) or global aphasia and apractognosia. Rupture of the anterior choroidal branch of the internal carotid artery produces hemianesthesia and hemiplegia on the contralateral side and homonymous hemianopsia or superior quadranopsia without cerebral edema. The relative small size of the anterior choroidal artery and its long subarachnoid course may account for its susceptibility to thrombosis. Absence of cerebral edema may be an important feature of this condition compared to occlusion of the middle cerebral artery.

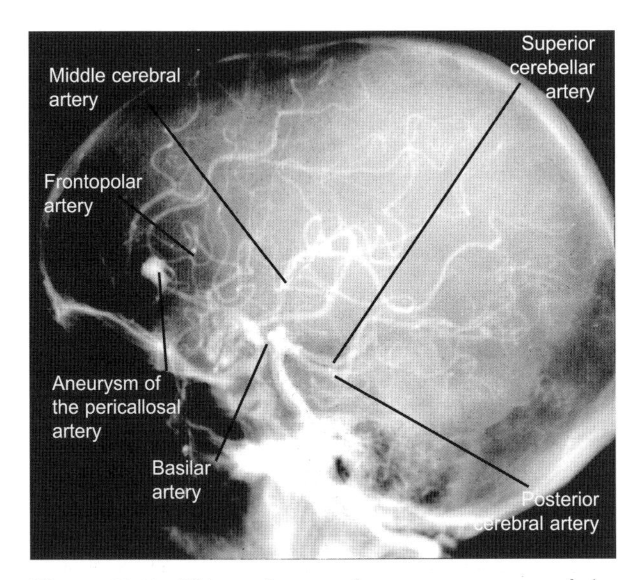

Figure 20.30 This angiogram shows an aneurysm of the pericallosal branch of the anterior cerebral artery. Note the middle cerebral artery and branches of the basilar artery

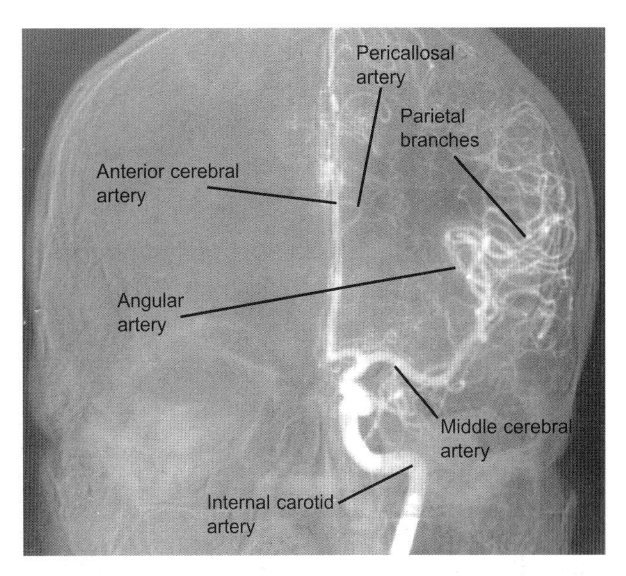

Figure 20.31 This internal carotid artery angiogram is shown as an overview of the distribution of the middle cerebral and anterior cerebral arteries

Table 20.3 Combined motor and sensory dysfunctions

Conditions	*Affected structures*	*Deficits*
Beck's syndrome	Spinothalamic and corticospinal tracts	Bilateral anesthesia and upper motor neuron palsy
Brown-Séquard syndrome	Anterolateral system, dorsal column, corticospinal tract, descending autonomic fibers, and neurons of the ventral and dorsal horns	Contralateral hemianesthesia of the body, ipsilateral loss of fine touch and proprioception, ipsilateral spastic palsy, signs of Horner's syndrome, and loss of all sensations, flaccid palsy, atrophy & areflexia on the side of the affected segment
Cauda equina syndrome	Lumbosacral roots	Pain in the involved dermatomes and flaccid palsy
Conus medullaris syndrome	Sacral segments	Bladder, bowel, and sexual disturbances, and anesthesia in the gluteal region
Ependymoma	Ependyma of sacral segments of spinal cord and fourth ventricle	Muscle paresis, incontinence, hydrocephalus, and paraplegia
Multiple sclerosis	Dorsal columns, corticospinal tracts, optic nerve, lateral spinothalamic tract, and ventral horn of the spinal cord	Locomotor ataxia, upper motor neuron palsy, paraesthesia or anesthesia, visual deficits, and incontinence
Spina bifida cystica (myeloschisis)	Lumbosacral roots and/or segments	Incontinence, sexual dysfunction, sensory and motor loss, and back pain
Pellagra	Anterior horn, cortical neurons, dorsal columns, and spinal nerves	Locomotor ataxia, cog-wheel rigidity, paresis, and subcortical motor signs
Refsum's disease	Dorsal columns, ventral horn, and spinal nerves	Locomotor ataxia, polyneuritis, visual deficits, and deafness
Subacute combined system degeneration	Dorsal columns and corticospinal tract	Impairment of proprioception and vibratory sensations, and signs of upper motor neuron palsy
Transection of the spinal cord	All ascending and descending tracts, dorsal and ventral horns	Loss of all sensations, muscle tone, and reflexes below the level of transection followed by stages of recovery
Avellis syndrome	Nucleus ambiguus, lateral spino-thalamic, and corticospinal tracts	Dysphagia, dysphonia, contralateral hemianesthesia, and hemiparesis
Medial medullary syndrome	Hypoglossal nucleus, medial lemniscus and corticospinal tract	Ipsilateral atrophy of lingual muscles, contralateral loss of proprioception and signs of spastic paralysis
Lateral medullary (Wallenberg's or PICA) syndrome	Nucleus ambiguus, spinal trigeminal nucleus, anterolateral system, descending sympathetic fibers	Dysphagia, ipsilateral loss of pain and temperature sensations from the face, contralateral loss of pain and temperature sensations from the body, Horner's syndrome
Anterior inferior cerebellar artery (AICA) syndrome	Facial motor nucleus, spinal lemniscus, spinal trigeminal tract, middle cerebellar peduncle, and vestibular nuclei	Ipsilateral facial paralysis, contralateral hemianesthesia of the body, ipsilateral hemianesthesia of face, and signs of cerebellar dysfunctions
Superior cerebellar artery syndrome	Lateral gaze center, anterolateral system, and middle cerebellar peduncle	Lateral gaze palsy, cerebellar dysfunctions, and contralateral hemianesthesia
Benedikt's syndrome	Superior cerebellar peduncle, oculomotor nerve, red nucleus, medial, spinal lemnisci, and possibly the corticospinal tract	Ataxia, oculomotor palsy, contralateral loss of position sense and hemianesthesia, and signs of cerebellar dysfunction. This syndrome may also exhibit hemiparesis

Suggested reading

1. Brandt J, Butters N. The neuropsychology of Huntington's disease. *Trends Neurosci* 1986;9:118–20
2. Chandler JD. Propranolol treatment of akathesia in Tourette's syndrome. *J Am Acad Child Adolesc Psychiatry* 1990;29:475–7
3. Chevalier G, Deniau JM. Disinhibition as a basic process in the expression of striatal functions. *Trends Neurosci* 1990;13:277–80
4. Fahn S. Fetal-tissue transplantation in Parkinson's disease. *N Engl J Med* 1992;327:1589–90
5. Groenewegen HJ, Berendse HW, Haber SN. Organization of the output of the ventral striatopallidal system in the rat: ventral pallidal afferents. *Neuroscience* 1993;57:113–42
6. Guridi J, Obeso JA. The role of the subthalamic nucleus in the origin of hemiballism and parkinsonism: new surgical perspectives. *Adv Neurol* 1997;74:235–47
7. Hallett M. Physiology of basal ganglia disorders: an overview. *Can J Neurol Sci* 1993;20:177–83
8. Hattori T. Conceptual history of the nigrostriatal dopamine system. *Neurosci Res* 1993;16:239–62
9. Lindvall O. Neural transplantation in Parkinson's disease. In Dunnett SB, Björklund A, eds. *Functional Neural Transplantation*. New York: Raven Press, 1994:101–38
10. Lindvall O, Sawle G, Widner H, *et al*. Evidence for long term survival and function of dopaminergic grafts in progressive Parkinson's disease. *Ann Neurol* 1994;35: 172–80
11. Monzillo PH, Sanvito WL, Da Costa AR. Cluster-tic syndrome: report of five new Cases. *Arq Neuropsiquiatr* 2000;58:518–21
12. Murphy TK, Goodman WK, Ayoub EM, *et al*. On defining Sydenham's chorea: where do we draw the line? *Biol Psychiatry* 2000;47:851–7
13. Oerlemans WG, Moll LC. Non-ketotic hyperglycemia in a young woman, presenting as hemiballism-hemichorea. *Acta Neurol Scand* 1999;100:411–14
14. Parent A, Hazrati LN. Anatomical aspects of information processing in primate basal ganglia. *Trends Neurosci* 1993;16:111–16
15. Robbins TW, Everitt BJ. Functions of dopamine in the dorsal and ventral striatum. *Semin Neurosci* 1992;4: 119–28
16. Selemon LD, Gottlieb JP, Goldman-Rakic PS. Islands on striosomes in the neostriatum of the rhesus monkey: nonequivalent compartments. *Neuroscience* 1994;58: 183–92
17. Wada Y, Kita Y, Yamamoto T, *et al*. Homocystinuria with generalized chorea and other movement disorders: a case report. *No To Shinkei* 2000;52:629–31
18. Webster KE. The functional anatomy of the basal ganglia. In Stem G, ed. *Parkinson's Disease*. London: Chapman and Hall Medical, 1990:356
19. Wexler NS. The Tiresias complex: Huntington's disease as a paradigm of testing for late-onset disorders. *FASEB J* 1992;6:2820–5
20. Yeterian EH, Pandya DN. Laminar origin of striatal and thalamic projections of the prefrontal cortex in rhesus monkeys. *Exp Brain Res* 1994;99:383–98

Extrapyramidal motor system 21

The extrapyramidal system comprises a group of structures that are concerned with stereotyped movements, complex motor activities, suppression of cortically induced movements, regulation of posture, and adjustment of muscle tone. It consists of the basal nuclei (ganglia), claustrum, substantia nigra, red nucleus, subthalamic nucleus, and the reticular formation. These neurons influence motor activity by projecting to specific thalamic nuclei en route to the motor and premotor cortices. The cortical influence on subcortical and spinal motor neurons is mediated via the corticofugal fibers. Therefore, projection of the basal nuclei, a component of the extrapyramidal system, to the motor cortex via the thalamus and thalamocortical fibers may account for the important role that this system plays in initiating and planning of movements.

Characteristics of the extrapyramidal system

The extrapyramidal motor system consists of the basal nuclei (ganglia), substantia nigra, red nucleus, subthalamic nucleus, and the reticular formation. The basal nuclei (Figures 21.1, 21.2 & 21.3, 21.4, 21.5 & 21.6) are a collection of subcortical nuclei embedded in the white matter of the cerebral hemispheres, which include the globus pallidus, caudate nucleus, and putamen that collectively form the corpus striatum. Some authors have extended the conventional definition of the basal nuclei to include the corpus striatum, subthalamic nucleus and the substantia nigra. These nuclei regulate stereotyped movements and mediate control of saccadic eye movements via reciprocal connections to the frontal eyefield and via their projections to the superior colliculus. They also coordinate orientation-associated memory and behavior via the bidirectional connections to the prefrontal and orbitofrontal cortices. This diverse origin and connections may account for the important role that the basal nuclei play in regulating motor activity and higher cognitive functions.

The caudate nucleus lies lateral to the thalamus, separated from it by the stria terminalis. It is comprised of the head, body, and tail regions. The head forms the floor of the anterior horn of the lateral ventricle, whereas the body is located in the floor of the central portion of the lateral ventricle, dorsolateral to the thalamus. It is separated from the thalamus by the stria terminalis and thalamostriate vein. The tail forms a curve running into the roof of the inferior horn of the lateral ventricle, terminating in the amygdala.

The putamen, the most lateral component of the corpus striatum, lies external to the globus pallidus and medial to the external capsule, joining rostrally the head of the caudate nucleus. The globus pallidus and putamen form the lentiform nucleus.

The caudate and putamen form the dorsal striatum (neostriatum), representing the afferent portion of the basal nuclei. They are derived from the telencephalon, maintaining identical cellular structures. In general, the dorsal striatum modulates complex motor responses. The dorsal striatum receives blood supply from the lateral striate (lenticulostriate) artery, recurrent artery of Heubner, and the anterior choroidal artery. The anterior cerebral artery provides blood to the head of the caudate and the adjacent part of the putamen via the medial striate artery (recurrent artery of Heubner).

Most striatal neurons contain γ-aminobutyric acid (GABA), enkephalin, substance P, as well as dopamine. Enkephalinergic neurons have D_2 dopamine receptors whereas substance P neurons have D_1 receptors. Somatostatin, acetylcholinesterase, choline acetyltransferase and pancreatic polypeptide containing neurons also exist. Glutamic acid decarboxylase (GAD) and serotonin concentrations are highest in the caudal striatum, whereas dopamine, acteylcholine, and substance P are highest in the rostral striatum. Triosomes, patches within the matrix of the caudate and less so in the putamen, contain enkephalin, dopamine, substance P and opiate receptors. The rest of the striatal matrix contains acetylcholine and somatostatin, receiving input from the intralaminar thalamic nuclei and superficial layers of the neocortex.

The nucleus accumbens septi and olfactory tubercle form the ventral striatum, which maintains significant role in memory and behavior.

The relatively small size of the anterior choroidal artery which supplies the tail of the caudate and its long subarachnoid course may account for its susceptibility to thrombosis. A hematoma that develops on the putamen as a result of hypertension or cerebrovascular malformation may quickly expand to involve the lateral ventricle and the adjacent internal capsule. Headache, vomiting, stiff neck, confusion and diminished consciousness, which are manifested in putaminal hemorrhage, are attributed to accumulation of blood in the lateral ventricle. Destruction of the internal capsule by this hematoma may also produce contralateral hemiparesis, contralateral conjugate gaze palsy, and Horner's syndrome.

Afferents of the dorsal striatum

The dorsal striatum receives corticostriate fibers from widespread areas of the cerebral cortex, thalamostriate fibers from the intralaminar thalamic nuclei, and nigrostriatal fibers from the substantia nigra (Figures 21.7 & 21.8).

• The corticostriate fibers originate from the entire neocortex and project to the ipsilateral striatum, utilizing the excitatory neurotransmitter glutamate. The inferior part of the caudate and adjacent ventral striatum receives specific projections from the orbitofrontal cortex. The caudate nucleus in particular receives input from the frontal eyefield and frontal association cortex. The putamen receives projections from the ipsilateral somatosensory, premotor, and motor cortices, and also from the contralateral motor cortex via the corpus callosum.

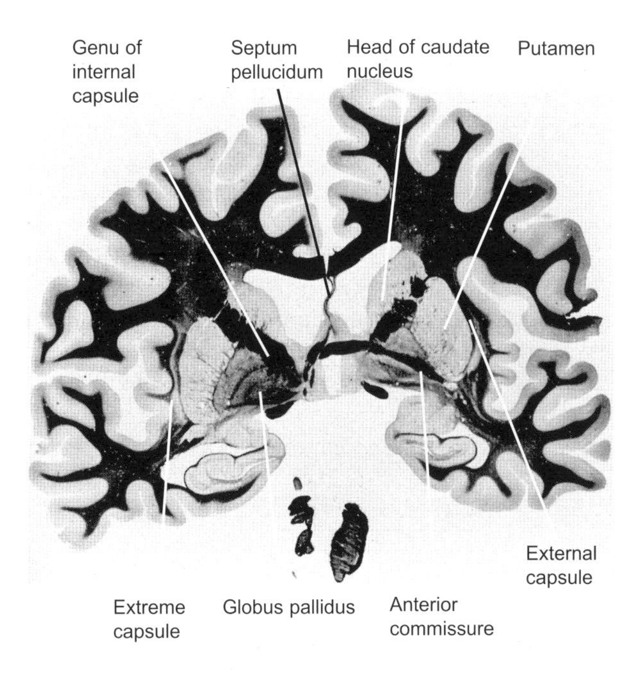

Figure 21.1 Coronal section of the brain. Observe the various components of the basal nuclei in relation to the internal capsule and the lateral ventricle

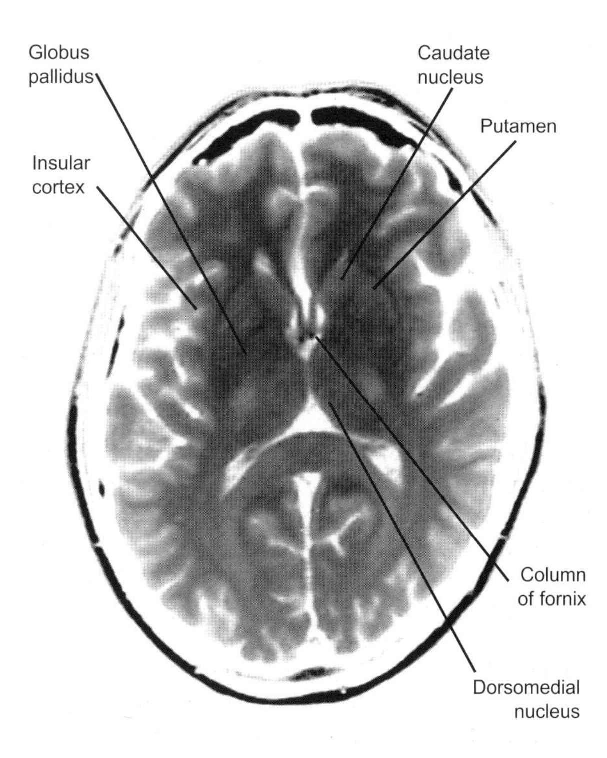

Figure 21.3 MRI scan showing a more elaborate structural organization of the basal nuclei

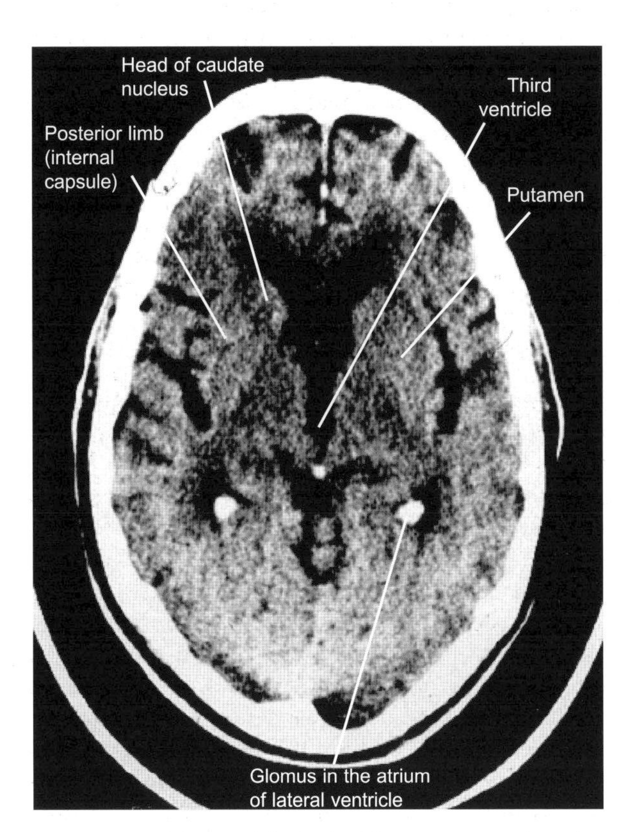

Figure 21.2 CT scan of the brain. The putamen and caudate nucleus are clearly visible

Additional projections to the caudate nucleus arise from the association areas of the frontal, parietal, and temporal cortices. The centromedian, an intralaminar nucleus of the thalamus, processes the striatal input and then projects back to the cortex (thalamostriate fibers), completing a closed feedback circuit. Through this circuit the basal nuclei affects not only the motor, but also the behavioral and cognitive functions.

• Nigrostriate fibers convey dopamine from the site of its production at the pars compacta of the substantia nigra to the striatum, exerting inhibitory influence upon the dorsal striatum through a direct path or possible excitatory effect via an indirect path. These afferents comprise the mesostriatal dopamine pathway. Some afferents to the striatum originate from the serotoninergic raphe nuclei of the midbrain (dorsal raphe nucleus B7), contributing to the high concentration of serotonin in the dorsal striatum. It also receives afferents from the adrenergic neurons of the locus ceruleus. There is additional input from the locus ceruleus (A6).

Efferent fibers from the dorsal striatum project to the globus pallidus and the reticular portion of the substantia nigra (striopallidal & strionigral fibers), delivering inhibitory GABA and excitatory substance P to the substantia nigra. Substance P may enhance the production and utilization of the neurotransmitter dopamine.

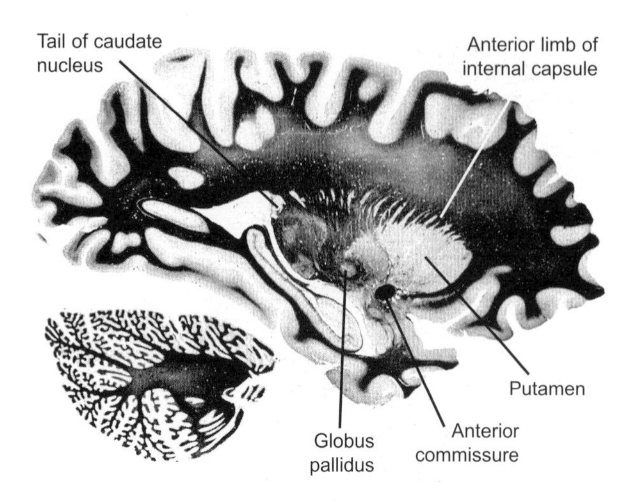

Figure 21.4 Midsagittal view of the brain. Observe the putamen, globus pallidus, and the fibers of the internal capsule

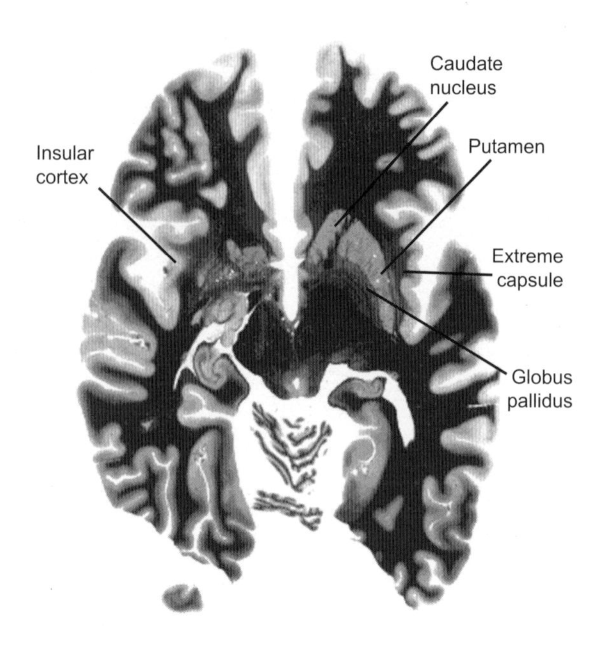

Figure 21.6 Horizontal section of the brain. Observe the continuation of the putamen with the caudate nucleus as well as the medial and lateral segments of the globus pallidus

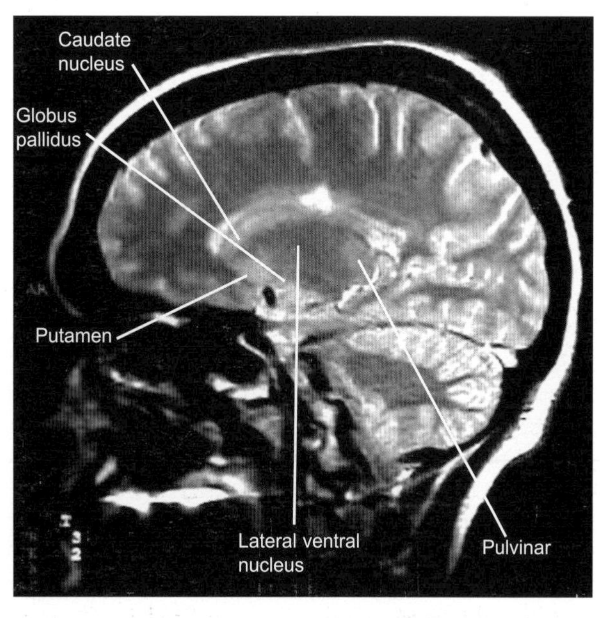

Figure 21.5 This MRI scan of the brain demonstrates the relationship of the basal nuclei to the thalamus

Receptor/ion channel interface and second messenger may determine the variation in the inhibitory and excitatory effects of GABA, substance P, and dopamine. The strionigral and striopallidal fibers project to both segments of the globus pallidus, delivering inhibitory GABA and excitatory substance P from the striatum. Due to their diverse connections, components of the striatum perform different functions; for example the putamen plays a role in motor activity, whereas the caudate nucleus is associated with cognitive tasks.

Afferents to the ventral striatum arise from the hippocampal gyrus and orbitofrontal, anterior cingulate, and temporal cortices. It also receives projections from the dorsal raphe (group B7), substantia nigra (medial part of pars compacta), paranigral nucleus (dopaminergic group A10), and the locus ceruleus (A6). The latter dopaminergic projections form the mesolimbic dopamine pathway, which also sends fibers to the entorhinal area, amygdala, septal area, and hippocampus. Efferents of the ventral striatum project to the ventral pallidum and the substantia nigra pars reticulata directly or via the subthalamic nucleus.

The globus pallidus (paleostriatum), derived from the diencephalon, is located lateral to the posterior limb of the external capsule and medial to the putamen (Figure 21.1, 21.4 & 21.5). The dorsal pallidum (pallidum), which refers to the globus pallidus, forms the efferent (output) portion of the basal nuclei. It consists of medial and lateral segments, separated by the internal medullary lamina. Both segments contain GABA and utilize acetylcholine as neurotransmitter. In particular, the medial segment exhibits similar cytological, morphological, and functional characteristics to the pars reticulata of the substantia nigra, and contains substance P. The lateral segment is comprised primarily of enkephalin.

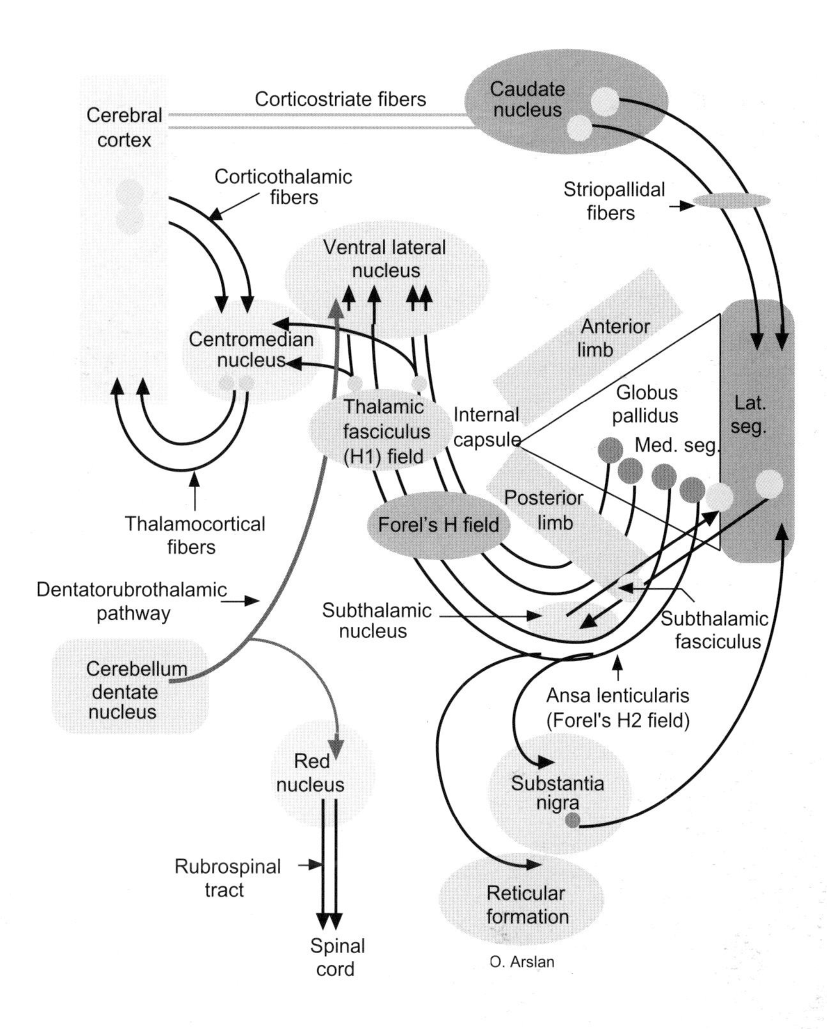

Figure 21.7 Afferents and efferents of the basal nuclei. Note the massive output from the medial segment (Med. seg.) of the globus pallidus to the ventral lateral nucleus of the thalamus. Lat. seg., lateral segment

The dorsal pallidum receives GABAergic fibers primarily from the striatum (striopallidal fibers) and to a lesser extent from the subthalamic nucleus. It is important to note that there are no direct spinal projections from the pallidum. The output from the medial segment of the dorsal pallidum is represented primarily by the lenticular fasciculus, ansa lenticularis, and thalamic fasciculus (Figures 21.7 & 21.8).

The lenticular fasciculus (Forel's field H2) traverses the posterior limb of the internal capsule and then joins the ansa lenticularis.

The ansa lenticularis (Figures 21.7 & 21.8) loops around the posterior limb of the internal capsule.

The lenticular fasciculus and ansa lenticularis merge rostral to the midbrain in Forel's field H, joining the dentatorubrothalamic fibers to form the thalamic fasciculus (Forel's field H1). The latter delivers impulses to the ipsilateral ventral lateral, ventral anterior, and the centromedian thalamic nuclei. A few fibers may project to the tegmentum of the midbrain as the pallidotegmental tract, constituting the principal link between the basal nuclei and the reticular formation. Some pallidal projections terminate in the habenular nucleus, thus establishing a link between the basal nuclei and the limbic system. The ventral lateral and ventral anterior thalamic nuclei influence, through the thalamocortical radiations, the ipsilateral motor and premotor cortices, respectively. The centromedian nucleus acts as a feedback loop, delivering generated impulses back to the striatum. The lateral segment of the globus pallidus projects to the subthalamic nucleus via the subthalamic fasciculus (Figure 21.7), utilizing the neurotransmitter GABA.

The ventral pallidum represents the area superolateral to the nucleus basalis of Mynert and caudal to the anterior perforated substance (substantia innominata), separated from the dorsal pallidum by the anterior commissure. Information received by the ventral pallidum, which originates from the ventral striatum, projects to the dorsomedial and midline thalamic nuclei and via thalamocortical fibers to the cingulate and prefrontal cortices, and hippocampal gyrus.

In summary, the corticostriate fibers activate the striatum to release the inhibitory GABA, which act upon the globus pallidus and substantia nigra. The striopallidal fibers block the release of inhibitory GABA from the globus pallidus, which acts upon the thalamus, resulting in activation of the thalamocortical fibers. This activation is thought to facilitate movement by exciting the motor cortex. Influences exerted by the subthalamic nucleus upon motor activity follows different logic. The corticostriate fibers produce excitation of the striatum and subsequent inhibition of the lateral segment of the globus pallidus, leading to disinhibition of the subthalamic nucleus and ultimate activation of the pallidal output.

The claustrum, a gray matter mass that resembles the thalamus, is located between the external and extreme capsules. It has temporal and insular parts, maintaining reciprocal connections with most areas of the isocortex. It receives retinotopically organized visual fibers from both eyes, as well as sensory fibers from the postcentral gyrus. The claustrum, superior colliculus, and pulvinar may mediate activities associated with visual attention.

The substantia nigra (Figures 21.6 & 21.9) represents the largest nucleus of the brainstem. It maintains reciprocal connections with the striatum. It shows degeneration in Parkinson's disease (see also the midbrain).

The red nucleus (Figures 21.7 & 21.9) occupies the center of the midbrain tegmentum. It is encircled by the superior cerebellar peduncle and is crossed by fibers of the oculomotor nerve en route to the interpeduncular fossa. The magnocellular part of this nucleus gives rise to the contralateral rubrospinal tract, a less distinct pathway in man that is thought to regulate flexor muscle tone. This nucleus receives contralateral cerebellar projections from the dentate, globose and emboliform nuclei, and ipsilateral rubro-olivary projection, contained in the central tegmental tract. It also projects to the ipsilateral motor cortex via the ventral lateral nucleus of the thalamus. Therefore, it is conceivable that the rubro-olivary tract may switch the control of movements from the corticospinal tract to the rubrospinal tract for programmed automation. In the same manner, the rubro-olivary tract may allow the corticospinal tract to intervene, in response to changes during on-going automated movements by the rubrospinal pathway. It delivers cerebellar and cerebral input to the upper three cervical spinal segments in man. Recovery of motor control (in monkeys) following corticospinal tract dysfunction may be attributed to the compensatory role that the rubro-olivary tract plays in re-routing the motor commands to the cerebellum and back to the rubrospinal tract.

A lesion of the subthalamic nucleus produces hemiballism, a rare disorder with sudden onset that affects the proximal part of the contralateral extremities. It is commonly seen in hypertensive individuals as a sequel to a vascular accident. It manifests forceful and uncontrollable movements commonly on the contralateral extremities, resembling a baseball pitcher's windup. This condition exacerbates under stressful conditions.

The subthalamic nucleus (Figures 21.7 & 21.9) is a biconvex nucleus which is separated by the zona incerta from the medially located internal capsule and ventrally positioned ventral thalamic nuclei. It integrates motor activities through its connections with the basal nuclei, substantia nigra and the tegmentum of the midbrain. It is thought to have an inhibitory influence upon the globus pallidus.

Reticular formation, another component of the extrapyramidal system, occupies the central core of the brainstem, consisting of the paramedian, medial, and lateral nuclear columns. It is concerned with the regulation of somatic and visceral (autonomic) motor activities and reflexes. It also modulates the electrocortical activities of the brain, mediates pain transmission, and regulates emotional expression.

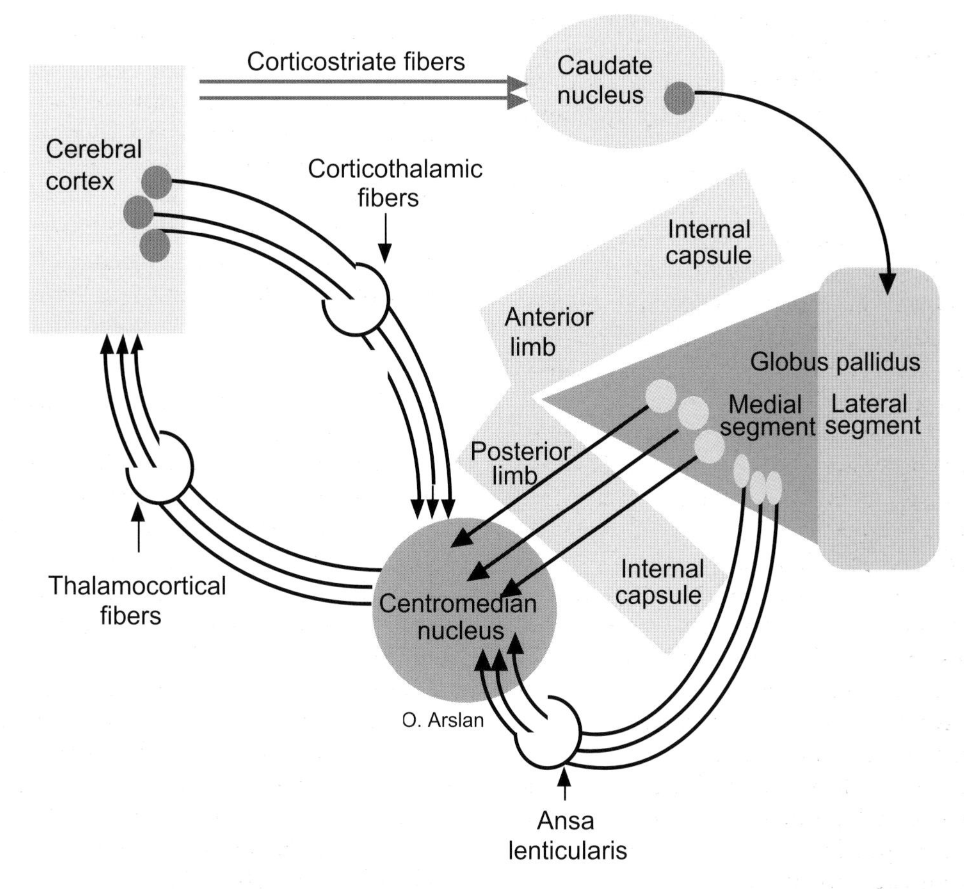

Figure 21.8 Feedback circuit formed by the efferents and afferents of the centromedian nucleus of the thalamus, which includes reciprocal connection with the cerebral cortex and projections from the medial segment of the globus pallidus

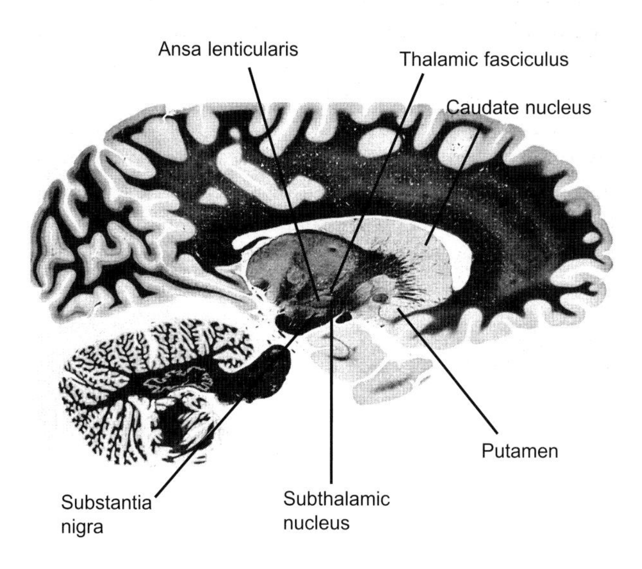

Figure 21.9 Section through the subthalamic area. Observe the subthalamic nucleus, lenticular and thalamic fasciculi, caudate nucleus, putamen, and substantia nigra

Figure 21.10 This is a depiction of the expressionless (masked) facies of an individual with Parkinsonism

Lesions of the extrapyramidal system may produce abnormal involuntary movements (dyskinesia) contralateral to the lesion side. Dyskinesia may be due to the excess of neural output and the imbalance in the dopamino-cholinergic-GABAergic systems. These motor disturbances can simply be explained as intact structures released from the inhibitory influences of the damaged structures associated with this motor system. Motor deficits may also be expressed as a reduction in movements (hypokinesia) and alteration in the residual contraction of the muscles (e.g. rigidity seen in Parkinson's disease). Dyskinesia may assume the forms of tremor, rigidity, choreiform movements, athetosis, dystonia, myoclonus, tardive dyskinesia, akathesia, and tics.

Tremor at rest is an involuntary rhythmic, sinusoidal, and alternating bilateral movement in a group of joints seen at rest that may disappear during voluntary action. It is rarely seen unilaterally and rarely involves the legs. Tremor may be aggravated under stress and in situations that require maintaining a constant posture. It often diminishes with alcohol intake and during sleep and voluntary movements.

Midbrain or rubral tremor is an extremely disabling condition, which may also be seen at rest, exhibiting exacerbation during movement. This form of tremor is seen in multiple sclerosis resulting from destruction of the superior cerebellar peduncle that surrounds the red nucleus. Tremor may also be observed in neonates as a result of low level of serum magnesium, epileptic seizures, and in individuals with hepatolenticular degeneration (Wilson's disease).

Rigidity refers to the generalized increase in the tone of all muscle groups (agonists and antagonists) to the same degree without significant change in reflex activity. In contrast, spasticity primarily affects the antigravity muscles. Rigidity may also be responsible for the fragmentation of a movement and for the slow regaining of the resting position subsequent to sudden release of passive resistance from the affected part. This is in contrast to the rebound phenomenon observed in cerebellar dysfunction. In rigidity, joints exhibit wax-like resistance to passive movements in all directions (during flexion and extension of the wrist or elbow). Conversely, spasticity exhibits resistance only in the initial stage of passive movement. Although rigidity is generally accompanied by tremor, congenital tremor observed in erythroblastosis fetalis (kernicterus) remains an exception.

Rigidity and tremor are pathognomic to Parkinson's disease (Figure 21.10), a slow progressive condition that develops as a consequence of encephalitis, manganese or carbon dioxide poisoning, trauma, neurosyphilis, cerebrovascular accidents, or cerebral arteriosclerosis. Phenothiazine and catecholamine-depleting drugs like reserpine, and metoclopramide may also produce characteristics of a reversible form of Parkinson's disease. The similarity between medication-induced Parkinsonism and Parkinson's disease is so great that the distinction is almost impossible.

True Parkinson's (paralysis agitans) disease is a familial disorder most frequently seen in males. In this disease, the melanin-containing cells of the pars compacta of the substantia nigra, nigrostriatal fibers, locus ceruleus (noradrenergic neurons), raphe nuclei (serotoninergic neurons), dorsal motor nucleus of the vagus, and the globus pallidus undergo degeneration and some depigmentation. Pathological examination of the pars compacta of the substantia nigra may reveal esinophilic inclusions or Lewy bodies. Deficits seen in Parkinson's disease are attributed to the abnormal reduction or depletion of dopamine in the striatum and the pars compacta of the substantia nigra and loss of its inhibitory effect. This is followed by overactivity of the cholinergic intrastriatal neurons and alteration in the output mechanism of the GABAergic neurons. It has been proposed that Parkinson's disease is an abnormal acceleration of the aging process. This is based on the fact that dopamine shows an increase in concentration very early in life, followed by a rapid decrease in the first two decades of life and then by prolonged decline of dopamine, a process that may even accelerate further by infection or toxicity.

The resting tremor is intermittent, slow (4–6 Hz) and shows characteristics of 'pill rolling' in which the cupped hand appears as though it was shaking pills. Tremor involves extension of the interphalangeal joints of the thumb and digits and flexion of the metacarpophalangeal joints. It usually begins unilaterally, most frequently involving the right hand and tends to disappear with action, intense concentration, and during sleep. It re-emerges as the extremity assumes a resting position. Rhythmic, regular, and involuntary contractions of the labial muscles without tongue movements (rabbit syndrome) may be seen as a manifestation of Parkinson's disease, or as a side-effect of antipsychotic drugs which respond to anticholinergic medications. The rigidity produces lead-pipe or cogwheel phenomenon (ratchet-like features) as a result of the gradual and alternate increase and decrease in muscle resistance to passive movement. Patients exhibit flexed and stooped posture which may be corrected upon command. A marked decrease or absence of movements (bradykinesia or akinesia) is manifested in small handwriting (micrographia). Akinesia may become severe, resulting in immobile posture. Movements may become slow to an extent that conveying food to the mouth may require an unusually long time. Patients exhibit difficulty in initiating or stopping movements, as well as difficulty in performing simple tasks, such as buttoning a shirt or eating.

Artificially-induced lesions in the globus pallidus and/or the ventral lateral nucleus of the thalamus may alleviate tremor and rigidity. Posteroventral pallidotomy re-emerged as a method of treatment developed by Leksell, and achieved good relief from tremor, rigidity and akinesia. However, central homonymous visual field deficit, facial weakness, and dysphasia are reported complications of this procedure. Innovative approaches to the treatment of Parkinson's disease included transplantation of dopamine producing adrenal medullary cells into the basal nuclei or ventricular system. In view of the fact that the brain is an immunologically privileged site, attempts to experimentally transplant mesencephalic fetal cells were carried out, though with no clear-cut success.

Habitual automatic movements become more difficult than usual, or they may assume the pattern of reflex activity. Thus, running may be easier than walking. Shuffling gait with short and rapid steps (festination) may also be seen in individuals with Parkinson's disease. Patients may also lack the associated movements during walking (i.e. arm-swing) and are unable to bend forward or flex the knee in attempt to sit on a chair (sitting en bloc), or stand up. Patients may have difficulty in maintaining the standing posture if a sudden move has occurred. Slow monotonous repetitive speech and masked expressionless face are also common symptoms of this disease. The head, shoulder, and body move together as a block when making turns, leading to entrapment in corners. Righting reflex, which involves smooth turns from the supine to prone position, is severely impaired. Intellect remains intact although some indications of slow thinking and response may be detected. Patients with Parkinsonism usually prefer cold weather and are easily kept warm. They may be comfortable with light cover even in cold weather. Emptying of the bladder may be inadequate, and constipation is a major and common complaint. Erectile function is impaired and blood pressure may be low. Patients commonly exhibit seborrheic dermatitis, which can be controlled with adequate hygiene. Tendon reflexes are usually unaffected and an abnormal extensor plantar reflex suggests a Parkinson-like syndrome. However, uninhibited glabellar reflex (Myerson sign), snout reflex, and palmomental reflexes are common to Parkinson's disease, even early in the course of this condition.

L-dopa, a precursor of dopamine, which permeates the blood–brain barrier, has been proven effective in ameliorating akinesia and rigidity, but is less effective in reducing tremor. Initially, one-fourth of the nigrostriatal neurons must be intact to convert L-dopa to dopamine. Dopamine agonists such as bromocriptine may be used as a supplement to L-dopa. L-dopa may be supplemented with medications such as amanthidine (Symmetrel) that enhance the release of dopamine and norepinephrine from presynaptic neurons. Tocopherol (vitamin E) is proven to have an important role in protecting dopamine from the effects of toxins and free radicals. Combined with Deprenyl, an antidepressant and antioxidant drug which blocks dopamine metabolism, tocopherol may prove to be of value in the treatment of this disease. Carbidopa, a decarboxylase inhibitor that does not cross the blood–brain barrier, is usually given in combination with L-dopa to raise the L-dopa concentration in the brain and minimize its side-effects.

Overdosage of L-dopa produces reversible choreiform movements of the extremities and perioral muscles. Since dopamine depletion is associated with excess of acteylcholine, anticholinergic medications are relatively effective in treating Parkinson's disease and drug-induced Parkinsonism and dystonia. It is important to bear in mind that anticholinergic therapy may accelerate the development of tardive dyskinesia and contribute to mental retardation. Neurosurgical procedures involving the contralateral thalamus (thalamotomy) used to be a successful method of treatment until the introduction of L-dopa. This surgical procedure also carried the risk of damaging the corticospinal and corticobulbar tracts within the posterior limb of the internal capsule, producing various forms of upper motor neuron palsies. Some patients developed dystonia and athetoid movements as complications of neurosurgical intervention.

Chorea is a brisk, rapid, graceful, and purposeless movement of short duration, which shows no rhythm and starts suddenly. These movements appear to be purposeful but in reality they do not serve any purpose. These involuntary movements are seen primarily in the distal appendicular musculature (e.g. hand) and appear to move from one part of the body to another in a random and continuous fashion. These movements may show exacerbation under stressful situations, persist as long as the patient is awake and may even continue during sleep. Twitching in the face or lip-smacking usually accompany the involuntary movements. Hypotonia, a decrease in the muscle tone or atonia, which occur at the end of each involuntary movement, may lead to a delay in the relaxation of the contracted muscles. Two classic diseases exhibit this type of movements, Sydenham's and Huntington's chorea.

Sydenham's chorea (St. Vitus' dance) is a rare reversible disorder which occurs in children between the age of 5–15 years (Figure 21.11), and commonly follows endocarditis as a complication of hemolytic streptococci-induced rheumatic fever. This disorder may last up to two years. It should be differentiated from dystonia, tics, and dyskinesia associated with drug withdrawal. Involuntary choreiform movements in this disorder are more rapid and may have a lightning character. Children develop unusual behavior and remain irritable.

Huntington's (adult) chorea (Figure 21.12) is a devastating progressive illness that eventually leads to death. It is an autosomal dominant disease and the mutated gene is located on chromosome number 4. The location of the mutated gene may allow prenatal diagnosis of this disease, eventual cloning, and replacement of the gene. Children of patients with this disease have a 50% chance of being affected. It is characterized by degeneration of the neostriatum and cerebral cortex. The reduction of glutamic acid decarboxylase, an enzyme responsible for the conversion of L-glutamic acid to GABA, is thought to be crucial in the pathogenesis of this disease. Changes in the levels of Substance P, enkephalin, dynorphin, and cholecystokinin may also be noticed. Mental deterioration, which includes paranoid delusions and eventual dementia, may occur prior to the onset of involuntary movements or immediately afterwards.

Symptoms typically appear late in the fourth or fifth decades after the affected individual has had children and passed on the gene. Unfortunately, one of the saddest travesties of this illness in the past was peoples' perception that afflicted patients were witches, creating the environment for their cruel and unfair treatment, torture, and sometimes execution. Atypical (Westphal subtype) Huntington's chorea may occur in children and occasionally in adults, exhibiting rigidity and akinesia, without choreiform movements.

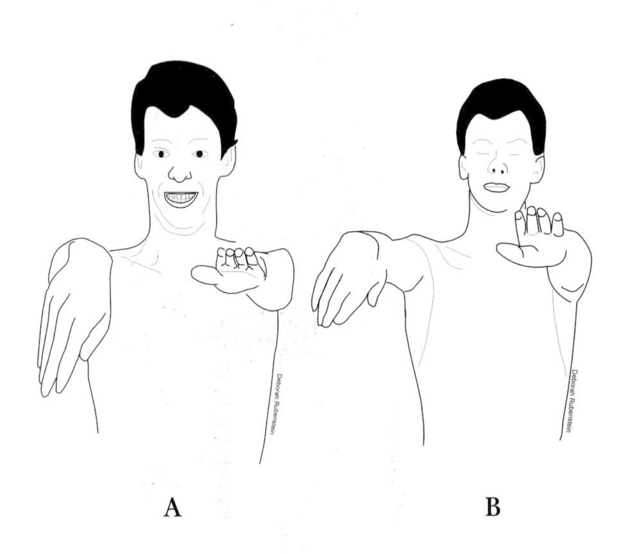

Figure 21.11 These two diagrams illustrate the choreiform movements associated with Sydenham's chorea. In (A) the patient's forearms are pronated and his right hand is flexed at the wrist, while he is grimacing. In (B) the same individual with eyes are closed and hypotonic right arm

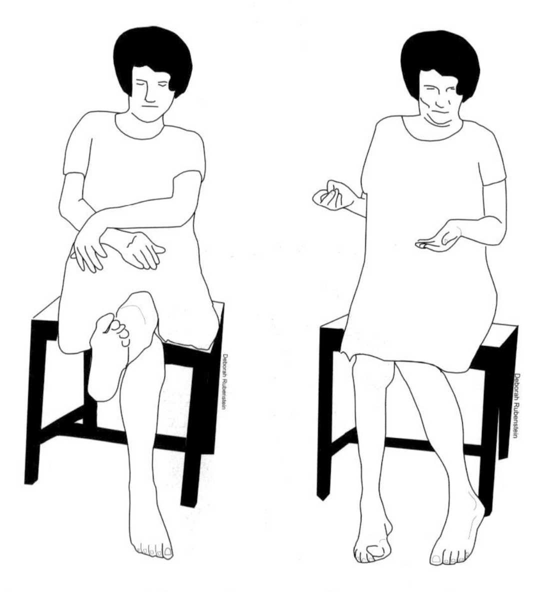

Figure 21.12 These diagrams shows manifestations of Huntington's chorea. Observe the bizarre posture, uncoordinated movements of the upper and lower extremities. Affected individuals are disoriented, but alert, exhibiting speech disorder and disorientation

Athetosis (Figure 21.13) is a slow, irregular, writhing, and worm-like movement involving the distal muscles of the extremities, neck, face, and tongue. These movements are cramp-like, characterized by hyperflexion or hyperextension of the joints of the hand with bizarre configuration accompanied by adduction and abduction of the shoulder joint. They alternate between extension–pronation and flexion–supination of the arm, flexion–extension of the fingers, and eversion–inversion of the foot. Sometimes dorsiflexion of the great toe may be erroneously diagnosed as Babinski's sign. Athetoid movements occur in birth trauma-related anoxia. They are also seen in kernicterus, a neurological condition caused by Rh or other blood incompatibility between the mother and the fetus. Kernicterus is associated with hemolysis, and accumulation of non-conjugated bilirubin in the brain.

Many patients with chorea may also exhibit athetoid (choreoathetoid) movements. Choreoathetoid movements are also seen in Lesch–Nyhan syndrome, a sex-linked disease that usually affects males and results from deficiency of hypoxanthine-guanine phosphoribosyl transferase. It is characterized by mental retardation, physical handicap, upper motor neuron palsy, and a striking phenomenon of self-mutilation.

Dystonia (torsion dystonia) is an idiopathic hereditary disorder characterized by contracture of the axial or appendicular musculature, resulting in severe torsion and deformed rigid posture (Figure 21.14). It is a physically impairing condition with no detectable impact on the mental status. Dystonia is also seen in hepatolenticular degeneration (Wilson's) and Huntington's diseases, and Lesch–Nyhan syndrome. In contrast to athetosis where the distal muscles are affected, dystonia primarily involves the proximal muscles. The twisted and fixed posture of a dystonic patient can be very striking and repulsive. It manifests itself in childhood or teenage years. Overactivity of the striatal dopaminergic system, which increases dopamine concentration, is thought to be a factor in this disorder. This hypothesis is based on the observation that the drugs which induce this condition block dopaminergic receptors with a resultant increase in dopamine concentration. It is furthermore based on the fact that administration of dopaminergic drugs exacerbates the associated movement disorders. Some patients may show minimal signs collectively known as writer's cramp, which is characterized by tonic contractions of the hand muscles during writing.

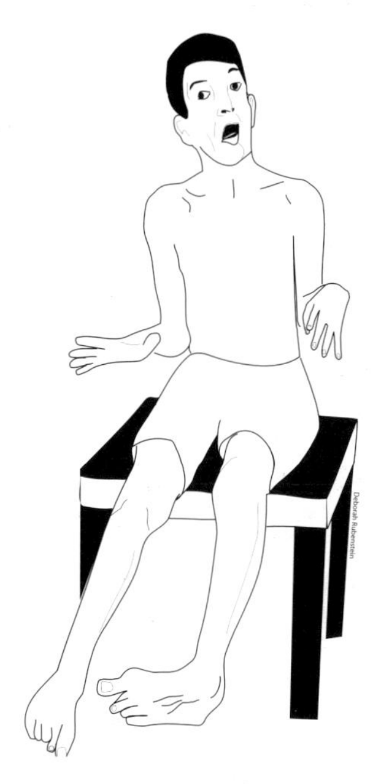

Figure 21.13 This drawing depicts an individual with athetosis affecting all muscles of the limbs, neck, and face. Note the flexion in the upper extremities and extension of the lower extremities

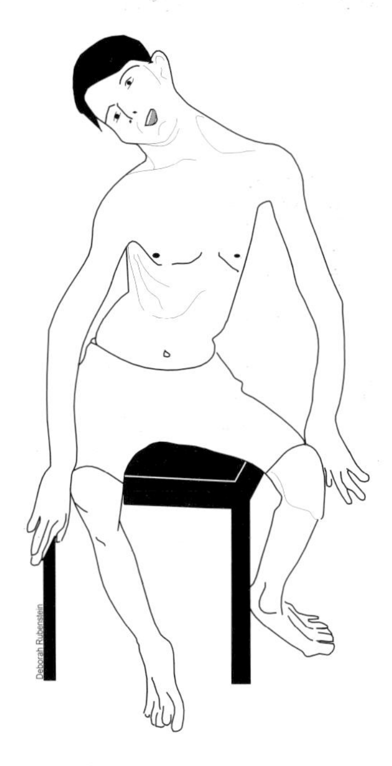

Figure 21.14 This is a depiction of a patient with torsion dystonia. Observe the torsions of the limbs, neck, and spine which are produced by the slow, yet powerful, involuntary movements

Myoclonus is a short, irregular and lightning-like contraction of a muscle or a group of muscles, which may be induced by voluntary action or by a stimulus from the examiner. It may or may not be synchronous or symmetrical, and may be focal or generalized. The repetitive fashion of the myoclonic jerks involving one or more sites is characteristic of this condition. Myoclonic movements may occur at the onset of sleep or upon awakening (nocturnal myoclonus or hypnic jerk). It is associated with the damage sustained by the cerebral cortex as a result of infectious diseases (i.e. Creutzfeldt–Jakob disease) or metabolic disorders such as toxemia or uremia. Seizures and dementia may accompany myoclonus. Degeneration of the inferior olivary nuclear complex and associated neuronal feedback circuit produces palatal myoclonus, rhythmic contractions of the soft palate which occur regularly and during sleep. Palatal myoclonus is not usually associated with mental disorder and generally results from infarction of the inferior olivary nucleus. Diaphragmatic myoclonus may involve the diaphragmatic muscles, producing constant hiccup.

Tardive dyskinesia is a very common symptom that develops in a great number of patients as side-effects of antipsychotic medications (e.g. phenothiazines) administered for a prolonged period of time. Choreoathetoid movements of the limbs and trunk, and tongue protrusion, particularly lip smacking and facial grimacing characterize this irreversible and persistent condition. Patients with this condition may exhibit loud labored respiration similar to that seen in Tourette's syndrome. Supersensitivity of the dopaminergic striatal neurons, depletion of GABA and glutamic acid decarboxylase may be attributed to the development of this condition. Several structures such as the substantia nigra and inferior olivary nucleus may show some degree of degeneration in this condition.

Akathesia is characterized by a state of restlessness and jitters and continuous compulsion to move in order to relieve the uncomfortable sensation in the leg and thigh. These movements may mimic those seen in Parkinson's disease and other disorders of the basal nuclei. It may also be seen as a consequence of usage of neuroleptics, resembling the dyskinesia produced by excessive L-dopa. Akathesia may also be seen in other conditions such as parietal lobe lesions or porta-caval encephalopathy. It has some resemblance to choreiform movements, but is restricted to the lower extremity and is generally accompanied by subjective restlessness that exceeds leg movements. The latter is an important factor that distinguishes akathesia from agitation. One form of this condition is tardive akathesia, which may mimic signs of manic depression or anxiety states. It is treated with opioid analgesic such as propoxyphene.

Wilson's disease (hepatolenticular degeneration) ia an autosomal recessive disorder of copper metabolism, which commonly occurs in the offspring of blood-related (consanguineous) marriages. This serious condition is caused by a low level of circulating copper binding α2 globulin (ceruloplasmin) and a high level of free copper in the serum which deposits in the basal nuclei, kidney, and liver. Deposition of copper ultimately produces degeneration in the putamen and liver (multilobular cirrhosis). Tremor, a manifestation of this disease, involves the hands and trunk. In the limbs, it resembles wing beating or flapping tremor. Rigidity, dysphagia, difficulty in speech, loss of facial expression, behavioral changes, and dementia are distinctive signs of this disease.

Deposition of the copper at the corneoscleral junction produces Kayser–Fleischer ring, a greenish-brown pigmentation in the Descemet's membrane of the cornea. Glomerular and tubular dysfunctions and glucosuria are also seen as a result of involvement of the urinary system. This disease is treated with D-penicillamine, a copper chelator. A low copper diet and potassium sulfide supplement, which retards copper absorption, is also needed. This disease has a prevalence of 1 in 200 000 individuals.

Tics are repetitive, rapid, irregular and non-rhythmic stereotyped movements or vocalizations that can be mimicked by the observer and often can be willfully held under control by the patient, usually at the expense of mounting inner tension. The inner tension is often relieved when the tic is allowed to occur. These movements usually involve simultaneously several muscle groups and may assume the form of a complex motor act. Simple tics may be in the forms of shoulder jerk, prolonged eye blink or head toss. Complex motor tics are seen as jumping, stamping, touching oneself. Vocal tics may be in the form of throat clearing, grunting, sniffing or vocalization of words or fractions of words. Patients may mimic words (echolalia) or movements (echopraxia or echokinesis). Tics may be intensified by excitement, anxiety, and exhaustion, and are suppressed by intense concentration. They may persist during sleep and may be accompanied by sleepwalking.

Tourette's syndrome is a familial (autosomal dominant) disorder, which begins as simple tics in childhood. It later advances into multiple motor and vocal tics. It affects boys three times more frequently than girls. Most cases develop by the age of 13. These stereotyped vocal tics may assume the form of throat clearing or barking noises and may evolve into unprovoked outbursts of coprolalia (utterance of obscene phrase). Most of the obscene words are fragmented, containing only fractions of the actual words. The latter phenomenon may be associated with indecent gestures (copropraxia) or lewd thoughts. CT scan and MRI do not show any abnormality, although an electroencephalogram may not be normal. Tics in this lifelong disease change in distribution and intensity and may show remission. Anxiety and tension associated with school years tends to aggravate this condition.

Some patients may develop obsessive–compulsive disorder (OCD) or attention-deficit hyperactivity disorder (ADHD) or both. Intelligence remains unaffected. Methylphenidate, a medication used for the treatment of ADHD, and cocaine which blocks the re-uptake of norepinephrine and dopamine, may predispose children for Tourette's syndrome. This syndrome may be treated by haloperidol, which suppresses vocal and motor tics, but it may also cause gynecomastia. Fluphenazine, pimozide, tetrabenazine, and clonidine are also commonly used.

Suggested reading

1. Bousser MG, Chiras J, Bories J, *et al.* Cerebral venous thrombosis – a review of 38 cases. *Stroke* 1985;16: 199–213
2. Crowell RM, Morawetz RB. The anterior communicating artery has significant branches. *Stroke* 1977;8: 272–3
3. Davis JM, Davis KR, Crowell RM. Subarachnoid hemorrhage secondary to ruptured intracranial aneurysm: prognostic significance of cranial CT. *Am J Roentgenol* 1980;134:711–15
4. Everhart DE, Harrison DW, Crews WD Jr. Hemispheric asymmetry as a function of handedness: perception of facial affect stimuli. *Percept Mot Skills* 1996;82:264–6
5. Flament D, Onsott D, Fu QG, Ebner TJ. Distance- and error-related discharge of cells in premotor cortex of rhesus monkeys. *Neurosci Lett* 1993;153:144–8
6. Fries W, Danek A, Scheidtmann K, Hamburger CS. Motor recovery following capsular stroke. Role of descending pathways from multiple motor areas. *Brain* 1993;116:369–82
7. Halsband U, Ito N, Tanji J, Freund H-J. The role of premotor cortex and the supplementary motor area in the temporal control of movement in man. *Brain* 1993;116:243–66
8. Horton JC, Chambers WA, Lyons SL, *et al.* Pregnancy and the risk of hemorrhage from cerebral arteriovenous malformations. *Neurosurgery* 1990;27:867–72
9. Kennedy PR. Corticospinal, rubrospinal and rubro-olivary projections: a unifying hypothesis. *Trends Neurosci* 1990;13:474–8
10. Kim SG, Ashe J, Hendrich K, *et al.* Functional magnetic resonance imaging of motor cortex: hemispheric asymmetry and handedness. *Science* 1993;267:615–17
11. Kristensen MO. Progressive supranuclear palsy – 10 years later. *Acta Neurol Sci* 1985;71:177–89
12. Mannen T, Iwata M, Toyokura Y, Magazhima K. Preservation of a certain motor neuron group of the sacral cord in amyotrophic lateral sclerosis: its clinical significance. *J Neurol Neurosurg Psychiatry* 1977;40: 464–9
13. Marcus JC. Flexor plantar responses in children with upper motor neuron lesions. *Arch Neurol* 1992;49: 1198–9
14. Matthews PBC. The human stretch reflex and the motor cortex. *Trends Neurosci* 1991;14:87–91
15. Morrison BM, Morrison JH. Amyotrophic lateral sclerosis associated with mutations in superoxide dismutase; a putative mechanism of degeneration. *Brain Res Rev* 1999;1:121–35
16. Porter R, Lemon RN. *Corticospinal Function and Voluntary Movements*. Oxford: Oxford University Press, 1993:428
17. Ropper AH, Fisher CM, Kleinman GM. Pyramidal infarction in the medulla: A cause of pure motor hemiplegia sparing the face. *Neurology* 1979;29:91–5
18. Strick PL. Anatomical analysis of ventrolateral thalamic

input to primate motor cortex. *J Neurophysiol* 1976;39: 1020–31
19. Tanji J. The supplementary motor area in the cerebral cortex. *Neurosci Res* 1994;19:251–68
20. Willoughby EW, Anderson NE. Lower cranial nerve motor function in unilateral vascular lesions of the cerebral hemisphere. *Br Med J* 1984;289:791–4

Index